OPTIMIZATION OF DISTRIBUTED
PARAMETER STRUCTURES : VOLUME I

NATO ADVANCED STUDY INSTITUTES SERIES

Proceedings of the Advanced Study Institute Programme, which aims at the dissemination of advanced knowledge and the formation of contacts among scientists from different countries.

The series is published by an international board of publishers in conjunction with NATO Scientific Affairs Division

A	Life Sciences	Plenum Publishing Corporation
B	Physics	London and New York
C	Mathematical and Physical Sciences	D. Reidel Publishing Company Dordrecht and Boston
D	Behavioural and Social Sciences	Sijthoff & Noordhoff International Publishers B.V.
E	Applied Sciences	Alphen aan den Rijn, The Netherlands and Rockville, Md., U.S.A.

Series E: Applied Sciences — No. 49

OPTIMIZATION OF DISTRIBUTED PARAMETER STRUCTURES - VOLUME I

edited by

Edward J. Haug
The University of Iowa,
College of Engineering
Iowa City, IA, 52242, U.S.A.

and

Jean Cea
University of Nice
Department of Mathematics
Nice, France

Sijthoff & Noordhoff 1981
Alphen aan den Rijn, The Netherlands
Rockville, Maryland, U.S.A.

Proceedings of the NATO Advanced Study Institute on
Optimization of Distributed Parameter Structural Systems
Iowa City, Iowa, U.S.A.
May 20 - June 4, 1980

ISBN-13: 978-94-009-8605-3   e-ISBN-13: 978-94-009-8603-9
DOI: 10.1007/978-94-009-8603-9

NATO-NSF ADVANCED STUDY INSTITUTE ON

OPTIMIZATION OF DISTRIBUTED PARAMETER STRUCTURES

Iowa City, Iowa, United States, 21 May - 4 June, 1980

Director : E. J. Haug, Materials Division, College of
Engineering, University of Iowa, Iowa City, Iowa

Co-director: J. Cea, Department of Mathematics, University of
Nice, Nice, FRANCE

Scientific Contents of the Advanced Study Institute

The Advanced Study Institute was organized to bring together
engineers and mathematicians working in optimization of distributed parameter structures. The principle attention of the Institute was focused on structures described by boundary-value problems, as opposed to finite element models of structures. The di-
versity of interest and specialization of participants ranged from
applications oriented engineers, through mathematically oriented
research engineers, to applied mathematicians working in fields
related to structural optimization. A key objective of the institute was to promote interaction by engineers and applied mathematicians who have, in the past, taken rather different approaches
to structural optimization . As part of this objective, the emerging field of shape optimal design was given a high degree of emphasis in the Institute, to provide a forum for the study of
mathematical techniques of shape optimization that have been applied to diverse fields of science and to consider their applicability for structural optimization.

The scientific program began with a review of distributed
parameter structural optimization literature (E. Haug) and a survey of design sensitivity analysis methods applicable to structural optimization (E. Haug). Reviews of optimality criterion methods (J.Taylor), variational methods for developing optimality criterion (E. Rozvany), and an analysis of singular problems in optimal design (E. Masur) were presented as an overview of optimality
criteria methods. Optimality criteria methods for design of columns for buckling, beams and shafts for vibration, and plates for
vibration were presented (N. Olhoff), as was optimization of
grids, shells, and arches (G. Rozvany). Mathematical aspects of
recently encountered optimal design problems with repeated eigenvalues were presented (B. Rousselet, and K. Choi). Design of
plates for minimum deflection and stress (N. Banichuk) and a survey of structural optimization methods under nonconservative
loading (R. Plaut) were presented. An in depth treatment of opti-

mization methods and their application for structures under earthquake loads was presented as a sequence of lectures in the program
(K. Pister, E. Polak, M. Bhatti). Numerical methods of optimal
remodeling (J. Taylor) and gradient projection techniques for
static and dynamic problems (E. Haug, J. Arora) were presented.
Approximately 40% of lecture time in the Institute was devoted to
presentation of basic mathematical ideas of shape optimazation,
with applications to numerious fields of applied physics (J. Cea,
B. Rousselet, and J. Zolesio). Applications of shape optimal design techniques to structural systems (N. Banichuk) and optimization of shape of elastic bodies contact (R. Benedict) were presented. In addition to the principle lectures of the institute,
participants gave contributed papers and lectures, many of which
are contained in these proceedings.

LECTURERS

J. Arora, University of Iowa, Iowa City, Iowa, U.S.A.
N. Banichuk, USSR Academy of Sciences, Moscow, USSR
R. Benedict, University of Iowa, Iowa City, Iowa, U.S.A.
M. Bhatti, University of Iowa, Iowa City, Iowa, U.S.A.
J. Cea, University of Nice, Nice, FRANCE
K. Choi, University of Iowa, Iowa City, Iowa, U.S.A.
E. Haug, University of Iowa, Iowa City, Iowa, U.S.A.
E. Masur, University of Illinois, Chicago, Illinois, U.S.A.
N. Olhoff, Technical University of Denmark, Lyngby, DENMARK
K. Pister, University of California, Berkeley, CA., U.S.A.
R. Plaut, Virginia Polytechnic Institute, Blacksburg, VA., U.S.A.
E. Polak, University of California, Berkeley, California, U.S.A.
B. Rousselet, University of Nice, Nice, FRANCE
G. Rozvany, Monash University, Clayton, AUSTRALIA
J. Taylor, University of Michigan, Ann Arbor, Michigan, U.S.A.
J. Zolesio, University of Nice, Nice, FRANCE

CONTRIBUTORS

C. Akkoc, Middle East Technical University, Gaziantep, TURKEY
H. Alper, Bogazici University, Istanbul, TURKEY
D. Anderson, University of Warwick, Coventry, ENGLAND
F. Cheng, University of Missouri-Rolla, Rolla, Missouri, U.S.A.
K. Cheng, Technical University of Denmark, Lyngby, DENMARK
Y. Chun, Villanova University, Villanova, Pensylvania, U.S.A.
C. Cinquini, Universita Di Pavia, Pavia, ITALY
J. Claudon, University of Tokoyo, Tokoyo, JAPAN
M. Delfour, Universite de Montreal, Montreal, CANADA
C. Fleury, Universite de Liege, Liege, BELGIUM
J. Kalker, Delft University of Technology, Delft, NETHERLANDS
P. Kirmser, Kansas State University, Manhattan, Kansas, U.S.A
R. Kohn, New York University, New York, New York, U.S.A.
V. Komkov, West Virginia University, Morgantown, WV., U.S.A.
J. Kruzelecki, Technical University of Cracow, Crakow, POLAND
B. Kwak, The Korea Advanced Institute of Science, Seoul, KOREA
E. Lightfoot, University of Oxford, Oxford, ENGLAND
C. Martin, University of Nebraska, Lincoln, Nebraska, U.S.A.
A. Morris, Royal Aircraft Establishment, Hampshire, ENGLAND
P. Pedersen, The Technical University of Denmark, Lyngby, DENMARK
B. Pierson, Iowa State University, Ames, Iowa, U.S.A.
C. Polizzotto, Universita di Palermo, Palermo, ITALY
G. Sacchi, Istituto di Scienza E Delle Costruzioni, Milano, ITALY
M. Saglam, Bogazici Universitesi, Istanbul, TURKEY
R. Sandstrom, University of Michigan, Ann Arbor, Michigan, U.S.A.
P. Sinha, University of Warwick, Coventry, ENGLAND
J. Sokolowski, Systems Research Institute, Warszawa, POLAND

CONTRIBUTORS (cont)

W. Spillers, Rensselaer Polytechnic Institute, Troy, N.Y., U.S.A.
M. Szata, Technical University of Wroclaw, Wroclaw, POLAND
B. Topping, The University of Edinburgh, Edinburgh, ENGLAND
A. Torkamani, University of Pittsburgh, Pittsburgh, Penn., U.S.A.
J. Whitesell, University of Michigan, Ann Arbor, Michigan, U.S.A.
K. Willmert, Clarkson College, Potsdam, New York, U.S.A.
H. Zehlein, Gesselschaft fur Kernforschung, Karlsruhe, GERMANY

PARTICIPANTS

A. Aly, University of Waterloo, Waterloo, Ontario, CANADA
E. Atimtay, Middle East Technical University, Ankara, TURKEY
A. Bilgutay, Middle East Technical University, Ankara, TURKEY
N. Carmichael, University of Warwick, Coventry, ENGLAND
R. Caron, University of Waterloo, Waterloo, Ontario, CANADA
A. Ghali, University of Calgary, Calgary, CANADA
J. Gierlinski, University of Southampton, Southampton, ENGLAND
D. Grierson, University of California, Los Angeles, Calif., U.S.A.
E. Johnson, Virginia Polytechnic Institute, Blacksburg, VA.,U.S.A.
A. Kildegaard, Aalborg University Center, Aalborg, DENMARK
O. Lev, Merritt Consulting Engineers, Redlands, California, U.S.A.
B. Lysik, Technical University of Wroclaw, Wroclaw, POLAND
J. McNabb, Bradley University, Peoria, Illinois, U.S.A.
C. Ng, David Taylor Res. & Dev. Center, Bethesda, MD, U.S.A.
T. Panzeca, Istituto di Scienza e Delle Costruzioni, Palermo,ITALY
M. Papadrakakis, Athens, GREECE
G. Payre, Universite de Sherbrooke, Sherbrooke, Quebec, CANADA
S. Tang, Ford Research Laboratory, Dearborn, Michigan, U.S.A.
G. Turvey, University of Lancaster, Lancaster, ENGLAND

TABLE OF CONTENTS

X

PREFACE

These proceedings contain lectures and contributed papers presented at the NATO-NSF Advanced Study Institute on Optimization of Distributed Parameter Structures (Iowa City, Iowa 21 May - 4 June, 1980). The institute was organized by E. Haug and J. Cea, with the enthusiastic help of leading contributors to the field of distributed parameter structural optimization. The principle contributor to this field during the past two decades, Professor William Prager, participated in planning for the Institute and helped to establish its technical direction. His death just prior to the Institute is a deep loss to the community of engineers and mathematicians in the field, to which he made pioneering contributions.

The proceedings are organized into seven parts, each addressing important problems and special considerations involving classes of structural optimization problems. The review paper presented first in the proceedings surveys contributions to the field, primarily during the decade 1970-1980. Part I of the proceedings addresses optimality criteria methods for analyzing and solving problems of distributed parameter structural optimization. Optimality criteria obtained using variational methods of mechanics, calculus of variation, optimal control theory, and abstract optimization theory are presented for numerous classes of structures; including beams, columns, plates, grids, shells, and arches. Optimality criteria and numerical methods based on these criteria represent a reasonably well developed field that is thoroughly covered in Part I. A special topic receiving considerable attention here is the emerging problem of occurrence of repeated eigenvalues in optimal designs, for both buckling and vibration. New results are presented and directions for future effort in this important class of problems are suggested.

Numerical methods for optimal remodeling and direct numerical optimization of structures are presented in Part II of the proceedings. The theme here is iterative optimization, primarily through linearization or other related approximations of the general nonlinear structural optimization problem. Numerical methods for vibration, buckling, displacement, and dynamic constraints are treated. Direct numerical methods for treating repeated eigenvalues are suggested and examples are presented.

Part III is devoted to optimization of structures under earthquake loads. The essentially nonlinear dynamic behavior of such structures is treated in detail and iterative optimization methods that are suitable for solution of such structural optimization problems are presented. A software system that has been used in earthquake structural optimization is presented.

Contributed papers on finite dimensional structural optimization are presented in Part IV. A survey of nonlinear programming methods and their application to large scale structures, using finite element methods, is presented. Numerous special applications found in papers presented in this part of the proceedings.

Part V is devoted to optimization of structures under nonconservative loading and other special problems of distributed parameter structural optimization. A survey of optimization under nonconservative loading is presented. Related problems in flutter optimization and other complex problems of distributed parameter structural optimization are presented.

Part VI is devoted to the shape optimal design problem. Since only limited work in the field of structural shape optimization has been done, lectures in this part of the proceedings focus on shape optimization examples from diverse fields of applied physics. Mathematical and numerical methods for solving such problems are presented to serve as a guide for future work in shape optimal design of structures. The coherence of the mathematical theory of shape optimization and results obtained for problems other than structures show great potential for development of this field in the future.

The final part of the proceedings, Part VII, presents a thorough treatment of design sensitivity analysis of structural systems. Recently developed, rigorous mathematical theories of design sensitivity analysis are presented and their application to structural optimization problems is illustrated. Results reported in this part of the proceedings are intended to serve as a foundation for future work in both optimality criteria and direct iterative methods for distributed parameter structural optimization.

The extent and variety of the lectures and papers presented in these proceedings illustrate the extensive contribution of numerous individuals in preparation and conduct of the Institute. The Institute directors wish to thank all contributors to these proceedings for their substantial effort. Special thanks go to T. Brannian for her efforts in the administrative planning and support of the Institute and for her typing of preliminary manuscripts. The dedicated and tireless support of R. Huff, K. Walters and A. Craven in typing the final copy of the entire manuscript is greatly appreciated. Thanks are also due to Dr. K.K. Choi and graduate assistants H. Lam, J. Hou, Y. Yoo, A. Shabana, A. Belegundo in proofreading of the final text. Finally, without the financial support of the NATO Office of Scientific Affairs and the U.S. National Science Foundation, the Institute and these proceedings would not have been possible. Their support is gratefully acknowledged by all concerned with the Institute.

December 1980 E. J. Haug

J. Cea

Part 1

OPTIMALITY CRITERIA METHODS FOR STRUCTURAL OPTIMIZATION

A REVIEW OF DISTRIBUTED PARAMETER STRUCTURAL OPTIMIZATION LITERATURE

Edward J. Haug

Materials Division, College of Engineering,
The University of Iowa, Iowa City, Iowa

INTRODUCTION

A multitude of papers have appeared since 1970 in the fields of optimal control, function space optimization, nonlinear programming, finite dimensional structural optimization, and to a lesser degree continuous material distribution in optimization of structural elements. The relationship among these fields in finite dimensional design problems (selection of a parameter vector in R^n) is becoming reasonably well understood and developed. In the area of infinite dimensional or distributed parameter optimization of structures (selection of a function of one or more design variables describing continuous distribution of material over a 1, 2, or 3 dimensional structure), the interrelationship of control theory, function space optimization, and structural optimization is far less clear. It is the purpose of this review to analyze the literature that has appeared since 1970 in the field of distributed parameter structural design, particularly as regards the methods employed and applications of these methods to specific classes of structural optimization problems.

A substantial literature on optimal design of elastic structures involving continuous distribution of a design variable over one or two space dimensions exists. Applications have generally been to test problems involving bars, shafts, beams, and plates of variable section. A variety of performance constraints have been placed on the design problems, which serve to segregate the classes of problems. Research in structural optimization is largely uncoupled from a substantial effort in mathematics and science on optimization of distributed parameter systems.

A brief review of books published on structural optimization, review articles that have appeared, and applicable distributed parameter optimization theory is given in this section. The detailed review of distributed parameter structural optimization literature is then organized according to problem type, as follows.

Section	Principal Focus
2	Buckling of columns, plates, and shells
3	Vibration of bars, beams, and plates
4	Deflection and compliance of structures
5	Dynamic response
6	Shape of two and three dimensional elements
7	Multi-purpose structures and special problems

This classification is not unique, but is consistent with categorizations of objectives and constraints that have been by authors of preceeding reviews. Papers are cited in the reference list in the order they are encountered in the problem classification used in this paper. They are referenced in later sections as they relate to another problem class.

It is intended that this review should be comprehensive for english language literature in the period 1970 to 1980. The author apologizes for any references that were inadvertently omitted, particularly as regards Soviet literature that has not yet been translated. Selected papers that appeared prior to 1970 are cited to provide correlation with previous work.

It should be noted that the massive literature on structural optimization by direct nonlinear programming is not included here. The criterion used was that methods based on nonlinear programming solution of a discretized model of the structure would not be included. Some literature is cited in which a finite dimensional design variable space is employed with a differential equation model of structural response. The literature on design of structures for plastic response has been completely ignored. Finally, the rather special literature on nonconservatively loaded structures has not been included. Even with this restriction in scope, a substantial literature has appeared in the 1970's.

Literature Reviews

The first comprehensive literature review on structural optimization appeared in 1963 by Wasiutynski and Brandt [1]. They present a comprehensive review of literature through approximately 1962, citing 234 references. Sheu and Prager [2] continue with a comprehensive review through approximately 1967, citing an additional 146 references. These reviews consider nonlinear programming and distributed parameter methods and include plastic structural optimization. They are the last comprehensive reviews that have appeared in the growing field of structural optimization.

In 1972, Pierson [3] presented a review on structural design under dynamic constraints, with emphasis on structural vibration, flutter, and transient dynamic response. In 1973, Niordson and Pedersen [4] gave a critical and analytical review of structural optimization literature from approximately 1968 to 1972, dealing with both mathematical programming and analytical methods of structural optimization. McIntosh [5] reviewed applications of optimal control techniques applied to structural optimization, through approximately 1974. Primary emphasis in this review is on structures that are modelled with ordinary differential equations and constraints on static, dynamic, and eigenvalue response.

In a sequence of reviews in the Shock and Vibration Digest, Olhoff [6] and Ranacharyulu and Done [7] review literature on dynamic structural optimization. Olhoff's reviews [6] focus primarily on vibration response, citing literature through 1976. The more recent review [7] treats both vibration and transient response constraints, through approximately 1978. Both of these reviews address numerical and analytical aspects of distributed parameter optimal structural design.

For reviews of related structural optimization problems, the reader is referred to a review by Venkayya [8], where primary emphasis is on finite dimensional structural optimization through 1977, and to a recent review by Weisshaar and Plaut [9] on optimization of structure under nonconservative loading.

Books

Prager [10] presented a series of lectures on structural optimization in 1974 that provides an insightful introduction to the field of distributed parameter structural optimization. His primary emphasis is on optimality criterion. The most substantial text on distributed parameter optimal design is the proceedings of a symposium held in Warsaw, Poland in 1973 [11], edited by Sawczuk and Mroz. This book contains a wealth of valuable papers, many of which will be discussed later in this review. In 1976, Rozvany [12] published a text on optimal design of flexural systems, with primary emphasis on design for plastic response, but with a chapter on elastic response optimization. Haug and Arora [13] present a text on numerical methods of structural optimization, with an introduction to distributed parameter optimization.

Finally, Banichuk [14] has very recently published a monograph (in Russian-English translation pending) that presents primarily analytical methods of elastic structural optimization.

Applicable Distributed Parameter Optimal Control Literature

The massive literature on distributed parameter optimal control is primarily oriented to classes of problems that cannot be related to the mainstream of structural design optimization. For example, much of the distributed parameter control literature is on linear control systems. The structural optimization problem is nonlinear, however, since the design characteristics to be selected appear in the coefficients of the otherwise linear differential equations. When taken as equations involving both state and design variables, the equations of solid mechanics are necessarily nonlinear. Further, most optimal control literature involves evolution equations (initial-value problems) and seeks a feedback, or time dependent solution of the optimization problem. The structural design optimization problem, however, may be viewed as a control problem in which the control is time independent, or open loop. This situation is of little interest to most researchers in the field of optimal control and is given very little attention. Finally, the massive literature on existence theory, sufficiency theory, identification theory, and much of the theory of necessary conditions is of little assistance in either theoretical or computational solution of structural design problems.

Robinson [15] surveyed the distributed parameter control literature, covering over two hundred papers during the period 1966 through 1969. In his review of open loop control techniques, which are most closely allied with the type of problem encountered in structures, Robinson observes that if the necessary conditions can be stated in integral equation form, then a number of Soviet generated techniques are available. However, if the equations are in differential equation form, some methods are available to obtain solutions, but not with any degree of generality. Robinson's closing remark is that the techniques developed apply

only to a very restricted class of problems and, in the general
case, the prospects of determining solutions from necessary
conditions are not good.

Of value in application of necessary conditions to problems
of distributed parameter optimal design is a paper by Lurie [16],
in which elliptic partial differential equations of the kind
encountered in structural optimization are reduced to first order
form and general necessary conditions are derived for nonlinear
optimization. The necessary conditions presented there have
served as the theoretical foundation for some of the structural
optimization work reviewed here. It may be noted that the even
ordered partial differential equations of conservative continuum
mechanics can generally be put in self-adjoint form. Taking
advantage of this fact, it is possible to obtain necessary
conditions by the same techniques employed by Lurie. The
resulting necessary conditions involve an adjoint variable that is
the solution of the same basic differential operator equation, but
with a different nonhomogeneous term.

As noted in the foregoing, the focus of optimal control
theory is almost always on determination of time dependent
functions, rather than purely space dependent design functions. A
notable exception is the formulation of Butkovskiy [17], in which
a more general problem requiring determination of two control
functions is defined, one depending on both time and space
variables, the other depending on only one variable, that can be
taken as a space variable. On page 44 of Reference [17], he
presents a maximum principle characterization of necessary
conditions for such problems, in which one of the necessary
conditions collapses the time variable through temporal
integration, allowing a direct statement of necessary conditions
in the design variables, of course involving integrals over time
of the state variable. This formulation of necessary conditions
has apparently not been exploited, to date. Butkovskiy then
proceeds to treat thermal problems, in which the equations of
state and necessary conditions can be obtained in integral
equation form. He then obtains analytical and numerical solutions
of such problems with major applications in the fields of thermal
systems. While this test is freely referenced in the structural
optimization literature, it is not clear that the techniques
presented therein, which are based on Soviet results prior to
1965, have been fully exploited.

In a recent review [18] and text [19] Ray and co-workers
address distributed parameter optimal control, with emphasis on
chemical engineering applications. While the primary emphasis is
on dynamic systems, some of the methods may be applicable to
distributed parameter structures.

The literature on distributed parameter optimization, since 1970, has taken a turn toward a more abstract formulation of the problem. Lions and co-workers have introduced function space techniques [20,21], for optimization of linear and nonlinear systems. Their necessary conditions generally result in variational inequalities, a field that has developed quite rapidly during the past decade. While the applications presented by Lions are rather abstract, the basic ideas employed should have considerable applicability in the field of mechanical system optimization. Full advantage of these techniques has clearly not been taken, to date.

A paper by McGlothin [22] continues the function space setting, but with somewhat less abstraction. A nonlinear dynamic system is treated in a control setting, for which the control variable is time dependent. The generality of the formulation, however, may allow application to structural problems.

Excellent books on theoretical and computational aspects of optimal control theory include the landmark texts of Pontryagin et.al. [23], Hestenes [24], Bryson and Ho [25], and Cea [26]. Pontryagin [23] and Hestenes [24] present very general necessary conditions of optimality for intial-value problems (ordinary differential equations) arising in optimal control theory. Bryson and Ho [25] present iterative numerical methods for direct solution of open loop optimal control problems. Their methods are extended for direct treatment of distributed parameter structural optimization problems by Haug and Arora [13]. Cea [26] presents necessary conditions of optimality and numerical methods that may be applicable for distributed parameter optimal design.

A different approach that has been pioneered by Soviet researchers such as Dubovitskii and Milyutin involves an abstract vector space setting, operator theory for representation of the state of the dynamic system, and a variety of formalisms for treating constraints. The abstraction ranges from purely set theoretic, cone ordering techniques for statement of constraints, to functional inequalities and utilization of Frechet differentiation theory. This theory is thoroughly treated in texts by Pshenichnyi [27] and Ioffe and Tihomirov [28]. These abstract methods for treatment of distributed parameter optimization problems have not been exploited to date. While the degree of abstraction is high, it appears that operator theoretic properties of continuum systems can be exploited to sharpen the abstract theory and obtain workable necessary conditions and, perhaps more important, a valid mathematical motivation for computational techniques.

Zolezzi [29] published an outstanding paper that treats elliptic state equations in which the coefficients of the differential equation, which is written in divergence form, are allowed to depend upon the control variable. This class of problems represents a valuable model of equations of continuum mechanics and allows treatment of the design problem. The motivation in Zolezzi's paper is stochastic control, so he also treats parabolic problems that are related to heat transfer problems of a continuum. This paper uses functional analysis methods of weak convergence and existence theory of partial differential equations to obtain a set of necessary conditions for the problem of minimizing an integral, subject to the above mentioned differential equations and boundary conditions. The formulation, however, does not include either functional or pointwise inequality constraints.

BUCKLING OF COLUMNS, PLATES, AND SHELLS

Column Buckling

Historically, one of the first optimal structural design problems addressed was treated by Lagrange in 1770 [30] and later by Clausen in 1849 [31]. For an accessible summary of these initial studies, the reader is referred to Reference [32] (Vol. I, pp. 66-67 and Vol. II, pp. 325-329, respectively). The first modern treatment of this problem, which sparked substantial interest in optimization by the mechanics community, was presented in an elegant paper by Keller [33]. Keller treated the problem of maximizing the fundamental buckling load for a pinned-pinned column, under the condition that the total volume of the column is specified. In his fundamental paper [33], Keller addresses both the question of optimum tapering of the column and selection of the optimum cross-sectional geometry. He employs a directional derivative approach to obtain necessary conditions of optimality and obtains closed form solutions. In a subsequent paper, Tadjbakhsh and Keller [34] treat a variety of boundary conditions for column optimization, using the analytical method Keller had earlier been presented [33]. In later related developments, Farsahd and Tadjbakhsh [35] and Gajewski and Zyczkowski [36] treat more general load conditions. It is interesting to note that since no lower bound on cross-sectional area or upper bound on stress is specified, zero cross-sections (singularities) occur in the designs obtained. In 1966, Keller and Niordson presented a similar analysis of column optimization, seeking the tallest-column that will remain stable under its own weight [37].

The earlier variational approach of References [33] and [34] was employed in this analysis. Barnes [38] recently extended an analysis of the optimum column shape problem of Keller, to obtain sufficient conditions of optimality for certain boundary conditions. Alblas [39] extended the basic Keller formulation of Reference [33] to treat column cross sections that are not convex or simply connected.

In a sequence of papers [40-43], Taylor and Prager develop a variational formulation of the problem of column optimization, employing stationarity of the Rayleigh quotient to obtain optimality criteria for fixed volume and maximum buckling loads, including lower-bounds on cross-sectional areas. They prove sufficiency conditions of optimality for sandwich columns. In Reference [43], Taylor considers prestress. In related developments, Huang and Sheu [44] and Frauenthal [45] treat optimization of clamped free columns, with a variety of cross-sections, including tubular. Popelar [46,47] has also employed potential and complementary energy approaches to obtain optimality criteria for column optimization. Hu and Kirmser [48] have presented a numerical method for solving optimality criteria of columns. Adali [49] treats a related problem in which optimal nonhomogeniety of material is determined.

There is a subtle difference in the technical approach to developing necessary conditions of optimality in References [33,34 and 37], where a directional derivative of the buckling eigenvalue with respect to design is calculated, and in References [40-45], where a first variation of the Rayleigh quotient is employed to calculate a Frechet derivative of the eigenvalue, which is then used with a Lagrange multiplier technique to obtain necessary conditions of optimality. In the second approach, at least a Gateaux derivative of the eigenvalue with respect to design is assumed to exist.

Haug [50] treats the problem of minimum weight design of a clamped-free column with a lower bound on the fundamental buckling load and an upper bound on stress. The problem is formulated and solved analytically using the Pontryagin maximum principle and numerically using a steepest descent method of optimal control theory. An alternate steepest descent numerical method is applied for solution of the same problem in Reference [51]. Hornbuckle and Boykin [52] demonstrated equivalence of the problems of minimum weight with an eigenvalue bound and maximum fundamental eigenvalue with weight bound. Dinkoff [53] has demonstrated that for a simply supported column, a design with linear taper is almost as efficient as the fully optimum design.

As a result of the zero cross- sections predicted for columns under various support conditions in References [33,34 a nd 37] several authors considered an alternate formulation of the optimum column problem [54-58], in which the location of the singular points was considered as a design variable. Masur [55] and Olhoff and Taylor [58] determine best location of connections and singularities, whereas Farshad [54] and Mroz and Rozvany [56,57] determine optimum support locations.

In a fundamental paper, Olhoff and Rasmussen [59] show that a repeated eigenvalue occurs for a clamped-clamped column at the optimum design, for certain values of a lower bound on the cross- sectional area. In this paper, the earlier result of Tadjbakhsh and Keller [34] is shown to be in error. They then present a variational formulation of the problem of maximum buckling load, using the first variations of the Rayleigh quotient, with constraints appended using Lagrange multipliers. It is suggested in this paper that the error made by Tadjbakhsh and Keller is one of requiring continuous first derivatives of the buckling mode, whereas at singular points that arise, the slope must in fact be allowed to be discontinuous. Olhoff and Rasmussen iteratively construct optimum designs for the clamped-clamped column and show that no singular value of the cross- section occurs, even if the lower bound cross- sectional area is taken as zero. They show that below a certain value of the lower bound on cross- sectional area, the lower bound constraint will not be active.

Banichuk [60] and Olhoff [61] and co-workers examine the nature of singularities of optimal columns and obtain transversality conditions that must hold at singular points. Komkov and Haug [62] present an argument, using nonlinear analysis of buckling, that indicates that zero cross-sections will generally not occur if a refined optimization formulation is employed. For a detailed technical discussion on column optimization, the reader is referred to Olhoff's paper [63] in these proceedings.

In very recent literature, Masur and Mroz [64-66] have developed an alternate form of optimality criteria for singular problems, such as those arising when repeated eignvalues occur in an optimization problem. They use an optimality criterion that requires that both eigenvalues cannot increase in any feasible direction of design modification. Using this necessary condition, they obtain optimality criterion that reduced to that of Olhoff and Rasmussen [59].

A simplified model of the clamped-clamped column, using discrete torsional springs to represent stiffness of the structure, is employed by Prager and Prager [67] to show that even with a simplified model, repeated eigenvalues occur at certain optimal designs. Prager employs a variable support stiffness that shows that the nature of the optimal design depends on the value of torsional rigidity of the support. A similar problem for the clamped free column, with a flexible support, is treated by Banichuk [68], but in this flexible support configuration no repreated eigenvalues occur. The occurance of repeated eigenvalues at optimum column designs for a variety of conservative and nonconservative loads has recently been developed by Blachut and Gajewski [69].

Haug, Choi, and Rousselet [70-74] use a directional differentiation theory and abstract optimization theory [27] to study structural optimization problems with repeated eigenvalues. They show [71-74] that a repeated eigenvalue is in general not Gateaux differentiable, much less Frechet differentiable. They demonstrate existence of a directional derivative and employ directional derivative ideas with modern optimization theory [27] to obtain a set of necessary conditions for optimal design problems [70,71] in which repeated eigenvalues do occur. Several examples are presented in References [70 and 71] to show that the clamped-clamped column is not the only structural optimization problem in which repeated eigenvalues occur at an optimum design. Examples are presented in Reference [71] to show that in general, erroneous results may be expected if Lagrange multiplier techniques are used formally to derive optimality criterion for problems in which repeated eigenvalues occur. Using directional derivative theory developed for repeated eigenvalues in Reference [74], it is shown [71] that for the clamped-clamped column problem treated by Olhoff and Rasmussen [59] and Prager [67], if the design is symmetric about the midpoint of the column and if the first two repeated eigenvalues correspond to symmetric and anti-symmetric modes, then the repeated eigenvalues are indeed Frechet differentiable. The variational form of optimality criterion, with Lagrange multipliers, used by Olhoff and Rasmussen [59] can then be justified, provided a certain set of eigenvectors corresponding to the repeated eigenvalue is used. As shown by other examples in Reference [71], however, this is not the general case. In fact, if symmetry of the design of the clamped-clamped column is destroyed by other constraints, then the Lagrange multiplier technique may be invalid. These results, supported by the analysis presented by Masur [66] indicate that great care is required in development of optimality criteria and for problems in which repeated eigenvalues occur.

As a final note on the column buckling problem, it is important to observe that other substantial technical difficulties may arise in design optimization in which repeated eigenvalues occur. It has been shown by Thompson and co-workers [75-78] that imperfection sensitivity may be associated with buckling of structures if repeated eigenvalues occur. They demonstrate that catastrophic failure may result if optimality criterion that imply repeated eigenvalues are adopted. Thus, to assure a safe design, it appears that future effort should be devoted to investigating post buckling behavior of an optimized design in which repeated eigenvalues occur. In order to assure imperfection insensitivity for practical design, additional constraints may be placed on the branching character of the buckling loads in post- buckling behavior. This complex subject will require great technical care and substantial development in optimal design theory. As indicated by the examples presented in Reference [71], occurrence of repeated eigenvalues at optimum designs is not limited to the clamped-clamped column. Vibration examples are also given that display this character and heuristic optimality criteria discussed by Thompson and Hunt [78] illustrate that in order to avoid catastrophic nonlinear response when repeated eigenvalues occur, future efforts should be devoted to adding constraints to the design problem to insure that imperfection sensitivity does not occur even in the presence of repeated eigenvalues. Important initial contributions to this problem have been made by Masur [79] and Roorda and Reis [80].

Other Buckling Problems

Only a few papers have appears in the literature concerning design of plates and shells for buckling behavior. Frauhenthal [81] treated the problem of optimization of an axisymmetic circular plate, subject to axisymmetric radial loading. Under these conditions, the governing differential equations reduce to ordinary differential equations and problems involving both sandwich and solid cross- section geometries are treated, including stress constraints on the pre- buckled structure. A somewhat more complicated problem of optimal thickness determination for a cylindrical shell that is loaded by external pressure is treated by Andreev and co-workers [82]. A Pontryagin type maximum principle is employed to obtain optimality criterion and an iterative algorithm is used to construct numerical solutions. Weight Optimization of a reinforced spherical shell under external pressure and the associated buckling constraint is treated by Manevich and Kaganov [83]. Approximations are employed to determine local and global buckling loads and elementary optimization methods are employed to construct solutions.

Zyczkowski and Kruzelecki [84] also formulate shell buckling optimal design problems involving external pressure and bending. They use a variational formulation to obtain optimality criterion and solve two example problems. Kruzelecki [85] extends these results for combined loading and variable shell cross sectional shapes and material distribution. Rikards [86] has recently shown convexity of selected optimization problems involving shells, which may shed light on sufficiency conditions.

Use of finite element approximations of stiffened shell structures for optimization against buckling has been made by Pappas and Allentuch [87,88]. Simitses and Sheinman [89] and Kunoo and Yang [90] have used related methods to treat shells with imperfections and stiffeners.

Finally, Tadjbakhsh and Farshad [91] formulate an arch optimal design problem in which the arch is presumed to support no bending moment or shear. A variational method is used to formulate optimality criteria and examples exhibiting singular cross sections are solved. Blachut and Gajewski [92] extend earlier work by Budiansky, Frauenthal, and Hutchinson [93] in optimization of Funicular arches. They use the Pontryagin Maximum Principle to derive optimality criteria and show that repeated eigenvalues occur for some optimal designs. Amazigo [94] has treated optimization of circular arches for snap-buckling.

A limited number of papers treat column buckling in conjunction with other structural performance constraints or other modes of structural deformation. These papers are discussed in Section 6 of this paper.

VIBRATION OF BARS, BEAMS, AND PLATES

Stimulated by the Keller papers on optimization of columns [33,34,37], an intense activity in optimization of structural elements for vibration was initiated by Niordson [95] and Turner [96]. In his fundamental paper [95], Niordson uses the directional derivative techniques associated with the Rayleigh quotient for vibration of a simply supported beam, to develop necessary conditions of optimality for distribution of a fixed amount of material to maximize fundamental frequency. He uses an iterative technique to solve a specific example problem, which exhibits singular sections. Turner [96], on the other hand, deals with minimization of total mass of a bar in axial vibration, leading to a second-order differential equation formulation for structural vibration. He uses a discrete element approximation method to construct numerical solutions.

Vibration of Bars

Taylor [97] reformulated the Turner problem of axial
vibration of a bar and employed a variational approach to obtain
necessary conditions for maximization of the fundamental
eigenvalue, subject to a constraint on the volume of material
available. The method employed by Taylor is closely related to
his development of optimality criterion for column buckling [40].
In a subsequent note, Taylor [98] extends his formulation for
axial vibration optimization to include a lower bound on
cross-sectional area. He uses the basic variational formulation
presented by himself and Prager [42], including lower bound
constraints, that is applicable to both buckling and natural
frequency problems. The axial vibration problem was studied by
Sheu [99], using piecewise uniform structural elements to
construct the rod. He derives optimality criterion and
numerically constructs solutions. Sippel and Warner [100] present
an extension of the axial vibration problem to a multi-element
structure with several bars connected in a series to a set of
discrete masses. They use the variational formulation of Prager
and Taylor [42,98] to obtain optimality criteria. They construct
optimal solutions with variable cross-section, within each of the
structural elements, and subsequently determine an optimum design
with constant cross-sections in each of the elements. They
conclude that the constant cross-section of the optimum design is
competitive with designs obtained by tapering the structural
members. Quite recently, Cardou and Warner [101] have developed a
more general and rigorous set of necessary conditions of
optimality for the axial vibration problem and have applied their
method to problems of beam and multi-element structural
deformation.

As a final note on the axial vibration problem, Miele and
co-workers [102] apply a gradient projection technique of optimal
control theory to numerically construct solutions of the
problem of mass minimization of an axial bar, subject to
constraints on the natural frequency and cross-section.

Vibration of Beams

Following the initial paper of Niordson on optimal vibration
of beams [95], Brach [103] treats the problem of maximization of
fundamental frequency for beams in which the moment of inertia of
the beam cross sectional area is proportional to the cross
sectional area. Using a variational formulation, Brach constructs
closed form solutions and finds that for certain boundary
conditions no extremal fundamental frequency may exist. Brach
[104] later shows that for this class of structures, the problems

of maximization of fundamental frequency with a given amount of material and the problem of minimization of mass with a constraint on natural frequency are not equivalent. The question of existence of solutions of optimization problems with eigenvalue constraints is similarly treated by Vepa [105]. In a related paper [106], Brach presents an alternate method of constructing approximately optimum designs, using a characteristic vibration shape technique and studies vibration of beams and bars.

McCart, Haug and Streeter [107] present a formulation for structural optimization with lower-bound constraints on beam cross-section and on natural frequency, using a gradient projection technique of optimal control theory. They apply the theory to optimization of a portal frame. Since the technique used does not necessarily control the fundamental frequency, numerical computation encounter difficulty when a shift of fundamental mode from antisymmetric to symmetric mode occurs. This behavior is now suspected to be similar to the coalessence of eigenvalues encountered in the buckling optimization problem presented by Olhoff and Rasmussen [59]. A similar portal frame application of the method of Cardou and Warner [101] was subsequently presented. The numerical method employed in Reference [107] was subsequently refined, using the gradient projection method of Reference [51], to develop a more effective computational technique that is applied to vibration problems in Reference [108]. A related numerical treatment of optimization of beams of minimum weight, subject to constraints on natural frequency and a lower bound on cross-sectional area, is presented by Pierson [109]. He employs a gradient projection technique of optimal control theory for beam optimization, with a variety of boundary conditions. Considerable numerical experience is presented. Kamat and Simitses [110] present a numerical method for natural frequency optimization of a beam on an elastic foundation. They use a finite element discretization and solve the problem using a nonlinear programming method. Foley and Citron [111] reduce the beam optimization prolem to a finite dimensional approximate problem, using Ritz expansion.

The optimality criterion method initially presented by Niordson for beam optimization [95] is extended by Karihaloo and Niordson [112,113] in treating the problem of maximizing fundamental frequency of beams, with a constraint on the amount of material available, but no lower bound on cross-sectional area. They treat both similar cross-sections and cross-sections of fixed width and height. Similar results are obtained by Seiranyan [114] for the problems in which no lower bound on cross section is imposed.

Warner and Vavrick [115] first treat the problem of structural optimization with constraints on several natural frequencies, using a variational formulation. Olhoff [116,117] addresses the problem of maximizing higher-order eigenfrequencies of beams, first without a lower bound on cross- section [116] and subsequently with lower bounds on cross- sectional area [117]. In a related development, Troitskii [118] seeks to maximize the difference between natural frequencies in the eigenspectrum associated with an optimum design. He develops optimality criteria, using an axial bar as an example.

Motivated by singular behavior of optimum designs with no lower bound on cross- sectional area, Mroz and Rozvany [119,120] seek the optimum placement of singularities for extremum eigenvalues. Singularities of optimum beams in vibration are addressed more comprehensively by Olhoff and Niordson [121] and in considerably more detail in Reference [122]. They show that there are numerous relationships between the problems of maximizing the nth natural frequency and optimal placement of singularities. Szelag and Mroz [123] recently treated a related class of beam optimization for vibration, with variable support locations. They show that, in such problems, repeated eigenvalues can occur.

As final observations on beam optimization for natural frequency, Kamat [124] uses a variational formulation of the beam optimization problem to investigate the effect of shear deformation and rotary inertia in optimum designs. In constructing solutions, he employs the finite element computational technique presented in Reference [110]. Gupta and Murthy [125] extended the beam vibration optimization problem to determine optimum non- homogeneous material distribution. More recently, Smirnov and Troitskii [126] have extended the conventional formulation to treat curved beams.

Torsional Vibration of Shafts

Weisshaar formulates a problem of optimization of a torsional shaft to minimize weight, subject to a condition that the lowest natural frequency is fixed [127]. Variable wall thickness of shaft is to be selected to minimize weight of the structure, subject to the frequency constraint. He employs necessary conditions of optimality given in References [128] and [129] and numerically constructs optimum designs. In recent papers, Vavrick and Warner [130,131] investigate the problem of minimum mass design with torsional frequency and thickess constraints, in considerable detail. They employ the Pontryagin maximum principle to obtain optimality criteria and construct pieced extremals that lead to optimum designs. They show that, for certain values of upper and lower bounds on thickness, discontinuous optimum designs can arise, even though there is no discontinuous forcing function or any other form of discontinuous input to the problem. They also show [131] that the problem of minimum mass with bounds on eigenvalues and the dual problem of maximization of eigenvalues for a given mass are equivalent. Elwany and Barr [132, 133] treat a similar class of torsional vibration problems, using a Lagrange multiplier, variational method to obtain optimality criteria.

Minimum weight design problems associated with shafts of high speed machinery and with turbine disks are treated in References [134] to [137]. Bending dynamics of shaft and turbine blades are incorporated in the optimization problem and bounds on natural frequency are imposed. Optimal control methods and methods of nonlinear programming are employed in obtaining solutions.

Vibration of Plates

The first treatment of variable thickness plate optimization for natural frequency response appears to have been presented by Olhoff [138] in extending the method of Niordson [95] to an axisymmetric plate for which the eigenvalue problem is an ordinary differential equation. In this paper, Olhoff obtains singular (zero cross-section) plates as optimum. The method of obtaining optimality criterion is the variational method presented earlier by Niordson [95].

From a totally different approach, Armand and Vitte [128] and Weisshaar [129] use optimality criterion of the Pontryagin maximum principle type for development of necessary conditions of optimality for a variety of plate geometries and forcing functions. The primary emphasis of their work is to develop optimality criterion for minimum weight design of structures with constraints on natural frequencies. For the general plate problem, this set of necessary conditions is terribly complex and reduces to a computationally feasible set of necessary conditions only for idealized problems. These results were extended by Armand [139,140] for treatment of shear panels and a variety of problems arising in aerospace applications. Selected problems are solved in closed form, but the complexity of necessary conditions does not lead to a general computational algorithm for more complex problems of plate bending. In a more recent paper Armand [141] presents a numerical method of constructing solutions of variable thickness plate problems for minimum weight, subject to eigenvalue constraints. In this work he shows that there is a tendancy for ribs to develop in a plate, to enchance stiffness and increase natural frequency.

Olhoff [142] extends his variational formulation and necessary conditions for the problem of maximizing the fundamental natural frequency of a rectanglular plate, subject to the condition that a given volume of material is available for a solid plate. In this paper, he writes necessary conditions and numerically constructs solutions that vary smoothly and have zero cross sections appearing. He observes that these designs should be considered as only local optimum, since one would expect ribs to develop to enhance stiffening of the structure, as noted by Armand [141]. Olhoff subsequently treats the problem of singularities and formation of stiffeners in optimal design of plates [143] using optimality criteria. In very recent work, Olhoff and co-workers [144-146], show that indeed ribs do develop in a natural way, if certain upper and lower bounds are placed on plate thickness. It is interesting to note that Vavrick and Warner [130] found the same basic criterion for occurrence of discontinuities in design of torsional vibrating shafts.

Lurie [147] shows that the Prager optimality criterion [42] is applicable to plate optimization, but that if one uses this formulation presuming considerable smoothness is the solution of the vibraton equations, only smooth designs will occur. He argues that the only way to treat problems with rib stiffeners is to model the structural response with explicit rib stiffeners attached to the variable thickness plate and to determine the optimal location of the stiffener and their geometric characteristics. Seiranyan [148] considers the problem of maximization of fundamental natural frequency for fixed volume of material for an axisymmetric circular plate and finds no ribs.

From a direct gradient direction approach, using function space derivatives of eigenvalues, Haug and co-workers [51,108,149] use a gradient projection computational technique to minimize the weight of plates with constraints on natural frequency and cross sectional thickness. As the cross sectional lower-bound is decreased, simply supported plates similar to those presented by Olhoff [142] are obtained. Use of the gradient projection algorithm has been compared with results obtained using discrete finite element formulation of the problem [150] and virtually identical results are obtained.

Several special problems concerning vibrating plates are treated in recent literature. Foley [151] uses the ε method for numerical optimization of a square plate, with constraints on one

or more natural frequencies. Rammerstorfer [152] determines
fields of initial stress in a plate to maximize the fundamental
frequency. Banichuk and Mironov [153] include the effect of
interaction of a plate and on ideal fluid in plate optimization.
Bert [154] and Rao and Singh [155] determine properties of
composite materials to optimize vibration characteristics of
plates. In a sequence of papers on optimization of plate
vibration. Carmichael [156,157,158] shows that optimality
criteria derived from a control formulation of the problem may be
singular.

Virtually no results exist on optimization of shell
structures for specified natural frequency. One paper that
presents an extension of beam optimization idea is to a class of
shells of revolution is Reference [159] where optimality criterion
are derived but no examples are solved.

DEFLECTION AND COMPLIANCE OF BEAMS AND PLATES

Beams with Deflection Constraints

In an early paper, Barnett [160] seeks to minimize the weight
of a beam, subject to the condition that deflection at a specified
point on the beam is bounded. He employs a Castigliano theorem to
characterize displacement at the given point and a variational
approach to obtain necessary conditon of optimality. Several
subsequent papers [161-164] treat the problem of minimum weight
design of beams with deflection constraints at specified points.
Dixon [161] applies Pontryagin's maximum principle for a
cantilevered beam with extreme displacement at its end. Huang and
Tang [162] use an extension of the Barnett technique with the
principle of virtual work, applying their necessary conditions to
both statically determinate and statically indeterminate beams.
Prager [163] employs the principles of stationary mutual potential
energy to obtain optimality criterion for beams with given
displacement, with displacement governed by applied load and
thermo-elastic considerations. Chern [164] treats an extension of
the Barnett problem by the same method, but allows the applied
load to depend on design. Cantu and Cinquini [165] present an
iterative method for beam optimization with prescribed
displacement or rotation at a given point.

The problem of minimum weight design of beams with a
constraint on the maximum deflection, occurring at an unknown
point, was first treated by Haug and Kirmser [166], using the

Pontryagin maximum principle to construct solutions by pieced extremals. Application to simply supported and cantilevered beams with fixed width and variable depth are presented, each having a smooth cross section. The same minimum weight design problem, with a constraint on the absolute value of displacement is treated numerically by a gradient projection method in Reference [108], with a load that leads to bicurvature. Results shown there indicate discontinuous cross sections at the optimum design, a result previously obtained Haug [167], with more difficult computations arising from the Pontryagin maximum principle. The same basic method, employing the Pontryagin maximum principle for optimization of structural elements is treated by DeSilva [168].

Shield and Prager [169] present a variational formulation of the problem of minimum weight design with constraints on deflection, using direct variational approach. They prove a principle of stationary mutual potential energy, associated with two kinematically admissible displacement states for the system. One of the displacement states is calculated to be associated with a unit load at the point where displacement is to be constrained. The principle then gives the displacement at this point, under a given load. This variational formulation of the displacement constraint allows the authors to extend the variational method of Prager [42] to the problem of optimization of structures with displacement constraints. Prager and co-workers [170,171] apply and extend the basic principle developed in Reference [169]. They demonstrate that the optimality criterion obtained through the stationary mutual complementary energy approach is sufficient for sandwich beams and necessary for other beam cross sections. In a subsequent paper, Huang [172] applies the Shield-Prager technique under a variety of loading conditions. Dafalias and Dupuis also apply the Shield-Prager method [173-175], using an iterative technique to locate the point at which maximum deflection occurs.

In independent developments, Bhargava and Duffin [176] and Simitses and Kotras [177] develop optimality criterion and solve example problems for cantilevered beams with a constraint on deflection at the end. Bhargava and Duffin [176] employ duality theory and variational techniques to develop a rigorous optimality criterion that is subsequently applied to a cantilevered beam on an elastic foundation, with fixed height and variable width. Simitses and Kotras [177] on the other hand, use dirac-delta functions to represent displacement at the given point and subsequently bound displacement and seek a minimum weight design. The same problem of minimum weight design of a beam on an elastic foundation, with displacement specified at a known point, is treated by Distefano and Todechini [178] using an invariant imbedding technique.

The problem of minimum weight design of beams with bounds on maximum deflection, occurring at an unknown point on the beam, is treated in References [179-181]. Huang [179] employs a Castigliano theorem to represent deflection at a specified point in terms of design and seeks to minimize weight, subject to the condition that the deflection constraint is satisfied at every point along the beam. He treats the point of maximum deflection as a variable in the formulation and determines it as part of the solution of the problem. Komkov and Coleman [180] use a similar argument, but with distribution theory to present deflection at selected points along the beam and subsequently impose the deflection constraint at all points. They develop optimality criterion for both beams and plates. Cinquini [181] uses a direct variational approach to adjoin the deflection constraint to the weight function that is to be minimized. He allows for discontinuities in adjoint variable and develops a set of necessary conditions for the minimum weight deflection problem that resemble those obtained for the same problem by the Pontryagin maximum principle in Reference [166]. He then carries out example calculations with statically determinate beams.

Armand [182] presents a general analysis of optimization of beams with equality and inequality constraints involving deflection and other response measures. Banichuk [183] formulates structural design problems with constraints on deflection over the entire beam as a min-max problem. He obtains optimality criteria for problems with deflection constraints and for other problems, using necessary condition of min-max optimization theory.

Plates with Deflection Constraints

Hegemier and Tang [184] treat the problem of plate optimization with bounds on displacement at several specified points. They employ a variational method to obtain optimality criterion and discretize the plate using finite element calculations to construct numerical solutions. Simply-supported and clamped sandwich plate examples are solved. Erbatur and Mengi [185,186] employ the principle of stationary mutual potential energy to derive optimality criteria for plates, with displacement at a given point constrained.

Using a gradient projection function space optimization algorithm, Haug and co-workers [51,108,149] solve the problem of minimum weight plate design, with constraints on deflection at all points in the plate and other response measures. Armand and Lodier [187] use a finite element formulation for optimization of a rectangular plate with a single deflection constraint. In their results they note development of apparent ribs in the design, to enchance stiffness. They also discuss the regularity of displacement functions that must be allowed in the finite element formulation to enable ribs to form in the iteratively obtained design. In an impressive sequence of papers, Banichuk and co-workers [188,191] develop a theory and computational method for minimum weight design of plates with a deflection constraint over the entire plate. Banichuk first [188] treats the design problem with a deflection constraint at a single point and obtains asymptotic estimates of the solution. He then [189] uses the min-max formulation of Reference [183] to obtain optimality criteria and a numerical method. More recently [190,191], he uses a p-norm estimate of maximum displacement and stress, hence reducing the min-max problem to an approximate minimum problem. Rectangular plates with a variety of constraints are optimized in Reference [191].

As a final note on plate design with deflection constraints, a paper investigating the influence of rib stiffeners is considered. Simitses [192] constructs a non-optimum circular plate with ribs and compares its displacement with that of a plate that is optimized for deflection. He finds that the ribbed design yields smaller deflection, so the non-ribbed optimum must be only a local optimum.

Beams and Plates with Compliance Constraints

Work done by the applied load system in undergoing displacement (compliance) is often treated as a measure of stiffness of the structure. This compliance measure is attractive from an optimization point of view, since it is a global measure of structural performance; i.e. it can be written as an integral over the entire element.

Prager and co-workers [193-196] make extensive use of compliance constraints in formulating and obtaining optimality criterion for numerous structural optimization problems. They obtain optimality criterion using the basic method in Prager's landmark paper with Taylor [42] and specifically treat problems of minimum weight with piecewise constant section properties [193] beams and frames [193-196], and Michell trusses [195]. Huang [197,198] uses the same basic approach for minimum weight design of beams and circular plates. A similar set of problems is addressed by Chern and Prager [199, 200,171] using optimality criteria associated with the compliance constraint. Save [201] formulates a more general problem, including many design variables and employs compliance constraints to relfect stiffness and deflection of the structure.

Masur, in a sequence of papers [202-204] develops a consistent optimization theory for optimization of beam and plate structures for compliance. He first [202] develops a rigorous set of optimality criteria for material distribution for minimum volume, with constraints on compliance. He then [203] treats the practical problem of selecting among available structural sections to optimize design with compliance constraints. Finally [204], he carefully studied the problem of maximizing stiffness (minimizing compliance), specifically studying the problem of a circular plate with uniform pressure. He derives optimality criterion and shows that hinges occur in the optimum design. He then analyzes the effect of discontinuities associated with discreteness and shows that irregularities may arise in the optimum design, even though no irregularity in the load is specified. Masur expands upon considerations of singularity and non-stationarity of optimal designs in Reference [66].

Mroz and Rozvany [205,56] treat optimization of a variety of structural elements with compliance and other global performance constraints. Mroz [205] shows that in some problems global theorems may be obtained, whereas in others only local conditions are possible. Mroz and Rozvany [56] then seek to determine optimum location of supports with constraints on compliance. Reiss [206] develops optimality criterion for maximum compliance associated with an axisymmetric plate. Mroz and co-workers [207,208,209] also treat compliance optimization problems in which load distribution is determined as the design variable.

Miscellaneous Problems

Chern and Prager [210] formulate a problem of optimal thickness of an axisymmetric rotating disk, with a bound on displacement of the edge. They use the Shield-Prager [169] principle of stationary mutual potential energy to derive necessary and sufficient conditions of optimality and construct numerical solutions. Chern [211] treats a related problem of a rod spinning about a line perpendicular to its axis, with thermal effects and a constraint on deflection at the end. Fuchs and Brull [212] formulate a problem of optimizing support location for a beam, with an objective minimizing maximum deflection. They then use a strain energy approximation for the min-max deflection problem and solve examples.

Rossow and Taylor [213] derive optimality criteria for minimum compliance of a plane elastic sheet and use finite element methods to calculate solutions. Mroz [214] treats composite material selection for minimum compliance design of structures, with application to plates.

DYNAMIC RESPONSE

Forced Steady State Oscillation

As an extension of structural optimization with bounds on natural frequencies, several authors have investigated design optimization of structures with harmonic steady-state input. Icerman [215] formulates a problem of minimizing virtual work of the applied load for steady-state forced oscillation of rods, beams, and trusses. He uses a variational formulation to obtain optimality criterion, from which he obtains analytical solutions for axial motion of a rod, bending of a beam, and response of a truss. Mroz [216] formulates a general problem of structural dynamic optimization under harmonic load. He also uses a dynamic compliance, which is the work done by the applied forces during one cycle of motion as his optimality criterion and develops a variational formulation of the problem and optimality criterion. His variational necessary condition is applicable to broad classes of structures and is proved to be both necessary and sufficient for sandwich structures. He applied the theory to sandwich and solid beams and plates.

Plaut [217] formulates the problem of minimum weight design of structures under periodic loading, subject to the constraint that deflection of a specified point of the structure is prescribed. He extends the principle of stationary mutual potential energy of Shield and Prager [169] and derives optimality criterion for elastic sandwich beams. Plaut subsequently [218] uses a Rayleigh Ritz technique to approximate optimal design of structures for for dynamic response.

Huang [219] expands on the periodic loading problem treated by Plaut [217], to include the effect of inertial force. He also employs the principle of stationary mutual potential energy and obtains optimality criterion. The development is applicable to general elastic structural systems. Rockenback [220] formulates a generalization of the problem treated by Plaut [217] to include deflection constraints under harmonic exitation for a specified range of exitation frequencies. He treats the problem as a nonlinear programming problem, using penalty function techniques to construct numerical solutions.

A related class of optimal design problems involves selection of support parameters to optimize dynamic response of a structure that is subjected to harmonic or statistically defined ground input motion. Typical treatments of such isolation problems may be found in References [221-223]. For a detailed review of literature regarding optimization of support substructures for dynamic isolation, References [224] and [225] may be consulted.

Transient Dynamic Response

Brach [226,227] appears to be the first to address transient response optimization of beams with suddenly applied load, seeking to minimize an integral measure of impulse response, subject to the constraint that a fixed amount of material is available. A gradient projection numerical method is employed to numerically solve the problem, after a discrete model of the beam is constructed.

Plaut [228] formulates a minimum weight beam design problem under general distributed transient load and subject to an upper-bound constraint on the deflection of the beam. He derives upper-bounds for displacement response to input and finds a design to minimize weight, subject to the condition that the upper bound does not exceed the deflection limit. He notes that the question of precision of the upper bound is unanswered and that conservative designs may in fact be generated. He gives applications to cantilever beam design optimization. A problem of similar level of generality is formulated by Pochtman [229], who seeks to minimize weight, subject to constraints on maximum displacement response over time and domain of the plate being considered. He treats general transient load and incorporates constraints on natural frequency. The design of the plate is characterized by a finite number of design parameters and a random search technique is used to numerically construct solutions.

In the first general computational approach to optimization of structures subject to transient dynamic response, Fox and Kapoor [230] present a technique based on an upper-bounding idea, similar to that employed by Plaut [228]. Fox and Kapoor use a finite element shock spectrum response estimate that gives upper bounds on peak displacement to shock loading. These bounds are stated in terms of eigenvalues and eigenfunctions of the finite element matrices, which are differentiated to obtain gradient information needed for feasible direction optimization. Two numerical examples involving truss frame structures are presented. The significance of this paper should not be under-estimated, since it represents the first step in development of a general approach to structural optimization with transient dynamic response constraints.

After a lapse of six years following the Fox and Kapoor paper [230], two papers presenting substantially more general approaches to the problem appeared almost simultaneously [231, 232]. Cassis and Schmit [231] seek to minimize weight, subject to constraints on stress and displacement at nodes within the finite element model of the structure. They employ approximation concepts to obtain expressions for derivatives of peak displacement and stress response with respect to design. They then employ an exterior penalty function, sequentially unconstrained minimization technique for numerical solution of design problems. They present examples involving frame structures with from three to twenty-one members. They also generate plots of feasible design regions for simplified structures that illustrate very complex constraint boundaries and, in some cases, disconnected feasible regions. Using an equivalent integral functional constraint formulation for inequality constraints on displacement and stress within the structure, Feng, Arora, and Haug [232] develop a formulation for optimization of structures under general transient dynamic response. They use a gradient projection technique and a finite element numerical analysis method for optimization of structures. No approximations are made in modeling of the structure, other than selecting the number of modes to be employed in eigenfunction expansion of the dynamic solution. Example problems involving up to seven truss frame elements are solved and numerical results presented. In subsequent discussion [233] Feng, Arora, and Haug treat example problems presented in Reference [231] using the gradient projection technique of Reference [232]. It is shown [233] that results obtained with the sequential unconstrained minimization technique of Reference [231] are at best local optima and substantially conservative.

The gradient projection technique employed in Reference [232] was extended to a distributed parameter function space formulation [234,235] and used to solve a variety of design optimization problems, including simply-supported beams, clamped beams, cantilevered beams, simply-supported plates, and clamped plates. The numerical algorithm is based on perturbation theory of differential operators and uses an ajoint variable method for design derivative calculation. Finite element methods were used for numerical analysis and the iterative optimization method tested for convergence. Numerical results obtained with the continuous formulation [234,235] compare quite favorably to the parallel approach in a finite development design space setting [232].

Generalizations of the methods of References [232], [234] and [235] are presented in References [236] and [237], with applications to structural and machine elements, vibration isolators, and vehicle suspension systems. A more rigorous and generally applicable development of distributed parameter structural optimization for dynamic response is presented, with applications, in Reference [238], based on design sensitivity analysis methods for operator equations of mechanics of References [73,74] and the gradient projection method of Reference [149].

Earthquake Structural Design

Kato, Nakamura, and Naraku [239] first addressed the explicit problem of optimization of shear building structures for earthquake resistance. They use ground motion as an exitation to a linear structural model with damping and predict dynamic response of each floor of the structure. Linear approximations are then made to obtain design derivatives of dynamic response and sequential linear programming is employed to optimize the design. Venkayya and Khot [240] present a technique for optimization of structures due to impulsive-type loading encountered in earthquake situations, as well as in aircraft structural design. They seek minimum weight with bounds on dynamic stiffness of the structure, defined by the Rayleigh quotient of the equations of motion, and bounds on stress. They use a displacement finite element model of the structure and develop and optimality criterion, using a Lagrange multiplier formulation, that implies a constant value of an energy measure throughout the structure if the structure responds in a single mode. They then extend this to an approximate technique for structures responding in a combination of modes. An iterative redesign algorithm is presented, based on constant distribution of their energy optimality criterion. They present applications of the technique to a wing span of an aircraft, modeled with 155 bar elements, and to a circular arch.

Cheng and Botkin [241] formulate a general problem of structural optimization for dynamic response using model analysis, obtaining an upper-bound for displacement and stress response. They differentiate through the equations of motion with respect to design parameters to obtain derivatives of the response measures with respect to design and employ a feasible direction algorithm for optimization. They present several examples of frame structural optimization associated with earthquake-type loading. In subsequent papers, Cheng and Srifuengfung [242,243] adopt the optimality criterion method presented by Venkayya [240] and calculate optimal designs of frame structures.

Ray, Pister, and co-workers [244-246] treat the earthquake structural design problem, using linear structural models and optimizing structural reliabilty. Feasible direction optimization methods are employed for selection of design parameters, under failure constraints associated with earthquake ground motion. In a more fundamental development, Ray, Pister, and Polak [247-248] develop a method for calculating derivatives of transient dynamic response of structures to earthquake inputs. Nonlinear behavior of structural response is incorporated in their analysis for multi- story frame structures. They present a numerical integration method for calculating explicit design derivatives of dynamic response, as needed by optimization algorithms. A specified earthquake isolation system for building structures is optimized by Bhatti, Pister, and Polak [249] for dynamic response constraints, using a feasible direction optimization method. Subsequently [250], they present a general-purpose optimization computer code capable of treating such problems, using a reliable general purpose optimization program. Bhatti [251] presents a detailed treatment of optimization of an earthquake energy absorbing device at the foundation of a tall building. The building itself is modeled by elastic structural response and the isolator is allowed to undergo plastic deformation, under earthquake shock loading. Performance constraints under small earthquake, requiring elastic response, and under larger earthquakes in which the structure is allowed to go plastic. A min-max optimization formulation is employed. Pister, Polak, and Bhatti [252-255] present a comprehensive treatment of dynamic structural optimization for earthquake application.

Miscellaneous Problems

Levy and Wolf [256] investigate the existence of a fully stressed design for dynamically loaded structures and show that such a design does exist for some statically determinate structures. Komkov and Coleman [257,258] investigate an optimal control approach to design sensitivity analysis and optimization of dynamically loaded structures. They investigate, from an analytical point of view, optimal damping of structural response. Thermann [259] treats a variety of dynamic structural optimization problems using the Pontryagin maximum principles. He treats problems with constraints on one and two natural frequencies and on response to harmonically varying input.

SHAPE OPTIMAL DESIGN

Numerical Methods for Shape Optimization

One of the first treatments of the general problem of selection of shape of a structure as the design variable is presented by Zienkiewicz and Campbell [260]. They formulate the shape optimal design problem using a finite element model of complex structures and treat the location of the nodal points of the finite element model as design variables. They then calculate derivatives of stiffness and load matrices with respect to design parameters and obtain derivatives of structural response measures and employ sequential linear programming for numerical solution. They present examples associated with dams and rotating turbine machinery. Ramakrishnan and Francavilla [261] employ a similar finite element formulation, but they use a penalty function method for numerical optimization. Francavilla, Ramakrishnan, and Zienkiewicz [262] employ the finite element method of References [260] and [261] for fillet optimization to minimize stress concentration. Schnack [263] and Oda [264] use a finite element formulation for stress calculation in the neighborhood of a stress concentration and iteratively modify the contour to minimize the peak stress.

More basic approaches for surface contouring to minimize stress concentration were initiated by Tvergaard in selecting the optimum shape of a fillet [265]. He employs a stress field model of the fillet, with a finite dimensional family of pertubations allowed in the boundary shape, defined in terms of coordinate parameters. He employs a variational analysis of the stress field equations to obtain derivatives of stress with respect to his parameters and uses sequential linear programming to iteratively construct an optimum design. Kristensen and Madsen [266] formulate a class of shape optimal design problems for planar solids, which generalizes the approach presented by Tvergaard [265]. They use orthogonal polynomials to locate the boundary of the body and treat the coefficients in these polynomials as design parameters. They employ a finite element model of the structural response to obtain derivatives of stress with respect to their design parameters and employ sequential linear programming to solve the optimization problem. They solve an elementary problem of the optimum shape of a hole in a biaxial stress field analytically. They numerically illustrate the method on more complex problems. Bhavikatti and Ramakishnan [267] present a refinement of the formulation of References [260], [261] and [262] for optimum design of fillets in flat and round tension bars. They also use a polynomial with coefficients taken as the design variables to characterize the shape of the fillet and a finite element model to calculate stress with the body. They investigate minimization of stress concentration factor, minimum volume design, and design for uniform stress distribution along the fillet boundary as optimality criterion. Derivatives of response measures with respect to design parameters are calculated using a finite element model. Sequential linear programming is employed for numerical solution.

A function space gradient projection method of optimal design of the shape of two-dimensional elastic bodies is presented by Chun and Haug [268,269], using design sensitivity analysis methods similar to those presented by Rousselet and Haug [270] and a gradient projection method of the kind presented in Reference [149]. The design objective in this work is weight minimization, with constraints on Von Mises yield stress and shear stress distribution on the boundary.

Optimality criteria have been developed for selected classes of shape optimal design problems, which have been used for constructing solutions. Kunar and Chan [271] use a fully stressed criteria and select geometrical variables to minimize weight. Dems and Mroz [272] present a quite general approach to shape optimal design. They use a boundary perturbation analysis to derive optimality criteria and a finite element numerical method to determine optimum boundaries.

Shape of Cross Section of Shafts in Torsion

The problem of optimization of cross sectional shape of torsion members is addressed by Henry [273]. He develops an analytical method for location of the boundary, in terms of a small number of parameters and iteratively selects these parameters to minimize weight, subject to constraints on cross sectional geometry and boundary stress. He treats the problem of a shaft with grooves that are required for keyways.

Banichuk [274,275] formulates a general problem of selecting the optimum shape of cross section for a nonhomogeneous shaft to maximize torsional stiffness, with a given amount of material available. He uses the fact that the functional minimized by the warping potential in a variational formulation of the boundary-value problem is proportional to the torsional stiffness of the shaft. He then takes variations of this functional with respect to both the warping function and boundary variation, using the material derivative idea of continuum mechanics, and obtains a necessary condition for optimum location of the boundary. He treats both simply-connected and multiply-connected cross sections. Kurshin and Onoprienko [276] treat the same problem of maximum torsional stiffness of a shaft with doubly-connected cross section, using a complex variable method to determine the optimum boundary.

Banichuk [277] subsequently presents an extension of the torsional stiffness maximization problem for rods, using optimal distribution on a given amount of stiffening material around the boundary. The method he employs is a direct extension of that used in References [274,275]. Gurvitch [278] presents an alternate analytical technique for optimizing the shape of an interior boundary that is associated with inhomogeneity in material, using a coordinate system associated with the warping function and obtaining necessary and sufficient conditions of optimality. Quite recently, Dems [279] uses the method of Reference [272] to formulate and numerically solve a variety of problems of shaft cross section shape optimization for torsional stiffness.

Shapes of Holes in Planar Solids

Neuber [280,281] and Cherpanov [282] treat the problem of finding a hole shape in a planar solid so as to make tangential normal stress acting on the boundary constant, under the assumption that this is a condition of optimality. Cherpanov [282] cites considerable earlier Soviet literature addressing the same design objective. Wheeler [283] investigates conditions under which constant tangential normal stress along the boundary of a hole, fillet, or notch is a valid optimality criterion for minimum peak stress. He develops criterion for axisymmetric torsion problems and plane problems involving holes, notches, and fillets. Bjorkman and Richards [284] treats the problem of hole shape in a plane elastic solid to minimize stress concentration on the hole contour.

Banichuk [285] formulates the problem of selecting hole shape in an infinite plane body that is in biaxial tension at infinity, to minimize peak stress in the vicinity of the hole. Using the maximum principle for harmonic functions, he proves that the optimum hole shape leads to constant tangential normal stress around the boundary of the hole, hence proving the optimality criterion employed in References [280-282] for this class of problems. In a related paper [286], Banichuk treats the problem of finding the optimum hole shape to minimize the maximum value of the second invariant of the stress tensor deviator over an entire plate, which is subjected to uniform bending at infinity. Again, using the maximum principle for harmonic functions, he shows that the maximum stress occurs at the hole boundary and that for certain classes of problems the hole boundary is uniformly stressed.

Miscellaneous Shape Optimal Design Problems

Banichuk and Karihaloo [287] seek the shape of the cross section in a cylindrical bar to minimize weight, subject to constraints on torsional stiffness and bending stiffness. They use a variational formulation of the torsion problem and a Lagrange multiplier technique to adjoin constraints to the cost function. Using a material derivative type calculation, the first variation of the augmented cost function is taken with respect to shape and optimality criterion are derived. Parbery and Karihaloo [288] use the same method to optimize hollow cylinders, with constraints on torsional and bending stiffness.

Cherkaev [289] presents a theoretical treatment of the problem of boundary shape selection to minimize volume of the structure, subject to a lower bound constraint on natural frequency. He develops a general necessary condition and shows that the variational formulation of Prager [171] can be applied to obtain the same result. As a final note, Durelli and co-workers [290,291] present an experimental method using photo-elasticity to find the optimum shape of a hole in a flat plate, under uni-axial load, to minimize stress concentration. Problems of mathematical physics also lead to shape determination for optimum response. For example, Hersch and Payne [292] treat membrane shape optimization for eigenvalue extremization.

Related Literature on Domain Optimization

Cea, Zolesio, and Rousselet [293-300] present techniques and applications, from fields other than structural optimization, for selecting optimum domain. They cite substantial literature in this general field and give examples that show potential for optimality criteria and direct numerical methods for shape optimization of structures.

MULTI-PURPOSE STRUCTURES AND MISCELLANEOUS OPTIMAL DESIGN PROBLEMS

Multi-purpose Structures

In a landmark paper [301] Prager and Shield formulate the problem of minimum weight design of sandwich structures, subject to several constraints. They treat a beam-tie, as an example, for specified traverse and longitudinal stiffness. A Lagrange multiplier technique is employed to obtain necessary conditions, which are solved for a simple problem. They discuss an extension of the technique to incorporate other constraints, such as buckling load, that can be represented by a global measure of structural response. Martin and Chern [302,303] employ the method of Reference [301] to develop optimality criterion for a structure under multiple loads and a bound on minimum cross section. They treat both continuous and piecewise uniform cross sections and construct solutions, using optimality criterion with Lagrange multipliers.

Karihaloo and Niordson [112] develop optimality criterion to maximize fundamental frequency of a vibrating beam, under constraints that the volume of material is fixed and that the buckling load is bounded from below. They use Lagrange multipliers to develop an optimality criterion and extend the earlier work of Niordson [95].

Sherman and Wang [304,305] treat the problem of volume minimization of an axisymmetric plate of varying thickness, subject to constraints that stress and deflection are bounded. They allow exponentially varying plate thickness, hence reducing the optimization problem to one of selecting a finite number of parameters through a nonlinear programming approach.

In a sequence of papers [306-311] Karihaloo, Parbery, and Wood treat multi- purpose tie- beams and beam-columns that are quite similar to problems treated in Reference [301]. They use Lagrange multipliers and a variational formulation of the constraints to obtain optimality criterion and present numerical solutions of test problems.

Haug and co-workers [51,108,149] use a function space gradient projection optimization technique for minimization of weight of beam, plate, and composite structures under a combined set of constraints on deflection, stress, natural frequency, and buckling loads. The gradients required in the gradient projection calculation are computed using adjoint design variable sensitivity methods of References [73,74,312,313]

Multiple failure criteria constraints, under the same loading system, are treated by several authors. Haug and Kirmser [166] use necessary conditions from optimal control theory to obtain pieced extremals for minimum weight design of a beam under both stress and displacement conditions. Similarly, Huang [314] extends the method of Prager and Taylor [42] to incorporate the effect of shear deformation in optimization of beams and columns for buckling and vibration.

In an important sequence of papers [315-318], Seiranyan and Gura analyze minimum weight design of beams and plates with multiple loading and multiple constraints, primarily vibration and frequency. They employ a Lagrange multiplier method for rigorous development of optimality criterion and carry out explicit solution of simple problems. In Reference [317], they use a quasi- optimal technique for approximating solutions of more complex problems. A distinguishing character of this work, as compared to much of the preceding work, is that it employs modern functional analysis techniques and abstract optimization theory in development of optimality criterion. In an important subsequent paper [319], Seiranyan rigorously demonstrates that most of the functionals arising in structural optimization, involving natural frequency, buckling, displacement, and compliance are homogeneous functionals. He employs functional analysis differentiation theory to rigorously obtain optimality criterion for multiple constraint types and gives a clear indication of a unifying direction that may be taken in future efforts and distributed parameter optimization. The method he employs is closely related to the abstract optimization theory employed by Choi, Haug, and Rousselet [71] in developing rigorous optimality criterion for general structural optimization problems involving repeated eigenvalues.

There appears to be great potential for organized development of optimality criterion for very general problems, using the method of References [71,319]. This feeling is strengthened by an approach concerning existence and stability of solution of optimal design problems by Velte and Villaggio [320]. They employ modern Sobolev space theory, which has become the foundation of modern theory of partial differential equations, to carefully analyze the existence, uniqueness, and stability of problems of structural optimization. Their study of optimization and of a rod shows that under relatively weak hypotheses, existence is guaranteed, but uniqueness and stability are problem dependent.

Special Problems

A problem of optimal determination of distribution of shear modulus in a nonhomogeneous material that makes up on elastic bar in torsion is investigated by Klosowicz and Lurie [321]. Variation in shear modulus is taken as the design variable and the objective is to maximize torsional rigidity, subject to the constraint that a given amount of material is available. They employ a variational method to obtain optimality criterion and discuss algorithms to obtain numerical solutions, but present no numerical results. Banichuk [322,323] treats a related problem of optimal anisotrophy of rods in torsion and subsequently [324] presents a more general formulation of optimal orientation of axes of anisotrophy in plane deformable solids. He formulates optimality criterion in which material properties play the role of design functions. Hirano [325] treats a similar problem seeking optimum fiber orientation in multi-layer composite materials, to maximize buckling load of a structure.

Banichuk [326,327] has also treated an interesting class of problems, in which he seeks to optimize some pointwise measure of structural performance, subject to the condition that constraints hold for an entire family of load systems. This problem [326] may be viewed as one with maximization over the load systems in the constraint. Alternatively [327], he treats problems in which the functional to be minimized is thè maximum value of a functional over all load classes. In this way, min-max problems associated with game theory arise. Optimality criterion or such problems are discussed, but solution methods are extremely difficult.

Aristov and Troitskii [328] address the problem of minimum weight design of beams and plates subject to stress constraints. They formulate the problem as a first order system of differential equations and apply necessary conditions of optimality from optimal control theory to obtain necessary conditions for the structural problem. They also account for discontinuities that may arise in the functional being treated.

Samsonov [329] treats the problem of optimum location of rib stiffeners in plates, with multiple failure criteria. The approach used is similar to shape optimal design, since the rib separates subdomains in which plate action occurs.

Solodovnikov [330] discusses the general problem of determining the shape of generating curves for shells of revolution to minimize weight, subject to constraints on stress and stability. He employs the shell equations to determine stress derivative and sequential linear programming to solve the optimization problem.

Olhoff and Taylor [331,332] have formulated a class of structural optimization problems called "structural remodeling". Such problems include most classical cost and constraint functions. The idea is to allow use of only a specified amount of material to vary, or remodel, the design to provide the greatest decrease in cost.

Battermman and Pavicic [333] treat minimum weight design of laminated, anisotropic shells of revolution, with shell thickness as the designn variable. Loading, fiber alignment, and material properties are specified. A membrane model of shell deformation is employed and constraints on stress are imposed in minimizing weight.

Taylor [334] and Rozvany [335] present generalizations of optimality criterion developed earlier by Prager and Taylor [42] and show that extended version of optimality criterion can often include new problems and models treated in earlier litreature.

The design problem of elastic body contour shape selection to minimize peak contact stress was first treated by Conry and Seireg [336]. They treat gear contour design using a nonlinear programming method. Haug and Kwak [337] treat a somewhat more general formulation, in which bounds on the allowable position of the surface are included. They use a sequential linear programming method for solution. Lukasiewicz [338] has addressed a similar problem, but uses constant contact stress as an optimality criteria and constructs solutions using the Hertz formulas for contact stress. Benedict [339] has developed a rigorous optimality criteria for contour optimization to minimize contact stress for general elastic bodies in contact.

Design of structural components as part of vibrators [340] or vibration isolators [341] presents a class of problems of future interest.

Grillages and archgrids represent an important class of structural optimization problems that have been studied extensively by Rozvany and co-workers [342-345]. For a more detailed technical review of this class of problems, the reader is referred to the paper by Rozvany [346] in these proceedings.

Optimization of shells for special purposes has been discussed in earlier sections of this review. General approaches to shell optimization are typified by papers of Solodovnikov [330] and Krivenkov [347].

As a final note, the potential for nonstandard optimization techniques, for multi-criteria optimization may be considered. Stadler [348-350] and Carmichael [351] apply the concept of Pareto optimality to multi-criteria structural optimization problems. The utility of this approach is yet to be determined, but it does offer an alternative to conventional formulations of optimal design problems that may be considered in the future.

38

References

1. Wasiutynski, Z., and Brandt, A. "The Present State of Knowledge In the Field of Optimum Design of Structures" Applied Mechanics Reviews, Vol. 16, No. 5, 1963, pp. 341-350

2. Sheu, C.Y., and Prager,W. "Recent Developments in Optimal Structural Design" Applied Mechanics Reviews, Vol. 21, No. 10, 1968, pp. 985-992

3. Pierson, B.L. "A Survey of Optimal Structural Design Under Dynamic Constraints" International Journal of Numerical Methods in Engineering, Vol. 4, 1972, pp, 491-499

4. Niordson, F.I., and Pedersen, P. "A Review of Optimal Structural Design" Proceedings, 13th International Congress on Theoretical and Applied Mechanics, Springer-Verlag, 1973

5. McIntosh, S.C., "Structural Optimization Via Optimal Control Techniques" Proceedings, ASME Structural Optimization Symposium, ASME, AMDT, New York, 1974

6. Olhoff, N. "A Survey of the Optimal Design of Vibrating Structural Elements, Part I: Theory" and "Part II: Applications" Shock and Vibration Digest, Vol. 8, No. 8, pp. 3-10, and No. 9, pp. 3-10

7. Rangacharyulu, M.A.V., and Done, G.T.S. "A Survey of Structural Optimization Under Dynamic Constraints" Shock and Vibration Digest, Vol. 11, No. 12, 1979, pp. 15-25

8. Venkayya, V.B. "Structural Optimization: A Review and Some Recommendations" International Journal of Numerical Methods in Engineering, Vol. 13, No. 2, 1978, pp. 203-228

9. Weisshaar, T.A. and Plaut, R.H. "Structural Optimization Under Nonconservative Loading" Optimization of Distributed Parameter Structures (Ed. E.J. Haug and J. Cea), Sijthoff & Noordhoff, Alphen aan den Rijn, Netherlands, 1980

10. Prager, W. Introduction to Structural Optimization, Courses and Lectures: International Centre for Mechanical Science, Vdine, No. 212, Springer-Verlag, Vienna, 1974

11. Sawczuk, A. and Mroz, Z. Optimization in Structural Design, Springer-Verlag, New York, 1975

12. Rozvany, G.I.N. Optimal Design of Flexural Systems, Pergamon Press, New York, 1976

13. Haug, E.J. and Arora, J.S. Applied Optimal Design, Wiley-Interscience, New York, 1979

14. Banichuk, N.V. Optimization of the Shapes of Elastic Bodies (in Russian), Nauka, Moscow, 1980

15. Robinson, A.C. "A Survey of Optimal Control of Distributed Parameter Systems" Automatica, Vol. 7, No. 3, 1971, pp. 371-388

16. Lurie, K.A. "The Mayer-Bolza Problem for Multiple Integrals: Some Optimum Problems for Elliptic Differential Equations Arising in Magnetohydrodynamics" in Topics in Optimization, (Ed. Leitmann), Academic Press, New York, 1967, pp. 147-197

17. Butkovskiy, A.G. Distributed Control Systems, American Elsevier Publishing Co., New York, 1969

18. Ray. W.H. "Some Recent Applications of Distributed Parameter Systems Theory - A Survey" Automatica, Vol. 14, 1978, pp. 281-287

19. Ray, W.H. and Lainiotis, D.G. Distributed Parameter Systems, Marcel Dekker, New York, 1978

20. Lions, J.L. Optimal Control of Systems Governed by Partial Differential Equations, Springer-Verlag, New York, 1971

21. Lions, J.L. Some Aspects of the Optimal Control of Distributed Parameter Systems, SIAM Series in Regional Conferences in Applied Mathematics, 1972

22. McGlothin, G.E. "Optimal Control of Distributed Parameter Systems with Penalites on Special Derivatives of the State" International Journal of Control, Vol. 24, No. 4, 1976, pp. 145-166

23. Pontryagin, L.S., Boltyanskii, V.G., Gamkrelidze, R.V. and Mishchenko. E.F. "The Mathematical Theory of Optimal Processes" Wiley-Interscience, New York, 1962

24. Hestenes, M.R., Calculus of Variations and Optimal Control Theory, Wiley, New York, 1966

25. Bryson, A.E. and Ho, Y.C. Applied Optimal Control, Wiley, New York, 1979

26. Cea, J. Optimization-Theory and Algorithms, Tata Institute, Bombay, 1978

27. Pshenichnyi, B.N. Necessary Conditions for an Extremum, Marcel Dekker, New York, 1971

28. Ioffe, A.D. and Tihomirov,V.M. Theory of Extremal Problems, North-Holland, Amsterdam, 1979

29. Zolezzi, T. "Necessary Conditions for Optimal Controls of Elliptic or Parabolic Problems" SIAM Journal of Control, Vol. 10, No. 4, 1972, pp. 594-607

30. Lagrange, J.L. "Sur la Figure des Colonnes" Miscellanea Taurinensia V, 1770-1773, p. 123

31. Clausen, T. "Uber die Form Architektonischer Saulen" Melanges Mathematiques et Astronomiques I, 1849-1853, pp. 279-294

32. Todhunter, I. and Pearson, K. A History of the Theory of Elasticity, Dover, New York, 1960

33. Keller, J.B. "The Shape of the Strongest Column" Archives of Rational Mechanics and Analysis, Vol. 5, 1960, pp. 275-285

34. Tadjbakhsh, I., Keller, J.B. "Strongest Columns and Isoperimetric Inequalities for Eigenvalues" Journal of Applied Mechanics, Vol. 9, 1962, pp. 159-164

35. Farshad, M. and Tadjbakhsh, I. "Optimum Shape of Columns with General Conservative End Loading" Journal of Optimization Theory and Applications, Vol. 11, No. 4, 1973, pp. 413-420

36. Gajewski, A. and Zyczkowski, M. "Optimal Design of Elastic Columns Subjected to the General Conservative Behavior of Loading" ZAMP, Vol. 21, No. 52, 1971, pp. 806-818

37. Keller, J.B. and Niordson, F.I. "The Tallest Column" Journal Math. & Mechanics, Vol. 16, 1966, pp. 433-446

38. Barnes, E.R. "The Shape of the Strongest Column and Some Related Extremal Eigenvalue Problems" Quarterly of Applied Mathematics, Vol. 35, 1977, pp. 393-409

39. Alblas, J.B. "Optimal Strength of a Compund Column" International Journal of Solids and Structures, Vol. 13, 1977, pp. 307-320

40. Taylor, J.E. "The Strongest Column: An Energy Approach" Journal of Applied Mechanics, Vol. 34, 1967, p. 486

41. Taylor, J.E. and Liu, C.Y. "Optimal Design of Columns" AIAA Journal, Vol. 6, 1968, pp. 1496-1502

42. Prager,W. and Taylor, J.E. "Problems of Optimal Structural Design" Journal of Applied Mechanics, Vol. 35, 1968, pp. 102-106

43. Taylor, J.E. "Optimal Prestress Against Buckling: An Energy Approach" International Journal of Solids and Structures, Vol. 7, No. 2, 1971, pp. 213-223

44. Huang, N.C. and Sheu, C.Y. "Optimal Design of an Elastic Column of Thin Walled Cross Section" Journal of Applied Mechanics, Vol. 35, No. 2, 1968, pp. 285-288

45. Frauenthal, J.C. "Constrained Optimal Design of Columns Against Buckling" Journal of Structural Mechanics, Vol. 1, 1972, pp. 79-89

46. Popelar, C.H. "Optimal Design of Beams Against Buckling: A Potential Engery Approach" Journal of Structural Mechanics, Vol. 4, 1976, pp. 181-196

47. Popelar, C.H. "Optimal Design of Structures Against Buckling: A Complementary Energy Approach" Journal of Structural Mechanics, Vol. 5, 1977, pp. 45-66

48. Hu, K.K. and Kirmser, P.G. "A Numerical Solution of A Nonlinear Differential-Integral Equation for the Optimal Shape of the Tallest Column" International Journal of Engineering Science, Vol. 18, 1980, pp. 333-339

49. Adali, S. "Optimal Shape And Non-Homogeneity of a Non-Uniformly Compressed Column" International Journal of Solids and Structures, Vol. 15, 1979, pp. 935-949

50. Haug, E.J. "Two Methods of Optimal Structural Design" Developments in Mechanics, Vol. 5 (Ed., Weiss, H.J., Young, D.F., Riley, W.F., and Rogge, T.R.), Iowa State University Press, 1969, pp. 847-860

51. Haug, E.J., Pan, K.C. and Streeter, T.D. "A Computational Method for Optimal Structural Design II: Continuous Problems" Numerical Methods in Engineering, Vol. 9, 1975, pp. 649-667

42

52. Hornbuckle, J.C. and Boykin, W.H. "Equivalence of a Constrained Minimum Weight and Maximum Column Buckling Load Problem with Solution" Journal of Applied Mechanics, Vol. 45 1978, pp. 159-164

53. Dinkoff, B., Levine, M. and Luus, R. "Optimum Linear Tapering In the Design of Columns" Journal of Applied Mechancis, Vol. 46, 1979, pp. 956-958

54. Farshad, M. "Optimum Shape of Continuous Columns" International Journal of Mechanical Science, Vol. 16, No. 8, 1974, pp. 597-602

55. Masur, E.F. "Optimal Placement of Available Sections in Structural Eigenvalue Problems" Journal of Optimization Theory and Applications, Vol. 15, 1975, pp. 69-84

56. Mroz, Z. and Rozvany, G.I.N. "Optimal Design of Structures with Variable Support Conditions" Optimization Theory Application, Vol. 15, 1975, pp. 85-101

57. Rozvany, G.I.N. and Mroz, Z. "Column Design: Optimization of Support Conditions and Segmentation" Journal of Structural Mechanics, Vol. 5, 1977, pp. 279-290

58. Olhoff, N. and Taylor, J.E. "Designing Continuous Columns for Minimum Total Cost of Material and Interior Supports" Journal of Structural Mechanics, Vol. 6, 1978, pp. 367-382

59. Olhoff, N. and Rasmussen, S.H. "On Single and Bimodal Optimum Buckling Loads of Clamped Columns" International Journal of Solids Structures, Vol. 13, 1977, pp. 605-614

60. Banichuk, N.V. and Karihaloo, B.L. "On the Solution of Optimization Problems with Singularities" International Journal of Solids Structures, Vol. 13, 1977, pp. 725-733

61. Olhoff, N. and Niordson, F.I. "Some Problems Concerning Singularities of Optimal Beams and Columns, ZAMM, Vol. 59, 1979, pp. T16-T26

62. Komkov, V. and Haug, E.J. "On the Optimum Shape of Columns" Optimization of Distributed Parameter Structures (Ed. E.J. Haug and J. Cea) Sijthoff & Noordhoff, Alphen aan den Rijn, Netherlands, 1980

63. Olhoff, N. "Optimization of Columns Against Buckling" Optimization of Distributed Parameter Structures (Ed. E.J. Haug and J. Cea) Sijthoff & Noordhoff, Alphen aan den Rijn, Netherlands, 1980

64. Masur, E.F. and Mroz, Z. "On Non-Stationary Optimality Conditions in Structural Design" International Journal of Solids Structures, Vol. 15, 1979, pp. 503-512

65. Masur, E.F. and Mroz, Z. " Singular Solutions in Structural Optimization Problems" in Proc. IUTAM Symposium on Variational Methods in the Mechanics of Solids (Ed. S. Nemat-Nasser) Northwestern University, Evanston, IL, USA, 1978, Pergamon Press (to appear)

66. Masur, E.F. "Singular Problems of Optimal Design" Optimization of Distributed Parameter Structures (Ed. E.J. Haug and J. Cea) Sijthoff & Noordhoff, Alphen aan den Rijn, Netherlands, 1980

67. Prager, S. and Prager W. "A Note on Optimal Design of Columns" International Journal of Mechanical Science, Vol. 21, 1979, pp. 249-251

68. Banichuk, N.V. "Optimizing the Stability of a Bar with Elastic Clamping" MTT (Mechanics and Solids), Vol. 4, 1974, pp. 150-154

69. Blachut, J. and Gajewski, A. "A Unified Approach to Optimal Design of Columns" Solid Mechanics Archives, December 1980

70. Haug, E.J. "Optimization of Distributed Parameter Structures with Repeated Eigenvalues", New Approaches to Nonlinear Problems in Dynamics (Ed. P.J. Holmes)

71. Choi, K.K., Haug, E.J. and Rousselet, B. "Optimization of Structures with Repeated Eigenvalues" Optimization of Distributed Parameter Structures (Ed. E.J. Haug and J. Cea) Sijthoff & Noordhoff, Alphen aan den Rijn, Netherlands, 1980

72. Rousselet, B. "Singular Dependence of Repeated Eigenvalues" Optimization of Distributed Parameter Structures (Ed. E.J. Haug and J. Cea) Sijthoff & Noordhoff, Alphen aan den Rijn, Netherlands, 1980

73. Haug, E.J. and Rousselet, B. "Design Sensitivity Analysis in Structural Mechanics I: Static Response Variations" Journal of Structural Mechanics, Vol. 8, No. 1, 1980, pp. 17-41

74. Haug, E.J. and Rousselet, B. "Design Sensitivity Analysis in Structural Mechanics II: Eigenvalue Variations" Journal of Structural Mechanics, Vol. 8, 1980, to appear

75. Thompson, J.M.T. and Lewis, G.M. "On the Optimum Design of Thin-walled Compression Members" Journal of Mechanics and Physics of Solids, Vol. 20, 1972, pp. 101-109

76. Thompson, J.M.T. "Optimization as a Generator of Structural Instability" International Journal of Mechanical Science, Vol. 14, No. 9, 1972, pp. 627-630

77. Thompson, J.M.T. and Supple, W.D. "Erosion of Optimum Design by Compound Branching Phenomena" Journal of Mechanics and Physics of Solids, Vol. 21, No. 3, 1972, pp. 135-144

78. Thompson, J.M.T. and Hunt, G.W. "Dangers of Structural Optimization" Engineering Optimization, Vol. 1, 1974, pp. 99-110

79. Masur, E.F. "Optimal Design of Symmetric Structures Against Postbuckling Collapse" International Journal of Solids and Structures, Vol. 14, 1978, pp. 319-326

80. Roorda, J. and Reis, A.J. "Nonlinear Interactive Buckling: Sensitivity and Optimality" Journal of Structural Mechanics, Vol. 5, 1977, pp. 207-232

81. Frauenthal, J.C. "Constrained Optimal Design of Circular Plates Against Buckling" Journal of Structural Mechanics, Vol. 1, No. 2, 1972, pp. 159-186

82. Andreev, L.V., Mossakovsii, V.I. and Obodan, N.I. "On Optimal Thickness of a Cylindrical Shell Loaded by External Pressure" PMM, Vol. 36, No. 4, 1972, pp. 677-685

83. Manevich, A.I. and Kaganov, M.Y. "Stability and Weight Optimization of Reinforced Spherical Shells Under External Pressure" Prikladnay Mekhanika, Vol. 9, No. 1, 1973, pp. 20-26

84. Zyczkowski, M. and Kruzelecki, J. "Optimal Design of Shells with Respect to Their Stability" Optimization in Structural Design (Ed. Sawczuk and Mroz), Springer -Verlag, New,York, 1975, pp. 229-247

85. Kruzelecki, J. "Optimization of Shells Under Combined Loadings via the Concept of Uniform Stability" Optimization of Distributed Parameter Structures (Ed. E.J. Haug and J. Cea) Sijthoff & Noordhoff, Alphen aan den Rijn, Netherlands, 1980

86. Rikards, R.B. "Convexity of Some Classes of Optimization Problems for Multilayer Shells Under Conditions of Stabilty and Vibration" MTT, Vol. 15, 1980, pp. 145-154

87. Pappas, M. and Allentuch, A. "Automated Optimal Design of Frame Reinforced, Submersible, Circular, Cylindrical Shells" Journal of Ship Research, Vol. 17, 1973, pp. 208-216

88. Pappas, M. and Allentuch,A. "Pressure Hull Optimization Using General Instability Equation Admitting More Than One Longitudinal Buckling Half-Wave" Journal of Ship Research, Vol. 19, 1975, pp.18-22

89. Simitses, G.J. and Sheinman, I. "Optimization of Geometrically Imperfect Stiffened Cylindrical Shells Under Axial Compression" Computers and Structures, Vol. 9, 1978, pp. 337-381

90. Kunoo,K. and Yang, T.Y. "Minimum Weight Design of Cylindrical Shell with Multiple Stiffener Sizes" AIAA Journal, Vol. 16, 1978, pp. 35-40

91. Tadjbakhsh, I. and Farshad, M. "On Conservatively Loaded Funicular Arches and Their Optimal Design" Optimization in Structural Design (Ed. Sawczuk and Mroz) Springer-Verlag, New York, 1975, pp. 215-228

92. Blachut, J. and Gajewski, A. "On Unimodal And Bimodal Optimal Design of Funicular Arches" International Journal of Solids and Structures, to appear, 1981

93. Budiansky, B., Frauenthal, J.C. and Hutchinson, J.W. "On Optimal Arches" Journal of Applied Mechanics, Vol. 36, 1969, pp. 880-882

94. Amazigo, J.C. "Optimal Shape of Shallow Circular Arches Against Snap-Buckling" Journal of Applied Mechanics, Vol. 45, 1978, pp. 591-594

95. Niordson, F.I. "On the Optimal Design of a Vibrating Beam" Quarterly of Applied Mechanics, Vol. 23, 1965, pp. 47-53

96. Turner, M.J. "Design of Minimum Mass Structures with Specified Natural Frequency" AIAA Journal, Vol. 5, No. 3, 1967, pp. 406-412

97. Taylor, J.E. "Minimum Mass Bar for Axial Vibration at Specified Natural Frequencies" AIAA Journal, Vol. 5, No. 10, 1967, pp. 1911-1913

98. Taylor, J.E. "Optimum Design of a Vibrating Bar with Specified Minimum Cross Section" AIAA Journal, Vol. 6, No. 7, 1968, pp. 1379-1381

99. Sheu, C.Y. "Elastic Minimum-weight Design for Specified Fundamental Frequency" International Journal Solids and Structures, Vol. 4, 1968, pp. 953-958

100. Sippel, D.L. and Warner, W.H. "Minimum-mass Design of Multi-element Structures Under a Frequency Constraint" AIAA Journal, Vol. II, No. 4, 1973, pp. 483-489

101. Cardou,A. and Warner, W.H. "Minimum-mass Design of Sandwich Structures with Frequency and Section Constraints" Journal of Optimization Theory and Applications, Vol. 14, No. 6, 1974, pp. 633-647

102. Miele, A., Mangiavacchi, A., Mohanty, B.P., and Wu, A.K., "Numerical Determination of Minimum Mass Structures with Specified Natural Frequencies" International Journal of Numerical Methods in Engineering, Vol. 13, No. 2, 1978, pp. 203-228

103. Brach, R.M. "On the Extremal Fundamental Frequencies of Vibrating Beams" International Journal of Solid Structures, Vol. 4, 1968, pp. 667-674

104. Brach, R.M. "On Optimal Design of Vibrating Structures" Journal of Optimization Theory and Applications, Vol. 11, 1973, pp. 662-667

105. Vepa, K. "On the Existence of Solutions to Optimization Problems with Eigenvalue Constraints" Quarterly of Applied Mechanics, Vol. 31, 1973-1974, pp. 329-341

106. Brach, R.M. "Optimized Design: Characteristic Vibration Shapes and Resonators" Journal of the Acoustic Society of America, Vol. 53, No. 1, 1973, pp. 113-119

107. McCart, B.R., Haug, E.J. and Streeter, T.D. "Optimal Design of Structures with Constraints on Natural Frequency" AIAA Journal, Vol. 8, No. 6, 1970, pp. 1012-1019

108. Haug, E.J., Arora, J.S. and Matsui, K. "A Steepest-Descent Method for Optimization of Mechanical Systems" Journal of Optimization Theory and Applications, Vol. 19, No. 3, 1976, pp. 401-424

109. Pierson, B.L. "An Optimal Control Approach to Minimum-Weight Vibrating Beam Design" Journal of Structural Mechanics, Vol. 5, 1977, pp. 147-178

110. Kamat, M.P. and Simitses, G.J. "Optimal Beam Frequencies by the Finite Element Displacement Method" International Journal of Solids and Structures, Vol. 9, 1973, pp. 415-429

111. Foley, M. and Citron, S.J. "A Simple Technique for the Minimum Mass Design of Continuous Structural Members" Journal of Applied Mechanics, Vol. 44, 1977, pp. 285-290

112. Karihaloo, B.L. and Niordson, F.I. "Optimum Design of Vibrating Beams Under Axial Compression" Archives of Mechanics, Vol. 24, 1972, pp. 1029-1037

113. Karihaloo, B.L. and Niordson, F.I. "Optimum Design of Vibrating Cantilevers" Journal of Optimization Theory and Applications, Vol. 11, 1973, pp. 638-654

114. Seiranyan, A.P. "Optimal Beam Design with Limitations on Natural Vibration Frequency and Buckling Load" MTT (Mechanics of Solids), Vol. 11, No. 1, 1976, pp. 147-152

115. Warner, W.H. and Vavrick, D.J. "Optimal Design in Axial Motion for Several Frequency Constraints" Journal of Optimization Theory and Applications, Vol. 15, No. 1, 1975, pp. 159-166

116. Olhoff, N. "Optimization of Vibrating Beams with Respect to Higher Order Natural Frequencies" Journal of Structural Mechanics, Vol. 4, 1976, pp. 87-122

117. Olhoff, N. "Maximizing Higher Order Eigenfrequencies of Beams with Constraints on the Design Geometry" Journal of Structural Mechanics, Vol. 5, 1977, pp. 107-134

118. Troitskii, V.A. "Optimization of Elastic Bars in the Presence of Free Vibrations" MTT (Mechanics of Solids) Vol. 11, No. 3, 1976, pp. 145-152

119. Mroz, A. and Rozvany, G.I.N. "Optimal Design of Structures with Variable Support Conditions" Journal of Optimization Theory Applications, Vol. 15, 1975, pp. 85-101

120. Rozvany, G.I.N. "Analytical Treatment of Some Extended Problems in Structural Optimization, Part I and II" Journal of Structural Mechanics, Vol. 3, 1974-75, pp. 359-402

121. Olhoff, N. and Niordson, F.I. "Some Problems Concerning Singularities of Optimal Beams and Columns" ZAMM, Vol. 59, 1979, pp. T16-T26

122. Olhoff, N. "Optimization of Transversely Vibrating Beams and Rotating Shafts" Optimization of Distributed Parameter Structures (Ed. E.J. Haug and J. Cea) Sijthoff & Noordhoff, Alphen aan den Rijn, Netherlands, 1980

123. Szelag, D. and Mroz, Z. "Optimal Design of Vibrating Beams with Unspecified Support Reactions" Computer Methods In Applied Mechanics and Engineering, Vol. 19, 1979, pp. 333-349

124. Kamat, M.P. "Effect of Shear Deformations and Rotary Inertia on Optimum Beam Frequencies" International Journal of Numerical Methods in Engineering, Vol. 9, 1975, pp. 51-62

125. Gupta, V.K. and Murthy, P.N. "Optimal Design of Uniform Non-Homogeneous Vibrating Beams" Journal of Sound and Vibration, Vol. 59, 1978, pp. 521-531

126. Smirnov, A.B. and Troitskii, V.A. "Optimization of Natural Vibrational Frequencies of Curvilinear Thin Elastic Rods" MTT, Vol. 14, 1979, pp. 162-168

127. Weisshaar, T.A. "Optimization of Simple Structures with Higher Mode Frequency Constraints, AIAA Journal, Vol. 10, No. 5, 1972, pp. 691-693

128. Armand, J.L. and Vitte, W.J. Foundations of Aeroelastic Optimization and Some Applications to Continuous Systems, Report No. SUDAAR-390, Department of Aeronautics and Astronautics, Stanford University, 1970

129. Weisshaar, T.A. An Application of Control Theory Methods to the Optimization of Structures Having Dynamic or Aeroelastic Constraints, Report No. SUDAAR 412, Department Aeronautics and Astronautics, Stanford University, 1970

130. Vavrick, D.J. and Warner, W.H. "Minimum Mass Design with Torsional Frequency and Thickness Constraints" Journal of Structural Mechanics, Vol. 6, no. 2, 1978, pp. 211-232

131. Vavrick D.J. and Warner, W.H. "Duality Among Optimal Design Problems for Torsional Vibration" Journal of Structural Mechanics, Vol. 6, No. 2, 1978, pp. 233-246

132. Elwany, M.H.S. and Barr, A.D.S. "Some Optimization Problems In Torsional Vibration" Journal of Sound and Vibration, Vol. 57, 1978, pp. 1-33

133. Elwany, M.H.S. and Barr, A.D.S. "Minimum Weight Design of Beams in Torsional Vibration with Several Frequency Constraints" Journal of Sound and Vibration, Vol. 62, 1979, pp. 411-425

134. Maday, C.J. "A Class of Minimum Weight Shafts" ASME Journal of Engineering for Industry, Vol. 96, No. 1, 1974, pp. 166-170

135. DeSilva, B.M.E. "Optimal Vibrational Modes of a Disc" Journal of Sound and Vibration, Vol. 21, No. 1, 1972, pp. 19-34

136. DeSilva, B.M.E. "Optimal Control Concepts in the Design of Turbine Discs and Blades" Shock and Vibration Digest, Vol. 7, 1975, pp. 63-76

137. Grinev, V.B. and Garal, Y.A. "Optimization of the Parameters of Rotating Rods" Soviet Applied Mechanics, Vol. 13, No. 9, 1975, pp. 389-393

138. Olhoff, N. "Optimal Design of Vibrating Circular Plates" International Journal of Solids Structures, Vol. 6, 1970, pp. 139-156

139. Armand, J-L. "Minimum-Mass Design of a Plate-Like Structure for Specified Fundamental Frequency" AIAA Journal, Vol. 9, No. 9, 1971, pp. 1739-1745

140. Armand, J-L. Applications of the Theory of Optimal Control of Distributed-Parameter Systems to Structural Optimization, NASA Contractor Report No. NASA CR-2044, NASA, Washington, D.C., 1972

50

141. Armand, J-L. "Numerical Solutions in Optimization of Structural Elements" Paper at First International Conference on Computational Methods in Nonlinear Mechanics, Austin, Texas, 1974

142. Olhoff, N. "Optimal Design of Vibrating Rectangular Plates" International Journal of Solids Structures, Vol. 10, 1974, pp. 93-109

143. Olhoff, N. "On Singularities, Local Optima and Formation of Stiffeners in Optimal Design of Plates" Optimization in Structural Design (Ed. Sawczuk and Mroz) Springer-Verlag, New York, 1975, pp. 82-103

144. Olhoff, N. "Optimal Design of Solid Elastic Plates" Optimization of Distributed Parameter Structures (Ed. E.J. Haug and J. Cea) Sijthoff & Noordhoff, Alphen aan den Rijn, Netherlands, 1980

145. Cheng, K-T. and Olhoff, N. An Investigation Concerning Optimal Design of Solid Elastic Plates, DCAMM-Rept. No. 174, The Danish Center for Applied Mathematics and Mechanics, 1980

146. Olhoff, N., Lurie, K.A., Cherkaev, A.V. and Fedorov, A.V. Sliding Regimes and Anisotropy in Optimal Design of Vibrating Axisymmetric Plates, DCAMM-Rept., The Danish Center for Applied Mathematics and Mechanics, 1980

147. Lurie, K.A. and Cherkaev, A.V. "Prager Theorem Application to Optimal Design of Thin Plates, MTT (Mechanics of Solids) Vol. 11, 1976, pp. 157-159

148. Seiranyan, A.P. "A Study of an Extremum in the Optimal Problem of a Vibrating Circular Plate" MTT (Mechanics of Solids) Vol. 13, No. 6, 1978, pp. 99-104

149. Haug, E.J. "A Gradient Projection Method for Structural Optimization" Optimization of Distributed Parameter Structures (Ed. E. J. Haug and J. Cea) Sijthoff & Noordhoff, Alphen aan den Rijn, Netherlands, 1980

150. Haug, E.J., Pan, K.C. and Streeter, T.D. "A Computational Method for Optimal Structural Design I: Piecewise Uniform Structures" International Journal of Numerical Methods in Engineering, Vol. 5, 1972, pp. 171-184

151. Foley, M.H. "A Minimim Mass Square Plate with Fixed Fundamental Frequency of Free Vibration" AIAA Journal, Vol. 16, 1978, pp. 1001-1004

152. Rammerstorfer, F.G. "Increase of the First Natural Frequency and Buckling Load of Plates by Optimal Field of Initial Stresses", Acta Mechanica, Vol. 27, 1977, pp. 217-238

153. Banichuck, N.V. and Mironov, A.A. "Optimization Problems for Plates Oscillating in an Ideal Fluid" PMM, Vol. 40, 1976, pp. 520-527

154. Bert, C.W. "Optimal Design of A Composite Material Plate to Maximize Its Fundamental Frequency" Journal of Sound and Vibration, Vol. 50, 1977, pp. 229-237

155. Rao, S.S. and Singh, K. "Optimum Design of Laminates with Natural Frequency Constraints" Journal of Sound and Vibration, Vol. 67, 1979, pp. 101-112

156. Carmichael, D. and Goh, B.S. "Optimal Vibrating Plates and a Distributed Parameter Singular Control Problem" International Journal of Control, Vol. 26, 1977, pp. 19-31

157. Carmichael D. "Singular Optimal Control Problems In the Design of Vibrating Structures" Journal of Sound and Vibration, Vol. 53, 1977, pp. 245-253

158. Carmichael, D. "Optimal Control in the Design of Material Continua" Archives of Mechanics, Vol. 30, 1978, pp. 743-755

159. Ainola, L.I. "On the Inverse Problem of Natural Vibrations of Elastic Shells" PMM, Vol. 35, No. 2, 1971, pp.358-364

160. Barnett, R.L. "Minimum Weight Design of Beams for Deflection" Journal of Engineering Mechanics, ASCE, Vol. 87, EM2, 1961, pp. 75-109

161. Dixon, L.C.W. "Pontryagin's Maximum Principle Applied to the Profile of a Beam" Journal of Royal Aeronautical Society, Vol. 71, 1967, pp. 513-515

162. Huang, N-C. and Tang, H.T. "Minimum-Weight Design of Elastic Sandwich Beams with Deflection Constraints" Journal of Optimization Theory and Applications, Vol. 4, No. 4, 1969,pp. 277-298

52

163. Prager, W. "Optimal Thermoelastic Design for Given
      Deflection" International Journal of Mechanical Science,
      Vol. 12, 1970, pp. 705-709

164. Chern, J-M. "Optimal Structural Design for Given Deflection
      in Presence of Body Forces" International Journal of Solids
      and Structures, Vol. 7, 1971, pp. 363-382

165. Cantu, E. and Cinquini, C. "Iterative Solutions for Problems
      of Optimal Elastic Design" Computer Methods In Applied
      Mechanics and Engineering, Vol. 20, 1979, pp. 257-266

166. Haug, E.J. and Kirmser, P.G. "Minimum Weight Design of Beams
      with Inequality Constraints on Stress and Deflection"
      Journal of Applied Mechanics, Vol. 34, 1967, pp. 999-1007

167. Haug, E.J. Minimum Weight Design of Beams with Inequality
      Constraints on Stress and Deflection, Ph. D. Thesis, Kansas
      State University, 1966

168. DeSilva, B.M.E. "Application of Pontryagin's Principle to a
      Minimum Weight Design Problem" Journal of Basic Engineering,
      Vol. 92, 1970, pp. 245-250

169. Shield, R.T. and Prager, W. "Optimal Structural Design for
      Given Deflection" ZAMP, Vol. 21, 1970, pp. 513-523

170. Prager, W. "Optimal Design of Statically Determinate Beams
      for Given Deflection" Journal of Mechanical Science, Vol. 13
      1971, p. 893

171. Prager, W. "Conditions for Structural Optimality" Computers
      and Structures, Vol. 2, 1972, pp. 833-840

172. Huang, N-C. "On Principle of Stationary Mutual Complementary
      Energy and Its Application to Structural Design" ZAMP, Vol.
      22, 1971, pp. 608-620

173. Dafalias, Y.F. and Dupuis, G. "Minimum-Weight Design of
      Continuous Beams Under Displacement and Stress Constraints"
      Journal of Optimization Theory and Applications, Vol. 9,
      No. 2, 1972, pp. 137-154

174. Dupuis, G. "Optimal Design of Statically Determinate Beams
      Subject to Displacement and Stress Constraints, AIAA
      Journal, Vol. 9, No. 5, 1971

175. Dupuis, G. "An Iterative Approach to Structural Optimization" International Journal for Numerical Methods in Engineering , Vol. 4, 1972, pp. 331-336

176. Bhargava, S. and Duffin, R.J. "Dual Extermum Principles Relating to Optimum Beam Design" Archives of Rational Mechanics and Analysis, Vol. 50, 1973, pp. 314-330

177. Simitses, G.J. and Kotras,T. "The Optimal Euler-Bernoulli Cantilever" ASCE Journal of Engineering Mechanics Division, Vol. 101, No EM6, 1975, pp. 922-929

178. Distefano, N. and Todeschini, R. "Invariant Imbedding and Optimum Beam Design with Displacement Constraints" International Journal of Solids and Structures, Vol. 8, No. 8, 1972, pp. 1073-1088

179. Huang, N.C. "Optimal Design of Elastic Beams for Minimum-Maximum Deflection" Journal of Applied Mechanics, Vol. 38, 1971, pp. 1078-1081

180. Komkov, V. and Coleman, N.P. "An Analytic Approach to Some Problems of Optimal Design of Beams and Plates" Archives of Mechanics", Vol. 27, No. 4, 1975, pp. 565-575

181. Cinquini, C. "Optimal Elastic Design for Prescribed Maximum Deflection" Journal of Structural Mechanics, Vol. 7, No. 1, 1979, pp. 21-34

182. Armand, J-L. "Applications of Optimal Control Theory to Structural Optimization: Analytical and Numerical Approach" Proceedings IUTAM Symposium on Optimization of Structural Design (Ed. A. Sawczuk and Z. Mroz) Springer-Verlag, 1975, pp. 15-39

183. Banichuk, N.V. "Game Problems in the Theory of Optimal Design" Optimization in Structural Design (Ed. A. Sawczuk and Z. Mroz) Springer-Verlag, New York 1975, pp. 111-121

184. Hegemier, G.A. and Tang, H.T. "A Variational Principle, The Finite Element Method, and Optimal Structural Design for Given Deflection" Optimization in Structural Design (Ed. A. Sawczuk and Z. Mroz) Springer-Verlag, New York, 1975, pp. 464-483

185. Erbatur, F. and Mengi, Y. "Optimal Design of Plates Under the Influence of Dead Weight and Surface Loading" Journal of Structural Mechanics, Vol. 5, 1977, pp. 345-356

54

186. Erbatur, F. and Mengi, Y. "On the Optimal Design of Plates for a Given Deflection" Journal of Optimization Theory and Applications, Vol. 21, 1977, pp. 103-110

187. Armand, J-L. and Lodier, B. "Optimal Design of Bending Elements" International Journal of Numerical Methods in Engineering, Vol. 13, 1978, pp. 373-384

188. Banichuk, N.V. "Optimal Elastic Plate Shapes in Bending Problems" MTT, Vol. 10, No. 5, 1975, pp. 151-158

189. Banichuk, N.V., Karelishvili, V.M. and Mironov, A.A. "Numerical Solution of Two-Dimensional Optimization Probelms for Elastic Plates", MTT, Vol. 12, No. 1, 1977, pp. 65-74

190. Banichuk, N.V., Karelishvili, V.M. and Mironov, A.A. "Optimization Problems with Local Performance Criteria in the Theory of Plate Bending" MTT, Vol. 13, No. 1, 1978, pp. 116-122

191. Banichuk, N.V. "Design of Plate for Minimum Deflection and Stress" Optimization of Distributed Parameter Structures (Ed. E.J. Haug and J. Cea) Sijthoff & Noordhoff, Alphen aan den Rijn, Netherlands, 1980

192. Simitses, G.J. "Optimal vs. the Stiffened Circular Plate" AIAA Journal, Vol. 11, No. 10, 1973, pp. 1409-1412

193. Sheu, C.Y. and Prager, W. "Minimum-weight Design with Piecewise Constant Specific Stiffness" Journal of Optimization Theory and Application, Vol. 2, 1968, pp. 179-189

194. Prager, W. "Optimality Criteria in Structural Design" Proceedings of National Academy of Sciences, Vol. 61, 1968, pp. 794-797

195. Hegemier, G.A. and Prager, W. "On Michell Trusses" International Journal of Mechanical Sciences, Vol. 11, 1969, p. 209

196. Martin, J.B. "The Optimal Design of Beams and Frames with Compliance" Internatioal Journal of Solids and Structures, Vol. 7, 1971, pp. 63-81

197. Huang, N-C. "Optimal Design of Elastic Structures for Maximum Stiffness" International Journal of Solids and Structures, Vol. 4, 1968, pp. 689-700

198. Huang, N.C. and Sheu, C.Y. "Optimal Design of Elastic Circular Sandwich Beams for Minimum Compliance" Journal of Applied Mechanics, Vol. 37, 1970, p. 569

199. Chern, J.M. and Prager, W. "Optimal Design for Prescribed Compliance Under Alternative Loads" Journal of Optimization Theory and Applications, Vol. 5, 1970, pp. 424-431

200. Chern, J.M. "Optimal Design of Beams for Alternative Loads and Constraints on generalized Compliance and Stiffness" International Journal of Mechanical Science, Vol. 13, No. 8, 1971, pp. 661-674

201. Save, M. "A General Criterion for Optimal Structural Design" Journal of Optimization Theory and Applications, Vol. 15, 1975, pp. 119-129

202. Masur E.F. "Optimum Stiffness and Strength of Elastic Structures" ASCE Journal of Engineering Mechanics Division, Vol. 95, No. EM5, 1970, pp. 621-640

203. Masur, E.F. "Optimal Structural Design for a Discrete Set of Available Structural Members" Computer Methods in Applied Mechanics and Engineering, Vol. 3, 1974, pp. 195-207

204. Masur, E.F. "Optimality in the Presence of Discreteness and Discontinuity" Optimization in Structural Design (Ed. A. Sawczuk and Z. Mroz), Springer-Verlag, 1975, pp. 441-453

205. Mroz, Z. "Multiparameter Optimal Design of Plates and Shells" Journal of Structural Mechanics, Vol.1, No. 3, 1973, pp. 371-392

206. Reiss, R. "Optimal Compliance Criterion for Axisymmetric Solid Plates" International Journal of Solids Structures, Vol. 12, 1976, pp. 319-329

207. Mroz, Z. and Garstecki, A. "Optimal Design of Structures with Unspecified Loading Distribution" Journal of Optimization Theory and Applications, Vol. 20, 1976, pp. 359-380

208. Szelag, D. and Mroz, Z. "Optimal Design of Elastic Beams with Unspecified Support Conditions" ZAMM, Vol. 58, 1978, pp. 501-510

209. Mroz, Z. "On Optimal Force Action and Reaction on Structures" Structural Control (Ed. H.H.E. Leipholz) North-Holland, 1980, pp. 523-544

56

210. Chern, J-M. and Prager, W. "Optimal Design of Rotating Disk for Given Radial Displacement of Edge" Journal of Optimization Theory and Applications, Vol. 6, No. 2, 1970, pp. 161-170

211. Chern, J-M. "Optimal Thermo-Elastic Deisgn for Given Deformation" Journal of Applied Mechanics, Vol. 38, No. 2, 1971, pp. 538-540

212. Fuchs, M.B. and Brull, M.A. "A New Strain Energy Theorem and Its Use in the Optimum Design of Continuous Beams" Computers and Structures, Vol. 10, 1979, pp. 647-657

213. Rossow, M.P. and Taylor, J.E. "A Finite Element Method for the Optimal Design of Variable Thickness Sheets" AIAA Journal, Vol. 11, 1973, pp. 1566-1569

214. Mroz, Z. "Optimal Design of Structures of Composite Materials" International Journal of Solids and Structures, Vol. 6, 1970, pp. 859-870

215. Icerman, L.J. "Optimal Structural Design for Given Dynamic Deflection" International Journal of Solids and Structures, Vol. 5, 1969, pp. 473-490

216. Mroz, Z. "Optimal Design of Structures Subject to Dynamic, Harmonically-Varying Loads," ZAMM, Vol. 50, 1970, pp. 303-309

217. Plaut, R.H. "Optimal Structural Design for Given Deflection Under Periodic Loading" Quarterly of Applied Mathematics, Vol. 29, 1971, pp. 315-318

218. Plaut, R.H. "Approximate Solutions to Some Static and Dynamic Optimal Structural Design Problems" Quarterly of Applied Mathematics, Vol. 31, 1973, pp. 535-539

219. Huang, N.C. "Minimum Weight Design of Vibrating Elastic Structures with Dynamic Deflection Constraint" Journal of Applied Mechanics, Vol. 43, 1976, pp. 171-180

220. Rockenback, P.C. Minimum-Mass Response-Constrained Design of Vibrating Sandwich Beams, Report R-604, Coordinated Science Laboratory, University of Illinois, Urbana, 1973

221. McNamara, R.J. "Turned Mass Dampers for Buildings" Journal of Structural Divison, ASCE, Vol. 103, No. ST9, 1977, pp. 1785-1798

222. Petersen, N.R. "Design of Large Scale Turned Mass Dampers" Preprint 3578, ASCE National Meeting, Boston, April, 1979

223. Wiesner, K.B. "Turned Mass Dampers to Reduce Building Wind Motion" Preprint 3510, ASCE National Meeting, Boston, April 1979

224. Rao, S.S. "Structural Optimization Under Shock and Vibration Environment" The Schock and Vibration Digest, Vol. 11, No. 2, 1979, pp. 3-12

225. Magne, R.W. "Optimization Techniques for Shock and Vibration Isolator Development" Shock and Vibration Digest, Vol. 11, No. 10, 1979, pp. 25-33

226. Brach, R.M. "Minimum Dynamic Response for a Class of Simply Supported Beam Shapes" International Journal Mechanical Science, Vol. 10, 1968, pp. 429-439

227. Brach, R.M. "Optimum Design of Beams for Sudden Loading" Journal of Engineering Mechanics Division, ASCE, Vol. 96, No. EM6, 1968, pp. 1395-1407

228. Plaut, R.H. "On Minimizing the Response of Structures to Dynamic Loading" ZAMP, Vol. 21, 1970, pp. 1004-1010

229. Pochtman, Y.M. "Optimization of Structures with Constraints on Dynamic and Frequency Characteristics" Doklady Akademii Nauk SSSR (in Russian) Vol. 203, No. 2, 1972, pp. 307-308, English Translation NASA TT F-14, 540, NASA, 1972

230. Fox, R.L. and Kapoor, M.P. "Structural Optimization in the Dynamics Response Regime: A Computational Approach" AIAA Journal, Vol. 8, No. 10, 1970, pp. 1798-1804

231. Cassis, J.H. and Schmit, L.A. "Optimum Design with Dynamic Constraints" Journal of Structural Division, ASCE, Vol. 102, No. ST10, 1976, pp. 2053-2071

232. Feng. T-T., Arora, J.S. and Haug, E.J. "Optimal Structural Design Under Dynamic Loads" International Journal of Numerical Methods in Engineering, Vol. 11, 1977, pp. 39-52

233. Arora, J.S. and Haug, E.J. "Optimum Structural Design with Dynamic Constraints" ASCE Journal of Structural Division, Vol. 103, No. ST10, 1977, pp. 2071-2074

58

234. Haug, E.J. and Feng, T-T. "Optimization of Distributed Parameter Structures under Dynamic Loads" Control and Dynamic Systems (Ed. C.T. Leandes), Vol. 13, 1977, pp. 207-246

235. Haug, E.J. and Feng, T-T. "Optimal Design of Dynamically Loaded Continuous Structures" International Journal of Numerical Methods in Engineering, Vol. 12, 1978, pp. 299-307

236. Haug, E.J., Arora, J.S. and Feng, T-T. "Sensitivity Analysis and Optimization of Structures for Dynamic Response", ASME Journal of Mechanical Design, Vol. 100, 1978, pp. 311-318

237. Hsiao, M.H., Haug, E.J. and Arora, J.S. "A State Space Method for Optimal Design of Vibration Isolators" ASME Journal of Mechanical Design, Vol. 101, 1979, pp. 309-314

238. Haug, E.J. and Arora, J.S. "Distributed Parameter Structural Optimization for Dynamic Response" Optimization of Distributed Parameter Structures (Ed. E.J. Haug and J. Cea) Sijthoff & Noordhoff, Alphen aan den Rijn, Netherlands, 1980

239. Kato, B., Nakamara, Y. and Anraku, H. "Optimum Earthquake Design of Shear Buildings" Journal of Engineering Mechanics Division, ASCE, Vol. 98, No. EM4, 1972, pp. 892-909

240. Venkayya, V.B. and Khot, N.S. "Design of Optimum Structures to Impulse Type Loading" AIAA Journal, Vol. 13, 1975, pp. 989-994

241. Cheng, F.Y. and Botkin, M.E., "Nonlinear Optimum Design of Dynamic Damped Frames" Journal of Structural Division, ASCE, Vol. 102, No. ST3, 1976, pp. 609-628

242. Cheng, F.Y. and Srifuengfung, D. "Earthquake Structural Design Based on Optimality Criterion" Sixth World Conference on Earthquake Engineering, Vol. 5, Earthquake Resistant Design, 1977

243. Cheng, F.Y. and Srifuengfung, D. "Optimal Structural Design for Simultaneous Multicomponent Static and Dynamic Input" International Journal of Numerical Methods in Engineering, Vol. 13, 1978, pp. 353-371

244. Ray, D., Pister, K.S. and Chopra, A.K., "Optimum Design of Earthquake-Resistant Shear Buildings" EERC 74-3, Earthquake Engineering Research Center, U. of Ca., Berkeley, January 1974

245. Vitiello, E., Pister, K.S. Applications of Reliability-Based Global Cost Optimization to Design of Earthquake-Resistant Structures, Report No. EERC 74-10, University of California, Berkeley, August 1974

246. Walker, N.D. and Pister, K.S. Study of a Method of Feasible Directions for Optimal Elastic Design of Framed Structures Subject to Earthquake Loading, EERC 75-39, Earthquake Engineering Research Center, Univ. of Calif., Berkeley, December 1975

247. Ray, D. Sensitivity Analysis for Hysteretic Dynamic Systems: Application to Earthquake Engineering, EERC 74-5, Earthquake Engineering Research Center, Univ. of Calif., Berkeley, April, 1974

248. Ray, D., Pister, K.S., and Polak, E. "Sensitivity Analysis for Hysteretic Dynamic Systems: Theory and Applications" Computer Methods in Applied Mechanics and Engineering, Vol. 14, 1978, pp. 179-208

249. Bhatti, M.A., Pister, K.S. and Polak, E. Optimal Design of an Earthquake Isolation Systems, Report No. EERC 78-22, Univ. of Calif., Berkeley, October, 1978

250. Bhatti, M.A., Polak, E. and Pister, K.S. OPTDYN-A General Purpose Optimization Program for Problems With or Without Dynamic Constraints, Report No. EERC 79-16, Univ. of Calif. Berkeley, July, 1979

251. Bhatti, M.A., Optimal Design of Localized Nonlinear Systems with Dual Performance Criteria Under Earthquake Excitations, Report No. EERC 79-15, Univ. of Calif., Berkeley, July, 1979

252. Pister, K.S. "Optimal Design of Structures Under Dynamic Loading" Optimization of Distributed Parameter Structures (Ed. E.J. Haug and J. Cea) Sijthoff & Noordhoff, Alphen aan den Rijn, Netherlands, 1980

253. Polak, E. "Algorithms for Optimal Design" Optimization of Distributed Parameter Structures (Ed. E.J. Haug and J. Cea) Sijthoff & Noordhoff, Alphen aan den Rijn, Netherlands, 1980

254. Bhatti, M.A., Essebo, T., Nye, W., Pister K.S., Polak, E., Sangiovanni-Vincentelli and Titz, A. "A Software System for Optimization-Based Interactive Computer-Aided Design" Optimization of Distributed Parameter Structures (Ed. E.J. Haug and J. Cea) Sijthoff & Noordhoff, Alphen aan den Rijn, Netherlands, 1980

255. Bhatti, M.A. and Pister, K.S. "Application of Optimal Design to Structures Subject to Earthquake Loading" Optimization of Distributed Parameter Structures, (Ed. E.J. Haug and J. Cea) Sijthoff & Noordhoff, Alphen aan den Rijn, Netherlands, 1980

256. Levy, H.J. and Wolf, B.M. "Fully Stressed Dynamically Loaded Structures" ASME Paper 74-WA/DE-19, ASME, New York, 1974

257. Komkov, V. Optimal Control Theory for the Damping of Elastic Vibrations of Simple Elastic Systems, Springer-Verlag, Berlin, 1972

258. Komkov, V. and Coleman, N. "Optimality of Design and Sensitivity Analysis of Beam Theory" International Journal of Control, Vol. 18, No. 4, 1973, pp. 731-740

259. Thermann, K. "Optimal Design Criteria of Dynamically Loaded Elastic Structures" Optimization in Structural Design (Ed. A. Sawczuk and Z. Mroz) Springer-Verlag, New York, 1975, pp. 152-167

260. Zienkiewicz, O.C. and Campbell, J.S. "Shape Optimization and Sequential Linear Programming" Optimum Structural Design (Ed. R.H. Gallagher and O.C. Zienkiewicz) Wiley, New York, 1973, pp. 109-126

261. Ramakrishnan, C.V. and Francavilla, A. "Structural Shape Optimization Using Penalty Functions", Journal of Structural Mechanics, Vol. 3, No. 4, 1975, pp. 403-432

262. Francavilla, A., Ramakrishnan, C.V. and Zienkiewicz, O.C. "Optimization of Shape to Minimize Stress Concentratrion" Journal of Strain Analysis, Vol. 10, 1975, pp. 63-70

263. Schnack, E. "An Optimization Procedure for Stress Concentrations by the Finite Element Technique" International Journal for Numerical Methods in Engineering, Vol. 14, 1979, pp. 115-124

264. Oda, J. "On A Technique to Obtain an Optimum Strength Shape by the Finite Element Method" Bulletin of the JSME, Vol. 20, 1977, pp. 160-167

265. Tvergaard, V. "On the Optimum Shape of a Fillet in a Flat Bar with Restrictions" Optimization in Structural Design (Ed. A. Sawczuk and Z. Mroz), Springer-Verlag, New York, 1975, pp. 181-195

266. Kristensen, E.S. and Madsen, N.F. "On the Optimum Shape of Fillets in Plates Subjected to Multiple In-Plane Loading Cases" International Journal of Numerical Methods in Engineering, Vol. 10, 1976, pp. 1007-1019

267. Bhavikatti, S.S. and Ramakrishnan, C.V. "Optimum Design of Fillets in Flat and Round Tensions Bars" ASME Paper, 77-DET-45, 1977

268. Chun, Y.W. and Haug, E.J. "Two Dimensional Shape Optimal Design" International Journal of Numerical Methods in Engineering, Vol. 13, 1978, pp. 311-336

269. Chun, Y.W. and Haug, E.J. "Shape Optimal Design of an Elastic Body of Revolution" Preprint No. 3516, ASCE Annual Meeting, Boston, April 1979

270. Rousselet, B. and Haug, E.J. "Design Sensitivity Analysis in Structural Mechanics III: Shape Variation" Optimization of Distributed Parameter Structures (Ed. E. J. Haug and J. Cea) Sijthoff & Noordhoff, Alphen aan den Rijn, Netherlands, 1980

271. Kunar, R.R. and Chan, A.S.L. "A Method for the Configurational Optimization of Structures" Computer Methods In Applied Mechanics and Engineering, Vol. 7, 1976, pp. 331-350

272. Dems, K. and Mroz, Z. "Multiparameter Structural Shape Optimization by the Finite Element Method" International Journal of Numerical Methods In Engineering, Vol. 13, 1978, pp. 247-263

273. Henry. A.S. The Analytic Design of Torsion Members, Ph.D. Thesis, University of Iowa, 1971

274. Banichuk, N.V. "Optimization of Elastic Bars in Torsion" International Journal of Solids and Structures, Vol. 12, 1976, pp. 275-286

275. Banichuk, N.V. "On a Variational Problem with Unknown Boundaries and the Determination of Optimal Shapes of Elastic Bodies" PMM, Vol. 39, No. 6, 1975, pp. 1037-1047

276. Kurshin, L.M. and Onoprienko, P.N. "Determination of the Shapes of Doubly-Connected Bar Sections of Maximum Torsional Stiffness" PMM, Vol. 40, No. 6, 1976, pp. 1020-1026

62

277. Banichuk,N.V. "On a Two Dimensional Optimization Problem in Elastic Bar Torsion Theory" Soviet Applied Mechanics, Vol. 11, No. 5, 1976, pl. 38-44

278. Gurvitch, E.L. "On Isoperimetric Problems for Domains with Partly Known Boundaries" Journal of Optimization Theory and Applications, Vol. 20, No. 1, 1976, pp. 65-79

279. Dems, K. "Multiparameter Shape Optimization of Elastic Bars In Torsion" International Journal for Numerical Methods in Engineering, Vol. 15, 1980, pp. 1517-1539

280. Neuber, H. "Der Zugbeanspruchte Flachstab Mit Optimalem Querschnittubergang" Forsch Ingenieurwesen, Vol. 35, 1969, pp. 29-30

281. Neuber, H. "Zur Optimierung Der Spannungskonzentration" in Continuum Mechanics and Related Problems of Analysis, Nauka, Moscow, 1972, pp. 375-380

282. Cherepanov, G.P. "Inverse Problems of the Plane Theory of Elasticity" PMM, Vol. 38, No. 6, 1974, pp. 963-979

283. Wheeler, L. "On the Role of Constant-Stress Surfaces in the Problem of Minimizing Elastic Stress Concentration" International Journal of Solids and Structures, Vol. 12, 1976, pp. 779-789

284. Bjorkman, G.S. and Richards, R. "Harmonic Holes - An Inverse Problem in Elasticity" Journal of Applied Mechanics, Vol. 43, 1976, pp. 414-418

285. Banichuk, N.V. "Optimality Conditions in the Problem of Seeking the Hole Shapes in Elastic Bodies" PMM, Vol. 41, No. 5, 1977, pp. 920-925

286. Banichuk, N.V. "Optimizing Hole Shape in Plates Working in Bending" Soviet Applied Mechanics, Vol. 12, No. 3, 1977, pp. 72-78

287. Banichuk, N.V. and Karihaloo, B.L. "Minimum Weight Design of Multi-Purpose Cylindrical Bars" International Journal of Solids and Structures, Vol. 12, 1976, pp.267-273

288. Parbery, R.D. and Karihaloo, B.L. "Minimum-Weight Design of Hollow Cylinders for Given Lower Bounds on Torsional and Flexural Rigidities" International Journal of Solids and Structures, Vol. 13, 1977, pp. 1271-1280

289. Cherkaev, A.V. "On the Question of Formulating the Problem of Optimal Design of Freely Oscillating Structures" PMM, Vol. 42, No. 1, 1978, pp. 194-197

290. Durelli, A.J. and Rajaiah, K. "Optimum Hole Shapes in Finite Plates Under Uniaxial Load" Journal of Applied Mechanics, Vol. 46, 1979, pp. 691-695

291. Durelli, A.J., Rajaiah, K., Hovanesian, J.D. and Hung, Y.Y., "General Method to Directly Design Stress-Wise Optimum Two-Dimensional Structures", Mechanics Research Communications, Vol. 6, 1979, pp. 159-165

292. Hersch, J. and Payne, L.E. "Extremal Principles and Isoperimetric Inequalities for Some Mixed Problems of Stekloff's Type" ZAMP, Vol. 19, 1968, pp. 802-817

293. Cea, J. "Solution of a Model Problem by Variational Methods and Examples of Problems of Shape Optimal Design" Optimization of Distributed Parameter Structures (Ed. E. J. Haug and J. Cea) Sijthoff & Noordhoff, Alphen aan den Rijn, Netherlands, 1980

294. Cea, J. "Definition of Boundaries for Shape Design" Optimization of Distributed Parameter Structures (Ed. E.J. Haug and J. Cea) Sijthoff & Noordhoff, Alphen aan den Rijn, Netherlands, 1980

295. Cea, J. "Continuous Steepest Descent in Hilbert Space and 'Domain Spaces'", Optimization of Distributed Parameter Structures (Ed. E.J. Haug and J. Cea) Sijthoff & Noordhoff, Alphen aan den Rijn, Netherlands, 1980

296. Zolesio, J.P. "The Material Derivative (Or Speed) Method for Shape Optimization" Optimization of Distributed Parameter Structures (Ed. E.J. Haug and J. Cea) Sijthoff & Noordhoff, Alphen aan den Rijn, Netherlands, 1980

297. Zolesio, J.P. "Speed Method in Several Examples" Optimization of Distributed Parameter Structures (Ed. E.J. Haug and J. Cea) Sijthoff & Noordhoff, Alphen aan den Rijn, Netherlands, 1980

298. Rousselet, B. "Implementation of Shape Optimal Design Algorithms" Optimization of Distributed Parameter Structures (Ed. E.J. Haug and J. Cea) Sijthoff & Noordhoff, Alphen aan den Rijn, Netherlands, 1980

299. Cea, J. "Other Methods in Shape Optimal Design" <u>Optimization of Distributed Parameter Structures</u> (Ed. E.J. Haug and J. Cea) Sijthoff & Noordhoff, Alphen aan den Rijn, Netherlands, 1980

300. Rousselet, B. "Dependence of Eigenvalues on Shape" <u>Optimization of Distributed Parameter Structures</u> (Ed. E.J. Haug and J. Cea) Sijthoff & Noordhoff, Alphen aan den Rijn, Netherlands, 1980

301. Prager, W. and Shield, R.T. "Optimal Design of Multi-Purpose Structures" <u>International Journal of Solids and Structures</u>, Vol. 4, 1968, pp. 469-475

302. Martin, J.B. "Optimal Design of Structures for Multipurpose Loading" <u>Journal of Optimization Theory and Applications</u>, Vol. 6, No. 1, 1970, pp. 22-40

303. Chern, J-M. and Martin, J.B. "The Multipurpose Optimal Design of Elastic Structure with Piecewise Uniform Cross Section" <u>ZAMP</u>, Vol. 22, No. 5, 1971, pp. 834-855

304. Sherman, Z. and Wang, P-C. "Volume Minimization of Thin Plates Subject to Constraints" <u>ASCE Journal of Engineering Mechanics Division</u>, Vol. 97, No. EM3, 1971, pp. 741-754

305. Sherman, Z. "Weight Minimization of Axisymmetric Clamped Plates Subject to Constraints" <u>International Journal of Solids and Structures</u>, Vol. 9, 1973, pp. 279-290

306. Karihaloo, B.L. "Optimal Design of Multi-Purpose Tie-Beams" <u>Journal of Optimization Theory and Applications</u>, Vol. 27, No. 3, 1979, pp 427-438

307. Karihaloo, B.L. and Parbery, R.D. "Optimal Design of Multi-Purpose Beam-Columns" <u>Journal of Optimization Theory and Applications</u>, Vol. 27, No. 3, pp. 439-448

308. Karihaloo, B.L. "Optimal Design of Multi-Purpose Structures" <u>Journal of Optimization Theory and Applications</u>, Vol. 27, No. 3, pp. 449-461

309. Karihaloo, B.L. and Wood, G.L. "Optimal Design of Multi-Purpose Sandwich Tie-Columns" <u>ASCE Journal of Engineering Mechanics Division</u>, Vol 105, No. EM3, 1979, pp. 465-469

310. Karihaloo, B.L. "Optimal Design of Multi-Purpose Tie Column of Solid Construction" <u>International Journal of Solids and Structures</u>, Vol. 15, 1979, pp. 103-109

311. Parbery, R.D. and Karihaloo, B.L. "Minimum-Weight Design of Thin-Walled Cylinders Subject to Flexural and Torsional Stiffness Constraints" Journal of Applied Mechanics, Vol. 47, 1980, pp. 106-110

312. Haug, E.J. and Komkov, V. "Sensitivity Analysis in Distributed-Parameter Mechanical Systems Optimization" Journal of Optimization Theory and Applications, Vol. 23, No. 3, 1977, pp. 445-464

313. Haug, E.J. and Arora, J.S. "Design Sensitivity Analysis of Elastic Mechanical Systems" Computer Methods in Applied Mechanics and Engineering, Vol. 15, 1978, pp. 35-62

314. Huang, N.C. "Effect Shear Deformation on Optimal Design of Elastic Beams" International Journal of Solids and Structures, Vol. 7, 1971, pp. 321-326

315. Seiranyan, A.P. "Elastic Plates and Beams of Minimum Weight with Several Types of Bending Loads" MTT, Vol. 8, No. 5, 1973, pp. 83-89

316. Seiranyan, A.P. "Optimal Beam Design with Limitations on Natural Vibration Frequency and Buckling Load" MTT, Vol. 11, No. 1, 1976, pp. 133-138

317. Seiranyan, A.P. "Quasioptimal Solutions of Optimal Design Probelms with Various Constraints" Soviet Applied Mechanics, Vol. 13, No. 6, 1977, pp. 544-550

318. Gura, N.M. and Seiranyan, A.P. "Optimum Circular Plate with Constraints on the Rigidity and Frequency of Natural Oscillations" MTT, Vol. 12, No. 1, 1977, pp. 129-136

319. Seiranyan, A.P. "Homogeneous Functionals and Structural Optimization Problems" International Journal of Solids and Structures, Vol. 15, 1979, pp. 749-759

320. Velte, W. and Villaggio, P. "Are the optimum Problems in Structural Design Well Posed?" Archives of Rational Mechanics and Analysis, to appear 1981

321. Klosowicz, B. and Lurie, K.A. "On the Optimal Nonhomogeneity of A Torsional Elastic Bar" Archives of Mechanics, Vol. 24, No. 2, 1971

322. Banichuk, N.V. "One Extremum Problem for a System with Distributed Parameters and Determination of the Optimal Properties of an Elastic Medium" DALK, AM SSSR, Vol. 242, No. 5, 1978

323. Banichuk, N.V. "Optimal Anisotropy of Rods in Torsion" MTT, No. 4, 1978

324. Banichuk, N.V. "Optimization of Anisotropic Properties of Deformable Media in Plane Problems of Elasticity" MTT, Vol. 14, No. 1, 1979, pp. 63-68

325. Hirano, Y. "Optimum Design of Laminated Plates Under Axial Compression" AIAA Journal, Vol. 17, No. 9, 1979, pp. 1017-1019

326. Banichuk, N.V. "Minimax Approach to Structural Optimization Problems" Journal of Optimization Theory and Applications, Vol. 20, No. 1, 1976, pp. 111-128

327. Banichuk, N.V. "On a Game Problem of Optimizing Elastic Bodies" Soviet Math Dakl, Vol. 226, No. 3, 1976, pp. 117-120

328. Aristov, M.V. and Troitskii, V.A. "Elastic Circular Plate of Minimum Weight" MTT, No. 3, 1975, pp. 153-156

329. Samsonov, A.M. "Optimum Location of Thin Elastic Rib on Elastic Plate" MTT, Vol. 13, No. 1, 1978, pp. 121-129

330. Solodovnikov, V.N. "Optimization of Elastic Shells of Revolution" PMM, Vol. 42, No. 3, 1978, pp. 511-520

331. Olhoff, N. and Taylor, J.E. "On Optimal Structural Remodelling" Journal of Optimization Theory and Applications Vol. 27, No. 4, 1979, pp. 571-581

332. Taylor, J.E. "Optimal Remodelling Theory and Applications" Optimization of Distributed Parameter Structures (Ed. E.J. Haug and J. Cea) Sijthoff & Noordhoff, Alphen aan den Rijn, Netherlands, 1980

333. Batterman, S.C. and Pavicic, N. "Optimum design of Fiber Reinforced Shells of Revolution" Optimization in Structural Design, Springer-Verlag, New York, 1975

334. Taylor, J.E. "On Variational Formulations for Structures Design Problems" Optimization in Structural Design (Ed. Sawczuk and Mroz), Springer-Verlag, New York, 1975, pp.60-67

335. Rozvany, G.I.N. "Analytical Treatment of Some Extended Problems in Structural Optimization" Journal of Structural Mechanics, Vol. 3, No. 4, 1974-75, pp. 359-385

336. Conry, T.F. and Seireg, A. "A Mathematical Programming Method for Design of Elastic Bodies in Contact" Journal of Applied Mechanics, Vol. 48, 1971, pp. 387-392

337. Haug, E.J. and Kwak,B.M. "Contact Stress Minimization by Contour Design" International Journal of Numerical Methods in Engineering, Vol. 12, 1978, pp. 917-930

338. Lukasiewicz, S.A. "Optimum Design in Junction and Contact Problems" Colloquium No. 110, Contact Problems and Load Transfer in Mechanical Assemblages, Linkoping, Sweden, 1978

339. Benedict, R.L. "Optimal Design for Elastic Bodies in Contact" Optimization of Distributed Parameter Structures (Ed. E.J. Haug and J. Cea) Sijthoff & Noordhoff, Alphen aan den Rijn, Netherlands, 1980

340. Youssef, N.A.N. and Popplewell, N. "A Theory of the Greatest Maximum Response of Linear Structures" Journal of Sound and Vibration, Vol. 56, 1978, pp. 21-33

341. Jacquot, R.G. "Optimal Dynamic Vibration Absorbers for General Beam Systems" Journal of Sound and Vibration, Vol. 60, 1978, pp. 535-542

342. Rozvany, G.I.N. "Grillages of Maximum Strength and Maximum Stiffness" International Journal of Mechanical Science, Vol. 14, 1972, pp. 651-666

343. Prager, W. and Rozvany, G.I.N. "Optimal Layout of Grillages" Journal of Structural Mechanics, Vol. 5, 1977, pp. 1-18

344. Hill, R.D. and Rozvany, G.I.N. "Optimal Beam Layouts: the Free Edge Paradox" Journal of Applied Mechanics, Vol. 44, 1977, pp. 696-700

345. Rozvany, G.I.N. "A New Class of Structural Optimization Problems: Optimal Archgirds" Computer Methods in Applied Mechanics and Engineering, Vol. 19, 1979, pp. 127-150

346. Rozvany, G.I.N. "Optimality Criteria for Grids, Shells and Arches" Optimization of Distributed Parameter Structures (Ed. E.J. Haug and J. Cea) Sijthoff & Noordhoff, Alphen aan den Rijn, Netherlands, 1980

68

347. Krivenkov, Iu.P. "Sufficient Conditions of Optimality in Linear Problems of the Mathematical Theory of Optimal Processes with Phase Constraints" PMM, Vol. 42, 1978, pp. 623-632

348. Stadler, W. "Natural Structural shapes of Shallow Arches" Journal of Applied Mechanics, Vol. 44, 1977, pp. 291-298

349. Stadler, W., "Uniform Shallow Arches of Minimum Weight and Minimum Maximum Deflection" Journal of Optimization Theory and Applications, Vol. 23, 1977, pp. 137-164.

350. Stadler, W. "Natural Structural Shapes (The Static Case), Quarterly Journal of Mechanics And Applied Mathematics, Vol. 31, 1978, 169-217

351. Carmichael, D.G. "Computation of Pareto Optima In Structural Design" International Journal of Numerical Methods in Engineering, Vol. 15, 1980, pp. 925-952

AUTHOR INDEX

Author	Reference Number
Adali, S.	49
Ainola, L.I.	159
Alblas, J.B.	39
Allentuch, A.	87, 88
Amazigo, J.C.	94
Andreev, L.V.	82
Anraku, H.	239
Aristov, M.V.	328
Armand, J.L.	128, 139, 140, 141, 182, 187
Arora, J.S.	13, 108, 232, 233, 236, 237, 238, 313
Banichuck, N.V.	14, 60, 68, 153, 183, 188, 189, 190, 191, 274, 275, 277, 285, 286, 287, 322, 323, 324, 326, 327
Barnes, E.R.	38
Barnett, R.L.	160
Barr, A.D.S.	132, 133
Batterman, S.C.	333
Benedict, R.L.	339
Bert, C.W.	154
Bhargava, S.	176
Bhatti, M.A.	249, 250, 251, 254, 255
Bhavikatti, S.S.	267
Bjorkman, G.S.	284
Blachut, J.	69, 92
Boltyanskii, V.G.	23
Botkin, M.E.,	241
Boykin, W.H.	52
Brach, R.M.	103, 104, 106, 226, 227
Brandt, A.	1
Brull, M.A.	212
Bryson, A.E.	25
Budiansky, B.	93
Butkovskiy, A.G.	17
Campbell, J.S.	260
Cantu, E.	165
Cardou,A.	101
Carmichael, D.	156, 157, 158, 351
Cassis, J.H.	231
Cea, J.	26, 293, 294, 295, 299
Chan, A.S.L.	271
Cheng, F.Y.	241, 242, 243,
Cheng, K.T.	145
Cherepanov, G.P.	282
Cherkaev, A.V.	146, 147, 289
Chern, J.M.	164, 199, 200, 210, 211, 303
Choi, K.K.	71

A REVIEW OF THE BASIS FOR OPTIMALITY CRITERIA METHODS

John E. Taylor

The University of Michigan, Ann Arbor, Michigan,
48109, USA

ABSTRACT

This paper presents a review of the basic optimality
criteria methods employed for structural optimization.

1. INTRODUCTION

In the jargon of literature in the field of structural
optimization, the label Optimality Condition is identified with
necessary conditions from the calculus of extremum problems.
Often it is simply the Euler-Lagrange equation or some reduced
or equivalent form of the equation that has this label attached
to it. Associated techniques for the computational solution of
optimal structural design problems are called Optimality Criteria
Methods. The objective in this review is to provide a partial
survey of optimality conditions, mainly in relation to the variety
of problem types and to approaches for problem formulation. The
focus is on ideas, methods, and interpretations. Detailed
developments are not reproduced here.

The author wishes to acknowledge all contributions of the
many research workers around the world who have participated in
the development of material related to the subject of this
article. The reader is referred to survey articles and tracts
cited in this section [1-9] and elsewhere in these proceedings
for lists of the extensive literature covering such contributions.
Explicit citations to the literature are limited in the present
text to a relatively few of the many sources related to the
particular material under consideration. Thus the coverage is

cursory rather than comprehensive, in reference both to specific content and to source.

2. REVIEW

Results reported by Michell [10] represent one of the earliest systematic developments in this field. Michell Truss is the name given to the layout of a curvi-linear, dense, fiber mesh predicted from the requirement that the strains for the loaded system have constant value throughout the field occupied by the mesh. This requirement comprises an optimality condition, or optimality criterion. It is a necessary condition for the minimization of total weight of the mesh, within a lower bound on fiber stress. Prager [11], among others [12,13], has made substantial contributions to the refinement and extension of the Michell theory.

Much more recently, Rozvany completed a very thorough study on the design of networks in flexure, or grillages [6]. The layouts and distributions predicted according to his theory represent, for fields subject to bending loads, the counterpart of what Michell designs are for fibers under direct stress. In both cases, the exploitation of optimality criteria leads to the prediction of a (generally unique) design, for specified loads and available supports. However the set of criteria for the flexure problem is rather more elaborate than the rules from Michell theory. The intricacies of this theory for optimal design of flexural systems are exposed in detail by Rozvany in Refs. 14 and 15.

The problem of optimal design relative to plastic collapse has received substantial treatment, especially during the period 1955 to 1970 [16-24]. For the unconstrained design of continuum structures, in which local stiffness depends linearly on the design variable, the optimality condition states that the local dissipation rate per unit volume of material has constant value over the optimal structure. The condition proves to be sufficient, as well as necessary for the prediction of minimum weight design. Similar results are established for a variety of types of design problems, e.g. in Refs. 25 and 26. Works of Wasiutynski on "Methods of Strength Design" presented for some cases the earliest exposition of optimality conditions (see, for example, references 42,108,109 cited in Ref. 1).

The design of Euler columns is considered next, as an example problem in optimal design for continuum elastic structures. On the bases that the change in eigensolution is smooth relative to change in design in the neighborhood of the optimal solution, Keller and Tadjbakhsh [27] derive the optimality condition for columns through a perturbation applied to the boundary value problem statement. An isoperimetric constraint is imposed, whereby the set of comparison designs is limited to those column shapes that have equal volume of structural material. According to the optimality conditions for this problem, the (Local) measure of maximum strain energy per unit length per unit cross-sectional area has constant value over the column. In the case of members with stiffness proportional to cross-sectional area, this measure equals strain energy per unit volume (specific strain energy).

Others subsequently derived the same results by application to different forms of statement for the variational problem. For example, Keller and Niordson [28] set up the problem in a form that reflects maximization of the eigenvalue, within the isoperimetric constraint on volume of material. Niordson and Olhoff, among others, used a similar form of problem statement to represent the optimal design problem for free-vibration of elastic structures [29-35]. The optimality criterion for dynamic problems states that; for the optimal design the sum of the unit strain energy plus unit kinetic energy is constant over the structure. This result is discussed along with many important refinements and applications by Olhoff in Refs. 36-38.

Optimality conditions for the column problem and for free vibrations were demonstrated independently using an energy formulation [39-41]. Martin [42] derived optimality criteria for a variety of problems, using Complementary Energy to model the mechanics. Prager's very broad contributions to the field of structural optimization also includes results on these problems (e.g. Refs. 9 and 43). Optimality criteria that are applicable for optimal design relative to dissipative, gyroscopic, or circulatory loads were presented by Plaut [44]. His derivation is facilitated by the introduction of an adjoint system corresponding to the equations of the original problem. Plaut furnishes the details in Ref. 45, as well as results for an interesting set of example problems.

For the situation of multipurpose design, the optimality condition expresses the requirement that the average unit energies over all, or some combination of the prescribed set of loads may dominate the design locally (see, for example Refs. 46 to 52).

Special considerations that must be made in order to identify singularities in the solutions for free-vibration design problems are described in detail by Olhoff [53]. Megarefs and Hodge [54] discuss examples that arose in frame design problems. Masur and Mroz [55] provide an elaboration on the statement of optimality conditions, as the means to handle problems where singularities may arise. Their results are also described in Ref. 56. Olhoff and Rasmussen [57] established the form of optimality condition that applies in column design for the case in which the response of the optimum structure is bimodal. This optimality condition is expressed in terms of the average over response modes of unit strain energy. Somewhat more general forms of optimality conditions that result from the consideration of Optimal Remodeling are discussed in Ref. 58.

A special issue of International Journal for Numerical Methods in Engineering [59] includes an outstanding collection of papers on optimality criteria methods themselves.

REFERENCES

1. Wasiutyński, Z. and Brandt, A., "The Present State of Knowledge in the Field of Optimum Design of Structures," Applied Mechanics Reviews, vol. 16, 1963, pp. 344-350.
2. Sheu, C.Y. and Prager, W., "Recent Developments in Optimal Structural Design," Applied Mechanics Reviews, vol. 21, 1968, pp. 985-992.
3. Niordson, F.I. and Pedersen, P., "A Review of Optimal Structural Design," Proc. 13th Int. Cong. Th. Appl. Mech., (ed. E. Becker and G.K. Mikhailov), Moscow, Springer-Verlag, 1973, pp. 264-278.
4. Barnett, Ralph L., "Survey of Optimum Structural Design," Experimental Mechanics, vol. 6, no. 12, 1966, pp. 19A-26A.
5. Reitman, M.I. and Shapiro, G.S., "Optimization of Structures under Dynamic Loads," (in Russian), The All-Union Symposium "On the Problems of Optimization in Mech. of Solid Deformable Bodies," Vilnius Civil Engineering Inst., Vilnius, USSR, 1974.
6. Rozvany, G.I.N., Optimal Design of Flexural Systems, Pergamon Press, 1976.
7. Olhoff, N., "A Survey of the Optimal Design of Vibrating Structural Elements," Shock Vib. Dig., vol. 8, No. 8, 1976, pp. 3-10 (Part I: Theory), vol. 8, No. 9, 1976, pp. 3-10 (Part II: Applications).
8. Prager, W., "Survey Paper: Optimization of Structural Design," J. Optimiz. Theory Appl., Vol. 6, No. 1, 1970, pp. 1-21.
9. Prager, W., Introduction to Structural Optimization, Courses and Lectures Int. Centre for Mech. Sci., Udine, No. 212, Springer-Verlag, Vienna, 1974.
10. Michell, A.G.M., "The Limits of Economy of Material in Frame Structures," Phil. Mag. S6, Vol. 8, No. 47, 1904, pp. 589-597.
11. Prager, W., "A Note on Discretized Michell Structures," Comp. Meth. Appl. Mech. Eng., Vol. 3, No. 3, 1974, pp. 349-355.

12. Hemp, W.S., "Studies in the Theory of Michell Structures," Proc. 11th Int. Congr. Appl. Mech., Munich, 1964, pp. 621-628.
13. Hegemier, G.A. and Prager, W., "On Michell Trusses," Int. J. Mech. Sci., Vol. 11, No. 2, 1969, pp. 209-215.
14. Rozvany, G.I.N., "Variational Methods and Optimality Criteria," Optimization of Distributed Parameter Structures, (Ed. E.J. Haug and J. Cea), Sijthoff & Nordhoff, Alphen aan den Rijn, Netherlands, 1980.
15. Rozvany, G.I.N., "Optimality Criteria for Grids, Shells, and Arches," Optimization of Distributed Parameter Structures, (Ed. E.J. Haug and J. Cea), Sijthoff & Nordhoff, Alphen aan den Rijn, Netherlands, 1980.
16. Drucker, D.C. and Shield, R.T., "Design for Minimum Weight," Proc. 9th Int. Congr. on Appl. Mech., Brussells, Book 5, 1956, pp. 212-222.
17. Drucker, D.C. and Shield, R.T., "Bounds on Minimum Weight Design," Quart. Appl. Math., Vol. 15, No. 3, 1957, pp. 269-281.
18. Prager, W., "Minimum Weight Design of Plates," De. Ingenieur, Vol. 67, 1955, pp. 0.141-0.142.
19. Mroz, Z., "The Load Carrying Capacity and Minimum Weight of Annular Plates," Rozpr. Inzyn, Vol. 4, 1958, pp. 605-625.
20. Prager, W. and Shield, R.T., "Minimum Weight Design of Circular Plates under Arbitrary Loading," Zeit. Ang. Math. Phys., Vol. 10, No. 4, 1959, pp. 421-426.
21. Mroz, Z., "On the Problems of Minimum Weight Design," Quart. Appl. Math., Vol. 19, No. 2, 1961, pp. 127-135.
22. Prager, W. and Shield, R.T., "A General Theory of Optimal Plastic Design," J. Appl. Mech., Vol. 34, No. 1, 1967, pp. 184-186.
23. Heyman, J., "On the Absolute Minimum Weight Design of Framed Structures," Quart. J. Mech. Appl. Math., Vol. 12, No. 3, 1959, pp. 314-324.
24. Hemp, W.S., Optimum Structures, Clarendon Press, 1973.
25. Prager, W. and Taylor, J.E., "Problems of Optimal Structural Design," J. Appl. Mech., Vol. 35, No. 1, 1968, pp. 102-106.
26. Save, M.A., "A Unified Formulation of the Theory of Optimal Plastic Design with Convex Cost Function," J. Struct. Mech., Vol. 1, No. 2, 1972, pp. 267-276.
27. Tadjbakhsh, I. and Keller, J.B., "Strongest Columns and Isoperimetric Inequalities for Eigenvalues," J. Appl. Mech., Vol. 9, 1962, pp. 159-164.
28. Keller, J.B., "The Tallest Column," J. Math. Mech., Vol. 16, No. 5, 1966, pp. 433-446.
29. Niordson, F.I., "On the Optimal Design of a Vibrating Beam," Quart. Appl. Mech., Vol. 23, 1954, pp. 47-53.
30. Olhoff, N., "Optimal Design of Vibrating Circular Plates," Int. J. Solids Structures, Vol. 6, 1970, pp. 139-156.
31. Olhoff, N., "Optimal Design of Vibrating Rectangular Plates," Int. J. Solids Structures, Vol. 10, 1974, pp. 93-109.

32. Sheu, C.Y., "Elastic Minimum Weight Design for Specified Fundamental Frequency," *Int. J. Solids Structures*, Vol. 4, 1968, pp. 953-958.

33. Brach, R.M., "On Optimal Design of Vibrating Structures," *J. Optimization Theory and Applications*, Vol. 11, 1973, pp. 662-667.

34. Cardou, A. and Warner, W.H., "Minimum-Mass Design of Sandwich Structures with Frequency and Section Constraints," *J. Optimization Theory and Applications*, Vol. 14, 1974, pp. 633-647.

35. Warner, W.H. and Vavrick, D.J., "Optimal Design in Axial Motion for Several Frequency Constraints," *J. Optimization Theory and Applications*, Vol. 15, 1975, pp. 157-166.

36. Olhoff, N., "Optimization of Columns Against Buckling," *Optimization of Distributed Parameter Structures* (Ed. E.J. Haug and J. Cea), Sijthoff & Nordhoff, Alphen aan den Rijn, Netherlands, 1980.

37. Olhoff, N., "Optimization of Transversly Vibrating Beams and Rotating Shafts," *Optimization and Distributed Parameter Structures*, (Ed. E.J. Haug and J. Cea), Sijthoff & Nordhoff, Alphen aan den Rijn, Netherlands, 1980.

38. Olhoff, N. and Cheng, K.-T., "Optimal Design of Solid Elastic Plates," *Optimization and Distributed Parameter Structures*, (Ed. E.J. Haug and J. Cea), Sijthoff & Nordhoff, Alphen aan den Rijn, Netherlands, 1980.

39. Taylor, J.E., "Minimum-Mass Bar for Axial Vibration at Specified Natural Frequency," *AIAA Journal*, Vol. 5, 1967, pp. 1911-1913.

40. Taylor, J.E., "The Strongest Column: An Energy Approach," *J. Appl. Mech.*, Vol. 34, 1967, p. 486.

41. Taylor, J.E. and Liu, C.Y., "Optimal Design of Columns," *AIAA Journal*, Vol. 6, 1968, pp. 1497-1502.

42. Martin, J.B., "The Optimal Design of Beams and Frames with Compliance Constraints," *Int. J. Solids Struct.*, Vol. 7, No. 1, 1971, pp. 63-81.

43. Prager, W., "Methods of Structural Optimization," *Int. Symp. Com. Meth. Appl. Sci. Eng.*, Springer-Verlag, Berlin, 1974.

44. Plaut, R., "On the Optimal Structural Design for a Non-Conservative Elastic Stability Problem," *J. Optimiz. Theory Appl.*, Vol. 7, No. 1, 1971, pp. 52-60.

45. Weisshaar, T.A. and Plaut, R.H., "Structural Optimization Under Nonconservative Loading," *Optimization and Distributed Parameter Structures*, (Ed. E.J. Haug and J. Cea), Sijthoff & Nordhoff, Alphen aan den Rijn, Netherlands, 1980.

46. Shield, R.T., "Optimum Design Methods for Multiple Loading," *Zei. Ang. Math. Phys.*, Vol. 14, No. 1, 1963, pp. 38-45.

47. Prager, W., "Optimum Plastic Design of a Portal Frame for Alternative Loads," *J. Appl. Mech.*, Vol. 34, No. 3, 1967, pp. 772-773.

48. Mayeda, R. and Prager, W., "Minimum-Weight Design of Beams for Multiple Loading," Int. J. Solids Struct., Vol. 3, No. 6, 1967, pp. 1001-1011.
49. Prager, W. and Shield, R.T., "Optimal Design of Multipurpose Structures," Int. J. Solids Struct., Vol. 4, No. 4, 1968, pp. 469-475.
50. Martin, J.B., "Optimal Design of Elastic Structures for Multi-purpose Loading," J. Optimiz. Theory Appl., Vol. 6, No. 1, 1970, pp. 22-40.
51. Chern, J.-M., "Optimal Design of Beams for Alternative Loads and Constraints on Generalized Compliance and Stiffness," Int. J. Mech. Sci., Vol. 13, No. 8, 1971, pp. 661-674.
52. Nagtegaal, J.C. and Prager, W., "Optimal Layout of a Truss for Alternative Loads," Int. J. Mech. Sci., Vol. 15, No. 7, 1973, pp. 583-592.
53. Olhoff, N., "On Singularities, Local Optima and Formation of Stiffeners in Optimal Design of Plates," Proc. IUTAM Symp. on Optimization in Structural Design (Ed. A. Sawczuk and Z. Mroz), Springer-Verlag, 1975, pp. 82-103.
54. Megarefs, G.J. and Hodge, P.G., "Singular Cases in the Optimum Design of Frames," Quart. Appl. Math., Vol. 21, No. 2, 1963, pp. 91-103.
55. Masur, E.F. and Mroz, Z., "Singular Solutions in Structural Optimization Problems," Proc. IUTAM Conf. on Var. Methods in Solid Mechanics, Springer-Verlag, to appear.
56. Masur, E.F., "Singular Problems of Optimal Design," Optimization of Distributed Parameter Structures, (Ed. E.J. Haug and J. Cea), Sijthoff & Nordhoff, Alphen aan den Rijn, Netherlands, 1980.
57. Olhoff, N. and Rasmussen, S.H., "On Single and Bimodal Optimum Buckling Loads of Clamped Columns," Int. J. Solids Structures, Vol. 13, 1977, pp. 605-614.
58. Taylor, J.E., "Optimal Remodelling Theory and Applications," Optimization of Distributed Parameter Structures, (Ed. E.J. Haug and J. Cea), Sijthoff & Nordhoff, Alphen aan den Rijn, Netherlands, 1980.
59. Int. J. Num. Meth. Engrg., Vol. 13, No. 2, 1978.

VARIATIONAL METHODS AND OPTIMALITY CRITERIA

George I.N. Rozvany

Faculty of Engineering, Monash University,
Clayton, Victoria, Australia

ABSTRACT

Optimality criteria for various classes of plastic and elastic structural design problems are reviewed and their use is illustrated with simple examples. It is shown that these criteria can be derived by using either variational methods or extremum principles of mechanics. The advantages of the so-called static-kinematic method are discussed.

1. INTRODUCTION

A few years ago, the term structural optimization was almost synonymous with the word optimism. Nearly all researchers in this field were convinced at the time that powerful new theories and the rapidly developing computer technology would soon enable the designer to discard all intuitive decisions, in view of the availability of direct techniques for systematically locating the most efficient solution. However, when the author attended a recent NASA symposium on future trends in computerized design,* it was already painfully obvious that there exists a general disillusionment with direct optimization methods among practising designers, although some limited optimization techniques based on repeated analysis and mathematical programming have been found useful in aircraft design. The roots of the present crisis can be traced back to the history of recent research in this field. During the

---

*Washington, D.C., November 1978.

last decade, research in structural optimization has polarized into two mainstreams of methodology which can be characterized as follows:

(1) Cost minimization through new analytical methods, employing calculus of variations, optimal control theory, and extremum principles of mechanics.

(2) Optimization using numerical methods (repeated analysis based on more and more efficient matrix operations, finite element techniques, and mathematical programming methods), making full use of ever increasing electronic computational capabilities.

A partial failure of numerical methods has been due to the following factors:

(1) Realistic design problems usually involve a very large number of parameters and the objective functions, as well as the constraints, are mostly nonlinear and nonconvex. Since the number of local minima is not known and is often extremely large, the global minimum can easily be overlooked, unless a vast amount of computational effort is devoted to the exercise (for a simple illustration see Ref. 1).

(2) It is known from analytical studies that the optimum solution is often nonunique, in terms of certain parameters. In such problems, a numerical method would provide only one value from an infinite set of equally optimum designs, although some other optimum design could be much preferable on practical grounds.

(3) Repeated analyses associated with redesign involve a considerably greater amount of computational effort than direct synthesis.

(4) The solution appears in a numerical form that provides very little insight into the real (mathematical) nature of the solution.

(5) In complex and highly nonlinear problems, the magnitude of error is difficult, often impossible to assess.

(6) After any change in the initial design constraints, much of the computation must be repeated, while analytical results are usually valid for an entire class of problems.

(7) Highly automated design processes can be dangerous in the hands of a half-educated designer, who may employ them

outside their range of validity and who is often unable to locate errors due to incorrect input.

In spite of the foregoing difficulties, numerical methods are preferred by designers because analytical techniques have the following limitations:

(1) For realistic design problems, analytical optimality criteria often yield a very large number of nonlinear equations. It is very time-consuming and often not feasible to solve these analytically, and numerical methods for finding all roots often prove unsatisfactory.

(2) Only highly idealized objective functions and design constraints are amenable to analytical solutions.

(3) Sophisticated analytical methods require a high degree of mathematical skill which the majority of designers do not possess.

In spite of the foregoing limitations, analytical methods have many advantages. As Prager [2] has pointed out, analytical study of optimal structures *furnishes information that is useful in testing the validity, accuracy and convergence of numerical methods and in assessing the efficiency of practical designs.* Although analytical methods have also been directed at discretized solutions more recently, they were applied originally to continuum type solutions in which (i) the cross-sectional properties vary continuously along the centerlines and (ii) the layout may consist of an infinite number of members at an infinitesimal spacing. As Prager says [3], *while this type of structure is not practical, it furnishes a lower bound on the structural weight of more realistic designs that is useful in the evaluation of their structural efficiency.*

The present difficulties could possibly be partially overcome by better cooperation between numerically and analytically inclined research groups. Particularly promising are some hybrid methods involving both analytical and numerical or computerized operations. Such developments include:

(1) Computer programs based entirely on analytical operations. While a very large number of long equations are often difficult if not impossible to handle manually, such expressions can be processed in a symbolic form on the computer. Such methods will be discussed in connection with optimal layout problems.

(2) Direct synthesis using finite elements and analytical optimality criteria [4,5]. These methods are based on the so-called static-kinematic approach, which is discussed in the next section.

2. THE STATIC-KINEMATIC METHOD OF STRUCTURAL OPTIMIZATION

Using Prager's terminology [6], the basic variables in structural mechanics are generalized stresses Q, strains q, loads p, and displacements u. A generalized stress can signify a local stress, such as normal stress σ, a shear stress τ, or a stress resultant that can be, for example, a bending moment M, a shear force V, a normal force N, or a torsional moment T. A generalized strain may represent again a local strain, e.g. longitudinal strain ε, shear strain γ, or a cross-sectional strain, which is some derivative of the displacement of an entire cross-section. The latter is based on the assumption that the cross-section remains plane and is free of distortions in certain directions. Such cross-sectional strains may represent a curvature κ or a twist θ of a bar, plate, or shell.

Generalized strains and stresses are referred to an entire cross-section in the case of one-dimensional and two-dimensional structures. In a one-dimensional structure, such as a beam, frame, truss, arch, or ring; the principal cross-sectional dimensions (width, depth) are small in comparison with the third dimension of the member (length) and thus the idealized behavior of the structure can be represented fully by specifying the generalized stresses and strains at all points of the centroidal axis. In two-dimensional structures, such as plates and shells, as well as in grid-like and shellgrid-like continua, it is sufficient to specify the generalized stresses and strains at any point of the middle surface or centroidal surface (grids). Although generalized strains in elastic structures are represented by traditional functions, the strain fields employed in optimization methods often contain impulse-like concentrations and hence they can only be described by generalized functions. Such strain-impulses are concentrated rotations (cusps) or discontinuities (steps) in the displacement field.

Generalized loads may consist of distributed loads $p(x)$, concentrated loads P, distributed couples $c(x)$, or concentrated couples C. Finally, generalized displacements u may have translational and rotational components.

A point of any potential centroidal axis or middle surface is specified by the spatial coordinates $x \in D$ where D is termed the structural domain.

The fundamental relations of structural mechanics are described in Fig. 2.1. Static continuity or equilibrium conditions express relations between loads and stresses, while kinematic continuity or compatibility conditions express relations between strains and displacements. The generalized strain-stress relations usually depend not only on material properties (e.g. modulus of elasticity E or modulus of rigidity G) but also on cross-sectional properties (e.g. moment of inertia or cross-sectional area). In addition, static and kinematic constraints (boundary conditions) are to be satisfied along some subset (S) of the structural domain D.

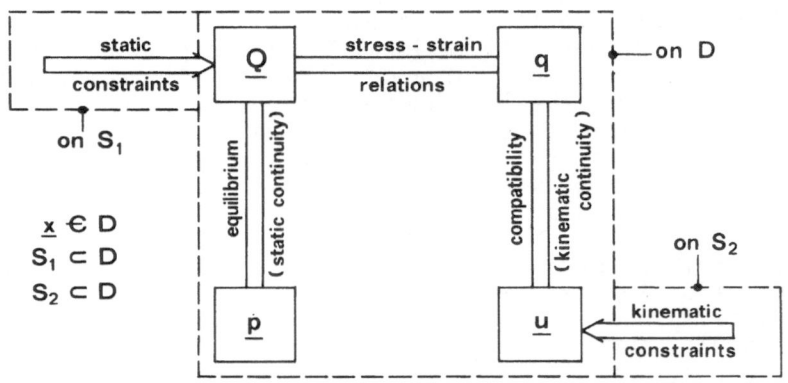

$Q(\underline{x})$: generalised stresses $q(\underline{x})$: generalised strains

$p(\underline{x})$: generalised loads $u(\underline{x})$: generalised displacements

Figure 2.1. Fundamental Relations of Structural Mechanics.

The basic variables and fundamental relations of structural mechanics are illustrated with a very simple example in Fig. 2.2. Considering a so-called Bernoulli-beam, the static continuity (equilibrium) condition is $M'' = -p$, where $M(x)$ is the bending moment and $p(x)$ is the external load. For one-dimensional structures, the notation $(\ )' = d(\ )/dx$ and $(\ )'' = d^2(\ )/dx^2$ is used. The kinematic continuity condition is $u'' = -\kappa$ and the generalized strain-stress relation (i.e. the moment-curvature relation) is $M = EI\kappa$, where E is the elastic modulus and I is the moment of inertia of the beam cross-section. At a free end S_1, there are only static constraints (end conditions) $M = M' = 0$. However, at a clamped end S_2, there are only kinematic constraints $u = u' = 0$. Beams may have, of course, mixed end conditions. At a simple end support S_3, $M = u = 0$ and at an end with a vertical guide S_4, $M' = u' = 0$. The latter two types of end conditions are partly kinematic constraints.

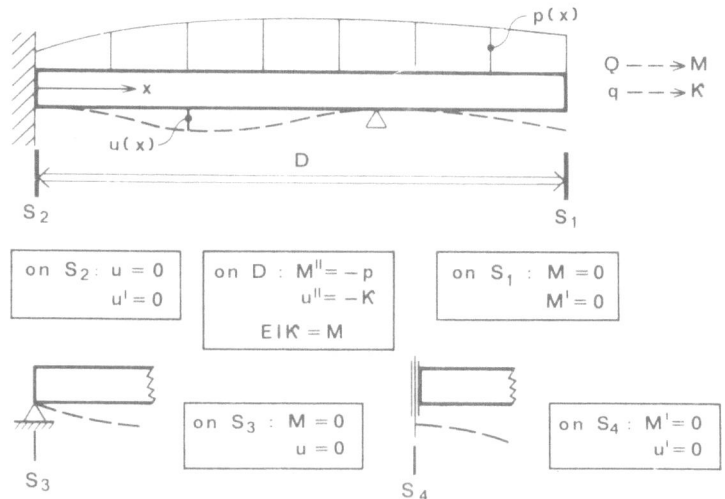

Figure 2.2. Example of Fundamental Relations: Bernoulli beam.

As indicated in Fig. 2.3, loads and stresses that satisfy static continuity and static constraints are said to be statically admissible and are denoted by $(\underline{p}^S, \underline{Q}^S)$, while kinematically admissible displacements and strains $(\underline{u}^K, \underline{q}^K)$ satisfy kinematic continuity and kinematic constraints. In structural analysis, a system must satisfy static and kinematic admissibility, as well as the generalized strain-stress relations based on the given cross-sectional geometry.

In structural design, the cross-sectional geometry is not given. Hence the generalized strain-stress relations are not known prior to the design procedure. For this reason, the solution is in general not unique in design problems. In optimal design, however, a further condition is introduced, namely a quantity called total cost Φ is required to take on a minimum value. In the so-called static-kinematic method of structural optimization, this cost minimality condition is converted into a special strain-stress relationship, as will be explained in detail in the next section. In addition to the optimal strain-stress relation, static and kinematic admissibility must be satisfied. The problem of optimal design is therefore, in effect, converted into a problem of analysis.

While the static-kinematic approach does not change the formal statement of the extremum conditions, the advantages of this method are psychological rather than mathematical. Engineers, in general, can visualize such physical analogies as moments or deflections more easily than abstract mathematical concepts.

Static and Kinematic Admissibility

$$\left.\begin{array}{l} \underline{p}^s, \underline{Q}^s \\ \underline{u}^k, \underline{q}^k \end{array}\right| \text{ satisfy } \left|\begin{array}{l} \text{static} \\ \text{kinematic} \end{array}\right| \text{ constraints } + \text{ continuity}$$

Φ : total cost
ψ : specific cost

Optimal Plastic Design

$$\min \Phi = \int_D \psi\left[\underline{Q}^s(\underline{x})\right] d\underline{x}$$

$\psi(\underline{Q})$: specific cost function

Prager — Shield Optimality Condition

$$\boxed{\text{on } D, \quad \underline{q}^k = \underline{G}\left[\psi(\underline{Q}^s)\right]}$$

$\underline{G}$: generalised gradient

Upper Bound :

$$\Phi^* = \int_D \psi\left[\underline{Q}^s(\underline{x})\right] d\underline{x} \geqslant \Phi_{\min}$$

Lower Bound (convex functions) :

$$\Phi^{**} = \int_D \left[\underline{u}^k \underline{p} - \widehat{\psi}(\underline{q}^k)\right] d\underline{x} \leqslant \Phi_{\min}$$

$\widehat{\psi}$: complementary cost

Figure 2.3. Optimal Plastic Design and the Prager-Shield Condition.

Because the general framework of finding complex optimal solutions requires a lot of intuition, such an analogy has been invaluable in solving such intricate problems as the optimal layout of beam systems. Moreover, there now exists a comprehensive set of general theorems for deriving directly the static-kinematic optimality criteria for almost all structural optimization problems, without resorting to such mathematical tools as the calculus of variations or optimal control-theory.

The idea of the static-kinematic method was first introduced, in the context of optimal plastic design, by Prager and Shield [7]. A large number of extensions of this theory have been proposed by the author and will be discussed in subsequent sections.

3.   THE PRAGER-SHIELD THEORY OF OPTIMAL PLASTIC DESIGN

Although other types of specific cost function may be adopted, it will be assumed here that the specific cost ψ per unit length, area, or volume at a point $\underline{x}$ depends only on the generalized stress $\underline{Q}(\underline{x})$ at that point, i.e., $\psi = \psi[\underline{Q}(\underline{x})]$. This functional relationship will be called the specific cost function. The total cost Φ can then be obtained by integration, $\Phi = \int_D \psi(Q)d\underline{x}$.

Plastic optimal design, based on the lower bound theorem of limit analysis, amounts to the minimization of such an integral, subject to statical admissibility of $\underline{Q}(\underline{x})$ on D, as is shown in Fig. 2.3.

A slightly reformulated version of the Prager-Shield condition is also given in Fig. 2.3. Expressed in words, this condition requires that one find a statically admissible stress field and a kinematically admissible strain field, such that the strains equal the generalized gradients of the specific cost function, with respect to the stresses adopted. While the generalized gradient is simply a set of partial derivatives for differentiable specific cost functions, its full meaning is explained in the next section. The Prager-Shield criterion is a necessary and sufficient condition for convex specific cost functions with linear equilibrium equations and a necessary condition for nonconvex problems. While the originators of this theory applied it to convex problems only [7], its usefulness for nonconvex problems has been demonstrated by the author [8, pp. 64-79].

Figure 2.3 also shows that an upper bound on the minimum total cost can be based on any statically admissible stress field and a lower bound can be derived from any kinematically admissible strain/displacement field. The concept of complementary cost $\hat{\psi}$ will be clarified in the next section.

Proofs of the Prager-Shield condition are discussed in Section 5.

4. THE GENERALIZED GRADIENT OPERATOR

The generalized gradient operator has been introduced in connection with Prager-Shield type optimality conditions, but it has obvious applications in other areas. Specific cost functions with a single variable are considered in Figure 4.1.

When the function is differentiable (smooth) then the gradient and the corresponding strain q become simply the first derivative of the specific cost function, $q = d\psi/dQ$. Naturally, the specific cost can be obtained by integration of the function $q(Q)$, as indicated in Fig. 4.1(a). Similarly, the so-called complementary cost $\hat{\psi}$ can be derived by integrating the function $Q(q)$. The latter is used in lower bound theorems (Fig. 2.3).

When the slope of a specific cost function is discontinuous at a point, the generalized gradient is a convex combination of the slopes (first derivatives) at either side of the cusp (Fig. 4.1(b)). This means that the gradient is nonunique at such points, which fact will have a number of important implications later.

Finally, when the specific cost function has a discontinuity (step), then the corresponding gradient becomes a unit impulse or Dirac measure, multiplied by the magnitude of the step (Fig. 4.1(c)).

90

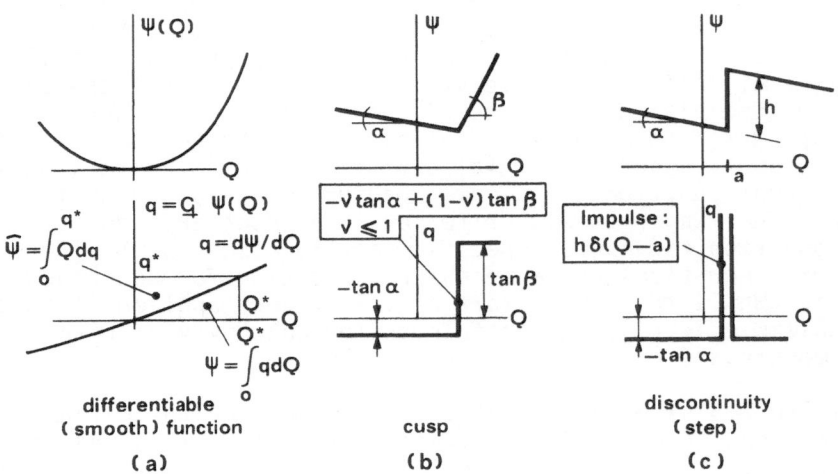

Figure 4.1. Generalized Gradients for Functions with One Variable.

The gradient of specific cost functions with several variables is illustrated in Fig. 4.2. For differentiable functions, the generalized gradient reduces to the conventional gradient of vector calculus, whose components are the first partial derivatives. In a graphical representation, such a gradient vector is normal to the cost contour (Fig. 4.2(a)).

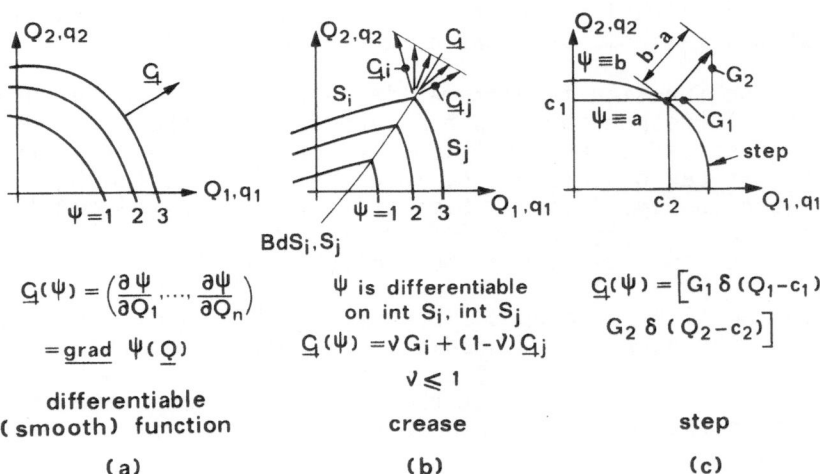

Figure 4.2. Generalized Gradients: Functions with Several Variables.

At a point of slope discontinuity or crease, the gradient is a convex combination of the gradients associated with the adjacent stress regimes (S_i and S_j in Fig. 4.2(b)), on whose interior the specific cost function is differentiable.

At a discontinuity or step in value of the specific cost function, all components of the gradient are impulses and the vectorial sum of the magnitude of such impulses equals the magnitude of the step (Fig. 4.2). Such discontinuous specific cost functions are of importance in design problems where the choice of available cross-sections is restricted to a finite number of sizes, as will be explained in the following. Impulse-type cost-gradients result in strain-impulses and discontinuities in the optimal displacement field. If the relevant strain is a curvature, then the strain impulse becomes a concentrated rotation (slope discontinuity in the deflection field). An impulse in shear strain causes a step in the displacement field. It has been shown by the author [8, p. 85] that the magnitude of such strain impulses depends on the step size h in the cost function and the slope of the relevant stress Q, with respect to the spatial co-ordinate involved, as can be seen in Fig. 4.3(a).

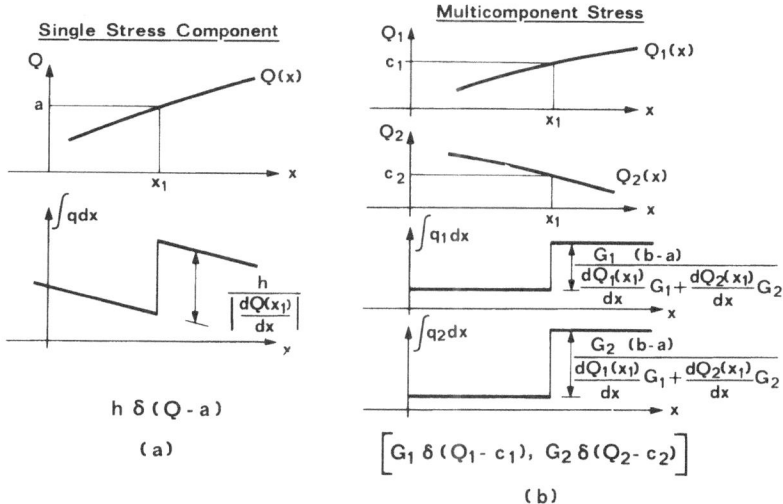

Figure 4.3. Strain Impulses in Pragerian Associated Displacement Fields.

If the specific cost depends on several stress components, then the components of the strain impulses are still proportional to the components G_1 and G_2 of the gradient vector normal to the step in the stress space, but their magnitude depends on the first

derivatives of the stress components with respect to the spatial
co-ordinate x(Fig. 4.3(b)). This result is derived by a limiting
process, as shown in Fig. 4.4. The variation of the stress components between $x = x_1$ and $x = x_2$ in a one-dimensional structure is
given in Fig. 4.4(b). It is shown in Fig. 4.4(a) that over the
same range of stress component values, the specific cost increases
from a constant level of $\psi = a$ to another constant level of $\psi = b$.
The sloping step occurs over a width of Δ and it becomes a real
step when Δ tends to zero. The equation giving Δ in Fig. 4.4 is
based on usual scalar multiplication and the strains given in
Fig. 4.4(c) are furnished by the Prager-Shield condition for the
sloping part of the cost surface. Then the simplifications shown
yield the previously given (Fig. 4.3(b)) integral values of the
strain impulse components (Fig. 4.4, bottom).

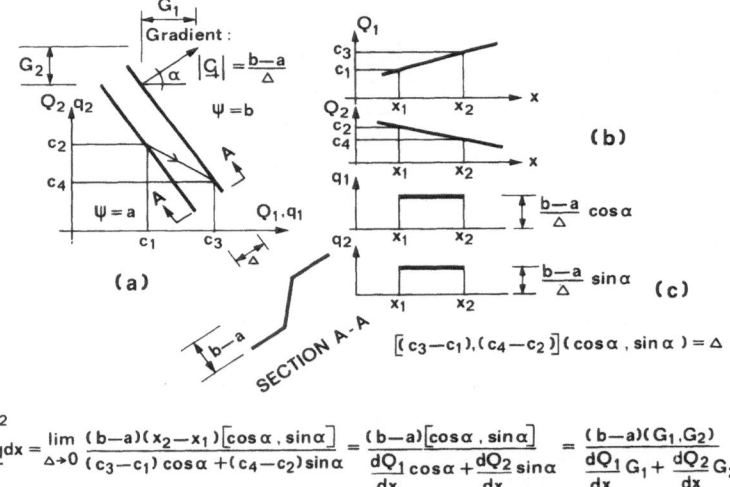

$$\int_{x_1}^{x_2} q\,dx = \lim_{\Delta \to 0} \frac{(b-a)(x_2-x_1)\left[\cos\alpha, \sin\alpha\right]}{(c_3-c_1)\cos\alpha + (c_4-c_2)\sin\alpha} = \frac{(b-a)\left[\cos\alpha, \sin\alpha\right]}{\frac{dQ_1}{dx}\cos\alpha + \frac{dQ_2}{dx}\sin\alpha} = \frac{(b-a)(G_1, G_2)}{\frac{dQ_1}{dx}G_1 + \frac{dQ_2}{dx}G_2}$$

Figure 4.4. Derivation of the Strain Impulse Equation.

5. PROOFS OF THE PRAGER-SHIELD CONDITION

The simplest proof of the Prager-Shield optimality criterion
is based on the complementary energy theorem for elastic structures. Suppose that the specific cost ψ of the plastic system to
be optimized equals the specific complementary energy of a non-
linearly elastic system, having the same boundary conditions and
loading as the plastic system. Be definition of the specific
complementary energy, $\partial\psi/\partial Q_i = q_i$ $(i = 1,\dots,n)$ gives the strain-
stress relation in the Prager-Shield condition. Furthermore,

while optimal plastic design requires statical admissibility, the complementary energy theorem states that minimality of the total complementary energy bi-implies kinematic admissibility. This means, in turn, that the total cost of the plastic system (and the total complementary energy of the associated elastic system) will be a minimum when the above strain-stress relation and kinematic admissibility are satisfied. Naturally, it also follows from the foregoing theorem, that the total cost (total associated complementary energy) of any system satisfying statical admissibility, but not kinematic admissibility, gives an upper bound on the minimum total cost (Fig. 2.3). This, however, is self-evident, because by definition, the cost of any admissible system is higher than that of the optimum cost.

The lower bound theorem in Fig. 2.3 follows readily from the minimum potential energy theorem. If ψ is the complementary energy of the associated elastic system, then the complementary cost $\hat{\psi}$ becomes equal to the strain energy of the same system. By the potential energy theorem, considering all kinematically admissible elastic systems, the total potential energy $\int_D (\hat{\psi} - \underline{u}\,\underline{p})dx$ is minimized by the system that also satisfies static admissibility. By simply reversing the sign of the quantities in the integrand, one gets maximization instead of minimization, which implies the lower bound theorem in Fig. 2.3.

The foregoing approach was used by Prager and Shield [7] for deriving their optimality criterion and by the author for proving the lower bound theorem in Fig. 2.3.

In order to prove more complex extensions of the Prager-Shield theory, it is convenient to combine the calculus of variations with known extremum principles of mechanics. This latter approach is demonstrated by again deriving the original Prager-Shield condition. For simplicity, a differentiable specific cost function $\psi(\underline{Q})$ is considered first. The equilibrium equations are denoted by

$$p_j = E_j(\underline{Q}) \tag{5.1}$$

where E_j is a term containing derivatives of the stress components Q_j. Using Lagrangian multipliers u_j, the extremum problem can be stated as

$$\min \, \Phi = \int_D \left\{ \psi + \sum_j u_j[p_j - E_j(Q)] \right\} d\underline{x} \quad , \tag{5.2}$$

The stationarity condition is expressed in the form of Euler-equations. First it is shown that the terms

$$\sum_j u_j \ E_j(\underline{Q}) \qquad\qquad (5.3)$$

give the term q_i^k for variations of Q_i. The principle of virtual forces [9] states that

$$\int_D \left(\sum_i q_i^k \ \delta Q_i^s - \sum_j u_j^k \ \delta p_j^s \right) d\underline{x} = 0 \qquad\qquad (5.4)$$

where (Q_i^s, p_j^s) and (q_i^k, u_j^k) need not be associated with the same elastic system. Replacing p_j^s with $E_j(\underline{Q}^s)$, on the basis of Eq. 5.1 and writing down the Euler equation for the variation of Q_i, one has

$$q_i^k = F_i \qquad\qquad (5.5)$$

where F_i is the term corresponding to Eq. 5.3 in the integrand. This means, that in any further derivation, one can always use the single term q_i^k in the Euler equation when dealing with the equilibrium equations of Eq. 5.3 in the integrand and consider variations of Q_i. The full set of Euler equations for Eq. 5.2 thus becomes

$$q_i^k = \partial\psi(\underline{Q}^s)/\partial Q_i \qquad\qquad (5.6)$$

which is the Prager-Shield condition for differentiable specific cost functions.

Specific cost functions with slope discontinuities can be dealt with easily by the following method: Since $\psi(\underline{Q})$ is differentiable on the interior of the stress regimes S_β (see Section 4), one may find a set of constituent cost functions $\psi_\beta(\underline{Q})$ such that

for $\quad \underline{Q} \in S_\beta: \ \psi_\beta(\underline{Q}) = \psi(\underline{Q})$

for $\quad \underline{Q} \notin S_\beta: \ \psi_\beta(\underline{Q}) \leq \psi(\underline{Q})$
$\qquad\qquad (5.7)$

The latter can be replaced by

$$\psi(\underline{Q}) = \psi_\beta(\underline{Q}) + s_\beta, \qquad\qquad (5.8)$$

where s_β is a non-negative slack function. Then $\psi(\underline{Q})$ in $\int_D \psi \ d\underline{x}$ can be represented by a series of constraints using Eq. 5.8 and Lagrangian multipliers. The resulting Euler equations then give (see Ref. 8, pp. 57-59) the Prager-Shield condition in terms of generalized derivatives.

Using the foregoing framework for deriving optimality criteria, a number of extensions of the Prager-Shield theory have been obtained which are discussed in the next section.

6. EXTENSIONS OF THE PRAGER-SHIELD THEORY

6.1 Reactions or Unspecified Forces of Non-Zero Cost

In most structural systems, some external forces are unspecified and depend on the design adopted. One such generalized force is a so-called reaction that is transmitted from a support to the structure. However, there exist other types of unspecified forces, e.g. a ballast whose weight is not prescribed and is to be optimized.

When the cost of a reaction is not taken into consideration or is zero, kinematical admissibility of the Prager-Shield displacement field requires that the displacement component corresponding to that reaction is zero. In practical problems, however, the cost of support usually does depend on the magnitude of the reaction(s) acting on it. The total cost of such a system is given at the top of Fig. 6.1, where the reaction (or unspecified force) is $\underline{R}(\underline{x})$, having up to six components (three couples and three forces) and acting on the subset S of the domain D. The extended optimality condition for this problem is also given in Fig. 6.1. The displacements $\underline{u}_k$ along supports, and hence the optimal displacement field, are fictitious. They are only an analogy for determining the optimal solution more easily, because the supports of non-zero cost may well be quite rigid in a mechanical sense. For supports of zero cost, on the other hand, the Prager-Shield displacement field does correspond to one possible velocity field at collapse.

6.2 Optimal Plastic Design for External Load and Body Forces

The effect of body forces, such as selfweight of long-span structures, can be significant in determining the optimal design. Denoting the body force per unit volume by $\underline{b}$, the optimality criterion for this problem is given in the middle of Fig. 6.1. When the specific cost is the selfweight itself and there is only a single deflection component (as in beams), then the multiplier in front of the gradient reduces to $(1 + u)$.

Heyman [10] proved some twenty years ago that the minimum weight of plastically designed beams with linear specific cost function $\psi = k|M|$ corresponds to a deflection field having constant curvature $\kappa = k$ or $\kappa = -k$, depending on the sign of the bending moments. The author has shown more recently [11] that in

Optimization of Reactions (Unspecified Generalised Forces)

$$\mathbb{I} = \int_D \psi(\underline{Q}) \, d\underline{x} + \int_s \Omega(\underline{R}) \, d\underline{x}, \quad S \subset D$$

$$\boxed{\text{on } D, \quad \underline{q}^k = \underline{\underline{G}}\left[\psi(\underline{Q}^s)\right]; \text{ on } S, \quad u^k = \underline{\underline{G}}\left[\Omega(\underline{R}^s)\right]}$$

Optimization for External Load and Body Forces

specific mass : $\psi(x)$, body forces $\underline{b}\ \psi(x)$

$$\boxed{\text{on } D, \quad \underline{q}^k = (1 + \underline{b} \cdot \underline{u}^k)\ \underline{\underline{G}}\left[\psi(\underline{Q}^s)\right]}$$

Optimization for Alternate Loads

Alternate loads : $\underline{p}_1, \cdots, \underline{p}_j, \cdots, \underline{p}_n$, Design cost $\overline{\psi} = \max_j \psi(\underline{Q}_j^s)$

$$\boxed{\begin{array}{l} \text{on } D, \quad \underline{q}_j^k = \lambda_j\ \underline{\underline{G}}\left[\psi(\underline{Q}_j^s)\right], \quad \lambda \geqslant 0 \\[2mm] \lambda_j > 0 \text{ only if } \overline{\psi} = \psi(\underline{Q}_j^s), \quad \sum_j \lambda_j = 1 \end{array}} \qquad \mathbb{I} = \int_D \overline{\psi} \, dx$$

For an infinite set (e.g. moving) loads: $\int_B \lambda(x,z) \, dz = 1$ for all $x \in D$

Figure 6.1. Extended Optimality Criteria in Plastic Design.

the presence of selfweight, the quadratic displacement functions of Heyman are replaced by simple trigonometric and hyperbolic relations (cos and cosh functions). The significance of self-weight in optimal design of long-span systems is discussed in a companion paper by the author.

6.3 Optimal Plastic Design for Alternate Loads

Most real structures are subjected to a variety of possible loading conditions. In plastic design, it is necessary to find a statically admissible stress field $\underline{Q}_j^s$ for each such alternate load and then to adopt the maximum value of the specified cost ψ, considering the cost requirements for each of the loading conditions, at each point of the structure.

This procedure and the corresponding optimality criteria are given at the bottom of Fig. 6.1. It is to be noted that only fully stressed cross sections can be subjected to non-zero strain. For problems in which there exist an infinite number of loading conditions associated with some parameter z defined on a set B, the summation in the last condition is replaced by an integral (Fig. 6.1). The problem of moving loads has been explored in several papers by Prager, Save, Shield and Lamblin (e.g. Refs. 12, 13, and 14).

6.4 Optimal Plastic Design: Prescribed Cost Distribution

Continuously varying cross-sections, along centerlines or middle surfaces, are often not economical to construct or the variation can be restricted for practical reasons to simple (e.g. linear) relationships. The optimal design of such systems is described at the top of Fig. 6.2. The structural domain D in such problems is divided into segments D_α and on each segment the cost variation is restricted to a given shape function γ_β, which is multiplied by an unknown constant. For piecewise prismatic systems, $\gamma_\beta = 1$ can be adopted. Optimality criteria for this class of problems are also given in Fig. 6.2. For a very simple class of beams, these conditions reduce to those of Foulkes [15], which were published some 25 years ago. Other applications are outlined in Ref. 8.

Optimal Plastic Design : Prescribed Cost Distribution

$D = U D_\alpha$ Int $D_i \cap$ Int $D_j = 0$ $(i \neq j)$ D_i : Segments

on D_α, $\overline{\Psi} = \Lambda_\alpha \, \mathring{\gamma}_\alpha (\underline{x})$, $\mathring{\gamma}_\alpha$: given shape functions

Λ_α : unknown constants, $\overline{\Psi} \geqslant \Psi (Q_j^s)$

$$\text{on } D, \quad \underline{q}_j^k = \lambda_j \, \underline{G}[\Psi (\underline{Q}_j^s)], \quad \lambda_j \geqslant 0$$
$$\lambda_j > 0 \text{ only if } \overline{\Psi} = \Psi (Q_j^s),$$
$$\int_{D_\alpha} \mathring{\gamma}_\alpha (\underline{x}) \, d\underline{x} = \sum_j \int_{D_\alpha} \lambda_j (\underline{x}) \, \mathring{\gamma}_\alpha (\underline{x}) \, d\underline{x}$$

Optimal Plastic Design : Cost of Connections (Joints)

$\Phi = \int_D \Psi [\underline{Q}^s(x)] \, dx + \eta_i [\underline{Q} (x_i)]$, δ : delta function

$$\text{on } D, \quad q^k = \underline{G}[\Psi (\underline{Q}^s)] + \sum_i \delta(x - x_i) \, \underline{G}\{\eta_i [\underline{Q}^s(x_i)]\}$$

Figure 6.2. Further Optimality Criteria in Plastic Design.

6.5 Allowance for the Cost of Connections

Structural connections (joints) are responsible for a significant proportion of the total cost of a structure. Optimality conditions allowing for connection cost are given at the bottom of Fig. 6.2. It will be seen that the connection cost depends only on the local stress value at a point x_i of the structure and the corresponding strain appears in the form of an impulse. While this theory was developed [8] for one-dimensional structures, it can be readily generalized to multi-dimensional systems.

6.6 Optimization of Segmentation

In Section 6.4 (Fig. 6.2, top), it was assumed that the boundaries of the segments are given. When the segment boundaries are also to be optimized, the foregoing conditions are still valid, but additional local conditions must be fulfilled at such boundaries. Although more general conditions can be easily derived, the conditions for optimal segmentation are given here in the context of a beam whose cost depends on the bending moment only. Then the optimality criteria are those given in Fig. 6.3 (left), where $\Delta\bar{\psi}_B$ is the step in the specific cost at the boundary B, V_B is the shear force, and θ is the relative rotation. Because the multipliers $\lambda_j(x)$ usually consist of impulse-type functions, such concentrated rotations occur in the displacement field when the cost distribution is prescribed. This is because, in general, $\bar{\psi} = \psi(Q)$ only at isolated points and hence (by the condition in Fig. 6.2, λ can only be non-zero locally at a few points.

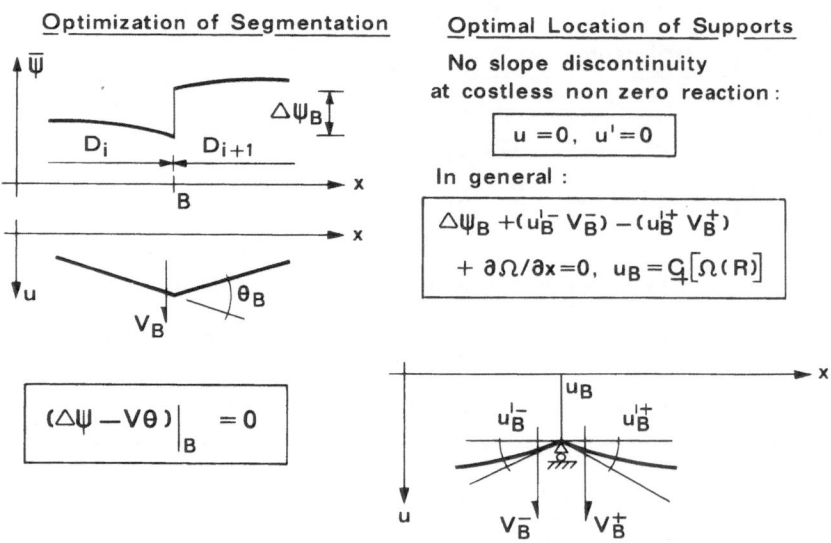

Figure 6.3. Local Optimality Criteria in Plastic Design.

6.7 Optimal Location of Supports

When the location of reactions (supports) is unspecified, further conditions must be fulfilled for optimality. If the reaction is costless and the beam deflection u has no slope discontinuity (cusp) at the support, then the condition for optimality is that both the deflection and slope must be zero at the

99

support (Fig. 6.3, right top). The general expression, which allows for non-zero support cost, slope discontinuity, and spatial dependency of the support cost, is given also in Fig. 6.3 (right bottom).

The relations in Sections 6.6 and 6.7 are based on transversality conditions of variational calculus. It is important to know that these are only necessary conditions for optimality, even for convex specific cost functions. All other extensions discussed so far (Section 6.1 to 6.5) are also sufficient conditions for convex cost functions and linear equilibrium constraints.

7. SPECIAL FEATURES OF HOMOGENEOUS SPECIFIC COST FUNCTIONS

Homogeneous cost functions are defined in Fig. 7.1. If the order of homogeneity is one, then these functions may be represented graphically by a cone and hence can be termed conical cost functions. An example of such a function is also given in Fig. 7.1.

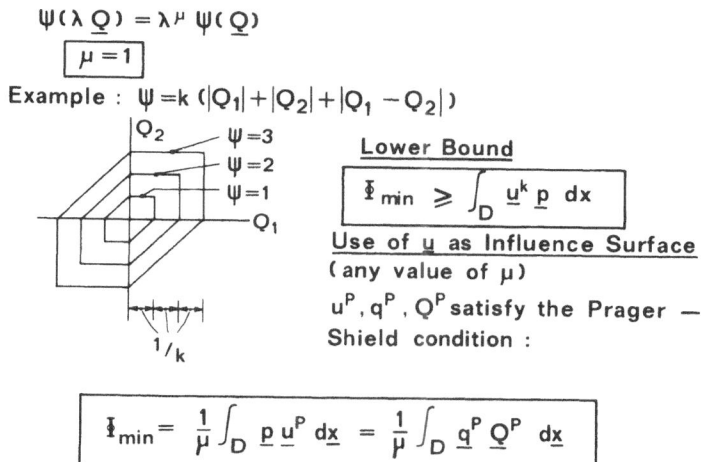

Figure 7.1. Theorems for Homogeneous Cost Functions of Order μ.

For conical cost functions, the complementary cost $\bar{\psi}$ takes on a zero value and thus the lower bound theorem in Fig. 2.3 takes on a much simpler form, as is shown in Fig. 7.1.

Another feature of homogeneous cost functions of any order is that the Pragerian displacement field u^P, satisfying the

Prager-Shield criterion and its extensions, can be used as an influence surface. The relevant expression is given in Fig. 7.1, together with an additional formula for the minimum total cost. This means that once the Pragerian displacement field is known, the minimum total cost can be determined by simply integrating the product of the loads and displacements. For conical cost functions, the Pragerian displacement field is often valid for an entire range of loading conditions. In that case, $\underline{u}^p$ can be used as an influence surface for optimum total cost.

8. OPTIMAL ELASTIC DESIGN

8.1 Deflection Constraints

Considering an elastic system, one may prescribe an upper limit on the deflection at a given point. An alternative, and more general formulation of deflection constraints is to specify upper limits on the weighted combination of deflections at various points of the structure (Fig. 8.1, top). The weighting of the deflections is given by factors $\bar{p}_i$ for local deflections and by functions $\bar{p}(x)$ for deflections along a finite subdomain. The optimality conditions for various subclasses of problems are also summarized in Fig. 8.1, in which all symbols with overbars are associated with $\bar{P}_i$ and $\bar{p}$, as if the latter were (virtual) external loads. When $\bar{P}_i$ and $\bar{p}$ are actually equal to the real external loads, then the problem reduces to design for a compliance constraint. While Fig. 8.1 considers only a Bernoulli beam, the same conditions can be readily derived for other systems.

$$\sum_i \bar{P}_i \, \underline{u}_i + \int_{\bar{D}} \bar{p}(\underline{x}) \, \underline{u}(\underline{x}) \, d\underline{x} \leq C \qquad \bar{D} \subseteq D$$

Bernoulli Beams :   S = stiffness

$$\text{on D, } \underline{G}\big[\psi(S)\big] = \lambda \, M^s / \bar{M}^s / S^2 = \bar{K}^k \, K^k \, \lambda$$

Prescribed Stiffness Distribution :

$$\text{on } D_\alpha, \ S(x) = \Lambda_\alpha \, \delta_\alpha(x)$$

$$\int_{D_\alpha} \delta_\alpha \, \underline{G}\big[\psi(S)\big] dx = \frac{\lambda}{\Lambda_\alpha^2} \int_{D_\alpha} (M^s \, \bar{M}^s / \delta_\alpha) \, dx$$

Support with Hinge, no cost discontinuity :

$$(\bar{u}'\bar{V} + \bar{u}'V)_{B^-} = (\bar{u}'\bar{V} + \bar{u}'V)_{B^+}$$

Figure 8.1. Optimal Elastic Design for a Deflection Constraint.

The number of papers in this field is too large to review here fully. Some more important contributions are due to Barnett [16], Haug and Kirmser [17], Prager and Taylor [18], and Masur [19, 20].

8.2 Elastic Design for Stress Constraints

The optimal stress design of fully stressed elastic beams was discussed by Taylor [21] and Masur [22]. Masur used the concept of energy density in the design fibers, which must be proportional to the gradient of the weight function. This concept can be shown to reduce to the Prager-Shield condition [7], which is equally valid for plastic and fully stressed elastic designs.

More recently [23, 24], it was found that in the elastic stress design of statically indeterminate beams, two types of regions occur: in fully stressed regions, the associated displacement field is governed by the usual cost gradient type (Prager-Shield) condition and in understressed regions, the optimal stiffness is determined by a deflection constraint, using the conditions given in Fig. 8.1. This is because all redundant supports can be replaced by (zero) displacement constraints, which will cover the design of some segments, while other segments are designed on the basis of the stress constraint. The existence of these two types of regions was pointed out, in a slightly different context, by Haug and Kirmser [17]. In effect, optimal elastic structures with stress constraints use their understressed segments for rectifying the kinematic inadmissibility of the fully stressed solution. The associated Pragerian displacement field for understressed regions is given by

$$-u'' = \lambda \bar{M}/S \tag{8.1}$$

in which the notation is the same as in Fig. 8.1.

The interesting aspect of this approach is the conclusion that elastic stress design can also be handled by the static kinematic approach and Pragerian associated displacements. However, in these problems one must satisfy simultaneously two types of kinematic admissibility, that of elastic displacements $u_{e\ell}$ and that of Pragerian associated displacements.

8.3 Elastic Design for Buckling

Conditions for the optimal design of elastic columns for a given buckling load are given in the framework of generalized gradients in Fig. 8.2. In addition, optimal support location requires zero lateral reaction at buckling. The extensive literature in this field is not reviewed here, because other numerous papers deal with the problem of buckling in detail.

$$\underline{G}\left[\Psi(S)\right] = \lambda\left[u''(x)\right]^2$$

$$\int_{D_\alpha} \bar{\delta}_\alpha(x)\ \underline{G}(S)dx = -\lambda\int_{D_\alpha} \bar{\delta}_\alpha(x)\left[u''(x)\right]^2\,dx$$

$$\left[\Psi(S) - \lambda S u''^2\right]_{B^-} = \left[\Psi(S) - \lambda S u''^2\right]_{B^+}$$

Figure 8.2. Optimal Elastic Design for a Given Buckling Load.

9. SIMPLE EXAMPLES ILLUSTRATING THE STATIC KINEMATIC METHOD

9.1 Optimal Plastic Design of a Frame for Combined Bending and Axial Force

Consider first the problem of transmitting a single point load P by two compression members of prescribed permissible stress to two equidistant supports (Fig. 9.1(a)) so that the total volume of the system is a minimum. This is a simple layout problem, which is discussed in greater detail in a companion paper. It will be found that the optimum slope of the bars is 45° and hence the optimal height of the frame is h = L/2.

One would like to know now if the same frame is still optimal if the bars can resist both bending moments and axial forces, where they are rigidly jointed at the top (Fig. 9.1(d)). In this problem, the centroidal geometry is fixed and one wants to determine only the optimal distribution of the generalized stresses. Using a rectangular cross section of variable width but constant depth, the yield condition has been derived by Prager [25] and the specific cost function and cost gradients by the author [8, pp. 90-91]. The specific cost, or weight per unit length, is given by

$$\psi = \left[\beta|M| + (\beta^2 M^2 + 4N^2)^{1/2}\right]\gamma/2\sigma_0 \qquad\qquad (9.1)$$

where $\beta = 4/d$ is a constant, d is the depth of the cross-section, γ is the specific weight of the arch material, and σ_0 is the yield stress. The corresponding cost contours are given in Fig. 9.1(b).

Solution by Static-Kinematic Method. In the current problem, the generalized stress vector consists of the bending moment and

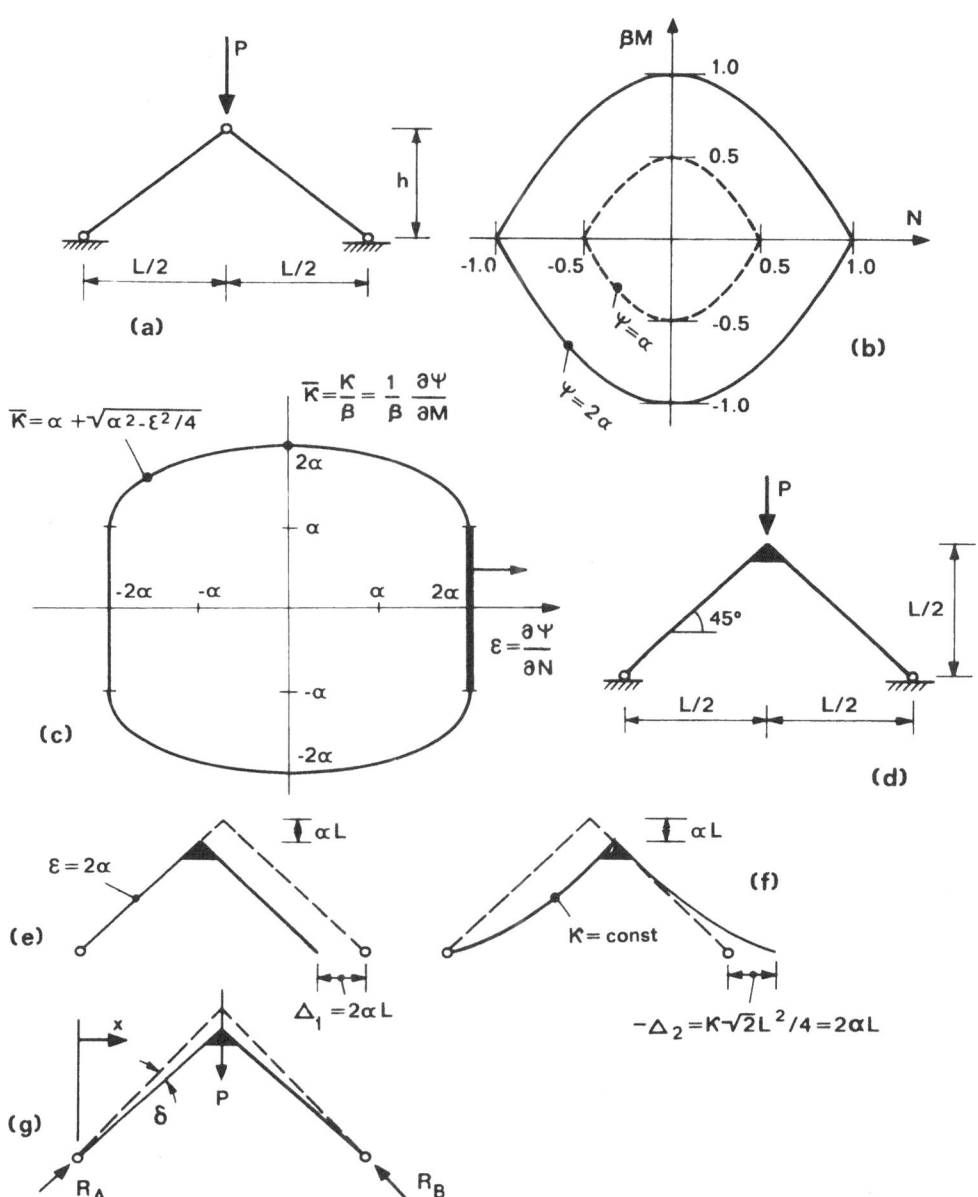

Figure 9.1.   Example: Frame Design for Bending and Compression.

axial force, $Q = (M, N)$ and the corresponding strain vector of the curvature and axial strain, $q = (\kappa, \varepsilon)$. Using the Prager-Shield optimality condition (Fig. 2.3), the strain-stress relations become

$$\varepsilon = 4\alpha N/(\beta^2 M^2 + 4N^2)^{1/2} \tag{9.2}$$

$$\kappa = \beta\alpha \left[\beta M/(\beta^2 M^2 + 4N^2)^{1/2} + \text{sgn } M \right] , \quad \text{for } M \neq 0 \tag{9.3}$$

$$|\kappa| \leq \alpha\beta , \quad \text{for } M = 0 \tag{9.4}$$

where $\alpha = \gamma/2\sigma_0$. Eliminating the parameter (M/N) from Eqs. 9.2 to 9.4, one gets the relationship for $\kappa(\varepsilon)$ indicated graphically in Fig. 9.1(c). In checking optimality of the design in which the moments are zero throughout ($M \equiv 0$), Eq. 9.2 gives an axial strain of $\varepsilon = 2\alpha$ and Eq. 9.4 gives a curvature $|\kappa| < \beta\alpha$. It is shown in Fig. 9.1(e) that the above axial stress causes a horizontal displacement of $\Delta_1 = 2\alpha L$ at the support. The maximum admissible curvature $\kappa = \beta\alpha$ would result in a horizontal displacement that is (Fig. 9.1(f)) equal to the product of the total rotation $\sqrt{2} \kappa L$ and the average lever arm $L/4$, giving $\beta\alpha\sqrt{2} L^2/4$. This implies that kinematic admissibility is satisfied only when

$$\beta L \geq 8/\sqrt{2} \qquad \text{or} \qquad L/d \geq \sqrt{2} \tag{9.5}$$

It is obvious that all realistic design problems would comply with this limitation. Hence, the flexureless solution can be adopted in all practical situations.

Derivation by Traditional Methods. The total cost of the foregoing solution is $\Phi_{opt} = 2\alpha PL$. Considering now a solution in which the reaction forces enclose an infinitesimal angle δ with the members (Fig. 9.1(g)), the generalized stresses become

$$N = \frac{P \cos \delta}{2 \sin (45° + \delta)}$$
$$\tag{9.6}$$
$$M = \frac{P \ x\sqrt{2} \sin \delta}{2 \sin (45° + \delta)}$$

After neglecting higher powers of δ, Eq. 9.6 reduces to

$$\left. \begin{array}{l} N = P\sqrt{2}(1 - \delta)/2 \\ \\ M = P(1 - \delta) \times \delta \end{array} \right\} \tag{9.7}$$

105

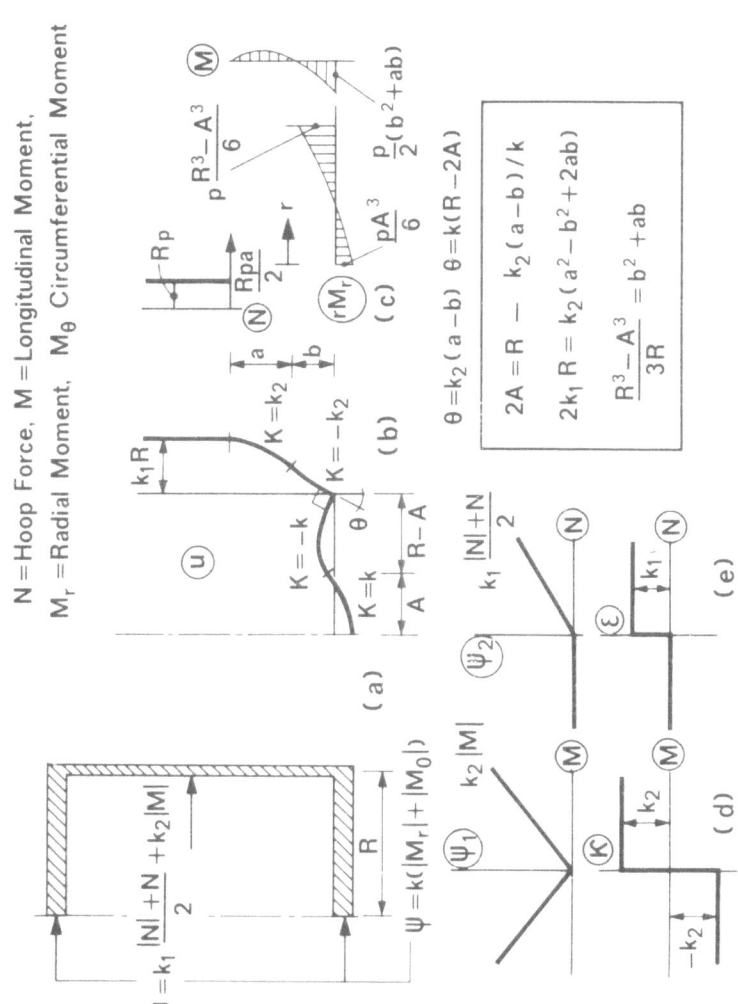

N = Hoop Force, M = Longitudinal Moment,
M_r = Radial Moment, M_θ Circumferential Moment

Figure 9.2. Example: tank design.

Over the top part of these diagrams, the entire load is resisted by hoop forces and the longitudinal moments are zero. After expressing the corner rotation θ from deformation of the shell and from that of the end plate, slope continuity furnishes the first framed equation in Fig. 9.2. The second equation follows from continuity of displacements at the end of the zone of hoop forces and the third follows from moment equilibrium at the corner. Optimal values of the parameters a, b, and A are given for various values of k_1, k_2, and k in Table 9.1.

TABLE 9.1   SOLUTION FOR CYLINDRICAL TANK

$k_1/k_2 R$	k/k_2	A/R	a/R	b/R
0.1	0.1	0.6676	0.3171	0.3506
0.1	0.5	0.5666	0.3197	0.3863
0.1	1.0	0.5369	0.3205	0.3942
0.2	0.1	0.2161	0.4490	0.3922
0.2	0.5	0.4232	0.4505	0.3737
0.2	1.0	0.4584	0.4511	0.3678

9.3   Optimal Design of a Beam with Unspecified Support Location

In Fig. 9.3 the position of the internal support C as well as the cross-sectional geometry of a beam is to be optimized for the specific cost function $\psi = k|M|$.

The moment diagram and the Pragerian associated displacement field are shown in Fig. 9.3. For optimal support location, $u'_C = 0$ at the internal support. The values of the shear forces V_C^- and V_C^+ at either side of the internal support and the distance of zero moment points (points of inflexion) b and d are determined from static conditions. Then, the ratio of b and d is obtained from kinematic conditions (i.e. similarity of the deflection diagrams for the two spans). Finally, the latter result is combined with statical relations and two equations are obtained for b. Since these two are contradictory, the solution given is clearly not optimum. This illustrates the earlier observation, that generalized cost gradients associated with a slope discontinuity (cusp) can be important in structural optimization. The real optimum solution is actually based on such gradients, as can be seen from

Then the new total cost becomes

$$\Phi_1 = P2\sqrt{2}\,\alpha \int_0^{L/2} \left\{ \beta(1-\delta)x\delta + \sqrt{\beta^2(1-\delta)^2 x^2 \delta^2 + 2(1-\delta)^2} \right\} dx$$

$$= P2\sqrt{2}\,\alpha(1-\delta)\left\{ \delta\beta L^2/8 + (\beta\delta/2)\left[x\sqrt{x^2 + 2/\beta^2\delta^2} \right.\right.$$

$$\left.\left. + (2/\beta^2\delta^2)\,\log\left(x + \sqrt{x^2 + 2/\beta^2\delta^2}\right)\right]\Big|_0^{L/2} \right\}$$

$$= P2\sqrt{2}\,\alpha(1-\delta)\left\{ (\delta\beta L^2/8) + (L/4)\sqrt{(\beta^2\delta^2 L^2/4)+2} \right.$$

$$\left. + \log\left[(L/2 + \sqrt{L^2/4 + 2/\beta^2\delta^2})/\sqrt{2/\beta^2\delta^2}\right]/\beta\delta \right\} \qquad (9.8)$$

Making use of the relations

$$\left.\begin{aligned} \sqrt{1+\Delta} &\simeq 1 + \Delta/2 + 0(\Delta^2) \\[2mm] \log(1+\Delta) &= \Delta + 0(\Delta^2) \end{aligned}\right\} \qquad (9.9)$$

a lengthy transformation, in which infinitesimals of higher order are neglected, yields the cost difference

$$\Delta\Phi = \Phi_1 - \Phi = P\alpha\delta\left[\beta L^2\sqrt{2}/4 - 2L\right] \qquad (9.10)$$

Then the minimality condition $\Delta\Phi \geq 0$ reduces to Eq. 9.5. It is left to the reader to decide, which one of the two methods is simpler.

9.2 Optimal Plastic Design of a Cylindrical Tank

Figure 9.2(a) shows a tank consisting of a cylindrical shell and two circular end plates. The specific cost functions representing the reinforcement volume in concrete plates and shells are also indicated. The tank is subject to constant internal pressure p and the total reinforcement volume is to be minimized.

Considering the cylindrical shell, the cost may be divided into two components ψ_1 and ψ_2 and then the Pragerian strain-stress requirements are given in Figs. 9.2(d) and 9.2(e). For the plates, the requirements for the two principal curvatures κ_θ and κ_r are similar to those given in Fig. 9.2(d).

The displacement fields and stress fields satisfying the Prager-Shield condition are given in Fig. 9.2(b) and 9.2(c).

108

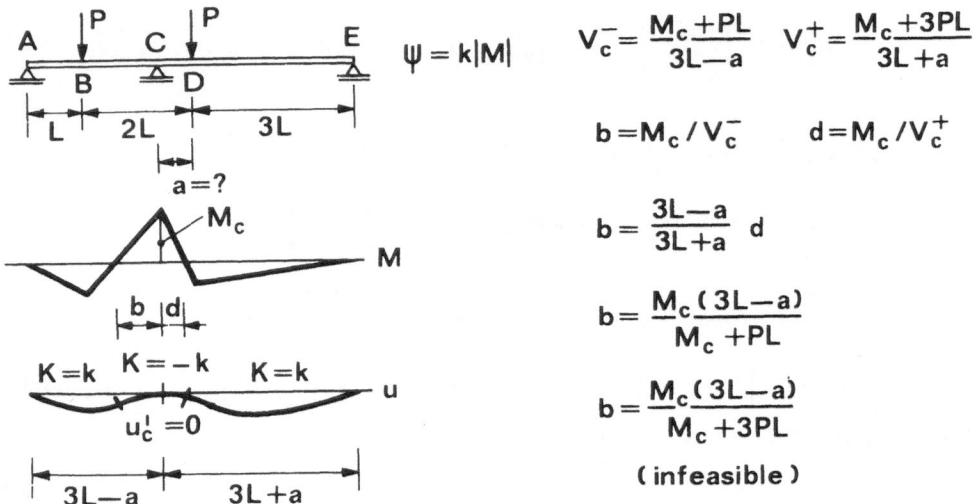

$$\psi = k|M| \qquad V_c^- = \frac{M_c + PL}{3L - a} \qquad V_c^+ = \frac{M_c + 3PL}{3L + a}$$

$$b = M_c / V_c^- \qquad d = M_c / V_c^+$$

$$b = \frac{3L - a}{3L + a} \, d$$

$$b = \frac{M_c(3L - a)}{M_c + PL}$$

$$b = \frac{M_c(3L - a)}{M_c + 3PL}$$

(infeasible)

Figure 9.3.   Beam Example: Infeasible Solution.

Fig. 9.4, in which the beam has a zero cross-section between D and E, but must still satisfy a constraint on the curvatures. In Fig. 9.4, x is expressed from both static and kinematic considerations and the optimal value of a is then calculated.

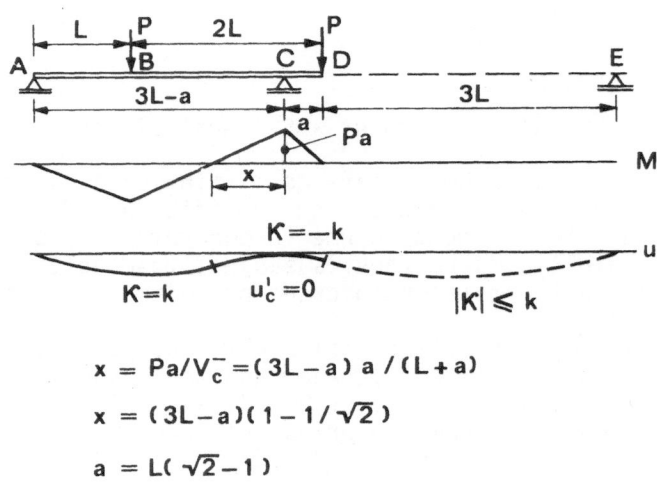

$$x = Pa/V_c^- = (3L - a) \, a / (L + a)$$

$$x = (3L - a)(1 - 1/\sqrt{2})$$

$$a = L(\sqrt{2} - 1)$$

Figure 9.4.   Beam Example: Correct Solution.

Another admissible, but nonoptimal solution has the support under the point load at D, positive moments with $\kappa = k$ from A to D, and zero moment with $\kappa = -k$ between D and E. Although the cost function is convex in this problem, the fact that the support location is unspecified makes the problem nonconvex and the solution potentially nonunique. It is, therefore, necessary to examine all solutions satisfying the Prager-Shield conditions.

10. CONCLUDING REMARKS

The aim of this paper was to demonstrate that general theorems and formulae are now available for deriving static-kinematic type optimality conditions for almost all conceivable problems in structural optimization. Although the original theory of Prager and Shield [7] was proposed in the context of optimal plastic design for convex cost functions, a single load condition, and continuous variation of the cross-sections, the idea of static-kinematic optimality criteria has since been extended to other fields of structural design.

A common feature of all methods reviewed here is that it is necessary to find a statically admissible stress-field and a kinematically admissible strain field that must then satisfy a special strain-stress relation given by a concise general formula containing the gradient of the specific cost function. Some additional local conditions are provided for optimizing such aspects as support location or segmentation.

In the current paper, only very simple examples were given, in order to illustrate the principles involved and to avoid lengthy and confusing algebraic or computational work. It is shown in a companion paper that these methods are suitable for handling highly complex design problems.

Finally it is to be remarked that although optimality criteria have been discussed here in the framework of the Prager-Shield theory, the same criteria could have been derived as generalizations of theories by Masur [22], Mroz [26], or Save [27].

REFERENCES

1. Prager, W. and Rozvany, G.I.N., "Plastic Design of Beams: Optimal Location of Supports and Steps in the Yield Moment," Int. J. Mech. Sci., Vol. 17, No. 10, 1975, pp. 627-631.
2. Prager, W. and Rozvany, G.I.N., "Optimal Layout of Grillages," J. Struct. Mech., Vol. 5, No. 1, 1977, pp. 1-18.

3. Rozvany, G.I.N. and Prager, W., "Optimal Design of Partially Discretized Grillages," J. Mech. Phys. Solids, Vol. 24, No. 2/3, 1976, pp. 125-136.

4. Thierauf, G., "A Method of Optimal Limit Design of Structures with Alternative Loads," Comp. Meth. Appl. Mech. Engg., Vol. 16, No. 2, 1978, pp. 135-156.

5. Pape, G., Eine Quadratische Approximation des Bewessungsroblem idealplastischer Tragwerke, Dr. Ing. Thesis, Essen Univ.,1979.

6. Prager, W., Introduction to Structural Optimization, Springer-Verlag, Vienna, 1974.

7. Prager, W. and Shield, R.T., "A General Theory of Optimal Plastic Design," J. Appl. Mech., Vol. 35, No. 1, 1967, pp. 184-186.

8. Rozvany, G.I.N., Optimal Design of Flexural System, Pergamon, Oxford, 1976.

9. Argyris, J.H. and Kelsey, S., Energy Theorems in Structural Analysis, Butterworth, London, 1960.

10. Heyman, J., "The Absolute Minimum Weight Design of Framed Structures," Quart. J. Mech. Appl. Math., Vol. 12, No. 3, 1959, pp. 314-324.

11. Rozvany, G.I.N., "Optimal Plastic Design: Allowance for Self-Weight," J. Engrg. Mech. Div. ASCE, Vol. 103, No. EM6, 1977, pp. 1165-1170.

12. Save, M.A. and Prager, W., "Minimum Weight Design of Beams Subjected to Fixed and Moving Loads," J. Mech. Phys. Solids, Vol. 11, No. 4, 1963, pp. 255-267.

13. Save, M.A. and Shield, R.T., "Minimum Weight Design of Sandwich Shells Subjected to Fixed and Moving Loads," Proc. 11th Int. Congr. Appl. Mech., Munich, 1964 (Springer, Berlin), pp. 341-349.

14. Lamblin, D.O. and Save, M.A., "Minimum Volume Plastic Design of Beams for Movable Loads," Meccanica, Vol. 6, No. 3, 1971, pp. 157-163.

15. Foulkes, J., "The Minimum Weight Design of Structural Frames," Proc. Roy. Soc., London, Ser. A, Vol. 223, No. 1155, 1954, pp. 482-494.

16. Barnett, R.L., "Minimum Weight Design of Beams for Deflection," J. Engrg. Mech. Div. Proc. ASCE, Vol. 87, No. EM1, 1961, pp. 75-109.

17. Haug, E.J. and Kirmser, P.G., "Minimum Weight Design of Beams with Inequality Constraints on Stress and Deflection," J. Appl. Mech., Vol. 34, No. 4, 1967, pp. 999-1004.

18. Prager, W. and Taylor, J.E., "Problems of Optimal Structural Design," J. Appl. Mech., Vol. 35, No. 1, 1968, pp. 102-106.

19. Masur, E.F., "Optimality in the Presence of Discreteness and Discontinuity," Proc. IUTAM Symp. on Optimiz. in Struct. Design, Springer-Verlag, Vienna, 1975, pp. 441-453.

20. Masur, E.F. "Optimal Design for a Discrete Set of Available Members," J. Comp. Meth. Appl. Mech. Engrg., Vol. 3, No. 3. 1974, pp. 195-207.

21. Taylor, J.E., "Maximum Strength Elastic Structural Design," J. Engrg. Mech. Div. ASCE, Vol. 95, No. EM3, 1969, pp. 653-664.
22. Masur, E.F., "Optimal Stiffness and Strength of Elastic Structures," J. Engrg. Mech. Div. ASCE, Vol. 96, No. EM5, 1970, pp. 621-640.
23. Rozvany, G.I.N., "Plastic Versus Elastic Strength Design," J. Engrg. Mech. Div. ASCE, Vol. 103, No. 1, 1977, pp. 210-214.
24. Rozvany, G.I.N., "Optimal Elastic Design for Stress Constraints," Computers and Struct., Vol. 8, No. 3, 1978, pp. 455-463.
25. Prager, W., Introduction to Plasticity, Addison-Wesley, Reading, 1959.
26. Mroz, Z., "Limit Analysis of Plastic Structures Subject to Boundary Variations," Arch. Mech. Stos., Vol. 15, No. 1, 1963, pp. 63-76.
27. Save, M., "A Unified Formulation of the Theory of Optimal Plastic Design with Convex Cost Function," J. Struct. Mech., Vol. 1, No. 2, 1972, pp. 247-276.

ACKNOWLEDGMENT

This lecture summarizes the author's joint work over many years with Professor W. Prager, who has also collaborated in the preparation of this text. Professor Prager's sudden death, soon after the completion of the manuscript, is a tragic loss to the world of science.

OPTIMALITY CRITERIA FOR GRIDS, SHELLS AND ARCHES

George I.N. Rozvany

Faculty of Engineering
Monash University
Clayton, Victoria, Australia

ABSTRACT

This paper is concerned with two major topics; (1) the theory
of optimal layouts and (2) the design of long-span surface structures for design-dependent loads. Optimal layout theory is based
on static-kinematic optimality criteria and Prager's concept of
basic structures. This method is illustrated with an example.
It is used for deriving Michell's criteria for least-weight
trusses and then the theory of optimal beam layouts (grillages)
is discussed in detail. Design dependent loads, such as self-
weight and skin-weight, play an important role in the design of
long-span surface structures. In this context, optimal shape
design of long-span grid-shells and rotational shells, as well
as one way arch systems, is discussed. Finally, optimization of
trusses, beam grids, and arches for alternative load conditions
is considered.

1.   INTRODUCTION

As Prager [1] has remarked,

Most of the literature on structural optimization is concerned with the optimal choice of cross-sectional dimensions. When the layout as well as the cross-sectional
dimensions are at the choice of the designer, structural
optimization becomes a much more challenging problem.

So far, only two structural layout problems have been investigated in detail. One study, the theory of least weight trusses,

dates back to the end of the last century with classical papers by Maxwell [2] and Michell [3]. As Prager [4] has noted,

Although the literature on Michell trusses is quite extensive, the mathematically similar theory of grillages of least weight was only developed during the last decade. Despite its late start, this theory has advanced farther than that of optimal trusses. In fact, grillages of least weight constitute the first class of plane structural systems for which the problem of optimal layout can be solved for almost all loadings and boundary conditions.

The first aim of this paper is to discuss optimal layout theory in general, and its rather successful application to minimum weight beam layouts. It is seen that the latter theory has been refined to the point where, in addition to being able to handle even the most complicated boundary shapes and load distributions, there exists now a computer algorithm that can generate, by purely analytical methods, and plot optimal beam layouts.

Because minimum weight trusses and grillages often consist of an infinity of densely space members, Prager [5] has termed them "truss-like continua" and "grillage-like continua." Naturally, the weight of these theoretical optimum solutions can be closely approximated by a layout consisting of a finite number of members (see e.g. Ref. 6). A very dense grillage may be regarded as an anisotropic plate. Naturally, the latter can be made much more efficient, in terms of weight, than an isotropic plate because its strength parameters are made dependent not only on location but also on direction. It therefore appears reasonable to consider the optimization of a similarly modified curved surface structure in which the shell is replaced by a dense system of intersecting arches. This new class of optimization problems deals with optimal arch grids or shell grids. So far, however, only the optimum shape of the arch grid surface has been studied, while the direction of the arch elements is assumed to be given [7,8]. The optimum layout of the arch-elements will be the subject of a forthcoming project. Nevertheless, the study of arch grids has lead to the conclusion that the optimal design of long-span systems for a given external load only is unrealistic, because such design-dependent loads as self-weight or skin-weight (weight at non-load bearing sheeting prescribed per unit surface area) influence the optimum solution significantly. Naturally, inclusion of design-dependent loads makes structural optimization mathematically more difficult, hence this field constitutes the second major topic of the current paper. As illustrations of the theory, shape optimization of one and two-way arch systems and of axially symmetric shells are considered. These systems are first optimized for a single load condition. The last section of the paper deals with optimization of trusses, grids, and arches for several alternate load conditions.

2. PRAGER'S APPROACH TO LAYOUT OPTIMIZATION

A very powerful method for deriving optimum structural layouts was proposed by Prager [5] in the context of trusses and grillages of minimum weight. To show the versatility of this technique, it is illustrated here with an example, in which a frame of unspecified geometry is subjected to both bending and compression.

2.1 Illustrative Example

In Fig. 2.1(a), a point load is to be transmitted to two equidistant hinges by a rigid frame consisting of two members. It has already been established in Ref. 9 that for the particular cross-section considered, the specific cost function is

$$\psi = \alpha[\beta|m| + (\beta^2 M^2 + 4N^2)^{1/2}] \tag{2.1}$$

and the corresponding cost gradients $\underline{q} = (\kappa, \varepsilon)$ are

$$\varepsilon = 4\alpha N/(\beta^2 M^2 + 4N^2)^{1/2} \tag{2.2}$$

$$\kappa = \alpha\beta[\beta M/(\beta^2 M^2 + 4N^2)^{1/2} + \operatorname{sgn} M] \ , \qquad \text{for } M \neq 0 \tag{2.3}$$

$$\kappa \leq \alpha\beta \ , \qquad \text{for } M = 0 \tag{2.4}$$

A graphical representation of these equations is given in Figures 9.1(a) and 9.1(b) of Ref. 9. One should add at this point, that for (M = 0, N = 0), the Prager-Shield condition [10] gives a cost gradient

$$\underline{G} = \lambda(\kappa, \varepsilon) \tag{2.5}$$

where κ and ε can have any value given by Eqs. 2.2 to 2.4 and $0 \leq \lambda \leq 1$. This follows from the fact that at the origin of the stress space, the cone representing the specific cost function has its vertex. Thus, the convex combination of the adjacent slopes give a highly non-unique gradient, as shown in Eq. 2.5. Graphically represented, the same gradient can be any point on or inside the cost gradient curve shown in Figure 9.1(c) of Ref. 9.

First consider a traditional method for solving this problem. Clearly, the independent variables are the height h of the frame

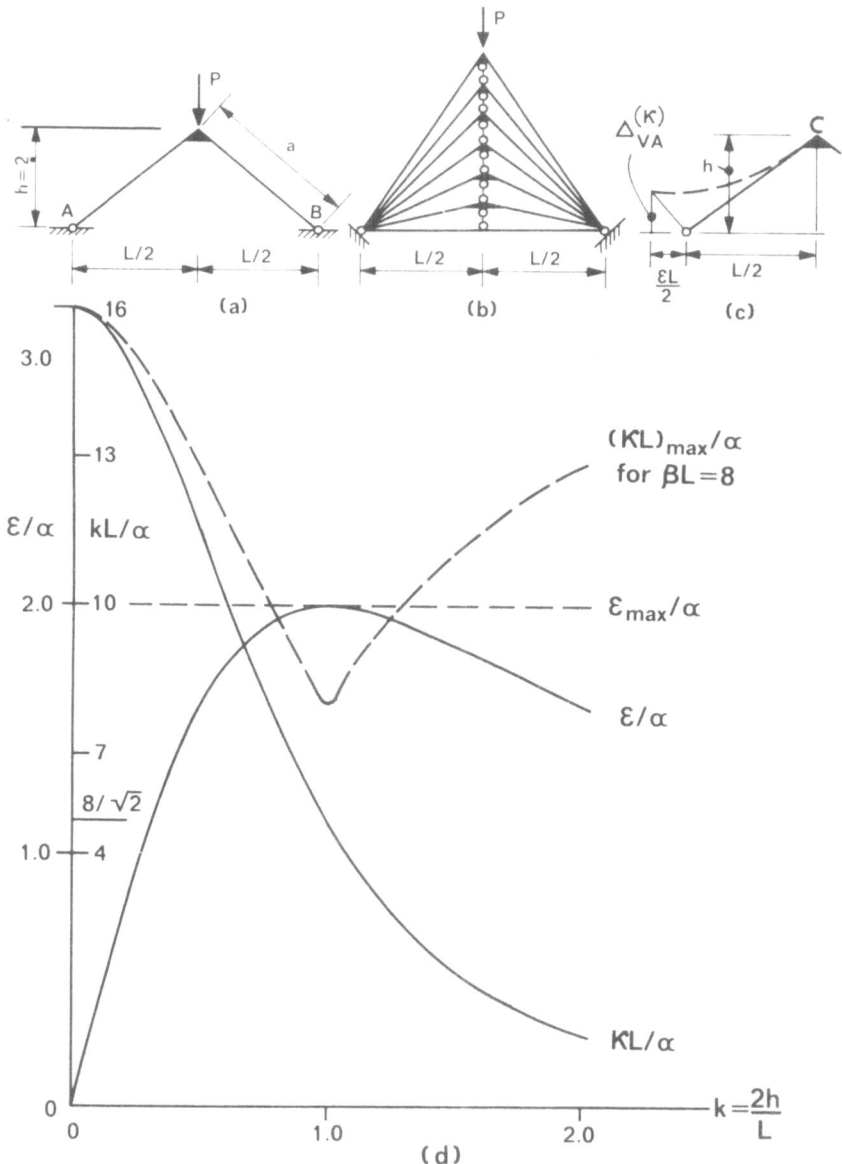

Figure 2.1   A Simple Layout Problem.

and the horizontal reaction H. Adopting a horizontal coordinate x, with its origin at support A in Fig. 2.1(a), the generalized stresses in this problem become

$$N = \frac{hP + HL}{2[h^2 + (L/2)^2]^{1/2}}$$

$$M = 2H \times h/L + Px/2$$
$\left. \right\}$ (2.6)

The values in Eq. 2.6 must be substituted into the specific cost function of Eq. 2.1 and then the integral

$$\Phi = 4[h^2 + (L/2)^2]^{1/2} \int_0^{L/2} \psi \, dx/L \tag{2.7}$$

evaluated. Finally, the minimum of Φ with respect to the independent variables H and h have to be found. This is rather difficult to do by analytical methods, because; (1) the specific cost function ψ in Eq. 2.1 is not everywhere differentiable and (2) the total cost Φ is a non-convex function of H and h. Nevertheless the above cost surface has been explored analytically and has confirmed results that are obtained herein by Prager's method.

Following Prager's idea, consider first a structure that consists of the centerlines of all admissible layouts (Fig. 2.1(b). Although a finite number of frames are indicated in Fig. 2.1(b), the system considered consists of an infinite number of frames, some of which are infinitely tall. Such an underlying system is referred to by Prager [5] as a basic structure. On the basis of set-theoretical terminology, it could also be called the universe of the layout problem. The total cost of such a system is

$$\Phi = \sum \int_D \psi \, ds \tag{2.8}$$

where s is the coordinate along the frame centerline, D is the total centerline of a given frame, and Σ signifies summation over all frames. It will be noted that the above problem is convex, because ψ in Eq. 2.1 is a convex function. One obvious advantage of Prager's layout theory is, therefore, that it conserves the convexity of the problem when ψ is convex.

In the basic structure (or universe), all (optimum) members of nonzero cross-section must satisfy the cost gradient condition as in Eqs. 2.2 to 2.4 and all (non-optimum) members of zero cost section must still be subjected to an inequality constraint, such as Eq. 2.5, on the strains. Kinematic admissibility then guarantees optimality, since the Prager-Shield condition is necessary and sufficient, if ψ is convex. Naturally, the generalized stresses must also be statically admissible. The vertical links connecting the frames in Fig. 2.1(b) are costless and (owing to the cost gradient condition) perfectly rigid. The vertical deflection at the apex of all frames therefore has the same value.

A kinematically admissible set of displacements is first established when both ε and κ are constant along a given frame. Assume first that one subjects a frame to an axial strain ε, while releasing one hinged support and thereby allowing horizontal displacements. Then, the apex of the frame moves downward by $\Delta_{VC}^{\varepsilon} = \varepsilon h$ and the released hinge moves horizontally by εL. In order to reestablish kinematic admissibility, the constant curvatures of each member must produce a horizontal displacement of $\varepsilon L/2$. It is explained in Fig. 2.1(c) that the same constant curvatures produce, on the basis of similar triangles, a vertical displacement of

$$(L/2h)(\varepsilon L/2) = \varepsilon L^2/4h \tag{2.9}$$

Hence the total vertical displacement at C is

$$\Delta_{VC} = \varepsilon h + \varepsilon L^2/4h \tag{2.10}$$

Next, it is established that a flexureless arch of unit slope

$$h = L/2 \tag{2.11}$$

is optimum for a certain range of βL values. For such an arch, Eq. 2.2 gives $\varepsilon = 2\alpha$ and hence Eqs. 2.10 to 2.12 give a vertical deflection at the apex of

$$\Delta_{VC} = 2\alpha L \tag{2.12}$$

which is also the deflection of any arbitrary arch at its apex. Denoting the slope $(2h/L)$ of any arch by k, Eqs. 2.10 and 2.12 furnish

$$(\varepsilon L/2)(k + 1/k) = 2\alpha L \tag{2.13}$$

and hence

$$\varepsilon = 4\alpha/(k + 1/k) \tag{2.14}$$

It is shown in Fig. 2.1(d) that ε in Eq. 2.14 has its maximum value of $\varepsilon = 2\alpha$ at k = 1 and then, by Eq. 2.2, only the arch of unit slope can resist purely axial forces in the optimum system.

For sufficiency, it is still necessary to show that all curvatures κ of the basic structure (i.e. all potential frames) satisfy the Prager-Shield condition (Eqs. 2.2 to 2.5, or

Figure 9.1(c) of Ref. 9. Kinematic admissibility requires equality of horizontal displacements due to axial and curvature effects in any arbitrary frame,

$$\epsilon L/2 = ah\kappa/2 \quad , \tag{2.15}$$

and hence

$$\kappa L = 4\alpha L^2/[ah(k + 1/k)] = 16\alpha/[k(1 + k^2)^{\frac{1}{2}}(k + 1/k)] \tag{2.16}$$

Such curvatures must be smaller than or equal to the ones given by Eqs. 2.2 to 2.4, which reduce to (see Figure 9.1(c) of Ref. 9

$$\kappa_{max} = \beta[\alpha + (\alpha^2 - \epsilon^2/4)^{\frac{1}{2}}] \tag{2.17}$$

For the particular case of $\beta L = 8$, Fig. 2.1(d) shows the curvature values given by Eqs. 2.16 and 2.17. It can be seen that for this particular β value, two solutions are equally optimum: for $k = 0$, (horizontal beam type member) $\epsilon = 0$ and $\kappa = 2\alpha\beta$ which by Eq. 2.2 or Fig. 9.1(c) of Ref. 9 admits only moments but no axial force; and for $k = 1$ (unit slope) $\epsilon = 2\alpha$ and $\kappa < \beta\alpha$, which admits a force without a moment. At $\beta L = 8$, the load can also be shared in an infinite number of combinations by the beam and the frame of unit slope, provided that the vertical tie is costless. All such combined solutions are equally optimum.

For $\beta L > 8$, the top curve $\kappa_{max} L$ moves higher than the one in Fig. 2.1(d) and hence only the solution with unit slope and purely axial force is optimum. For $\beta L < 8$, all three curves move downward in proportion to βL, but ϵ_{max} remains unchanged. Hence only the horizontal (beam) solution, with bending but no axial force, is optimum. As pointed out earlier, these conclusions have been verified by a conventional method that requires considerably lengthier calculations.

2.2 General Procedure

Although in the above example a few intuitive decisions have been made with a view to simplifying the procedure, Prager's optimal layout theory can be used in a systematic fashion, as will be seen in subsequent sections. Essentially, the following operations are being employed:

(1) Set up the basic structure (universe) consisting of all potential structural members.

(2) Determine the strain-stress relation from the Prager-Shield condition for (optimum) members of nonzero cross section.

(3) Derive the strain inequality conditions from the cost gradients for (non-optimum) members of zero cross section.

(4) Consider a kinematically admissible strain distribution and scale it in such a manner that, along lines of maximum strain, the Prager-Shield condition for nonzero members is satisfied. The term maximum strain must here be interpreted in terms of the cost gradient surface and does not necessarily correspond to the maximum value of a single strain component.

(5) Check if, along all non-optimum members, the strain inequality condition is satisfied.

(6) Check if the strains in the optimum members admit a statically admissible solution of generalized stresses.

In the example of Section 2.1, the last step was trivial, because in the two solutions furnished by Fig. 2.1(d), one (a beam) was required to resist only bending and the other (the frame) only axial forces. Both were clearly statically admissible.

2.3 Trusses of Minimum Weight

Prager's layout theory can be applied readily to least weight trusses, for which the specific cost function is

$$\psi = k|N| \tag{2.18}$$

where k is a constant and N is the member force. The Prager-Shield condition then furnishes the following strain-stress relations:

$$\varepsilon = k \text{ sgn } N \quad , \qquad \text{for } N \neq 0 \tag{2.19}$$

$$|\varepsilon| \leq k \quad , \qquad \text{for } N = 0 \tag{2.20}$$

Considering a plane truss, the basic structure consists of potential members running in all possible directions at all points. However, it is well known that at a point of a plane strain system, the maximum and minimum strains occur in the two so-called principal directions, which are orthogonal to each other. Because

Eqs. 2.19 and 2.20 do not admit a greater strain than k, one or both principal directions must correspond to $|\varepsilon| = k$, if nonzero forces are to be admitted at the point considered. This conclusion is naturally identical to the conditions proposed originally by Michell [3]. Depending on the signs of the principal strains, various geometrical properties of least weight trusses can be derived. This subject will not be discussed here in detail, because relatively little progress has been made in the theory of least weight trusses in recent years. However, Prager has developed a rather ingeneous method for optimizing trusses with a finite number of joints [11-13].

3. THE THEORY OF OPTIMAL BEAM LAYOUT
   (MINIMUM WEIGHT GRILLAGES)

One of the most sophisticated theories in structural optimization deals with optimal layout of beam systems (grillages), whose cross-sectional area ψ is proportional to the absolute value of the bending moment M in the beam,

$$\psi = k|M| \tag{3.1}$$

where k is a given constant. Such a specific cost function can be used for rectangular beams of given depth but variable width or for beams consisting of two thin flanges, neglecting the cost of the web. Because the Pragerian associated displacement field is identical to the elastic displacement field, the optimal solution is equally valid for plastic and elastic grillages. Moreover, since the specific cost function is a homogeneous function of degree one, the solution also minimizes the product of loads and displacements $\int_D p\ u\ d\underline{x}$ (see Figure 7.1 in Ref. 9), ensuring maximum average stiffness. The same solutions minimize the reinforcement in concrete slabs of constant thickness.

Naturally, it is impossible to discuss in a single paper such an extensive field in detail. Already in 1976, a major part of a book [14] and a 130 page review article in the Advances in Applied Mechanics [15] was devoted by the author to the theory of optimal load transmission by flexure. Therefore, only some of the fundamental features of this development will be reviewed here, briefly.

3.1 Transformation into a Geometrical Problem
    via Optimality Criteria

One can again usefully employ Prager's concept of basic structures, which in this case consist of potential beams running in all directions at any point of the structural domain D. For the specific cost function in Eq. 2.1, the Prager-Shield condition [10] furnishes the stress-strain relations

$$\kappa = k \text{ sgn } M \quad , \qquad \text{for } M \neq 0 \qquad\qquad (3.2)$$

$$|\kappa| \leq k \quad , \qquad \text{for } M = 0 \qquad\qquad (3.3)$$

where κ is the beam curvature. Considering the Pragerian associated displacement field at a point, the maximum and minimum curvatures occur again in the principal directions, at right angles. Since Eqs. 3.2 and 3.3 set the maximum absolute value of the curvatures at k, the condition $|\kappa| = k$ for nonzero cross section can only be fulfilled in one or both principal directions (unless the curvature is the same in all directions). The consequence of the above conclusion is that only the types of optimum regions shown in Fig. 3.1 are admissible in the optimum solution. In R-regions, only one principal curvature has the absolute value k, $(R^+: \kappa_1 = k, |\kappa_2| < k \text{ or } R^-: |\kappa_1| < k, \kappa_2 = -k)$ and beams run only in the corresponding direction. In S-regions the curvatures are the same in all directions and have the value k $(S^+: \kappa_1 = \kappa_2 = k$ or $S^-: \kappa_1 = \kappa_2 = -k)$, and thus beams can run in all directions. In T-regions the two principal curvatures are of opposite sign $(\kappa_1 = -\kappa_2 = k)$ and beams can be oriented only in the two principal directions. In addition, in all regions the beams must have a moment of the same sign as that of the curvature of the associated displacement field.

The relatively complicated variational problem is now reduced to a simple geometrical one. One must simply find a displacement field consisting of the optimum regions in Fig. 3.1, so that the deflected surface is smoothly jointed at region boundaries and satisfies the kinematic boundary conditions. Such conditions are zero displacement (u = 0) along simple supports and zero displacement as well as slope (u = $\partial u/\partial x$ = $\partial u/\partial y$ = 0) along clamped edges.

3.2 A Simple Example

The proposed procedure is illustrated with an example. Two forces (P_1 and P_2) in Fig. 3.2(a) are to be transmitted by a system of beams to given simple supports along a square boundary (ABCD). The solution must consist of smoothly jointed regions of the type shown in Fig. 3.1 and must have a zero deflection (u = 0) along the supports ABCD. It can be easily checked that deflections given in Fig. 3.2(b) indeed satisfy the conditions of

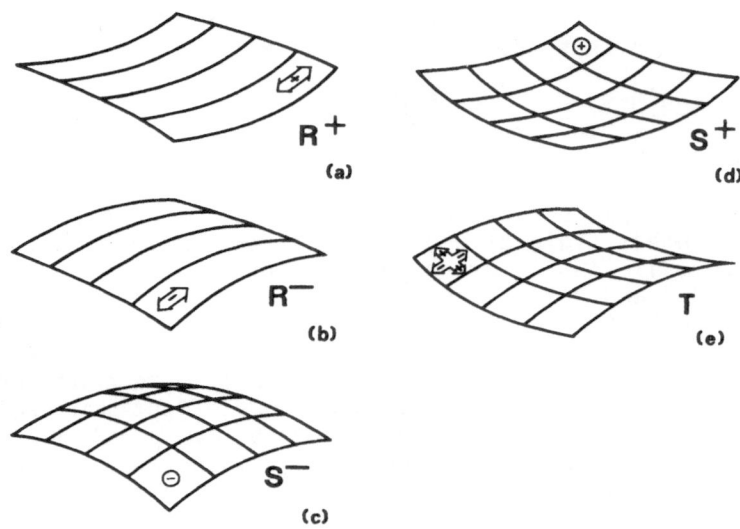

Figure 3.1   Optimal Regions in Grillages.

optimality.   Over the central region, $\kappa_x = -\partial^2 u/\partial x^2 = \kappa_y = \partial^2 u/\partial y^2$
$= k$ and hence one has an S^+-region.   In the corner regions,
$\kappa_x = \overset{-}{+} k$ and $\kappa_y = \pm k$ and therefore they consist of T-regions.
A simplified representation of this solution, using the symbols
introduced in Fig. 3.1 is given in Fig. 3.2(c) and an oblique
view in Fig. 3.2(d).   It can be readily checked that the boundary
and continuity conditions are satisfied by the displacement in
Fig. 3.2(b).

Bearing in mind that one can use beams in positive bending
in any direction in the central S^+-region, the solution in Fig.
3.2(c) allows a number of equally optimum beam layouts, two of
which are indicated in Figs. 3.2(e) and 3.2(f).   In the latter,
beams shown in broken lines are in negative bending.   The above
solution was found independently by Morley [16] and the author.
Because of the difficulty in finding region patterns that satisfy
all boundary and continuity conditions, no significant progress
was the made for several years and Morley even suggested [16]
that such a solution may not exist for built-in corners. Finally,
a better insight into patterns of optimum regions was achieved by
observing certain topological features of these layouts, in
simple examples.

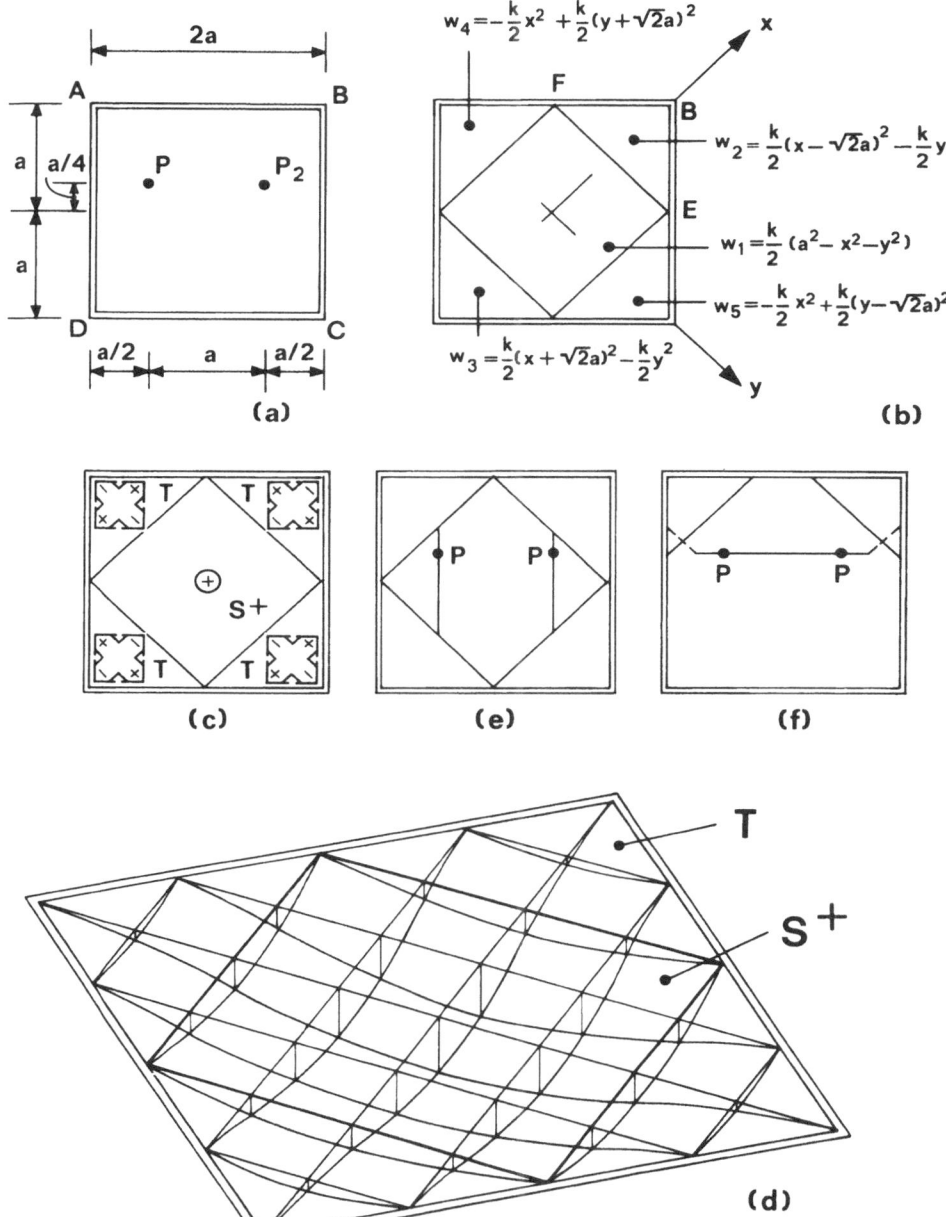

Figure 3.2   Optimal Beam Layout: A Simple Example.

124

3.3 The Anatomy of Optimum Beam Layouts

Consider first the optimum grillage layout in Fig. 3.3, in
which the beams are built-in along the boundary and the point P.
One can observe that the solution consists of two subsets:
(1) Junctions consisting of S^+-regions (shaded areas) and
(2) Branches, which either connect two junctions or a junction
with a boundary. Moreover, it has been found that centerlines
(dash-dot lines) of such branches may intersect the boundary only
at points where the boundary curvature takes on a locally maximum
value. The centerlines themselves represent a locally maximal
deflection along a line normal to the centerline. By first
locating all potential starting points of branches, a systematic
method for finding the beam layout for any combination of simply
supported and built in boundaries was developed in 1973.

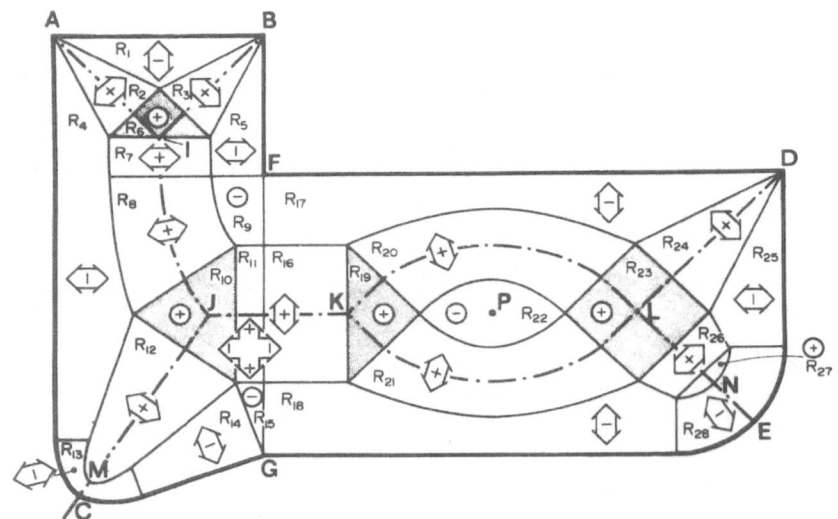

Figure 3.3   Branches and Junctions in Optimum Layouts.

The centerline divides each branch into two half-branches.
or fields, and it has been found that only four types of fields
are associated with clamped and simply supported boundaries. A
field is more restricted geometrically than an optimum region by
itself, because it takes the relevant boundary conditions into
consideration. So-called β-fields (Fig. 3.4(a)) occur along built
in boundaries and consist of a beam system of negative curvature
(R^--region) along the boundary and a R^+-region along the centerline. Along simply supported boundaries, however, three types
of fields may occur. In an α-field (Fig. 3.4(b)), which is the
most common, beams have simply a positive curvature throughout
(R^+-region). If the boundary is concave from the outside and

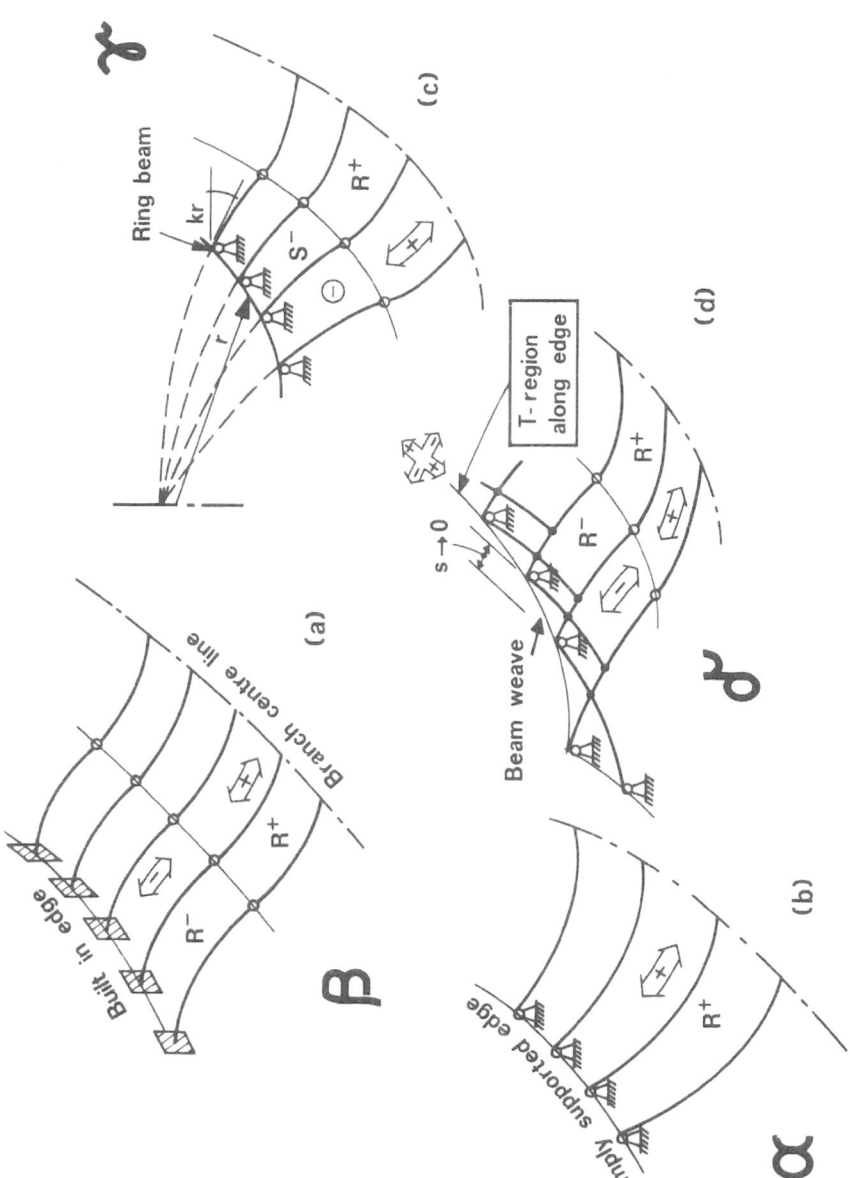

Figure 3.4 Optimum Fields Associated with Simple and Clamped Supports.

its radius is sufficiently small, then it is more economical to have a concentrated ringbeam along the simple supports (see the γ-field in Fig. 3.6(c) of Section 3.4), which can balance negative bending moments in the radial direction. This possibility is not only an abstract theory, but has been verified experimentally in the context of annular reinforced concrete slabs with concentrated reinforcement along the inner edge.

The limiting case, beyond which a γ-field is more economical than an α-field, has been derived by Prager's static-kinematic method and then checked by purely static calculations. A sharp corner in a simply supported boundary can be regarded as a clamped point (see Ref. 15, p. 298). In the vicinity of such a corner, it is more economical to transmit a negative support moment to the corner by a so-called beam-weave, a term suggested by Prager. The resulting δ-field is shown in Fig. 3.4(d), in which the curvature conditions along the edge (locally) correspond to that of a T-region. In a beam-weave, beams of infinitesimal length are subjected to positive moments and the end of long beams to negative moments.

Depending on the types of fields they contain, branches bounded by simple supports and/or clamped edges may be of ten different types, namely $\alpha\alpha$, $\alpha\beta$, $\alpha\gamma$, $\alpha\delta$, $\beta\beta$, $\beta\gamma$, $\beta\delta$, $\gamma\gamma$, $\gamma\delta$, and $\delta\delta$ branches. In Fig. 3.3, all branches are of $\beta\beta$ type.

The rather intricate geometrical properties of all types of branches have been fully explored [14,15]. In addition to finding general equations for the centerlines and region boundaries in various types of branches, simple geometrical constructions have been developed for deriving such curves, which also promote a better insight into the intrinsic features of the solution. These methods are also used in computer programs for analytically generating optimum layouts. As an example, the construction of some branches is given in Fig. 3.5 (see Ref. 15).

3.4   The Free Edge Paradox

In addition to the foregoing types of branches, the general solution has been obtained for internal simple supports, over which the grillage is continuous (e.g. Ref. 14, p. 219). By 1974, the foregoing theory could handle all boundary conditions, with the exception of free edges, which defied all attempts to find a solution. The main difficulty was that often only positive moments (i.e. R^+-regions) are statically admissible between free edges and simple supports. However, a positive moment in the neighborhood of a free edge requires an upward force at the free end (Fig. 3.6(a)) and there is no external support to provide such a force.

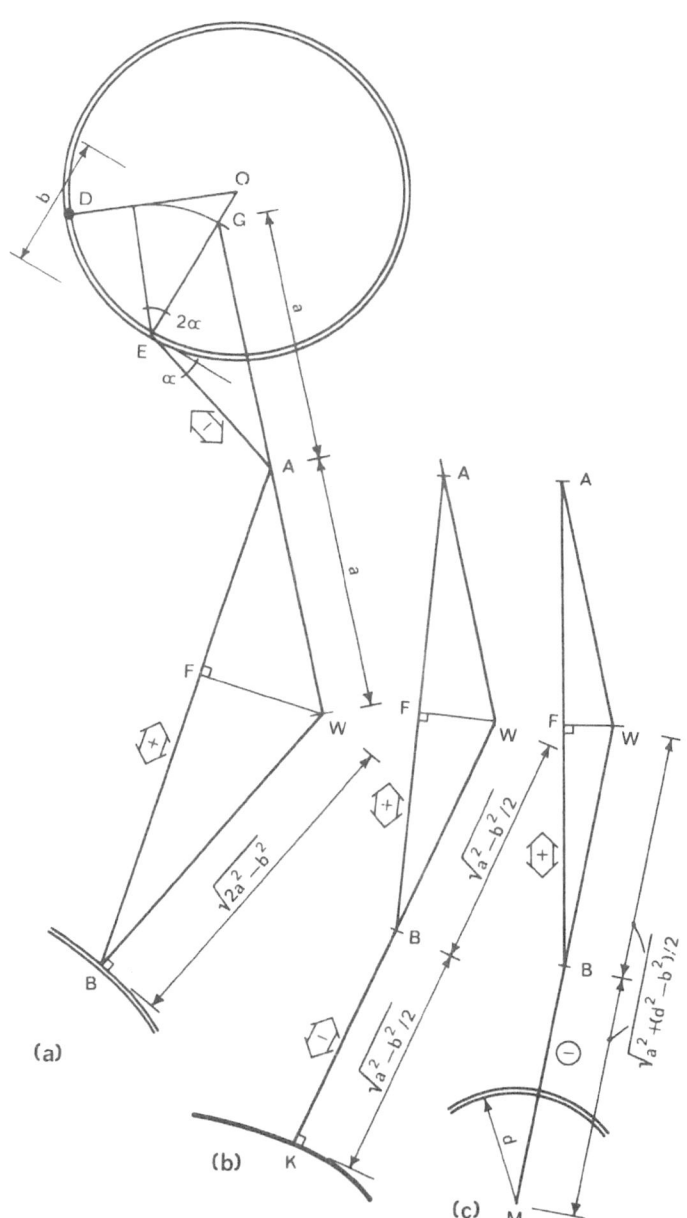

(a)

(b)

(c)

Figure 3.5   Construction of $\alpha\delta$, $\beta\delta$, and $\gamma\delta$ Branches.

128

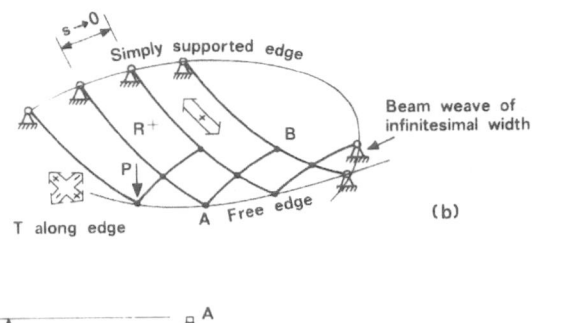

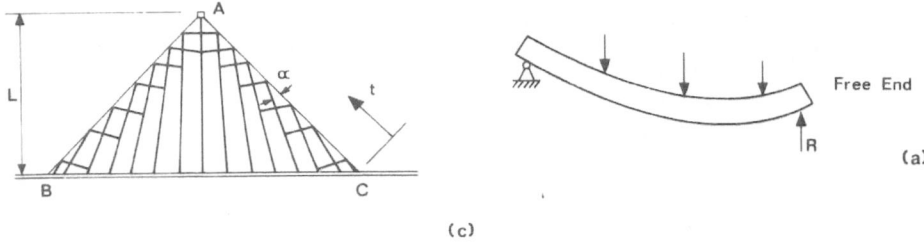

(c)

Figure 3.6   The Problem of Free Edges.

Finally, the free edge paradox was solved by Hill and the author [16]. Prager has derived the same solution independently [4]. It was found that, in general, a beam weave occurs only along a free edge in optimal layouts. One can see from Fig. 3.6(b) that each long beam is supported at its end A by a short beam AB, which is balanced by the adjacent long beams. Because in these problems the condition $\kappa_1 = -\kappa_2 = k$ (as in

T-regions) can only be satisfied along a line coinciding with the free edge; in the theoretical optimum solution, the length of the short beams (Fig. 3.6(b)) shrinks to an infinitesimal value. For a particular problem, with simple point support at A, simple line support along BC, and free edges along AB and AC, the optimum solution is given in Fig. 3.6(c). It is interesting that the optimum angle α between the beams and the free edge is given by the simple analytical expression

$$t = L(1 + 1/\sin 2\alpha)^{\frac{1}{2}} e^{(1-\cot \alpha)/2} \tag{3.4}$$

The solution given in Fig. 3.6(c) is valid for any downward symmetric load distribution. The same solution has been verified by numerically optimizing (using nonlinear programming) the layout of a finite number of beams for the same problem. The Prager-efficiency (i.e. theoretical optimal weight/weight of a discrete solution) for the 2, 6 and 10 beam solutions was found to be 96.3, 97.9, and 98.5 per cent [17]. Moreover, the beam directions in the discrete 10-beam solution were very similar to those in the absolute optimal solution (Fig. 3.6(c)).

Since the foregoing developments, the optimal beam layout for free edges has been determined in a number of other combinations. While Fig. 3.7(a) shows the load transmission of a single point load, in principle along a free edge, Figs. 3.7(b), 3.7(c), and 3.7(d) indicate only the optimum long beams for free edges of various orientations. In these three diagrams, respectively, the relation between α and t is given by

$$t/L = \tan 2\alpha/\tan 2\alpha_0 \tag{3.5}$$

$$t/\sqrt{2}\, L = e^{\frac{1}{2}(\cot \alpha_0 - \cot \alpha)}[(1 + \mathrm{cosec}\ 2\alpha)/(1 + \mathrm{cosec}\ 2\alpha_0)]^{\frac{1}{2}} \tag{3.6}$$

$$t/L = \ln[\tan(\alpha + \pi/4)/\tan(\alpha_0 + \pi/4)] - \cot \alpha + \cot \alpha_0 \tag{3.7}$$

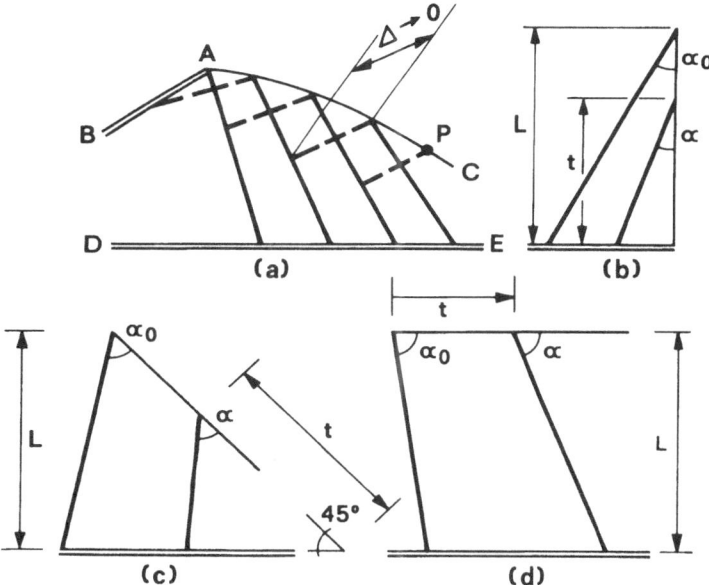

Figure 3.7   Some Optimum Beam Layouts with Free Edges.

3.5 Beam Supported Edges

In practical design problems, beams along free edges are often deeper than interior beams of the system. It is therefore reasonable to investigate the case in which the specific cost function is $\psi = k|M|$ for interior beams and $\psi = k_1|M|$, with $k_1 < k$, for edge beams. In the context of concrete slabs, a similar formulation was introduced by Lowe and Melchers [18]. In Fig. 3.8, two edges (BD) are beam-supported and the third (DD) has a rigid simple support. There is also a rigid simple point support at B. In all three cases, $k = 1$ is adopted. For $k_1 = 1/3$ and $k_1 = 1/2$, the Lowe-Melchers formula [18] gives correctly the result in Figs. 3.8(a) and 3.8(b), because for this problem it reduces to (after correcting a printing error)

$$(k_1 + 1)\tan^2 \bar{\alpha}/2 + (k_1 - 1)\tan \bar{\alpha} + (k_1 - 1)/2 = 0 \qquad (3.8)$$

However, the same formula would give an inadmissible result for $k_1 = 2/3$, since its central portion would not be kinematically admissible. It turns out that the outer regions are governed by the Lowe-Melchers formula and the layout in the inner region by Eq. 3.6 herein.

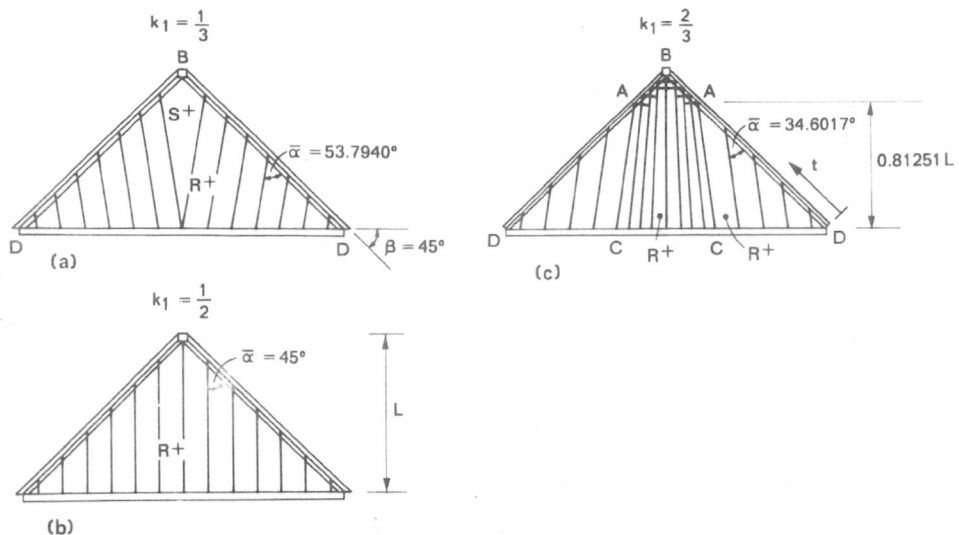

Figure 3.8 Optimal Triangular Grillages with Beam-Supported Edges.

It is interesting to note that over the segments AB, no edge beam is used, in spite of its relatively lower cost, because

a beam weave proves cheaper. The result in Fig. 3.8(c) also
independently confirms the validity of the present treatment of
free edges, because Eq. 3.6 gives exactly the same direction for
AC as the corrected Lowe-Melchers formula. Moreover, the curvature increases along the line BA and reaches exactly $\kappa = k_1 = 2/3$
at A, where the edge beam takes over.

3.6 The Effect of Nonuniform Depth

If the depth h of a grillage of a reinforced plate varies,
i.e. $h = h(\underline{x})$, then the curvature condition also changes with
location and it can be expressed as

$$\kappa_i = k \text{ sgn } M_i/h(\underline{x}) \quad , \qquad \text{for } M_i \neq 0 \tag{3.9}$$

In a recent paper [19], it was claimed that for an axisymmetric plate with polar coordinates (r,θ), the condition for only
circumferential reinforcement (beams) is that the thickness
increases monotonically with the radius. This is because a condition of only circumferential reinforcement, on the basis of
the Prager-Shield theory [10], is

$$\kappa_\theta \gtrless \kappa_r$$

or

$$u'(r)/r \geq u'' \tag{3.10}$$

which is only possible if $\kappa_\theta = u'/r$ decreases and, by Eq. 3.9
$h(r)$ increases with the radius.

It has since been found that the foregoing condition is not
entirely correct, because κ_r often takes on a negative value and
therefore the criterion

$$|\kappa_\theta| \geq |\kappa_r| \tag{3.11}$$

is more appropriate. This means that, in addition to the lower
limit $u'' - u'/r = 0$ implying $h(r) = $ const, the upper limit
$u'' + u'/r = 0$ must also be observed for only circumferential
reinforcement in the optimum solution. A simple example of this
limiting case is $h(r) = ar^2$, with $u'/r = k/ar^2$ and $u'' = -k/ar^2$.
Moreover, the problem with $h(r) = ar^3$, $u'/r = 1/ar^3$, and
$u'' = -2/ar^3$ clearly satisfies the condition of monotonically
increasing thickness variation claimed in Ref. 19, but fails to
satisfy Eq. 3.11. The condition in Ref. 19 also breaks down for
stepwise variation of the thickness, as is shown in Fig. 3.9.
The reason for this is that in a region with $\kappa_\theta \gtrless \kappa_r$, a step in

132

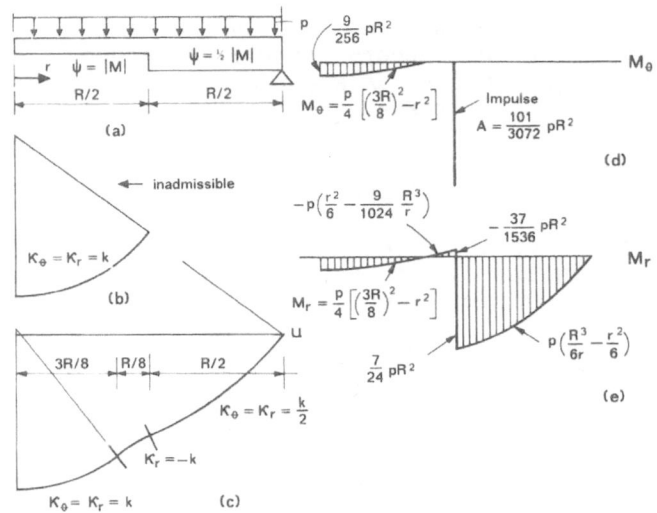

Figure 3.9   Optimum Solution for a Slab of Varying Depth.

the thickness would require a step in the curvature $\kappa_\theta(r) = u'/r$.
Since the slope $u'(r)$ must be continuous, such a sudden change
in the value of κ_θ is not possible.   This case is illustrated in
Fig. 3.9 in which the depth of a uniformly loaded circular plate
over the outer half radius is twice the depth value for the inner
half.   Whereas the criterion in Ref. 19 would give $\kappa_\theta = \kappa_r = k$
for the inner region (Fig. 3.9(b)), this would be inadmissible
at $r = R/2$, owing to the requirement $\kappa_\theta = k/2$ for $r > R/2$.   The
correct solution is shown in Figs. 3.9(c), 3.9(d), and 3.9(e)
and requires a concentrated band of reinforcement at $r = R/2$.

The foregoing solution is verified by successive approximations in Fig. 3.10, in which three discrete beam layouts give
cost values of 23.5, 21.81 and 14.54 for a single unit point load.
The absolute optimum cost can be evaluated by using the deflection surface in Fig. 3.9(c) as an influence surface.   The deflection value at the unit point load is 14.5 (for $R = 8$), which is
clearly just slightly better than the best discrete solution in
Fig. 3.10.

3.7   Other Recent Developments

Many solutions given by the foregoing theory are somewhat
unrealistic, because they contain high shear concentrations and
the cost of shear forces is neglected.   In order to rectify

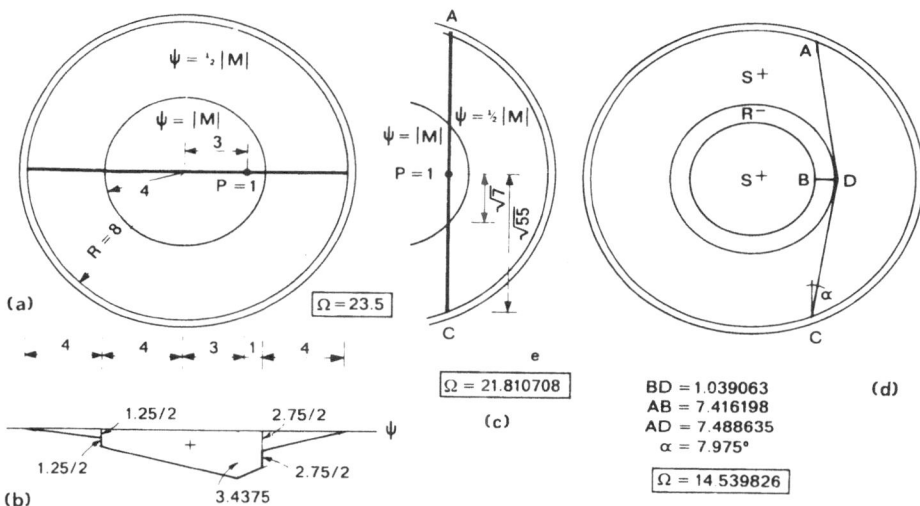

Figure 3.10   Discrete Solutions Approximating
the Solution in Fig. 3.9.

this situation, a more refined theory has been developed [20],
in which the beam specific cost function takes the form

$$\psi = k|M| + k_1|V| \tag{3.12}$$

where k and k_1 are constants and V is the shear force in the
beam. Other refinements include allowance for the cost of supports and connections, as well as self-weight of grillages.

4. OPTIMIZATION FOR DESIGN-DEPENDENT LOADS

The author came across the problem of design-dependent loads
the first time, when investigating optimal design of long span
surface structures, under a government-sponsored research contract in Stuttgart, Germany two years ago. Because this project
involved arch grids (grid shells) and cable networks, the optimization of these structures for a given external load must first
be considered.

4.1 Optimum Arch Grids: Given External Load

The optimization of arch grids for a given external load
[7] was based on a number of simplifying assumptions. It was

stipulated that all arches were to resist axial compression only under the single load system considered. Moreover, no allowance was made for elastic compatibility, nor for instability. The resulting solution thus represents optimum plastic design for a given vertical load system $p(x,y)$. In addition, the design was restricted to arches whose horizontal projection was parallel to the rectangular coordinates (x,y) contained in a horizontal reference plane.

With a view of facilitating the mathematical treatment, it was first assumed that at a point (x,y) of the reference plane, the elevation of the x-arch and y-arch is not necessarily the same, but the two arch systems are connected by vertical ties, whose weight is disregarded. The total weight of the arches was to be minimized, within a permissible stress constraint. It was then found that the optimum arch grid satisfies the following three conditions:

(1) Zero bending, which was actually assumed without proof of optimality, is ensured if each arch centerline has the shape of a scaled moment diagram for a beam that has the same span and loading as the arch considered. In other words, each arch must form a funicular curve for the share of load it is resisting and the loads on the x and y arch systems must add up to the total load at all points $[p_x(x,y) + p_y(x,y) = p(x,y)]$.

(2) The unit mean square slope condition requires, for an x arch for example, that

$$\int_{x_1}^{x_2} \left[\frac{dz_x}{dx}\right]^2 \frac{dx}{L_x} = 1.0 \tag{4.1}$$

where L_x is the horizontal arch length, z_x is the arch elevation, and (x_1,x_2) are the end-coordinates of the arch.

(3) The equal elevation condition stipulates that the elevation of the x and y arches at any point (x,y) of the reference plane is the same

$$z_x(x,y) \equiv z_y(x,y) \tag{4.2}$$

This means that one has a single layer of arches in the optimum solution, although initially two separate layers of arches were permitted.

Using the foregoing three optimality conditions, which have been derived by both variational methods and the Prager-Shield condition [10], the optimum shape of arch grids has been determined for various boundary shapes.

4.2 Optimal Arch Grids: Allowance for Self-Weight

It has been shown [8] that the optimization of arch grids for combined external load and self-weight can still be based on the three criteria in Section 4.1, except that the unit mean square slope condition is replaced by (for any x-arch, for example)

$$\left[\int_{x_1}^{x_2} z^2 e^{2zc} dx \right] \bigg/ \left[\int_{x_1}^{x_2} e^{2zc} dx \right] = 1.0 \tag{4.3}$$

where $c = \gamma/\sigma_0$, γ is the specific weight of the arch material, and σ_0 is the permissible stress. This modified criterion is termed the weighted unit mean square slope condition.

The three types of optimality criteria again yield the solution directly. However, owing to the computational complexity of this problem, another two methods have been used for checking the results. In one method, the archgrid was discretized, differential calculus was used to derive optimality conditions, and then particular examples were optimized. In the third method, nonlinear programming was applied directly to specific examples. All three methods have produced excellent agreement, although in the second two, equal elevations were not stipulated. Naturally, the elevations of the x-arches and y-arches turned out to be equal in the latter solutions. In Figs. 4.1 and 4.2, cross sections and isometric views of square optimized archgrids for $\alpha = 0$ and $\alpha = 0.8$ are compared, where $\alpha = cL$, L is the span of all arches, and c has been defined above. The solutions for higher α-values differe even more from the solution in which the self-weight is neglected ($\alpha = 0$). An interesting, but understandable feature of the solutions [7] is the nonconvexity of the arches in the vicinity of the edges.

Research into archgrids was proposed by Prager and the first paper on this topic was written jointly by the author and Prager [21]. The work on optimal archgrids with self-weight was carried out jointly by H Nakamura and the author [8].

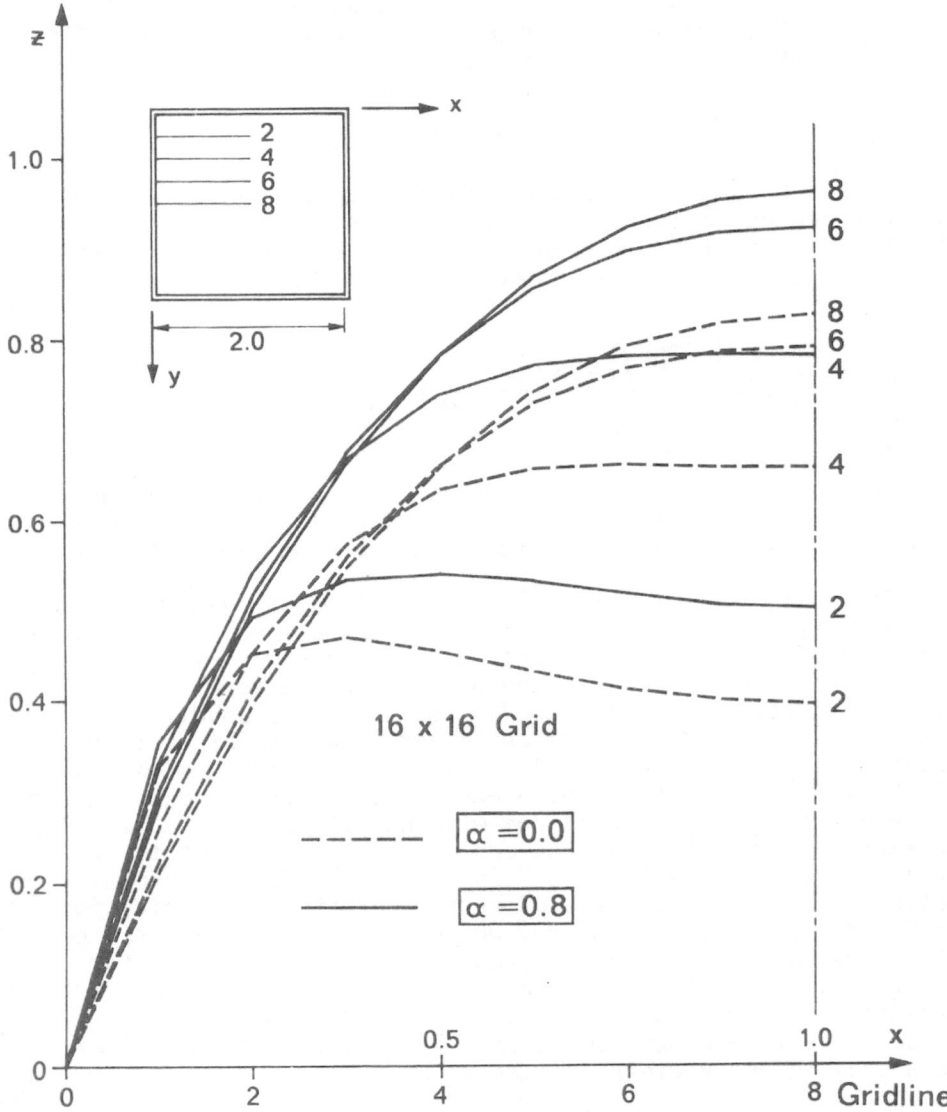

Figure 4.1   Comparison of Optimal Archgrids With
and Without Self-Weight.

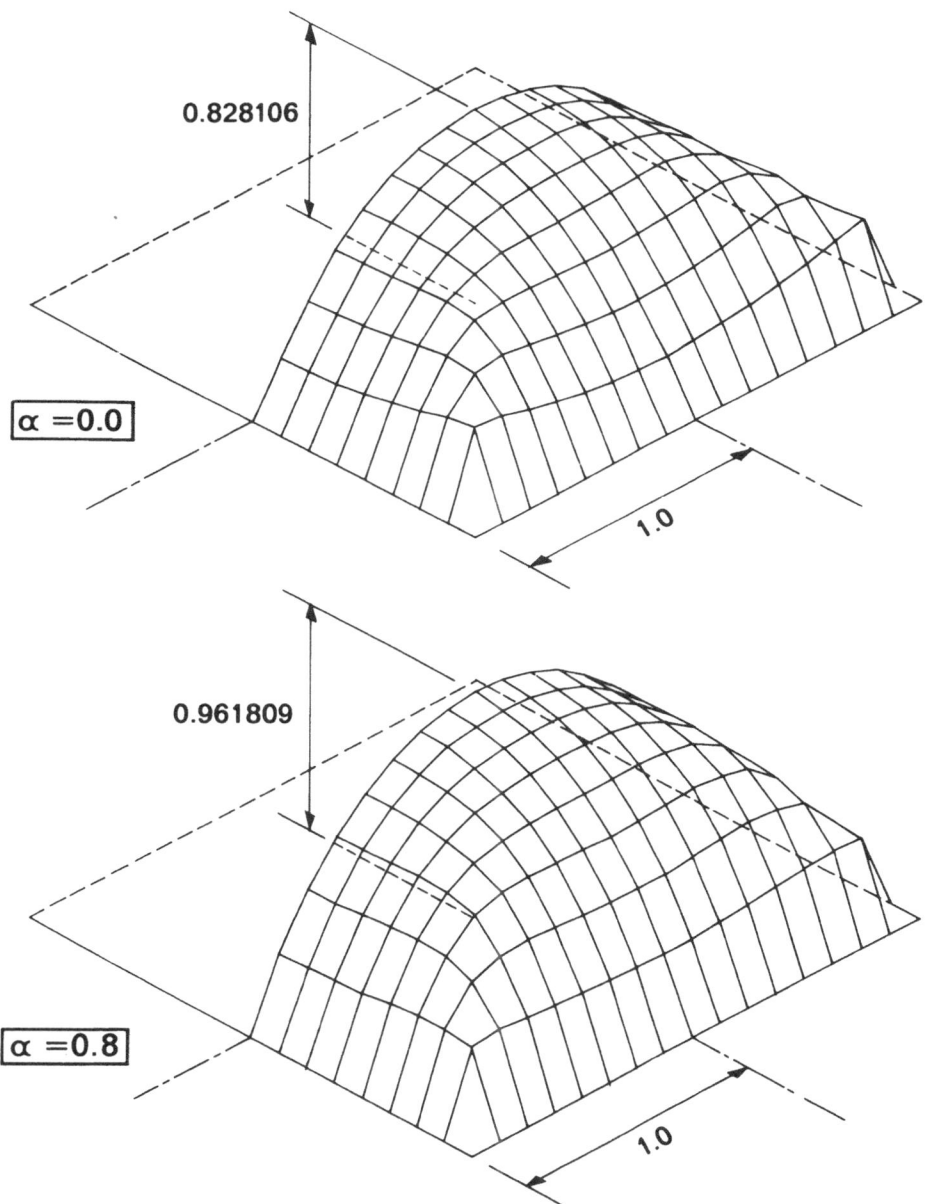

Figure 4.2   Comparison of Optimal Archgrids With
and Without Self-Weight.

4.3 One-Way Arch Systems: Allowance for Self-Weight and Skin-Weight

With a view to checking the results obtained for archgrids, and in order to get a better insight into optimization for design-dependent loads, a number of closed form analytical solutions were derived for one-way arch systems. This project was again initiated by Prager who also derived the first solution (self-weight without external load).

The loading conditions considered in this study are shown in Fig. 4.3. The external load p is given and often assumed to be uniformly distributed. The weight of non-load bearing roof sheeting r (or skin-weight) is prescribed per unit surface area, but does depend on the slope when expressed in terms of horizontal unit surface area. Finally the self-weight s depends both on cross-sectional area and slope. Assuming a flexureless arch of constant compressive stress throughout, the governing differential equations are

Self-weight only: $\qquad\qquad y'' - c(1 + y'^2) = 0 \qquad\qquad$ (4.4)

Self-weight + External Load: $\quad y'' - c(1 + y'^2) - p/H = 0 \qquad$ (4.5)

Self-weight + Skin-weight: $\quad y'' - c(1 + y'^2) - r(1 + y'^2)^{\frac{1}{2}}/H = 0$

$$\qquad\qquad\qquad\qquad\qquad\qquad\qquad\qquad (4.6)$$

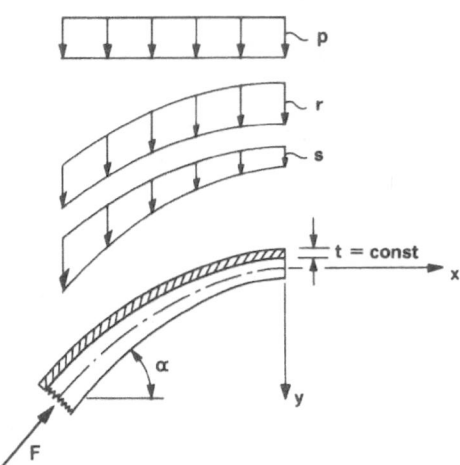

Figure 4.3   Types of Loads Considered in Arch Optimization.

As mentioned earlier, closed form analytical solutions, although lengthy, have been obtained for the above problems considering various types of distributed and point loads [21]. The optimum shape of the centerline was derived first, then the total cost was determined, in terms of the horizontal reaction H, by integration and finally differentiation with respect to H yielded the optimum value of the latter. The results were then checked by using the weighted unit mean square slope condition of Eq. 4.3 for self-weight problems. Figure 4.4 shows, for example, the optimum value of the horizontal reactions for symmetric, partial uniformly distributed loads and self-weight, in terms of a non-dimensional parameter $\bar{\beta}$. Figure 4.5 gives the optimum arch shape for the relatively simple case of a full uniformly distributed load (a = 1.0 in Fig. 4.4) over the entire span. Only a certain part of this curve is used, depending on the magnitude of the nondimensional span $\alpha = \gamma L/\sigma_0$.

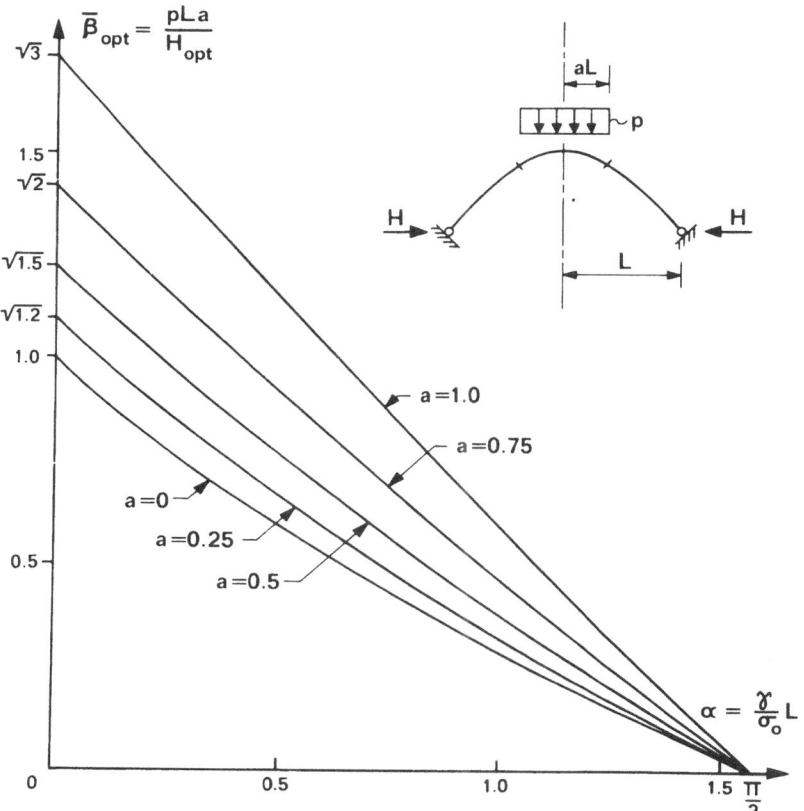

Figure 4.4   Optimum Values of the Horizontal Reaction for Arches with Partial Symmetric Uniform Load and Self-Weight.

140

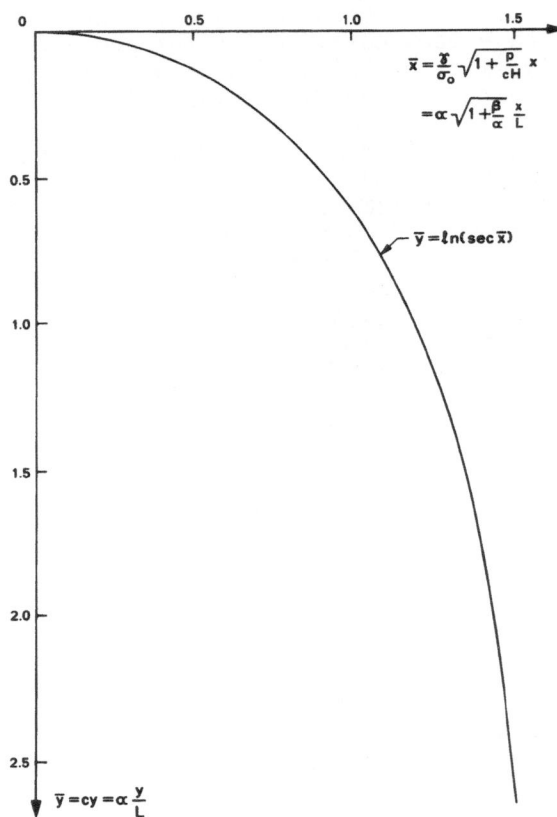

Figure 4.5   Optimal Arch Shape for Full
Distributed Load and Self-Weight.

When a = 0 in Fig. 4.4, then the load shrinks to a single
point load P at the center. For this simple case, the optimum
horizontal reaction H is given by the short expression [21]

$$H = P \cos \alpha / (1 - \sin \alpha) \qquad (4.7)$$

The optimum shape of half-arches for various values of the
nondimensional span α is given in Fig. 4.6. It has been found
[21] that $\alpha = \pi/2$ is the limiting span for any type of external
load on an arch. Beyond this span, the structure is not capable
of carrying its own weight. At the limiting span value, the
optimum arch becomes infinitely tall and, at slightly lower span
values, all optimum arches have a very similar shape, irrespective of the type of loading applied. This can be seen in
Fig. 4.6 (curve for α = 1.5) and Fig. 4.5.

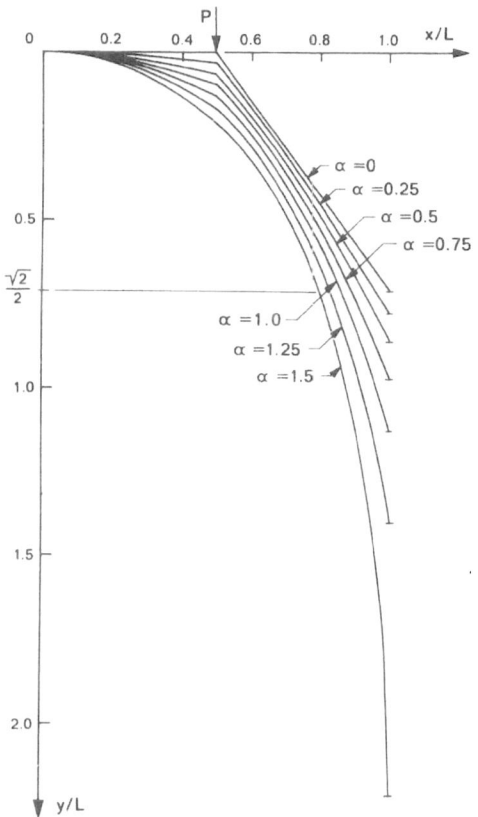

Figure 4.6   Optimum Arch Shapes for Two Point Loads.

Finally, the cost sensitivity of the design, in terms of the horizontal reaction H, is shown in Fig. 4.7 for various span values. It can be seen that the design cost is very sensitive to H, at high values of the nondimensional span α.

Significant contributions were made to this project by C.M. Wang and R.D. Hill, while H. Nakamura verified some of the above results by numerical methods used for archgrids [8].

4.4   Optimum Rotational Shells Subject to Self-Weight

Ziegler [22] discussed the design of fully stressed spherical domes for self-weight only, considering a Tresca yield condition (Fig. 4.8). As in the case of an arch that is fully stressed under its self-weight, Ziegler's solution is

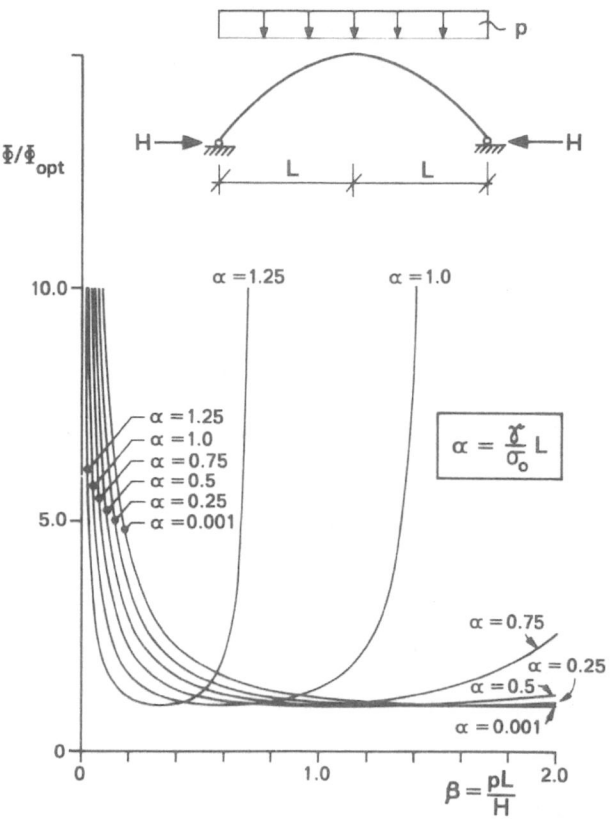

Figure 4.7   Cost Sensitivity in Terms of H.

indeterminate, because the total cost depends linearly on the
shell thickness adopted at the vertex. The extension of Ziegler's
problem to self-weight plus skin-weight (non-load bearing roof-
sheeting of given weight per unit area of middle surface) was
suggested by Prager and is described in a joint paper by Prager
and the author [23]. Ziegler [22] has tried two fully stressed
solutions corresponding to sides AB and AC in Fig. 4.8 and has
found that the latter was more economical. The same conclusion
was reached [23] for the extended problem. Side AC of the
Tresca-hexagon furnishes the governing differential equation

$$t' \sin \phi + t(2 \cos \phi - \rho) = q \rho \tag{4.8}$$

where t is the nondimensional thickness, ρ is the nondimensional
radius of the middle surface, ϕ is the meridian angle starting at
the apex and q is the ratio of the skin-weight and shell-weight

per unit area at the apex. The differential equation in Eq. 4.8
was solved by generating the Taylor coefficients of $t(\phi)$ by an
automatic process on the computer, using a recursion formula. The
thickness variation for various q-values is given in Fig. 4.9.
The broken line in Fig. 4.9 shows the limit of validity of Eq.
4.8, corresponding to point C in Fig. 4.8.

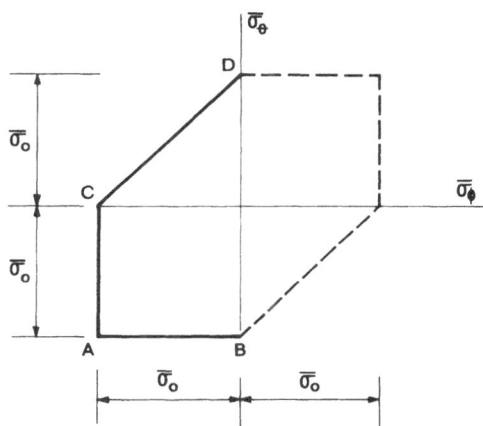

Figure 4.8   Tresca Yield Condition Used in Dome-Design.

In addition to the foregoing thickness variation, the following simple relation has been found [23]:

$$\rho = 2/(q + 1) \tag{4.9}$$

This means that the total weight (including skin) W and structural weight (without skin) W* can be readily determined for a
given nondimensional base radius r, as a function of the apex
thickness q adopted. The results of such calculations are given
for two base radius values in Fig. 4.10. Moreover, for given
values of the base radius r, the optimum values of the total
weight W, maximum meridian angle ϕ_1, maximum thickness t_1, apex
thickness q, and ratio α of the principal stresses at the base
have been determined and are given in Fig. 4.11.

The foregoing solutions have been restricted to side AC of
the Tresca hexagon (Fig. 4.8). Since the completion of Ref. 23,
H. Nakamura and the author extended this project to the next
Tresca side (CD in Fig. 4.8), which applies to optimization of
domes with a base radius greater than 0.7 (nondimensional). The
extended thickness variations are shown analytically, for the
limiting q-values only, in Fig. 4.12, but they have been determined for intermediate values also. In addition to the foregoing

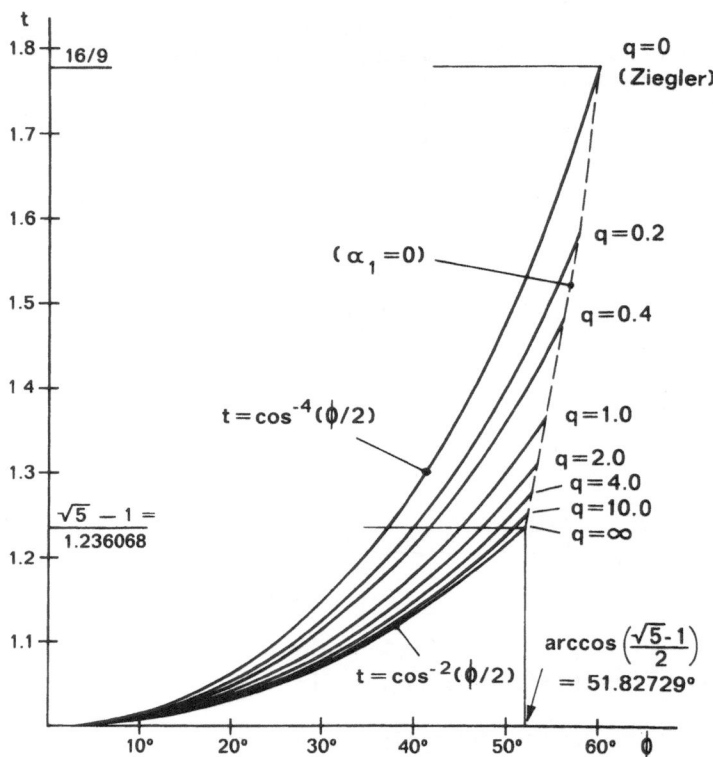

Figure 4.9   Thickness Variation in Fully Stressed Domes.

investigation, similar solutions have been obtained more recently
for two other loading conditions; uniform vertical load and uniform outside pressure, both with self-weight. The three types of
loading conditions investigated are summarized in Fig. 4.13.

Owing to Professor Prager's sudden death, his paper on
spherical domes was to be, tragically, his last publication.
However, he intended to write a second paper with the author on
this topic, in which the optimum shape of domes would have been
discussed. This investigation has since been completed by M. Dow,
H. Nakamura, and the author. In addition to the spherical shape,
the optimum quadratic meridian curve and then the shape corresponding to point A in Fig. 4.8 (equal principal stresses throughout) have been considered. The latter is likely to be the absolute optimum. The optimum spherical and quadratic solutions are
compared for r = 0.7 in Fig. 4.13. The absolute optimum differed
only slightly from the latter, with a height of 0.381667. The
nondimensional weights for the three solutions are; W = 2.370057,
2.194171, and 2.186050. It is important to note that the

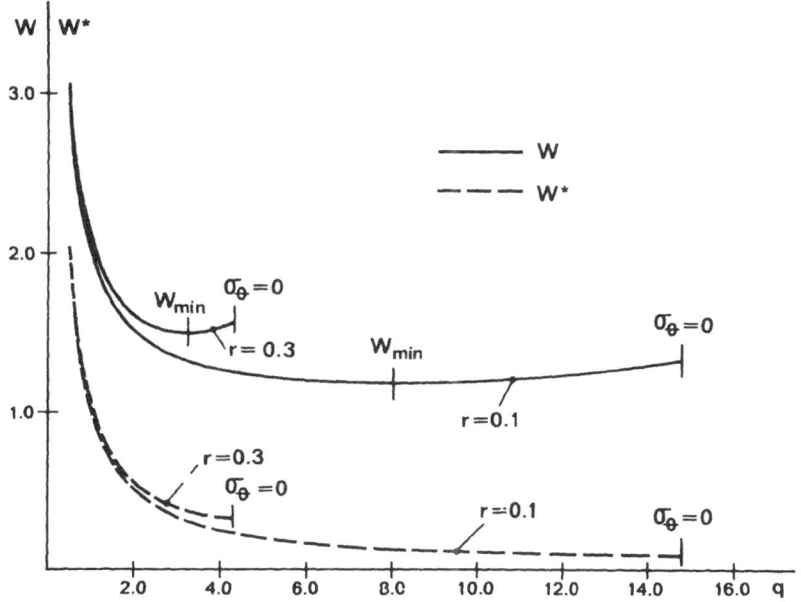

Figure 4.10 Total Weight W and Structural Weight W* as Functions of q.

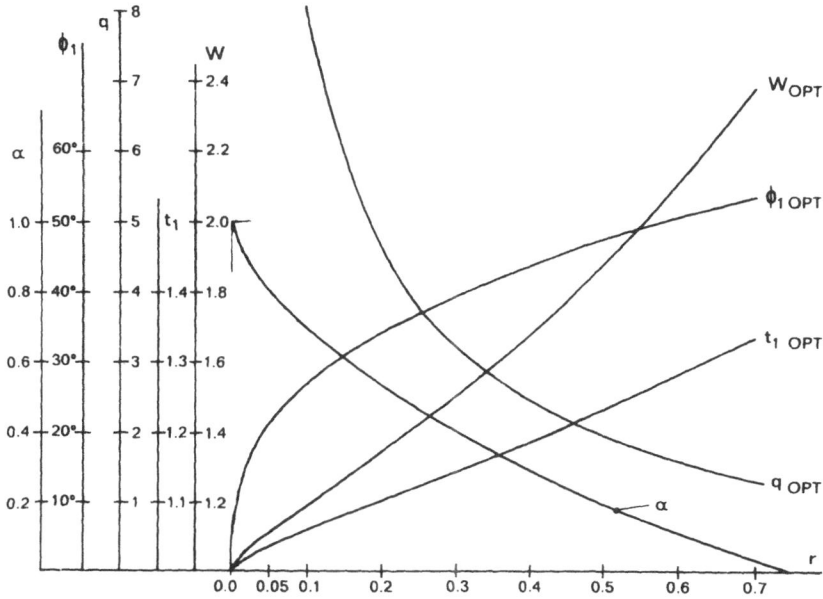

Figure 4.11 Optimum Values of Parameters for Domes: Skin and Self-Weight.

146

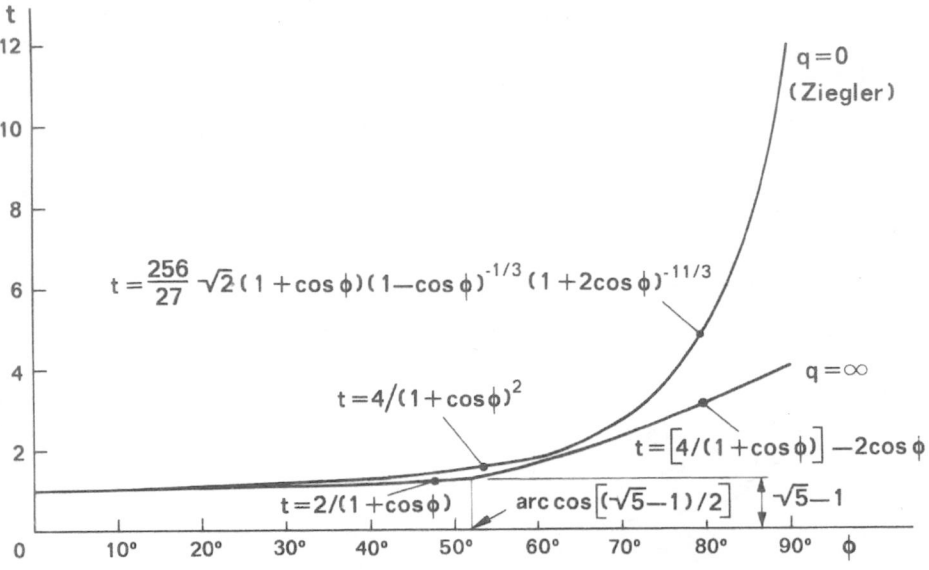

Figure 4.12   Extended Thickness Distribution for Spherical Domes.

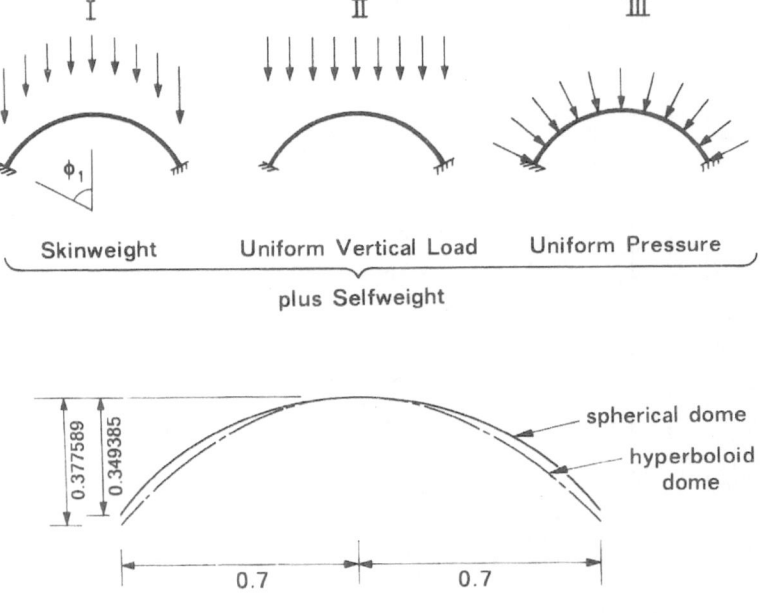

Figure 4.13   Other Problems Investigated More Recently.

nondimensional radius r equals the product of the base radius and the specific weight of the dome material, divided by the yield stress.

5. OPTIMAL DESIGN FOR ALTERNATE LOADS

Optimality criteria for alternate loads were summarized in Figure 6.1 of Ref. 9. In addition to these, very useful super-position principles are available. For two load conditions, these principles were obtained independently by Hemp [24], Prager and Nagtegaal [25] as well as Spillers and Lev [26] and then they were extended to an arbitrary number of load conditions by Hill and the author [27]. In all these methods, instead of optimizing the system for n alternate loading conditions, the same system is optimized for n single load conditions, separately, and then the strength parameters of the n designs are superimposed, which is considerably easier than the traditional method. Hill and the author have applied this method to Michell frames and grillages, subjected to four alternate load conditions.

More recently, M. Dow, C.M. Wang and the author have inves-tigated the optimum shape of arches for alternate load conditions. The foregoing superposition principle is inapplicable here, if the arch is to consist of a single curved member. It has been found, surprisingly, that even for two alternate point loads, the optimum arch centerline contains curved segments. Along segment CE in Fig. 5.1, for example, the specific costs for the two load conditions are equal. In this problem, the specific cost function is a weighted sum of the absolute values of the moment and axial force on the arch cross section.

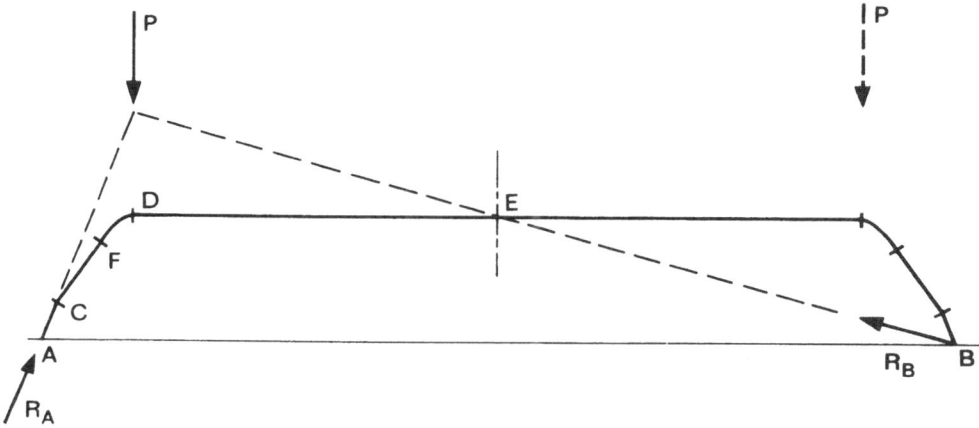

Figure 5.1   Optimum Arch Centerline for Two Alternate Point Loads.

148

6.   CONCLUDING REMARKS

One of the most intriguing developments in structural optimization has been the development of a computer program [6] that
generates and plots optimum beam layouts for any boundary condition specified in the input. Such a program could be used
readily by a nonspecialist designer, who wants to get some idea
about the most efficient beam layout for a given design problem.
Figure 5.2, for example, shows a well-known beam layout designed
by L. Nervi. The absolute optimum layout for the same support
conditions is shown in Fig. 5.3. The latter minimizes the sum of
the absolute values of the principal moments for the entire system,
whereas Nervi's design can be shown to minimize the sum of the
squares of the moments. For a more complex boundary shape, such
as the one shown in Fig. 5.4, construction of the optimum beam
layout by manual methods would require a considerable amount of
time, while the computer time required for the same is only a few
seconds. The program is extremely efficient, because it carries
out all operations analytically. Such symbolic processing of
optimization problems could open up entirely new avenues in structural design by electronic computers.

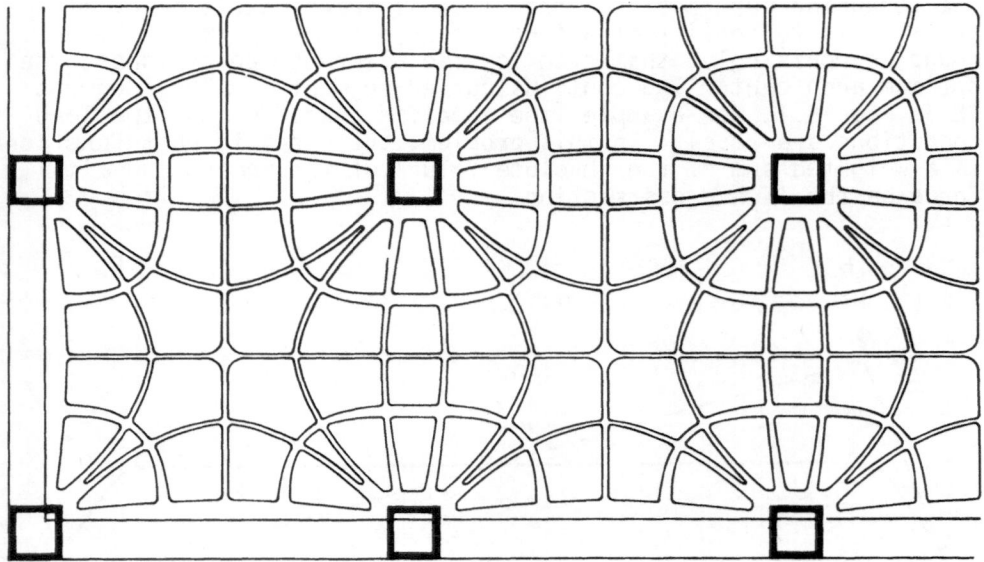

Figures 5.2   A Well-Known Beam Layout by Nervi.

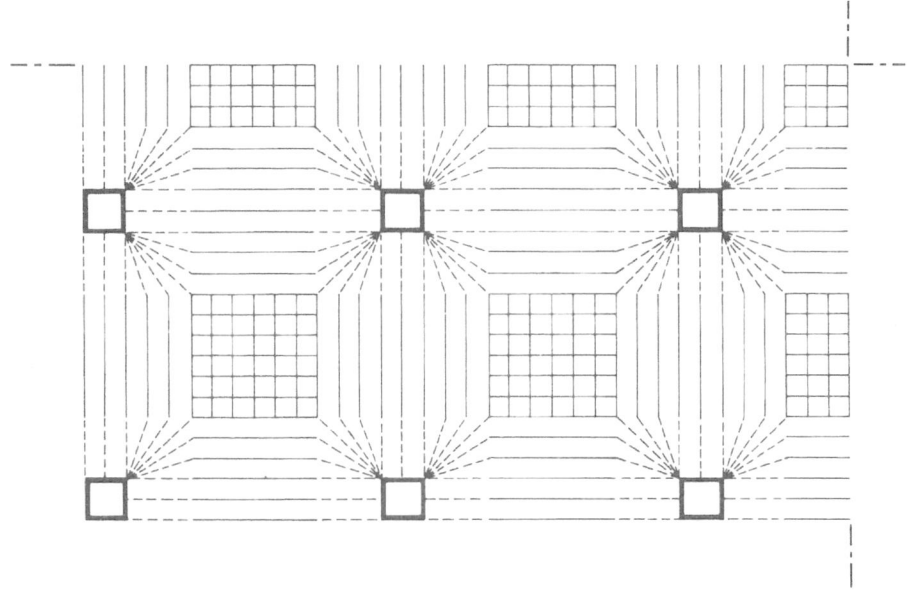

Figure 5.3   Optimum Beam Layout for the Support
              Conditions in   Fig. 5.2.

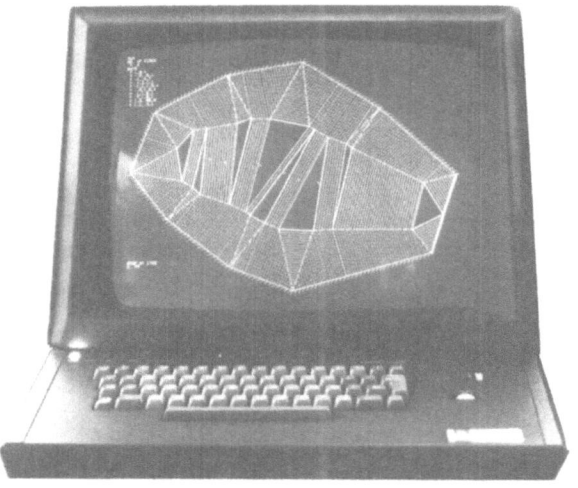

Figure 5.4   Optimum Beam Layout Generated and Plotted by
              Purely Analytical Methods on the Computer.

150

REFERENCES

1. Prager, W., and Rozvany, G.I.N., "Optimization of Structural Geometry," Proc. Conf. Dynamical Systems, Academic Press, New York, 1977, pp. 265-294.
2. Maxwell, G., "On Reciprocal Figures, Frames, and Diagrams of Forces," Sci. Papers II, Cambridge Univ. Press, 1890, pp. 175-177.
3. Michell, A.G.M., "The Limits of Economy of Material in Frame Structures," Phil. Mag., Vol. 8, 1904, pp. 589-597.
4. Prager, W., and Rozvany, G.I.N., "Optimal Layout of Grillages," J. Struct. Mech., Vol. 5, 1977, pp. 1-18.
5. Prager, W., Introduction to Structural Optimization, Springer-Verlag, Vienna, 1974.
6. Hill, R.D., and Rozvany, G.I.N., "A Computer Program for Deriving Analytically and Plotting Optimal Beam Layouts," Comp. Struct., Vol. 10, 1979, pp. 295-300.
7. Rozvany, G.I.N., and Prager, W., "A New Class of Optimization Problems, Optimal Archgrids," Comp. Meth. Appl. Mech. Engg., Vol. 19, 1979, pp. 127-150.
8. Rozvany, G.I.N., Nakamura, H., and Kuhnell, B.T., "Optimal Archgrids, Allowance for Selfweight," Comp. Meth. Appl. Mech. Engr., to appear 1980.
9. Rozvany, G.I.N., "Variational Methods and Optimality Criteria," Optimization of Distributed Parameter Structures (Eds. E.J. Haug and J. Cea), Sijthoff & Noordhoff, Alphen aan den Rihn, Netherlands, 1980.
10. Prager, W., and Shield, R.T., "A General Theory of Optimal Plastic Design," J. Appl. Mech., Vol. 34, 1967, pp. 184-186.
11. Prager, W., "Nearly Optimal Design of Trusses," Computers and Structures, Vol. 8, 1978, pp. 451-454.
12. Prager, W., "Optimal Layout of Trusses with Finite Numbers of Joints," J. Mech. Phys. Solids, Vol. 26, 1978, pp. 241-250.
13. Prager, W., "Optimal Layout of Cantilever Trusses," J. Optimiz. Theory Appl., Vol. 23, 1977, pp. 111-117.
14. Rozvany, G.I.N., Optimal Design of Flexural Systems, Pergamon, Oxford, 1976.
15. Rozvany, G.I.N., and Hill, R.D., "General Theory of Optimal Force Transmission of Flexure," Advances in Appl. Mech., Vol. 16, 1976, pp. 183-308.
16. Morley, C.T., "The Minimum Reinforcement in Concrete Slabs," Int. J. Mech. Sci., Vol. 8, 1966, pp. 305-320.
17. Hill, R.D., and Rozvany, G.I.N., "Optimal Beam Layouts: The Free Edge Paradox," J. Appl. Mech., Vol. 44, 1977, pp. 696-700.
18. Lowe, P.G., and Melchers, R.E., "On the Theory of Optimal Edge Beam Supported Fibre Reinforced Plates," Int. J. Struct. Mech., Vol. 16, 1974, pp. 627-641.
19. Melchers, R.E., "Minimum Reinforcement in Nonuniform Plates," J. Eng. Mech. Div. ASCE, Vol. 102, 1976, pp. 943-956.

20. Rozvany, G.I.N., "Optimal Beam Layouts: Allowance for Cost of Shear," Comp. Meth. Appl. Mech. Engg., Vol. 19, 1979, pp. 49-58.
21. Hill, R.D., Rozvany, G.I.N., Wang, C.M., and Leong, K.H., "Optimization, Spanning Capacity and Cost Sensitivity of Fully Stressed Arches," J. Struct. Mech., Vol. 7, 1979, pp. 375-410.
22. Ziegler, H., "Kuppeln Gleicher Festigheit," Ing.-Arch., Vol. 26, 1958, pp. 378-382.
23. Prager, W., and Rozvany, G.I.N., "Optimal Spherical Cupola of Uniform Strength," Ing.-Arch., to appear 1980.
24. Hemp, W.S., Optimum Structures, Clarendon, Oxford, 1973.
25. Nagtegall, J.C., and Prager, W., "Optimal Layouts of a Truss for Alternative Loads," Int. J. Mech. Sci., Vol. 15, 1973, pp. 383-392.
26. Spillers, W.R., and Lev, O., "Design for Two Loading Conditions," Int. J. Solids Struct., Vol. 7, 1971, pp. 1261-1267.
27. Rozvany, G.I.N., and Hill, R.D., "Optimal Plastic Design: Superposition Principles and Bounds on the Minimum Cost," Comp. Meth. Appl. Mech. Engg., Vol. 13, 1978, pp. 151-173.

152

OPTIMIZATION OF COLUMNS AGAINST BUCKLING

Niels Olhoff

Department of Solid Mechanics, The Technical University
of Denmark, DK-2800 Lyngby, Denmark

ABSTRACT

This paper presents formulations and optimality criteria for
a variety of column optimization problems. Column material
distribution and interior support location are treated as design
variables. Constraints are imposed on cross section and material
volume. Both single and bimodal formulations are treated.

1.   INTRODUCTION

As early as 1770, optimal design of elastic columns against
buckling was considered by Lagrange [1]. Unfortunately, however,
his solutions are incorrect, due to computational errors. The
problem was repeated by Clausen [2] in 1851 and, in a more general
form that has provided inspiration for contemporary research in
the area, by Keller [3] in 1960. Keller determined the shape of
a simply supported, elastic column of given length and material
volume that maximizes the buckling load in Ref. 3, and subsequently considered columns with other sets of classical boundary
conditions, together with Tadjbakhsh in Ref. 4. In Refs. 3 and 4,
the governing differential equations of optimality were
established by variational analysis of the equations of
equilibrium and the boundary conditions. Alternate variational
formulations, based on well known energy principles of structural
mechanics, were later presented by Keller and Niordson [5] and
Taylor [6], but only geometrically unconstrained formulations for
optimal design were considered.

In 1968, Taylor and Liu [7] demonstrated an important extension of the formulation used in Ref. 6, to accomodate constraints
on the minimum cross-sectional area of the column. Such constraints are equivalent to constraints on the maximum prebuckling
stress, and were since taken into account by Prager and Taylor [8],
Frauenthal [9], and Olhoff and Rasmussen [10] in investigations
of different aspects pertaining to optimal design of elastic
columns with continuously varying cross-section. Optimization of
columns with different segments of uniform cross-sections has been
considered by Masur [11,12], and optimal design with support
locations as additional design variables is dealt with by Mróz
and Rozvany [13,14], Prager and Rozvany [15], Szelag and Mróz
[16], and Olhoff and Taylor [17].

2. GEOMETRICALLY UNCONSTRAINED, SINGLE MODE FORMULATION FOR OPTIMAL DESIGN

Consider a thin, straight, elastic column that has volume V,
length L, and Young's Modulus E and is subjected to an axial compressive force, the value of which is P at buckling. The cross-
section of the column is permitted to vary along the column axis,
according to the relation

$$I(x) = cA^p(x) \qquad (2.1)$$

where the second area moment of inertia is $I(x)$ and the cross-
sectional area is $A(x)$. In Eq. 2.1, c and p are positive constants, which are assumed to be given. The relationship models
beams of solid rectangular cross-section of fixed width and
variable height for $p = 3$, solid cross-sections of geometrically
similar shape for $p = 2$, and solid rectangular sections of fixed
height and variable width for $p = 1$. The case of $p = 1$ also
includes constant-width sandwich cross-sections with lightweight
cores of uniform height and zero stiffness covered by two identical, thin face sheets of variable thickness.

The columns considered are of Bernoulli-Euler type with
given support conditions, flexible supports being excluded for
brevity. Introducing a dimensionless cross-sectional area $\alpha(x)$
and buckling load λ by

$$\alpha(x) = A(x)L/V \qquad (2.2)$$

$$\lambda = \frac{PL^{p+2}}{EcV^p} \qquad (2.3)$$

where the coordinate x is nondimensionalized by means of L, the
lateral deflection $y(x)$ at buckling is governed by an eigenvalue

problem consisting of the differential equation

$$(\alpha^p y'')'' = - \lambda y'' \qquad (2.4)$$

and given, homogeneous boundary conditions at the column ends $x = 0$ and $x = 1$.

A convenient expression for the eigenvalue λ, actually the Rayleigh quotient

$$\lambda = \frac{\int_0^1 \alpha^p y''^2 dx}{\int_0^1 y'^2 dx} \qquad (2.5)$$

is obtained if one multiplies the differential equation of Eq. 2.4 by $y(x)$ and invokes the boundary conditions through two integrations by parts. The property of λ being stationary at the actual deflection $y(x)$, among all other kinematically admissible deflection functions, is well known. To be kinematically admissible, a deflection $y(x)$ must be continuous, satisfy the kinematic boundary conditions, and have continuous slope y' except possibly at sections of vanishing bending stiffness.

The optimization problem consists of determining the column shape that maximizes the fundamental buckling load P for given values of V, L, E and c. In dimensionless terms, this problem is equivalent to determining the design variable $\alpha(x)$ so that the eigenvalue λ given by Eq. 2.5 is maximized, subject to the constraint

$$\int_0^1 \alpha \, dx = 1 \qquad (2.6)$$

Note from Eq. 2.3 that the non-dimensional solution to the problem will at the same time minimize V for given values of P, L, E, and c. It is convenient to normalize the deflection $y(x)$ such that the denominator in Eq. 2.5 is set equal to unity, i.e. so that

$$\int_0^1 y'^2 dx = 1 \qquad (2.7)$$

Assume now that λ is a simple eigenvalue. Then, the cross-sectional area function $\alpha(x)$ and corresponding deflection $y(x)$ that maximize λ subject to the constraints of Eqs. 2.6 and 2.7 are identified with stationarity of the functional

$$\lambda^* = \int_0^1 \alpha^p y''^2 dx - \beta p \left\{ \int_0^1 \alpha dx - 1 \right\} - \kappa \left\{ \int_0^1 y'^2 dx - 1 \right\} , \qquad (2.8)$$

where the constants β and κ are Lagrangian multipliers.

The Euler-Lagrange equation expressing stationarity of λ^* with respect to arbitrary admissible variation of α is

$$\alpha^{p-1} y''^2 = \beta \qquad (2.9)$$

which is termed the optimality condition of the problem [3-9], and is necessary for optimality. If one multiplies this equation by α, integrates over the interval $0 \leq x \leq 1$, and takes constraints of Eqs. 2.6 and 2.7 into account, he easily finds that

$$\beta = \lambda \qquad (2.10)$$

Thus, for cases of $p = 3$ or $p = 2$, Eq. 2.9 gives the optimum cross-sectional area function $\alpha(x)$ in the form

$$\alpha(x) = \left(\frac{\lambda}{y''^2} \right)^{\frac{1}{p-1}} , \quad \text{if} \quad p = 3,2 \qquad (2.11)$$

while, for $p = 1$, Eq. 2.9 states that the curvature $y''(x)$ of the deflection is constant throughout, except for possible sign shifts,

$$y''(x) = \pm \sqrt{\lambda} , \quad \text{if} \quad p = 1 \qquad (2.12)$$

The condition of stationarity of λ^* of Eq. 2.8, with respect to variation of $y(x)$, requires that $\kappa = \lambda$ and re-establishes the Euler buckling differential equation of Eq. 2.4, together with natural boundary conditions. If one assumes an interior point of discontinuity, the variation of y gives conditions of continuity of the bending moment $m(x) = \alpha^p y''$ and of the function $(\alpha^p y'')' + \lambda y'$,

$$\langle \alpha^p y'' \rangle = 0 \qquad (2.13)$$

$$\langle (\alpha^p y'')' + \lambda y' \rangle = 0 \qquad (2.14)$$

at all points free from kinematic constraints (column supports). Here the notation $\langle \cdot \rangle$ indicates the jump discontinuity of the argument. In fact, the function $(\alpha^p y'')' + \lambda y'$ attains a constant value at any point x, $x_j < x < x_{j+1}$, lies between adjacent points

x_j and x_{j+1} of kinematic constraints. Thus, if one integrates Eq. 2.4 in an interval $x_j < x < x_{j+1}$, the shear force $t(x)$ takes the form

$$t(x) = -(\alpha^p y'')' = t_0 + \lambda y' \qquad (2.15)$$

where t_0 is a constant of integration. The constant $-t_0$ identifies $(\alpha^p y'')' + \lambda y'$ and physically represents the force component of the stress resultants in the buckled column in the direction perpendicular to the x-axis. The shear force $t(x)$ is perpendicular to the deflected beam axis.

One is now able to conclude that, for a given value of p, the three differential and integral equations of Eqs. 2.4, 2.5, and 2.9, with $\beta = \lambda$, together with the normalization condition of Eq. 2.7, continuity conditions of Eqs. 2.13 and 2.14, and given boundary conditions for y constitute the mathematical formulation of the dimensionless optimization problem. The unknowns to be determined are $\alpha(x)$, $y(x)$, and λ.

For p = 1, the problem is seen to be quite simple, so solutions can be obtained analytically [6,8,18,19]. In cases p = 2 and p = 3, where the problem becomes non-linear in $y(x)$, it is still possible to apply analytical methods of solution provided that the boundary conditions are sufficiently simple [2-4,20]. Otherwise, numerical methods are available [5,9,10,14,16,21,22].

Optimal shapes $\pm\sqrt{\alpha}$ and corresponding buckling modes y are as indicated in Fig. 2.1 for (a) clamped-free, (b) simply supported-simply supported, and (c) clamped-simply supported columns with p = 2. The buckling loads of the optimum solutions in Fig. 2.1 are increased by (a) 1/3, (b) 1/3, and (c) 35.1%, when compared with the buckling loads of correspondingly supported uniform columns of the same volume, length, and material [3,4].

3. SINGULAR BEHAVIOR AT POINTS OF ZERO BENDING MOMENT

Consider now the behavior of a solution to the geometrically unconstrained, single mode formulation of Section 2 at a point $x = x_i$ of zero bending moment $m = \alpha^p y''$. For the case p = 1, zero bending moment must imply, in view of Eq. 2.12, that α also vanishes. For p = 2 and p = 3, Eq. 2.11 and the relation $m = \alpha^p y''$ show that the cross-section α again vanishes and that y'' tends to infinity at a point $x = x_i$ of vanishing bending moment m.

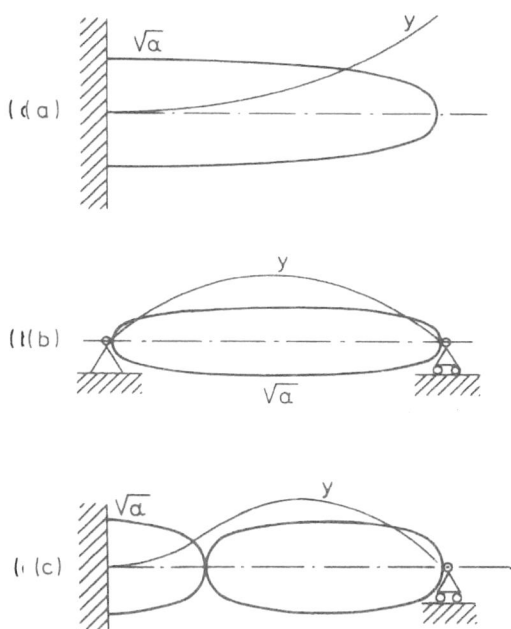

Figure 2.1 Optimum Columns

Now, if the point $x = x_i$ of $m = \alpha = 0$ is an interior point in the interval for x, then a discontinuity of the slope y' of the deflection is possible at that point and, in view of Eq. 2.14, such behavior would then imply a discontinuity of the shear force $t(x) = -(\alpha^P y'')'$.

Analyzing the functional behavior predicted by Eqs. 2.4 and 2.9 in the right and left hand vicinities of an inner singular point $x = x_i$, by means of power series expansions [23], one finds that the leading terms in the expansions of y and α satisfying continuity of y and $(\alpha^P y'')' + \lambda y'$, are given by

$$y(x) = \begin{cases} a_0 + a_1^+(x-x_i) + b^+(x-x_i)^r + \ldots, & x \geq x_i \\ a_0 + a_1^-(x_i-x) + b^-(x_i-x)^r + \ldots, & x \leq x_i \end{cases} \tag{3.1}$$

and

158

$$\alpha(x) = \begin{cases} g^+(x-x_i)^s + \dots , & x \geq x_i \\ \\ g^-(x_i-x)^s + \dots , & x \leq x_i , \end{cases} \tag{3.2}$$

for $p = 1$, 2, and 3, provided that $(\alpha^p y'')' + \lambda y'$ is non-vanishing at $x = x_i$. The power r in Eq. 3.1 and the leading power s in Eq. 3.2 are given by

$$\left. \begin{array}{l} p = 1 : r = 2 \quad , \ s = 1 \\ \\ p = 2 : r = 5/3 , \ s = 2/3 \\ \\ p = 3 : r = 3/2 , \ s = 1/2 \end{array} \right\} \tag{3.3}$$

For $p = 1$, higher powers than $r = 2$ are not present in Eq. 3.1 and the coefficients b^+ and b^- are equal to $\sqrt{\lambda}/2$ or $-\sqrt{\lambda}/2$. For $p = 2$ and 3, the power r in Eq. 3.1 is the leading non-integer power of the series, and is seen from Eq. 3.3 to cause singularity of y''. The coefficients g^+ and g^- in Eq. 3.2 are both positive and slopes $\alpha'(x_i^+)$ and $\alpha'(x_i^-)$ of the cross-sectional area function are therefore of opposite signs. These slopes are finite for $p = 1$, but are infinite for $p = 2$ and $p = 3$.

The functional behavior given by Eqs. 3.1, 3.2, and 3.3 for $p = 1$, 2, and 3 at an arbitrary inner point $x = x_i$ of zero bending moment is associated with a possible slope jump ($a_i^+ \neq a_i^-$ in general) and a jump of shear force $t(x) = -(\alpha^p y'')'$, even though the deflection $y(x)$ and the function $(\alpha^p y'')' + \lambda y'$ are continuous. Therefore, the point $x = x_i$ corresponds, physically and kinematically, to an inner hinge of the optimum column.

It should finally be noted that Eqs. 3.1 and 3.2, as special cases (with $a_0 = 0$), express the behavior of y and α in the vicinity of a simply supported column end point $x = x_i = 0$ (or $x = x_i = 1$) for $x \geq x_i = 0$ (or $x \leq x_i = 1$).

4. CONDITIONS FOR OPTIMALLY PLACED INNER HINGES

In problems of optimizing statically determinate columns by means of a single mode, geometrically unconstrained formulation, the locations of singular points of zero bending moment are known beforehand. Thus, simply supported or free end points are pre-determined to be singular. However, in a priori statically in-determinate problems of the type mentioned, singularities may occur at inner points. The positions of such points can be pre-scribed for a particular column to be optimized, while in other problems one may consider the locations $x = x_i$, $i = 1,\ldots,S$ of the singularities (hinges) to be additional design variables. Problems of the latter type were considered for the first time by Masur [12]. To derive specific conditions for optimum locations of the points $x = x_i$, one may follow Refs. 19 and 23 and demand stationarity of the functional λ^* given by Eq. 2.8, with respect to arbitrary admissible variation δx_i, $i = 1,\ldots,S$. First, however, one integrates the first term on the right hand side of Eq. 2.8 by parts and, for convenience, defines $x_0 = 0$ and $x_{S+1} = 1$, so that λ^* takes the form

$$\lambda^* = [\alpha^p y'' y']_0^1 - \sum_{i=1}^{S} <\alpha^p y'' y'>_{x_i} - \sum_{i=1}^{S+1} \int_{x_{i-1}^+}^{x_i^-} (\alpha^p y'')' y' dx$$

$$- \beta p \left\{ \sum_{i=1}^{S+1} \int_{x_{i-1}^+}^{x_i^-} \alpha dx - 1 \right\} - \kappa \left\{ \sum_{i=1}^{S+1} \int_{x_{i-1}^+}^{x_i^-} y'^2 dx - 1 \right\} \tag{4.1}$$

where symbols $[\ ]_0^1$ and $< >_{x_i}$ are to be interpreted according to $[z]_0^1 = z(1) - z(0)$ and $<z>_{x_i} = z(x_i^+) - z(x_i^-)$. The terms involving these symbols in Eq. 4.1 all vanish, due to the natural boundary conditions and the conditions $m(x_i) = \alpha^p y''(x_i) = 0$, $i = 1,\ldots,S$, respectively. Then, stationarity of λ^*, subject to variations $\delta x_i^+ = \delta x_i^- = \delta x_i$, $i = 1,\ldots,S$, is expressed by

$$\delta_{x_i} \lambda^* = \sum_{i=1}^{S} \left\{ \left[(\alpha^p y'')' y' + \kappa y'^2 + \beta p \alpha \right]_{x=x_i^+} \right.$$

$$\left. - \left[(\alpha^p y'')' y' + \kappa y'^2 + \beta p \alpha \right]_{x=x_i^-} \right\} \delta x_i = 0 \tag{4.2}$$

One has $\kappa = \lambda$, $\alpha(x_i) = 0$, and assuming that no column support is placed at the points $x = x_i$, $i = 1,\ldots,S$, the function $(\alpha^p y'')' + \lambda y'$ will be continuous at these points (see Eq. 2.14). Since the variations δx_i, $i = 1,\ldots,S$, are arbitrary, one obtains the following necessary conditions for optimally placed inner points $x = x_i$ of zero bending moment:

$$\left(\left(\alpha^p y''\right)' + \lambda y'\right)_{x=x_i} <y'>_{x_i} = 0 , \quad i = 1,\ldots,S \qquad (4.3)$$

Here, the relation $\left(\alpha^p y''\right)' + \lambda y' = 0$ cannot be used as a condition for an optimum position x_i, because this relation would then also hold for any other point of the particular interval between adjacent supports (see the previous discussion), implying that any of these points would designate an optimum location.

Consequently, the result following from Eq. 4.3 is

$$<y'>_{x_i} = 0 \quad i = 1,\ldots,S \qquad (4.4)$$

In view of its derivation, the result can be interpreted as follows: The optimal location $x = x_i$ of an inner hinge is such that there is a fundamental mode $y(x)$ that has continuous slope at the point $x = x_i$ and that satisfies Eqs. 2.4 to 2.7, 2.9, 2.13, and 2.14.

4.1 Alternative Condition at Optimally Placed x_i If the Optimum Buckling Eigenvalue is Multiple

The non-dimensional buckling load λ of a beam with inner hinges is generally not a simple eigenvalue, but may be associated with a multiplicity of modes. If this is the case for the optimum eigenvalue, then the symbol $y(x)$ used above may identify any of these modes.

Following Ref. 19, one can now consider a particular point $x = x_i$, which is the optimal location of an inner hinge. For this point, a condition is presented here that constitutes an alternative to Eq. 4.4, if the optimum buckling eigenvalue λ is multiple.

Assume that the column has at least one additional hinge at a point $x = x_{i+1}$, which is either an inner point or an end point of the column. Assume also that no column supports or hinges are placed between the points x_i and x_{i+1}. Furthermore, without loss of generality, let $x_i < x_{i+1}$.

Denote by y_1 the mode that satisfies Eq. 4.4 at a particular x_i, thereby ensuring optimum location of the point. For this mode, in view of Eq. 2.14,

$$(\alpha^P y_1'')'_{x_i^+} = (\alpha^P y_1'')'_{x_i^-} \tag{4.5}$$

Note that $y_1(x)$ and the optimum cross-sectional area function $\alpha(x)$ and buckling load λ must satisfy Eqs. 2.4 to 2.7. However, if y_1 satisfies these equations, along with α and λ, then Eqs. 2.4 to 2.7, 2.9 and 2.13 are satisfied by the same α and λ, together with the continuous function y_2 given by

$$\left.\begin{array}{ll} y_2''(x) \equiv -y_1''(x) \,, & x_i < x < x_{i+1} \\[2mm] y_2(x) \equiv y_1(x) \,, & 0 \le x \le x_i \text{ and } x_{i+1} \le x \le 1 \end{array}\right\} \tag{4.6}$$

which indicates that λ is a multiple eigenvalue. It follows from Eq. 4.6 that $(\alpha^P y_2'')'_{x_i^+} = -(\alpha^P y_1'')'_{x_i^+}$ and $(\alpha^P y_2'')'_{x_i^-} = (\alpha^P y_1'')'_{x_i^-}$, so that in view of Eq. 4.5,

$$(\alpha^P y_2'')'_{x_i^+} = -(\alpha^P y_2'')'_{x_i^-} \tag{4.7}$$

In order to be a buckling mode, y_2 now only needs to satisfy Eq. 2.14. Expressing Eq. 2.14 in terms of y_2 at $x = x_i$ and using Eq. 4.7 one gets $2(\alpha^P y_2'')'_{x_i^+} + \lambda y_2'(x_i^+) - \lambda y_2'(x_i^-) = 0$ and, by means of Eq. 2.15 one arrives at the condition

$$y_2'(x_i^+) = -y_2'(x_i^-) - 2\frac{t_0}{\lambda} \tag{4.8}$$

for the mode y_2 at an optimally placed x_i.

Equation 4.8 is a mixed geometrical-static condition. The interpretation of the constant t_0 has been given in the foregoing. In order to convert Eq. 4.8 into a purely geometrical condition, consider Fig. 4.1, which indicates the modes y_1 and y_2 and shows the components in the x and y directions of the stress resultants at the hinge locations $x = x_i$ and $x = x_{i+1}$. For reasons of equilibrium, the direction of the resultant of λ and t_0 must coincide with the chord of the hinged beam segment. This implies

162

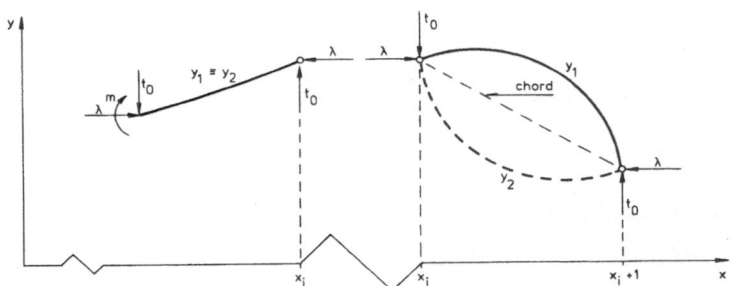

Figure 4.1   Geometry of Alternative Modes

that the ratio t_0/λ may be expressed as $t_0/\lambda = (y_2(x_i) - y_2(x_{i+1}))/(x_{i+1} - x_i)$, and Eq. 4.8 may hereby be translated into the geometrical condition

$$y_2'(x_i^+) = - y_2'(x_i^-) - 2 \frac{y_2(x_i) - y_2(x_{i+1})}{x_{i+1} - x_i} \tag{4.9}$$

at optimally placed x_i.

Since $y_1(x_i) = y_2(x_i)$ and $y_1'(x_i^+) = y_1'(x_i^-) = y_2'(x_i^-)$, it is easily seen how Eq. 4.9 may be formulated in words: The slopes of y_1 and y_2, relative to the chord, are equal and opposite at $x = x_i^+$.

In fact, since one has $y_2'' \equiv -y_1''$ for $x_i < x < x_{i+1}$, the modes y_1 and y_2 are symmetric with respect to the chord in this interval and the slopes of y_1 and y_2, relative to the chord, are therefore equal and opposite at any point of the interval.

4.2   Complete Conditions for Optimally Placed Inner Hinges

In a single mode formulation of optimal design for maximum buckling load, where the resulting optimum buckling load λ is not simple, the single mode $y(x)$ can only identify one of the possible types of behavior exemplified above by y_1 and y_2 at the optimum location $x = x_i$ of an inner hinge.

Consequently, for such problems, the condition for optimum location $x = x_i$ of an inner hinge is that either

$$<y'>_{x_i} = 0 \tag{4.10}$$

or

$$y'(x_i^+) = -y'(x_i^-) - 2 \frac{y(x_i) - y(x_{i+1})}{x_{i+1} - x_i} \tag{4.11}$$

In Eq. 4.10, x_{i+1} denotes the position of an adjacent inner hinge or hinged end point of the beam, and Eq. 4.11 presumes that no beam supports or additional hinges are placed between the points x_i and x_{i+1}.

4.3 Discussion

The buckling load of the clamped-simply supported optimum column shown in Fig. 2.1, originally determined in Ref. 4, is actually bimodal. For this type of column, one obtains the same optimum design independently of whether he uses Eq. 4.10 or Eq. 4.11 to govern the location of the inner hinge of zero bending moment. However, as is illustrated next, it is necessary to pay full attention to both conditions in other other geometrically unconstrained problems.

Figure 4.2 shows the geometrically unconstrained optimum design of a doubly clamped column (p = 2) with two inner hinges [10,19]. This design is also bimodal, but in this case, the optimum position of the left hand hinge is governed by Eq. 4.11, while the position of the right hand hinge is governed by Eq. 4.10. If, for example, Eq. 4.10 were used for both inner hinges, their positions would change, and a slightly different design be obtained. This design would maximize the second buckling eigen-value (with a symmetric mode), see Ref. 10, but it would have a much lower fundamental Euler buckling eigenvalue than the design

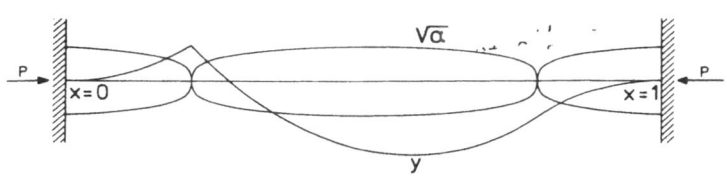

Figure 4.2   Optimum Clamped-Clamped Column with Inner Hinges

shown in Fig. 4.2, and hence would not be optimum in the sense of maximizing the buckling load. In fact, the doubly clamped column solution published in Ref. 4 is subject to this mistake.

5. BIMODAL FORMULATION FOR OPTIMAL DESIGN

In this section a problem is considered in which a bimodal rather than a single mode formulation is necessary, in order to arrive at the correct optimum design. Such a bimodal formulation is presented.

The example problem considered consists in maximizing the Euler buckling eigenvalue λ of Eq. 2.3, for a doubly clamped, solid elastic column (p = 2) of given volume and length. This problem was first considered in Ref. 4, but as pointed out in Ref. 10, an erroneous solution was arrived at. The design shown in Fig. 4.2 constitutes the correct solution, within the premises of a single mode formulation of the problem, and it replaces the design from Ref. 4, which also was obtained on the basis of a single mode formulation.

However, the design shown in Fig. 4.2 is only optimum within the class of doubly clamped columns with two inner hinges. It is quite obvious that the column would obtain a greater Euler buckling load for the same volume, length, and material, if it were made to buckle in a symmetric fundamental buckling mode with a continuous slope throughout. This could easily be achieved by restributing the given material slightly so that the hinges became locked. The problem is, however, that the field equations of the geometrically unconstrained column do predict zero cross-section (singular behavior) at points of vanishing bending moment, and two such points are necessarily present in a clamped-clamped column whenever a single mode formulation is used.

This clearly indicates that the single mode formulation is inadequate for the problem under consideration, which motivated a reformulation of the problem in Ref. 10, leading to a new optimality condition that does not necessarily lead to vanishing cross-sections at points of zero bending moment.

The initial formulation of the optimization problem of Section 2 is now reformulated and expanded. Following Ref. 10, consider the possibility that the dimensionless fundamental buckling load λ of Eq. 2.3 for the optimum column is a double eigenvalue, i.e. that

$$\lambda = \int_0^1 \alpha^p y_i''^2 dx \qquad i = 1,2 \tag{5.1}$$

where $y_1(x)$ and $y_2(x)$ denote the two buckling modes corresponding to the optimum λ. When λ is double, y_1, and y_2 need not be mutually orthogonal, but the form of Eqs. 5.1 presupposes the mode normalization

$$\int_0^1 y_i'^2 dx = 1 \quad i = 1,2 \tag{5.2}$$

The condition of given volume for the column is again expressed by

$$\int_0^1 \alpha dx = 1 \tag{5.3}$$

To formulate the problem in more generality, consider a geometric minimum constraint for the design variable $\alpha(x)$, namely require that $\alpha(x) \geq \bar{\alpha}$ throughout, assuming the minimum allowable value $\bar{\alpha}$ ($0 \leq \bar{\alpha} < 1$) to be given. This minimum constraint can be expressed by means of a real slack variable $g(x)$, in the standard fashion

$$g^2(x) = \alpha(x) - \bar{\alpha} \tag{5.4}$$

Applying a variational formulation of the optimization problem, one considers the functional

$$\Lambda^* = \int_0^1 \alpha^P y_1''^2 dx - \gamma \left\{ \int_0^1 \alpha^P y_1''^2 dx - \int_0^1 \alpha^P y_2''^2 dx \right\}$$

$$- \sum_{i=1}^2 n_i \left\{ \int_0^1 y_i'^2 dx - 1 \right\} - \beta p \left\{ \int_0^1 \alpha dx - 1 \right\}$$

$$- p \int_0^1 \mu(x) \{ g^2 - \alpha + \bar{\alpha} \} dx \tag{5.5}$$

appending to the Rayleigh quotient based upon y_1, the assumption of the eigenvalue being double (see Eqs. 5.1), and the constraints of Eqs. 5.2, 5.3 and 5.4 by means of Lagrangian multipliers γ, n_i ($i = 1,2$), β and $\mu(x)$, respectively. The governing equations, in addition to the constraint equations of the optimization problem are now obtained as the Euler-Lagrange equations, following from the stationarity of the functional Λ^* with respect to variation of $\alpha(x)$, $g(x)$, $y_1(x)$, and $y_2(x)$, respectively.

The stationarity of Λ^* for arbitrary admissible variations of y_i, $i = 1,2$, yields the well known buckling differential equation for y_i, $i = 1,2$,

$$(\alpha^p y_i'')'' = -\lambda y_i'' \quad i = 1,2 \tag{5.6}$$

and the natural boundary conditions, after one has identified the Lagrangian multipliers $n_i = \lambda$, $i = 1,2$, by means of Eq. 5.1.

Variation of $\alpha(x)$ and $g(x)$ leads to the necessary conditions

$$\alpha^{p-1} \left\{ (1 - \gamma)y_1''^2 + \gamma y_2''^2 \right\} + \mu(x) = \beta \tag{5.7}$$

and

$$\mu(x)g(x) = 0 \tag{5.8}$$

respectively. Equation 5.7 is the bimodal optimality condition, which, with $\mu(x) = 0$, may be compared with the optimality condition of Eq. 2.9 for a geometrically unconstrained, single mode, formulation. Now, to remove explicit appearance of the functions $\mu(x)$ and $g(x)$, denote by x_u and x_c the unions of subintervals in which $g(x) \neq 0$ and $g(x) = 0$, respectively, noting that x_u and x_c make up the entire interval $0 \leq x \leq 1$.

For $x \in x_u$, $\alpha(x) > \bar{\alpha}$, i.e. the cross-sectional area is unconstrained, and, moreover, $\mu(x) = 0$, which reduces Eq. 5.7. These results follow directly from Eqs. 5.4 and 5.8. For $x \in x_c$, $\alpha(x) \equiv \bar{\alpha}$, i.e. the cross-sectional is constrained. Consequently, Eqs. 5.7 and 5.8 may be replaced by the formula

$$\alpha(x) = \begin{cases} \left(\dfrac{\beta}{(1 - \gamma)y_1''^2 + \gamma y_2''^2} \right)^{\frac{1}{p-1}} & (\text{if} > \bar{\alpha}), \quad x \in x_u \\[4mm] \bar{\alpha} & , \quad x \in x_c \end{cases} \tag{5.9}$$

for $p = 2$ or $p = 3$. For the case $p = 1$, one may replace Eqs. 5.7 and 5.8 by the equations

$$\left. \begin{array}{ll} (1 - \gamma)y_1''^2 + \gamma y_2''^2 = \beta, & (\text{if } \alpha > \bar{\alpha}), \quad x \in x_u \\[3mm] \alpha(x) = \bar{\alpha} & , \quad x \in x_c \end{array} \right\} \tag{5.10}$$

As in the case of Eq. 2.12, note that for p = 1, the optimality condition for the geometrically unconstrained sub-interval(s) does not contain the design variable α, which must therefore be determined from the buckling differential equation. A method of solution for p = 1 is presented in Ref. 24. The case of p = 1 will not be considered further here.

For exemplification, follow Ref. 10 and derive convenient expressions for the Lagrangian multipliers β and γ and the optimum buckling eigenvalue λ, for the more complex cases of p = 3 and p = 2. First, substitute Eq. 5.9 into the volume constraint of Eq. 5.3, thereby obtaining an explicit expression for β,

$$
\beta = \left[\frac{1 - \bar{\alpha} \int_{x_c} dx}{\int_{x_u} \frac{dx}{\{(1 - \gamma)y_1''^2 + \gamma y_2''^2\}^{1/(p-1)}}} \right]^{p-1}
\tag{5.11}
$$

for p = 2 or 3. Then, subtracting the two equations in Eq. 5.1, substituting α(x) from Eq. 5.9 and using Eq. 5.11, one finds the following implicit equation for γ:

$$
\int_{x_u} \frac{y_1''^2 - y_2''^2}{\{(1 - \gamma)y_1''^2 + \gamma y_2''^2\}^{p/(p-1)}} dx +
$$

$$
\bar{\alpha}^p \left[\frac{\int_{x_u} \frac{dx}{(1 - \gamma)y_1''^2 + \gamma y_2''^2}^{1/(p-1)}}{1 - \bar{\alpha} \int_{x_c} dx} \right]^p \int_{x_c} (y_1''^2 - y_2''^2) dx = 0
\tag{5.12}
$$

Finally, substitution of Eqs. 5.9 and 5.11 into the first of Eqs. 5.1 gives an explicit expression for λ,

$$
\lambda = \bar{\alpha}^p \int_{x_c} y_1''^2 dx
$$

$$
+ \left[\frac{1 - \bar{\alpha} \int_{x_c} dx}{\int_{x_u} \frac{dx}{\{(1 - \gamma)y_1''^2 + \gamma y_2''^2\}^{1/(p-1)}}} \right]^p \times
$$

$$\times \int_{x_u} \frac{y_1''^2}{\{(1 - \gamma)y_1''^2 + \gamma y_2''^2\}^{p/(p-1)}} \, dx \tag{5.13}$$

Equations 5.2, 5.6, 5.9, 5.11, 5.12, and 5.13 comprise the complete set of necessary equations governing the bimodal optimal design problem, for $p = 3$ and $p = 2$. They constitute a strongly coupled, non-linear integro-differential eigenvalue problem. The unknowns to be determined are the optimum buckling eigenvalue λ, the optimum column cross-sectional area function $\alpha(x)$ (and thereby the sub-intervals x_u and x_c), the eigenfunctions y_1 and y_2, and the Lagrangian multipliers β and γ. The solution depends in general on the minimum constraint $\bar{\alpha}$, which is the only specified quantity in the non-dimensional formulation.

A method of numerical solution, based on successive iterations, is presented in Ref. 10.

5.1 Discussion

The bimodal formulation for optimal design described above contains geometrically unconstrained optimization and/or single mode optimization as special cases. The principal advantage of the new formulation is that while the optimality condition of Eq. 2.9 of the single mode formulation predicts formation of hinges at points of zero bending moment in a geometrically unconstrained formulation of optimal design, the bimodal optimality condition of Eq. 5.7 (where $\mu(x) \equiv 0$ for geometrically unconstrained optimization) does not necessarily lead to zero cross-section and formation of hinges at points of zero bending moment.

Special cases of the extended problem formulation are easily obtained. Note first that one may remove the geometric minimum constraint from the formulation by setting $\mu(x) \equiv 0$ in Eq. 5.5, and that the governing equations for the resulting geometrically unconstrained optimization problem for $p = 3$ or $p = 2$ are then obtained as the special case of Eqs. 5.2, 5.6, 5.9, 5.11, 5.12, and 5.13, associated with $\bar{\alpha} = 0$ (implying vanishing of the sub-interval(s) x_c).

One may also reduce the bimodal formulation to the single mode formulation by specifying $\gamma = 0$ and $\eta_2 = 0$ in the functional Λ^* of Eq. 5.5. This leads to a subset of the governing equations in which Eqs. 5.1, 5.2, and 5.6 are only to be considered for $i = 1$, and where Eq. 5.12 drops out. Since $\gamma = 0$, y_2 and its derivatives vanish from all equations. Note that the equations for the geometrically unconstrained, single mode buckling load optimization problem thus obtainable as a special case, are presented in Section 2.

Being based on the functional Λ^* of Eq. 5.5, the bimodal formulation outlined above has a particularly valuable property. It has the ability of automatically handling a problem in which the optimal buckling load is, in fact, a simple eigenvalue. Clearly, such behavior would manifest itself by the functions y_1 and y_2 becoming identical, thereby leaving the Lagrangian multiplier γ undetermined (see Eq. 5.5). The reason why such behavior is not excluded in the formulation is that no condition of linear independence or mutual orthogonality is imposed on y_1 and y_2.

Hence, the general formulation, by solution, directly provides the answer to the a priori question; is the optimum column, subject to a given value of $\bar{\alpha}$, associated with a simple or a double fundamental buckling load λ.

In the following, a problem is discussed in which it is necessary to adopt a bimodal formulation in order to arrive at the correct optimum design. Other examples are discussed in Refs. 24, 25 and 26. It is evident that the necessity of applying a bimodal, or multimodal, formulation in optimal design is increasing with the degree of statical indeterminacy and complexity of the structure to be optimized.

5.2 Bimodal Optimization of a Doubly Clamped p = 2 Column [10]

Figure 5.1 illustrates optimum designs, $\pm \sqrt{\alpha}$, and associated fundamental single or double modes corresponding to selected values of a geometric minimum constraint $\bar{\alpha}$ on the cross-sectional area α of a doubly clamped, p = 2 column. In Fig. 5.1(a), $\bar{\alpha} = 0.7$ and $\lambda = 48.690$ is simple. In Fig. 5.1(b), $\bar{\alpha} = 0.4$ and $\lambda = 51.775$ is simple. In Fig. 5.1(c), $\bar{\alpha} = 0.25$ and the optimum buckling load $\lambda = 52.349$ is bimodal. Figure 5.1(d) shows the optimum solution corresponding to any value of $\bar{\alpha}$ belonging to the interval $0 \leq \bar{\alpha} < 0.226$, where the constraint is no longer active in the design. The corresponding optimum buckling load $\lambda = 52.3563$ is bimodal.

In Fig. 5.2, curve ABCD is based on a number of solutions and shows λ as a function of the geometric minimum constraint $\bar{\alpha}$. For $0.280 < \bar{\alpha} \leq 1$, the optimal designs are associated with a simple fundamental eigenvalue λ, given by curve CD. Curve CE shows the second order eigenvalues λ_2 of the simple optimum eigenvalue designs behind curve CD. At point C, the two curves are seen to coalesce at the value 0.280 for $\bar{\alpha}$, and for $0 \leq \bar{\alpha} \leq 0.280$, the optimum designs are associated with a bimodal fundamental eigenvalue, cf. modes y_1 and y_2 in Figs. 5.1(c) and 5.1(d). All the designs obtained are symmetrical (this was not assumed in the solution procedure), and purely symmetrical and antisymmetrical linear combinations of double modes y_1 and y_2 can be constructed.

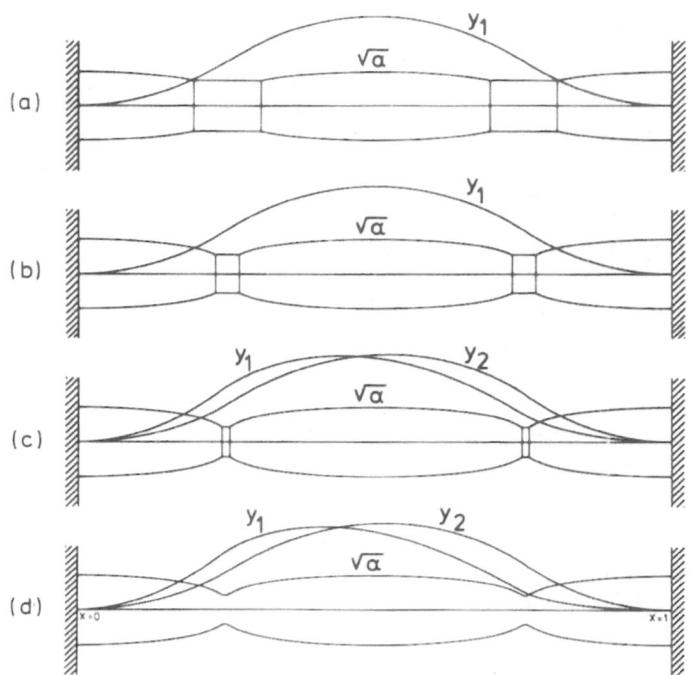

Figure 5.1   Optimum Doubly Clamped Columns for Different Values
of the Cross-Sectional Area Constraint $\bar{\alpha}$

As is shown by curve DCB of Fig. 5.2, the optimum buckling
eigenvalue λ increases with decreasing constraint, for
$0.226 \leq \bar{\alpha} \leq 1$, and for these values of $\bar{\alpha}$ the constraint is active
in the optimum designs, cf. Figs. 5.1(a), 5.1(b) and 5.1(c).
However, for values of $\bar{\alpha}$ belonging to the interval $0 \leq \bar{\alpha} \leq 0.226$,
the minimum constraint is inactive in the optimum design and the
associated bimodal fundamental buckling eigenvalue is constant,
cf. AB in Fig. 5.2.   For these values of $\bar{\alpha}$, $0 \leq \bar{\alpha} \leq 0.226$, the
optimum design, see Fig. 5.1(d), is the same, and it has finite
variable cross-section throughout, with a minimum magnitude of
$\bar{\alpha} = 0.226$.

This $\bar{\alpha}$ independent bimodal optimum design in Fig. 5.1(d) is
the solution to the geometrically unconstrained optimization
problem for a doubly clamped column, with p = 2.   Its fundamental,
double buckling eigenvalue λ is 32.62% higher than the fundamental
eigenvalue of a corresponding uniform column of the same volume,
length, and material.   The bimodal optimum design replaces not
only the solution arrived at in Ref. 4, but also the geomtrically
unconstrained, candidate design in Fig. 4.2.

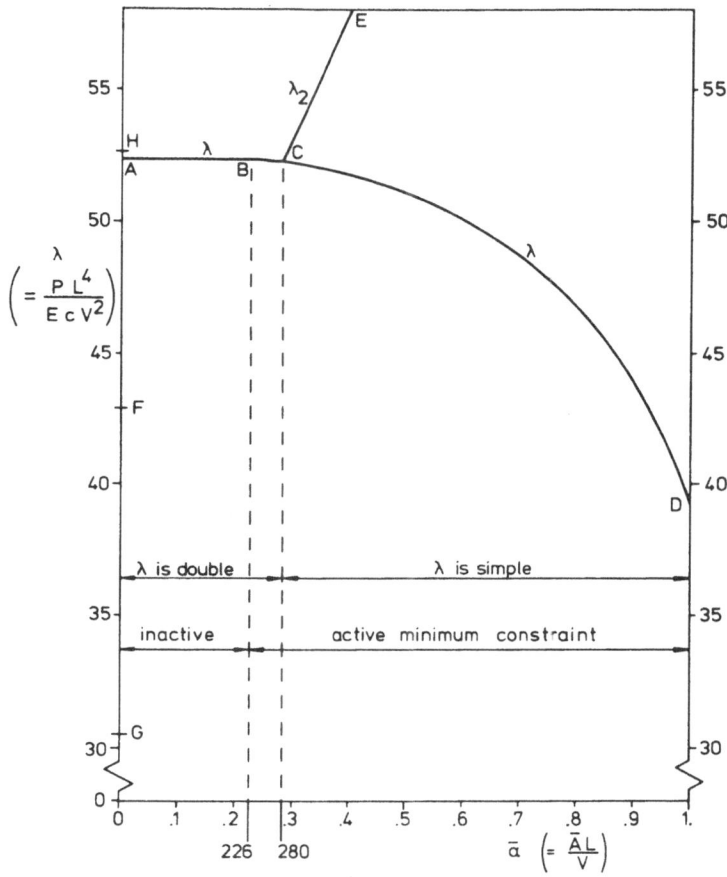

Figure 5.2   Optimum Buckling Load vs. $\bar{\alpha}$

The result provides a noteworthy example of a statically indeterminate solution to a geometrically unconstrained, one-dimensional, single purpose, structural optimization problem. It constitutes an obvious exception to the general rule that solutions to the broad class of all such problems will be statically determinate.

6.   SINGLE MODE FORMULATION OF GEOMETRICALLY CONSTRAINED OPTIMAL DESIGN

The mathematical formulation of the problem of optimizing a column with a prescribed minimum constraint $\alpha \geq \bar{\alpha}$ on the cross-sectional area α, assuming the optimum fundamental buckling load

λ to be a simple eigenvalue, is obtained as a special case of the bimodal formulation in Section 5 by setting $\gamma = 0$ and $\eta_2 = 0$, which implies that y_2 and its derivatives drop out of the problem. Denoting y_1 by y, the governing equations of the problem are identified with stationarity of the simplified form F

$$F = \int_0^1 \alpha^p y''^2 dx - \eta \left\{ \int_0^1 y'^2 dx - 1 \right\} - \beta p \left\{ \int_0^1 \alpha dx - 1 \right\}$$

$$- p \int_0^1 \mu(x) \{g^2 - \alpha + \bar{\alpha}\} dx \tag{6.1}$$

of the functional Λ^* in Eq. 5.5. From Eqs. 5.2, 5.6, 5.9, 5.11 and 5.13, the set of governing equations for problems associated with $p = 2$ or $p = 3$ is easily found to be

$$(\alpha^p y'')'' = -\lambda y'' \tag{6.2}$$

$$\int_0^1 y'^2 dx = 1 \tag{6.3}$$

$$\alpha(x) = \begin{cases} \left(\dfrac{\beta}{y''^2} \right)^{\frac{1}{p-1}} & (\text{if} > \bar{\alpha}) \ , \quad x \in x_u \\[3mm] \bar{\alpha} & , \quad x \in x_c \end{cases} \tag{6.4}$$

$$\beta = \left[\frac{1 - \bar{\alpha} \displaystyle\int_{x_c} dx}{\displaystyle\int_{x_c} \frac{dx}{(y'')^{2/(p-1)}}} \right]^{p-1} \tag{6.5}$$

$$\lambda = \bar{\alpha} \int_{x_c} y''^2 dx + \left[\frac{1 - \bar{\alpha} \displaystyle\int_{x_c} dx}{\displaystyle\int_{x_u} \frac{dx}{(y'')^{2/(p-1)}}} \right]^p \int_{x_u} \frac{dx}{(y'')^{2/(p-1)}} \tag{6.6}$$

This type of column problem (associated with $p = 2$ or $p = 3$) has been solved for different boundary conditions in Refs. 9, 17, 22,

27, and 28. Note that the equations above reduce to those presented in Section 2 for single mode, geometrically unconstrained,
optimal design, if one sets $\bar{a}=0$, which makes subdomain(s) x_c
vanish.

The set of equations corresponding to Eqs. 6.2 to 6.6, for
the case p = 1, is easily derived and is not dealt with here.
Geometrically constrained optimization problems with p = 1 were
first considered and solved in the significant papers of Refs.
6, 7, and 8.

In the papers cited in this section, the a priori assumption
of the optimum buckling eigenvalue being simple is undoubtedly
correct for the examples considered. When a single mode formulation is used, the assumption should always - and can easily - be
checked a posteori.

7. CONDITIONS FOR OPTIMALLY PLACED INNER SUPPORTS

Consider finally the possibility that positions $x = x_i$,
i = 1,...,M of a given number $M \geq 1$ of inner simple supports are
design variables, in addition to the cross-sectional area function
α. Assuming the optimum buckling eigenvalue λ to be single-modal,
one derives specific conditions for optimal locations of the
supports, on the basis of the functional F of Eq. 6.1.

Necessary conditions of optimality are derived in a similar
way as in Ref. 17. One first rewrites Eq. 6.1 so that the support
locations $x = x_i$ and the shear force appear explicitly. Sub-
dividing the interval $0 \leq x \leq 1$, defining $0 = x_0^+$ and $1 = x_{M+1}^-$, and
integrating $\alpha^p y''^2$ once by parts, F defined by Eq. 6.1 takes the
form

$$F = - \sum_{i=1}^{M+1} \int_{x_{i-1}^+}^{x_i^-} \left\{ (\alpha^p y'')' y' + n y'^2 + \beta p \alpha + p \mu \{g^2 - \alpha + \bar{a}\} \right\} dx + n + \beta p$$

$$(7.1)$$

The Dirichlet boundary terms vanish at x = 0 and x = 1, due to
the types of boundary conditions assumed, and they cancel at
$x = x_i$, i = 1,..., M, because the bending moment $\alpha^p y''$ and the
slope y' are continuous at the inner supports (the continuity of
y' follows from the constraint on α, which prevents formation of
inner hinges).

Stationarity of F for admissible variations $\delta x_i^+ = \delta x_i^- = \delta x_i$,
i = 1,...,M, is expressed by

$$\delta_{x_i} F = \sum_{i=1}^{M} \left\{ \left[(\alpha^P y'')' y' + \eta y'^2 + \beta p\alpha + p\mu(g^2 - \alpha + \bar{\alpha}) \right]_{x=x_i^+} \right.$$

$$\left. - \left[(\alpha^P y'')' y' + \eta y'^2 + \beta p\alpha + p\mu(g^2 - \alpha + \bar{\alpha}) \right]_{x=x_i^-} \right\} \delta x_i = 0$$

$$(7.2)$$

Recall now from Section 5 the definition of g in Eq. 5.4, and that $\eta = \lambda$. Moreover, it is easily seen that α is continuous throughout. For arbitrary variations δx_i, $i = 1,\ldots,M$, one therefore has

$$\left\langle \left\{ (\alpha^P y'')' + \lambda y' \right\} y' \right\rangle_{x_i} = 0 \quad i = 1,\ldots,M \tag{7.3}$$

Since y' is continuous and, in general, $y'(x_i) \neq 0$, Eqs. 7.3 reduce to the necessary conditions

$$\left\langle (\alpha^P y'')' + \lambda y' \right\rangle_{x_i} = 0 \quad i=1,\ldots,M \tag{7.4}$$

which, in view of the continuity of y' further reduce to the following conditions for optimum locations $x = x_i$, $i = 1,\ldots,M$, of interior supports:

$$\left\langle (\alpha^P y'')' \right\rangle_{x_i} = 0 \quad i = 1,\ldots,M \tag{7.5}$$

Equations 7.5 state that the optimum locations of M inner simple supports are such that the shear force is continuous at the supports and Eqs. 7.4 state that reactions at optimally placed supports are zero. These conditions were originally derived in Ref. 13, in different contexts for both elastic and plastic design. The conditions for plastic design are also given in Ref. 15. Solutions of column optimization problems with interior simple supports as additional design variables are available in Refs. 14, 16, and 17. Figure 7.1 shows an example from Ref. 17 of an optimally design column which has clamped-simply supported ends and one optimally placed interior simple supports.

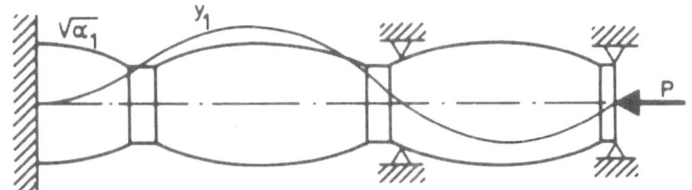

Figure 7.1   Optimum Column with Interior Support

REFERENCES

1.   Lagrange, J.L., "Sur la Figure des Colonnes," Miscellanea
     Taurinensia V, 1770-1773, p. 123.
2.   Clausen, T., "Über die Form Architektonischer Säulen,"
     Melanges Mathematiques et Astronomiques I, 1849-1853, pp.
     279-294.
3.   Keller, J.B., "The Shape of the Strongest Column," Arch.
     Rational Mech. Anal., Vol. 5, 1960, pp. 275-285.
4.   Tadjbakhsh, I., and Keller, J.B., "Strongest Columns and Iso-
     perimetric Inequalities for Eigenvalues," J. Appl. Mech.,
     Vol. 9, 1962, pp. 159-164.
5.   Keller, J.B. and Niordson, F.I., "The Tallest Column,"
     J. Math. Mech., Vol. 16, 1966, pp. 433-446.
6.   Taylor, J.E., "The Strongest Column: An Energy Approach,"
     J. Appl. Mech., Vol. 34, 1967, p 486.
7.   Taylor, J.E. and Liu, C.Y., "Optimal Design of Columns,"
     AIAA Journal,Vol. 6, 1968, pp. 1497-1502.
8.   Prager, W. and Taylor, J.E., "Problems of Optimal Structural
     Design," J. Appl. Mech., Vol. 35, 1968, pp. 102-106.
9.   Frauenthal, J.C. "Constrained Optimal Design of Columns
     Against Buckling," J. Struct. Mech., Vol. 1, 1972, pp. 79-89.
10. Olhoff, N. and Rasmussen, S.H., "On Single and Bimodal
     Optimum Buckling Loads of Clamped Columns," Int. J. Solids
     Structures, Vol. 13, 1977, pp. 605-614.
11. Masur, E.F., "Optimal Placement of Available Sections in
     Structural Eigenvalue Problems," J. Optimization Theory Appl.,
     Vol. 15, 1975, pp. 69-84.
12. Masur, E.F., "Optimality in the Presence of Discreteness and
     Discontinuity," Proc. IUTAM Symposium on Optimization in
     Structural Design (eds. A. Sawczuk and Z. Mróz), Springer-
     Verlag, Berlin, 1975, pp. 441-453.
13. Mróz, Z. and Rozvany, G.I.N., "Optimal Design of Structures
     With Variable Support Conditions," J. Optimization Theory
     Appl., Vol. 15, 1975, pp. 85-101.
14. Mróz, Z. and Rozvany, G.I.N., "Column Design: Optimization
     of Support Conditions and Segmentation," J. Struct. Mech.,
     Vol. 5, 1977, pp. 279-290.

15. Prager, W. and Rozvany, G.I.N., "Plastic Design of Beams: Optimal Locations of Supports and Steps in Yield Moment," Int. J. Mech. Sci., Vol. 17, 1975, pp. 627-631.

16. Szelag, D. and Mróz, Z., Optimal Design of Elastic Beams with Unspecified Support Conditions, Report of the Institute of Fundamental Technical Research, Warsaw, Poland, 1977.

17. Olhoff, N. and Taylor, J.E., "Designing Continuous Columns for Minimum Total Cost of Material and Interior Supports," J. Struct. Mech., Vol. 6, 1978, pp. 367-382.

18. Banichuk, N.V., and Karihaloo, B.L., "On the Solution of Optimization Problems with Singularities", Int. J. Solids Structures, Vol. 13, 1977, pp. 725-733.

19. Olhoff, N. and Niordson, F.I., "Some Problems Concerning Singularities of Optimal Beams and Columns," ZAMM, Vol. 59, 1979, pp. T16-T26.

20. Banichuk, N.V., "Optimizing the Stability of a Bar with Elastic Clamping," MTT (Mechanics of Solids), Vol. 9, 1974, pp. 150-154.

21. Haug, E.J. and Arora, J.S., Applied Optimal Design of Mechanical and Structural Systems, Wiley, 1979.

22. Rasmussen, S.H., "On the Optimal Shape of an Elastic-Plastic Column," J. Struct. Mech., Vol. 4, 1976, pp. 307-320.

23. Olhoff, N., "Optimization of Vibrating Beams with Respect to Higher Order Natural Frequencies," J. Struct. Mech., Vol. 4, 1976, pp. 87-122.

24. Masur, E.F. and Mróz, Z., "Singular Solutions in Structural Optimization Problems", Proc. IUTAM Symposium on Variational Methods in the Mechanics of Solids (ed. S. Nemat-Nasser), Pergamon Press, to appear.

25. Masur, E.F., and Mróz, Z., "On Non-Stationary Optimality Conditions in Structural Design," Int. J. Solids Structures, Vol. 15, 1979, pp. 503-512.

26. Prager, S. and Prager, W., "A Note on Optimal Design of Columns," Int. J. Mech. Sci., Vol. 21, 1979, pp. 249-251.

27. Simitses, G.J., Kamat, M.P., and Smith, C.V. Jr., "Strongest Column by the Finite Element Displacement Method," AIAA Journal, Vol. 11, 1973, pp. 1231-1232.

28. Murthy, K.N. and Christiano, P., "Design of Least Weight Structures for Prescribed Buckling Load," AIAA Journal, Vol. 12, 1973, pp. 1773-1775.

OPTIMIZATION OF TRANSVERSELY VIBRATING BEAMS AND ROTATING SHAFTS

Niels Olhoff

Department of Solid Mechanics
The Technical University of Denmark
DK-2800 Lyngby, Denmark

ABSTRACT

This paper formulates and solves optimal design problems for beams and rotating shafts involving vibration characteristics. Both fundamental and higher order natural frequencies are optimized or constrained.

1. INTRODUCTION

Optimization of dynamically loaded structures is a comparatively new field of research, which was opened by the significant paper [1] published by Niordson in 1965. Since then, the field has been characterized by considerable activity, and many interesting and useful results have been obtained. Survey papers by Pierson [2], Reitman and Shapiro [3], Rao [4] and the author [5] provide keys to the literature of the field.

This paper considers problems of determining the distribution of structural material of transversely vibrating beams or of rotating circular shafts, such that maximum values of natural frequencies or critical speeds are obtained, for a prescribed amount of material and a given length of the beam or shaft. Equivalently, the volume of structural material is minimized, for a specified vibration frequency or critical speed. The practical significance of such problems is that they provide designs of minimum weight (or cost) of material, subject to constraints on beam vibrational resonance due to external excitation of given frequency or frequency range, and on failure due to whirling instability at service speeds of rotating shafts.

In Section 2 the problem of optimizing with respect to the fundamental frequency of free, transverse vibrations of a beam, or with respect to the first critical speed of a rotating shaft is formulated. The governing equations for geometrically constrained problems of these types are derived in Section 3, and are specialized to geometrically unconstrained optimal design in Section 4. The significance of the types of problems considered in these sections is that they yield minimum material volume designs against resonance or whirling instability, respectively, subject to all external excitation frequencies or service speeds within a large range from zero up to the particular fundamental eigenvalue.

Section 5 deals with optimization with respect to a particular higher order eigenvalue (vibration frequency or critical speed) of given order [6,7]. This type of optimization produces a considerable gap between the subject eigenvalue and the adjacent lower order eigenvalue and offers even more competitive designs for problems of vibration or whirling instability for which external excitation frequencies or service speeds are confined within an interval of finite upper and lower limits. Optimum designs obtained in this way are significant even when the ranges of the external excitation frequencies or the service speeds include small frequencies at which resonance or whirling instability cannot be accepted. Hence, such phenomena can be removed, while preserving the optimum characteristics of the original higher order optimum design, for an extra cost corresponding to the appropriate number of optimally placed inner supports [6,7].

2.   OPTIMAL DESIGN WITH RESPECT TO THE FUNDAMENTAL NATURAL FREQUENCY

Consider an elastic Bernoulli-Euler beam of length L and structural volume V, which vibrates at its fundamental angular frequency ω of free transverse vibration. The beam is made of a material with Young's modulus E and the mass density ρ. It has variable but similarly oriented cross-sections, with the relationship

$$I = c \, A^p \qquad\qquad\qquad (2.1)$$

between the area moment of inertia I and the area A. Attention is restricted to the cases of $p = 2$ (geometrically similar, solid cross-sections) and $p = 3$ (solid cross-sections of fixed width and variable height), because the case of $p = 1$ (sandwich cross-sections) is usually degenerate for the types of problems to be considered [5,8-10]. The constant c and the value of p ($p = 2$ or $p = 3$) are assumed to be given in the following. A number of given nonstructural masses Q_k, $k = 1,\ldots,K$, are assumed to be attached to a beam at specified points $X = X_k$, where X denotes the beam coordinate.

Introducing dimensionless quantities,

$$x = X/L \quad , \qquad 0 \le x \le 1 \tag{2.2}$$

$$\alpha(x) = A(x)L/V \tag{2.3}$$

$$q_k = \frac{Q_k}{\rho V} \quad , \qquad x = x_k \quad , \qquad k = 1,\ldots,K \tag{2.4}$$

$$\lambda = \omega^2 \frac{\rho L^{3+p}}{E c V^{p-1}} \tag{2.5}$$

i.e. coordinate x, cross-sectional area function $\alpha(x)$, nonstructural masses q_k, and eigenvalue λ, respectively, the nondimensional Rayleigh quotient associated with the fundamental vibration mode $y(x)$ has the form

$$\lambda = \int_0^1 \alpha^p {y''}^2 \, dx \tag{2.6}$$

provided that no flexible supports are present and that the vibration mode $y(x)$ is normalized by

$$\int_0^1 \alpha y^2 \, dx + \sum_{k=1}^K q_k y^2(x_k) = 1 \tag{2.7}$$

The optimization problem consists in determining the cross-sectional area function $\alpha(x)$, for a beam having given homogeneous boundary conditions and carrying a set of prescribed nonstructural masses q_k, such that a maximum value is obtained for the fundamental eigenvalue λ. The length and the structural volume of the beam are assumed to be specified, i.e., α is subject to the constraint

$$\int_0^1 \alpha \, dx = 1 \tag{2.8}$$

cf. Eqs. 2.2 and 2.3.

Consider geometrically constained optimization of the beam, assuming that $\alpha(x)$ may nowhere be less than an a priori specified value $\bar{\alpha}$, belonging to the interval $0 \le \bar{\alpha} < 1$, the limiting values of which correspond to geometrically unconstrained and fully constrained (i.e. uniform) design, respectively. The geometric minimum constraint is conveniently expressed by means of the real slack variable $g(x)$ as

$$g^2(x) = \alpha(x) - \bar{\alpha} \tag{2.9}$$

A few aspects of the optimization problem outlined above should be noted at this point. By means of Eq. 2.5 one easily sees that for $p > 1$ and given values of ρ, E, L, c, and $\bar{\alpha}$, a design α that is associated with a maximum value of λ will maximize ω for a fixed V and at the same time minimize V for a fixed value of ω. The formulation has additional practical significance, however. For $p = 2$, it also governs the problem of maximizing the first critical speed ω for fixed V (or minimizing V for fixed first critical speed) of a rotating circular shaft, when gyroscopic effects are excluded. In this case q_k may be considered as masses of nonstructural circular disks on the shaft.

3. GOVERNING EQUATIONS FOR GEOMETRICALLY CONSTRAINED OPTIMAL DESIGN

A variational formulation of the optimization problem stated above is employed, using the functional

$$\lambda^* = \int_0^1 \alpha^p y''^2 \, dx - \eta \left\{ \int_0^1 \alpha y^2 \, dx + \sum_{k=1}^K q_k y^2(x_k) - 1 \right\}$$

$$- \beta \left\{ \int_0^1 \alpha \, dx - 1 \right\} - \int_0^1 \mu(x)\{g^2 - \alpha + \bar{\alpha}\} dx \tag{3.1}$$

which constitutes an augmented form of the Rayleigh quotient of Eq. 2.6, where the constraints of Eqs. 2.7, 2.8, and 2.9 are adjoined by means of the Lagrangian multipliers η, β, and $\mu(x)$, respectively. Together with Rayleigh's minimum principle, these Lagrangian multipliers permit variations of λ^* with respect to y, α, and $g(x)$ to be taken independently.

Hence one directly obtains the Euler-Lagrange equations

$$\left(\alpha^p y''\right)'' = \eta \alpha y \tag{3.2}$$

$$p\alpha^{p-1} y''^2 - \eta y^2 = \beta - \mu(x) \tag{3.3}$$

$$g(x)\mu(x) = 0 \tag{3.4}$$

In addition of Eq. 3.2, the stationarity of λ^* with respect to y also produces (1) the natural boundary conditions, (2) the condition of continuity of the bending moment $m = \alpha^p y''$, except at points of prescribed y', and (3) the condition of continuity of the shear force $t(x) = -(\alpha^p y'')'$, except at points of prescribed y_n and at the points $x = x_k$, $k = 1,\ldots,K$, with attached nonstructural masses q_k. At the latter points the jumps of t are identified as $\langle t \rangle_{x_k} = -\eta q_k y(x_k)$. Now multiplying through Eq. 3.2 by y and integrating by parts twice, with the boundary conditions and the jump conditions taken into account, yields an equation which, when compared with Eqs. 2.6 and 2.7, gives

$$\eta = \lambda \tag{3.5}$$

Hence, Eq. 3.2 takes the well-known form

$$\left(\alpha^p y''\right)'' = \lambda \alpha y \tag{3.6}$$

Equation 3.3 follows from the variation of the design variable $\alpha(x)$ and is identified as the so-called optimality condition (which is necessary for optimality). To obtain a convenient formula for $\alpha(x)$, one combines Eq. 3.3 with Eq. 3.4 and the defining Eq. 2.9 for $g(x)$. Noting that either $g(x) \neq 0$, implying that $\alpha(x) > \bar{\alpha}$ and $\mu(x) = 0$ (which reduces Eq. 3.3), or $g(x) = 0$, which is equivalent to $\alpha(x) = \bar{\alpha}$, one has

$$\alpha(x) = \begin{cases} \left[\dfrac{\beta + \lambda y^2}{p y''^2}\right]^{\frac{1}{p-1}}, & (\text{if} > \bar{\alpha}), \quad x \in x_u \\[3ex] \bar{\alpha}, & x \in x_c \end{cases} \tag{3.7}$$

where x_u and x_c denote subintervals in which the optimum cross-sectional area function $\alpha(x)$ is unconstrained or constrained, respectively. Clearly, the subintervals x_u and x_c together occupy the entire interval $0 \leq x \leq 1$.

The optimization problem is governed by Eqs. 2.6, 2.7, 2.8, 3.6, and 3.7, together with the continuity and jump conditions noted in the foregoing. This constitutes a coupled, nonlinear, integro-differential eigenvalue problem in dimensionless form, where the unknowns to be determined are the optimum fundamental eigenvalue λ, the associated normalized vibration mode y, the optimum cross-sectional area function α (which includes the determination of the subdomains x_u and x_c), and the Lagrangian multiplier β. The value of the minimum allowable cross-sectional area $\bar{a}$, and the number K, the locations x_k, k = 1,...,K, and the magnitudes q_k of the attached nonstructural masses, respectively, are given quantities for a particular problem.

Due to the nonlinearity and coupling of the governing equations, closed form solutions cannot be expected for p = 2 and p = 3, so numerical solution procedures must be applied. Such procedures are available in Ref. 6, 11, and 12. References 6 and 12 present results for geometrically constrained cantilevers with p = 2, with and without nonstructural masses, respectively. Reference 11 offers results for p = 2 and p = 3, with various other boundary conditions.

Figure 3.1 shows results obtained in Ref. 6, namely cantilever beams of geometrically similar cross-sections (p = 2) (or for rotating circular cantilever shafts), optimized with respect to the fundamental natural transverse vibration frequency (or first critical speed) $\omega = \omega_1$. The beams are illustrated by optimum shapes $\pm\sqrt{\alpha_1}$, where $\alpha_1(x) = \alpha(x) = A(x)L/V$, and the solutions in Figs. 3.1(a) and 3.1(c) and 3.1(b) and 3.1(d), respectively, correspond to minimum constraints $\bar{\alpha}$ = 0.05 and 0.5. The dimensionless nonstructural tip mass in Figs. 3.1(c) and 3.1(d) is $q_1 = Q_1/\rho V = 0.1$. The fundamental frequencies ω_1 of the optimal designs are increased by (a) 279%, (b) 88%, (c) 81% and (d) 57%, respectively, in comparison with those corresponding to uniform designs of the same volume, length, material, and, for (c) and (d), tip mass.

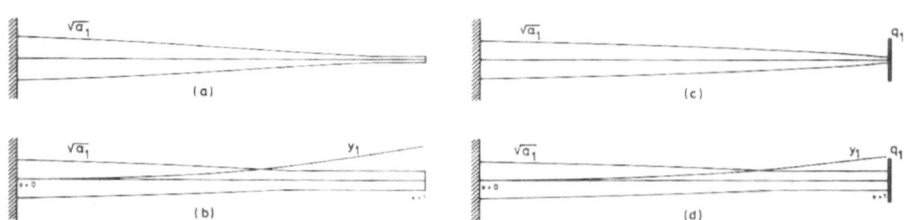

Figure 3.1. Optimum Vibrating Cantilever Beams

Figure 3.2 shows the square root $\sqrt{\lambda_1}$ of the fundamental eigenvalue $\lambda = \lambda_1$ and the square root $\sqrt{\lambda_2}$ of next eigenvalue λ_2, as functions of the dimensionless cross-sectional area constraint $\sqrt{\bar{\alpha}} = \sqrt{\bar{A}L/V}$, for optimum cantilevers of the type shown in Fig. 3.1. Note that $\bar{\alpha}$ = 0 and $\bar{\alpha}$ = 1 correspond to geometrically unconstrained and fully constrained (uniform) designs, respectively, and that

the square roots of the eigenvalues are proportional to the first
and second vibration frequencies (critical speeds) ω_1 and ω_2,
respectively, for given beam volume, length, and material. The
solid curves in Fig. 3.2 represent optimum $\lambda = \lambda_1$ designs without
nonstructural mass, see for example Figs. 3.1(a) and 3.1(b), while
other curves are for optimum $\lambda = \lambda_1$ designs with a dimensionless
tip mass $q_1 = Q_1/\rho V$, see e.g. Figs. 3.1(c) and 3.1(d).

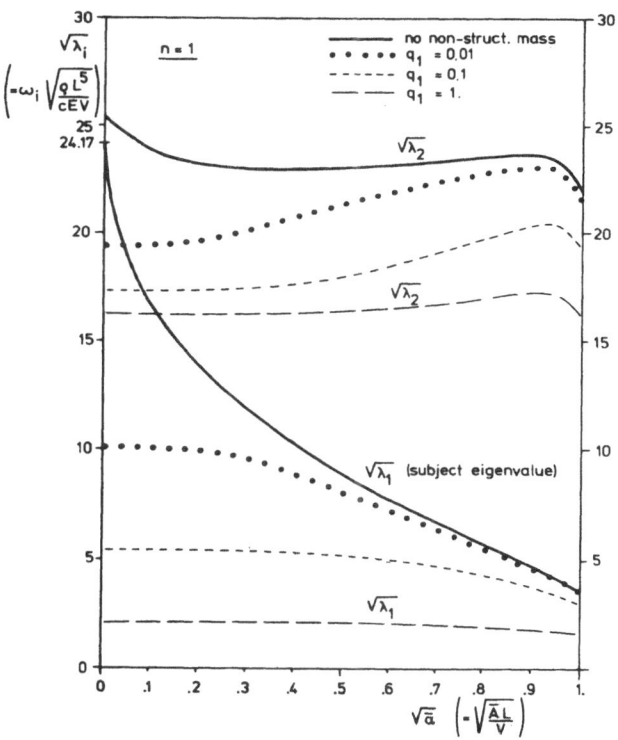

Figure 3.2. $\sqrt{\lambda_1}$ and $\sqrt{\lambda_2}$ For Optimum λ_1 Cantilevers vs. $\sqrt{\bar{\alpha}}$

Figure 3.2 clearly illustrates that geometrically unconstrained designs are associated with maximum obtainable merits in
comparison with corresponding geometrically constrained designs.
Although geometrically constrained optimum designs are preferable
in practice, it is obvious that knowledge of the maximum obtainable efficiency (and the associated unconstrained design) is of
both theoretical and practical importance.

184

4. GEOMETRICALLY UNCONSTRAINED OPTIMIZATION

Consider now the case in which no geometric constraint is specified for the cross-sectional area function $\alpha(x)$ in the process of optimization. This constitutes a special case of the formulation considered in Section 3, and corresponds to setting $\mu(x) \equiv 0$ in Eq. 3.1, or to set $\bar{\alpha} = 0$ and the interval x_u equal to the entire interval $0 < x < 1$. Doing this, the system of Eqs. 2.6, 2.7, 2.8, 3.6 and 3.7 reduces to the following system for geometrically unconstrained optimization:

$$\lambda = \int_0^1 \alpha^p y''^2 \, dx \tag{4.1}$$

$$\int_0^1 \alpha y^2 \, dx + \sum_{k=1}^K q_k y^2(x_k) = 1 \tag{4.2}$$

$$\int_0^1 \alpha dx = 1 \tag{4.3}$$

$$(\alpha^p y'')'' = \lambda \alpha y \tag{4.4}$$

$$\alpha(x) = \left[\frac{\beta + \lambda y^2}{py''^2} \right]^{\frac{1}{p-1}} \tag{4.5}$$

Here the optimality condition of Eq. 4.5 is valid in the entire interval $0 \leq x \leq 1$. If one writes this equation in the form $p\alpha^{p-1} y'' - \lambda y^2 = \beta$, multiplies it by α, integrates over the interval, and uses Eqs. 4.1 to 4.3, one finds that

$$\beta = \lambda \left[p - 1 + \sum_{k=1}^K q_k y^2(x_k) \right] \tag{4.6}$$

i.e., Lagrangian multiplier β is always positive for problems with $p > 1$.

Geometrically unconstrained solutions obtained numerically by successive iterations are available in Refs. 1, 7, 10, 13, and 14. Reference 10 presents optimum $p = 2$ and $p = 3$ cantilevers, with and without tip mass, and $p = 2$ solutions without nonstructural mass are available in Refs. 1 and 7 for other boundary conditions.

Figure 4.1 shows examples of geometrically unconstrained optimum designs of p = 2 beams, with respect to the fundamental natural vibration frequency $\omega = \omega_1$, when no nonstructural masses are considered. The design in Fig. 4.1(a) is the solution for a simply supported beam [1], Fig. 4.1(b) shows the optimum canti-lever design [10], and Fig. 4.1(c) illustrates the optimum design of a doubly clamped beam [7]. The design in Fig. 4.1(b) may be compared with the constrained designs in Fig. 3.1(a) and 3.1(b). It should be noted that its optimal characteristics (maximum ω_1 for a given V and L, or minimum V for given ω_1 and L) are repre-sented by the value indicated for the solid $\sqrt{\lambda}_1$ curve at $\sqrt{\bar{\alpha}} = 0$ in Fig. 3.2. It is also worth mentioning that the design in Fig. 4.1(b) is at the same time the optimum design of a clamped-simply supported beam [7].

The fundamental frequencies ω_1 of the optimum solutions in Fig. 4.1 are increased by (a) 6.6%, (b) 588%, and (c) 332%, respectively, when compared with the frequencies of corresponding uniform beams of the same volume, length, and material. Comparing the design of Fig. 4.1(b) with a uniform clamped-simply supported beam, its fundamental frequency ω_1 is increased by 57%.

4.1 Types of Singular Behavior

In problems of optimizing transversely vibrating Bernoulli-Euler beams without geometric constraint, there may occur two different types of singular behavior, both of which are associated with zero bending moment $m(x) = \alpha^p y''$ and cross-section α. In one, called Type I, the shear force is finite, while in the other, called Type II, the shear force vanishes at the singularity.

The singular behavior can be determined analytically by expanding a solution of Eqs. 4.4 and 4.5 in power series (see

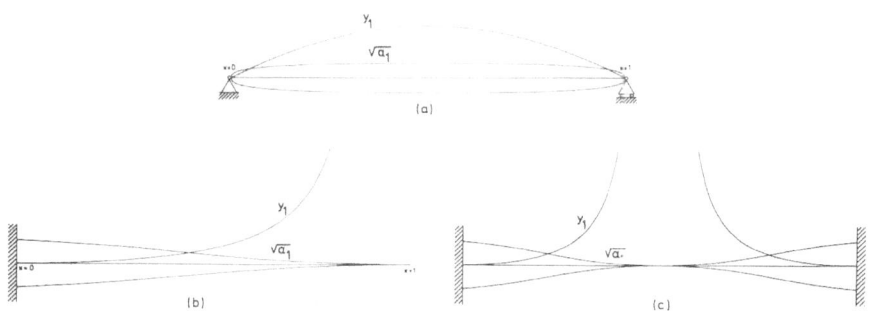

Figure 4.1.   Optimum Geometrically Unconstrained Beams

e.g. Refs. 1, 7 and 15) near the singular point $x = x_i$, which may either be an inner point or an end point of the beam. In the following, let $x = x_i$ be an arbitrary inner singular point (singular boundary points are special cases). The leading terms of the power series expansions at a Type I singularity for the deflection $y(x)$ and the design variable $\alpha(x)$ are given by [7,15]

$$
y(x) = \begin{cases} a_0 + a_1^+(x - x_i) + b_1^+(x - x_i)^r + \dots \, , & x \geq x_i \\[2mm] a_0 + a_1^-(x_i - x) + b_1^-(x_i - x)^r + \dots \, , & x \leq x_i \end{cases} \tag{4.7}
$$

which reflects continuity of y, and

$$
\alpha(x) = \begin{cases} g_1^+(x - x_i)^s + \dots \, , & x \geq x_i \\[2mm] g_1^-(x_i - x)^s + \dots \, , & x \leq x_i \end{cases} \tag{4.8}
$$

respectively, where a_0, a_1^+, a_1^-, b_1^+, b_1^-, g_1^+ (>0), and g_1^- (>0) are coefficients. Depending on the value of p, the powers r and s in Eqs. 4.7 and 4.8 attain the following values:

$$
\left. \begin{array}{r} r = 5/3 \\ s = 2/3 \end{array} \right\} \quad , \quad p = 2 \atop \left. \begin{array}{r} r = 3/2 \\ s = 1/2 \end{array} \right\} \quad , \quad p = 3 \tag{4.9}
$$

In Eq. 4.7, r is the leading noninteger power of the series, which implies that the second and higher order derivatives of y are unbounded in the neighborhood of the singularity.

The finite shear force $t(x) = -(\alpha^p y'')'$ at a Type I singularity ensures nonzero values of the coefficient b_1^+ and b_1^- in Eq. 4.7, as well as of the coefficients g_1^+ and g_1^- in Eq. 4.8. With the exception of problems in which an external transverse force acts at the point $x = x_i$ (and in buckling problems), the shear force $t(x) = -(\alpha^p y'')'$ will be continuous at $x = x_i$. Therefore, $b_1^+ = -b_1^-$ in Eq. 4.7 and $g_1^+ = g_1^-$ in Eq. 4.8. Clearly the behavior of the beam is characterized kinematically and physically by a hinge at the location of a Type I singularity.

Consider now singularities of Type II in which,the behavior
is associated with zero shear force $t(x) = -(\alpha^p y'')$, in addition
to zero bending moment $m(x) = \alpha^p y''$. At such an interior singularity, the deflection y as well as the slope y' may exhibit discontinuities. The point therefore corresponds to an inner separation of the beam.

Type II singularities are not found in the same broad class
of problems as is the case with Type I singularities. In optimal
design for maximum buckling load for example, inner Type II singularities are not admitted by the governing equations, which
seems obvious from physical grounds.

Now, investigating Eqs. 4.4 and 4.5 for optimum, geometrically
uncontrained transversely vibrating beams, by means of power
series expansions, the leading terms of the deflection $y(x)$ and
design variable $\alpha(x)$ are found to be [7,15]

$$y(x) = \begin{cases} d^+(x - x_i)^n + \dots \ , & x > x_i \\ \\ d^-(x_i - x)^n + \dots \ , & x < x_i \end{cases} \tag{4.10}$$

and

$$\alpha(x) = g(x - x_i)^s + \dots \ , \qquad x \geq x_i \ , \qquad x \leq x_i \tag{4.11}$$

respectively, in the vicinity of an arbitrary inner point $x = x_i$
with a Type II singularity. In Eqs. 4.10 and 4.11, d^+, d^-, and g
are nonvanishing coefficients (g is positive) and the powers n and
s of the leading terms attain the following values:

$$\left.\begin{array}{c} \left.\begin{array}{c} n = -2 \\ s = 4 \end{array}\right\} \ , \qquad p = 2 \\ \\ \left.\begin{array}{c} n = -1 \\ s = 2 \end{array}\right\} \ , \qquad p = 3 \end{array}\right\} \tag{4.12}$$

The type of singular behavior is seen to be quite significant, in that the deflection $y(x)$ and all its derivatives tend
toward infinity at the point $x = x_i$ where the bending moment,
shear force, and beam cross-section vanish.

4.2 Conditions for Optimally Placed Inner Type I Singularities (hinges) and Type II Singularities (separations)

Conditions for optimum positions of Type I and II singularities can now be derived as conditions of stationarity of the geometrically unconstrained form of the functional λ^* in Eq. 3.1, with $\mu \equiv 0$ [15]. This form of the functional, which leads to Eqs. 4.1 to 4.5 for geometrically unconstrained design, is to be stationary with respect to variation of the positions $x = x_i$, $i = 1,\ldots,S$ of the singular points. Integrating the term $\alpha^p y''^2$ by parts, noting that $\eta = \lambda$ and $\alpha^p y'' y'(x_i) = 0$ for both Type I and Type II singularities, and proceeding as in Section 4 of Ref. 16, one finds the following necessary conditions at optimally placed points $x = x_i$, $i = 1,\ldots,S$, of zero bending moment:

$$\left\langle (\alpha^p y'')' y' + \lambda \alpha y^2 \right\rangle_{x_i} = 0 \quad , \qquad i = 1,\ldots,S \qquad (4.13)$$

At an inner Type I singularity (hinge), one has $\alpha(x_i) = 0$ and $\alpha y^2(x_i) = 0$, and the shear force $-(\alpha^p y'')'$ is finite. Since, in general, the shear force will also be continuous at the point, one obtains the condition

$$\left\langle y' \right\rangle_{x_i} = 0 \qquad (4.14)$$

That is, the slope y' is continuous at an optimally placed inner Type I singularity in a vibrating beam. This result implies that $a^+ = -a^-$ in Eq. 4.7, if $x = x_i$ is optimally placed. This result was first derived by Masur [17].

Now, if a particular x_i is the location of an inner Type II singularity (beam separation), one has $(\alpha^p y'')'_{x=x_i} = \alpha^p y''(x_i) = \alpha(x_i) = 0$, but for $p = 2$ and $p = 3$, the terms $(\alpha^p y'')' y'$ and $\lambda \alpha y^2$ in Eq. 4.13 precisely attain finite values at the two sides of the point $x = x_i$. By using Eqs. 4.10 and 4.11, with values of n and s given by Eq. 4.12, Eq. 4.13 yields [15]

$$d^+ = \pm d^- \qquad (4.15)$$

That is, the coefficients of the leading singular terms in the expansion of the vibration mode $y(x)$ are equal in magnitude (possibly with opposite signs) at an optimally placed inner Type II singularity. By considering the case $p = 2$ and assuming d^+ and d^- to have equal signs, the condition $d^+ = d^-$ was first derived in Ref 7.

In view of Eqs. 4.10 and 4.11 where n = -2 for p = 2 and n = -1 for p = 3, Eqs. 4.15 imply that, in the vicinity of the singular point, the vibration mode is either symmetrical with respect to the line $x = x_i$, or symmetrical with respect to the point $x = x_i$, if x_i is the optimal location of an inner Type II singularity. Note that an optimally placed inner Type II singularity is found in Fig. 4.1(c).

5. OPTIMIZATION WITH RESPECT TO HIGHER ORDER NATURAL FREQUENCIES

Consider now the problem of maximizing a particular higher order natural frequency ω_n of given order n (n > 1) for a transversely vibrating Bernoulli-Euler beam of prescribed structural volume, length, and material. This problem is governed by a set of equations consisting of Eqs. 2.6 to 2.8, 3.6 and 3.7 for geometrically constrained optimal design, or Eqs. 4.1 to 4.5 for unconstrained design, with λ, $\alpha(x)$, and $y(x)$ subscripted as λ_n, $\alpha_n(x)$, and $y_n(x)$ (indicating reference to the given order n of the subject frequency), and the additional equations [6,7]

$$\int_0^1 \alpha_n y_n y_j \, dx + \sum_{k=1}^{K} q_k y_n(x_k) y_j(x_k) = 0 \, , \quad j = 1,\ldots,n-1 \quad (5.1)$$

$$(\alpha_n^p y_j'')'' = \lambda_j \alpha_n y_j \, , \qquad j = 1,\ldots,n-1 \quad (5.2)$$

Equations 5.1 are conditions of orthogonality of $y_n(x)$ and the lower modes $y_j(x)$, j = 1,...,n-1. Equations 5.2 constitute, together with the boundary conditions, n-1 eigenvalue problems for the lower modes $y_j(x)$ of the optimum design α_n.

For p = 2, a solution α_n to this problem is at the same time the optimum design associated with a maximum value of a higher order critical angular speed of revolution ω_n for a rotating circular shaft, when gyroscopic effects are excluded.

A further significance of the type of problem under consideration, is that a geometrically unconstrained solution α_n to the problem coincidently constitutes the optimum design for the problem of maximizing the difference between two adjacent natural frequencies (or critical speeds) ω_n and ω_{n-1} for given volume and length of the beam (or rotating shaft) [7]. It is not surprising, therefore, that geometrically constrained solutions also exhibit large gaps between two adjacent frequencies [6] and that these gaps are very close to the maximum obtainable gaps. The point is that the geometrically constrained problem of maximum, single, higher order natural frequency is simpler to solve than the

problem formulated in terms of maximum difference between two
adjacent natural frequencies. A formulation of the latter problem
is available in Ref. 18.

Thus, the type of problem considered here is directly related
to practical design against resonance of beams due to external
excitation and whirling instability of rotating shafts. If the
structure is designed such that the external excitation frequency
or the service speed is isolated in a broad gap between two consecutive higher order natural frequencies or higher order critical speeds, considerable weight savings is possible, compared with
designs in which the excitation frequency or service speed is
placed between zero and a high value of the fundamental natural
frequency or first critical speed.

Another direct and very important advantage of considering
optimization (geometrically constrained as well as unconstrained)
with respect to a single, higher order natural frequency (or
critical speed) of given order n is that the resulting optimum
design is, at the same time, the optimal design for the problem
of optimizing with respect to the fundamental natural frequency
(or first critical speed), assuming the positions of n-1 available
interior supports to be design variables, in addition to the
cross-sectional area distribution [6,7]. According to Mróz and
Rozvany [19] and Rozvany [20] , and as was derived in Section 7
of Ref. 16, zero support reaction is a necessary condition for
optimum location of an interior simple support. This implies [6,7]
that the optimum positions for available interior supports in a
problem of optimal design with respect to the fundamental frequency, are simply identified with the n-1 nodal points of the
vibration mode $y_n(x)$ of the higher order frequency optimal design.

Figure 5.1 illustrates, as an example, the geometrically
unconstrained design with optimal positions of four available
inner supports that maximize the fundamental frequency of a transversely vibrating, p = 2, beam with clamped and free end points.
The design is determined by optimizing the beam without inner
supports with respect to its fifth eigenfrequency. The inner
supports are subsequently optimally placed at the four nodal
points of the corresponding mode.

5.1 Results of Geometrically Constrained Optimal Design

Consider now some examples of geometrically constrained
solutions for transversely vibrating Bernoulli-Euler cantilever
beams (p = 2) and rotating, cantilevered circular shafts from
Ref. 6. Results for Timoshenko beams are available in Ref. 21.

Figure 5.1.   Optimum Design and Interior Support Placement
for a Vibrating Beam

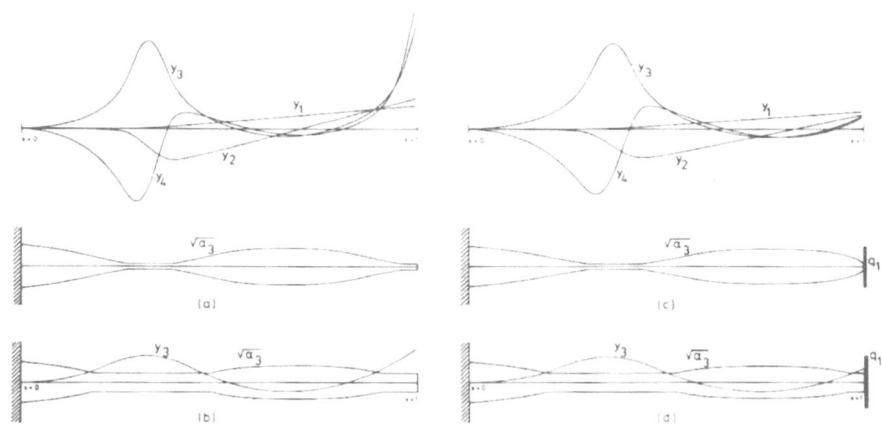

Figure 5.2.   Constrained Cantilever Beams Optimized for
Third Natural Frequency

Figure 5.2 shows cantilever beams optimized with respect to
the third natural frequency ω_3. The optimum designs are associated with p = 2 and are gometrically constrained with $\bar{\alpha} = \bar{A}L/V$
= 0.05 for Figs. 5.2(a) and 5.2(c), and $\bar{\alpha}$ = 0.5 for Figs. 5.2(b)
and 5.2(d), respectively. Designs in Figs. 5.2(c) and 5.2(d) are
equipped with a dimensionless tip mass, $q_1 = Q_1/\rho V = 0.1$. The
first four vibration modes of the optimum designs of Figs. 5.2(a)
and 5.2(c) are also shown in Fig. 5.2. The natural frequencies
ω_3 of the optimum beams are increased by (a) 129%, (b) 39%,
(c) 82%, and (d) 28%, respectively, when compared with the same
frequency of uniform beams of the same volume, length, material,
and, for Figs. 5.2(c) and 5.2(d), tip mass. The frequency differences $\omega_3 - \omega_2$ of the optimum beams are (a) 228%, (b) 53%, (c) 156%,
and (d) 41% higher than the corresponding frequency differences
of the uniform beams.

192

Figure 5.3 summarizes, for n = 3, the results for a number of
optimum solutions of the types shown in Fig. 5.2. Adopting the
concept of given volume V and length L, the square roots of the
eigenvalues in Fig. 5.3 directly represent the natural frequencies
of the optimum designs associated with n = 3.   The results in
Fig. 5.3 may be compared with corresponding results for n = 1 in
Fig. 3.2, noting especially the different ordinate scales and the
different type of behavior of the subject eigenfrequencies of the
mass-carrying beams, substantiating advantages of optimizing with
respect to higher order eigenfrequencies.   Similarly, the optimum
n = 3 designs in Fig. 5.2 may be compared with the n = 1 designs
in Fig. 3.1.

As is illustrated for n = 3 in Fig. 5.3, one finds for any
value of n > 1 that both the optimum natural frequency ω_n and the
distance between the consecutive natural frequencies ω_n and ω_{n-1}
increase with decreasing geometric constraint.   In fact, the
absolute values of the differences between ω_n and ω_{n-1} increase

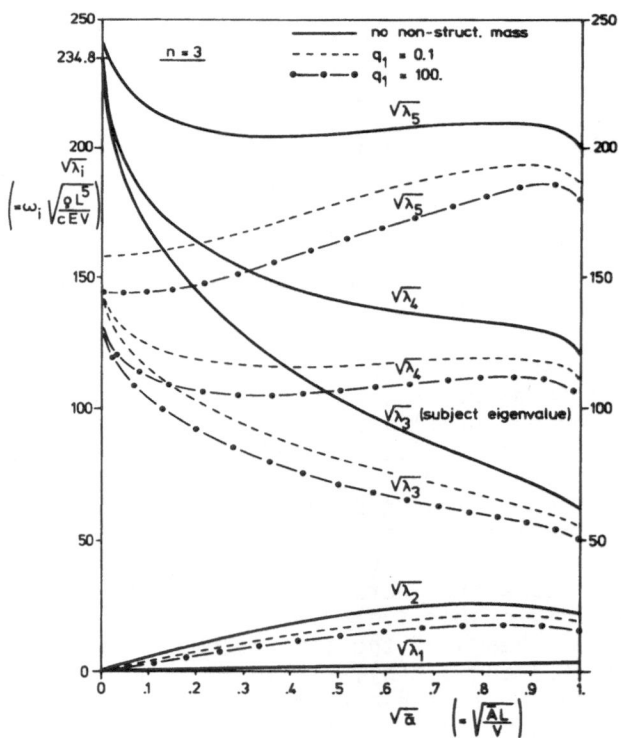

Figure 5.3. Square Root Eigenvalues of Optimal λ_3 Cantilevers
vs. $\sqrt{\bar{\alpha}}$.

with increasing n, irrespective of whether the beams are equipped with nonstructural mass or not.

If one optimizes cantilevers for n > 3, not only is the subject frequency ω_n increased, but also the closest subsequent natural frequencies ω_{n+1}, ω_{n+2}, etc. are pushed upwards. The subspectrum consisting of these natural frequencies becomes significantly condensed for optimum solutions associated with small geometric constraint. In the limiting case of geometrically unconstrained optimization, the subject frequency may be increased to the extent that it coalesces with one or more of the subsequent natural frequencies [6], as shown in Fig. 5.3. For the problems considered in Ref. 6, coalescence of the subject eigenvalue with one or more higher order eigenvalues is always found to take place in the limiting case of geometrically unconstrained design ($\bar{a} = 0$). This implies the advantage that it is not necessary to apply a bi- or multimodal formulation in order to obtain the correct optimal design.

Another characteristic feature of optimizing with respect to a higher order natural frequency ω_n, n > 2, is that all the lower natural frequencies ω_j, j = 1,...,n-1, are kept small by this process. Further, they tend toward a multiple zero eigenvalue as the geometric minimum constraint tends towards zero, as shown in Fig. 5.3.

5.2 Geometrically Unconstrained Optimal Design

Geometrically unconstrained solutions are important because they constitute limiting designs for corresponding geometrically unconstrained designs, and their associated optimum eigenvalues constitute upper bound values for corresponding eigenvalues of all similar beams, with and without nonstructural mass. Some geometrically unconstrained optimum designs obtained in [7] for transversely vibrating Bernoulli-Euler beams with p = 2 are now considered.

Solutions are determined numerically in Ref. 7, on the basis of a formal integration of the geometrically unconstrained formulation for optimal design, with possible types of singular behavior appropriately allowed in the numerical solution procedure. The positions as well as the Types (I and II) of the singularities are additional design variables in the optimization process. The optimum solutions are, in fact, determined via a path through a class of geometrically unconstrained, suboptimal solutions.

Figure 5.4 provides an illustration of this in the case of optimizing a cantilever for n = 3. The solutions in Fig. 5.4(a) are both suboptimal solutions. The dashed solution has inner

194

Type I singularities at x = 0.218 and x = 0.418 and the eigenvalue
λ_3 is $\lambda_3 = 1.77 \cdot 10^4$. For the solid curves in Fig. 5.4(a),
Type I singularities are placed at x = 0.268 and x = 0.368, and
the eigenvalue is increased to $\lambda_3 = 2.87 \cdot 10^4$. Figures 5.4(b)
shows the resulting optimum solution with $\lambda_3 = 5.511 \cdot 10^4$. This
solution has an inner Type II singularity, which is optimally
placed at x = 0.321, and the optimum frequency ω_3 of the design is
increased by 280%, in comparison with the corresponding frequency
of a uniform cantilever of the same volume, length, and material.
It should be noted that it is the design in Fig. 5.4(b) that lies
behind the result for $\bar{\alpha} = 0$ in Fig. 5.3 and that it may be compared to corresponding constrained optimum designs in Figs. 5.2(a)
and 5.2(b).

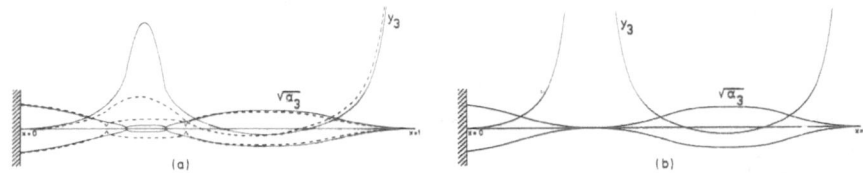

Figure 5.4.   Unconstrained Cantilever Beams Optimized for
              Third Natural Frequency

Each Type I and Type II singularity introduces, respectively,
one and two degrees of kinematic freedom to a geometrically unconstrained beam. When optimizing with respect to the n'th natural
frequency ω_n (n > 1), it turns out [7] that the optimum design
possesses exactly n - 1 degrees of kinematic freedom to perform
rigid body motions. Thus, all natural frequencies lower than the
subject natural frequency of a geometrically unconstrained optimum
design correspond to rigid body motion[*] and attain zero value,
cf. Fig. 5.3, with $\bar{\alpha} = 0$.

This clearly implies that a geometrically unconstrained solution to the problem of optimizing the n'th natural frequency ω_n
(n > 1) is coincidently the solution to the problem of optimizing
the difference between the two adjacent frequencies ω_n and ω_{n-1}.

The conditions of Section 4.2 for optimal location of the
two possible types of inner singularities are also valid for

---

[*]
Note in Fig. 5.2 the tendency of the lower vibration modes
y_1 and y_2 to become rigid body motions in the limiting case of
zero constraint. These modes indicate the two degrees of kinematic freedom associated with an inner Type II singularity,
namely jumps in both deflection and slope.

optimization with respect to higher order natural frequencies.
These conditions are illustrated by optimally placed inner Type I
singularities in Figs. 5.5(a), 5.5(b), and 5.7(b) and by optimally
placed innter Type II singularities in Figs. 5.4(b), 5.5(b), and
5.6(b). Figures 5.5(a) and 5.5(b) show optimum cantilevers for
n = 2 and n = 4, respectively. Figures 5.6(a) and 5.6(b) show
optimum simply supported beams for n = 2 and n = 3, respectively,
and Figs. 5.7(a) and 5.7(b) illustrate free beams optimized for

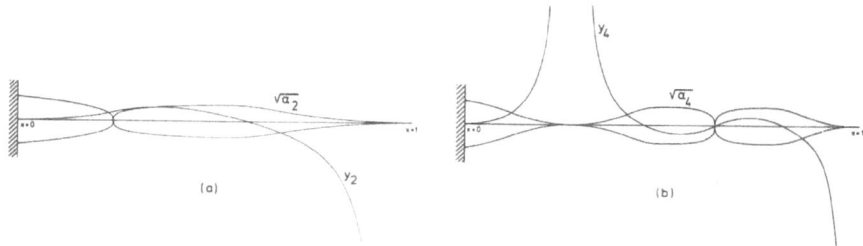

Figure 5.5. Optimum Cantilevers for n = 2 and n = 4

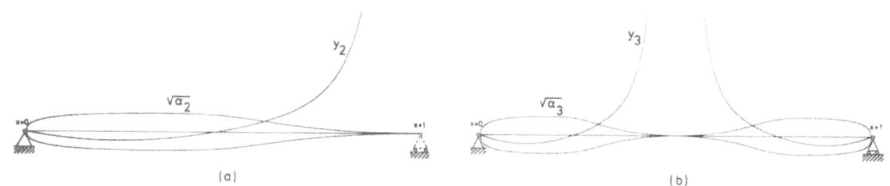

Figure 5.6. Optimum Simply Supported Beams for n = 2 and n = 3

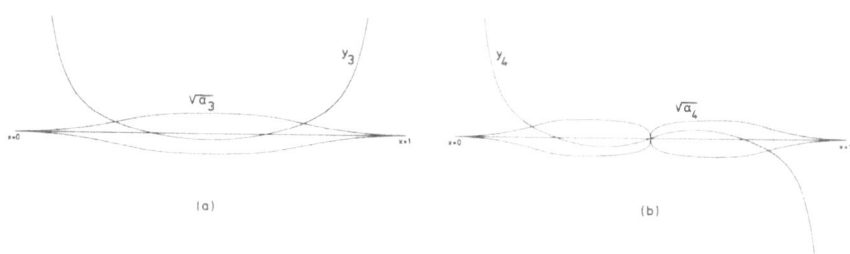

Figure 5.7. Optimum Free-Free Beams for n = 3 and n = 4

n = 3 and n = 4, respectively. The first two natural frequencies
of free beams correspond to rigid body motion, and are always zero.

The study in Ref. 7 revealed the notable feature that two
Type I singularities may coalesce, thereby forming one Type II
singularity. Figure 5.4 provides an illustration of such coalescence at an inner point. Note also that the inner beam separation
formed at the resulting Type II singularity of the optimum n = 3
cantilever in Fig. 5.4(b) in fact divides this beam into a scaled
optimum cantilever corresponding to n = 1, see Fig. 4.1(b), and a
scaled optimum free beam corresponding to n = 3, as shown in
Fig. 5.7(a). The coalescence may also take place at a beam end
point, e.g. the Type II singularity of the simply supported,
optimum n = 2 beam in Fig. 5.6(a) results from an original inner
Type I singularity coalescing with an a priori singularity of
Type I at the beam end.

The formation of Type II singularities is found to contribute
considerably to high subject eigenvalues. Singularities of this
type consequently play a predominant role in geometrically unconstrained optimal design. Thus, no more than one inner Type I
singularity is found in any of the designs [7].

In fact when optimizing beams of a given type of end conditions, it is only necessary to apply the numerical solution procedure for values of n up to a particular value, N, where the first
Type II singularity occurs in the corresponding optimum design.
As discussed and outlined in Ref. 7, the inner Type II singularities open up the possibility of determining the optimum designs
associated with higher values of n, simply by assembling optimum
beam elements obtained numerically for small values of n. A
so-called method of scaled beam elements is developed in Ref. 7,
by means of which optimum designs subject to any higher value of
n are easily determined for beams with any combination of clamped,
simply supported, and free end conditions.

Some geometrically unconstrained, p = 2 designs corresponding
to small values of n, determined numerically in Ref. 7 for the
combinations of end conditions mentioned, are shown in Figs. 4.1,
5.4(b), 5.5, 5.6, and 5.7 of the present paper. Their optimum
eigenvalues λ_n are all given in Ref. 7. For higher values of n,
the compositions of optimum, p = 2 beams with the sets of end
conditions considered are indicated in Table 5.1 (from Ref. 7).
Their associated optimum eigenvalues λ_n, n = N + 2j and N + 2j + 1,
with j = 0,1,..., are given by

$$\lambda_{N+2j} = \left[\lambda_N^{1/4} + j \cdot 10.4048\right]^4 , \qquad j = 0,1,... \qquad (5.3)$$

$$\lambda_{N+2j+1} = \left[\lambda_{N+1}^{1/4} + j \cdot 10.4048\right]^4 , \qquad j = 0,1,... \qquad (5.4)$$

TABLE 5.1 COMPOSITIONS AND DATA FOR OPTIMUM, p = 2 BEAMS CORRESPONDING TO n ≥ N (from Ref. 7)

Beam type	Free beam	S. supp.-free beam	S. supp. beam	Cantilever	Clamped-s. supp. beam	Doubly clamped beam
N	5	4	3	3	2	1
λ_N	$1.875\cdot10^5$	$8.611\cdot10^4$	$3.274\cdot10^4$	$5.511\cdot10^4$	$1.837\cdot10^4$	$9.350\cdot10^3$
λ_{N+1}	$3.239\cdot10^5$	$1.875\cdot10^5$	$8.611\cdot10^4$	$1.138\cdot10^5$	$5.511\cdot10^4$	$\lambda_{N+3} = 2.939\cdot10^5$
Optimal beam composition for n=N+2j , j=0,1,...	$\dfrac{cc..cc}{j}$	$\dfrac{bc..cc}{j}$	$\dfrac{bc..cb}{j}$	$\dfrac{ac..cc}{j}$	$\dfrac{ac..cb}{j}$	$\dfrac{ac..ca}{j}$
Optimal beam composition for n=N+2j+1 , j=0,1,...	$\dfrac{cc..cC}{j}$	$\dfrac{Bb..cc}{j}$	$\dfrac{bc..cB}{j}$	$\dfrac{ac..cC}{j}$	$\dfrac{ac..cB}{j}$	j=0 : Aa; j≥1 : $\dfrac{aCc..ca}{j-1}$

Optimal beam elements

$\lambda^a = 5.844\cdot10^2$ $\lambda^A = 3.540\cdot10^3$ $\lambda^b = 2.046\cdot10^3$ $\lambda^B = 1.172\cdot10^4$ $\lambda^c = 1.172\cdot10^4$ $\lambda^c = 1.172\cdot10^4$ $\lambda^C = 3.274\cdot10^4$

a   A   b   B   c   c   C

and appropriate values of N, λ_N, and λ_{N+1} are available in Table 5.1. Equations 5.3 and 5.4 hold for the beam types included in Table 5.1, with the single exception that for the doubly clamped beam, $\lambda_2 = 2.545 \cdot 10^4$ and its optimum eigenvalues λ_{N+2j+1}, j = 1,2,..., are given by

$$\lambda_{N+2j+1} = \left[\lambda_{N+3}^{1/4} + (j-1) \cdot 10.4048 \right]^4 , \quad j = 1,2,... \quad (5.5)$$

instead of Eq. 5.4.

The individual volumes v^i and lengths ℓ^i of the longitudinally and transversely scaled optimum elements a, A, b, B, c, and C (see Table 5.1), meeting overall structural volume V and length L, are determined by Ref. 7 as

$$\frac{v^i}{V} = \frac{\ell^i}{L} = \left[\frac{\lambda^i}{\lambda_n} \right]^{1/4} \quad (5.6)$$

where superscript i denote lower or upper case letter of a particular beam element and where the values of λ^i for individual elements are given in Table 5.1.

REFERENCE

1. Niordson, F.I., "On the Optimal Design of a Vibrating Beam," Quart. Appl. Mech., Vol. 23, 1965, pp. 47-53.
2. Pierson, B.L., "A Survey of Optimal Structural Design under Dynamic Constraints," Int. J. Num. Meth. Engrg., Vol. 4, 1972, pp. 491-499.
3. Reitman, M.I. and Shapiro, G.S., "Optimization of Structures under Dynamic Loads (in Russian)," The All-Union Symposium On the Problems of Optimization in Mech. of Solid Deformable Bodies, Vilnius Civil Engineering Inst., Vilnius, USSR, June 1974.
4. Rao, S.S., "Optimum Design of Structures under Shock and Vibration Environment," Shock Vib. Dig., Vol.7, No. 12, 1975, pp. 61-70.
5. Olhoff, N., "A Survey of the Optimal Design of Vibrating Structural Elements," Shock Vib. Dig., Vol. 8, No. 8, 1976, pp. 3-10 (Part I: Theory); Vol. 8, No. 9, 1976, pp. 3-10 (Part II: Applications).
6. Olhoff, N., "Maximizing Higher Order Eigenfrequencies of Beams with Constraints on the Design Geometry," J. Struct. Mech., Vol. 5, 1977, pp. 107-134.

7. Olhoff, N., "Optimization of Vibrating Beams with Respect to Higher Order Natural Frequencies," J. Struct. Mech., Vol. 4, 1976, pp. 87-122.
8. Brach, R.M., "On the Extremal Fundamental Frequenices of Vibrating Beams," Int. J. Solid Structures, Vol. 4, 1968, pp. 667-674.
9. Brach, R.M., "On the Optimal Design of Vibrating Structures," J. Optimization Theory and Appl., Vol. 11, 1973, pp. 662-667.
10. Karihaloo, B.L. and Niordson, F.I., "Optimum Design of Vibrating Cantilevers," J. Optimization Theory and Appl., Vol. 11, 1973, pp. 638-654.
11. Kamat, M.P. and Simitses, G.J., "Optimal Beam Frequencies by the Finite Element Displacement Method," Int. J. Solids Structures, Vol. 9, 1973, pp. 415-429.
12. Vepa, K., "On the Existence of Solutions to Optimization Problems with Eigenvalue Constraints," Quart. Appl. Mech., Vol. 31, 1973-1974, pp. 329-341.
13. Karihaloo, B.L. and Niordson, F.I., "Optimum Design of Vibrating Beams under Axial Compression," Archives of Mechanics, Vol. 24, 1972, pp. 1029-1037.
14. Seiranyan, A.P., "Optimal Beam Design with Limitations on Natural Vibration Frequency and Buckling Load (in Russian)," MTT (Mechanics of Solids), Vol. 11, No. 1, 1976, pp. 147-152.
15. Olhoff, N. and Niordson, F.I., "Some Problems Concerning Singularities of Optimal Beams and Columns," ZAMM, Vol. 59, 1979, pp. T16-T26.
16. Olhoff, N., "Optimization of Columns Against Buckling," Optimization of Distributed Parameter Structures (Eds. E.J. Haug and J. Cea), Sijthoff & Nordhoff, Alphen aan den Rijn, Netherlands, 1980.
17. Masur, E.F., "Written discussion to paper by N. Olhoff: On Singularities, Local Optima and Formation of Stiffeners in Optimal Design of Plates," Proc. IUTAM Symposium on Optimization in Structural Design (Ed. A. Sawczuk and Z. Mróz), Springer-Verlag, 1975, pp. 82-103.
18. Troitskii, V.A., "Optimization of Elastic Bars in the Presence of Free Vibrations (in Russian)," MTT (Mechanics of Solids), Vol. 11, No. 3, 1976, pp. 145-152.
19. Mróz, Z. and Rozvany, G.I.N., "Optimal Design of Structures with Variable Support Conditions," J. Optimization Theory Appl., Vol. 15, 1975, pp. 85-101.
20. Rozvany, G.I.N., "Analytical Treatment of Some Extended Problems in Structural Optimization, Parts I and II," J. Struct. Mech., Vol. 3, 1974-75, pp. 359-402.
21. Pierson, B.L., "An Optimal Control Approach to Minimum-Weight Vibrating Beam Design," J. Struct. Mech., Vol. 5, 1977 pp. 147-178.

SINGULAR PROBLEMS OF OPTIMAL DESIGN

Ernest F. Masur*

Department of Materials Engineering, University of
Illinois (Chicago Circle), Chicago, Illinois 60680

ABSTRACT

Although a structure whose design is to be optimized is
continuous, the result of the optimization process may be discontinuous or singular. Four cases are discussed: discontinuous
deformations at points (or regions) of vanishing stiffness, discontinuous design specifications (e.g. the placing of cover
plates), multiple roots of eigenvalue problems, and the imposition
of multiple constraints. It is demonstrated that the probability
of nonanalyticity increases with increasing complexity of the
structure.

1. INTRODUCTION

This paper deals with continuous structural members, such as
beams, plates, shells, etc. It is reasonable to expect that the
problem of optimizing the design of sub structural elements should
lead to solutions that are continuous and analytical. Nevertheless, there are a number of cases of common concern in which the
solution becomes singular. Specifically, these cases include the
development of discontinuities in the deformation field at points
of vanishing cross section (or of hinges), the effect of prescribed design discontinuities, the presence of multiple eigenvalues, and the imposition of multiple constraints.

---

*Currently on leave with the U.S. National Science Foundation.

The discussion here is cast into the framework of generalized stress vectors and associated generalized strain vectors e. The compliance density matrix K is introduced, as a function of the design parameter h, with e = Ks. The cost density is given by C, which is again a function of h. Suppose the problem is to minimize the deflection at a given point P by choosing a suitable design for given total cost. This is equivalent to finding the minimum of

$$J = \int_D s^T K(h)\bar{s} \, dV - b^2 \int_D C(h)dV, \tag{1.1}$$

in which $\bar{s}$ represents the stress vector in the structure associated with a unit load at point P, b^2 is a Lagrangian multiplier, and D is the physical domain over which the structure is distributed. Note that the problem of minimizing the total compliance is included in Eq. 1.1 by setting $\bar{s}$ = s.

If the design h is varied by an amount $\dot{h}$, then the variation of J is given by

$$\dot{J} = \int_D \dot{s}^T K\bar{s} \, dV + \int_D \left(s^T \frac{dK}{dh} \bar{s} - b^2 \frac{dC}{dh} \right) \dot{h} \, dV$$

$$+ \int_D s^T K\dot{\bar{s}} \, dV \tag{1.2}$$

If there are no singularities, the first integral represents the internal work of a self-equilibrated stress system over a compatible strain field and therefore vanishes. The third integral vanishes for the same reason. If J is to be a minimum, that is, for

$$\dot{J} \geq 0 \tag{1.3}$$

it follows by the usual argument of the calculus of variations that

$$s^T \frac{dK}{dh} \bar{s} - b^2 \frac{dC}{dh} = 0, \qquad x \in D \tag{1.4}$$

provided $\dot{h}$ is unconstrained. If there is a constraint on h, e.g. $0 \leq h_0 = h < h_1$, then the equality in Eq. 1.4 is replaced by the appropriate inequality. Equation 1.4 is a well-known optimality condition and implies, for minimum compliance and given volume, that the average stress energy density in the critical fibers must be constant [1,2,3].

Note, for future reference, that without significant loss of generality one may set $C(0) = K(0) = 0$. Also note that for $h > 0$, $C > 0$ and K is positive definite. Finally, one may introduce the displacement vector u through

$$\left. \begin{array}{l} e = \ell(u) \\ \bar{e} = \ell(\bar{u}) \end{array} \right\} \tag{1.5}$$

in which ℓ is a linear operator and the barred quantities are once again associated with the case of a unit load at the point at which the deflection is to be minimized.

2. DISCONTINUOUS DEFORMATION

The first type of singularity considered is a discontinuity in the deformation, which may occur over some interior boundary B_C (identified by the coordinates x_C) if the design is unconstrained. Assume that $h = 0$ over B_C. Then

$$s_N = \bar{s}_N = 0 \tag{2.1}$$

over the same region, in which s_N and $\bar{s}_N$ represent, respectively, the normal components of the stress vectors s and $\bar{s}$ across B_C. It is noted, moreover, that on B_C the displacements u and $\bar{u}$ are not necessarily continuous, their discontinuities being expressed by [u] and [$\bar{u}$]. For example, in the case of the bending of beams, B_C may be represented by a hinge and the deflection curve may then exhibit a discontinuity in the slope θ.

In the case of discontinuous deformation, the vanishing of the first and last integrals in Eq. 1.2 is replaced by

$$\int_D \dot{s}^T K \bar{s} \, dV = - \int_{B_C} [\bar{u}]^T \dot{s}_N \, dS \tag{2.2}$$

$$\int_D s^T K \dot{\bar{s}} \, dV = - \int_{B_C} [u]^T \dot{\bar{s}}_N \, dS \tag{2.3}$$

It is noted that although s_N vanishes on B_C, the same may not be true of $\dot{s}_N$. In fact, if B_C is not prescribed and is allowed to move with a change in the design, then

$$s_N + \dot{s}_N = 0 \quad \text{on} \quad B_C + \dot{B}_C \ (x_C + \dot{x}_C) \tag{2.4}$$

or, after linearization,

$$\dot{s}_N + (s_N \nabla)\dot{x}_C = 0 \ , \quad x \in x_C \tag{2.5}$$

$$\dot{\bar{s}}_N + (\bar{s}_N \nabla)\dot{x}_C = 0 \ , \quad x \in x_C \tag{2.6}$$

where $(g\nabla) \equiv \left[\dfrac{\partial g_i}{\partial x_j}\right]$ is the Jacobian of a vector function $g = [g_1(x), \ldots, g_m(x)]^T$.

In view of Eqs. 1.2, 1.4, 2.2, 2.5, and 2.6,

$$\dot{J} = \int_{B_C} \left\{ [\bar{u}]^T (s_N \nabla) + [u]^T (\bar{s}_N \nabla) \right\} \dot{x}_C \ dS \tag{2.7}$$

and for unconstrained x_C, Eq. 1.3 is satisfied provided

$$[\bar{u}]^T (s_N \nabla) + [u]^T (\bar{s}_N \nabla) = 0 \ , \quad \text{on} \quad B_C \tag{2.8}$$

Once again, for constrained x_C the equality sign in Eq. 2.8 has to be replaced by the appropriate inequality [3,4].

If instead of the deflection at some point, the compliance is to be minimized, then $\bar{s}$ is replaced by s, and Eq. 2.8 becomes

$$[u]^T (s_N \nabla) = 0 \ , \quad \text{on} \quad B_C \tag{2.9}$$

which implies, except in special cases, that u is continuous. The exceptional cases have been discussed in Refs. 5 and 6.

Consider, for example, the case of a sandwich beam of given volume V and of length L, which is fixed at both ends and subjected to a uniform load p, and whose central deflection is to be minimized. The optimality condition corresponding to Eq. 2.8 and governing the location of the hinges is then

$$\frac{dM}{dx} [\bar{\theta}] + \frac{d\bar{M}}{dx} [\theta] = 0 \ ,$$

in which M and $\bar{M}$ are the bending moments in the beam due to the actual load and a unit load at the center of the beam, respectively. As has been shown in Ref. 4, for the optimum design hinges develop very near the quarter points and the discontinuity in the slopes at these hinges is negligible. The same conclusion was reached previously [7], through other means and without the use of the optimality condition given above.

A problem leading to less trivial results is posed by optimal design of a sandwich beam that is simply supported at one end and fixed at the other, which is once again subjected to a uniform load. The beam, which is of depth 2H and which consists of two cover plates of thickness t (to be designed), is to exhibit the smallest possible rotation θ_0 at the simply supported end. The solution of this problem [4] leads to a rotation at the simply supported end given by 0.0167 pL^4/EH^2V, which represents a reduction of almost one fourth, compared with its prismatic counterpart. Of interest here is the fact that at the hinge at x_C = 0.685 L, there is a discontinuity in the slope that is given by [θ_C] = -0.0018 pL^4/EH^2V, which is of the same order of magnitude as the average slope of the beam. If the compliance is to be minimized, i.e. if $\bar{M}$ is replaced by M, then there is no discontinuity in the slope. The same applies to plates, as was pointed out in Refs. 5 and 6.

3. DESIGN DISCONTINUITY

As another example of singular behavior, consider the case in which a certain design discontinuity is prescribed and the object is to place the discontinuity in such a way as to optimize the structure with respect to some given criterion. For example, the discontinuity may be in the nature of the addition of a cover plate of prescribed thickness but unknown location. It may also include, for example, the case of a structure that is to be built out of two different materials, the location of the interface being the objective of the optimization process.

Once again, designate the internal boundary (to be found) by B_C and note that both K and C may be discontinuous on B_C. In this case, Eq. 1.2 is replaced by

$$\dot{J} = \int_D \dot{s}^T K \bar{s}\, dV + \int_D \left(s^T \frac{dK}{dh} \bar{s} - b^2 \frac{dC}{dh} \right) \dot{h}\, dV$$

$$+ \int_D s^T K \dot{\bar{s}}\, dV - \int_{B_C} [s^T K \bar{s} - b^2 C]\, \dot{x}_{CN}\, dS \qquad (3.1)$$

where $\dot{x}_{CN}$ represents the normal component of the displacement of the internal boundary.

An equation of the type of Eq. 3.1 was first presented in Ref. 4. As has recently been pointed out in Ref. 8, however, it is important to note, but was overlooked in Ref. 4, that although s_N and $\bar{s}_N$ are continuous across the internal interface, their variations $\dot{s}_N$ and $\dot{\bar{s}}_N$ need not be continuous. As a result, the first and third integrals in Eq. 1.2 once again do not vanish, but are given by

$$\int_D \dot{s}^T K \bar{s} \, dV = - \int_{B_C} \bar{u}^T [\dot{s}_N] \, dS \tag{3.2}$$

$$\int_D s^T K \dot{\bar{s}} \, dV = - \int_{B_C} u^T [\dot{\bar{s}}_N] \, dS \tag{3.3}$$

Moreover, through an argument that is similar to the one employed in establishing Eqs. 2.5 and 2.6, one obtains

$$[\dot{s}_N] + (s_N \nabla) \dot{x}_C = 0 \,, \quad \text{on} \quad B_C \tag{3.4}$$

$$[\dot{\bar{s}}_N] + (\bar{s}_N \nabla) \dot{x}_C = 0 \,, \quad \text{on} \quad B_C \tag{3.5}$$

When this is substituted in Eqs. 3.1, 3.2, and 3.3, after considerable manipulation that is far from trivial and that follows a scheme first employed in Ref. 8, one obtains

$$\dot{j} = \int_D \left\{ \dot{s}^T K \bar{s} + \left(s^T \frac{dK}{dh} \bar{s} - b^2 \frac{dC}{dh} \right) \dot{h} + s^T K \dot{\bar{s}} \right\} dV$$

$$+ \int_{B_C} \left\{ [\bar{s}_T]^T (Ks)_T - s_N{}^T [K\bar{s}]_N + b^2 [C] \right\} \dot{x}_{CN} \, ds$$

$$= \int_C \left\{ \dot{s}^T K \bar{s} + \left(s^T \frac{dK}{dh} \bar{s} - b^2 \frac{dC}{dh} \right) \dot{h} + s^T K \dot{\bar{s}} \right\} dV$$

$$+ \int_{B_C} \left\{ [s_T]^T (K\bar{s})_T - \bar{s}_N{}^T [Ks]_N + b^2 [C] \right\} \dot{x}_{CN} \, dS \tag{3.6}$$

In Eq. 3.6, the subscripts T and N refer, respectively, to components tangential and normal to the interface. Use has been made of the fact that only the tangential components of the stresses and the normal components of the strains are discontinuous.

Once again, if $\dot{x}_{CN}$ is arbitrary then Eqs. 1.3 and 3.6

$$[\bar{s}_T]^T (Ks)_T - s_N^T [K\bar{s}]_N = [s_T]^T (K\bar{s})_T - \bar{s}_N^T [Ks]_N$$

$$= b^2 [C], \quad x \in x_C \tag{3.7}$$

If the compliance is to be minimized this reduces to

$$[s_T]^T (Ks)_T - s_N^T [K_s]_N = -b^2 [C], \quad x \in x_C \tag{3.8}$$

which was first established in Ref. 8.

A typical problem of the type discussed here is the case of a circular sandwich plate of depth 2H, with each face consisting of n cover plates of constant thickness and radius r_i. The problem is radially symmetric, it being assumed that the total thickness of the cover plates is t_i between the limits $r_{i-1} < r < r_i$, $i = 1,\ldots,n$, with $r_n = R$. The applied load p is constant and the problem is to find the radii r_i so as to make the total compliance a minimum. This problem is governed by

$$J = \frac{\pi}{2EH^2} \sum_{i=1}^{n} \frac{1}{t_i} \int_{r_{i-1}}^{r_i} (M_r^2 - 2\nu M_r M_t + M_t^2) r dr$$

$$- b^2 \pi \sum_{i=1}^{n} t_i (r_i^2 - r_{i-1}^2)$$

and Eq. 3.8 becomes

$$\left[\frac{1}{t} (M_r^2 - M_t)^2 \right]_{r=r_i} = -2b^2 EH^2 (t_{i+1} - t_i), \quad i = 1,\ldots,n-1$$

which can also be obtained directly for this specific problem. The analysis has been carried out in Ref. 8, under the assumption that all individual cover plates are of the same thickness, and the results are given in Fig. 3.1, in which w and q represent compliance and volume, respectively, in nondimensional form.

It is noted that the analysis in Ref. 8 follows the same procedure as the earlier analysis [4] of the same problem. In Ref. 4, however, the effects of the discontinuity of the variations of the normal stresses, i.e. Eqs. 3.4 and 3.5, were not

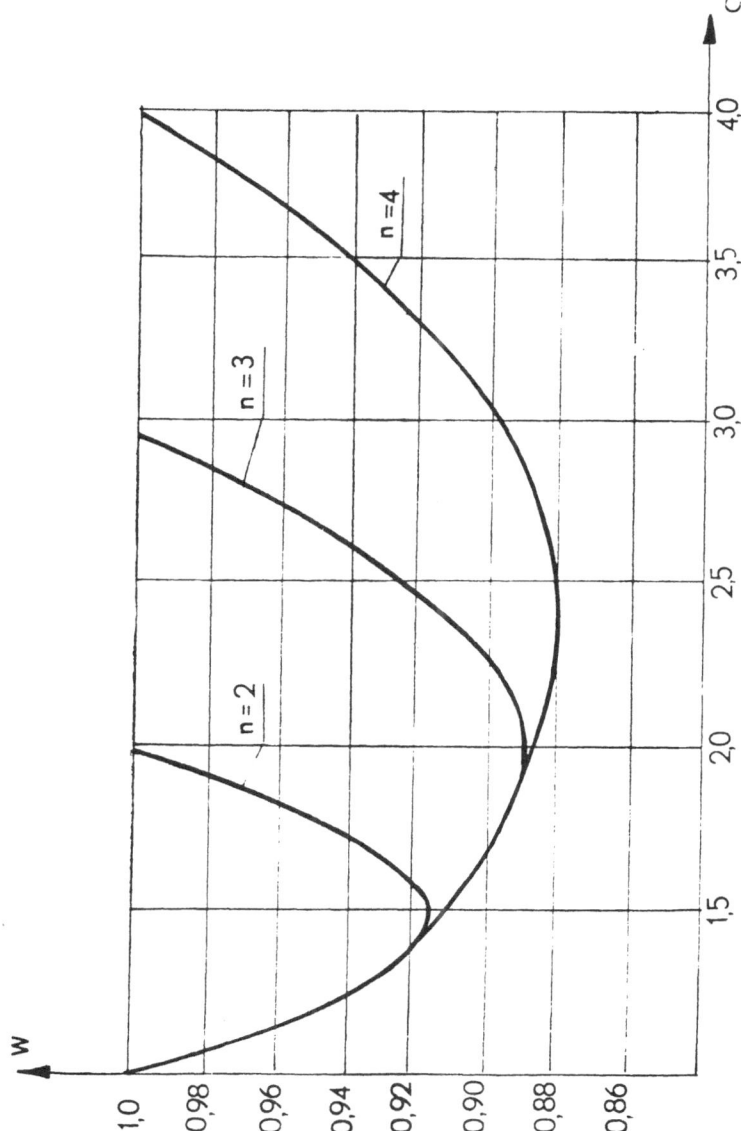

Figure 3.1 Compliance W vs. Volume q in a Circular Plate with n Cover Plates.

included and the results were therefore erroneous. On the other hand, the results carried out on equivalent beam problems or column buckling problems in Refs. 9 and 10 are correct, since in this case the normal stresses (i.e. the bending moments) are not only continuous, but have continuous derivatives at the points of discontinuities and their variations, by Eqs. 3.4 and 3.5, are therefore also continuous.

4.   EIGENVALUES

Singularities of a totally different type may occur in the case of eigenvalue problems. Suppose a structure of prescribed cost is to be designed in such a way that its buckling load (or smallest eigenvalue) is to be as large as possible. Similar problems occur in connection with vibrations, shaft instability, flutter, etc. In such problems, the optimization process may lead either to single eigenvalues or to multiple eigenvalues. In the first case the problem is ordinarily analytical, or station- ary, and in the second case it usually involves a type of singularity. The fact that a simple eigenvalue problem, in this case the problem of a column fixed at both ends, may lead to multiple eigenvalues was first discovered in Ref. 11.

If the optimum structure exhibits a simple smallest eigen- value $\lambda_1 < \tau_2 \leqq \ldots$, then the optimality condition is

$$\dot{\lambda}_1 \leqq 0 \tag{4.1}$$

which must hold for all admissible design variations $\dot{h}$ for which the cost is unchanged. If there are multiple eigenvalues $\lambda_1 = \lambda_2 \ldots = \lambda_n < \lambda_{n+1}$, then it is required that, for all variations $\dot{h}$ for which the cost is unchanged,

$$\min_i \dot{\lambda}_i \leqq 0, \quad i = 1,2,\ldots,n \tag{4.2}$$

If $\dot{h}$ is unconstrained, that is, if the admissibility of $\dot{h}$ implies also the admissibility of $-\dot{h}$, then the inequality of Eq. 4.1 may be removed, and the solution is therefore stationary. The same is not true in the case of Eq. 4.2. In particular, for unconstrained variation $\dot{h}$, Eq. 4.2 is equivalent to

$$\dot{\lambda}_1 \, \dot{\lambda}_2 \leqq 0 \tag{4.3}$$

for the case of n = 2. The retention of the inequality in Eqs. 4.2 and 4.3 is the cause of the singular character of the multi- ple eigenvalue solution.

A typical linear eigenvalue problem is governed by the quadratic form

$$P(u,\lambda,h) = \int_D \{Q_2(u,h) - \lambda W_2(u)\} \, dV \tag{4.4}$$

in which both Q_2 and W_2 are positive definite in u. Equation 4.4 represents a buckling problem. In the case of vibrations, W_2 may also be a function of the design parameter h. An extension of the discussion to this case is straightforward. Note that the eigenvalues λ_i and associated eigenmodes u_i are obtained by setting the first variation of P with respect to u equal to zero and by solving the resulting characteristic equation.

If the admissible modes are normalized by means of

$$\int_D W_2(u_i) \, dV = 1 \tag{4.5}$$

then for a simple eigenvalue

$$\dot{\lambda}_1 = \int_D \frac{\partial Q_2}{\partial h} (u_1,h)\dot{h} \, dV \tag{4.6}$$

and, for given cost, Eq. 4.1 corresponds to the optimality condition

$$\frac{\partial Q_2}{\partial h} (u_1,h) = b^2 \frac{dC}{dh} , \qquad x \in D \tag{4.7}$$

Now let the optimum design correspond to a dual eigenvalue $\lambda_1 = \lambda_2$. Then, because of the linear homogeneous nature of the eigenvalue problem, the associated eigenmode is not defined uniquely, but represents a linear combination of two solutions, say u_1^* and u_2^*, of the characteristic equation. Assume, without loss of generality, that u_1^* and u_2^* have been normalized in the sense of

$$\int_D Q_{11}(u_1^*,u_2^*,h)dV = \int_D W_{11}(u_1^*,u_2^*)dV = 0 \tag{4.8}$$

where Q_{11} and W_{11} are symmetric bilinear forms associated with the eigenvalue problem, such that $Q_2(u,h) = Q_{11}(u,u,h)$ and $W_2(u) = W_{11}(u,u)$.

It has been shown in Ref. 12 that the uncertainty regarding the eigenmode may be removed if the design variation $\overset{\circ}{h}$ is prescribed. That is, for a given $\overset{\circ}{h}$, one may take

$$\left.\begin{array}{l} u_1 = u_1^* \cos \phi + u_2^* \sin \phi \\[8pt] u_2 = -u_1^* \sin \phi + u_2^* \cos \phi \end{array}\right\} \qquad (4.9)$$

in which

$$\left.\begin{array}{l} \tan 2\phi = \dfrac{2\dot{\lambda}_{12}^*}{\dot{\lambda}_{11}^* - \dot{\lambda}_{22}^*} \\[16pt] \dot{\lambda}_{ij}^* = \displaystyle\int_D \dfrac{\partial Q_{11}}{\partial h} (u_i^*, u_j^*, h)\overset{\circ}{h} \ dV, \quad i,j = 1,2 \end{array}\right\} \qquad (4.10)$$

and hence

$$\dot{\lambda}_1 \dot{\lambda}_2 = \tfrac{1}{2}(\dot{\lambda}_{11}^* + \dot{\lambda}_{22}^*) \pm \tfrac{1}{2}\sqrt{(\dot{\lambda}_{11}^* - \dot{\lambda}_{22}^*)^2 + 4\dot{\lambda}_{12}^{*2}} \qquad (4.11)$$

where $\dot{\lambda}_i$ are obtained from Eq. 4.6, using u_i of Eq. 4.9 to obtain $\dot{\lambda}_i$. If Eq. 4.11 is substituted in Eq. 4.3, then for unconstrained design variations $\overset{\circ}{h}$, the latter is satisfied provided there is a constant $\gamma \in [0,1]$ such that [12]

$$(1 - \gamma) \frac{\partial Q_2}{\partial h} (u_1^*, h) + \gamma \frac{\partial Q_2}{\partial h} (u_2^*, h) = b^2 \frac{dC}{dh}, \quad x \in D \qquad (4.12)$$

Note that $\dot{\lambda}_1$ and $\dot{\lambda}_2$ given in Eq. 4.11 represent the eigenvalues of the 2 x 2 matrix $[\dot{\lambda}_{ij}^*]$. An extension to the case of multiple eigenvalues beyond n = 2 follows an analogous procedure (mutatis mutandis) and has recently been formulated in Refs. 13 and 14, which of course, is based on Eq. 4.2 rather than on Eq. 4.3.

An interesting example of startling simplicity is the case of the buckling of a column that is fixed at both ends. This problem was first studied in Ref. 15, but it has recently been shown in Ref. 11 that the earlier analysis, which was based on the concept of a simple eigenvalue, is in error and has to be replaced by a dual eigenvalue solution.

For this problem, the two quadratic forms are given by

$$Q_2(u,h) = EI(h)u''^2$$
$$W_2(u) = u'^2$$

(4.13)

where $' \equiv \frac{d}{dx}$, $C(h) = h$, and Eq. 4.7 then corresponds to

$$E \frac{dI}{dh} u''^2 = b^2, \quad x \in [0,L]$$

(4.14)

On the other hand, Eq. 4.12, with the asterisk deleted, corresponds to

$$E \frac{dI}{dh} \left[(1 - \gamma)u_1''^2 + \gamma u_2''^2 \right] = b^2, \quad x \in [0,L], \gamma \in [0,1]$$

(4.15)

subject to the normality conditions of Eqs. 4.8. Note that these restrictions, as well as the restriction on γ, are required if Eq. 4.12 is to be sufficient for optimality [12].

The solution of Eq. 4.14, as given in Ref. 15 is shown in Fig. 4.1. It was noted in Ref. 11 that this solution, though

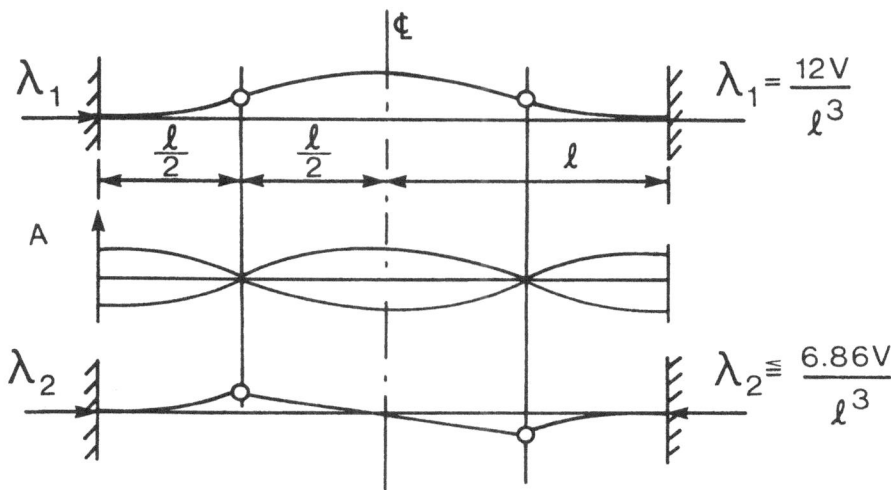

Figure 4.1   "Optimum" Fixed-Fixed Column According to Ref. 15 ("Optimum" Buckling Mode, Design, and Actual Buckling Mode).

stationary, is not optimum, since the mode shown in the top figure represents the second, rather than the first, buckling mode. A column designed as shown in Fig. 4.1 will instead buckle into the mode shown at the bottom of the figure, under a buckling load that is substantially smaller. In other words, λ_1 has become λ_2, and the process has led to optimizing the second eigenvalue, at considerable expense to the lowest eigenvalue. The correct solution, which is based on Eq. 4.15 and which was obtained in Ref. 11 for various values of h_0, is shown in Fig. 4.2 and contains no hinges. A schematic description of these relations is shown in Fig. 4.3.

On single and bimodal optimum buckling loads of clamped columns

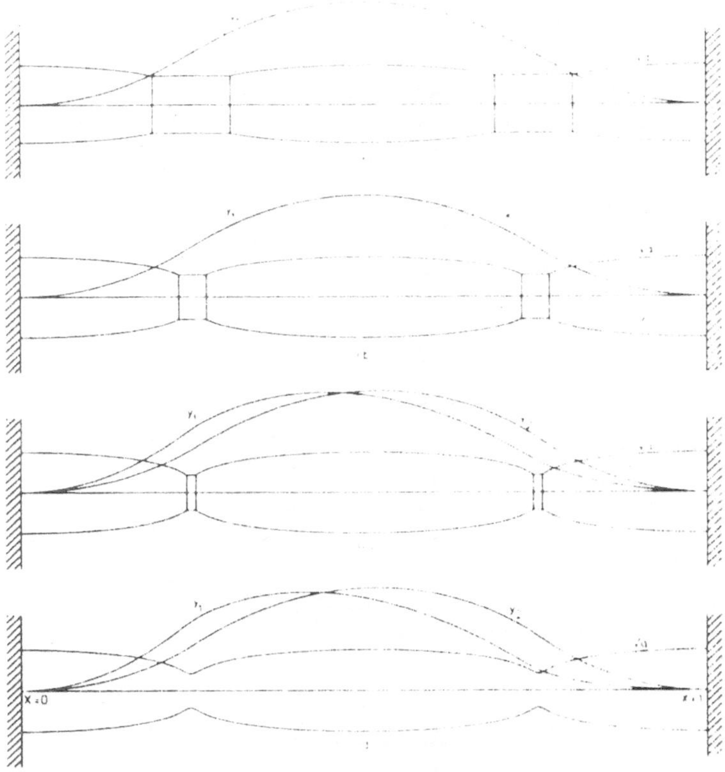

Figure 4.2   Optimum Designs and Associated Buckling Modes for Fixed-Fixed Column, According to Ref. 11 (for Various Minimum Design Constraints h_0).

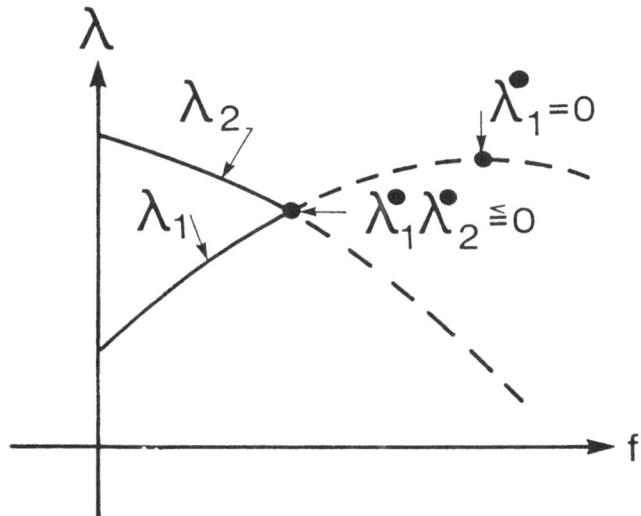

Figure 4.3   Schematic Picture of $\lambda_1 - \lambda_2$ Relationship.

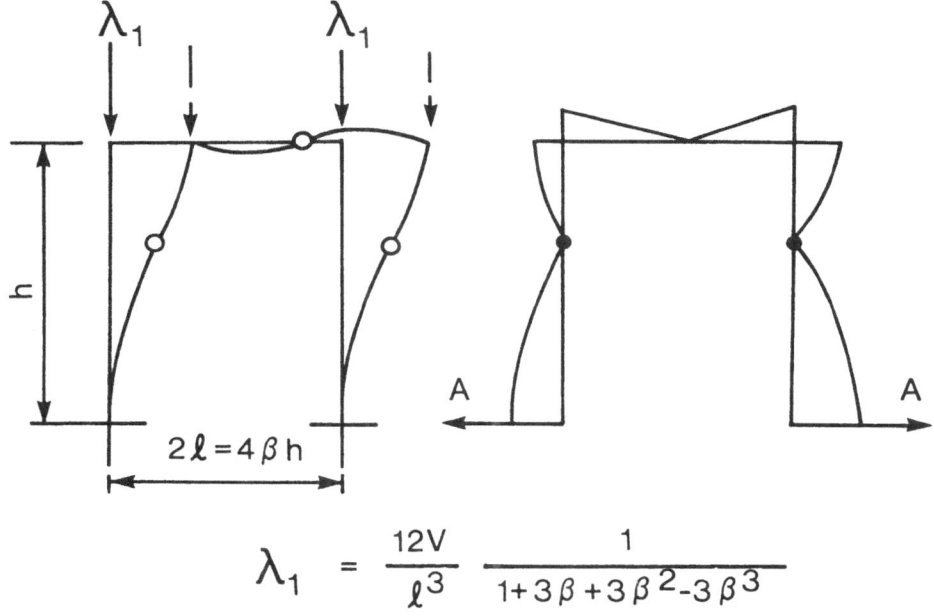

$$\lambda_1 = \frac{12V}{\ell^3} \; \frac{1}{1 + 3\beta + 3\beta^2 - 3\beta^3}$$

Figure 4.4   "Optimum" Design of Portal Frame Against Sidesway.

214

The case of multiple eigenvalues arising out of the process of optimization is by no means confined to the simple case of a fixed-fixed column. On the contrary, it becomes increasingly common for increasing structural complexity. For example, Fig. 4.4 shows the case of a portal frame that has been designed in such a way as to maximize the load λ that will produce buckling by sidesway. The optimum structure exhibits hinges in the two columns and at the center of the beam. However, a structure so designed will actually buckle into the symmetric pattern shown in Fig. 4.5, and under a load that is considerably smaller than the load associated with sidesway. Once again, the process of optimization based on the presumption of a simple buckling load has led to a crossing of the two eigenvalues and the optimum solution is instead to be found at the point where, as in Fig. 4.3, the two curves intersect. Further complexity in the structure is likely to introduce an even higher order of multiplicity in the eigenvalue solutions.

This section is concluded by considering optimal design of a ring that is subjected to constant radial pressure. As in all cases discussed previously, the straightforward application of the calculus of variations leads to a design that exhibits hinges. In the present case there are four hinges, placed at

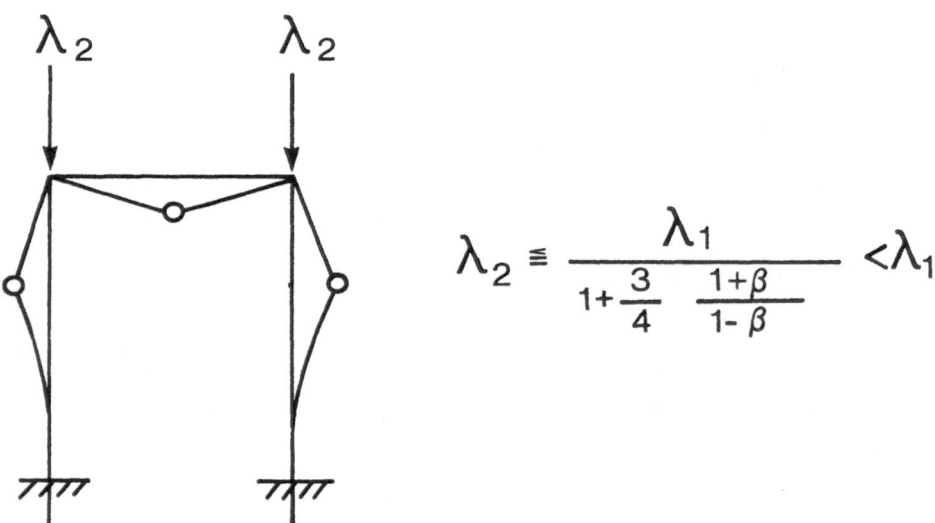

$$\lambda_2 \equiv \frac{\lambda_1}{1 + \dfrac{3}{4}\,\dfrac{1+\beta}{1-\beta}} < \lambda_1$$

Figure 4.5   Actual Buckling Mode of Frame Shown in Fig. 4.4.

equal distances, and the resulting "structure" is a mechanism with one degree of freedom. Once again, the process has led to an optimum solution of λ_2, for which λ_1 has shrunk to zero. The correct optimum design is therefore bimodal and singular.

5. MULTIPLE CONSTRAINTS AND CONCLUDING REMARKS

It is interesting to observe that the problems covered in Section 4 are of the max-min type, that is, they are concerned with determination of the largest possible value of the smallest element of a set. The opposite case, namely to design a structure with a view toward minimizing some maximum value leads to at least superficially analogous results. It occurs when optimization takes place subject to multiple constraints and has been covered widely in the literature. Representative examples are given in Refs. 16 and 17, while a fundamentally different approach, based on the properties of homogeneous functions has recently been introduced in Ref. 18.

Consider, for example, the problem of designing a structure of given cost, in such a way that the maximum displacement at any point is to be as small as possible. If the point of maximum displacement, as yet unknown, is identified by the position vector ξ, and if the stress field in the body due to a unit force at that point is identified by $\bar{s}(x,\xi)$ then with ξ itself being varied, Eq. 1.2 is now replaced by

$$\dot{J} = \int_D \{\dot{s}^T K \bar{s} + (s^T \frac{dK}{dh} \bar{s} - b^2 \frac{dC}{dh})\dot{h} + s^T K \dot{\bar{s}}\} \, dV$$

$$+ \int_D s^T K(\bar{s}\nabla_\xi)\dot{\xi} dV \tag{5.1}$$

in which ∇_ξ identifies the gradient with respect to ξ. If $\dot{\xi}$ is arbitrary, the optimality equations derived previously are now augmented by the condition

$$\int_D s^T K (\bar{s}\nabla_\xi) \, dV = 0 \tag{5.2}$$

It is of interest to note that the left side of Eq. 5.2 represents the displacement in the structure due to a set of dipoles applied at the point ξ.

If the problem is to minimize the maximum deflection in a beam, Eq. 5.2 becomes

$$\int_0^{\ell} \frac{M \frac{\partial \bar{M}}{\partial \xi}}{EI} \, dx = 0 \tag{5.3}$$

The left side in Eq. 5.3 now represents the deflection at ξ due to a unit couple at ξ, and Eq. 5.3 therefore constitutes the (not unreasonable) condition that the slope at the point of maximum deflection vanish. By differentiating Eq. 5.3 once again with respect to ξ, it can easily be shown, again not unexpectedly, that a local maximum occurs if $M(\xi) > 0$. The problem introduced here has been discussed in Ref. 19, but by using an alternate method that does not employ the optimality condition of Eq. 5.3.

As was already pointed out in Ref. 1, optimality problems with deflection constraints often have no solution if there is a reversal of the sign of the bending moment M. This is easy to demonstrate on a mathematical basis. From a physical point of view, this means that the maximum positive deflection can be made as small as desired, at the price, however, of increasing the maximum negative deflection.

The impasse can be resolved if the problem is restated to the effect that the maximum absolute value of the deflection is to be minimized. In this case, both positive and negative deflections (occurring, respectively, at ξ_1 and ξ_2) must be considered and, if both constraints are active, the solution is governed by the following equations:

$$\left[(1 - \zeta)\bar{M}(x,\xi_1) + \bar{M}(x,\xi_2) \right] \frac{M}{EI^2} \frac{dI}{dC} = b^2 \tag{5.4}$$

$$\int \frac{M \frac{\partial \bar{M}(x,\xi_i)}{\partial \xi_i}}{EI} \, dx = 0 \, , \quad i = 1,2 \tag{5.5}$$

In addition, it is required that $M(\xi_1)$ and $M(\xi_2)$ have opposite signs. The foregoing set of equations, covering non-stationary conditions of optimality, is typical of a large class of problems involving more than one constraint condition.

REFERENCES

1. Barnett, R.L., "Minimum Weight Design of Beams for Deflection," J. Engr. Mech. Div. ASCE, Vol. 87, No. EM2, 1961, pp. 75-109.

2. Masur, E.F., "Optimum Stiffness and Strength of Elastic Structures," J. Engr. Mech. Div, ASCE, Vol. 96, 1970, pp. 621-640.

3. Rozvany, G.I.N., Optimal Design of Flexural Systems, Pergamon Press, 1976.

4. Masur, E.F., "Optimality in the Presence of Discreteness and Discontinuity," Proc. IUTAM Symposium on Optimization in Structural Design, (Ed. A. Sawczuk and Z. Mroz), Springer-Verlag, 1975, pp. 441-453.

5. Olhoff, N., "On Singularities Local Optima and Formulation of Stiffeners in Optimal Design of Plates," Ibid., pp. 82-103. See also, discussion by E. F. Masur, Ibid.

6. Olhoff, N., "Optimization of Vibrating Beams with Respect to Higher Order Natural Frequencies," J. Struc. Mech., Vol. 4, 1976, pp. 87-122.

7. Huang, N.C., "On Principle of Stationary Mutual Complementary Energy and Its Application to Optimal Structural Design," J. Appl. Math. Phys. (ZAMP), Vol. 22, 1971, pp. 608-620.

8. Dems, K. and Mroz, Z., "Optimal Shape Design of Multi-Composite Structures," J. Struc. Mech., Vol. 8, 1980, No. 3, to appear, 1980.

9. Masur, E.F., "Optimal Structural Design for a Discrete Set of Available Structural Members," Comp. Methods in Appl. Mech. and Engr., Vol. 3, 1974, pp. 195-207.

10. Masur, E.F., "Optimal Placement of Available Sections in Structural Eigenvalue Problems," J. Optim. Theory and Appl., Vol. 15, 1975, pp. 69-84.

11. Olhoff, N. and Rasmussen, S.H., "On Single and Bimodal Optimum Buckling Loads of Clamped Columns," Int. J. Solids Struc., Vol. 13, 1977, pp. 605-614.

12. Masur, E.F. and Mroz, Z., "Singular Solutions in Structural Optimization Problems," IUTAM Conference on Variational Methods in Solid Mechanics, Springer-Verlag, to appear.

13. Haug, E.J., "Optimization of Distributed Parameter Structures with Repeated Eigenvalues," New Approaches to Nonlinear Problems in Dynamics (ed. P.J. Holmes), SIAM, to appear

14. Haug, E.J. and Rousselet, B., "Design Sensitivity Analysis in Structural Mechanics II: Eigenvalue Variations," J. Struc. Mech., Vol. 8, No. 2, to appear, 1980.

15. Tadjbakhsh, I. and Keller, J.B., "Strongest Columns and Isoperimetric Inequalities for Eigenvalues," J. Appl. Mech. Vol. 29, 1962, pp. 159-164.

16. Prager, W. and Shield, R.T., "Optimal Design of Multi-Purpose Structures," Int. J. Solids Struct., Vol. 4, 1968, pp. 469-475.

17. Olhoff, N. and Taylor, J.E., "On Optimal Structural Remodeling," J. Optim. Theory and Appl., Vol. 27, 1979, pp. 571-582.

218

18. Segranian, A.P., "Homogeneous Functionals and Structural Optimization Problems," Int. J. Solids Struct., Vol. 15, 1979, pp. 749-759.
19. Huang, N.C., "Optimal Design of Elastic Beams for Minimum-Maximum Deflection," J. Appl. Mech., Vol. 38, 1971, pp. 1078-1081.

OPTIMIZATION OF STRUCTURES WITH REPEATED EIGENVALUES[*]

Kyung K. Choi and Edward J. Haug

Materials Division, College of Engineering, The
University of Iowa, Iowa City, Iowa 52242

ABSTRACT

Structural optimization problems in which multiple eigenvalues may occur are formulated and analyzed. Several optimal
design problems are presented that exhibit a repeated eigenvalue
at an optimum design. Elementary examples show that for some
values of defining parameters for these problems, repeated eigenvalues occur and for other values only simple eigenvalues occur.
The examples show that repeated eigenvalues at an optimum design
can occur and must be accounted for in theory and numerical
methods of structural optimization. Recent results concerning
differentiability of eigenvalues of structures and vector space
optimization theory are used to develop necessary conditions of
optimality. Results are applied to show the validity of the
necessary conditions. A formal Lagrange multiplier method is
also applied to show that the formal Lagrange multiplier method
is not valid in general. Finite dimensional structural design
problems are first analyzed. Quasi-differentiability of the
eigenvalue is proved and useful necessary conditions are derived.
Methods and results are then extended to distributed parameter
structures, for which response and design are described by functions. While the analysis in function spaces is more technical
than in the case of the matrix problem, results very similar to

[*]Research supported by National Science Foundation Project No.
ENG 77-19967. This paper expands on parts of a joint report by
the authors and B. Rousselet. The authors wish to acknowledge
Rousselet's suggestions that contributed to results presented
herein.

220

those obtained for finite dimensional problems are derived.
Results are applied to the buckling of a clamped-clamped column
to study necessary conditions.

1.   INTRODUCTION

        The purpose of this paper is to formulate and analyze struc-
tural optimization problems in which multiple eigenvalues may
occur.   It has recently been shown by Olhoff, Rasmussen, Masur,
and Mroz [1,2] that in a certain clamped column that is optimized
for maximum buckling load, a repeated eigenvalue may occur.   In a
recent paper [3] Prager has shown that repeated eigenvalues can
be predicted for the optimum column, using a finite dimensional
model.   In this paper, optimality conditions are derived and
applied to construct and analyze optimum designs.

        Elementary examples presented in Section 2 show that for
some values of defining parameters for these problems, repeated
eigenvalues occur and for other values only simple eigenvalues
occur.   It is not, therefore, clear a-priori whether repeated
eigenvalues will occur.   The examples do show that repeated
eigenvalues at an optimum design can occur and must be accounted
for in theory and numerical methods of structural optimization.

        Recent results concerning differentiability of eigenvalues of
structures presented in Refs. 4 and 5 and vector space optimization theory [6] are used to develop necessary conditions of optimality.   In Section 3, optimality conditions are derived using
the directional differentiability of the eigenvalues [5].
Results are applied to examples given in Section 2 to show the
validity of the necessary conditions.   A formal Lagrange multiplier method is also applied to an example to show that the formal
Lagrange multiplier method is not valid in general.   In Section
4, finite dimensional structural design problems are analyzed.
Here, response and design of the structure are specified by a
finite number of parameters.   Matrix eigenvalue problems arise,
in which the matrices depend on design variables.   In this
section, quasi-differentiability of the eigenvalue is proved and
useful necessary conditions are derived.   In Section 5 the
methods and results used in Section 4 are extended to distributed
parameter structures, for which response and design are described
by functions.   Differential operators arise in the eigenvalue
problem, which depend on design.   While the analysis in function
spaces is more technical than in the case of the matrix problem,
results very similar to those presented in Section 4 are obtained.
Results of this section are applied to the buckling of a clamped-
clamped column [1] to study necessary conditions.

2. EXAMPLES OF STRUCTURAL OPTIMIZATION PROBLEMS WITH MULTIPLE EIGENVALUES

The purpose of this section is to formulate and analyze elementary examples of optimization problems that exhibit multiple eigenvalues. Some of the examples are simple enough that they can be solved in closed form. These examples are used in later sections to test methods and theories developed.

2.1 Simple Spring-Mass Optimal Design Problem with Repeated Natural Frequencies

Consider the spring-mass system shown in Fig. 2.1. The eigenvalue problem for small amplitude vibration of the rigid body is simply derived as

$$\begin{bmatrix} 4k_1 + k_2 & k_2 \\ k_2 & 4k_1 + k_2 \end{bmatrix} y = \zeta \begin{bmatrix} 2 & 1 \\ 1 & 2 \end{bmatrix} y \tag{2.1}$$

where $\zeta = \dfrac{2\omega^2 m}{3}$, m is the mass of the bar, $I = mL^2/12$ is the polar moment of inertia and horizontal motion of the bar is ignored.

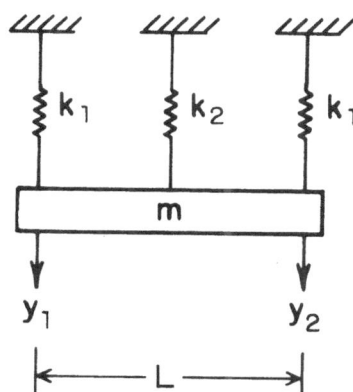

Figure 2.1   Two-degree of freedom spring-mass system.

The optimal design objective is to find design variables k_1 and k_2 to minimize weight of the spring supports, which is presumed to be of the form

$$\psi_0 = c_1 k_1 + c_2 k_2 \tag{2.2}$$

where c_1 and c_2 are known constants, subject to constraints that the eigenvalues be no lower than $\zeta_0 > 0$ and the spring constants are nonnegative. In inequality constraint form, this is

$$\left. \begin{aligned} \psi_1 &= \zeta_0 - \zeta_1 \leq 0 \\[1em] \psi_2 &= \zeta_0 - \zeta_2 \leq 0 \\[1em] \psi_3 &= -k_1 \leq 0 \\[1em] \psi_4 &= -k_2 \leq 0 \end{aligned} \right\} \tag{2.3}$$

The eigenvalues of Eq. 2.1 are $\zeta_1 = \dfrac{4k_1 + 2k_2}{3}$ and $\zeta_2 = 4k_1$, which gives $\omega_1^2 = \dfrac{2k_1 + k_2}{m}$ and $\omega_2^2 = \dfrac{6k_1}{m}$. Thus, the constraints of Eq. 2.3 becomes

$$\left. \begin{aligned} \psi_1 &= \zeta_0 - \frac{4k_1 + 2k_2}{3} \leq 0 \\[1em] \psi_2 &= \zeta_0 - 4k_1 \leq 0 \\[1em] \psi_3 &= -k_1 \leq 0 \\[1em] \psi_4 &= -k_2 \leq 0 \end{aligned} \right\} \tag{2.4}$$

Note that Eqs. 2.2 and 2.4 form a linear programming problem. The feasible set is shown graphically in Fig. 2.2. Note that the slope of the line connecting points A and B in Fig. 2.2 is -2. The level lines of the cost function of Eq. 2.2 are straight, with slope equal to $-c_1/c_2$. The cost function decreases as level lines of the cost function move to the lower left. Thus, it is clear that point A (repeated eigenvalue) is the optimum design if $c_1/c_2 \geq 2$ and point B (simple eigenvalue) is the optimum design if $c_1/c_2 \leq 2$.

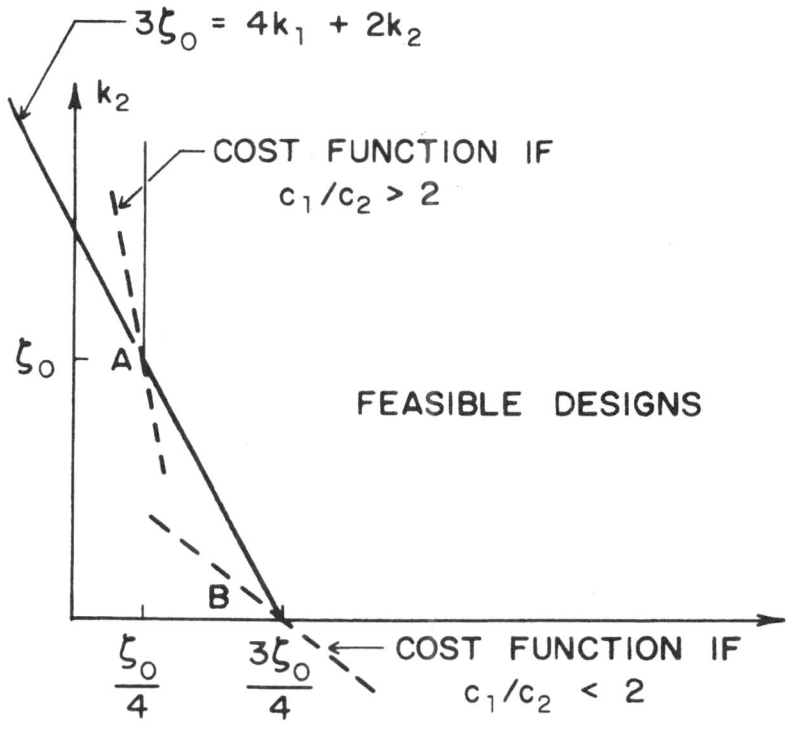

Figure 2.2   Feasible Region in Design Space for 2 DOF System

This result is quite interesting, since for certain values of parameters in the problem a repeated eigenvalue (A) occurs at the optimum design and for other values of the design parameters, only a simple eigenvalue (B) occurs at the optimum design.   It is expected that this uncertainty as to whether repeated roots arise will also occur in more complex optimal design problems.

2.2   A Column Problem With Repeated Eigenvalues

In Ref. 3, Prager presents an elegant analysis of a higher dimensional column buckling problem that exhibits repeated eigenvalues at an optimum point.   His analysis and results for a column with elastically clamped ends are summarized here.

Rotation of the end sections of Fig. 2.3 by an angle θ_0 is opposed by a clamping moment $M_0 = b_0\theta_0$, where b_0 is a given constant.   The cases $b_0 = 0$ and $b_0 = \infty$ correspond to pin-supported or rigidly clamped ends, respectively. By localizing the symmetric bending stiffness of the column in a finite number of elastic hinges that are connected by rigid segments, a structure is obtained whose deformation is specified by a finite number of displacement coordinates, rather than a function of the distance measured along the column. To keep the number of unknowns small, consider the column shown in Fig. 2.3, which has five rigid segments of length L and six elastic hinges, the hinges at the ends of the column having the given bending stiffness b_0. Because

the boundary conditions at the two ends are identical, the bending stiffnesses of the optimum design will be symmetric with respect to the center of the column and the buckling modes will be symmetric or antisymmetric with respect to this center. A column design is specified by the bending stiffness b_1 of hinges 1 and 4 and b_2 of hinges 2 and 3 in Fig. 2.3. A buckling mode that is known to be symmetric or antisymmetric is specified by the deflection y_1 of nodes 1 and 4 and y_2 of nodes 2 and 3. Upward deflections regarded as positive.

At the left end, the column is subject to the buckling load P, a reaction R, and a clamping moment M_0 (Fig. 2.3). The bending moment at hinge i is

$$M_i = M_0 - iLR - Py_i , \qquad i = 0,1,2,3,4 \qquad (2.5)$$

where $y_0 = 0$. If θ_i denotes the relative rotation of the segments meeting at hinge i, considered positive if counterclockwise rotation of the segment to the right of i exceeds that of the segment to the left of i,

$$M_i = b_i \theta_i = b_i(y_{i+1} - 2y_i + y_{i-1})/L , \quad i = 0,1,2,3,4 \quad (2.6)$$

where $y_{-1} = y_0 = 0$.

It is convenient to introduce a reference stiffness b* and define the dimensionless variables

$$\overline{P} = PL/b^* , \quad \overline{R} = RL/b^* , \quad \overline{M}_i = M_i/b^* , \quad \overline{y}_i = y_i/L \qquad (2.7)$$

Note that with these dimensionless variables, Eqs. 2.5 and 2.6 yield

$$M_0 - iR - Py_i = b_i(y_{i+1} - 2y_i + y_{i-1}), \quad i = 0,1,2,3,4 \quad (2.8)$$

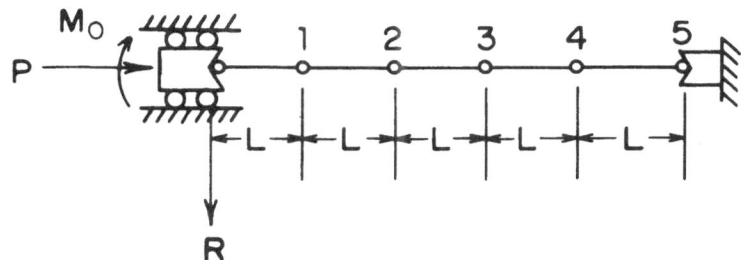

Figure 2.3   Elastically Supported Column

For a symmetric buckling mode, $y_3 = y_2$ and $R = 0$.   For $i = 0,1,2$, Eq. 2.8 yields

$$\left. \begin{array}{l} M_0 = b_0 y_1 \\[2mm] M_0 - P_s y_1 = b_1(y_2 - 2y_1) \\[2mm] M_0 - P_s y_2 = b_2(-y_2 + y_1) \end{array} \right\} \qquad (2.9)$$

where P_s is the buckling load of the symmetric mode.   When M_0 from the first of these equations is substituted into the other two, linear homogeneous equations for y_1 and y_2 are obtained that admit a nontrivial solution only if

$$P_s^2 - (b_0 + 2b_1 + b_2)P_s + b_0(b_1 + b_2) + b_1 b_2 = 0 \qquad (2.10)$$

The smaller root of this equation is the buckling load, if buckling is symmetric.

Assume that the cost of the design b_1, b_2 is fixed as

$$b_1 + b_2 = 1 \qquad (2.11)$$

and find a design that has the greatest buckling load.   In view of Eq. 2.11, Eq. 2.10 reduces to

$$P_s^2 - (1 + b_0 + b_1)P_s + b_0 + b_1(1 - b_1) = 0 \qquad (2.12)$$

For an antisymmetric buckling mode, $y_3 = -y_2$ and $R = 2M_0/(5L)$, because bending moment and deflection vanish at the center of the column. Proceeding as above, one obtains the quadratic equation for the buckling load P_a of the antisymmetric mode

$$P_a^2 - (3 + 0.6b_0 - b_1)P_a + b_0(1.8 - 1.6b_1) + 5b_1(1 - b_1) = 0$$

$$(2.13)$$

The smaller root of this equation is the antisymmetric buckling load.

In Fig. 2.4, P_s and P_a are shown as functions of b_1, for fixed values of b_0. To indicate important features, consider the case $b_0 = 0.25$, for which the variation of the buckling load is shown by the line ABCD. The arcs AB and CD correspond to antisymmetric buckling, while the arc BC corresponds to symmetric buckling. At point B in Fig. 2.4, that is for $b_1 = 0.122$, both symmetric and antisymmetric buckling is possible and the buckling load is a double eigenvalue. A similar observation applies to point C, that is to $b_1 = 0.953$. The arc, BC, however, has its greatest ordinate at point F, so the optimum design for $b_0 = 0.25$ corresponds to $b_1 = 0.300$, which buckles in a symmetric mode, the buckling load being a simple eigenvalue.

Along the curve OBE in Figure 2.4, the loads P_s and P_a for symmetric and antisymmetric buckling have the same value. The maxima of the arcs for symmetric buckling lie on the straight line EFG with the equation $P = 1 - 2b_1$, which intersects the curve OBE at the point H corresponding to $b_0 = 0.526$, $b_1 = 0.190$. For $b_0 < 0.526$, the optimum design buckles symmetrically, the buckling load being a single eigenvalue. For $b_0 > 0.526$, however, the buckling load is a double eigenvalue and the optimal design may buckle in a symmetric or an antisymmetric mode or in any linear combination of the two. Note that the maximum of the buckling load is or is not analytical, depending on whether the buckling load is a single or double eigenvalue. In a slightly different context, this has already been pointed out by Masur and Mroz [2].

In many structures, it is natural to express the bending stiffness as square of the design variable; i.e.,

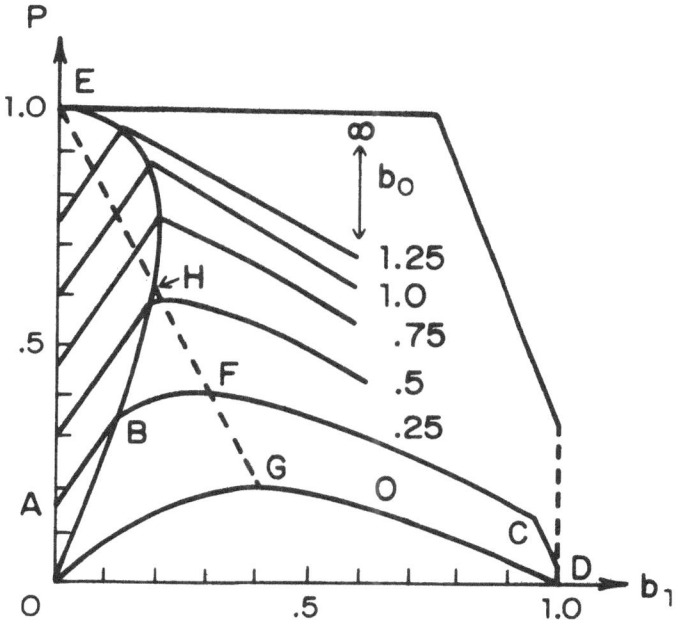

Figure 2.4   Buckling Loads for Optimum Columns

$$M_i = b_i^2 \, \theta_i \, , \quad i = 0,1,2,3,4 \tag{2.14}$$

If one proceeds as before, using the bending moments of Eq. 2.14, he has the characteristic equation for the symmetric mode as

$$P_s^2 - (b_0^2 + 2b_1^2 + b_2^2)P_s + b_0^2(b_1^2 + b_2^2) + b_1^2 b_2^2 = 0 \tag{2.15}$$

whose smaller root is the symmetric buckling load.   Also the characteristic equation for the antisymmetric mode is

$$P_a^2 - (0.6b_0^2 + 2b_1^2 + 3b_2^2)P_a + b_0^2(0.2b_1^2 + 1.8b_2^2)$$

$$+ 5b_1^2 b_2^2 = 0 \tag{2.16}$$

where the smaller root is the antisymmetric buckling load.   If one assumes that the cost of the design is fixed as in Eq. 2.11, then Eqs. 2.15 and 2.16 can be expressed in terms of b_0 and b_1.

In Fig. 2.5, P_s and P_a are shown as functions of b_1, for fixed values of b_0. Note that the maximum point is a double eigenvalue if b_0 is larger than the value of b_0 that corresponds to the point C in Fig. 2.5. Unlike the previous case, local maxima occur as double eigenvalues for b_0 larger than the b_0 that corresponds to point D. Another local maximum occurs as a simple eigenvalue for b_0 larger than the b_0 that corresponds to point E. At the left-hand ordinate of Fig. 2.5, if $b_0 > 1.29099$ then $P = P_s = 1$, which is simple eigenvalue. If $b_0 = 1.29099$, then $P = P_s = P_a = 1$, which is double eigenvalue. For the case $b_0 < 1.29099$, $P = P_a = 0.6b_0^2$, which is simple eigenvalue.

The governing eigenvalue equation for the column of Fig. 2.3, with stiffness of Eq. 2.14, can be obtained as

$$K(b)y = \zeta Dy \tag{2.17}$$

where

$$K(b) = \begin{bmatrix} b_0^2+4b_1^2+b_2^2 & -2b_1^2-2b_2^2 & b_2^2 & 0 \\ -2b_1^2-2b_2^2 & b_1^2+4b_2^2+b_3^2 & -2b_2^2-2b_3^2 & b_3^2 \\ b_2^2 & -2b_2^2-2b_3^2 & b_2^2+4b_3^2+b_4^2 & -2b_3^2-2b_4^2 \\ 0 & b_3^2 & -2b_3^2-2b_4^2 & b_0^2+b_3^2+4b_4^2 \end{bmatrix} \tag{2.18}$$

and

$$D = \begin{bmatrix} 2 & -1 & 0 & 0 \\ -1 & 2 & -1 & 0 \\ 0 & -1 & 2 & -1 \\ 0 & 0 & -1 & 2 \end{bmatrix} \tag{2.19}$$

where $\zeta = P$. Thus, the general design problem can be formulated analytically as; find $b \in R^4$ that minimizes

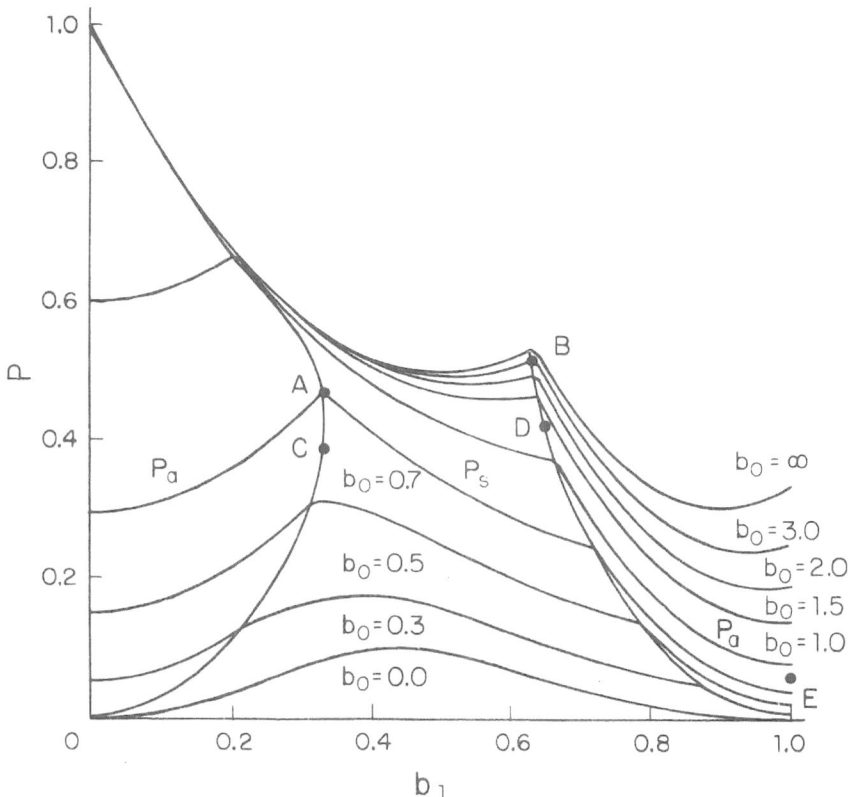

Figure 2.5   Buckling Loads For Optimum Columns

$$\psi_0 = \sum_{i=1}^{4} b_i \tag{2.20}$$

subject to eigenvalue constraints

$$\psi_i = \zeta_0 - \zeta_i \leq 0 , \quad i=1,2 \tag{2.21}$$

and design variable constraints

$$\phi_i = c_i - b_i \leq 0 , \quad i=1,2,3,4 \tag{2.22}$$

230

where ζ_1 and ζ_2 are two smallest eigenvalues of Eq. 2.17 and c_i are given constants.

2.3 A Clamped-Clamped Distributed Column with Repeated Eigenvalues

The clamped-clamped column of Fig. 2.6 was studied as an optimal design problem by Tadjbakhsh and Keller [8]. They sought a cross-sectional area $u(x)$, where moment of inertia of the cross section is $I = \alpha u^2$, to maximize the fundamental buckling load for a given volume of material in the column.

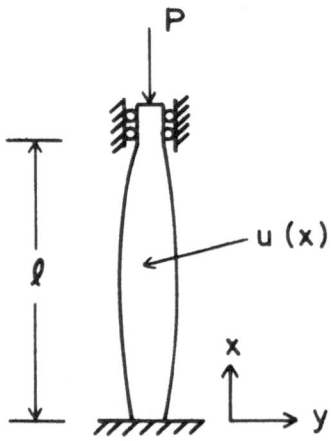

Figure 2.6   Column

Thus, the problem is to find $u \in L^\infty(0,1)$ that minimizes

$$\psi_0(u) = -\zeta(u) \tag{2.23}$$

subject to the volume constraint

$$\psi_1(u) = \int_0^\ell u\,dx - V = 0 \tag{2.24}$$

Here the buckling load ζ is the smallest eigenvalue of

$$Ay \equiv \left(E\alpha u^2(x)y'' \right)'' = -\zeta y'' \equiv \zeta By \\ y(0) = y'(0) = y(\ell) = y'(\ell) = 0 \Bigg\}$$ (2.25)

where $' \equiv \dfrac{d}{dx}$, E and α are material and geometrical constants, and

$$\phi \equiv c - u(x) \leq 0 , \quad \text{a.e. in } [0,\ell]$$ (2.26)

It has recently been shown by Olhoff and Rasmussen [1] that one should expect that the fundamental mode is bimodal at the optimum design, for some values of c > 0 in Eq. 2.26.

This prototype structural optimization problem is shown in Refs. 4 and 5 to be typical of structural optimization problems involving plates and planar elasticity, where the operators A and B are elliptic differential operators. Since $u \in L^{\infty}(0,\ell)$, it is clear that distributional solutions of the operator equation of Eq. 2.25 are required. In variational notation, this is

$$a_u(y,v) \equiv \int_0^\ell E\alpha u^2 y''v''dx = \zeta \int_0^\ell y'v'dx \equiv \zeta b_u(y,v)$$ (2.27)

with $y,v \in H_0^2(0,\ell)$

3. NECESSARY CONDITIONS OF OPTIMALITY

3.1 Directional Derivatives of Eigenvalues

For broad classes of structures, including the examples of Section 2, it is shown in Ref. 4 that the operators are strongly elliptic, the bilinear forms a_u and b_u of Eq. 2.27 are Fréchet differentiable with respect to u, and their differentials in a direction h, $a_{u,h}^{(1)}$ and $b_{u,h}^{(1)}$, are relatively bounded by a_u and b_u, respectively. It is shown in Ref. 5 that if the eigenvalue ζ is simple, with eigenfunction y normalized by $b_u(y,y) = 1$, then ζ is Fréchet differentiable with respect to u, and its Fréchet differential in direction h is

$$\zeta_{u,h}^{(1)} = a_{u,h}^{(1)}(y,y) - \zeta b_{u,h}^{(1)}(y,y)$$ (3.1)

which, in the case of the column of Section 2.3, is

$$\zeta_{u,h}^{(1)} = \int_0^\ell 2E\alpha u h(y'')^2 \, dx \qquad (3.2)$$

If, on the other hand, $\zeta = \zeta_1 = \ldots = \zeta_m$ is an m-fold eigenvalue, with eigenfunctions $y^i(x)$ orthonormalized so that $b_u(y^i,y^j) = \delta_{ij}$, then it is shown in Ref. 5 that ζ_i, as functions of u, have at most directional differentials $\zeta_{u,h,i}^{(1)}$, which are eigenvalues of the matrix

$$M(u,h) = \left[a_{u,h}^{(1)}(y^i,y^j) - \zeta b_{u,h}^{(1)}(y^i,y^j) \right] \qquad (3.3)$$

Since the eigenvalues of M depend on h, they define the bifurcation of the repeated eigenvalue ζ into possibly simple eigenvalues as u is changed to u + h, for h small.

In the case of the column with a pair of repeated eigenvalues and b_u-orthonormal eigenfunctions y^1 and y^2,

$$M = 2E\alpha \begin{bmatrix} \int_0^\ell uh(y^{1''})^2 dx & \int_0^\ell uh y^{1''} y^{2''} \, dx \\[2ex] \int_0^\ell uh y^{1''} y^{2''} dx & \int_0^\ell uh(y^{2''})^2 \, dx \end{bmatrix} \qquad (3.4)$$

3.2 Necessary Conditions of Optimality

In the case of a simple eigenvalue, the cost functional of Eq. 2.23 is differentiable and one may define a convex cone K of feasible directions, for example, in the case of the column the constraint of Eq. 2.26, with

$$K = \{h \in L^\infty : h = \gamma(u - u_0) \text{ for u such that}$$

$$c - u \leq 0 \text{ and } \gamma > 0\} \qquad (3.5)$$

Now, an optimality condition for u_0 to be optimum is that [6, p. 83]

$$\lambda_0 \psi_{u_0,h,0}^{(1)} + \lambda_1 \psi_{u_0,h,1} \geq 0 \qquad (3.6)$$

for some λ_0 and λ_1, not both zero, and all $h \in K$. In the case of the column, this is

$$\int_0^\ell [-\lambda_0 2 E\alpha u(y'')^2 + \lambda_1] \, hdx \geq 0 \qquad (3.7)$$

The situation is more complicated in case of a repeated eigenvalue. The directional derivative of $\zeta(u)$ is the smallest eigenvalue of the matrix of Eq. 3.3. But this is just

$$\zeta_{u_0,h}^{(1)} = \min_{\substack{\xi \in R^m \\ \xi^T \xi = 1}} \xi^T M(u_0,h) \, \xi \qquad (3.8)$$

For fixed ξ, $\xi^T M \xi$ is linear in h, since M is linear in h, so $\min_\xi \xi^T M \xi$ is concave. Thus, the directional derivative $\psi_{u_0,h,0}^{(1)}$ of Eq. 2.23 is convex and an optimality condition for the repeated eigenvalue problem is, from Eq. 3.6,

$$- \lambda_0 \min_{\substack{\xi \in R^m \\ \xi^T \xi = 1}} \sum_{i,j=1}^m \left[\xi_i \xi_j \left(a_{u_0,h}^{(1)}(y^i,y^j) - \zeta \, b_{u_0,h}^{(1)}(y^i,y^j) \right) \right]$$

$$+ \lambda_1 \psi_{u_0,h,1}^{(1)} \geq 0 \qquad (3.9)$$

for all $h \in K$. Since for fixed h, $a_{u_0,h}^{(1)}$ and $b_{u_0,h}^{(1)}$ are bilinear forms on $H_0^2(0,\ell)$, if one defines $z = \sum_{i=1}^m \xi_i y^i$, then

$$\sum_{i,j=1}^m \left[\xi_i \xi_j \left(a_{u_0,h}^{(1)}(y^i,y^j) - \zeta \, b_{u_0,h}^{(1)}(y^i,y^j) \right) \right]$$

$$= a_{u_0,h}^{(1)}(z,z) - \zeta \, b_{u_0,h}^{(1)}(z,z) \qquad (3.10)$$

Further, the space of eigenfunctions associated with $\zeta(u_0)$ is

$$N_u = \{z \in H_0^2(0,\ell) : z = \sum_{i=1}^m \xi_i y^i, \; \xi_i \text{ real}\}.$$ The subset of N_u such

that $\xi^T \xi = 1$ may now be characterized by the condition

$$b_u(z,z) = \sum_{i,j=1}^m \xi_i \xi_j \, b_u(y^i, y^j) = \sum_{i,j=1}^m \xi_i \xi_j \, \delta_{ij} = \xi^T \xi = 1$$

since the y^i were selected to be b_u-orthonormal.   Thus, Eq. 3.9
may be written

$$-\lambda_0 \; \min_{\substack{z \in N_u \\ b_u(z,z)=1}} \left(a_{u_0,h}^{(1)}(z,z) - \zeta \, b_{u_0,h}^{(1)}(z,z) \right) + \lambda_1 \psi_{u_0,h,1}^{(1)} \geq 0 \quad (3.11)$$

for all $h \in K$.

In the case of the column with a double eigenvalue, Eq. 3.9
is

$$-\lambda_0 \; 2E\alpha \min_{\substack{\xi \in R^2 \\ \xi^T \xi = 1}} \int_0^\ell u_0(\xi_1 y^{1''} + \xi_2 y^{2''})^2 \, h \, dx + \lambda_1 \int_0^\ell h \, dx \geq 0 \quad (3.12)$$

for all $h \in K$, where y^1 and y^2 is any b_u-orthonormal pair of eigenfunctions of Eq. 2.27 associated with $\zeta(u_0)$.   Use of this condition for determination of an optimum design u appears to be nontrivial.

3.3 Optimality Conditions for the Simple Spring-Mass System

It is of interest to verify the optimality criterion of Eq.
3.6 for the simple spring-mass system of Section 2.1, at the
solution with a repeated eigenvalue; $k_1 = \zeta_0/4$, $k_2 = \zeta_0$.   The
minimum in the right-hand side of Eq. 3.8 is the smaller of the
eigenvalues of the matrix M which are $4h_1/3 + 2h_2/3$ and $4h_1$, so
Eq. 3.6 is, using the differential of ψ_0 from Eq. 2.2 and $\lambda_0 = 1$
(see Section 4.5),

$$c_1 h_1 + c_2 h_2 - \lambda_1 \min\{4h_1/3 + 2h_2/3, \, 4h_1\} \geq 0 \qquad (3.13)$$

Consider first the case $4h_1 = 4h_1/3 + 2h_2/3$; i.e., $h_2 = 4h_1$. Then Eq. 3.13 is

$$c_1 h_1 + 4c_2 h_1 \geq \lambda_1 4h_1$$

for all $h_1 \in R^1$. Thus, equality must hold and

$$\lambda_1 = c_1/4 + c_2 \geq 0 \tag{3.14}$$

Consider second the case $4h_1 > 4h_1/3 + 2h_2/3$; i.e., $h_2 < 4h_1$. Then Eq. 3.13 is

$$c_1 h_1 + c_2 h_2 - \lambda_1 (4h_1/3 + 2h_2/3) \geq 0$$

Using λ_1 from Eq. 3.14, this may be written

$$(c_1 - 2c_2)h_2 \leq (c_1 - 2c_2)4h_1$$

Since this must hold for all h with $h_2 < 4h_1$, it is necessary that $c_1 - 2c_2 \geq 0$, or

$$c_1/c_2 \geq 2 \tag{3.15}$$

Finally, consider the case $4h_1 < 4h_1/3 + 2h_2/3$; i.e., $h_2 > 4h_1$. Then Eq. 3.13, with λ_1 from Eq. 3.14, becomes

$$c_2 h_2 \geq c_2 (4h_1)$$

Since this must hold for all h with $h_2 > 4h_1$, it is necessary that

$$c_2 \geq 0 \tag{3.16}$$

The necessary conditions thus reduce to the requirements of Eqs. 3.15 and 3.16, which simply require that $c_i \geq 0$, $i = 1,2$, and Eq. 3.15, which is precisely the condition determined in the Section 2.1 of this paper. Thus, the necessary condition of Eq. 3.11 is sharp for this problem; i.e., it is both necessary and sufficient (see Section 4.4) for an optimum design.

3.4 Formal Lagrange Multiplier Analysis

<u>Lagrange Multiplier Analysis of the Two Degree of Freedom Spring-Mass Example.</u>   The foregoing optimality conditions, specifically Eq. 3.6, may seem to be unnecessarily complicated. This viewpoint is reinforced by the well-known optimization results obtained by Prager, Taylor, and Niordson, using variational formulations with Lagrange multipliers in problems with a simple eigenvalue [9,10].   To evaluate the Lagrange multiplier method in the spirit of this foregoing work, consider the elementary two degree of freedom spring-mass vibration problem of Section 2.1.

The idea of the formal Lagrange multiplier method is to use the Rayleigh quotient characterization of the eigenvalues.   That is, minimize

$$\psi_0(k) = c_1 k_1 + c_2 k_2 \tag{3.17}$$

subject to the constraints

$$\zeta_0 - \zeta_1 = \zeta_0 - y^{1^T} \begin{bmatrix} 4k_1 + k_2 & k_2 \\ \\ k_2 & 4k_1 + k_2 \end{bmatrix} y^1 \leq 0 \tag{3.18}$$

$$\zeta_0 - \zeta_2 = \zeta_0 - y^{2^T} \begin{bmatrix} 4k_1 + k_2 & k_2 \\ \\ k_2 & 4k_1 + k_2 \end{bmatrix} y^2 \leq 0 \tag{3.19}$$

$$y^{i^T} \begin{bmatrix} 2 & 1 \\ \\ 1 & 2 \end{bmatrix} y^i = 1 , \qquad i = 1,2 \tag{3.20}$$

$$-k_1 \leq 0 , \qquad -k_2 \leq 0 \tag{3.21}$$

where y^1 and y^2 are eigenvectors corresponding to the eigenvalues ζ_1 and ζ_2.

The variational formulation is stated by defining Lagrange multipliers $\gamma_i \geq 0$, n_i, and $\mu_i \geq 0$, $i = 1,2$ and

$$L = c_1 k_1 + c_2 k_2 + \gamma_1(\zeta_0 - {y^1}^T K y^1) + \gamma_2(\zeta_0 - {y^2}^T K y^2)$$

$$+ n_1({y^1}^T M y^1 - 1) + n_2({y^2}^T M y^2 - 1)$$

$$+ \mu_1(-k_1) + \mu_2(-k_2) \tag{3.22}$$

The Kuhn-Tucker necessary conditions of optimality for the problem of Eqs. 3.17 to 3.21 are

$$\frac{\partial L}{\partial k} = \left(\frac{\partial L}{\partial k_1}, \frac{\partial L}{\partial k_2}\right) = 0 \tag{3.23}$$

or

$$c_1 - \gamma_1 \left({y^1}^T \begin{bmatrix} 4 & 0 \\ 0 & 4 \end{bmatrix} y^1\right) - \gamma_2 \left({y^2}^T \begin{bmatrix} 4 & 0 \\ 0 & 4 \end{bmatrix} y^2\right) - \mu_1 = 0 \tag{3.24}$$

$$c_2 - \gamma_1 \left({y^1}^T \begin{bmatrix} 1 & 1 \\ 1 & 1 \end{bmatrix} y^1\right) - \gamma_2 \left({y^2}^T \begin{bmatrix} 1 & 1 \\ 1 & 1 \end{bmatrix} y^2\right) - \mu_2 = 0 \tag{3.25}$$

and

$$\gamma_1(\zeta_0 - {y^1}^T K y^1) = 0$$

$$\left. \right\} \tag{3.26}$$

$$\gamma_2(\zeta_0 - {y^2}^T K y^2) = 0$$

$$\mu_1(-k_1) = 0$$

$$\left. \right\} \tag{3.27}$$

$$\mu_2(-k_2) = 0$$

Consider the known optimum design $k_0 = [\zeta_0/4, \zeta_0]^T$, where a double eigenvalue occurs. Then from Eq. 3.27, $\mu_1 = \mu_2 = 0$ and the eigenvectors y^i are solutions of the equation

$$(K - \zeta_0 M)\Big|_{k=k_0} y^i = \begin{bmatrix} 0 & 0 \\ 0 & 0 \end{bmatrix} y^i = 0 , \quad i = 1,2 \tag{3.28}$$

Hence any nonzero vector y is an eigenvector. Eigenvectors may now be selected and substituted into Eqs. 3.24 and 3.25, which must of course be satisfied.

With the M-normal eigenvectors

$$y^1 = \frac{1}{\sqrt{2}} \begin{bmatrix} 1 \\ 0 \end{bmatrix} , \quad y^2 = \frac{1}{\sqrt{2}} \begin{bmatrix} 0 \\ 1 \end{bmatrix} , \tag{3.29}$$

Eqs. 3.24 and 3.25, with $\mu_i = 0$, $i = 1,2$, yield

$$\left. \begin{aligned} c_1 &= 2\gamma_1 + 2\gamma_2 \geq 0 \\ c_2 &= \frac{1}{2}\gamma_1 + \frac{1}{2}\gamma_2 \geq 0 \end{aligned} \right\} \tag{3.30}$$

Thus by observation, the necessary condition is only satisfied if $c_1 = 4c_2 \geq 0$. But from Section 2.1, it is known that

$k_0 = \begin{bmatrix} \frac{\zeta_0}{4} , \zeta_0 \end{bmatrix}^T$ is an optimum design as long as $c_1 \geq 2c_2 \geq 0$. Thus the supposed necessary conditions of Eq. 3.24 and 3.25 are in fact not necessary at all.

Consider another set of eigenvectors, which are M-orthonormal,

$$y^1 = \frac{1}{\sqrt{2}} \begin{bmatrix} 1 \\ 0 \end{bmatrix} , \quad y^2 = \frac{1}{\sqrt{6}} \begin{bmatrix} 1 \\ -2 \end{bmatrix} \tag{3.31}$$

with these eigenvectors, Eqs. 3.24 and 3.25 are

$$\left. \begin{aligned} c_1 &= 2\gamma_1 + \frac{10}{3}\gamma_2 \geq 0 \\ c_2 &= \frac{1}{2}\gamma_1 + \frac{1}{6}\gamma_2 \geq 0 \end{aligned} \right\} \tag{3.32}$$

From Eqs. 3.32, one has

$$\gamma_1 = \frac{1}{8} (-c_1 + 20c_2) \geq 0$$

$$\gamma_2 = \frac{1}{8} (3c_1 - 12c_2) \geq 0$$

which give

$$20c_2 \geq c_1 \geq 4c_2 \geq 0 \tag{3.33}$$

These are also not valid necessary conditions.

Try now the set of eigenvectors, which are M-orthonormal and diagonalize the matrix $M(k_0, h)$ for all h,

$$y^1 = \frac{1}{\sqrt{6}} \begin{bmatrix} 1 \\ 1 \end{bmatrix}, \qquad y^2 = \frac{1}{\sqrt{2}} \begin{bmatrix} 1 \\ -1 \end{bmatrix} \tag{3.34}$$

By direct calculation in Eq. 3.3,

$$M_{11}(k_0, h) = \frac{4}{3} h_1 + \frac{2}{3} h_2 \tag{3.35}$$

$$M_{22}(k_0, h) = 4h_1 \tag{3.36}$$

and

$$M_{12}(k_0, h) = 0 \tag{3.37}$$

Substituting y^1 and y^2 of Eq. 3.34 into Eqs. 3.24 and 3.25, one obtains

$$\left. \begin{aligned} c_1 &= \frac{4}{3} \gamma_1 + 4\gamma_2 \geq 0 \\ c_2 &= \frac{2}{3} \gamma_1 \geq 0 \end{aligned} \right\} \tag{3.38}$$

or

$$\left. \begin{aligned} \gamma_1 &= \frac{3}{2} c_1 \geq 0 \\ \gamma_2 &= \frac{1}{2} (c_1 - 2c_2) \geq 0 \end{aligned} \right\} \tag{3.39}$$

This gives the correct result $c_1 \geq 2c_2 \geq 0$. Hence if one uses a pair of eigenvectors that diagonalize $M(k_0,h)$ for all h, then the variational formulation may be valid. However, in general, one can not expect that there exists a pair of eigenvectors which diagonalize $M(k_0,h)$ for all h.

To find out "what went wrong" in the formal Lagrange multiplier method, note that with y^i of Eq. 3.29, the value of the Rayleigh quotients are

$$\left.\begin{aligned} f_1(k) &\equiv {y^1}^T K(k)y^1 = \tfrac{1}{2}(4k_1 + k_2) \\[2mm] f_2(k) &\equiv {y^2}^T K(k)y^2 = \tfrac{1}{2}(4k_1 + k_2) \end{aligned}\right\} \tag{3.40}$$

For $k_0 = \left[\dfrac{\lambda_0}{4}, \lambda_0\right]^T$, $f_1 = f_2 = \zeta_0$ and the gradients of the Rayleigh quotients are, from Eq. 3.40,

$$\nabla f_1 = \nabla f_2 = \begin{bmatrix} 2 \\ \frac{1}{2} \end{bmatrix} \tag{3.41}$$

From the algebraic solution of ζ_i, one has $\zeta_1 = \tfrac{2}{3}(2k_1 + k_2)$ and $\zeta_2 = 4k_1$. Thus the gradients of the actual eigenvalues are

$$\nabla\zeta_1 = \begin{bmatrix} \frac{4}{3} \\ \frac{2}{3} \end{bmatrix}, \quad \nabla\zeta_2 = \begin{bmatrix} 4 \\ 0 \end{bmatrix} \tag{3.42}$$

Thus, $\nabla\zeta_i \neq \nabla f_i$, so the wrong derivatives were used in the Kuhn-Tucker optimality criteria. It is clear that one must be very careful in his analysis of derivatives of repeated eigenvalues.

To see why this difficulty arises, review the engineering perturbation analysis of the eigenvalue problem, which starts with

$$A(k) \ y^i = \zeta_i B(k) \ y^i \ , \quad y^{i^T} B(k) \ y^i = 1 \tag{3.43}$$

and premultiply by a constant vector v^T to obtain

$$v^T \ A(k) \ y^i = \zeta_i v^T \ B(k) \ y^i \tag{3.44}$$

Denoting variation with an over-dot and treating k, y, and ζ_i as independent, one has

$$v^T \ \dot{A} \ y^i + v^T \ A \ \dot{y}^i = \dot{\zeta}_i \ v^T \ B \ y^i + \zeta_i \ v^T \ \dot{B} \ y^i + \zeta_i \ v^T \ B \ \dot{y}^i \tag{3.45}$$

Rewriting, using symmetry of A and B, and setting $v = y^j$, one has

$$y^{j^T} \ \dot{A} \ y^i - \zeta_i y^{j^T} \ \dot{B} \ y^i + \dot{y}^{i^T}(A \ y^j - \zeta_i B \ y^j) = \dot{\zeta}_i y^{j^T} \ B \ y^i \tag{3.46}$$

Putting i = j in Eq. 3.46 and using Eq. 3.43, one has

$$\dot{\zeta}_i = y^{i^T} \ \dot{A} \ y^i - \zeta_i y^{i^T} \ \dot{B} \ y^i \ , \quad i = 1,2 \tag{3.47}$$

But, the examples treated above show that this is "wrong."

To see what goes wrong at repeated eigenvalues, consider the two degree of freedom example of Section 2.1. The eigenvalue ζ_0 is repeated if $k_1 = \dfrac{\zeta_0}{4}$ and $k_2 = \zeta_0$. If one perturbs k to $k_1 = \dfrac{\zeta_0}{4} + h_1$, $k_2 = \zeta_0 + h_2$, the eigenvalue equation of Eq. 3.43 is

$$\begin{bmatrix} 2\zeta_0 + 4h_1 + h_2 & \zeta_0 + h_2 \\ \zeta_0 + h_2 & 2\zeta_0 + 4h_1 + h_2 \end{bmatrix} y_h = \zeta_h \begin{bmatrix} 2 & 1 \\ 1 & 2 \end{bmatrix} y_h \tag{3.48}$$

where the subscript h denotes dependence on h. The solution is $\zeta_h^1 = \zeta_0 + \dfrac{4h_1 + 2h_2}{3}$, $\zeta_h^2 = \zeta_0 + 4h_1$. If $4h_1 \neq h_2$, $\zeta_h^1 \neq \zeta_h^2$ the M-orthonormal eigenvectors are

242

$$y_h^1 = \frac{1}{\sqrt{6}} \begin{bmatrix} 1 \\ 1 \end{bmatrix}, \qquad y_h^2 = \frac{1}{\sqrt{2}} \begin{bmatrix} 1 \\ -1 \end{bmatrix} \qquad (3.49)$$

which happen to be independent of h.   If on the other hand
$4h_1 = h_2$, anything is an eigenvector;   e.g.,

$$\bar{y}^1 = \frac{1}{\sqrt{2}} \begin{bmatrix} 1 \\ 0 \end{bmatrix}, \qquad \bar{y}^2 = \frac{1}{\sqrt{2}} \begin{bmatrix} 0 \\ 1 \end{bmatrix} \qquad (3.50)$$

To see what has happened geometrically, Fig. 3.1 shows the
eigenvectors just calculated.   The perturbation of design from
k_0, which gives a repeated eigenvalue, transforms $\bar{y}^i$ of Eq. 3.50
to y_h^i of Eq. 3.49.   Clearly, this transformation is not continuous, so the perturbation analysis of Eq. 3.45 is meaningless at
the repeated eigenvalue

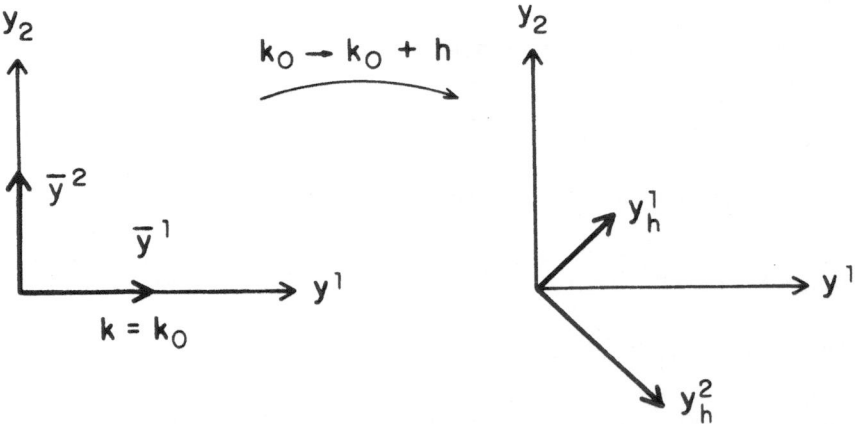

Figure 3.1   Transformation of Eigenspace with $h_2 \neq 4h_1$

To be more precise, let $h_1 = \varepsilon$ and $h_2 = \varepsilon$.   Then

$$||\dot{y}^1|| = ||\bar{y}^1 - y_h^1|| = \left|\left|\frac{1}{\sqrt{6}} \begin{bmatrix} \sqrt{3} - 1 \\ -1 \end{bmatrix} \right|\right| = \left[\frac{5 - 2\sqrt{3}}{6}\right]^{\frac{1}{2}}$$

$$||\dot{y}^2|| = ||\overline{\dot{y}}^2 - y_h^2|| = \left|\left|\frac{1}{\sqrt{2}}\begin{bmatrix}-1\\2\end{bmatrix}\right|\right| = \sqrt{\frac{5}{2}}$$

Since this is true for any $\varepsilon > 0$, arbitrarily small, $||\dot{y}^i||$ do not approach zero as $||h|| \to 0$.

Thus the linearization of

$$v^T A y = \zeta v^T By$$

to

$$v^T \dot{A} y + v^T A \dot{y} = \dot{\zeta} v^T By + \zeta v^T \dot{B} y + \zeta v^T B \dot{y}$$

is meaningless, since $y(h)$ is not even continuous, much less differentiable.

Lagrange Multiplier Analysis of the Column Problem with Repeated Eigenvalues. The formal Lagrange multiplier method is now applied to the column problem of Section 2.2 using Rayleigh quotient characterization of the eigenvalues. It is shown that the formal Lagrange multiplier method again fails to give the correct necessary conditions, unless one uses the eigenvector pairs that diagonalize the $M(b^0,h)$ matrix for all variations h, where b^0 is the optimum point. However, as noted in the previous example, this is not to say that one can find such eigenvectors in general problems.

The problem is to minimize

$$\psi_0 = \sum_{i=1}^{4} b_i \tag{3.51}$$

subject to the constraints

$$\zeta_0 - \zeta_i = \zeta_0 - y^{i^T} K(b) y^i \leq 0 , \quad i = 1,2 \tag{3.52}$$

$$y^{i^T} D y^i = 1 , \quad i = 1,2 \tag{3.53}$$

$$- b_i \leq 0 , \quad i = 1,2,3,4 \tag{3.54}$$

where $K(b)$ and D are given in Eqs. 2.18 and 2.19, respectively, and y^1 and y^2 are eigenvectors corresponding to the eigenvalues ζ_1 and ζ_2.

The Lagrange multiplier method is given by defining multipliers $\gamma_i \geq 0$, n_i, $i=1,2$ and $\mu_i \geq 0$, $i=1,2,3,4$ and

$$L = \sum_{i=1}^{4} b_i + \gamma_1 (\zeta_0 - y^{1^T} K y^1) + \gamma_2 (\zeta_0 - y^{2^T} K y^2)$$

$$+ n_1 (y^{1^T} D y^1 - 1) + n_2 (y^{2^T} D y^2 - 1)$$

$$- \sum_{i=1}^{4} \mu_i b_i \tag{3.55}$$

the Kuhn-Tucker necessary conditions are

$$\frac{\partial L}{\partial b} = \left[\frac{\partial L}{\partial b_1} , \frac{\partial L}{\partial b_2} , \frac{\partial L}{\partial b_3} , \frac{\partial L}{\partial b_4} \right] = 0 \tag{3.56}$$

or

$$1 - \gamma_1 y^{1^T} \begin{bmatrix} 8b_1 & -4b_1 & 0 & 0 \\ -4b_1 & 2b_1 & 0 & 0 \\ 0 & 0 & 0 & 0 \\ 0 & 0 & 0 & 0 \end{bmatrix} y^1 - \gamma_2 y^{2^T} \begin{bmatrix} 8b_1 & -4b_1 & 0 & 0 \\ -4b_1 & 2b_1 & 0 & 0 \\ 0 & 0 & 0 & 0 \\ 0 & 0 & 0 & 0 \end{bmatrix} y^2 - \mu_1 = 0 \tag{3.57}$$

$$1 - \gamma_1 y^{1^T} \begin{bmatrix} 2b_2 & -4b_2 & 2b_2 & 0 \\ -4b_2 & 8b_2 & -4b_2 & 0 \\ 2b_2 & -4b_2 & 2b_2 & 0 \\ 0 & 0 & 0 & 0 \end{bmatrix} y^1 - \gamma_2 y^{2^T} \begin{bmatrix} 2b_2 & -4b_2 & 2b_2 & 0 \\ -4b_2 & 8b_2 & -4b_2 & 0 \\ 2b_2 & -4b_2 & 2b_2 & 0 \\ 0 & 0 & 0 & 0 \end{bmatrix} y^2 - \mu_2 = 0 \tag{3.58}$$

$$1-\gamma_1 y^{1^T}\begin{bmatrix} 0 & 0 & 0 & 0 \\ 0 & 2b_3 & -4b_3 & 2b_3 \\ 0 & -4b_3 & 8b_3 & -4b_3 \\ 0 & 2b_3 & -4b_3 & 2b_3 \end{bmatrix} y^1 - \gamma_2 y^{2^T}\begin{bmatrix} 0 & 0 & 0 & 0 \\ 0 & 2b_3 & -4b_3 & 2b_3 \\ 0 & -4b_3 & 8b_3 & -4b_3 \\ 0 & 2b_3 & -4b_3 & 2b_3 \end{bmatrix} y^2 - \mu_3 = 0 \tag{3.59}$$

$$1-\gamma_1 y^{1^T}\begin{bmatrix} 0 & 0 & 0 & 0 \\ 0 & 0 & 0 & 0 \\ 0 & 0 & 2b_4 & -4b_4 \\ 0 & 0 & -4b_4 & 8b_4 \end{bmatrix} y^1 - \gamma_2 y^{2^T}\begin{bmatrix} 0 & 0 & 0 & 0 \\ 0 & 0 & 0 & 0 \\ 0 & 0 & 2b_4 & -4b_4 \\ 0 & 0 & -4b_4 & 8b_4 \end{bmatrix} y^2 - \mu_4 = 0 \tag{3.60}$$

and

$$\gamma_i \left(\zeta_0 - y^{i^T} K y^i \right) = 0 , \quad i=1,2 \tag{3.61}$$

$$- \mu_i b_i = 0 , \quad i=1,2,3,4 \tag{3.62}$$

Consider the case $b_0 = 0.7$ in Fig. 2.5, where point A is the optimum design. Since the design variables are not active, $\mu_i = 0$, $i=1,2,3,4$. Now Eqs. 3.57 through 3.60 are four equations with two unknowns.

If y^1 and y^2 are symmetric and antisymmetric respectively, then Eqs. 3.57 and 3.60 are identical and Eqs. 3.58 and 3.59 are identical, because the optimum design b^0 is symmetric. Hence one has two equations with two unknowns, which can be solved. One can easily verify that these two eigenvectors diagonalize the $M(b^0,h)$ matrix for all symmetric variations h. The reason that one can consider only symmetric variations h is because the optimum point b^0 is symmetric (see Section 5.4).

However, if y^1 and y^2 are not symmetric and antisymmetric, then Eqs. 3.57 through 3.60 may be inconsistent. In fact, for point A one has $b_1^0 = b_4^0 = 0.32956$ and $b_2^0 = b_3^0 = 0.67044$. With this optimum design, the Eispack eigenvalue analysis package [16] gives two D-orthonormal eigenvectors

$$y^1 = \begin{bmatrix} -0.55935 \\ -0.24855 \\ 0.18679 \\ 0.53133 \end{bmatrix}, \quad y^2 = \begin{bmatrix} 0.43471 \\ 0.98852 \\ 1.00203 \\ 0.46855 \end{bmatrix} \tag{3.63}$$

Substituting these two eigenvectors into Eqs. 3.57 through 3.60, one can easily verify that Eqs. 3.57 through 3.60 are inconsistent.

3.5 An Alternate Treatment of the Two Degree of Freedom Spring-Mass Example

An alternate and equivalent formulation of the two degree of freedom spring-mass example of Section 2.1 is to choose k_1 and k_2 to maximize the fundamental eigenvalue, subject to the condition that cost is fixed. That is, maximize

$$\overline{\psi}_0 = \min_{i=1,2} \varsigma_i \tag{3.64}$$

subject to the conditions

$$\left. \begin{array}{l} \overline{\psi}_1 = c_1 k_1 + c_2 k_2 = c \\[2mm] k_i \geq 0, \qquad i = 1,2 \end{array} \right\} \tag{3.65}$$

A necessary condition for this problem is obtained from results presented by Masur and Mroz in Ref. 2. They state the necessary condition

$$\varsigma_{k_0,h,1}^{(1)} \times \varsigma_{k_0,h,2}^{(2)} \leq 0 \tag{3.66}$$

for all design variations h that are consistent with constraints. At the known optimum design $k_1 = \varsigma_0/4$ and $k_2 = \varsigma_0$, both k_i are positive, so h need only satisfy

$$c_1 h_1 + c_2 h_2 = 0 \tag{3.67}$$

To use the necessary condition of Eq. 3.66, $\varsigma_{k_0,h,i}^{(1)}$ must be calculated. But $\varsigma_{k_0,h,i}^{(1)}$ are eigenvalues of the matrix $M(k_0,h)$, which were given in section 3.3 as $4h_1/3 + 2h_2/3$ and $4h_1$. Using this result, one has the necessary condition

$$\left(\frac{4}{3} h_1 + \frac{2}{3} h_2\right) 4h_1 \leq 0 \tag{3.68}$$

for all h_i satisfying Eq. 3.67. Solving Eq. 3.67 for h_2 and substituting into Eq. 3.68, one has

$$\left(\frac{4}{3} - \frac{2}{3} \frac{c_1}{c_2}\right) 4h_1^2 \leq 0 \tag{3.69}$$

which is just

$$\frac{c_1}{c_2} \geq 2 \tag{3.70}$$

which is the correct result, as shown in Section 2.1.

Since the product of all eigenvalues of a matrix is equal to the determinant of the matrix, the necessary condition of Eq. 3.66 is equivalent to

$$\det M(k_0,h) \leq 0 \tag{3.71}$$

For all variations h that satisfy Eq. 3.67. The determinant of a matrix is rotation invariant; i.e., $\det M(k_0,h) = \det R(\theta) M(k_0,h) R(\theta)^T$, where $R(\theta)$ is any orthogonal matrix. Thus, for given M-orthonormal eigenvectors y^1 and y^2 of Eq. 3.31 for the spring-mass example, if one uses the following eigenvectors (which are also M-orthonormal)

$$\tilde{y}^1 = \cos \theta\, y^1 + \sin \theta\, y^2$$
$$\tilde{y}^2 = -\sin \theta\, y^1 + \cos \theta\, y^2 \tag{3.72}$$

for the evaluation of the matrix $M(h)$, then the result will be the same as Eq. 3.71.

Thus the necessary condition of Ref. 2, which is based on only existence of directional derivatives of the repeated eigenvalues, yields correct results. This observation strengthens the case made in Section 3.3 and 3.4 that in treating optimal design problems in which repeated eigenvalues arise, one must use methods that rely only on existence of directional derivatives of the repeated eigenvalues. As shown in Section 3.4, use of formal Lagrange multiplier methods that rely on differentiability of functions arising in the optimization problem may yield incorrect results.

4. DERIVATION OF NECESSARY CONDITIONS OF OPTIMALITY FOR DISCRETE SYSTEMS

4.1 Introduction

In Section 3 necessary conditions of optimality were derived, using the fact that m-fold eigenvalues have directional derivatives [5]. However use of those necessary conditions for determination of an optimum design appears to be difficult (e.g., Eq. 3.12). In this section, it is shown that the smallest eigenvalue of buckling and vibration problems is quasi-differentiable [6, p. 68]. The quasi-differentiability of the smallest eigenvalue enables one to derive more useful necessary conditions.

In this section, the structure to be optimized is first approximated by finite element methods. The eigenvalue problem that arises in vibration or buckling of structures is then of the form

$$A(u)y = \varsigma B(u)y \qquad (4.1)$$

where $u \in R^n$, $y \in R^k$, and $A(u)$ and $B(u)$ are k x k matrices. In the problems considered here, $A(u)$ and $B(u)$ are symmetric and positive definite.

The statement of the problem is as follows; find $u \in R^n$ that minimizes the cost function $\psi_0(u)$, subject to Eq. 4.1 and the constraints

$$\psi_i(u) = \varsigma_0 - \varsigma(u) \leq 0 \qquad (4.2)$$

$$u \in Q = \{u \in R^n : c_i - u_i \leq 0, c_i > 0, i = 1,2,\ldots,n\} \qquad (4.3)$$

Here it is assumed that the cost function $\psi_0(u)$ is differentiable and $\varsigma(u)$ in Eq. 4.2 is the smallest eigenvalue of Eq. 4.1. It is also assumed that the smallest eigenvalue has multiplicity $m \geq 1$ at the optimum point $u_0 \in Q$.

4.2 Quasi-Differentiable Functionals

It is shown in Ref. 6 (p. 42) that if $\psi(u)$ is a real convex functional on a Banach space U, then the directional differential

$$\psi_{u_0,h}^{(1)} = \lim_{t\to+0} \frac{\psi(u_0 + th) - \psi(u_0)}{t} \tag{4.4}$$

exists for all u_0 and h, and is given by

$$\psi_{u_0,h}^{(1)} = \sup_{m\,\in\,\partial\psi(u_0)} m(h) \tag{4.5}$$

where $\partial\psi(u_0)$ is the subdifferential of ψ at u_0, which is defined as

$$\partial\psi(u_0) = \{m \in U^* : \psi(u) - \psi(u_0) \geq m(u - u_0) \text{ for all } u \in U\} \tag{4.6}$$

Moreover, $\partial\psi(u_0)$ is non-empty, convex, weak* closed, and bounded in U^* [6, p. 39].

A real functional $\psi(u)$ on a Banach space U is said to be quasi-differentiable at a point u_0 if there exists a convex, weak* closed, and bounded set $\partial\psi(u_0) \subset U^*$ such that Eq. 4.5 holds. Note that if $\psi(u)$ is quasi-differentiable, then its directional derivative $\psi_{u_0,h}^{(1)}$ is convex in h, since it is the supremum of linear functionals. The class of quasi-differentiable functionals contains all Fréchet and Gateaux differentiable functionals, because in these cases $\partial\psi(u_0)$ consists of a single element, the Fréchet or Gateaux differential. Also, Eq. 4.5 shows that all convex functionals are quasi-differentiable.

If $\psi_1(u)$ in Eq. 4.2 is convex in u, then it is quasi-differentiable at $u_0 \in Q$. Hence it is natural to ask whether $\psi_1(u)$ of Eq. 4.2 is convex or not. Of course the convexity of $\psi_1(u)$ depends on the operators A(u) and B(u) of each given problem. Since A(u) and B(u) of the problems considered here are positive definite, one can find the smallest eigenvalue as

$$\zeta(u) = \inf_{\substack{y \in D_A \\ y \neq 0}} \frac{(A(u)y,y)}{(B(u)y,y)} \tag{4.7}$$

Since the supremum of convex functionals is convex [11, p. 168]; if, for fixed $y \in D_A$, $-(A(u)y,y)/(B(u)y,y)$ is convex in u, then $\psi_1(u)$ is convex. For example, in the spring-mass system of Section 2.1, $-(A(k)y,y)/(B(k)y,y)$ is linear in k, so $\psi_1(k)$ in Eq. 4.2 is convex. However $\psi_1(u)$ is not always convex in u. A counter example is the column problem of Section 2.2. Figure 2.5 clearly shows that $\psi_1(b)$ is not convex for certain values of b_0. Another counter example can be constructed using the clamped-clamped column problem of Section 2.3, where

$$\zeta(u) = \inf_{\substack{y \in D_A \\ y \neq 0}} \frac{\int_0^\ell E\alpha u^2 y''^2 dx}{\int_0^\ell y'^2 dx} \tag{4.8}$$

Since for fixed $x \in [0,\ell]$, $u^2(x)$ is convex, one has

$$[\gamma_1 u(x) + \gamma_2 v(x)]^2 \leq \gamma_1 u^2(x) + \gamma_2 v^2(x) \tag{4.9}$$

for all $\gamma_1, \gamma_2 \geq 0$ $\gamma_1 + \gamma_2 = 1$. Thus

$$-\frac{\int_0^\ell E\alpha(\gamma_1 u + \gamma_2 v)^2 y''^2 dx}{\int_0^\ell y'^2 dx} \geq -\frac{\int_0^\ell E\alpha(\gamma_1 u^2 + \gamma_2 v^2) y''^2 dx}{\int_0^\ell y'^2 dx} \tag{4.10}$$

which is concave. More specifically, let $v = 2u$ and let $\gamma_1 = \gamma_2 = 1/2$. Then,

$$\psi_1(\gamma_1 u + \gamma_2 v) = \psi_1(\tfrac{3}{2} u) = \zeta_0 - \frac{9}{4} \inf_{\substack{y \in D_A \\ y \neq 0}} \frac{\int_0^\ell E\alpha u^2 y''^2 dx}{\int_0^\ell y'^2 dx} \tag{4.11}$$

and

$$\gamma_1 \psi_1(u) + \gamma_2 \psi_1(v) = \frac{1}{2} \psi_1(u) + \frac{1}{2} \psi_1(2u)$$

$$= \zeta_0 - \frac{5}{2} \inf_{\substack{y \in D_A \\ y \neq 0}} \frac{\int_0^\ell E\alpha u^2 y''^2 \, dx}{\int_0^\ell y'^2 \, dx} \tag{4.12}$$

so $\psi_1(\gamma_1 u + \gamma_2 v) \geq \gamma_1 \psi_1(u) + \gamma_2 \psi_1(v)$ and ψ_1 is not convex.

4.3 Quasi-Differentiability of the Eigenvalue

In this section, it is shown that $\psi_1(u)$ in Eq. 4.2 is quasi-differentiable at $u_0 \in Q$. To do so, first note that $\psi_1(u)$ is directionally differentiable at u_0 and $\psi_{u_0,h,1}^{(1)} = -\zeta_{u_0,h}^{(1)}$ where $\zeta_{u_0,h}^{(1)}$ is the smallest eigenvalue of the matrix [5]

$$M(u_0,h) = \left[\left(A_{u_0,h}^{(1)} y^i, y^j \right) - \zeta(u_0) \left(B_{u_0,h}^{(1)} y^i, y^j \right) \right] \tag{4.13}$$

where $\{y^i\}_{i=1,2,\ldots,m}$ is any $B(u)$-orthonormal basis of the eigenspace associated with $\zeta(u_0)$. Using the Rayleigh quotient, one has

$$\psi_{u_0,h,1}^{(1)} = \sup_{\substack{\xi \in R^m \\ ||\xi||=1}} (-M(u_0,h)\xi, \xi) \tag{4.14}$$

which is convex in h, since it is the supremum of linear functionals.

To show that $\psi_1(u)$ is quasi-differentiable at u_0, one has to find a convex, closed, and bounded set $\partial\psi_1(u_0) \subset R^n$ such that

$$\psi_{u_0,h,1}^{(1)} = \sup_{m \in \partial\psi_1(u_0)} (m,h) \tag{4.15}$$

Let $N = \{\xi \in R^m : ||\xi|| = 1\}$. As a functional of $\xi \in N$ and $\gamma \in R^n$, $(-M(u_0,h)\xi, \xi)$ in Eq. 4.14 has the Fréchet differential with respect to h, $(-M(u_0,e_i)\xi,\xi)_{i=1,2,\ldots,n}$, and satisfies the hypothesis of theorem 3.5 of Ref. 6, where N is compact and the e_i are vectors in R^n with all zero elements except the ith, which is one. Thus $\psi_{u_0,h,1}^{(1)}$, as a functional of h, is quasi-differentiable at h = 0. Moreover, the subdifferential of $\psi_{u_0,h,1}^{(1)}$ at h = 0 is

$$\partial\psi_{u_0,h,1}^{(1)}(0) = \{m \in R^n : (m,h) = \int_N (-M(u_0,h)\xi,\xi)\, d\mu(\xi)$$

$$= \int_N \sum_{i=1}^n (-M(u_0,e_i)\xi,\xi)\, h_i\, d\mu(\xi)\} \qquad (4.16)$$

where $d\mu(\xi)$ denotes a non-negative measure whose total variation on N is one; i.e., $\int_N d\mu(\xi) = 1$. Also, $\partial\psi_{u_0,h,1}^{(1)}(0)$ is convex, closed, and bounded (see proof of theorem 3.5 of Ref. 6) in R^n.

Now,

$$(m,h) = \int_N (-M(u_0,h)\xi,\xi)\, d\mu(\xi) \le \sup_{\xi \in N} (-M(u_0,h)\xi,\xi) \qquad (4.17)$$

on the other hand, by taking $d\mu(\xi)$ to be a Dirac measure at $\xi \in N$,

$$\psi_{u_0,h,1}^{(1)} = \sup_{\xi \in N} (-M(u_0,h)\xi,\xi) = \sup_{m \in \partial\psi_{u_0,h,1}^{(1)}} (m,h) \qquad (4.18)$$

Hence $\psi_1(u)$ is quasi-differentiable at u_0, with $\partial\psi_1(u_0) = \partial\psi_{u_0,h,1}^{(1)}(0)$. Note that for a general non-convex functional $\psi(u)$ with a directional derivative, the subdifferential of the functional at u_0 can be defined to be $\partial\psi_{u_0,h}^{(1)}(0)$ [11, p. 196].

4.4 Necessary Conditions of Optimality

Necessary conditions of optimality are now derived by evaluating the derivatives of the cost functional and constraint functionals. In order to apply theorem 4.1 of Ref. 6, one has to construct a convex cone K_Q, such that for any $h \in K_Q$, $u_\lambda = u_0 + \lambda h$ [6, p. 88] belongs to Q for sufficiently small $\lambda > 0$. Indeed, define

$$K_Q = \{h \in R^n : h = \gamma(u - u_0), u \in Q, \gamma > 0\} \tag{4.19}$$

Then $u_\lambda = u_0 + \lambda\gamma(u - u_0) = (1 - \lambda\gamma) u_0 + \lambda\gamma u \in Q$ for $\lambda \in [0, 1/\gamma]$, because Q is a convex set. Now by theorem 4.1 of Ref. 6, the necessary condition is that there exist $\lambda_0 \geq 0$ and $\lambda_1 \geq 0$, not both zero, such that

$$\lambda_0 \sum_{i=1}^{n} \frac{\partial \psi_0}{\partial u_i} (u_0) h_i - \lambda_1 \zeta_{u_0,h}^{(1)}$$

$$= \lambda_0 \sum_{i=1}^{n} \frac{\partial \psi_0}{\partial u_i} (u_0) h_i - \lambda_1 \inf_{\xi \in N} (M(u_0,h)\xi,\xi) \geq 0 \tag{4.20}$$

for all $h \in K_Q$ and $\lambda_1(\zeta_0 - \zeta(u_0)) = 0$, which is the complementary slackness condition. It should be noted that the above expression is not linear in h. Since $\zeta_{u_0,h}^{(1)}$ is positively homogeneous, by Eq. 4.18; i.e., $\zeta_{u_0,th}^{(1)} = t\zeta_{u_0,h}^{(1)}$ for all $t \geq 0$, the necessary condition of Eq. 4.20 is equivalent (be definition of K_Q) to

$$\lambda_0 \sum_{i=1}^{n} \frac{\partial \psi_0}{\partial u_i} (u_0) \left(u_i - u_{0_i}\right) - \lambda_1 \inf_{\xi \in N} (M(u_0,u - u_0)\xi,\xi) \geq 0 \tag{4.21}$$

for all $u \in Q$.

Since $\psi_1(u)$ is quasi-differentiable, Eq. 4.18 holds and the necessary condition of Eq. 4.20 is equivalent to

$$\lambda_0 \sum_{i=1}^{n} \frac{\partial \psi_0}{\partial u_i} (u_0) h_i + \lambda_1 \sup_{m \in \partial \psi_{u_0,h,1}^{(1)}(0)} (m,h) \geq 0 \tag{4.22}$$

for all $h \in K_Q$. Because $\partial \psi_{u_0, h, 1}^{(1)} (0)$ is bounded, by lemma 4.2 of Ref. 6, the necessary condition of Eq. 4.22 is equivalent to the existence of an $m \in \partial \psi_{u_0, h, 1}^{(1)} (0)$ (that is, existence of a measure $d\mu(\xi)$) such that

$$\lambda_0 \sum_{i=1}^{n} \frac{\partial \psi_0}{\partial u_i} (u_0) \, h_i - \lambda_1 \sum_{i=1}^{n} \int_N (M(u_0, e_i)\xi, \xi) \, d\mu(\xi) \, h_i \geq 0 \tag{4.23}$$

for all $h \in K_Q$. That is, there exists an $f \in R^n$ with components

$$f_i = \lambda_0 \frac{\partial \psi_0}{\partial u_i} (u_0) - \lambda_1 \int_N (M(u_0, e_i)\xi, \xi) \, d\mu(\xi) \tag{4.24}$$

that is an element of the dual cone K_Q^* of K_Q, where

$$K_Q^* = \{f \in R^n : (f, h) \geq 0 \text{ for all } h \in K_Q\} \tag{4.25}$$

and the necessary condition of Eq. 4.23 is equivalent to, by definition of K_Q,

$$(f, u - u_0) = \sum_{i=1}^{n} f_i \left(u_i - u_{0_i}\right) \geq 0 \tag{4.26}$$

for all $u \in Q$.

It is to be shown that the necessary condition of Eq. 4.26 is equivalent to

$$f_i = \lambda_0 \frac{\partial \psi_0}{\partial u_i} (u_0) - \lambda_1 \int_N (M(u_0, e_i)\xi, \xi) \, d\mu(\xi) \geq 0, \text{ if } i \in I(u_0)$$

$$f_j = \lambda_0 \frac{\partial \psi_0}{\partial u_j} (u_0) - \lambda_1 \int_N (M(u_0, e_j)\xi, \xi) \, d\mu(\xi) = 0, \text{ if } j \notin I(u_0) \tag{4.27}$$

where $I(u_0) = \{i : 1 \leq i \leq n, u_{0i} = c_i\}$.

Indeed, if $u_{0i} = c_i$, then let $u_j = u_{0j}$ for all $j \neq i$. Then by Eq. 4.26, $f_i(u_i - c_i) \geq 0$ where $u_i \geq c_i$, so $f_i \geq 0$. On the other hand, if $u_{0j} > c_j$, then let $u_i = u_{0i}$ for all $i \neq j$. By Eq. 4.26, $f_j(u_j - u_{0j}) \geq 0$. But since $u_{0j} > c_j$, $u_j - u_{0j}$ may take positive or negative values, so $f_j = 0$. Conversely, if the f_i's satisfy Eq. 4.27, then $\sum_{i=1}^{n} f_i(u_i - u_{0i}) \geq 0$, so Eq. 4.26 is satisfied.

Now if the design variable constraint is everywhere active before the eigenvalue constraint is active, then by the complementary slackness condition, $\lambda_1 = 0$ and the system is optimized. Hence one can assume that the design variable constraint is not everywhere active. Thus λ_1 will not be zero. If it were, λ_0 will be zero in Eq. 4.27, unless $\partial\psi_0/\partial u_i$ $(u_0) \geq 0$ for all $i \in I(u_0)$ and $\partial\psi_0/\partial u_j$ $(u_0) = 0$ for all $j \notin I(u_0)$. But if $\partial\psi_0/\partial u_j$ $(u_0) = 0$ for all $j \notin I(u_0)$, then one can eliminate the u_j's from the design variable vector and, in this case, the design variable constraint is everywhere active. Hence, one assumes $\partial\psi_0/\partial u_j$ $(u_0) \neq 0$ for all $j \notin I(u_0)$.

The next question is "when can one let $\lambda_0 = 1$?" Note in Eq. 4.20, if $\lambda_0 = 0$ then $-\lambda_1 \zeta_{u_0,h}^{(1)} \geq 0$ and $\zeta_{u_0,h}^{(1)} \leq 0$ for all $h \in K_Q$. Hence, of one shows that there exists an $h \in K_Q$ such that $\zeta_{u_0,h}^{(1)} > 0$, then one can set $\lambda_0 = 1 > 0$. Using $I(u_0)$, K_Q in Eq. 4.19 can be expressed as

$$K_Q = \{h \in R^n : h_i \geq 0 \text{ for } i \in I(u_0), h_j \in R \text{ for } j \notin I(u_0)\}$$
(4.28)

Define a subset of K_Q as

$$\overline{K} = \{h \in R^n : h_i = 0 \text{ for } i \in I(u_0), h_j \in R \text{ for } j \notin I(u_0)\}$$
(4.29)

Then, $\overline{K}$ is a subspace of R^n. In case design constraint is nowhere active, $I(u_0) = \emptyset$, and $K_Q = \overline{K} = R^n$. Consider two cases;

(A) If $M(u_0,e_j) = 0$ for any $j \notin I(u_0)$, then $\lambda_0 = 0$ by Eq. 4.27.

(B) If $M(u_0,e_j) \neq 0$ for any $j \notin I(u_0)$, then there is an h $\overline{K} \in K_Q$ such that $\zeta^{(1)}_{u_0,h} \neq 0$, which in turn implies that there exists an $h \in \overline{K} \subset K_Q$ such that $\zeta^{(1)}_{u_0,h} > 0$. Hence one can let $\lambda_0 = 1$.

To prove (B), first suppose $\zeta^{(1)}_{u_0,h} = 0$ for all $h \in \overline{K}$. Then by Eq. 4.14,

$$\inf_{\xi \in N} (M(u_0,h)\xi,\xi) = 0 \qquad (4.30)$$

for all $h \in \overline{K}$. Hence by substituting $-h$ for h, one has

$$\sup_{\xi \in N} (M(u_0,h)\xi,\xi) = 0 \qquad (4.31)$$

for all $h \in \overline{K}$. Thus, by Eqs. 4.30 and 4.31, the eigenvalues of the matrix $M(u_0,h)$ are all zero for all $h \in \overline{K}$. But since $M(u_0,h)$ is symmetric, this implies $M(u_0,h) = 0$, for all $h \in \overline{K}$. But then by taking $h = e_j$ for each $j \notin I(u_0)$, $M(u_0,e_j) = 0$ for all $j \notin I(u_0)$, which is a contradiction. Hence there is a $h \in \overline{K}$ such that $\zeta^{(1)}_{u_0,h} \neq 0$. Now since $\zeta^{(1)}_{u_0,h}$ is concave in h, by a corollary of Ref. 6 (p. 38),

$$\alpha(t) = \frac{\zeta^{(1)}_{u_0,th} - \zeta^{(1)}_{u_0,0}}{t} , \quad t \neq 0 \qquad (4.32)$$

is a nonincreasing function. In Eq. 4.32, if $t > 0$, then $\alpha(t) = \zeta^{(1)}_{u_0,h}$ and if $t < 0$, then $\alpha(t) = -\zeta^{(1)}_{u_0,-h}$. Thus, $\zeta^{(1)}_{u_0,h} \leq -\zeta^{(1)}_{u_0,-h}$. Hence if $\zeta^{(1)}_{u_0,h} \geq 0$, then $\zeta^{(1)}_{u_0,-h} \leq 0$. Now suppose that there is no $h \in \overline{K}$ such that $\zeta^{(1)}_{u_0,h} > 0$. Then for all

$h \in \overline{K}$, $\zeta^{(1)}_{u_0,h} = 0$ which is a contradiction.

From (A) and (B), if $M(u_0,e_j) = 0$ for any $j \notin I(u_0)$, then $M(u_0,e_j) = 0$ for all $j \notin I(u_0)$ and if $M(u_0,e_j) \neq 0$ for any $j \notin I(u_0)$, then $M(u_0,e_j) \neq 0$ for all $j \notin I(u_0)$. Moreover, if case (B) holds, then there exists an $h \in K_Q$ such that $\zeta^{(1)}_{u_0,h} > 0$. Hence there exists a $u \in Q$ such that $\psi_1(u) < 0$ which is the Slater condition.

In section 4.2 it was shown that the convexity of $\psi_1(u)$ depends on the operators $A(u)$ and $B(u)$ of the given problem. Suppose case (B) holds, then if $\psi_1(u)$ of a given problem is convex, the necessary condition of Eq. 4.27 is also a sufficient condition of optimality [11, p. 69], whereas if $\psi_1(u)$ is not convex, it may not be a sufficient condition.

However, even if $\psi_1(u)$ is not convex, one can show that up to first order approximation, it is impossible to have an improved design that satisfies the necessary condition. Indeed, if one wants to improve the design $u_0 \in Q$, then one is looking for a change δu in design that satisfies $\sum_{i=1}^{n} \frac{\partial \psi_0}{\partial u_i}(u_0) \delta u_i < 0$, while $\zeta^{(1)}_{u_0,\delta u} \geq 0$, so the eigenvalue constraint is not violated and $\delta u_i \geq 0$ for $i \in I(u_0)$. From Eq. 4.18,

$$\zeta^{(1)}_{u_0,\delta u} = \inf_{d\mu} \int \sum_{i=1}^{n} (M(u_0,e_i)\xi,\xi) \, d\mu(\xi) \, \delta u_i \geq 0 \tag{4.33}$$

so

$$\sum_{i \in I(u_0)} \lambda_1 \int_N (M(u_0,e_i)\xi,\xi) d\mu(\xi)\delta u_i$$

$$+ \sum_{j \notin I(u_0)} \lambda_1 \int_N (M(u_0,e_j)\xi,\xi) d\mu(\xi)\delta u_j \geq 0$$

which implies, using Eq. 4.27 with $\lambda_0 = 1$,

$$\sum_{i \,\in\, I(u_0)} \frac{\partial \psi_0}{\partial u_i}(u_0)\delta u_i + \sum_{j \,\neq\, I(u_0)} \frac{\partial \psi_0}{\partial u_j}(u_0)\delta u_j$$

$$= \sum_{i=1}^{n} \frac{\partial \psi_0}{\partial u_i}(u_0)\delta u_i \geq 0$$

which is a contradiction.

4.5 Simple Spring-Mass Example

It is of interest to verify the necessary condition of optimality of Eq. 4.27 for the simple spring-mass example of Section 2.1 at the optimum point, $k_0 = \left[\frac{\zeta_0}{4}, \zeta_0\right]^T$. Using M-orthonormal eigenvectors in Eq. 3.31,

$$(k_0,e_1) = \begin{bmatrix} 2 & \frac{2}{\sqrt{3}} \\ \frac{2}{\sqrt{3}} & \frac{10}{3} \end{bmatrix}, \quad (k_0,e_2) = \begin{bmatrix} \frac{1}{2} & -\frac{1}{2\sqrt{3}} \\ -\frac{1}{2\sqrt{3}} & \frac{1}{6} \end{bmatrix} \qquad (4.34)$$

Hence, $M(k_0,e_1) \neq 0$ and $M(k_0,e_2) \neq 0$, so one can let $\lambda_0 = 1$, which justifies Eq. 3.13.

Now the necessary condition of Eq. 4.27 is

$$\left. \begin{array}{l} c_1 - \lambda_1 \displaystyle\int_N \xi^T M(k_0,e_1)\xi d\mu(\xi) = 0 \\[2ex] c_2 - \lambda_1 \displaystyle\int_N \xi^T M(k_0,e_2)\xi d\mu(\xi) = 0 \end{array} \right\} \qquad (4.35)$$

A comparison of the necessary condition of Eq. 4.35 and the formal Lagrange multiplier result of Eqs. 3.24 and 3.25 gives an interesting observation. Since $\displaystyle\int_N d\mu(\xi) = 1$, from Eq. 4.35,

$$c_1 - 2c_2 = \lambda_1 \int_N \xi^T \begin{bmatrix} 1 & \sqrt{3} \\ \sqrt{3} & 3 \end{bmatrix} \xi \, d\mu(\xi)$$

$$\geq \lambda_1 \inf_{\xi \in N} \xi^T \begin{bmatrix} 1 & \sqrt{3} \\ \sqrt{3} & 3 \end{bmatrix} \xi \geq 0 \tag{4.36}$$

On the other hand, if one uses the eigenvector pair in Eq. 3.34, which diagonalizes $M(k_0,h)$ for all h, then,

$$M(k_0,e_1) = \begin{bmatrix} \dfrac{4}{3} & 0 \\ 0 & 4 \end{bmatrix}, \quad M(k_0,e_2) = \begin{bmatrix} \dfrac{2}{3} & 0 \\ 0 & 0 \end{bmatrix} \tag{4.37}$$

and the necessary condition of Eq. 4.35 is

$$\left.\begin{aligned} c_1 - \lambda_1 \left[\frac{4}{3} \int_N (\xi_1)^2 \, d\mu(\xi) + 4 \int_N (\xi_2)^2 \, d\mu(\xi) \right] &= 0 \\[2mm] c_2 - \lambda_1 \left[\frac{2}{3} \int_N (\xi_1)^2 \, d\mu(\xi) \right] &= 0 \end{aligned}\right\} \tag{4.38}$$

By letting $\int_N (\xi_1)^2 \, d\mu(\xi) = \alpha_1 \geq 0$ and $\int_N (\xi_2)^2 \, d\mu(\xi) = \alpha_2 \geq 0$, where $\alpha_1 + \alpha_2 = 1$, Eq. 4.38 is

$$\left.\begin{aligned} c_1 &= \lambda_1 \left(\frac{4}{3} \alpha_1 + 4 \alpha_2 \right) \geq 0 \\[2mm] c_2 &= \lambda_1 \left(\frac{2}{3} \alpha_1 \right) \geq 0 \end{aligned}\right\} \tag{4.39}$$

which is the same as Eq. 3.38. The validity of the Lagrange multiplier method in this case is due to the fact that m-fold eigenvalues are Fréchet differentiable if there exist m eigenvectors that diagonalize the matrix $M(u_0,h)$ for all h. However, in this case, one must use those m eigenvectors to have a valid variational formula.

5. DERIVATION OF NECESSARY CONDITIONS OF OPTIMALITY FOR DISTRIBUTED SYSTEMS

5.1 Introduction

In this section, methods of Section 4 are extended to distributed parameter systems, for which design and response variables of the system are functions. As in Section 4, it is to be shown that the smallest eigenvalue of a structural system is quasi-differentiable and using this quasi-differentiability, derive pointwise necessary conditions.

In this section the state equations are given in variational form, since the design variable u is in $L^\infty(\Omega)$ [5]. The general form of variational state equation is

$$a_u(y,v) = \varsigma b_u(y,v) \text{ for all } v \in V(\Omega) \tag{5.1}$$

where the physical domain Ω is a bounded open subset of R^n, $V(\Omega)$ is a Hilbert space that is dense in $L^2(\Omega)$, $a_u(y,v)$ is a strongly elliptic bilinear form on $V(\Omega)$, and $b_u(y,v)$ is a bilinear form on $L^2(\Omega)$.

The statement of the problem is as follows; find $u \in L^\infty(\Omega)$ that minimizes the cost function

$$\psi_0(u) = \int_\Omega F(u(x)) \, dx \tag{5.2}$$

subject to Eq. 5.1 and

$$\psi_1(u) = \varsigma_0 - \varsigma(u) \leq 0 \tag{5.3}$$

$$u \in Q = \{u(x) \in L^\infty(\Omega) : c - u(x) \leq 0 \text{ a.e. in } \Omega, c > 0\} \tag{5.4}$$

Here it is assumed that $\varsigma(u)$ is the smallest eigenvalue of Eq. 5.1 and that the smallest eigenvalue has multiplicity m at the optimum design $u_0 \in Q$. It is also assumed that $F(u)$ in Eq. 5.2 is continuous in u, differentiable with respect to u, and $F_u(u)$ is continuous in u. Then, $\psi_0(u)$ is Frechet differentiable and

$$\psi_{u_0,h,0}^{(1)} = \int_\Omega F_u(u_0(x))\; h(x)\; dx \qquad (5.5)$$

Indeed, by the mean value theorem,

$$\psi_0(u_0 + h) - \psi_0(u_0) = \int_\Omega [F(u_0(x) + h(x)) - F(u_0(x))]dx$$

$$= \int_\Omega F_u(u_0(x))h(x)\,dx + \int_\Omega [F_u(u_0(x) + \alpha(x)h(x))$$

$$- F_u(u_0(x))]h(x)\,dx$$

where $0 \le \alpha(x) \le 1$. But $F_u(u)$ is continuous on R, hence uniformly continuous on $S \subset R$ where $S = \{u : |u| \le ||u_0||_{L^\infty} + ||h||_{L^\infty}\}$, i.e., for every $\varepsilon > 0$, there exists a $\delta > 0$ such that

$$|F_u(u_1) - F_u(u_2)| \le \varepsilon \text{ for } u_1, u_2 \in S, \; |u_1 - u_2| < \delta$$

Since, $u_0(x) \in S$ for each x, $u_0(x) + \alpha(x)h(x) \in S$ for all $x \in \Omega$. Thus

$$|F_u(u_0(x) + \alpha(x)h(x)) - F_u(u_0(x))| \le \varepsilon$$

for all $||h||_{L^\infty} \le \delta$. Hence,

$$\sup_{x \in \Omega} |F_u(u_0(x) + \alpha(x)h(x)) - F_u(u_0(x))| \to 0$$

as $||h||_{L^\infty} \to 0$ and

$$|\int_\Omega [F_u(u_0(x) + \alpha(x)h(x)) - F_u(u_0(x))]\; h(x)\; dx|$$

$$\le \sup_{x \in \Omega} |F_u(u_0(x) + \alpha(x)h(x)) - F_u(u_0(x))|\; ||h||_{L^\infty} \int_\Omega 1\; dx$$

$$= K \sup_{x \in \Omega} |F_u(u_0(x) + \alpha(x)h(x)) - F_u(u_0(x))|\; ||h||_{L^\infty}$$

$$= o(||h||_{L^\infty}).$$

5.2 Quasi-Differentiability of the Eigenvalue

In this section, it is to be shown that $\psi_1(u)$ in Eq. 5.3 is quasi-differentiable at $u_0 \in Q$. To show this, first note that $\psi_1(u)$ is directionally differentiable at u_0 and $\psi_{u_0,h,1}^{(1)} = -\zeta_{u_0,h}^{(1)}$, where $\zeta_{u_0,h}^{(1)}$ is the smallest eigenvalue of the matrix [5]

$$M(u_0,h) = \left[a_{u_0,h}^{(1)} (y^i,y^j) - \zeta(u_0) b_{u_0,h}^{(1)} (y^i,y^j) \right] \tag{5.6}$$

where $\{y^i\}_{i=1,2,\ldots,m}$ is any b_u-orthonormal basis of the eigen-space of $\zeta(u_0)$. Using the Rayleigh quotient,

$$\psi_{u_0,h,1}^{(1)} = \sup_{\substack{\xi \in R^m \\ ||\xi||=1}} (-M(u_0,h)\xi,\xi) \tag{5.7}$$

which is convex in h, being the supremum of linear functionals.

To prove that $\psi_1(u)$ is quasi-differentiable at u_0, one has to find a convex, weak* closed, and bounded set $\partial\psi_1(u_0) \subset (L^\infty(\Omega))^*$ such that

$$\psi_{u_0,h,1}^{(1)} = \sup_{m \in \partial\psi_1(u_0)} (m,h) \tag{5.8}$$

Let $N = \{\xi \in R^m : ||\xi|| = 1\}$. Then as a functional of $\xi \in N$ and $h \in L^\infty(\Omega)$, $(-M(u_0,h)\xi,\xi)$ in Eq. 5.7 satisfies the hypothesis of theorem 3.5 of Ref. 6, with Fréchet differential with respect to h equal to $(-M(u_0,\cdot)\xi,\xi)$, where N is compact. Thus $\psi_{u_0,h,1}^{(1)}$, as a functional of h, is quasi-differentiable at h = 0. Moreover, the subdifferential of $\psi_{u_0,h,1}^{(1)}$ at h = 0 is

$$\partial\psi_{u_0,h,1}^{(1)} (0) = \{m \in (L^\infty(\Omega))^* : (m,h) = \int_N (-M(u_0,h)\xi,\xi)d\mu(\xi) \} \tag{5.9}$$

where $d\mu(\xi)$ denotes a non-negative measure whose total variation on N is equal to one; i.e., $\int_N d\mu(\xi) = 1$. Also, $\partial\psi_{u_0,h,1}^{(1)}(0)$ is convex, weak* closed, and bounded (see proof of theorem 3.5 of Ref. 6).

Now

$$(m,h) = \int_N (-M(u_0,h)\xi,\xi) \, d\mu(\xi) \leq \sup_{\xi \in N} (-M(u_0,h)\xi,\xi) \qquad (5.10)$$

On the other hand, by taking $d\mu(\xi)$ to be a Dirac measure at $\xi \in N$,

$$\psi_{u_0,h,1}^{(1)} = \sup_{\xi \in N} (-M(u_0,h)\xi,\xi) = \sup_{m \in \partial\psi_{u_0,h,1}^{(1)}(0)} (m,h) \qquad (5.11)$$

Hence $\psi_1(u)$ is quasi-differentiable at u_0, with $\partial\psi_1(u_0) = \partial\psi_{u_0,h,1}^{(1)}(0)$. In fact, as in the discrete case, $\partial\psi_{u_0,h,1}^{(1)}(0)$ can be defined to be the subdifferential of the functional $\psi_1(u)$ at u_0.

5.3 Necessary Conditions

As in the discrete case, in order to apply theorem 4.1 of Ref. 6, one has first to construct a convex cone K_Q such that for any $h \in K_Q$, $u_\lambda = u_0 + \lambda h$ [6, p. 88] belongs to Q, for all $\lambda > 0$ sufficiently small. As in the discrete case, one can set

$$K_Q = \{h \in L^\infty(\Omega) : h = \gamma(u(x) - u_0(x)), \, u \in Q, \, \gamma > 0\} \qquad (5.12)$$

This is true, since $u_\lambda = u_0 + \lambda\gamma(u - u_0) = (1 - \lambda\gamma) u_0 + \lambda\gamma u \in Q$ for $\lambda \in [0, 1/\gamma]$ and Q is a convex set. By theorem 4.1 of Ref. 6, the necessary condition of optimality is that there exist $\lambda_0 \geq 0$ and $\lambda_1 \geq 0$, not both zero, such that

$$\lambda_0 \int_\Omega F_u(u_0(x)) \, h(x) \, dx - \lambda_1 \, \zeta_{u_0,h}^{(1)}$$

$$= \lambda_0 \int_\Omega F_u(u_0(x)) \, h(x) \, dx - \lambda_1 \inf_{\xi \in N} (M(u_0,h)\xi,\xi) \geq 0 \quad (5.13)$$

for all $h \in K_Q$. Moreover, $\lambda_1(\zeta_0 - \zeta(u_0)) = 0$, which is the complementary slackness condition. Note that the above expression is not linear in h. Here $\zeta_{u_0,h}^{(1)}$ is positively homogeneous by

Eq. 5.11; i.e., $\zeta_{u_0,th}^{(1)} = t\zeta_{u_0,h}^{(1)}$ for all $t \geq 0$. Hence, by the

definition of K_Q, the necessary condition of Eq. 5.13 is equivalent to

$$\lambda_0 \int_\Omega F_u(u_0(x)) \, [u(x) - u_0(x)] \, dx - \lambda_1 \inf_{\xi \in N} (M(u_0,u-u_0)\xi,\xi) \geq 0$$
$$(5.14)$$

for all $u \in Q$.

By Eq. 5.11, the necessary condition of Eq. 5.13 is equivalent to

$$\lambda_0 \int_\Omega F_u(u_0(x)) h(x) \, dx + \lambda_1 \sup_{m \in \partial\psi_{u_0,h,1}^{(1)}(0)} (m,h) \geq 0 \quad (5.15)$$

for all $h \in K_Q$. Since $\partial\psi_{u_0,h,1}^{(1)}(0)$ is bounded, by lemma 4.2 of

Ref. 6, the above condition is equivalent to the existence of an $m \in \partial\psi_{u_0,h,1}^{(1)}(0)$ (i.e., existence of a measure $d\mu(\xi)$) such that $f \in (L^\infty(\Omega))^*$ defined by

$$(f,\overline{u}) = \lambda_0 \int_\Omega F_u(u_0(x)) \, \overline{u}(x) \, dx + \lambda_1(m,\overline{u})$$

$$= \lambda_0 \int_\Omega F_u(u_0(x)) \, \overline{u}(x) \, dx - \lambda_1 \int_N (M(u_0,\overline{u})\xi,\xi) \, d\mu(\xi) \quad (5.16)$$

for all $\bar{u} \in L^\infty(\Omega)$, is an element of dual cone K_Q^*. That is,

$$(f,h) = \lambda_0 \int_\Omega F_u(u_0(x)) h(x) dx$$

$$- \lambda_1 \int_N (M(u_0,h)\xi,\xi) d\mu(\xi) \geq 0 \qquad (5.17)$$

for all $h \in K_Q$. By definition of K_Q, the necessary condition of Eq. 5.17 is equivalent to

$$(f,u - u_0) = \lambda_0 \int_\Omega F_u(u_0(x)) [u(x) - u_0(x)] dx$$

$$- \lambda_1 \int_N (M(u_0,u - u_0)\xi,\xi) d\mu(\xi) \geq 0 \qquad (5.18)$$

for all $u \in Q$. Let $\bar{M}_{ij}(u_0,\bar{u})$ denote the integrands of the elements of the matrix $M(u_0,\bar{u})$; i.e.,

$$M_{ij}(u_0,\bar{u}) = \int_\Omega \bar{M}_{ij}(u_0,\bar{u}) dx \qquad (5.19)$$

Then Eq. 5.16 can be expressed as

$$(f,\bar{u}) = \lambda_0 \int_\Omega F_u(u_0(x)) \bar{u}(x) dx - \lambda_1 \int_N \int_\Omega (\bar{M}(u_0,\bar{u})\xi,\xi) dx\, d\mu(\xi)$$

$$= \lambda_0 \int_\Omega F_u(u_0(x)) \bar{u}(x) dx - \lambda_1 \int_\Omega \bar{u}(x) \int_N (\bar{M}(u_0,1)\xi,\xi) d\mu(\xi) dx$$

$$= \int_\Omega \left[\lambda_0 F_u(u_0(x)) - \lambda_1 \int_N (\bar{M}(u_0,1)\xi,\xi) d\mu(\xi) \right] \bar{u}(x) dx \qquad (5.20)$$

where Fubini's theorem [14, p. 150] is used in the second equality. Thus

$$f(x) = \lambda_0 F_u(u_0(x)) - \lambda_1 \int_N (\overline{M}(u_0,1)\xi,\xi) \, d\mu(\xi) \tag{5.21}$$

and $f \in L^1(\Omega)$. By same method used in example 10.5 of Ref. 11 (in Ref. 11, $\Omega \subset R$ is considered which can be extended to $\Omega \subset R^n$), the necessary condition of Eq. 5.18 is equivalent to

$$\left(\lambda_0 F_u(u_0(x)) - \lambda_1 \int_N (\overline{M}(u_0,1)\xi,\xi) \, d\mu(x)\right)[u(x) - u_0(x)] \geq 0 \tag{5.22}$$

for all $u \in Q$ and almost all $x \in \Omega$.

It is to be shown that the necessary condition of Eq. 5.22 is equivalent to

$$f(x) = \lambda_0 F_u(u_0(x)) - \lambda_1 \int_N (\overline{M}(u_0,1)\xi,\xi) d\mu(\xi) \geq 0 \text{ a.e. in } \Omega/I(u_0)$$

$$f(x) = \lambda_0 F_u(u_0(x)) - \lambda_1 \int_N (\overline{M}(u_0,1)\xi,\xi) d\mu(\xi) = 0 \text{ a.e. in } I(u_0) \tag{5.23}$$

where $I(u_0) = \{x \in \Omega: \ u_0(x) > c \text{ a.e.}\}$, which is the set the design variable constraint is not active almost everywhere.

Indeed, if $f(x)$ satisfies Eq. 5.23, then $f(x)[u(x) - u_0(x)] \geq 0$ for all $u \in Q$ and a.e. in Ω. Conversely, if $u_0(x) = c$ a.e. in the subset $\Omega/I(u_0)$ with strictly positive measure, then $u(x) - u_0(x)$ is non-negative a.e. in this set and $f(x) \geq 0$ a.e. in $\Omega/I(u_0)$, by Eq. 5.22. On the other hand, if $u_0(x) > c$ a.e. in the subset $I(u_0)$ with strictly positive measure, then $f(x) = 0$ a.e. in $I(u_0)$, by Eq. 5.22, because $u(x) - u_0(x)$ may take a positive or a negative value in this set.

Now if the design variable constraint is active almost everywhere, before the eigenvalue constraint is active, then by the complementary slackness condition $\lambda_1 = 0$ and the system is optimized. Hence, one can assume that the design variable constraint is not almost everywhere active. As in the discrete case, assume $F_u(u_0(x)) \neq 0$ a.e. in $I(u_0)$. Then λ_1 will not be zero, because otherwise, λ_0 will be zero in Eq. 5.23, which is a contradiction.

As in the discrete case, it is to be determined when one can let $\lambda_0 = 1$. In Eq. 5.13, if $\lambda_0 = 0$, then $-\lambda_1 \zeta_{u_0,h}^{(1)} \geq 0$ and $\zeta_{u_0,h}^{(1)} \leq 0$ for all $h \in K_Q$. Hence, if one shows that there exists an $h \in K_Q$ such that $\zeta_{u_0,h}^{(1)} > 0$, then one can let $\lambda_0 = 1 > 0$. The set K_Q in Eq. 5.12 can be written as

$$K_Q = \{h \in L^\infty(\Omega) : h(x) \geq 0 \text{ a.e. in } \Omega/I(u_0) \text{ and}$$

$$h(x) \in R \text{ a.e. in } I(u_0)\} \tag{5.24}$$

Define a subset of K_Q as

$$\overline{K} = \{h \in L^\infty(\Omega) : h(x) = 0 \text{ a.e. in } \Omega/I(u_0) \text{ and}$$

$$h(x) \in R \text{ a.e. in } I(u_0)\} \tag{5.25}$$

which is a subspace of $L^\infty(\Omega)$. Note that if the design variable constraint is not active almost everywhere, then $I(u_0) = \Omega$ and $K_Q = \overline{K} = L^\infty(\Omega)$.

Consider two cases;

(A) If $\overline{M}(u_0,1) = 0$ for any subset of $I(u_0)$ with strictly positive measure, then $\lambda_0 = 0$ by Eq. 5.23.

(B) If $\overline{M}(u_0,1) \neq 0$ for any subset of $I(u_0)$ with strictly positive measure, then there is an $h \in \overline{K} \subset K_Q$ such that $\zeta_{u_0,h}^{(1)} \neq 0$, which in turn implies that there exists an $h \in \overline{K} \subset K_Q$ such that $\zeta_{u_0,h}^{(1)} > 0$. Hence one can let $\lambda_0 = 1$.

To prove (B), first suppose $\zeta_{u_0,h}^{(1)} = 0$ for all $h \in \overline{K}$. Then by Eq. 5.7,

$$\inf_{\xi \in N} (M(u_0,h)\xi,\xi) = 0 \qquad (5.26)$$

for all $h \in \overline{K}$. Hence by substituting $-h$ for h, one has

$$\sup_{\xi \in N} (M(u_0,h)\xi,\xi) = 0 \qquad (5.27)$$

for all $h \in \overline{K}$. By Eqs. 5.26 and 5.27, the eigenvalues of the matrix $M(u_0,h)$ are all zero for all $h \in \overline{K}$. But since $M(u_0,h)$ is symmetric, this implies

$$M(u_0,h) = \int_\Omega \overline{M}(u_0,1)h(x)\,dx = \int_{I(u_0)} \overline{M}(u_0,1)h(x)\,dx = 0 \quad (5.28)$$

for all $h \in \overline{K}$. Here, $\overline{M}_{ij}(u_0,1) \in L^1(I(u_0))$, so by the corollary 4.2.6 of Ref. 13, $\overline{M}(u_0,1) = 0$ a.e. in $I(u_0)$, which is a contradiction. By the same method as in the discrete case, if $\zeta_{u_0,h}^{(1)} \geq 0$, then $\zeta_{u_0,-h}^{(1)} \leq 0$ for all $h \in \overline{K}$. Suppose there is no $h \in \overline{K}$ such that $\zeta_{u_0,h}^{(1)} > 0$, then for all $h \in \overline{K}$, $\zeta_{u_0,h}^{(1)} = 0$, which is a contradiction.

From (A) and (B), either $\overline{M}(u_0,1) = 0$ a.e. in $I(u_0)$ or $\overline{M}(u_0,1) \neq 0$ a.e. in $I(u_0)$. Moreover, if case (B) holds, then there exists an $h \in K_Q$ such that $\zeta_{u_0,h}^{(1)} > 0$. Hence there exists a $u \in Q$ such that $\psi_1(u) < 0$, which is the Slater condition.

When case (B) holds, if $\psi_1(u)$ of Eq. 5.3 is convex, then the necessary condition of Eq. 5.23 is also a sufficient condition of optimality [11, p. 69], whereas if $\psi_1(u)$ is not convex, it may not be sufficient.

Even if $\psi_1(u)$ is not convex, if case (B) holds, then one can prove that it is not possible, up to first order approximation, to have a better design than u_0, which satisfies the necessary condition.

Indeed, to improve the design u_0, one is looking for a change δu in design that satisfies $\int_\Omega F_u(u_0(x))\delta u(x)dx < 0$, with $\zeta^{(1)}_{u_0,\delta u} \geq 0$, that does not violate the eigenvalue constraint and $\delta u(x) \geq 0$ a.e. in $\Omega/I(u_0)$. From Eq. 5.11,

$$\zeta^{(1)}_{u_0,\delta u} = \inf_{d\mu} \int_N (M(u_0,\delta u)\xi,\xi) \; d\mu(\xi) \geq 0 \tag{5.29}$$

so by Fubini's theorem [14, p. 150],

$$\lambda_1 \int_{\Omega/I(u_0)} \delta u(x) \int_N (\overline{M}(u_0,1)\xi,\xi)d\mu(\xi)dx$$

$$+ \lambda_1 \int_{I(u_0)} \delta u(x) \int_N (\overline{M}(u_0,1)\xi,\xi)d\mu(\xi)dx \geq 0$$

which implies, using Eq. 5.23 with $\lambda_0 = 1$,

$$\int_{\Omega/I(u_0)} F_u(u_0(x))\delta u(x)dx + \int_{I(u_0)} F_u(u_0(x))\delta u(x)dx$$

$$= \int_\Omega F_u(u_0(x))\delta u(x)dx \geq 0$$

which is a contradiction.

5.4 Clamped-Clamped Column Example

Necessary Conditions. In this section, the buckling problem of the clamped-clamped column of Section 2.3 [1,8] is analyzed, using the necessary conditions developed in the previous sections. Instead of looking for $u \in L^\infty(0,\ell)$, as in Section 2.3, u will be restricted to be in $c(0,\ell)$, as in Refs. 1 and 8. Introducing the dimensionless variables

$$\overline{u}(x) = \frac{u(x)\ell}{V}, \quad \overline{\zeta} = \frac{3\ell^4}{E\alpha V^2} \tag{5.30}$$

where the coordinate x is non-dimensionalized by ℓ, the column problem of Section 2.3 becomes; find $u \in c(0,1)$ that minimizes

$$\psi_0(u) = -\zeta(u) \tag{5.31}$$

subject to the volume constraint

$$\psi_1(u) = \int_0^1 u(x)\,dx - 1 = 0 \tag{5.32}$$

and design variable constraint

$$Q = \{u \in c(0,1) : c \leq u(x), \ c > 0\} \tag{5.33}$$

where $\zeta(u)$ is the smallest eigenvalue of

$$\left. \begin{array}{l} (u^2(x)y'')'' = -\zeta y'' \\ y(0) = y'(0) = y(1) = y'(1) = 0 \end{array} \right\} \tag{5.34}$$

In variational notation, Eq. 5.34 is

$$a_u(y,v) \equiv \int_0^1 u^2(x)y''v''\,dx = \zeta \int_0^1 y'v'\,dx = \zeta b_u(y,v) \tag{5.35}$$

It is assumed that at the optimum design $u_0(x)$, the eigenvalue $\zeta(u_0)$ is double. Note that Eq. 5.32 is an equality constraint and $\psi_1(u)$ is a linear functional. Thus, a corollary of Ref. 6 (p. 88) applies and one can set

$$K_Q = \{h \in c(0,1) : h = \gamma(u(x) - u_0(x)), \ u \in Q, \ \gamma > 0\} \tag{5.36}$$

and the same method used in Section 5.3 applies. Hence, the necessary condition is that there exist $\lambda_0 \geq 0$ and λ_1, not both zero, such that

$$-\lambda_0 \zeta_{u_0,h}^{(1)} + \lambda_1 \int_0^1 h(x)\,dx = -\lambda_0 \inf_{\xi \in N} (M(u_0,h)\xi,\xi) + \lambda_1 \int_0^1 h(x)\,dx$$

$$= \lambda_0 \sup_{m \in \partial\psi_{u_0,h,0}^{(1)}(0)} (m,h) + \lambda_1 \int_0^1 h(x)\,dx \geq 0 \tag{5.37}$$

for all $h \in K_0$, where $\partial\psi_{u_0,h,0}^{(1)}(0)$ is the subdifferential of the functional $\psi_0(u)$ at u_0. By proceeding in same way as in Section 5.3, one gets the equivalent pointwise necessary conditions

$$\left.\begin{array}{l} -\lambda_0 \int_N (\overline{M}(u_0,1)\xi,\xi)\,d\mu(\xi) + \lambda_1 \geq 0 \quad \text{in } [0,1]/I(u_0) \\[2ex] -\lambda_0 \int_N (\overline{M}(u_0,1)\xi,\xi)\,d\mu(\xi) + \lambda_1 = 0 \quad \text{in } I(u_0) \end{array}\right\} \tag{5.38}$$

where $I(u_0) = \{x \in [0,1] : u_0(x) > c\}$, $\xi_1^2 + \xi_2^2 = 1$, and $\overline{M}(u_0,1)$ is the integrand of the matrix $M(u_0,1)$,

$$\overline{M}(u_0,1) = 2u_0(x) \begin{bmatrix} (y^{1''})^2 & y^{1''}y^{2''} \\[2ex] y^{1''}y^{2''} & (y^{2''})^2 \end{bmatrix} \tag{5.39}$$

If $0 < c < 1$, then the design variable constraint will not be active everywhere and λ_0 will not be zero. If it were zero, λ_1 will be zero by Eq. 5.38, which is a contradiction. Hence, one can set $\lambda_0 = 1 > 0$. Also, since $\overline{M}(u_0,1) \neq 0$ for any subset of $I(u_0)$, $\lambda_1 \neq 0$ (see Section 5.3). In fact, since $\overline{M}(u_0,1)$ is positive semi-definite for each x, $\lambda_1 > 0$ by Eq. 5.38.

Define a function $\nu(x)$ such that $\nu(x) \geq 0$ if $u_0(x) = c$ and $\nu(x) = 0$ if $u_0(x) > c$. Then the necessary condition of Eq. 5.38 becomes

$$\int_N (\overline{M}(u_0,1)\xi,\xi)\,d\mu(\xi) - \lambda_1 + \nu(x) = 0 \tag{5.40}$$

The necessary condition of Eq. 5.40 is different from the result of Olhoff and Rasmussen [1]. Note that

$$\int_N (M(u_0,h)\xi,\xi)d\mu(\xi) = \int_N \xi^T \begin{bmatrix} M_{11}(u_0,h) & 0 \\ 0 & M_{22}(u_0,h) \end{bmatrix} \xi \; d\mu(\xi)$$

$$= M_{11}(u_0,h) \int_N \xi_1^2 \; d\mu(\xi) + M_{22}(u_0,h) \int_N \xi_2^2 \; d\mu(\xi) \qquad (5.41)$$

for all $h \in K_Q$ if and only if $M_{12}(u_0,h) = M_{21}(u_0,h) = 0$, for all $h \in K_Q$. Let $\int_N \xi_1^2 \; d\mu(\xi) = 1 - \gamma$, $0 \leq \gamma \leq 1$. Then $\int_N \xi_2^2 d\mu(\xi) = \int_N (1 - \xi_1^2) \; d\mu(\xi) = \gamma$. Thus, the necessary condition of Eq. 5.37 is equivalent to

$$-\lambda_0 \left[(1 - \gamma) M_{11}(u_0,h) + \gamma M_{22}(u_0,h) \right] + \lambda_1 \int_0^1 h(x)dx \geq 0 \qquad (5.42)$$

for all $h \in K_Q$ if and only if $M_{12}(u_0,h) = 0$ for all $h \in K_Q$. By the same procedure used in Section 5.3, the functional necessary condition of Eq. 5.42 is equivalent to the pointwise necessary condition (with $\lambda_0 = 1$)

$$2u_0(x) \left[(1 - \gamma)(y^{1''})^2 + \gamma(y^{2''})^2 \right] - \lambda_1 + \nu(x) = 0 \qquad (5.43)$$

where $\nu(x) \geq 0$ if $u_0(x) = c$ and $\nu(x) = 0$ if $u_0(x) > c$. Equation 5.43 is the result of Ref. 1. Hence the necessary condition of Eq. 5.40 is equivalent to Eq. 5.43 if and only if

$$M_{12}(u_0,h) = \int_0^1 2u_0(x) \; y^{1''}y^{2''} h(x) \; dx = 0 \qquad (5.44)$$

for all $h \in K_Q$.

Since the boundary conditions at both ends are identical and the design variable constraint is symmetric with respect to the center of the column, the area of the optimum design will be symmetric. Hence, one can restrict the variations $h \in K_Q$ to be symmetric. Also, at the optimum, one can choose two b_u-ortho-normal eigenfunctions, which give the same eigenvalue, so that one is symmetric and the other is antisymmetric. Then $u_0(x)y^{1''}y^{2''} h(x)$ is antisymmetric and Eq. 5.44 is satisfied, for all symmetric $h \in K_Q$. Hence in this case the result of Ref. 1 is obtained.

Alternate Necessary Conditions. There is an alternate way to derive necessary conditions, which was obtained by Masur and Mroz [2] (see Section 3.5). Their necessary condition is that

$$\zeta_{u_0,h,1}^{(1)} \times \zeta_{u_0,h,2}^{(2)} \leq 0 \tag{5.45}$$

for all variations h that are consistent with constraints; i.e., for all $h \in K_Q$ such that $\int_0^1 h(x)dx = 0$, where $\zeta_{u_0,h,i}^{(1)}$ are eigen-values of the matrix $M(u_0,h)$. Since the product of all eigen-values of a matrix is equal to the determinant of the matrix, the necessary condition of Eq. 5.45 is equivalent to

$$\det M(u_0,h) \leq 0 \tag{5.46}$$

for all $h \in K_Q$ such that $\int_0^1 h(x)dx = 0$. For notational conven-ience, let

$$M(u_0,h) = \begin{bmatrix} \int_0^1 2u_0(x)h(x)\left(y^{1''}\right)^2 dx & \int_0^1 2u_0(x)h(x) y^{1''}y^{2''} dx \\ \int_0^1 2u_0(x)h(x) y^{1''}y^{2''} dx & \int_0^1 2u_0(x)h(x)\left(y^{2''}\right)^2 dx \end{bmatrix}$$

$$= \begin{bmatrix} a(h) & c(h) \\ c(h) & b(h) \end{bmatrix} \tag{5.47}$$

Then, the necessary condition of Eq. 5.46 is

$$a(h)b(h) - c^2(h) \leq 0 \tag{5.48}$$

for all $h \in K_Q$ such that $\int_0^1 h(x)\,dx = 0$.

It is to be shown that the necessary condition of Eq. 5.48 is equivalent to the necessary condition of Eq. 5.37.

Indeed, since $\inf_{\xi \in N} (M(u_0,h)\xi,\xi)$ is the smallest eigenvalue of the matrix $M(u_0,h)$, the necessary condition of Eq. 5.37 can be written as follows: there exist $\lambda_0 \geq 0$ and λ_1, not both zero, such that

$$\lambda_0 \psi_{u_0,h,0}^{(1)} + \lambda_1 \psi_{u_0,h,1}^{(1)}$$

$$= -\lambda_0 \frac{a + b - \sqrt{(a+b)^2 - 4(ab - c^2)}}{2} + \lambda_1 \int_0^1 h(x)\,dx \geq 0 \tag{5.49}$$

for all $h \in K_Q$. Suppose Eq. 5.49 holds, then let $h \in K_Q$ be such that $\int_0^1 h(x)\,dx = 0$. Then, from Eq. 5.49

$$- \frac{a + b - \sqrt{(a+b)^2 - 4(ab - c^2)}}{2} \geq 0$$

or

$$a(h)\,b(h) - c^2(h) \leq 0 \tag{5.50}$$

for all $h \in K_Q$ such that $\int_0^1 h(x)\,dx = 0$. Conversely, assume Eq. 5.48 holds. Then

$$\psi_{u_0,h,0}^{(1)} = - \frac{a + b - \sqrt{(a+b)^2 - 4(ab - c^2)}}{2} \geq 0 \tag{5.51}$$

for all $h \in K_Q$ such that $\psi_{u_0,h,1}^{(1)} = \int_0^1 h(x)\,dx = 0$. Let

$$\widetilde{K} = \{\xi \in R^2 : \xi_i = \psi^{(1)}_{u_0,h,i}, \ h \in K_Q, \ i = 0,1\}$$

$$(5.52)$$

$$P = \{n \in R^2 : n_0 < 0, \ n_1 = 0\}$$

Then Eq. 5.51 implies that the convex hull of $\widetilde{K}$ and the set P have empty intersection. To show this, suppose the assertion is false. Then there exists a $\overline{\xi} \in \text{co}(\widetilde{K})$ such that $\overline{\xi}_0 < 0$ and $\overline{\xi}_1 = 0$. Since $\overline{\xi} \in \text{co}\,(\widetilde{K})$, there exist elements $\xi^j \in \widetilde{K}$ such that

$$\overline{\xi} = \sum_j \alpha_j \ \xi^j$$

$$(5.53)$$

where $\alpha_j \geq 0$ and $\sum_j \alpha_j = 1$. Since $\xi^j \in \widetilde{K}$, then for some $h^j \in K_Q$,

$$\xi_0^j = \psi^{(1)}_{u_0,h^j,0}$$

$$\left.\vphantom{\begin{matrix}a\\a\end{matrix}}\right\}$$

$$(5.54)$$

$$\xi_1^j = \psi^{(1)}_{u_0,h^j,1}$$

Let $\overline{h} = \sum_j \alpha_j \ h^j$. Then $\overline{h} \in K_Q$, since K_Q is a convex cone. Thus

$$\psi^{(1)}_{u_0,\overline{h},0} \leq \sum_j \alpha_j \ \psi^{(1)}_{u_0,h^j,0} = \sum_j \alpha_j \ \xi_0^j = \overline{\xi}_0 < 0$$

$$\left.\vphantom{\begin{matrix}a\\a\\a\end{matrix}}\right\}$$

$$(5.55)$$

$$\psi^{(1)}_{u_0,\overline{h},1} = \sum_j \alpha_j \ \psi^{(1)}_{u_0,h^j,1} = \sum_j \alpha_j \ \xi_1^j = \overline{\xi}_1 = 0$$

because $\psi^{(1)}_{u_0,h,0}$ is convex and $\psi^{(1)}_{u_0,h,1}$ is linear in h. But Eq. 5.55 is a contradiction to Eq. 5.51. Since $\text{co}(\widetilde{K})$ and P are convex sets in R^2, they can be separated [6, p. 24]. Thus, there exists a nonzero vector $[\lambda_0, \lambda_1]^T$ such that

$$\sum_{i=0}^{1} \lambda_i \ \xi_i \geq 0 \geq \sum_{i=0}^{1} \lambda_i \ n_i$$

$$(5.56)$$

276

for all $\xi \in \tilde{K}$ and $\eta \in P$.   Since $\xi \in \tilde{K}$ can be expressed as
$\xi_i = \psi^{(1)}_{u_0,h,i}$, $h \in K_Q$ and one has

$$\lambda_0 \psi^{(1)}_{u_0,h,0} + \lambda_1 \psi^{(1)}_{u_0,h,1} \geq 0 \tag{5.57}$$

for all $h \in K_Q$.   Also, by the right-hand side Eq. 5.56, one has
$\lambda_0 \geq 0$.

REFERENCES

1. Olhoff, N. and Rasmussen, S.H., "On Single and Bimodal Optimum Buckling Loads of Clamped Columns," Int. J. Solids and Structures, Vol. 13, 1977, pp. 605-614.
2. Masur, E.F. and Mroz, Z., "Singular Solutions in Structural Optimization Problems," Proceedings, IUTAM Symposium On Variational Methods In The Mechanics Of Solids, (Ed. S. Nemat-Nasser), Pergamon Press, New York, to appear 1980.
3. Prager, S. and Prager, W., "A Note on Optimal Design of Columns," Int. J. Mech. Sci., Vol. 21, 1979, pp. 249-251.
4. Haug, E.J. and Rousselet, B., "Design Sensitivity Analysis in Structural Mechanics I: Static Response Variations," J. Str. Mech., Vol. 8, No. 1, 1980, pp. 17-41.
5. Haug, E.J. and Rousselet, B., "Design Sensitivity Analysis in Structural Mechanics II: Eigenvalue Variations," J. Str. Mech., Vol. 8, No. 2, 1980.
6. Pschenichnyi, B.N., Necessary Conditions for an Extremum, Marcel Dekker, New York, 1971.
7. Haug, E.J. and Arora, J.S., Applied Optimal Design, Wiley-Interscience, New York, 1979.
8. Tadjbakhsh, I. and Keller, J.B., "Strongest Columns and Iso-parametric Inequalities for Eigenvalues," J. of Applied Mechanics, Vol. 29, 1962, pp. 159-164.
9. Prager, W. and Taylor, J.E., "Problems of Optimal Structural Design," J. Applied Mechanics, Vol. 35, No. 1, 1968, pp. 102-106.
10. Niordson, F.I., "On the Optimal Design of a Vibrating Beam," Quarterly Applied Math., Vol. 23, No. 1, 1965, pp. 47-53.
11. Ioffe, A.D. and Tihomirov, V.M., Theory of Extremal Problems, North-Holland Publishing Co., New York, 1979.
12. Girsanov, I.V., Lecture Notes in Economics and Mathematical Systems, Springer-Verlag, New York, 1972.
13. Larsen, R., Functional Analysis, Marcel Dekker, New York, 1973.

14. Rudin, W., _Real and Complex Analysis_, McGraw-Hill, Inc.,
New York, 1974.
15. Coddington, E.A. and Levinson, N., _Theory of Ordinary
Differential Equations_, McGraw-Hill, Inc., New York, 1955.
16. Garbow, B.S., Boyle, J.M., Dongarra, J.J., and Moler, C.B.,
Lecture Notes in Computer Science, Springer-Verlag, New York,
1977.

OPTIMAL DESIGN OF SOLID ELASTIC PLATES

Niels Olhoff and Keng-Tung Cheng[*]

Department of Solid Mechanics, The Technical University
of Denmark, DK-2800 Lyngby, Denmark

ABSTRACT

This paper presents formulations, necessary conditions of
optimality, and solutions of problems of optimal design of solid
elastic plates. It is shown that only local optima should be
expected if one models the plate as an isotropic solid.
Further, results depend on the discretization used. An anisotropic formulation is then presented, which alleviates some of
these difficulties.

1. INTRODUCTION

While optimal design of elastic sandwich plates has not
presented particularly unexpected results, the field of optimal
design of solid, elastic plates has been concerned with a number
of questions [1 to 4], all of which are not yet fully clarified.

Optimization problems for thin, solid, elastic plates were
first considered in geometrically unconstrained form, i.e. without specification of constraints on the plate thickness function.
Numerical solutions of this type of problem are available in Refs.
5 to 9. The thickness distribution of these solutions are all
fairly smooth, although it can be argued in different ways [1,2,
4,7,8,10,11] that plate designs with integral stiffeners will
generally be associated with more nearly optimum characteristics.

---

[*]Visiting from the Department of Solid Mechanics, The Dalian
Engineering Institute, Dalian, The People's Republic of China.

In fact, the advantage of direct use of stiffeners for reinforcement has been demonstrated in different contexts in Refs. 12 to 16. Therefore, smooth solutions obtained from geometrically unconstrained formulations for optimal design must be considered as local optima [1,2,4,8,10,17].

Contrary to plates of sandwich type [18 to 20], the optimality condition for solid, elastic plates is only a stationary condition, i.e., it is necessary, but not sufficient for global optimality [1,2,21]. In fact, it is shown in Ref. 4 that a geometrically unconstrained, solid plate optimization problem may possess several local optima. It is also found that the characteristics of these local optimal designs increase rapidly with the number of stiffeners contained in the designs, but that a global optimal solution does not exist, when no constraints are prescribed for the plate thickness function.

Optimization of solid elastic plates, in which both maximum and minimum constraints are specified on the thickness function, is performed in Refs. 3, 11, and 22 to 25. Consideration of such constraints in the optimal design formulation seems to make the problem well posed, if the ratio between the maximum and minimum allowable thicknesses is sufficiently small. However, if larger values of this ratio are prescribed, it is found in the recent investigation of Ref. 25 that the global optimal design should be sought within a class of plates that has a very large number of very thin stiffeners (see also Ref. 8). Subsequent studies [26, 27] have shown that problems with a large ratio between the maximum and minimum allowable thicknesses become well formulated if the concentration (or density) of stiffeners is used as the design variable and the plate bending rigidity is considered to be anisotropic, rather than isotropic.

As a typical solid elastic plate optimization problem, the problem of minimizing plate volume subject to prescribed compliance (or integral stiffness) is considered in the following. The load on the plate, the plate domain, boundary conditions, and material, are assumed to be given. To some extent, this problem is less complicated than designing with respect to other behavioral constraints, e.g. prescribed vibration frequency or buckling load, but it contains all the significant features that are inherent in optimal design of solid, elastic plates.

The discussion in the following, which has implications for a number of similar two-dimensional optimization problems, is mainly focused on recent results obtained in Refs. 25 to 27.

2. OPTIMAL DESIGN FORMULATION ON THE BASIS OF ISOTROPIC PLATE BENDING RIGIDITY

Consider a thin, solid elastic plate of variable thickness h, whose mid-plane occupies a given domain Ω with boundary ω in the x^1, x^2 - plane of some three-dimensional coordinate system x^i, the x^3 axis of which is perpendicular to the x^1, x^2 - plane. The bending rigidity of the plate is assumed to be isotropic, i.e. independent of orientation, and is given by the scalar plate bending rigidity function $D(x^\alpha)$,

$$D = \frac{Eh^3}{12(1 - \nu^2)} \tag{2.1}$$

which is cubic in the plate thickness function $h(x^\alpha)$ for solid, elastic plates. The constants E and ν denote Young's modulus and Poisson's ratio, respectively, of the plate material.

The plate is acted on by a given transverse, static loading of intensity $p(x^\alpha)$, whereby the equation of equilibrium in the x^3-direction can be written in the general tensor form [28,29]

$$d_\alpha d_\beta M^{\alpha\beta} = p \tag{2.2}$$

Here and in the following, Greek indices take values 1 and 2 only, and repeated indices imply summation. The symbol d_α denotes the operator of covariant differentiation with respect to the independent variable x^α and $M^{\alpha\beta}$ is the contravariant, second order moment tensor

$$M^{\alpha\beta} = D \left\{ (1 - \nu)d^\alpha d^\beta w + \nu a^{\alpha\beta} d_\gamma d^\gamma w \right\} \tag{2.3}$$

where the scalar function $w(x^\alpha)$ is the plate deflection in the x^3-direction, $a^{\alpha\beta}$ is the contravariant metric tensor for the plate mid-plane coordinates x^α, and $d^\alpha = a^{\alpha\gamma}d_\gamma$ is the operator of contravariant differentiation.

Substituting Eq. 2.3 into Eq. 2.2, one obtains the following fourth order, partial differential equation for the deflection $w(x^\alpha)$ of a plate of variable thickness $h(x^\alpha)$

$$d_\alpha d_\beta \left[D \left\{ (1 - \nu)d^\alpha d^\beta w + \nu a^{\alpha\beta} d_\gamma d^\gamma w \right\} \right] = p \tag{2.4}$$

The boundary conditions at the edge ω of the plate are given by

$$w = 0 \qquad \Big\}$$
$$M^{\alpha\beta}n_\alpha n_\beta = 0 \quad \Big\} \tag{2.5}$$

at a simply supported edge,

$$w = 0 \qquad \Big\}$$
$$n_\alpha d^\alpha w = 0 \quad \Big\} \tag{2.6}$$

at a clamped edge, and

$$M^{\alpha\beta}n_\alpha n_\beta = 0 \qquad \qquad \Big\}$$
$$-n_\alpha d_\beta M^{\alpha\beta} - \frac{\partial(M^{\alpha\beta}t_\alpha n_\beta)}{\partial\omega} = 0 \quad \Big\} \tag{2.7}$$

at a free edge. Here, n_α and t_α denote the outward unit normal vector and the unit tangential vector in the x^α plane to the curve ω. The expression $n_\alpha d^\alpha w$ in the second of Eqs. 2.6 identifies the scalar slope of the deflection w normal to ω and $M^{\alpha\beta}n_\alpha n_\beta$ in the second of Eqs. 2.5 and the first of Eqs. 2.7 is the effective bending moment per unit length of ω. The expression on the left-hand side of the second of Eqs. 2.7 is the effective Kirchhoff shear force, per unit length of ω.

Together with any of the sets of boundary conditions of Eqs. 2.5, 2.6, and 2.7 the plate differential equation of Eq. 2.4 constitutes a self-adjoint boundary-value problem.

2.1 The Optimization Problem

Using the plate thickness function $h(x^\alpha)$ as the design variable, the problem of minimizing the plate volume V is now considered, where

$$V = \int_\Omega h \, d\Omega \tag{2.8}$$

subject to a given value Π of the compliance

$$\Pi = \int_{\Omega} pw \, d\Omega \tag{2.9}$$

of the plate, i.e. the work done by the applied forces. The transverse load distribution $p(x^{\alpha})$, the domain Ω, and the material constants E and ν for the elastic plate are assumed to be given. The problem of minimizing the plate volume for given compliance is equivalent to the dual problem of minimizing the compliance, or maximizing the integral plate stiffness, for given plate volume.

Consider now a geometrically constrained formulation of the optimization problem, i.e. assume that a maximum and minimum allowable value h_{max} and h_{min}, respectively, are specified for the design variable $h(x^{\alpha})$, such that $h_{max} \geq h \geq h_{min}$ throughout. These inequality constraints can be expressed as the equality constraints

$$h_{max} - h = \sigma^2 \tag{2.10}$$

$$h - h_{min} = \tau^2 \tag{2.11}$$

where the real functions $\sigma(x^{\alpha})$ and $\tau(x^{\alpha})$ are slack variables.

One may now apply a variational formulation of the optimization problem, and construct the functional V^*,

$$V^* = \int_{\Omega} h \, d\Omega + \Gamma \left[\int_{\Omega} pw \, d\Omega - \Pi \right]$$

$$- \int_{\Omega} \eta \left[d_{\alpha} d_{\beta} \left[D \left\{ (1 - \nu) d^{\alpha} d^{\beta} w + \nu a^{\alpha\beta} d_{\gamma} d^{\gamma} w \right\} \right] - p \right] d\Omega$$

$$- \int_{\Omega} \lambda \left[h - h_{max} + \sigma^2 \right] d\Omega - \int_{\Omega} \kappa \left[h_{min} - h + \tau^2 \right] d\Omega \tag{2.12}$$

where the compliance constraint of Eq. 2.9, the differential constraint of Eq. 2.4, and the geometric constraints of Eqs. 2.10 and 2.11 are adjoined to the functional V of Eq. 2.8 by means of Lagrangian multipliers Γ, $\eta(x^{\alpha})$, $\lambda(x^{\alpha})$, and $\kappa(x^{\alpha})$, respectively.

The set of necessary governing equations for this optimization problem can now be derived as the Euler-Lagrange equations
expressing stationarity of the functional V^* in Eq. 2.12, with
respect to its variables.

Proceeding similarly as in Ref. 25, where the dual formulation is considered, one finds that the condition of stationarity
of V^* with respect to variation of w leads to the result that the
Lagrangian multiplier function $\eta(x^\alpha)$, the so-called adjoint
variable, is proportional to w, with the Lagrangian multiplier Γ
as proportionality factor, i.e.

$$\eta(x^\alpha) \equiv \Gamma w(x^\alpha) \qquad (2.13)$$

This result is intimately connected with the self-adjointness of
the boundary value problem consisting of Eq. 2.4 and the boundary
conditions of Eqs. 2.5, 2.6, or 2.7.

Following the derivation in Ref. 25, writing the third
integral on the right-hand side of Eq. 2.12 in terms of $M^{\alpha\beta}$ given
by Eq. 2.3, and eliminating η by means of Eq. 2.13, one obtains
the condition of stationarity of V^* with respect to variation of
h, in the form

$$\int_\Omega \left\{ \Gamma \frac{dD}{dh} \frac{1}{D^2(1-\nu^2)} \left[(1+\nu)M^{\alpha\beta}M_{\alpha\beta} - \nu M^\alpha_\alpha M^\beta_\beta \right] - 1 + \lambda - \kappa \right\} \delta h d\Omega = 0$$

$$(2.14)$$

after using the divergence theorem, the boundary conditions, and
the moment-curvature relations. Since the variation δh is
arbitrary, one obtains from Eq. 2.14 the following necessary
condition of optimality, upon using Eq. 2.1 and redefining the
Lagrangian multiplier Γ by a simple scaling factor:

$$\Gamma h^{-4} \left[(1+\nu)M^{\alpha\beta}M_{\alpha\beta} - \nu M^\alpha_\alpha M^\beta_\beta \right] = 1 - \lambda + \kappa \qquad (2.15)$$

The variation of V^* with respect to σ and τ, respectively,
gives the switching equations

$$\left. \begin{array}{l} \lambda\sigma = 0 \\[1em] \kappa\tau = 0 \end{array} \right\} \qquad (2.16)$$

Comparing these equations with the defining equations of Eqs.
2.10 and 2.11 for the slack variables $\sigma(x^\alpha)$ and $\tau(x^\alpha)$, one is
able to deduce the following results by studying possible

combinations of σ and τ : If $\sigma \neq 0$ and $\tau \neq 0$, then $\lambda = \kappa = 0$ and $h_{max} < h < h_{min}$, i.e. the thickness function is unconstrained at the point considered. If $\sigma \neq 0$ and $\tau = 0$, then $\kappa = 0$ and $h = h_{min}$, i.e. the thickness is constrained from below. If $\sigma = 0$ and $\tau \neq 0$, then $\lambda = 0$ and $h = h_{max}$, i.e. h is constrained from above. The combination of $\sigma = \tau = 0$ is not possible since it was assumed that $h_{max} > h_{min}$.

The foregoing results enable one to eliminate the functions λ, κ, σ, and τ, and instead introduce the unions Ω_{ca}, Ω_u, and Ω_{cb} of subdomains of $\Omega(\Omega = \Omega_{ca} \cup \Omega_u \cup \Omega_{cb})$ in which the plate thickness $h(x^\alpha)$ is constrained from above, unconstrained, and constrained from below, respectively. Introducing the short-hand notation g for the scalar function

$$g(x^\alpha) = (1 + \nu)M^{\alpha\beta}M_{\alpha\beta} - \nu M^\alpha_\alpha M^\beta_\beta \qquad (2.17)$$

one can write the following convenient formula for the plate thickness function h:

$$h(x^\alpha) = \begin{cases} h_{max}, & \text{if } \left(\Gamma g(x^\alpha)\right)^{1/4} \geq h_{max}, \text{ i.e. } x^\alpha \in \Omega_{ca} \\ \left(\Gamma g(x^\alpha)\right)^{1/4}, & x^\alpha \in \Omega_u \qquad (2.18) \\ h_{min}, & \text{if } \left(\Gamma g(x^\alpha)\right)^{1/4} \leq h_{min}, \text{ i.e. } x^\alpha \in \Omega_{cb} \end{cases}$$

where the expression for h in the unconstrained subdomain Ω_u (with $\lambda = \kappa = 0$) follows from Eqs. 2.15 and 2.17.

To determine a suitable expression for the Lagrangian multiplier Γ, one first multiplies Eq. 2.2 by w and integrates over the domain Ω. Using the divergence theorem and the boundary conditions, one finds by means of Eq. 2.9 that the given compliance can be expressed alternatively as

$$\Pi = \int_\Omega M^{\alpha\beta} d_\alpha d_\beta w \, d\Omega$$

In view of the curvature-moment relationships [28,29],

$$d_\alpha d_\beta w = \frac{1}{D(1 - \nu^2)} \left[(1 + \nu)M_{\alpha\beta} - \nu a_{\alpha\beta}M^\gamma_\gamma \right]$$

and the defining equations of Eqs. 2.1 and 2.17 for $D(x^\alpha)$ and $g(x^\alpha)$, one thus has

$$\Pi = \frac{12}{E} \int_\Omega h^{-3} g(x^\alpha)d\Omega \qquad (2.19)$$

Multiplying Eq. 2.15 by h and integrating over Ω_u (where $\lambda = \kappa = 0$), replacing the left-hand side integral by integrals over Ω, Ω_{ca}, and Ω_{cb}, and using Eq. 2.19 Γ may be given by

$$\Gamma = \frac{\displaystyle\int_{\Omega_u} h\, d\Omega}{\displaystyle\frac{E}{12} \Pi - h_{max}^{-3} \int_{\Omega_{ca}} g(x^\alpha)d\Omega - h_{min}^{-3} \int_{\Omega_{cb}} g(x^\alpha)d\Omega} \qquad (2.20)$$

To summarize, the complete system of necessary conditions for the optimization problem consists of Eqs. 2.1, 2.3, 2.4, 2.8, 2.17, 2.18, and 2.20, together with a particular set of boundary conditions of Eq. 2.5, 2.6, or 2.7. This system of equations constitutes a coupled, non-linear, integro-partial differential boundary-value problem with unknown interior boundaries. Closed form solutions cannot be expected. The primary unknowns of the problem are the minimum volume V, the deflection $w(x^\alpha)$, and the thickness function $h(x^\alpha)$ of the optimum plate, which in turn require determination of the Lagrangian multiplier Γ and the sub-domains Ω_{ca}, Ω_u, and Ω_{cb}.

An efficient and quite general numerical solution procedure for this optimization problem (strictly speaking: the dual problem) is available in the recent paper of Ref. 25, where a number of solutions are presented for rectangular plates and axisymmetric, annular plates with various boundary conditions. Some of these results will be discussed briefly in the following.

2.2 Rectangular Plates

Choose x^i as a Cartesian coordinate system xyz. In terms of physical bending moments M_{xx} and M_{yy} and the physical twisting moment M_{xy}, Eqs. 2.3 then take the form

$$M_{xx} = D(w_{,xx} + \nu w_{,yy})$$

$$M_{yy} = D(w_{,yy} + \nu w_{,xx})$$

$$M_{xy} = D(1 - \nu)w_{,xy}$$

$$\left.\begin{array}{c}\end{array}\right\} \qquad (2.21)$$

where commas indicate partial differentiation with respect to succeeding coordinates. The plate differential equation of Eq. 2.4 attains the well known form [30]

$$(Dw_{,xx})_{,xx} + (Dw_{,yy})_{,yy} + \nu(Dw_{,xx})_{,yy} + \nu(Dw_{,yy})_{,xx}$$

$$+ 2(1 - \nu)(Dw_{,xy})_{,xy} = p \qquad (2.22)$$

in Cartesian coordinates, and the optimality condition of Eq. 2.15 takes the following form [25]

$$\Gamma h^{-4} \left[(M_{xx} + M_{yy})^2 + 2(1 + \nu)(M_{xy}^2 - M_{xx}M_{yy}) \right] = 1 \qquad (2.23)$$

in the geometrically unconstrained subdomain Ω_u. If one substitutes the moment expressions of Eq. 2.21, with D given in Eq. 2.1, then the optimality condition of Eq. 2.23 is seen to be precisely the same as is derived in Ref. 23.

Consider now some thickness distributions obtained in Ref. 25, for the dual problem of minimizing the compliance for given material volume of square, solid elastic plates that are acted on by a uniformly distributed static loading.

Figure 2.1 illustrates a full, square plate with clamped edges. The solution is obtained subject to a comparatively small ratio, $h_{max}/h_{min} = 1.5$, between the maximum and minimum thickness constraints. The compliance of the plate is 70.7% of the compliance of a uniform, clamped plate with the same volume, side lengths, material ($\nu = 0.25$), and loading. Thickness variation of the plate in Fig. 2.1 is seen to be smooth, and it is quantitatively very similar to solutions obtained in Ref. 23, for slightly different problems with small h_{max}/h_{min} ratios.

For problems associated with moderate to large h_{max}/h_{min} ratios, it is found in Ref. 25 that different local optimal designs can be obtained by starting the iterative numerical solution procedure with different initial thickness functions. In fact, a different optimization problem studied in Ref. 4 has earlier been found to possess a number of local optima.

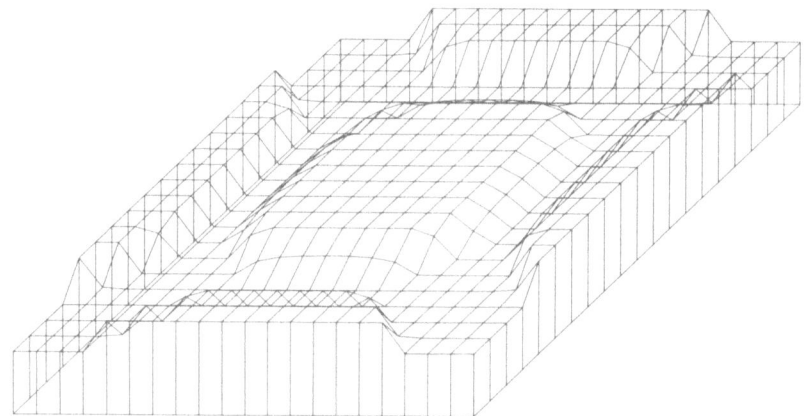

Figure 2.1   Optimum Rectangular Plate, With h_{max}/h_{min} = 1.5

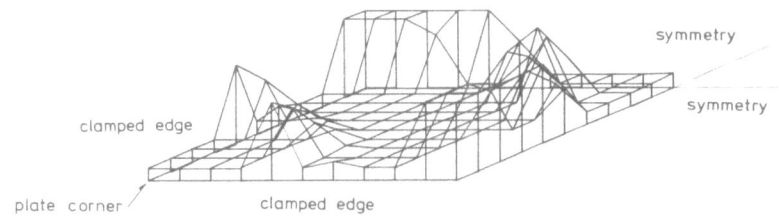

Figure 2.2   Optimum Rectangular Plate, With h_{max}/h_{min} = 6

Figure 2.2 shows a stationary, local optimum thickness distribution over a quarter of a clamped, square plate, obtained in
Ref. 25.   Again, the plate is acted on by uniformly distributed
static loading, but in this example, a comparatively large ratio
between the maximum and minimum thickness constraints is prescribed, namely h_{max}/h_{min} = 6.   Now, the compliance of the
optimized plate is only 21.8% of the compliance of a corresponding
uniform plate of the same volume and the solution is seen to be
equipped with significant integral stiffeners.   These stiffeners
are formed automatically by the optimization.   Except for the
compliance and thickness constraints, no constraints, e.g.,
conditions concerning location, number or width of the stiffeners,
are imposed.

However, the numerical solution shown in Fig. 2.2 cannot be
conceived as the final answer to the problem considered, because
it turns out [25] that solutions with large h_{max}/h_{min} ratios are

288

significantly dependent on the fineness of the grid used in the numerical solution procedure. The result in Fig. 2.2 is obtained on the basis of a 10 × 10 grid of equally spaced points for the quarter plate. By precisely solving the same problem as in Fig. 2.2, but using a 20 × 20 grid of equally spaced points for the quarter plate, the result shown in Figure 2.3 was obtained [25]. In this stationary, local optimum solution, more integral stiffeners are formed and the compliance is now 20.4% of the compliance of the uniform reference plate of the same volume.

For sufficiently large h_{max}/h_{min} ratios it is thus found in Ref. 25 that the plate solutions are equipped with more and more integral stiffeners and that the plate compliance decreases, if finer and finer grids are used in the numerical solution procedure. A possible limiting design could not be found by increasing the grid fineness to the limit of the capacity of the available computer in Ref. 25, and probably cannot be found at all. Thus, the dual formulation considered in Ref. 25, and hence also the present formulation of the optimization problem, does not seem to possess global optimal solutions if h_{max}/h_{min} is large.

2.3 Axisymmetric Plates

Following Ref. 25, one may now use a polar coordinate system r,θ as a convenient reference frame for circular, axisymmetric plates. Choosing $r = x^1$ and $\theta = x^2$, the metric tensor for the plate mid-plane is

$$a^{\alpha\beta} = \begin{Bmatrix} 1 & 0 \\ 0 & r^{-2} \end{Bmatrix} \quad , \quad a^{\beta}_{\alpha} = \begin{Bmatrix} 1 & 0 \\ 0 & 1 \end{Bmatrix}, \quad a_{\alpha\beta} = \begin{Bmatrix} 1 & 0 \\ 0 & r^2 \end{Bmatrix} \quad (2.24)$$

which implies that the only non-vanishing Christoffel symbols are

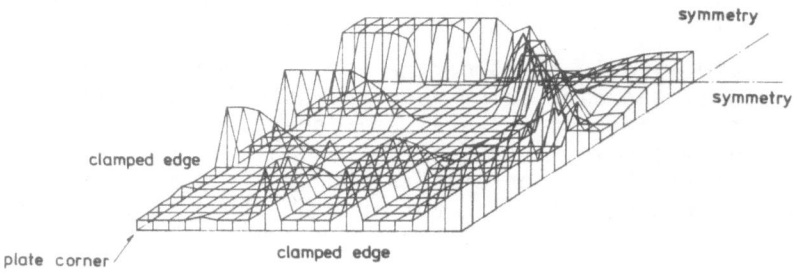

Figure 2.3   Optimum Rectangular Plate, With h_{max}/h_{min} = 6 and Fine Grid

$$\left\{\begin{matrix} 2 \\ 1\,2 \end{matrix}\right\} = \left\{\begin{matrix} 2 \\ 2\,1 \end{matrix}\right\} = r^{-1} \quad , \quad \left\{\begin{matrix} 1 \\ 2\,2 \end{matrix}\right\} = -r \tag{2.25}$$

and the contravariant, mixed, and covariant components of the moment tensor in Eq. 2.3 are expressed in terms of the deflection w and the scalar plate bending rigidity D as

$$\left.\begin{aligned}
M^{11} &= M_1^1 = M_{11} = D\left(w_{,rr} + \frac{\nu}{r} w_{,r} + \frac{\nu}{r^2} w_{,\theta\theta}\right) \\[2mm]
M^{22} &= r^{-2}M_2^2 = r^{-4} M_{22} = D\, r^{-2}\left(\nu w_{,rr} + \frac{1}{r} w_{,r} + \frac{1}{r^2} w_{,\theta\theta}\right) \\[2mm]
M^{12} &= M^{21} = M_1^2 = r^{-2} M_2^1 = r^{-2} M_{12} = r^{-2} M_{21} \\[2mm]
&= (1 - \nu)D\, r^{-1}\left(\frac{1}{r} w_{,\theta}\right)_{,r}
\end{aligned}\right\} \tag{2.26}$$

Along a curve r = const, the Kirchhoff shear force in the second of Eqs. 2.7 has the form $-M^{11}_{,r} + r M^{22} - r^{-1} M^{11} - 2M^{12}_{,\theta}$ and corresponds physically to the radial (Kirchhoff) shear force Q_r per unit length. By means of Eq. 2.26, this quantity can be expressed by w and D as

$$Q_r = -D_{,r}\left(w_{,rr} + \frac{\nu}{r} w_{,r} + \frac{\nu}{r^2} w_{,\theta\theta}\right)$$

$$\quad - D\left(w_{,rr} + \frac{1}{r} w_{,r} + \frac{1}{r^2} w_{,\theta\theta}\right)_{,r}$$

$$\quad - (1 - \nu)D\, \frac{1}{r}\left(\frac{1}{r} w_{,\theta\theta}\right)_{,r}$$

$$\quad - 2(1 - \nu)D_{,\theta}\, \frac{1}{r}\left(\frac{1}{r} w_{,\theta}\right)_{,r} \tag{2.27}$$

The physical moment components, i.e. radial bending moment M_{rr}, tangential bending moment $M_{\theta\theta}$, and twisting moment $M_{r\theta} = M_{\theta r}$ (all per unit length) are defined as follows, in terms of the mixed moment tensor components, and can be expressed by w and D as:

$$M_{rr} = M_1^1 = D\left(w,_{rr} + \frac{\nu}{r}\,w,_r + \frac{\nu}{r^2}\,w,_{\theta\theta}\right)$$

$$M_{\theta\theta} = M_2^2 = D\left(\nu\,w,_{rr} + \frac{1}{r}\,w,_r + \frac{1}{r^2}\,w,_{\theta\theta}\right)$$
$$\left. \right\} \qquad (2.28)$$

$$M_{r\theta} = M_{\theta r}\sqrt{M_1^2 M_2^1} = (1 - \nu)D\left(\frac{1}{r}\,w,_\theta\right),_r$$

Assume from now on that the plate is rotationally symmetric, i.e. that its thickness h and bending rigidity D depend only on the distance r from the symmetry axis, implying h = h(r) and D = D(r). Furthermore, attention is limited to load distribution functions p(r,θ) of the special type

$$p(r,\theta) = f(r)\cos n\theta \qquad (2.29)$$

where f is a given function that only depends on r and where n is a given integer. Equation 2.29 thus models a rotationally symmetric load distribution for n = 0, whereas Eq. 2.29 for n ≠ 0 models a load p(r,θ) that has the trace f(r) for θ = 0 and varies harmonically with θ in the circumferential direction. Assuming the boundary conditions to be homogeneous, the plate deflection function w(r,θ) then has the simple form

$$w(r,\theta) = v(r)\cos n\theta \qquad (2.30)$$

The assumptions introduced here offer the mathematical simplification that the governing non-linear partial differential equations of the optimization problem reduce to ordinary differential equations, after a separation of variables. This means, in turn, that much less computer space and time are required for the numerical solution procedure. These simplifications do not impede further study of plate stiffener formation, since the rotationally symmetric plate possesses the possibility of increasing its stiffness against circumferentially varying loads p(r,θ) by forming concentric, circumferential stiffeners that may effectively counteract the circumferential curvatures of the deflection function w(r,θ).

Substituting Eq. 2.26, with w given by Eq. 2.30 into the plate equilibrium equation of Eq. 2.2, cosnθ factors out and one obtains an ordinary differential equation for the plate ridigity function D(r) and the θ-independent part v(r) of the deflection function. After some manipulation, this equation can be written in the comparatively compact form

$$\left\{ r \left[D \left(v'' + \frac{1}{r} v' - \frac{n^2}{r^2} v \right)' + D' \left(v'' + \frac{v}{r} v' - \frac{vn^2}{r^2} v \right) \right] \right\}'$$

$$- D \frac{n^2}{r} \left(v'' + \frac{1}{r} v' - \frac{n^2}{r^2} v \right) - (1 - v)D'n^2 \left(\frac{1}{r} v \right)' = f(r)r \quad (2.31)$$

where primes denote differentiation with respect to r. Equation 2.31 replaces Eq. 2.4 in the formulation of the new optimization problem.

For deflection functions in the form of Eq. 2.30, the stress resultants in Eqs. 2.27 and 2.28 reduce to

$$\left. \begin{aligned} Q_r &= q_r \cos n\theta \quad, \quad M_{rr} = m_{rr} \cos n\theta \\ M_{\theta\theta} &= m_{\theta\theta} \cos n\theta \quad, \quad M_{r\theta} = m_{r\theta} \sin n\theta \end{aligned} \right\} \quad (2.32)$$

where the θ-independent parts of the stress resultants are given by

$$\left. \begin{aligned} q_r &= -D' \left(v'' + \frac{v}{r} v' - \frac{vn^2}{r^2} v \right) - D \left(v'' + \frac{1}{r} v' - \frac{n^2}{r^2} v \right)' \\ &\quad + (1 - v)D \frac{n^2}{r} \left(\frac{1}{r} v \right)' \\ m_{rr} &= D \left(v'' + \frac{v}{r} v' - \frac{vn^2}{r^2} v \right) \\ m_{\theta\theta} &= D \left(vv'' + \frac{1}{r} v' - \frac{n^2}{r^2} v \right) \\ m_{r\theta} &= -(1 - v)D \left(\frac{n}{r} v \right)' \end{aligned} \right\} \quad (2.33)$$

In order to establish the optimality condition, one may now substitute Eq. 2.28, with M_{rr}, $M_{\theta\theta}$, and $M_{r\theta}$ given by Eqs. 2.32, into Eq. 2.14 and perform a separate integration over θ in the interval $0 \le \theta \le 2\pi$, thereby ruling out θ-dependence. Upon applying the usual argument from the calculus of variations, using Eq. 2.1, and scaling the Lagrangian multiplier Γ by a multiplying factor, one obtains the optimality condition in the form

$$\Gamma h^{-4} \left[(m_{rr} + m_{\theta\theta})^2 + 2(1 + \nu)(m_{r\theta}^2 - m_{rr} m_{\theta\theta}) \right] = 1 - \lambda + \kappa \quad (2.34)$$

instead of Eq. 2.15. The equations replacing Eqs. 2.18 and 2.20 for the present, essentially one-dimensional problem, are easily derived [25].

The numerical solution procedure used to obtain the results shown in Figs. 2.1, 2.2, and 2.3, is based on a finite difference discretization of the problem [25], where the plate thickness function can only attain a single value at each grid point. However, the results associated with a large h_{max}/h_{min} ratio in Figs. 2.2 and 2.3 clearly indicate that the plate thickness function of the optimized plate may tend to exhibit finite jumps or discontinuities, over interior curves in the plate domain.

In order to facilitate such behavior, a finite element discretization is adopted for the solution of the axisymmetric plate optimization problem in Ref. 25. In this discretization, the plate is subdivided into a number of concentric, equidistant ring-elements. Each individual element has a constant thickness, but the thickness function of the plate may exhibit finite jumps between neighboring elements of different thicknesses.

Reference 25 presents a number of numerical results obtained in this way for axisymmetric annular plates acted on by load distributions of Eq. 2.29, with f constant and with a given integer value assigned to the circumferential wave number n of the load and deflection. Figure 2.4 shows results obtained in Ref. 25 for the dual problem of minimizing the compliance for given plate volume. The plate is axisymmetric and annular with clamped inner and outer radii, the inner radius being one-fifth of the outer radius. The ratio between the maximum and minimum thickness constraints is taken to be large, $h_{max}/h_{min} = 5$, and the load wave number is n = 4. This problem is solved by using (a) 150, (b) 200, (c) 250, and (d) 300 equally spaced elements to cover the interval of the radial distance from the inner to the outer plate radius. Each corresponding numerical solution is illustrated by a radial section through the plate in Figs. 2.4(a) to 2.4(d), respectively.

Figures 2.4(a) to 2.4(d), illustrate that the number of stiffeners increase rapidly, as the number of elements is increased. At the same time, the plate compliance is decreased. The stiffeners are seen to become thinner and thinner as the number of finite elements increases, the widths of most in fact being equal to the width of one finite element only. These results clearly indicate that for a large value of the ratio

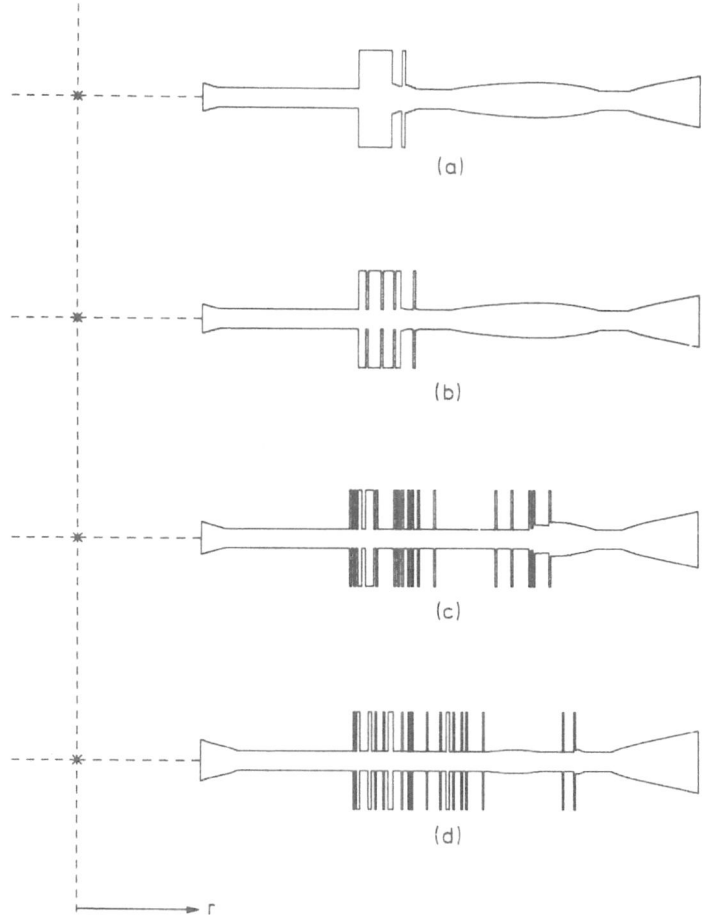

Figure 2.4   Annular Plate Sections

h_{max}/h_{min}, a possible limiting design cannot be obtained by continuously increasing the number of finite elements. This implies that a possible global optimal plate design does not seem to exist, within the present formulation of the optimization problem, where the thickness functions considered are only admitted to have a finite number of discontinuities (jumps). Consequently, in order to be able to determine possible global optimal designs associated with large h_{max}/h_{min} ratios, it is necessary to change the traditional formulation for optimal design.

3. OPTIMAL DESIGN FORMULATION FOR ANISOTROPIC, AXISYMMETRIC PLATES WITH INTEGRAL STIFFENERS

While a number of local optimum designs are found for plate optimization problems associated with a large h_{max}/h_{min} ratio, a possible global optimum design does not seem to exist within the class of thickness functions considered in the traditional formulation outlined in Section 2. The same must be expected [25] for corresponding formulations of optimal design for prescribed maximum deflection, fundamental frequency, buckling load, etc.

The results of Ref. 25 clearly indicate that the global optimum design associated with a sufficiently large h_{max}/h_{min} ratio may be a plate which, at least in some regions, is equipped with an infinite number of infinitely thin stiffeners. This means that the global optimum solution should be viewed as a limiting solution, from the point of view of practical design.

In order to determine such a limiting, global optimum design by means of continuum theory, one must change the optimal design formulation so as to use the concentration (or density) of stiffeners as the design variable. Moreover, seeking the design of a plate that is inherently anisotropic, it is necessary to consider the plate bending rigidity to be a tensor [26], rather than a scalar function.

3.1 The Anisotropic Plate Model

Consider a solid, axisymmetric plate made of linearly elastic material, with Young's modulus E and Poisson's ratio ν. Figure 3.1 shows a radial section through a small element of the plate. The plate element has the extent Δr in the radial direction, and consists of a finite number of successive, concentric, cylindrical layers of material disposed perpendicular to the plate mid-plane. These layers have given, constant heights h_{max} and h_{min}, respectively, as shown in Figure 3.1. Denote now by Δd_i the radial extent of the i-th layer of height h_{max}, and by Δc_i the radial extent of the neighboring layer of height h_{min}. The values Δd_i as well as the values Δc_i, i = 1,2,..., within the element may all be different from each other.

Define the concentration (or density) of cylindrical layers of heights h_{max} (i.e. plate stiffeners) within the plate element as

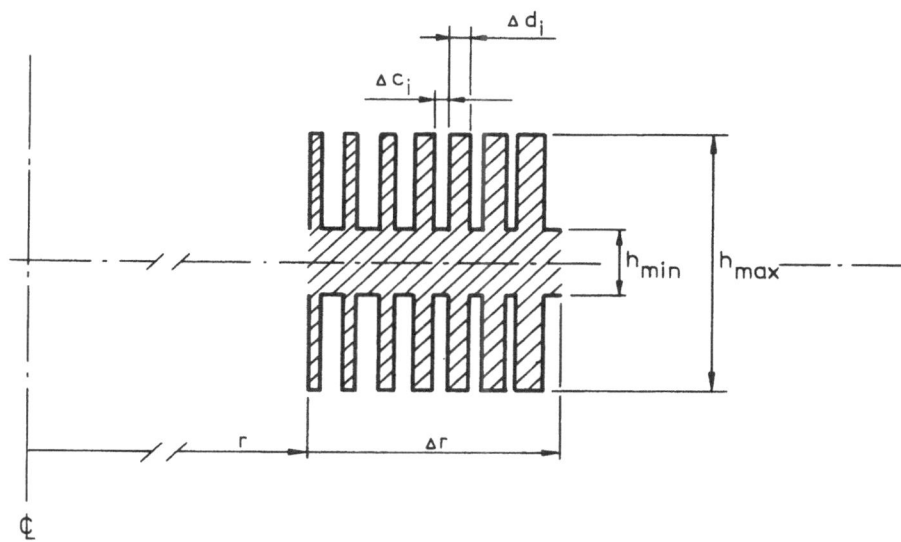

Figure 3.1   Axisymmetric Plate Element With Ribs

$$\mu = \frac{\sum\limits_{i} \Delta d_i}{\Delta r} \ , \ 0 \le \mu \le 1 \tag{3.1}$$

where Δr is given by

$$\Delta r = \sum_{i} (\Delta d_i + \Delta c_i) \tag{3.2}$$

and is infinitesimal. With $\mu = \mu(r)$, the total volume V of the anisotropic axisymmetric plate is given by

$$V = 2\pi \int_{R_*}^{R} \{\mu h_{max} + (1 - \mu)h_{min}\} r dr \tag{3.3}$$

where R_* and R are the inner and outer plate radii, respectively ($R_* = 0$ for a full plate).

For reasons of brevity the derivation of the smear-out equations is not repeated in the following. Instead, refer to Refs. 26 and 27, where two different paths are taken. The expression for the averaged radial Kirchhoff shear force and the relationships between averaged physical moment and curvature components for the anisotropic plate, which is in fact cylindrically orthotropic, can be written in the form

$$Q_r = -M_{rr,r} - \frac{2}{r} M_{r\theta,\theta} - \frac{1}{r}(M_{rr} - M_{\theta\theta})$$

$$M_{rr} = D_r \left[w_{,rr} + \nu_\theta \left(\frac{1}{r} w_{,r} + \frac{1}{r^2} w_{,\theta\theta} \right) \right]$$

$$M_{\theta\theta} = D_\theta \left(\nu_r w_{,rr} + \frac{1}{r} w_{,r} + \frac{1}{r^2} w_{,\theta\theta} \right)$$

$$M_{r\theta} = D_{r\theta} \left(\frac{1}{r} w_{,\theta} \right)_{,r}$$

$$(3.4)$$

These equations may be compared with Eqs. 2.28 for isotropic plates. In terms of $\mu = \mu(r)$, the elastic moduli D_r, D_θ, $D_{r\theta}$, ν_r, and ν_θ , with $D_r \nu_\theta = D_\theta \nu_r$, are given by

$$D_r = D_{//}$$

$$D_\theta = (1 - \nu^2)D_\perp + \nu^2 D_{//}$$

$$D_{r\theta} = (1 - \nu)D_\perp$$

$$\nu_\theta = \nu$$

$$\nu_r = \nu \frac{D_{//}}{(1 - \nu^2)D_\perp + \nu^2 D_{//}}$$

$$(3.5)$$

where $D_{//}$ and $D_\perp$ are defined as

$$D_{//} = \left[\frac{\mu}{D_{max}} + \frac{1 - \mu}{D_{min}} \right]^{-1}$$

$$D_\perp = \mu D_{max} + (1 - \mu)D_{min}$$

$$(3.6)$$

Here, D_{max} and D_{min} are given by h_{max} and h_{min}, respectively, via Eq. 2.1. It is worth noting at this point that Eqs. 3.4 reduce to their isotropic plate counterparts of Eq. 2.28 with the scalar, isotropic bending rigidity $D = D_{max}$ for $\mu = 1$ and $D = D_{min}$ for $\mu = 0$, respectively.

Consider now the compliance Π of the plate. First, one establishes an alternate expression for Π in Eq. 2.9. To do this, multiply the general equilibrium equation of Eq. 2.2 by w, integrate over the plate domain Ω, use the divergence theorem twice, and the particular set of boundary conditions of Eqs. 2.5, 2.6, or 2.7 to obtain

$$\Pi = \int_{\Omega} pwd\Omega = \int_{\Omega} M^{\alpha\beta} d_{\alpha} d_{\beta} wd\Omega \tag{3.7}$$

With polar coordinates $r = x^1$ and $\theta = x^2$, the contravariant moment components are given in terms of physical moment components as

$$M^{11} = M_{rr} \quad , \quad M^{12} = r^{-1} M_{r\theta} \quad , \quad M^{22} = r^{-2} M_{\theta\theta} \tag{3.8}$$

and the covariant curvatures $d_{\alpha} d_{\beta} w$ are given by

$$d_1 d_1 w = w_{,rr} \quad , \quad d_1 d_2 w = w_{,r\theta} - \frac{1}{r} w_{,\theta} \quad , \quad d_2 d_2 w = w_{,\theta\theta} + r w_{,r} \tag{3.9}$$

Using Eqs. 3.4 to 3.6, 3.8 and 3.9, one easily finds that

$$\int_{\Omega} M^{\alpha\beta} d_{\alpha} d_{\beta} wd\Omega = \int_{\Omega} \left\{ D_{//} \left(w_{,rr} + \frac{\nu}{r} w_{,r} + \frac{\nu}{r^2} w_{,\theta\theta} \right)^2 \right.$$
$$\left. + D_{\perp}(1 - \nu) \left(2 \left(\frac{1}{r} w \right)_{,r\theta}^{\prime 2} + (1 + \nu) \left(\frac{1}{r} w_{,r} + \frac{1}{r^2} w_{,\theta\theta} \right)^2 \right) \right\} d\Omega \tag{3.10}$$

Assuming load distributions and corresponding deflections in the form

$$\left. \begin{array}{l} p(r,\theta) = f(r)\cos n\theta \\ w(r,\theta) = v(r)\cos n\theta \end{array} \right\} \tag{3.11}$$

where n is an integer (cf. Eqs. 2.29 and 2.30), Eq. 3.7 becomes

$$c\Pi = \int_{R_*}^{R} fvrdr = \int_{R_*}^{R} \left\{ D_{//} \left(v'' + \frac{\nu}{r} v' - \frac{\nu n^2}{r^2} v \right)^2 \right.$$
$$\left. + D_{\perp}(1 - \nu) \left(2n^2 \left(\frac{1}{r} v \right)^{\prime 2} + (1 + \nu)\left(\frac{1}{r} v' - \frac{n^2}{r^2} v \right)^2 \right) \right\} rdr \tag{3.12}$$

where $c = (2\pi)^{-1}$ for $n = 0$ and $c = \pi^{-1}$ for $n > 0$.

3.2 Variational Formulation

Using the concentration $\mu(r)$ ($0 \leq \mu \leq 1$) of integral, circumferential plate stiffeners as the design variable, the objective is now to minimize the total plate volume V of Eq. 3.3, subject to given compliance Π in Eq. 3.12. The plate loading of Eq. 3.11, the plate radii R_* and R, the boundary conditions, the maximum and minimum thicknesses h_{max} and h_{min}, respectively, and the material constants E and ν are assumed to be given.

The optimum solution is associated with stationarity of the functional V^* defined by

$$V^* = \int_{R_*}^{R} \left\{ \mu h_{max} + (1 - \mu) h_{min} \right\} r\,dr + \Gamma \left[\int_{R_*}^{R} fvr\,dr - c\Pi \right]$$

$$- \Lambda \int_{R_*}^{R} \left\{ D_{//} \left(v'' + \frac{\nu}{r} v' - \frac{\nu n^2}{r^2} v \right)^2 \right.$$

$$+ D_{\perp}(1 - \nu) \left(2n^2 \left(\frac{1}{r} v \right)'^2 + (1 + \nu)\left(\frac{1}{r} v' - \frac{n^2}{r^2} v \right)^2 \right) - \left. fv \right\} r\,dr$$

$$- \int_{R_*}^{R} \lambda[\mu - 1 + \sigma^2]r\,dr + \int_{R_*}^{R} \kappa[\mu - \tau^2]r\,dr \tag{3.13}$$

where Γ, Λ, $\lambda(r)$, and $\kappa(r)$ are Lagrangian multipliers and $\sigma(r)$ and $\tau(r)$ are slack variables.

It follows from the path taken in the foregoing and the form of the functional V^* in Eq. 3.13 that the condition of stationarity of V^* for arbitrary admissible variation v will lead to a special form of the general plate equilibrium equation of Eq. 2.2, corresponding to loadings and deflections of the type given in Eq. 3.11. The form of this equilibrium equation (plate equation) is

$$(r\,q_r)' + \frac{n^2}{r} m_{\theta\theta} - 2\frac{n}{r} m_{r\theta} = -f(r)r \tag{3.14}$$

where the θ-independent parts q_r, m_{rr}, $m_{\theta\theta}$, and $m_{r\theta}$ are defined as in Eqs. 2.32, but now on the basis of the effective radial

(Kirchhoff) shear force Q_r, bending moments M_{rr} and $M_{\theta\theta}$, and twisting moment $M_{r\theta}$, given by Eqs. 3.4 for the anisotropic plate. The functions q_r, m_{rr}, $m_{\theta\theta}$ and $m_{r\theta}$ are now expressed in terms of $v(r)$ and the anisotropic plate bending rigidities $D_{//}(r)$ and $D_{\perp}(r)$, as follows:

$$q_r = -m'_{rr} - 2 \frac{n}{r} m_{r\theta} - \frac{1}{r}(m_{rr} - m_{\theta\theta})$$

$$m_{rr} = D_{//}\left(v'' + \frac{\nu}{r} v' - \frac{\nu n^2}{r^2} v\right)$$

$$m_{\theta\theta} = D_{//}\nu v'' + \left[(1 - \nu^2)D_{\perp} + \nu^2 D_{//}\right]\left(\frac{1}{r} v' - \frac{n^2}{r^2} v\right) \tag{3.15}$$

$$m_{r\theta} = -(1 - \nu)D_{\perp}\left(\frac{n}{r} v\right)'$$

Compare this with Eqs. 2.33. Substituting Eqs. 3.15, with $D_{//}$ and $D_{\perp}$ given by Eq. 3.6 into Eqs. 3.14, one has a form of the anisotropic plate equation, where the design variable $\mu(r)$ appears explicitly.

The optimality equation for this problem expresses stationarity of the functional V^* in Eq. 3.13, with respect to variation of $\mu(x)$. By means of Eq. 3.6, it is easily found to take the form

$$\Lambda(D_{max} - D_{min})\left\{\frac{1}{D_{max}D_{min}}\left[\frac{\mu}{D_{max}} + \frac{1 - \mu}{D_{min}}\right]^{-2}\left(v'' + \frac{\nu}{r} v' - \frac{\nu n^2}{r^2} v\right)^2\right.$$

$$\left. + (1 - \nu)\left(2n^2\left(\frac{1}{r} v\right)'^2 + (1 + \nu)\left(\frac{1}{r} v' - \frac{n^2}{r^2} v\right)^2\right)\right\}$$

$$= h_{max} - h_{min} + \lambda(r) - \kappa(r) \tag{3.16}$$

where the design variable μ appears explicitly. In geometrically unconstrained subintervals, one has $\lambda(r) = \kappa(r) = 0$, which simplifies Eq. 3.16. By means of the second of Eqs. 3.15 and the first of Eqs. 3.6, the optimality condition of Eq. 3.16 can be written in the shorter form

300

$$\Lambda(D_{max} - D_{min}) \left\{ \frac{m_{rr}^2}{D_{max}D_{min}} + (1 - \nu) \left(2n^2 \left(\frac{1}{r} v \right)'^2 \right. \right.$$

$$\left. \left. + (1 + \nu) \left(\frac{1}{r} v' - \frac{n^2}{r^2} v \right)^2 \right) \right\} = h_{max} - h_{min} \qquad (3.17)$$

for geometrically unconstrained sub-intervals, cf. Refs. 26 and 27.

3.3 Example and Discussion

The governing equations for the optimum, anisotropic plate can be solved numerically by methods described in Refs. 26 and 27. Figure 3.2 shows as an example the result of solving the new optimal design formulation outlined in Sections 3.1 and 3.2 for the case of an annular, axisymmetric plate with clamped inner and outer boundaries. The inner plate radius is one-fifth of the outer radius, and the ratio between the maximum and minimum plate thicknesses, is $h_{max}/h_{min} = 5$. The load wave number n is equal to 4, the radial load function f(r) is taken to be constant, and Poisson's ratio of the plate material is $\nu = 0.25$. All these data are precisely equal to those specified for the plates shown in Fig. 2.4. Taking also the total plate volume to be the same as before, and considering the problem of minimizing the compliance for given plate volume, one obtains on the basis of the new formulation the optimal solution illustrated in Fig. 3.2.

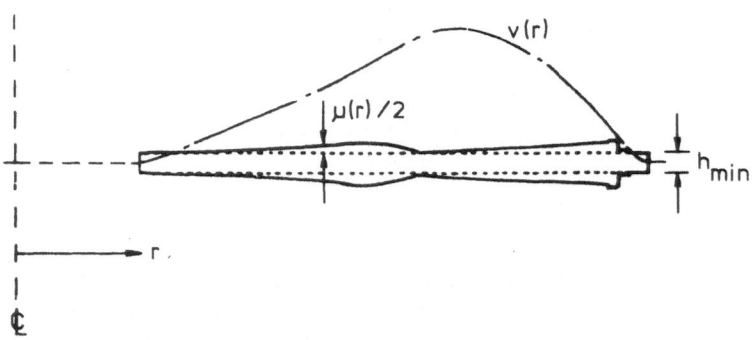

Figure 3.2   Optimum Distribution $\mu(r)$ of Integral Stiffeners on Clamped Annular Plate with $h_{max}/h_{min} = 5$ and n = 4

In Figure 3.2 the distance between the two dotted lines represents the minimum plate thickness h_{min}, and the sum of the vertical distances from the dotted lines to the solid curves indicates the concentration $\mu(r)$ of integral stiffeners (with total height $h_{max} - h_{min}$, cf. Fig. 3.1). The dashed-dotted curve above the plate in Fig. 3.2 shows the θ-independent part $v(r)$ of the deflection function.

The compliance of the integrally stiffened plate in Fig. 3.2 is found to be 41.5% of the compliance of a uniform, solid reference plate with the same volume, material, inner and outer radii, boundary conditions, and loading. This result, and the associated design in Fig. 3.2, are obtained by using 50 elements in the numerical solution procedure, but it is characteristic that only absolutely negligible changes are found if any other reasonable number of elements is used. The low compliance index, which cannot be reached by solving the corresponding traditional formulation of the example problem (cf. the results given in the text related to Fig. 2.4), and the aforementioned invariance of the compliance and the design with respect to the number of elements applied in the numerical solution, clearly indicates the superiority of the new optimal design formulation for problems with large h_{max}/h_{min} values.

From numerical results obtained in Refs. 26 and 27, it is obvious that the new formulation is superior to the traditional formulation and provides a regularization of axisymmetric plate optimization problems associated with large h_{max}/h_{min} values. Specific conditions for the superiority of the new formulation are derived in Ref. 27, on the basis of analytical considerations.

REFERENCES

1. Lurie, K.A., and Cherkaev, A.V., "Prager Theorem Application to Optimal Design of Thin Plates (in Russian)," MTT (Mechanics of Solids), Vol. 11, 1976, pp. 157-159.
2. Reiss, R., "Optimal Compliance Criterion for Axisymmetric Solid Plates," Int. J. Solids Structures, Vol. 12, 1976, pp. 319-329.
3. Armand, J.-L., "Numerical Solutions in Optimization of Structural Elements," Int. Conf. on Computational Methods in Nonlinear Mechanics, Austin, Texas, USA, 1974.
4. Olhoff, N., "On Singularities, Local Optima and Formation of Stiffeners in Optimal Design of Plates," Proc. IUTAM Symposium on Optimization in Structural Design (ed. A. Sawczuk and Z. Mróz), Springer-Verlag, 1975, pp. 82-103.

5. Huang, N.C., "Optimal Design of Elastic Structures for Maximum Stiffness," _Int. J. Solids Structures_, Vol. 4, 1968, pp. 689-700.

6. Olhoff, N., "Optimal Design of Vibrating Circular Plates," _Int. J. Solids Structures_, Vol. 6, 1970, pp. 139-156.

7. Frauenthal, J.C., "Constrained Optimal Design of Circular Plates Against Buckling," _J. Struct. Mech._, Vol. 1, 1972, pp. 159-186.

8. Olhoff, N., "Optimal Design of Vibrating Rectangular Plates," _Int. J. Solids Structures_, Vol. 10, 1974, pp. 93-109.

9. Gura, N.M. and Seyranian, A.P., "Optimum Circular Plate with Constraints on the Rigidity and Frequency of Natural Osciala- tions (in Russian)," _MTT (Mechanics of Solids)_, Vol. 12, 1977, pp. 138-145.

10. Simitses, G.J., "Optimal vs. Stiffened Circular Plate," _AIAA Journal_, Vol. 11, 1973, pp. 1409-1410.

11. Armand, J.-L. and Lodier, B., "Optimal Design of Bending Elements," _Int. J. Num. Meth. Engrg._, Vol. 13, 1978, pp. 373-384.

12. Nash, W.A., "Effects of a Concentric Reinforcing Ring on Stiffness and Strength of A Circular Plate," _J. Appl. Mech._, Vol. 15, 1949, pp. 25-29.

13. Megarefs, G.J., "Method of Minimal Weight Design of Axi- symmetric Plates," _Proc. ASCE_, Vol. 92, EM6, 1966, pp. 79-99.

14. Brotchie, J.F., "Method for Minimal Design of Axisymmetric Plates," _Proc. ASCE_, Vol. 93, EM5, 1967, pp. 173-175.

15. Kozlowski, W. and Mróz, Z., "Optimal Design of Solid Plates," _Int. J. Solids Structures_, Vol. 5, 1969, pp. 781-794.

16. Samsonov, A.M., "Optimum Location of Thin Elastic Rib on Elastic Plate (in Russian)," _MTT (Mechanics of Solids)_, Vol. 13, 1978, pp. 132-138.

17. Seyranian, A.P., "A Study of an Extremum in the Optimal Problem of a Vibrating Circular Plate (in Russian)," _MTT Mechanics of Solids)_, Vol. 13, 1978, pp. 113-118.

18. Armand, J.-L., _Application of the Theory of Optimal Control of Distributed-Parameter Systems to Structural Optimization_, NASA CR-2044, 1972.

19. Masur, E.F., "Optimality in the Presence of Discreteness and Discontinuity," _Proc. IUTAM Symposium on Optimization in Structural Design (ed. A. Sawczuk and Z. Mróz)_, Springer- Verlag, 1975, pp. 441-453.

20. Hegemier, G.A. and Tang, H.T., "A Variational Principle, the Finite Element Method, and Optimal Structural Design for Given Deflection," _Proc. IUTAM Symposium on Optimization in Structural Design (ed. A. Sawczuk and Z. Mróz)_, Springer- Verlag, 1975, pp. 464-483.

21. Mróz, Z., "Multi-Parameter Optimal Design of Plates and Shells," _J. Struct. Mech._, Vol. 1, 1973, pp. 371-392.

22. Banichuk, N.V., "Optimal Plate Shapes in Bending Problems (in Russian)," MTT (Mechanics of Solids), Vol. 10, 1975, pp. 180-188.

23. Banichuk, N.V., Kartvelishvili, V.M. and Mironov, A.A., "Numerical Solution of Two-dimensional Optimization Problems for Elastic Plates (in Russian)," MTT (Mechanics of Solids), Vol. 12, 1977, pp. 68-78.

24. Banichuk, N.V., Kartvelishvili, V.M., and Mironov, A.A., "Optimization Problems with Local Performance Criteria in the Theory of Plate Bending (in Russian)," MTT (Mechanics of Solids), Vol. 13, 1978, pp. 124-131.

25. Cheng, K.-T., and Olhoff, N., An Investigation Concerning Optimal Design of Solid Elastic Plates, DCAMM-Rept. No. 174, The Danish Center for Applied Mathematics and Mechanics, 1980. (To appear in Int. J. Solids Structures.)

26. Olhoff, N., Lurie, K.A., Cherkaev, A.V., and Fedorov, A.V., Sliding Regimes and Anisotropy in Optimal Design of Vibrating Axisymmetric Plates, DCAMM-Rept. No. 192, The Danish Center for Applied Mathematics and Mechanics, 1980.

27. Cheng, K.-T., On Some New Optimal Design Formulations for Plates, DCAMM-Rept. No. 189, The Danish Center for Applied Mathematics and Mechanics, 1980.

28. Flügge, W., Tensor Analysis and Continuum Mechanics, Springer-Verlag, 1972.

29. Niordson, F., Introduction to Shell Theory, Dept. of Solid Mechanics, The Technical University of Denmark, 1980.

30. Timoshenko, S. and Woinowsky-Krieger, S., Theory of Plates and Shells, McGraw-Hill, 1959.

304

ON SOME NEW OPTIMAL DESIGN FORMULATIONS FOR PLATES

Keng-Tung Cheng*

Department of Solid Mechanics, The Technical University
of Denmark, Lyngby, Denmark   DK-2800

ABSTRACT

In this paper, minimum compliance design problems of thin,
solid elastic plates are discussed. Indications provided by
numerical results obtained in a recent paper are extended and a
new plate model is proposed, in which a uniform plate is equipped
with an infinite number of infinitely thin integral stiffeners.
On the basis of this plate model, numerical and analytical investigations of a new formulation of the minimum compliance design problem are carried out. It is demonstrated that the new
formulation, under certain conditions, is superior to the corresponding traditional formulation for determining a possible
global optimal design of solid plates. To gain a deeper under-
standing of special features involved in optimal design of solid
plates, sandwich plates and beam stiffened plates are also studied.

1.   INTRODUCTION

In a recent paper [1] by Olhoff and the author, a series of
numerical results for minimum compliance design problems of thin,
solid elastic plates with given material volume and constraints
on the thickness variation are presented. Results for both
annular and rectangular plates show a clear tendency of an optimum
design with integral stiffeners. The optimum thickness distributions for annular plates further indicate that under some

---

*Visiting from the Department of Solid Mechanics, The Dalian
 Engineering Institute, Dalian, The People's Republic of China.

conditions the global optimum design may be a plate, which, at
least in some subregions, is equipped with an infinite number of
infinitely thin stiffeners.

However, the numerical investigations in Ref. 1 cannot be
used to conclude that densely stiffened plates are superior to
smooth ones in general. One obvious reason is that it is impossible to use an infinite number of elements in numerical calculations, so as to obtain compliances for plates with an infinite
number of discontinuities. Moreover, using finite, but very high
numbers of elements in the calculations, the results would be
encumbered with inevitable numerical errors, as more and more
thickness jumps emerge in the design. It follows from considerations of this kind that it would be worthwhile to investigate the
possibility of establishing some theorems, by means of which
smooth plates and plates with infinite numbers of stiffeners can
be compared and evaluated.

The present paper mainly deals with minimum compliance design
problems for three different types of plates and outlines significant features of problems in which the design variable is allowed
to have an infinite number of discontinuities. As an alternative
to the traditional formulation (see, e.g. Ref. 2), which usually
leads to smooth designs, a new plate model is developed in this
paper, in which the thickness is allowed to have an infinite
number of discontinuities and the density of infinitely thin
integral plate stiffeners is used as the design variable. Furthermore, appropriate bending rigidity moduli are applied in the
corresponding, inherently anisotropic plate model.

This new plate model is used for solid plates and certain
conditions under which the new plate model is superior to the
corresponding smooth one are derived. This superiority is demonstrated by a series of numerical examples. For sandwich plates,
however, it is shown that the new model can never be better than
the corresponding smooth one.

In addition to solid plates and sandwich plates, so-called
beam stiffened plates are considered, in which circumferential
stiffeners are circular beams (or rings) in annular plates. The
minimum compliance design problem for this type of plates is
formulated and necessary conditions for optimality are derived.
Several numerical examples are then presented to compare this
type of plates with others. For any smooth distribution of the
design variable of this model, it is shown that there will always
exist a corresponding integrally stiffened design with the same
material distribution and the same compliance.

For simplicity, this paper only deals with annular plates,
whose thickness distributions and boundary conditions are axially

symmetric. As is in Ref. 1, the applied loads are assumed to vary harmonically in the circumferential direction.

2. SOLID PLATES

2.1 Formulation for Smooth Plates

In polar coordinates r and θ, if one considers only load distribution functions $p(r,\theta)$ of the special type

$$p(r,\theta) = f(r)\cos n\theta \qquad (2.1)$$

where n is a given integer, for an annular plate with axisymmetric thickness distribution and boundary conditions, the minimum compliance design problem associated with given material volume and prescribed constraints for the thicknesses $h(r)$ can be formulated as follows:

Find the design variable $h(r)$ to minimize

$$\phi = \int_\Omega f(r)w(r)r\,dr$$

subject to the constraints

$$\int_\Omega h(r)r\,dr = 1$$

$$h_{min} \leq h(r) \leq h_{max} , \qquad r \in \Omega\{r|R_i \leq r \leq 1\}$$

$$(2.2)$$

Here $w(r)$ is the θ-independent factor of the deflection $W(r,\theta) = w(r)\cos n\theta$ of the plate, under the given load $p(r,\theta)$ of Eq. 2.1 and specific boundary conditions. All variables are dimensionless, and the inner and outer plate radii are R_i and 1, respectively.

On the basis of energy principles of linear elasticity, the compliance Φ can also be expressed in terms of the specific strain energy ϕ_A or complementary energy ϕ_C, i.e.

$$\Phi = \int_\Omega \phi_A r\,dr = \int_\Omega D\left[\kappa_{rr}^2 + \kappa_{\theta\theta}^2 + 2\nu\kappa_{rr}\kappa_{\theta\theta} + 2(1-\nu)\kappa_{r\theta}^2\right]r\,dr \qquad (2.3)$$

or

$$\Phi = \int_\Omega \Phi_c r dr$$

$$= \int_\Omega \frac{1}{D(1-\nu^2)} \left[M_{rr}^2 + M_{\theta\theta}^2 - 2\nu M_{rr} M_{\theta\theta} + 2(1+\nu) M_{r\theta}^2 \right] r dr \qquad (2.4)$$

Here,

$$D = h^3 \qquad (2.5)$$

and ν is Poisson's ratio; κ_{rr}, $\kappa_{\theta\theta}$, and $\kappa_{r\theta}$ are the θ-independent factors of radial, circumferential, and twisting curvatures, respectively; and M_{rr}, $M_{\theta\theta}$, and $M_{r\theta}$ are the θ-independent factors of corresponding moments. The curvature-deflection relations are

$$\left. \begin{aligned} \kappa_{rr} &= w'' \\[2mm] \kappa_{\theta\theta} &= \frac{w'}{r} - \frac{n^2 w}{r^2} \\[2mm] \kappa_{r\theta} &= - \left(\frac{nw}{r} \right)' \end{aligned} \right\} \qquad (2.6)$$

and the moment-curvature relations are

$$\left. \begin{aligned} M_{rr} &= -D(\kappa_{rr} + \nu\kappa_{\theta\theta}) \\[2mm] M_{\theta\theta} &= -D(\kappa_{\theta\theta} + \nu\kappa_{rr}) \\[2mm] M_{r\theta} &= -D(1 - \nu)\kappa_{r\theta} \end{aligned} \right\} \qquad (2.7)$$

where the primes denote differentiation with respect to r.

The necessary condition for optimality, which can be derived in different ways [1,2], is

$$\frac{\partial D}{\partial h} \left[\kappa_{rr}^2 + \kappa_{\theta\theta}^2 + 2\nu\kappa_{rr}\kappa_{\theta\theta} + 2(1 - \nu)\kappa_{r\theta}^2 \right] = \Lambda - \alpha + \beta \qquad (2.8)$$

or

$$\frac{1}{D^2} \frac{\partial D}{\partial h} \left[M_{rr}^2 + M_{\theta\theta}^2 - 2\nu M_{rr} M_{\theta\theta} + 2(1+\nu) M_{r\theta}^2 \right] = \Lambda^* - \alpha^* + \beta^* \qquad (2.9)$$

where Λ and Λ^* are constants and α, α^*, β, and β^* are functions that are equal to zero at all points r where the thickness constraints are inactive [1].

2.2 A New Plate Model

Numerical results presented in Ref. 1 clearly show the tendency that the optimum design under some conditions may be a plate, which at least in some of its subregions is stiffened with infinitely thin stiffeners. Motivated by this indication, a new plate model is proposed here.

The new plate model consists of a uniform part of thickness h_{min}, to which an infinite number of integral stiffeners in the circumferential direction are attached, as shown in Fig. 2.1. The stiffeners have rectangular cross-sections of fixed height $(h_{max}-h_{min})$ and infinitesimal width. It is further assumed that all the stiffeners are placed symmetrically with respect to the mid-plane of the plate. The design variable is chosen as the density b(r) of stiffeners defined by

$$b(r) = \lim_{\Delta r \to 0} \frac{\sum_i \Delta d_i}{\Delta r} = 1 - \lim_{\Delta r \to 0} \frac{\sum_i \Delta c_i}{\Delta r} \qquad (2.10)$$

where Δd_i is the width of the i-th stiffener and Δc_i is the distance between the i-th stiffener and the (i + 1)-st stiffener. All Δd_i, Δc_i, and Δr shown in Fig. 2.1 are infinitesimal.

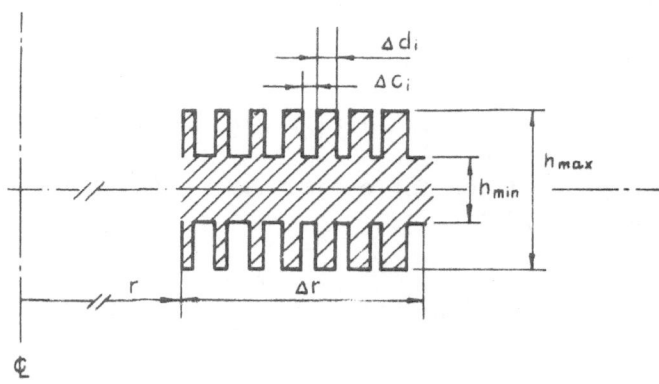

Figure 2.1 New Plate Model.

2.3 Specific Strain Energy for the New Model

The total strain energy in the small region Δr can be calculated as a sum of all contributions. As in Ref. 1, all the stiffeners are assumed to be parts of the plate. Thus, the contribution from the i-th stiffener is

$$\Delta\Phi_i = D_{max}\left[\kappa_{rr,i}^2 + \kappa_{\theta\theta,i}^2 + 2\nu\kappa_{rr,i}\kappa_{\theta\theta,i} + 2(1-\nu)\kappa_{r\theta,i}^2\right]\Delta d_i r_i \tag{2.11}$$

and the contribution from the i-th section between the i-th and (i + 1)-st stiffener is

$$\Delta\Phi_{\bar{i}} = D_{min}\left[\kappa_{rr,\bar{i}}^2 + \kappa_{\theta\theta,\bar{i}}^2 + 2\nu\kappa_{rr,\bar{i}}\kappa_{\theta\theta,\bar{i}} + 2(1-\nu)\kappa_{r\theta,\bar{i}}^2\right]\Delta c_i \bar{r}_i \tag{2.12}$$

where

$$\left.\begin{array}{l} D_{max} = h_{max}^3 \\[2ex] D_{min} = h_{min}^3 \end{array}\right\} \tag{2.13}$$

According to its definition, the specific strain energy is a limit of the average strain energy, i.e.

$$\Phi_A = \lim_{\Delta r \to 0} \frac{\Sigma\Delta\Phi_i + \Sigma\Delta\Phi_{\bar{i}}}{\Sigma\Delta d_i r_i + \Sigma\Delta c_i r_{\bar{i}}} \tag{2.14}$$

Based on Eqs. 2.11 and 2.12, the limiting process indicated above cannot be performed, because the thickness jumps cause discontinuities in the radial curvature κ_{rr}. In fact, continuity conditions at interfaces between adjacent subregions only provide continuity of deflection w, rotation w', radial moment M_{rr}, and radial shear force Q_r. Taking Eq. 2.6 into consideration, curvatures $\kappa_{\theta\theta}$ and $\kappa_{r\theta}$ are also continuous. Therefore, alternative expressions for $\Delta\Phi_i$ and $\Delta\Phi_{\bar{i}}$ in terms of continuous quantities, can be written as

$$\Delta\Phi_i = \left[D_{max}(1-\nu^2)\kappa_{\theta\theta,i}^2 + 2D_{max}(1-\nu)\kappa_{r\theta,i}^2 + \frac{M_{rr,i}^2}{D_{max}}\right]\Delta d_i r_i \tag{2.15}$$

$$\Delta\Phi_{\bar{i}} = \left[D_{min}(1-\nu^2)\kappa^2_{\theta\theta,\bar{i}} + 2D_{min}(1-\nu)\kappa^2_{r\theta,\bar{i}} + \frac{M^2_{rr,\bar{i}}}{D_{min}} \right]\Delta c_i r_{\bar{i}} \quad (2.16)$$

Substituting from Eqs. 2.15 and 2.16 into Eq. 2.14 and completing the limiting process, one obtains

$$\Phi_A = \left[D_{min} + b(r)(D_{max} - D_{min}) \right]\left[(1-\nu^2)\kappa^2_{\theta\theta} + 2(1-\nu)\kappa^2_{r\theta} \right]$$

$$+ M^2_{rr}\left[\frac{b(r)}{D_{max}} + \frac{1 - b(r)}{D_{min}} \right] \quad (2.17)$$

The radial curvature κ_{rr} for plates is, in general, defined as

$$\kappa_{rr} = \frac{d^2w}{dr^2} = \lim_{\Delta r \to 0}\left(\left.\frac{dw}{dr}\right|_{r+\Delta r} - \left.\frac{dw}{dr}\right|_r \right)/\Delta r \quad (2.18)$$

Then, the radial curvature κ_{rr} for the new plate model can be expressed as

$$\kappa_{rr} = \lim_{\Delta r \to 0} \frac{\Sigma\kappa_{rr,i}\Delta d_i + \Sigma\kappa_{rr,\bar{i}}\Delta c_i}{\Delta r} \quad (2.19)$$

Expressing $\kappa_{rr,i}$ and $\kappa_{rr,\bar{i}}$ by means of $M_{rr,i}$, $\kappa_{\theta\theta,i}$,..., and substituting into Eq. 2.19, the continuity of M_{rr} and $\kappa_{\theta\theta}$ leads to

$$M_{rr} = \left[\frac{b(r)}{D_{max}} + \frac{1 - b(r)}{D_{min}} \right]^{-1} (\kappa_{rr} + \nu\kappa_{\theta\theta}) \quad (2.20)$$

Resubstituting Eq. 2.20 into Eq. 2.17, one has the specific strain energy

$$\Phi_A = \left[D_{min} + b(D_{max} - D_{min}) \right]\left[\kappa^2_{rr} + \kappa^2_{\theta\theta} + 2\nu\kappa_{rr}\kappa_{\theta\theta} + 2(1-\nu)\kappa^2_{r\theta} \right]$$

$$+ b(b-1)(\kappa_{rr} + \nu\kappa_{\theta\theta})^2 (D_{max} - D_{min})^2/\left[bD_{min} + (1-b)D_{max} \right] \quad (2.21)$$

2.4 Minimum Compliance Design Formulation for the New Plate Model

Having obtained the specific strain energy, it is straight-forward to derive the moment-curvature relations for the new model. After introducing the convenient short-hand notation

$$
\left.\begin{aligned}
D_r &= \frac{D_{max}D_{min}}{bD_{min} + (1-b)D_{max}} \\[2mm]
D_{r\theta} &= bD_{max} + (1-b)D_{min} \\[2mm]
D_\theta &= (1 - \nu^2)D_{r\theta} + \nu^2 D_r \\[2mm]
\nu_r &= \frac{D_r \nu}{D_\theta}
\end{aligned}\right\} \tag{2.22}
$$

the moment-curvature relations can be written as

$$
\left.\begin{aligned}
M_{rr} &= -D_r(\kappa_{rr} + \nu\kappa_{\theta\theta}) \\[2mm]
M_{\theta\theta} &= -D_\theta(\kappa_{\theta\theta} + \nu_r\kappa_{rr}) \\[2mm]
M_{r\theta} &= -D_{r\theta}(1 - \nu)\kappa_{r\theta}
\end{aligned}\right\} \tag{2.23}
$$

or

$$
\left.\begin{aligned}
\kappa_{rr} &= \frac{-\nu(M_{rr} - \nu_r M_{\theta\theta})}{\nu_r D_{r\theta}(1 - \nu^2)} \\[3mm]
\kappa_{\theta\theta} &= \frac{-(M_{\theta\theta} - \nu M_{rr})}{D_{r\theta}(1 - \nu^2)} \\[3mm]
\kappa_{r\theta} &= \frac{-M_{r\theta}}{D_{r\theta}(1 - \nu)}
\end{aligned}\right\} \tag{2.24}
$$

In the above manipulations the following identity

$$D_{r\theta} = D_r \left[1 + \frac{b(1-b)(D_{max} - D_{min})^2}{D_{max} D_{min}} \right] \tag{2.25}$$

has frequently been used.

By studying the moment-curvature relations, it is noticed that the new model is nothing but a plate made of cylindrical anisotropic material [3], for which the governing equations are available in Refs. 3 and 4.

For the linearly elastic materials considered here, the specific complementary energy Φ_c is easily obtained by substituting Eq. 2.24 into Eq. 2.21, to obtain

$$\Phi_c = \frac{1}{D_{r\theta}(1-\nu^2)} \left[M_{rr}^2 - 2\nu M_{rr} M_{\theta\theta} + M_{\theta\theta}^2 + 2(1+\nu)M_{r\theta}^2 \right]$$

$$+ \frac{D_{r\theta} - D_r}{D_r D_{r\theta}} M_{rr}^2 \tag{2.26}$$

One may now formulate the minimum compliance design problem for the new plate model as follows:

With $b(r)$ as the design variable, minimize

$$\Phi = \int_\Omega f(r)w(r)r dr$$

subject to the volume constraint

$$\int_\Omega \left[bh_{max} + (1 - b)h_{min} \right] r dr = 1 \tag{2.27}$$

and the density constraints

$$0 \le b(r) \le 1$$

where $w(r)$ is the θ-independent factor of deflection of the plate, under given load $p(r,\theta) = f(r)\cos n\theta$ and certain boundary conditions.

Applying the Lagrange multiplier method [1], the necessary condition for optimality is found to be

$$(D_{max}-D_{min}) \left[(1-\nu^2)\kappa_{\theta\theta}^2 + 2(1-\nu)\kappa_{r\theta}^2\right] - M_{rr}^2 \left[\frac{1}{D_{max}} - \frac{1}{D_{min}}\right]$$

$$= \Lambda(h_{max} - h_{min}) - \alpha + \beta \qquad (2.28)$$

where Λ is a constant and α and β satisfy

$\alpha = 0$ and $\beta = 0$, when $0 < b(r) < 1$

$\alpha \neq 0$ and $\beta = 0$, when $b(r) = 0$

$\alpha = 0$ and $\beta \neq 0$, when $b(r) = 1$

Several numerical solutions are presented in Section 2.7.

2.5 Min-Min Problems

In order to make a theoretical comparison between solid plates and corresponding new model plates, cast the above formulations of Eqs. 2.2 and 2.27 into min-min problems [6]. This is done mainly on the basis of the minimum complementary energy principle, which can be stated as follows [5]:

For the statically admissible moment field associated with an actual displacement field, the total complementary energy attains a minimum value, when compared to values resulting from any other statically admissible moment field.

As is well known, the compliance of a structure of linear elastic material equals the total complementary energy of the structure, when it is in equilibrium. Therefore, the minimum compliance design problem can easily be reformulated as a min-min problem [2], i.e.

$$\underset{\text{admissible designs}}{\text{Min}} \left\{ \underset{\text{admissible moments}}{\text{Min}} \left\{ \int_\Omega \phi_c \, rdr \right\} \right\} \qquad (2.29)$$

The expressions for ϕ_c of solid plates and the new plates are given in Eqs. 2.4 and 2.26, respectively. The so-called admissible designs are conceived as designs that satisfy the given constraints for the volume and for the design variable. The admissible moment field is to satisfy the statical boundary conditions and equilibrium equations, in terms of moments. For both solid plates and new model plates, the set of admissible moment fields will be completely the same if the same loads and statical boundary conditions are considered. In other words, a statically admissible moment field for a solid plate is also an admissible one for the new model plate, and vice versa, if they have the same R_i, applied loads, and statical boundary conditions. Furthermore, keeping the statical boundary conditions and external loads unchanged, the actual moment field for a solid plate is certainly an admissible moment field for the new model plate.

2.6 Superiority of the New Model

The following theorem establishes the conditions under which a new model plate is superior to the corresponding smooth one.

Theorem 2.1: Suppose that in a smooth, solid plate optimum design there exists a subregion $\Omega_s \{r: a \leq r \leq d\}$ in which $h_{min} < h < h_{max}$ and in which the actual moments satisfy

$$\frac{M_{rr}^2}{[M_{rr}^2 - 2\nu M_{rr} M_{\theta\theta} + M_{\theta\theta}^2 + 2(1+\nu)M_{r\theta}^2]} < \frac{(\alpha + \beta + 1)\beta^3}{(1-\nu^2)(\beta^2 + \beta + 1)^2 \alpha^3} \qquad (2.30)$$

with $\alpha = h/h_{min}$ and $\beta = h_{max}/h_{min}$. Then, the compliance is further decreased if the material distribution is kept unchanged but a new model section is constructed instead of the original model in Ω_s.

Proof: Let ϕ^s denote the compliance of the original smooth design. By means of its actual moments M_{rr}, $M_{\theta\theta}$, and $M_{r\theta}$, Eq. 2.4 leads to

$$\phi^s = \int_\Omega \frac{1}{D(1-\nu^2)} \left[M_{rr}^2 - 2\nu M_{rr} M_{\theta\theta} + M_{\theta\theta}^2 + 2(1+\nu)M_{r\theta}^2 \right] r dr \qquad (2.31)$$

Now change the smooth plate section into the new model section and obtain a new plate. During that change, the material distribution is kept unchanged, i.e. the density b(r) of stiffeners in Ω_s satisfies

$$h_{min} + b(r)(h_{max} - h_{min}) = h(r) \left.\begin{array}{c}\\\\\\\end{array}\right\} r \in \Omega_s \qquad (2.32)$$
$$0 < b(r) < 1$$

Taking the moments M_{rr}, $M_{\theta\theta}$, and $M_{r\theta}$ as statically admissible for the new plate, the integral in Eq. 2.29 can be calculated as

$$\Phi^A = \int_{\Omega_s} \left\{ \frac{1}{D_{r\theta}(1 - \nu^2)} \left[M_{rr}^2 - 2\nu M_{rr} M_{\theta\theta} + M_{\theta\theta}^2 + 2(1+\nu) M_{r\theta}^2 \right] \right.$$

$$+ \frac{D_{r\theta} - D_r}{D_r D_{r\theta}} M_{rr}^2 \left. \right\} r dr$$

$$+ \int_{\Omega_1} \frac{1}{D(1 - \nu^2)} \left[M_{rr}^2 - 2\nu M_{rr} M_{\theta\theta} + M_{\theta\theta}^2 + 2(1+\nu) M_{r\theta}^2 \right] r dr \quad (2.33)$$

where

$$\Omega_1 = \Omega - \Omega_s \qquad (2.34)$$

According to 2.29, it turns out that

$$\Phi^N \leq \Phi^A , \qquad (2.35)$$

where Φ^N denotes the actual compliance for the new plate. Based on Eqs. 2.31 and 2.33, the difference between Φ^A and Φ^S is given by

$$\Phi^A - \Phi^S = \int_{\Omega_s} \left\{ \frac{(D_{r\theta}^{-1} - D^{-1})}{(1 - \nu^2)} \left[M_{rr}^2 - 2\nu M_{rr} M_{\theta\theta} + M_{\theta\theta}^2 + 2(1+\nu) M_{r\theta}^2 \right] \right.$$

$$+ (D_r^{-1} - D_{r\theta}^{-1}) M_{rr}^2 \left. \right\} r dr \qquad (2.36)$$

which will certainly be less than zero if the integrand on the right-hand side is less than zero everywhere in Ω_s. The latter condition is identical with the inequality,

$$\frac{M_{rr}^2}{[M_{rr}^2 - 2\nu M_{rr}M_{\theta\theta} + M_{\theta\theta}^2 + 2(1+\nu)M_{r\theta}^2]} < \frac{1}{1-\nu^2}\left(\frac{1}{D} - \frac{1}{D_{r\theta}}\right)\frac{D_r D_{r\theta}}{(D_{r\theta} - D_r)} \qquad (2.37)$$

Solving $b(r)$ from Eq. 2.32 and inserting it into Eqs. 2.22 and 2.37, some algebraic manipulations yield

$$\left(\frac{1}{D} - \frac{1}{D_{r\theta}}\right)\frac{D_r D_{r\theta}}{(D_{r\theta} - D_r)} = \frac{(\alpha + \beta + 1)\beta^3}{(\beta^2 + \beta + 1)^2 \alpha^3} \qquad (2.38)$$

Comparing Eq. 2.38 with Eq. 2.30, one concludes that if Eq. 2.30 is satisfied everywhere in Ω_s, then

$$\Phi^A < \Phi^S \qquad (2.39)$$

Taking Eq. 2.35 into consideration, one further concludes that if the conditions of Theorem 2.1 hold, then the inequality

$$\Phi^N < \Phi^S \qquad (2.40)$$

will hold. However, Eq. 2.40 simply implies superiority of the new model over the original smooth model, i.e. the latter cannot constitute a global optimal solution and the proof is complete.

In order to discuss this result, denote the expression on the right-hand side of Eq. 2.30 by μ_{cr}. Figure 2.2 shows a set of curves of μ_{cr} versus α, for different values of β. With a smooth (or partially smooth) local optimum design for a solid plate and its actual moments in hand, the curves in Fig. 2.2 can serve to determine whether there exists a new model plate that is superior to the smooth one. This can be done as follows. First, using the available thicknesses and moments, calculate

$$\alpha = \frac{h}{h_{min}}$$

$$\mu = \frac{M_{rr}^2}{[M_{rr}^2 - 2\nu M_{rr}M_{\theta\theta} + M_{\theta\theta}^2 + 2(1+\nu)M_{r\theta}^2]} \qquad (2.41)$$

at each point of subregions where $h_{min} < h < h_{max}$. Then, points

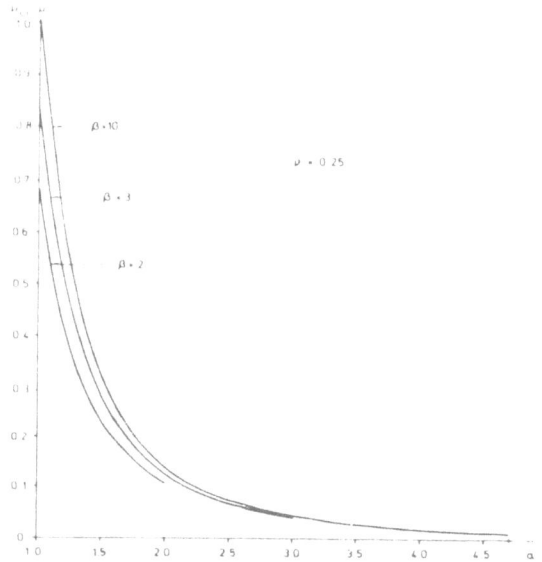

Figure 2.2 μ_{cr} vs α.

with coordinates (α, μ) are plotted in Fig. 2.2. If some of these points lie below the curve corresponding to the given β, it can be concluded that there exists a new model plate that is better than the smooth model.

Two interesting features should be emphasized here. First, for any finite $\beta > 0$, points with $M_{rr} = 0$ are always below the curves. This means that along curves with $M_{rr} = 0$, it is always worthwhile to construct a new model section, instead of the original smooth one, no matter how small β ($\beta > 0$) is. Second, for large values of β, the new model plate can be still better than the corresponding smooth plate, although M_{rr}^2 is rather large in comparison with

$$M_{rr}^2 - 2\nu M_{rr} M_{\theta\theta} + M_{\theta\theta}^2 + 2(1 + \nu)M_{r\theta}^2$$

as long as the condition of Eq. 2.30 holds. This statement is consistent with the result in Ref. 6, where Reiss proved that the compliance can be made arbitrarily small for a simply supported solid circular plate loaded by any axisymmetric loading, unless this includes a concentrated load at the plate center.

Finally, it should be emphasized that the condition in Theorem 2.1 is only a sufficient condition. For points lying above the curves in Fig. 2.2, there still exist the possibility that a new model plate may be superior to the corresponding smooth one.

2.7 Example

Since the above results hold for full circular plates as well, consider an example of a full circular plate with unit radius and uniformly distributed edge moments M_0, applied to the outer edge. For this example, see Ref. 6, a uniform thickness distribution constitutes a local optimal design, as may easily be checked by solving the plate equation and boundary conditions, and then substituting its solutions

$$
\left.
\begin{aligned}
M_{rr} &= M_{\theta\theta} = M_0 \\
w &= \frac{M_0}{2h^3(1 + \nu)} (1 - r^2)
\end{aligned}
\right\} \tag{2.42}
$$

into the necessary condition of Eq. 2.7 or 2.8, for optimality. The corresponding compliance,

$$
\Phi = \frac{M_0^2}{h^3(1 + \nu)} \tag{2.43}
$$

is determined as the work done by the edge moments M_0.

Turn now to the new model, i.e. a full circular plate consisting of a uniform part of thickness h_{min}, to which is attached a set of uniformly distributed stiffeners of height $(h_{max} - h_{min})$ and density

$$
b(r) = (h - h_{min})/(h_{max} - h_{min}) = (\alpha - 1)/(\beta - 1)
$$

where h_{max} and h_{min} are all given

Referring to the theory of anisotropic plates [3,4], the solution

$$
w = \frac{M_0}{D_r(m + \nu)(m + 1)} (1 - r^{m+1})
$$

$$M_{rr} = M_0 r^{m-1} \Big\}$$ (2.44)

$$M_{\theta\theta} = \frac{M_0 \nu}{(m + \nu)} \left(m + \frac{1}{\nu_r} \right) r^{m-1}$$

for the new plate enables one to obtain the compliance

$$\Phi_N = \frac{M_0^2}{D_r (m + \nu)}$$ (2.45)

where

$$m^2 = \frac{D_\theta}{D_r} = [b + (1 - b)\beta^{-3}][b + (1 - b)\beta^3] (1 - \nu^2) + \nu^2$$ (2.46)

Although the new plate has uniformly distributed stiffeners, which is not an optimum choice for this type of plate, its compliance Φ_N is already lower than the corresponding Φ in Eq. 2.43 when the conditions in Thm. 2.1 hold. The three curves in Fig. 2.3 show the dependence of Φ/Φ_N on h_{min}, for different values of

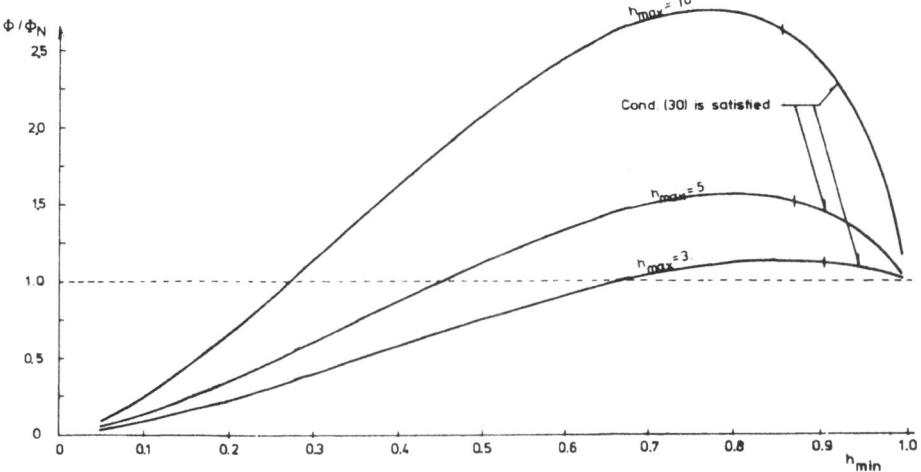

Figure 2.3 Φ/Φ_N vs h_{min}

h_{max}. Notice that the fact that the conditions in Thm. 2.1 are only sufficient conditions is directly implied in Fig. 2.3.

2.8 Numerical Results and Discussion

The minimum compliance design problem for the new model plate is now solved numerically. The general iterative scheme developed in Ref. 1 is slightly revised and used to obtain all the numerical results presented in the following.

To compare with numerical results for solid plates in Ref. 1, the plate to be optimized has the same Poisson ratio $\nu = 0.25$ and the same geometrical size as in Ref. 1, i.e. the inner plate radius is taken to be one-fifth of the outer plate radius, the ratio between the thickness constraints is $h_{max}/h_{min} = 5$, and h_{min} is given via the ratio $h_u/h_{min} = 1.6579$, where h_u is the solid plate thickness corresponding to a uniform distribution of the available volume over the plate area. Figures 2.4(a) to 2.4(i) illustrate numerical solutions for annular plates, using the new model with nine combinations of clamped, simply supported, and free inner and outer plate edges. Each solution is illustrated by a radial section through the plate, together with the θ-independent part $w(r)$ of the deflection function. All results are associated with the load wave number $n = 4$, and 50 finite elements are used in the numerical calculations. In Table 2.1, the corresponding compliances are listed for the nine combinations of boundary conditions. As is in Ref. 1, one states the compliances as fractions of the compliance $\Phi_{u,s}$ of a corresponding uniform plate that has the same loading, boundary conditions, total volume, inner and outer plate radii, and that is made of the same material. The results in the second column are the minimum compliances for solid plates quoted from Fig. 5 in Ref. 1. The third column shows compliances for new model plates, which consist of uniform plates of thickness h_{min} and uniform density distributions $b(r)$ of integral stiffeners. Comparisons of the numerical results in the second and third column give a clear demonstration of the superiority of the new model plate. In the last column the minimum compliances for the new model plates are shown, which correspond to the optimum designs shown in Figs. 2.4(a) to 2.4(i). Referring to the results in the third column, it is obvious that for the new model plates themselves, one gains very little by the optimization. This may be explained by the fact that the new type of structure, i.e. the new model plate, is itself an optimum design in the present case.

321

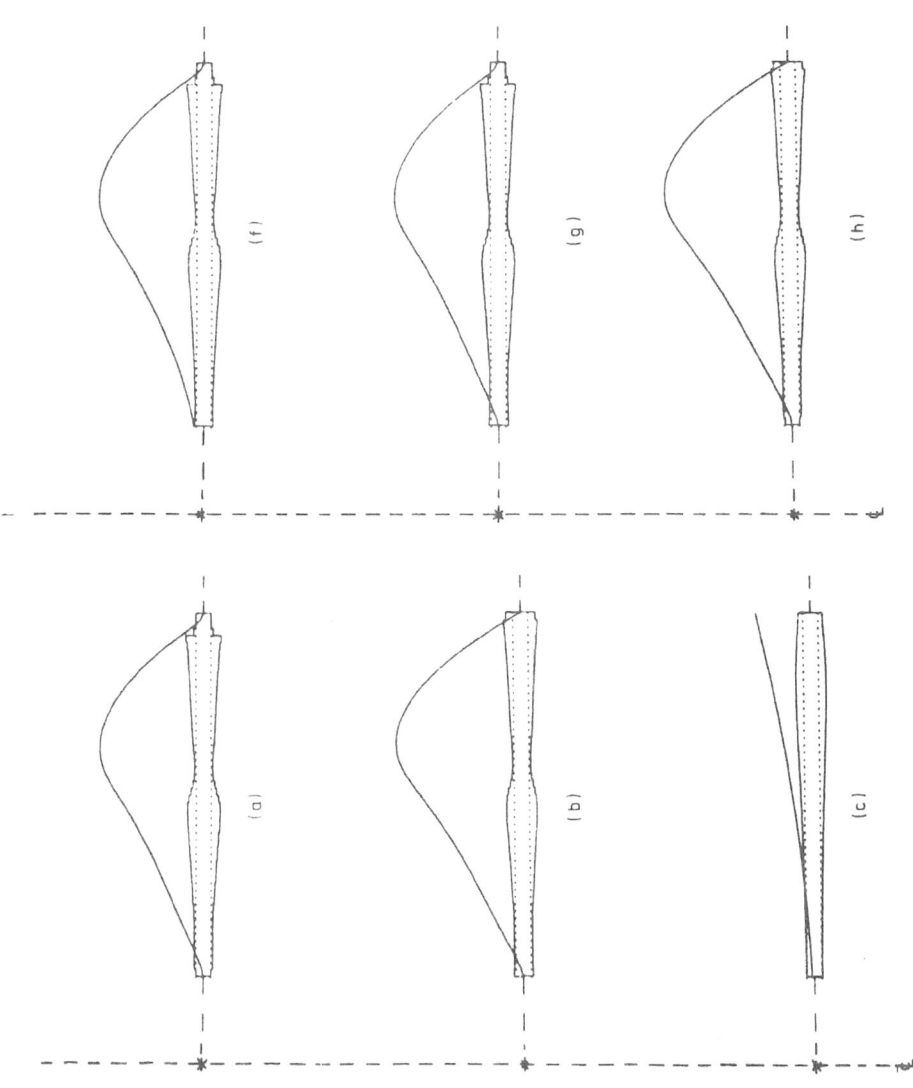

Figure 2.4 Results of Optimizing Annular, New Model Plates with Different Boundary Conditions

322

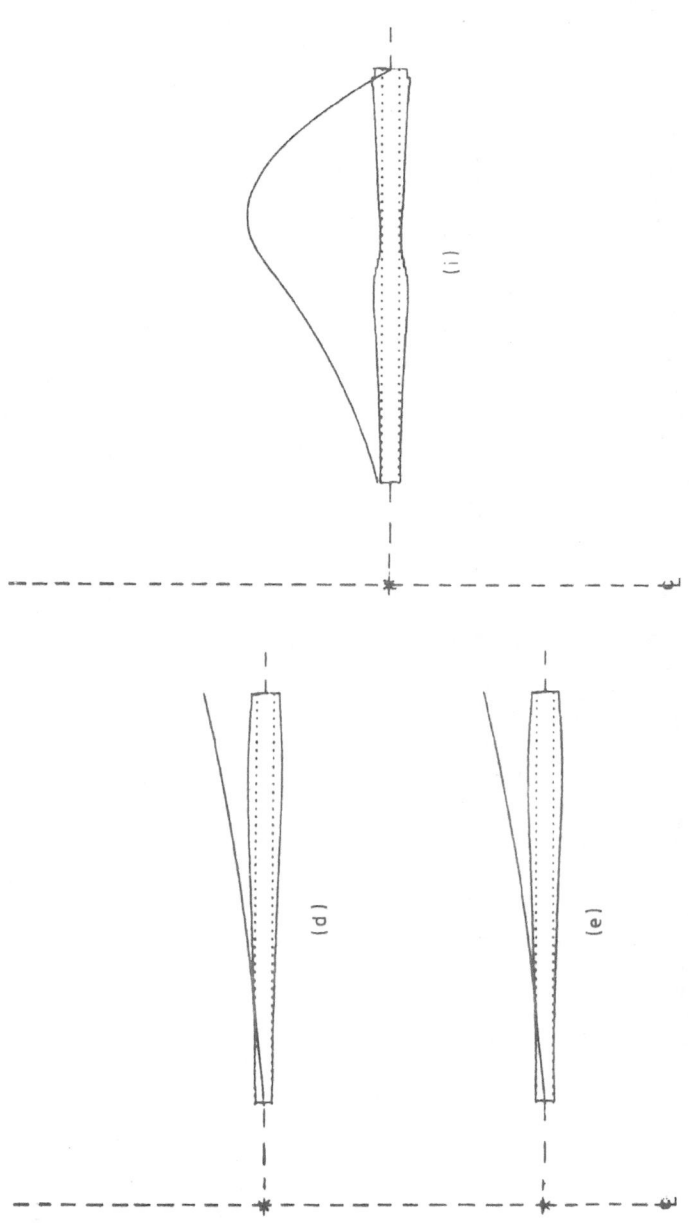

Figure 2.4 (cont'). Dotted lines show radial sections of their uniform parts. Densities of stiffeners are visualized by the distances between solid curves and dotted lines. In each picture the r-independent part w of the deflection function is also indicated by a solid curve. (a) Clamped-clamped plate. (b) Simply supported-simply supported plate. (c) Free-free plate. (d) Clamped-free plate. (e) Simply supported-free plate. (f) Free-clamped plate. (g) Simply supported-clamped plate. (h) Clamped-simply supported plate. (i) Free-simply supported plate, n = 4, 50 elements. Their compliances are given in Table 2.1

TABLE 2.1.   COMPARISON BETWEEN SOLID PLATES AND NEW PLATE MODELS.

Boundary conditions at inner and outer edges are indicated in the
first column: c - clamped, s - simply supported, f - free. Subscripts s, n, m, and u refers to solid plates, new model plates,
minimum compliance design, and uniform design, respectively.

B.C.	$\Phi_{m,s}/\Phi_{u,s}$	$\Phi_{u,n}/\Phi_{u,s}$	$\Phi_{m,n}/\Phi_{u,s}$
c - c	0.536	0.444	0.415
s - s	0.605	0.324	0.302
f - f	0.265	0.215	0.198
c - f	0.251	0.217	0.198
s - f	0.256	0.215	0.197
f - c	0.617	0.430	0.404
s - c	0.564	0.434	0.407
c - s	0.584	0.330	0.307
f - s	0.645	0.323	0.302

The above results should not lead to an impression that the
new model plates are always superior to the corresponding solid
plates. Numerical results reveal that there exist situations in
which the optimum design for the new model plate is inferior to
the corresponding uniform, solid, plate. Take the clamped-clamped
plate as an example, and optimize it, subject to the load wave
number n = 0.   Results for h_{max}/h_{min} = 5 and h_{max}/h_{min} = 20 are
shown in Figs. 2.5(a) and 2.5(b), respectively. Stating the compliances as a fraction of the compliance $\Phi_{u,s}$ as before, the
compliance ratio for the optimum new model plate is equal to
1.016 for h_{max}/h_{min} = 5, which implies that the optimum new model
plate is inferior to the corresponding uniform, solid plate.
However, with h_{max}/h_{min} = 20, one finds $\Phi_{m,n}/\Phi_{u,s}$ = 0.290, i.e.
the new model plate is certainly superior.

It should finally be mentioned that the new model plates
overestimate the stiffness of the physical structure, particularly
when the density b(r) is very small. As is well known, a very thin
stiffener should be considered as a thin curved beam, whose strains
are described by the theory of small deformations of thin curved
beams [7].

324

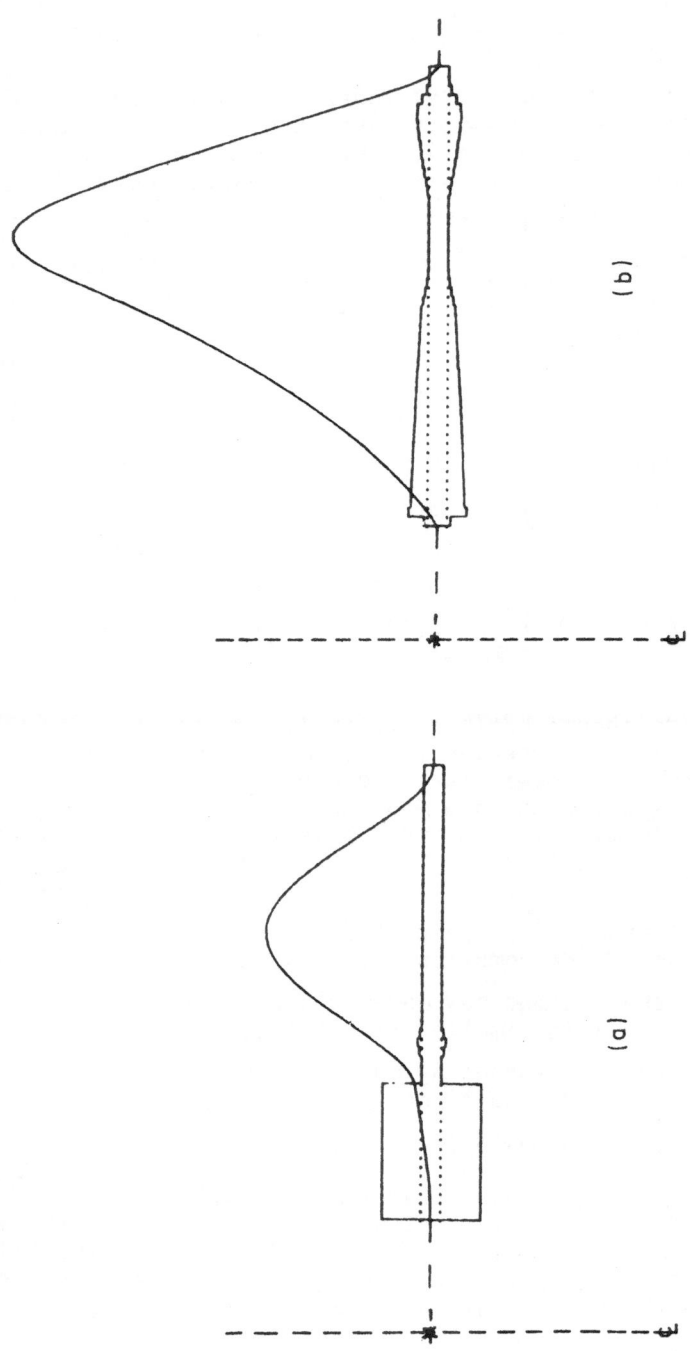

Figure 2.5 Influence of Ratio h_{max}/h_{min} on the Comparisons Between New Model Plates and Solid Plates.

Axial symmetric uniform loads. $R_i = 0.2$. Both inner and outer plate edges are clamped. (a) $h_{max}/h_{min} = 5$, $\Phi_{m,n}/\Phi_{u,s} = 1.016$. (b) $h_{max}/h_{min} = 20$, $\Phi_{m,n}/\Phi_{u,s} = 0.290$. The deflection curve in (b) is amplified 10 times

3.   SANDWICH PLATES

If one defines

$$D = h, \quad D_{max} = h_{max}, \quad D_{min} = h_{min} \tag{3.1}$$

the formulation in Section 2.1 represents the minimum compliance design problem for sandwich plates, with fixed uniform core of zero bending stiffness covered by thin face sheets of variable thickness $h(r)$ [8]. Using the algorithm in Ref. 1, with a minor modification, numerical calculations can easily be performed for sandwich plates.

Having studied solid plates, one may now ask what happens if the thickness of the face sheets is allowed to have an infinite number of discontinuities, i.e. will the new model also be superior to the smooth model for sandwich plates?

Figure 3.1 illustrates the application of the new model. At all junctions where the thickness of face sheets has jumps, all the quantities behave in the same way as in solid plates. Therefore, the derivations in Sections 2.3 and 2.4 are valid as well. However, the proof of Thm. 2.1 breaks down, due to the fact that

$$D = h = bh_{max} + (1 - b)h_{min} = D_{r\theta}$$

By noticing that

$$D_{r\theta} - D_r \geq 0$$

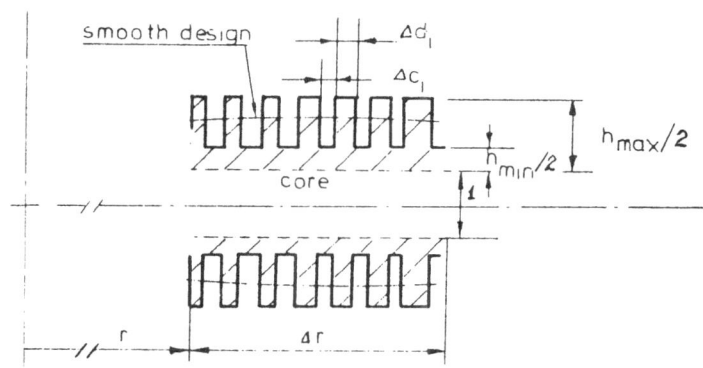

Figure 3.1   New Model for Sandwich Plates

one can conclude from Eq. 2.36 that

$$\phi^A \geq \phi^S$$

and that no comparison between ϕ^N and ϕ^S can be made. Therefore, a different path is followed, leading to the following theorem.

Theorem 3.1: For the minimum compliance design problem of sandwich plates, the new model with an infinite number of discontinuities of the face sheet thicknesses can only be inferior to the smooth plate of the same material distribution.

Proof: Assume that a smooth optimum design $h(r)$ of a sandwich plate is available and denote its compliance by ϕ^S. Keeping the material distribution unchanged within a small subregion Ω_s, change the smooth plate section into a new model section with density given by Eq. 2.32. Afterwards, solve the new plate and calculate its actual moments M_{rr}^*, $M_{r\theta}^*$, $M_{\theta\theta}^*$, and compliance ϕ^N. According to the argument advanced in Section 2.6, the moments are also statically admissible for the original smooth plate, if the applied loads and boundary conditions are not changed. Based on these moments, the integral in Eq. 2.29 provides an upper bound on ϕ^S, i.e.

$$\phi^A = \int_\Omega \frac{1}{D(1 - \nu^2)}$$

$$\times \left[M_{rr}^{*2} + M_{\theta\theta}^{*2} - 2\nu M_{rr}^* M_{\theta\theta}^* + 2(1 + \nu)M_{r\theta}^{*2} \right] rdr \geq \phi^S \qquad (3.2)$$

On the other hand, ϕ^N is given by

$$\phi^N = \int_{\Omega_s} \left\{ \frac{1}{D_{r\theta}(1 - \nu^2)} \right.$$

$$\times \left[M_{rr}^{*2} + M_{\theta\theta}^{*2} - 2\nu M_{rr}^* M_{\theta\theta}^* + 2(1 + \nu)M_{r\theta}^{*2} \right] + \frac{D_{r\theta} - D_r}{D_r D_\theta} M_{rr}^{*2} \left. \right\} rdr$$

$$+ \int_{\Omega_1} \frac{1}{D(1 - \nu^2)}$$

$$\times \left[M_{rr}^{*2} + M_{\theta\theta}^{*2} - 2\nu M_{rr}^* M_{\theta\theta}^* + 2(1 + \nu)M_{r\theta}^{*2} \right] rdr \qquad (3.3)$$

Taking $D_{r\theta} = D$ into consideration, one finds

$$\Phi^N = \Phi^A + \int_{\Omega_s} \frac{D_{r\theta} - D_r}{D_r D_\theta} M_{rr}^{*2} \, rdr \tag{3.4}$$

and therefore

$$\Phi^N \geq \Phi^A \tag{3.5}$$

where equality only holds if the integrand vanishes completely in Ω_s, i.e. if M_{rr}^* is equal to zero everywhere in Ω_s.

Relating Eq. 3.5 to Eq. 3.2, Thm 3.1 has been proved, i.e. $\Phi^N \geq \Phi^S$.

4. BEAM STIFFENED PLATES

Instead of considering integral stiffeners, as in Section 2, consider stiffeners of the kind mentioned at the end of Section 2.8, namely thin curved beams. Due to the fact that all stiffeners are infinitely thin, their twisting rigidities will be neglected [7].

The total strain energy consists of two parts, one of which is stored in the uniform plate of thickness h_{min} and the other in the stiffeners of height $(h_{max} - h_{min})$ and density $b(r)$. Therefore,

$$\Phi = \int_\Omega h_{min}^3 \left[\kappa_{rr}^2 + \kappa_{\theta\theta}^2 + 2\nu\kappa_{rr}\kappa_{\theta\theta} + 2(1 - \nu)\kappa_{r\theta}^2 \right] rdr$$

$$+ \int_\Omega b(r) \, (h_{max}^3 - h_{min}^3)\kappa_{\theta\theta}^2 \, rdr \tag{4.1}$$

The moment-curvature relation derived on the basis of Eq. 4.1 turns out to be the same as Eqs. 2.23 and 2.24, with the exception that one has to redefine

$$\left. \begin{array}{l} D_r = D_{min} = h_{min}^3 \\[2mm] D_{r\theta} = D_{min} \end{array} \right\} \tag{4.2}$$

$$D_\theta = D_{min} + b(r) (h^3_{max} - h^3_{min}) \Bigg\}$$

With a given material volume and h_{min} and h_{max}, the formulation for minimizing the compliance can still be stated as in Eq. 2.27. Of course, the deflection w has to satisfy a new governing equation, but this is available in Ref. 3. Again, adopting the Lagrange multiplier method, one finds that the necessary condition for optimality takes the following simple form:

$$\kappa^2_{\theta\theta} = \Lambda - \alpha + \beta$$

where

$$
\left.
\begin{array}{l}
\alpha \neq 0 \quad \text{and} \quad \beta = 0, \quad \text{if} \quad b = 0 \\[6pt]
\alpha = 0 \quad \text{and} \quad \beta \neq 0, \quad \text{if} \quad b = 1 \\[6pt]
\alpha = 0 \quad \text{and} \quad \beta = 0, \quad \text{if} \quad 0 < b < 1
\end{array}
\right\}
\tag{4.3}
$$

and Λ is a constant.

Using the algorithm given in Ref. 1, the optimality condition of Eq. 4.3, the governing plate equation, and suitable boundary conditions are easily solved numerically. For comparison, several results are presented.

It is intuitively clear that the beam stiffened plates considered here must be more compliant than the corresponding integrally stiffened plates proposed in Section 2. Again, take the clamped-clamped plate as an example and optimize it, subject to $n = 4$ and $h_{max}/h_{min} = 5$. The optimum design and its deflection are shown in Fig. 4.1. Stating the compliance $\Phi_{m,b}$ in the same way as before, the minimum compliance is now

$$\Phi_{m,b}/\Phi_{u,s} = 0.585.$$

Comparing this with the compliance 0.415 for the optimum integrally stiffened plate, the latter is obviously superior to the beam stiffened plate.

As is in Section 2, the ratio h_{max}/h_{min} and load wave number n strongly influence the comparison between solid plates and beam stiffened plates. With the same h_{max}, h_{min}, and R_i as in Fig. 2.4, taking load wave numbers $n = 0$, 1, and 2, the minimum

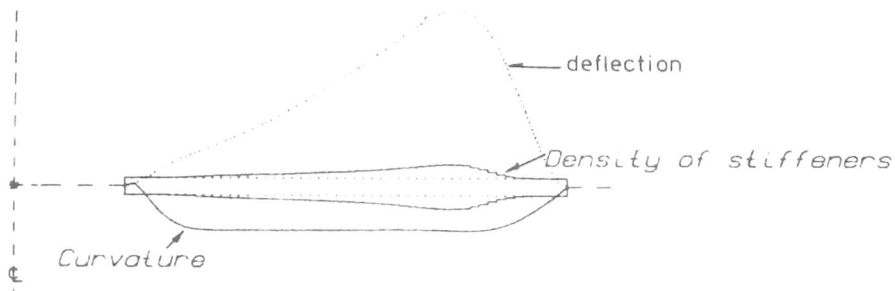

Figure 4.1   Results of Optimizing Annular, Beam Stiffened Plate.
$n = 4$, $h_{max}/h_{min} = 5$, $R_i = 0.2$, $\Phi_{m,b}/\Phi_{u,s} = 0.585$

compliances for this type of structure are even larger than the
ones provided by solid plates of uniform thickness (see Fig. 4.2).

Consider now the case in which the design variable is allowed
to have an infinite number of discontinuities. Then, Fig. 2.1 can
still be used to visualize the new model. Of course, the discontinuous function now represents the density of stiffeners. One
must also write b_{max} and b_{min}, instead of h_{max} and h_{min} in Fig.
2.1. For the new model, the design variable is

$$\mu(r) = \lim_{\Delta r \to 0} \frac{\Sigma \Delta d_i}{\Delta r}$$

and the volume element is now given by

$$dV = \{h_{min} + (h_{max} - h_{min})[\mu b_{max} + (1 - \mu)b_{min}]\}rdr \qquad (4.4)$$

Since one has already set $b_{min} = 0.0$ and $b_{max} = 1.0$, the constraint on material volume becomes

$$\int_{\Omega} \left[h_{min} + \mu(h_{max} - h_{min}) \right] rdr = 1 \qquad (4.5)$$

The same routine as for both solid and sandwich plates is
now briefly reviewed for this problem. Differences emerge as soon
as one reaches the continuity conditions. The requirement of
geometrical continuity ensures that the deflection w, rotation w',
and circumferential curvature $\kappa_{\theta\theta}$ must be continuous everywhere.
Due to the equation

330

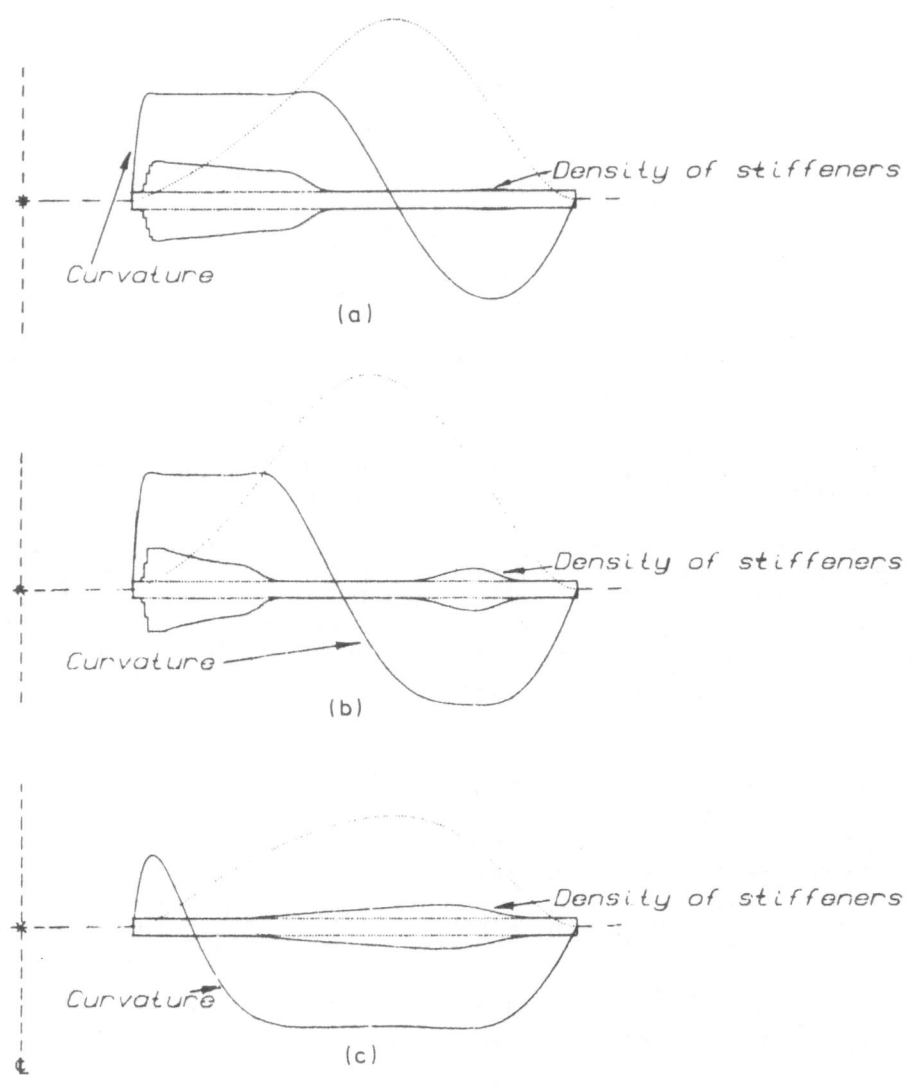

Figure 4.2   Influence of Different Load Wave Numbers n on the
             Comparisons Between Solid and Beam Stiffened Plates

(a) n = 0, $\Phi_{m,b}/\Phi_{u,s}$ = 1.425.   (b) n = 1, $\Phi_{m,b}/\Phi_{u,s}$ = 1.885.
(c) n = 2, $\Phi_{m,b}/\Phi_{u,s}$ = 1.451.   The curvatures refer to the circumferential curvatures $\kappa_{\theta\theta}$.

$$M_{rr} = -D_r(\kappa_{rr} + \nu\kappa_{\theta\theta}) = -h^3_{min}(\kappa_{rr} + \nu\kappa_{\theta\theta})$$

the continuity of M_{rr} implies continuous radial curvature κ_{rr} at interfaces. Therefore, one can directly perform the summation and limiting process behind Eq. 4.1 without any transformation. If one lets the material distribution remain unchanged, which leads to b = d, both the smooth design and the new design will have exactly the same expression for Φ. What has been established may be summarized as follows.

Theorem 4.1: For a beam stiffened plate with a smooth density distribution b(r), the new structure will have the same compliance as the smooth one, where b(r) is changed to be a function with an infinite number of discontinuities (as in Fig. 2.1) and the material distribution, applied load, and boundary conditions are unchanged.

Finally, it is emphasized that the beam stiffened plates considered here will underestimate the stiffness of the physical structure if the density b(r) is close to 1. Hence, for b = 1, the physical structure ought to become a solid plate with $h = h_{max}$, but due to the assumption that each individual stiffener behaves as a beam, the structure cannot withstand any radial or twisting moments.

5.   CONCLUSIONS

In this paper, both analytical and numerical investigations clearly show special features of optimizing solid plates. To determine a possible global optimal design for solid plates, the new formulation developed in Section 2 should be considered. Similar conclusions [9] are to be expected in optimization of solid plates with respect to maximizing the fundamental frequency, buckling load, minimizing the largest deflection, etc.

REFERENCES

1.   Cheng, K.-T. and Olhoff, N., An Investigation Concerning Optimal Design of Solid Elastic Plates, DCAMM-Report No. 174, The Danish Center for Applied Mathematics and Mechanics, 1980. (To appear in Int. J. Solids & Structures).
2.   Olhoff, N., Optimal Design for Vibrating Circular Plates, Int. J. Solids Structures, Vol. 6, 1970, pp. 139-156.
3.   Ambartzumjan, S.A., Theory of Anisotropic Plates (in Russian), Nauka Press, Moscow, 1967.
4.   Lekhnitskii, S.G., Anisotropic Plates, Gordon and Breach Science Publishers, New York, 1968.

332

5. Leipholz, H.H.E., Six Lectures on Variational Principles in Structural Engineering, Solid Mechanics Division, University of Waterloo, Waterloo, Ontario, Canada, 1978.

6. Reiss, R., "Optimal Compliance Criterion for Axisymmetric Solid Plates," Int. J. Solids Structures, Vol. 12, 1976, pp. 319-329.

7. Savin, G.N. and Fleishman, N.P., Rib-Reinforced Plates and Shells, (in Russian), Naukova Dumka, Kiev, 1964.

8. Armand, J.-L., Applications of the Theory of Optimal Control of Distributed-Parameter Systems to Structural Optimization, NASA, CR-2044, 1972.

9. Olhoff, N., Lurie, K.A., Cherkaev, A.V. and Fedorov, A.V., Sliding Regimes and Anisotropy in Optimal Design of Vibrating Axisymmetric Plates, DCAMM-Report, No. 192, The Danish Center for Applied Mathematics and Mechanics, 1980.

ACKNOWLEGMENT

The author wishes to thank Profs. Frithiof I. Niordson and Niels Olhoff for valuable discussions.

DESIGN OF PLATES FOR MINIMUM DEFLECTION AND STRESS

Nick V. Banichuk

Institute of Problems of Mechanics, USSR Academy of Sciences, Moscow, USSR

ABSTRACT

In this paper, optimal plate design problems are solved for various performance criterions. Optimal design problems of elastic plates having maximum bending stiffness are first examined. Optimality conditions are derived and boundary-value problems that serve to determine the optimum solutions are formulated. Analytical and numerical solutions are obtained and presented. Design of plates for minimum deflection is then considered and an optimization method is described. The method is based on the reduction of problems with a local functional to problems with an integral performance criterion. The method is also applied to the problems of stress intensity minimization. The present paper includes the results of Refs. 1 to 4.

1. PERFORMANCE CRITERIA-ELASTIC STRAIN POTENTIAL ENERGY

1.1 Optimality Criteria Static Bending Of An Elastic Three-Layer Plate

The problem of static bending of an elastic three-layer plate is first examined. The plate is simply supported along the contour Γ bounding the region Ω of the xy-plane and is loaded at the point $O(x_0, y_0)$ by the concentrated load P.

The thickness of the outer reinforcing layers of the plate is variable and, consequently, its cylindrical stiffness D is variable. The basic equations of bending of variable-thickness elastic plates are presented, for example, in Ref. 5. One may represent D in the form D = Kh, where the structural variable $\frac{1}{2}h(x,y)$ denotes the thickness of the outer reinforcing layers of the plate and $K = EH^2/4(1-\nu^2)$, where ν is the Poisson ratio, E is the Young's modulus of the plate material, and H is the constant middle-layer thickness. The volume of the material of the load-carrying layers is assumed given, which leads to the isoperimetric condition imposed on the outer layers thickness distribution

$$\iint_\Omega h\,dxdy = V \qquad (1.1)$$

In the studies made in the following, the magnitude of the force P is considered given. The deflection $w_0 = w(x_0,y_0)$ at the point $O(x_0,y_0)$ is

$$w_0 = \frac{K}{P} \iint_\Omega hU\,dxdy \qquad (1.2)$$

where

$$U = [(w_{xx}+w_{yy})^2 - 2(1-\nu)(w_{xx}w_{yy}-w_{xy}^2)]$$

and w_{xx}, w_{xy}, and w_{yy} are partial derivatives of the deflection function $w(x,y)$ with respect to the variables indicated by the subscripts. Equation 1.2 expresses the equality of the work done by the force on the displacement w_0 to the potential energy accumulated by the plate in the deformation process.

When the plate shape, i.e. the function $h(x,y)$, is varied, the value of the deflection w_0 at the point O will change. The optimization problem involves finding the function $h(x,y)$, satisfying the isoperimetric condition of Eq. 1.1 and minimizing the functional w_0, i.e.

$$w_{0*} = \min_h w_0 = \min_h \frac{K}{P} \int_\Omega hU\,dxdy \qquad (1.3)$$

For regions Ω whose contour Γ is symmetric relative to some point $O_* \in \Omega$ (ellipse, equilateral n-gon, rectangle, etc.) it is most natural to colocate the point O with O_* and evaluate the plate stiffness by the magnitude of the deflection at the point O.

The necessary minimum condition for the functional w_0, under the constraint of Eq. 1.1 and the equation of plate bending can be obtained, using the usual variational technique, as

$$U = \lambda^2 \qquad\qquad (1.4)$$

Here λ^2 denotes the Lagrange multiplier corresponding to the isoperimetric condition of Eq. 1.1.

It is not difficult to show, by direct arguments, that the condition of Eq. 1.4 is not only a necessary, but also a sufficient optimum condition. To do this, one examines two distributions of thicknesses, h_* and h, satisfying Eq. 1.1. Let w_* and w denote the actual realized deflections corresponding to the distributions of the thicknesses h_* and h. Assume that the functions h_* and w_* are connected by the optimality condition of Eq. 1.4, while the functions h and w are not subject to this requirement. Since for the distribution of the thicknesses h, the function w minimizes the integral of Eq. 1.2, the following inequality holds

$$K \iint\limits_\Omega hU dx dy - 2Pw_0 \leq K \iint\limits_\Omega h\, U_* dx dy - 2Pw_0$$

Using this inequality and making the following estimates

$$w_0 - w_{0*} = -\frac{1}{P}\left[K \iint\limits_\Omega hU dx dy - 2Pw_0 - K \iint\limits_\Omega h_* U_* dx dy \right.$$

$$\left. + 2Pw_{0*} \right] \geq -\frac{K}{P} \iint\limits_\Omega U_*(h-h_*) dx dy = 0$$

one obtains $w_0 \geq w_{0*}$. Consequently, the functional w_0 reaches a minimum for the functions h_* and w_*, subject to the condition of Eq. 1.4 (the equilibrium equations and the boundary and isoperimetric conditions are assumed satisfied). Thus, Eq. 1.4 is not only a necessary, but also a sufficient optimality condition.

In the previous considerations, the method proposed in Ref. 6 was used.

1.2 Solution of Plate Optimality Criteria

The general problem of finding the optimum distributions of h, w, and the quantity w_{0*} breaks down into two boundary-value problems for second-order partial differential equations. The optimum deflection distribution can, in principle, be found as a solution of the boundary-value problem of Eq.1.4, with the condition $(w)_\Gamma = 0$. Introducing the new variable $u = w/\lambda$, one writes this boundary-value problem in the form

$$\left.\begin{array}{l} (u_{xx}+u_{yy})^2 - 2(1-\nu)(u_{xx}u_{yy}-u_{xy}^2) = 1, \ (x,y) \in \Omega \\[2mm] u = 0 \ , \ \ (x,y) \in \Gamma \end{array}\right\} \tag{1.5}$$

The constant λ is defined by the isoperimetric constraint of Eq. 1.1, as

$$\lambda = \frac{Pu(x_0,y_0)}{KV} \tag{1.6}$$

The magnitude of the deflection w_{0*}, whose calculation does not require knowledge of the optimum thickness distribution $h(x,y)$, can be found from the following formula, derived from Eqs. 1.3, 1.4, and 1.6:

$$w_{0*} = \frac{K}{P}\iint\limits_{\Omega} h_* U_* dxdy = \frac{Pu^2(x_0,y_0)}{KV} \tag{1.7}$$

Note that the solution of Eq. 1.5 is independent of the position of the point 0 and is determined completely by the shape of the region Ω. This makes it possible, after solving Eq. 1.5 once for a given shape and determining the function $u(x,y)$, to obtain (by simple scaling in terms of u) the optimum deflection distribution and the value of the quantity w_{0*} being minimized, for different variants of location of the point 0 in Ω.

After finding the optimum deflection distribution $w(x,y)$ and substituting it into the plate bending equation and boundary conditions, one obtains a boundary-value problem for h. The function $h(x,y)$ is then found as the solution of this boundary-value problem.

One may now study the asymptotic behavior of the optimum solution near the contour Γ. Assume the contour Γ to be smooth and introduce the local orthotropic coordinate system $\xi\eta$, with coordinate origin at an arbitrary point Q of the contour Γ, directing the η axis along the tangent to the contour Γ and the ξ axis into the region Ω. Near the point Q, Eq. 1.5 is represented asymptotically in the form $u_{\xi\xi}^2 = 1$ and the optimum deflection distribution, as follows from this equation and the boundary condition of Eq. 1.5, is given by the quadratic function $u = \frac{1}{2}\xi^2 + a_0\xi$, where a_0 is an arbitrary constant. Using the asymptotic representation $u_{\xi\xi}^2 = 1$ and zero moment condition on Γ, one obtains $(h)_\Gamma = 0$. To determine the behavior of h near Γ, one can use the asymptotic behavior of the function u and write the asymptotic representation of the equation of the plate bending for small $\xi : h_{\xi\xi} = 0$. Hence, accounting for the boundary condition $(h)_\Gamma = 0$, one obtains $h = a_1\xi$, where a_1 is the asymptotic behavior constant. Thus, in determining the function h from the solution of the boundary-value problem of plate bending, one must use the boundary condition $(h)_\Gamma = 0$.

1.3 Optimum Circular And Elliptic Plates

One may now examine the problem of optimal design of a circular plate of radius R. The plate is simply supported along the contour and loaded at the center by the concentrated force P. The problem is solved in the polar coordinate system $r\phi$, located with the system origin $r = 0$ at the center of the plate. Using the axial symmetry property, one seeks the deflection and thickness distributions as functions of the single independent variable r, i.e. $w = w(r)$ and $h = h(r)$, where $0 \leq r \leq R$. In the selected coordinate system, the Eqs. 1.5 and 1.6 that serve to determine the deflection $w = \lambda u$ of the optimum plate take the form

$$\left(u_{rr} + \frac{u_r}{r}\right)^2 - \frac{2(1-\nu)}{r} u_r u_{rr} = 1 \qquad (1.8)$$

with the conditions

$$u(R) = 0, \quad u_r(0) = 0, \quad \lambda = \frac{w(0)}{u(0)}$$

The condition $u_r(0) = 0$ follows directly from Eq. 1.8. In fact, assuming the contrary, i.e. that $u_r(0) \neq 0$ and representing

Eq. 1.8 in the form

$$(u_{rr} + \nu u_r/r)^2 + u_r^2(1-\nu^2)/r^2 = 1$$

one finds that for r tending to zero, the left side of the equation increases without bound and thereby for $u_r(0) \neq 0$, this equation cannot be satisfied.

The solution of Eq. 1.8, satisfying the boundary conditions below Eq. 1.8, has the form $u = (R^2-r^2)/2\sqrt{2(1+\nu)}$ and the constant λ is $\lambda = PR^2/2\sqrt{2(1+\nu)}$ KV. Thus, the optimum deflection distribution is described by a quadratic function in the variable r,

$$w_* = \frac{PR^2(R^2-r^2)}{8(1+\nu)KV}$$

$$w_{0*} = \left(w_*\right)_{r=0} = \frac{PR^4}{8(1+\nu)KV}$$

Note that for a circular plate with constant reinforcing-layer thickness,

$$w_0 = PR^4(3+\nu)/16(1+\nu)KV$$

Comparing the value of the maximum deflection for optimum plate with the corresponding value of w_0 for the constant-thickness plate, one finds that the relative gain in deflection as a result of optimal distribution of the thickness is equal to $(w_0-w_{0*})/w_0 = 1-2/(3+\nu)$, which varies from 33% to 56%, depending on the values taken by the Poisson ratio ν, $0 < \nu < 0.5$.

To determine the optimum thickness distribution, one writes the bending equation in polar coordinates,

$$\left[h_r(rw_{rr}+\nu w_r) + h(rw_{rrr}+w_{rr}-\frac{w_r}{r})\right]_r = P\delta(r) \qquad (1.9)$$

The conditions for selecting the integration constant of Eq. 1.9 have the form

$$h(R) = 0$$

$$2\pi \int_0^R hrdr = V$$

$$(1.10)$$

Substituting into Eq. 1.9 in place of w, the function w(r) found and solving Eqs. 1.9 and 1.10 for h(r), one obtains

$$h = -\frac{4V}{R^2} \ell n\left(\frac{r}{R}\right)$$

The optimum thickness distribution has a singularity at the point of force application. An analogous singularity was identified in Ref. 7, in studying the optimal form h(r) of a circular plate in the framework of limit plastic design theory.

The optimum solution found cannot be used directly for practical purposes. However it can be used for theoretical estimation of the optimization possibilities and for constructing quasi-optimum solutions that are close to the found optimum solution, but do not have singularities.

Finally, consider the two-dimensional bending problem for an elliptic plate loaded at the point $O(x_0, y_0)$ by the force P and simply-supported along the contour $x^2/a^2 + y^2/b^2 = 1$. It is not difficult to verify that, for this form of the region Ω, the solution of the boundary-value problem of Eq. 1.5 has the form

$$u = \frac{a^2 b^2}{2\sqrt{a^4 + b^4 + 2\nu a^2 b^2}}\left(1 - \frac{x^2}{a^2} - \frac{y^2}{b^2}\right)$$

$$\lambda = \frac{a^2 b^2 P}{\sqrt{a^4 + b^4 + 2\nu a^2 b^2}\ 2KV}$$

and the optimum deflection distribution and the value of the functional are given by

$$w_* = \frac{Pa^4 b^4}{4KV\sqrt{a^4 + b^4 + 2\nu a^2 b^2}}\left(1 - \frac{x^2}{a^2} - \frac{y^2}{b^2}\right)$$

$$w_{0*} = \frac{Pa^4 b^4}{4KV\sqrt{a^4 + b^4 + 2\nu a^2 b^2}}$$

These formulas for w_* and w_{0*} can now be used to estimate elliptical plate effectiveness.

2. OPTIMIZATION PROBLEMS WITH LOCAL PERFORMANCE CRITERIA IN THE THEORY OF PLATE BENDING

2.1 Formulation Of The Problem

Consider the more complicated optimization problem for a solid elastic plate, clamped along contour Γ in the xy plane and loaded by the transverse forces $q = q(x,y)$. The contour Γ bounds the region Ω whose area is S. The volume of the plate is V. In the undeformed state, the plate middle surface coincides with Ω. On a part Γ_1 of Γ the plate is simply supported and on the other part Γ_2 it is rigidly clamped ($\Gamma = \Gamma_1 + \Gamma_2$). Let $h = h(x,y)$ represent the distribution of plate thickness. To simplify notation, one introduces the dimensionless variables

$$x' = \frac{x}{\sqrt{S}}, \ y' = \frac{y}{\sqrt{S}}, \ w' = \frac{w}{\sqrt{S}}, \ h' = \frac{hS}{V}, \ q = \frac{12(1-\nu^2)S^{9/2}q}{EV^3}$$

The discussion is carried out in dimensionless variables, so the primes on the dimensionless variables are omitted. In the usual notation, the equilibrium equation and the boundary conditions are written in the following form (see Ref. 5, for example):

$$Lw \equiv [h^3(w_{xx} + \nu w_{yy})]_{xx} + [h^3(w_{yy} + \nu w_{xx})]_{yy} + 2(1-\nu)(h^3 w_{xy})_{xy} = q$$

$$(w)_\Gamma = 0, \ \left(\frac{\partial w}{\partial n}\right)_{\Gamma_2} = 0, \ \left(h^3\left[\Delta w - \frac{1-\nu}{R}\frac{\partial w}{\partial n}\right]\right)_{\Gamma_1} = 0$$

(2.1)

where $\partial w/\partial n$, R, and Δ represent the derivative of the function w with respect to the external normal to the contour, the radius of curvature of the contour, and the Laplace operator, respectively.

For the given problem, the plate rigidity is evaluated in terms of the maximum deflection $J = \max_{xy}|w|$. The functional J depends on the distribution of plate thickness, i.e. $J = J(h)$.

The optimization problem consists of finding a distribution $h(x,y)$, among all continuous thickness distributions, that satisfies the conditions

$$h_{min} \leq h \leq h_{max}$$

$$\iint_\Omega h\,dx\,dy = 1$$

$$\left.\phantom{\begin{array}{c}1\\1\\1\\1\end{array}}\right\}$$ (2.2)

such that the functional $J(h)$ is minimized, i.e. the maximum plate deflection is minimized,

$$J_* = \min_h J = \min_h \max_{xy} |w| \qquad (2.3)$$

The problem of Eqs. 2.1 to 2.3 is a minimax, optimization problem with a local performance criterion. Here Eq. 2.1 are differential constraints, the first constraints of Eq. 2.2 are minimum and maximum values of the control function, and the second constraint of Eq. 2.2 is an isoperimetric condition.

Consider the constraints of Eq. 2.2 imposed on the control function. Direct allowance for the inequalities of Eq. 2.2 in optimization problems necessitates determination of the lines for which the function $h(x,y)$ reaches the constraints. This leads to certain difficulties in solution of the problem. Thus, one henceforth makes use of the auxiliary control function

$$h = \alpha + \beta\sin\phi \qquad (2.4)$$

where

$$\alpha = \frac{1}{2}(h_{max} + h_{min})$$

$$\beta = \frac{1}{2}(h_{max} - h_{min})$$

By introducing the auxiliary control function ϕ, one is able to eliminate inequalities in Eq. 2.2 from consideration. In fact, by substituting $h = \alpha + \beta\sin\phi$ into Eq. 2.2, one sees that the inequalities are satisfied for any value of ϕ. The sole restriction on ϕ is the isoperimetric condition of Eq. 2.2, which from Eq. 2.4 may be written as

$$\iint_\Omega \sin\phi\,dx\,dy = \gamma \qquad (2.5)$$

where

$$\gamma = \beta^{-1}(1 - \alpha) \quad .$$

In contrast to the integral functionals that are ordinarily considered in optimization problems for elastic bodies, such as elastic strain energy and fundamental frequency of plate vibrations, the functional $J(h) = \max_{xy}|w|$ to be minimized here is not determined by the entire realization of the state function, but only by its values at certain maximum points, which are not known in advance. The local nature of this functional is associated with fundamental difficulties that arise when one solves the problem given by Eqs. 2.1 and 2.3 to 2.5.

2.2 Transformation Of Local Performance Criteria To Integral Form

Consider a method that makes it possible to reduce investigation of the problem of optimization with a local performance criterion to a problem with an integral functional. The functional to be minimized takes the form of a norm in the space of the continuous functions C, i.e.

$$J = ||w||_C = \sup_{\Omega}|w(x)|$$

This norm is connected with the norm in the space L_p of functions that are integrable with degree p. As is known from functional analysis, for any p the inequality

$$||w||_{L_p} \leq ||w||_C$$

holds for all $w \in C$, while

$$\lim_{p \to \infty}||w||_{L_p} = ||w||_C$$

Taking into account the properties indicated, in addition to the initial problem of Eqs. 2.1 and 2.3 to 2.5 (denoted here as Problem 1), consider the minimization with respect to ϕ of the integral functional

$$J_{p*} = \min_{\phi} J_p = \min_{\phi} \left(\iint_{\Omega}|w|^p dxdy\right)^{\frac{1}{p}} \tag{2.6}$$

under the conditions of Eqs. 2.1, 2.4, and 2.5. This problem is called Problem 2. Henceforth p is taken to be even. In this case the absolute-value symbol in Eq. 2.6 may be omitted. For sufficiently large p, the solutions of Problems 1 and 2 will not be very different with $J_{p*} < J_*$.

2.3 Optimality Conditions

One may now obtain the optimality conditions for Problem 2. To do this, write an expression for the first variation of the functional to be minimized,

$$\delta J_p = \left(||w||_{L_p}\right)^{1-p} \iint_\Omega w^{p-1} \delta w \, dxdy$$

One now wishes to express the variation δw of the function w in the integrand in terms of the variation $\delta\phi$ of the control function. To do this, introduce the function v, which satisfies the boundary conditions of Eq. 2.1. One may also write an equation in variations corresponding to the differential equation in Eq. 2.1. Multiplying the expression on the left side of this equation by v and then integrating the product over the region Ω, one has

$$\iint_\Omega vL \delta w \, dxdy + 3\iint_\Omega v[(h^2 w_{xx}\delta h)_{xx} + (h^2 w_{yy}\delta h)_{yy}$$

$$+ v(h^2 w_{xx}\delta h)_{yy} + v(h^2 w_{yy}\delta h)_{xx} + 2(1-v)(h^2 w_{xy}\delta h)_{xy}]dxdy = 0$$

Integrating by parts and taking into account the boundary conditions of Eq. 2.1 for the functions w and v, one obtains

$$\iint_\Omega \delta w Lv \, dxdy + 3\iint_\Omega \delta h h^2[w_{xx}v_{xx} + w_{yy}v_{yy} + v(w_{xx}v_{yy}+w_{yy}v_{xx})$$

$$+ 2(1-v)w_{xy}v_{xy}]dxdy = 0$$

Taking this equality into account, one may represent the expression for δJ_p as

$$\delta J_p = \iint_\Omega \delta w\left[\left(\frac{w}{||w||_{L_p}}\right)^{p-1} - Lv\right]dxdy$$

$$- 3\beta\iint_\Omega \cos\phi\,\delta\phi(\alpha+\beta\sin\phi)^2[w_{xx}v_{xx} + w_{yy}v_{yy}$$

(Equation continued on next page)

$$+ \nu(w_{xx}v_{yy}+w_{yy}v_{xx}) + 2(1-\nu)w_{xy}v_{xy}]dxdy$$

One may determine the conjugate function v as a solution of the boundary-value problem

$$Lv = \Phi_p \equiv \left(\frac{w}{||w||_{L_p}}\right)^{p-1}$$

$$(v)_\Gamma = 0, \quad \left(\frac{\partial v}{\partial n}\right)_{\Gamma_2} = 0, \quad \left(h^3\left[\Delta v - \frac{1-\nu}{R}\right]\right)_{\Gamma_1} = 0 \qquad (2.7)$$

The first integral on the right side of the expression for δJ_p vanishes when the function v is defined in this way. As a result, one arrives at the desired formula for the first variation of the functional J_p, explicitly in terms of $\delta\phi$,

$$\delta J_p = - 3\beta \iint_\Omega \Lambda\cos\phi\,\delta\phi\,dxdy \qquad (2.8)$$

where

$$\Lambda \equiv (\alpha+\beta\sin\phi)^2[\Delta w\Delta v - (1-\nu)(w_{xx}v_{yy}+w_{yy}v_{xx}-2w_{xy}v_{xy})]$$

It follows from Eq. 2.5 that any variation of the control function $\delta\phi$ that occurs in the integrand of Eq. 2.8 must satisfy the equation

$$\iint_\Omega \cos\phi\,\delta\phi\,dxdy = 0 \qquad (2.9)$$

The necessary condition for minimization of J_p thus has the form $\delta J_p = 0$, for all $\delta\phi$ satisfying Eq. 2.9. From this and from Eqs. 2.8 and 2.9, one obtains the optimality condition for problem 2 as

$$(\Lambda - C)\cos\phi = 0 \qquad (2.10)$$

where C is a constant.

For prescribed p, the optimization Problem 2 reduces to solution of boundary-value problems of Eqs. 2.1 and 2.7 for the function w and the conjugate variable v, the optimality condition of Eq. 2.10 for the control function ϕ, and the isoperimetric condition of Eq. 2.5 for the unknown constant C. Thus, in seeking the optimum plate thickness distribution (in the sense of Problem 2), one has a boundary-value problem that is nonlinear, owing to the nonlinearity of Eqs. 2.7 and 2.10.

The solution of Problem 2 depends on p, since this parameter occurs on the right side of Eq. 2.7 for the conjugate variable. The remaining relationships of Eqs. 2.1, 2.5, and 2.10 do not depend explicitly on p. In particular, the optimality condition of Eq. 2.10 is independent of p and it will therefore have the same form in the limiting case $p = \infty$ as well. As a consequence, Eq. 2.10 is also the optimality condition for Problem 1 and one may obtain the system of equations determining the solution of Problem 1 by calculating the limit of the expression written on the right side of Eq. 2.7 as $p \to \infty$. This limit is calculated in Ref. 3, where it is shown that the limit of the function $\phi_p(x,y)$, as $p \to \infty$, is the delta function

$$\lim_{p \to \infty} \phi_p(x,y) = \delta(x-\xi,y-\eta)$$

Assume that (x,y) is an arbitrary point of the region Ω and that the maximum of the function $|w|$ over the region Ω is reached at the point $(\xi,\eta) \in \Omega$. As a consequence, the equation for the conjugate-variable v in Problem 1 is written as

$$Lv = \delta(x-\xi,y-\eta) \tag{2.11}$$

For the case in which a maximum of $|w|$ is realized at several isolated points simultaneously, the expression on the right side of Eq. 2.11 is replaced by a sum of δ-functions.

Certain difficulties are involved in numerical solution to the boundary-value problem of Eqs. 2.1, 2.5, 2.10, and 2.11, owing to the presence of a delta function on the right side of Eq. 2.11 and the fact that the $\xi\eta$ coordinates are not known in advance, but are determined in the course of the solution. Thus, instead of directly solving Problem 1, on the basis of the equations, indicated, a method is proposed for constructing quasi-optimal solutions. By this is meant approximate solutions of Problem 2 approach solutions of Problem 1 when $p \to \infty$.

2.4 A Numerical Method For Problem 2

Consider now a numerical method for solving Problem 2 and finding the optimum thickness distribution. To construct the computational algorithm, one must have an expression for the variation $\delta\phi$ that reduces the functional J_p, i.e. $\delta J_p < 0$, and that satisfies Eq. 2.9, i.e. that produces no change in the volume V. It may be shown (see Ref. 2 that the variation $\delta\phi$ determined by the formulas

$$\left.\begin{aligned}
\delta\phi &= \tau\psi \\[2em]
\psi &= \cos\phi\left[\Lambda - \frac{\iint\limits_{\Omega}\Lambda\cos^2\phi\,dxdy}{\iint\limits_{\Omega}\cos^2\phi\,dxdy}\right]
\end{aligned}\right\} \tag{2.12}$$

meets these requirements. In Eq. 2.12, τ is used to represent a positive constant (the gradient step).

The proposed computational algorithm consists of successive approximation to the optimum solution. The algorithm is made up of steps. Consider the computational operations performed at one step, numbered k+1(k=0,1,...). On the basis of the calculations carried out at the preceding step, one takes the functions $\phi^k(x,y)$, $w^k(x,y)$, $v^k(x,y)$, and the corresponding value of the functional J_p^k to have been found. At the (k+1) -st step one first determines the new values of the control function $\phi^{k+1}(x,y)$. To do this the gradient projection method is used, i.e.

$$\phi^{k+1} = \phi^k + \tau\psi^k$$

The function ψ^k is calculated from Eq. 2.12, where $\phi = \phi^k$, $\Lambda = \Lambda^k = \Lambda(w^k,v^k,\phi^k)$. From Eq. 2.4, with $\phi = \phi^{k+1}$, one finds the distribution of thickness $h = h^{k+1}(x,y)$. Then, for this thickness distribution one uses the local-variation method, with variable variation steps, to solve the variational problems of minimization of the functionals

$$I_w = \iint\limits_{\Omega}h^3[w_{xx}^2+w_{yy}^2+2\nu w_{xx}w_{yy}+2(1-\nu)w_{xy}^2]dxdy-2\iint\limits_{\Omega}qwdxdy$$

$$I_v = \iint\limits_{\Omega}h^3[v_{xx}^2+v_{yy}^2+2\nu v_{xx}v_{yy}+2(1-\nu)v_{xy}^2]dxdy-2\iint\limits_{\Omega}\phi_p vdxdy$$

and determines the deflection $w = w^{k+1}(x,y)$ and the conjugate function $v = v^{k+1}(x,y)$ (in the (k+1)-st approximation). The functionals I_w and I_v are minimized on the class of functions w and v satisfying the first two boundary conditions of Eq. 2.1 and 2.7. The third boundary condition is natural for the functionals I_w and I_v, so it need not be satisfied in advance.

2.5 Symmetric Cases

The relations of Eqs. 2.1, 2.5, 2.10, and 2.11, corresponding to Problem 1 ($p = \infty$), are convenient if the location of maximum deflection point is known beforehand. Consider the problem of Eqs. 2.1, 2.3, and 2.5 for the case of a symmetrical region Ω and boundary conditions of Eq. 2.1. Symmetry here is with respect to the origin of coordinate (0,0). The load is supposed to be concentrated and applied to the plate at the point (0,0), i.e. $q = P\delta(x,y)$, where P is the value of the force.

The equilibrium equation has the form

$$L(h)w = P\delta(x,y) \qquad (2.13)$$

so the maximum plate deflection is achieved at the point (0,0), for any symmetrical thickness distribution. Consequently, $\xi = 0$, $\eta = 0$ in Eq. 2.11. Thus, the conjugate function v of Problem 1 satisfies the equation

$$L(h) = \delta(x,y)$$

and the same boundary conditions as for the deflection function. Hence

$$w = Pv$$

and one need only solve one boundary value problem at any algorithm step, instead of two boundary value problems for the general nonsymmetrical case. Indeed, this relation provides the opportunity to exclude the function v from consideration and to reduce the problem to that of determining the deflection function w.

To determine the deflection distribution w, the following method is used. Suppose that the deflection value w_0 at the point (0,0) is given. Then the deflection distribution can be found from the following variational principle

$$P_0 = \frac{1}{w_0} \min_w \iint_\Omega (\alpha + \beta \sin\phi)^3 [(\Delta w)^2 - 2(w_{xx}w_{yy} - w_{xy}^2)] dx dy \qquad (2.15)$$

The minimum in Eq. 2.15 is determined among the functions satisfying the boundary conditions of Eq. 2.1 and the constraint $w(0,0) = w_0$. Let $w'(x,y)$ denote a function for which the minimum in Eq. 2.15 is achieved. It is evident that the solution of the boundary value problem for Eq. 2.13, with the boundary conditions of Eq. 2.1 is recalculated from $w'(x,y)$, according to the formulae

$$\left. \begin{aligned} w &= \frac{P}{P_0} w' \\[2ex] J &= ||w||_C = \frac{P}{P_0} w_o \end{aligned} \right\} \qquad (2.16)$$

Thus, the problem of obtaining the deflection distribution and the value of the optimized functional reduces to the solution of the variational problem of Eq. 2.15 and calculation of w and J, according to Eq. 2.16. In other aspects, the procedure for numerical solution of the optimization problem is the same as in the general case.

This algorithm has been implemented as a computer program. It was used to calculate the optimum thickness distribution for square plates (Ω: $-1/2 \leq x \leq 1/2$, $-1/2 \leq y \leq 1/2$), for various boundary conditions and values of the parameters h_{min} and h_{max}. The load q was taken to be concentrated $q = \delta(x,y)$, i.e. $P = 1$. The cases of rigid and hinged attachment of the edges of the plates were considered. In each case, the existing symmetry was used and the problem was solved in the square $0 \leq x \leq 1/2$, $0 \leq y \leq 1/2$, comprising a quarter of region Ω.

The optimum plate shapes shown in Figs. 2.1 to 2.3 (symmetrical relative to the neutral surface) correspond to the case $h_{min} = 0.8$ and $h_{max} = 1.2$. For the results shown, the initial thickness distribution was assumed to be constant $h^0(x,y) = 1$.

Results are first described for a built-in plate. Figure 2.1 shows the thickness distribution found via the foregoing calculations. Calculations show that the plate material is concentrated in the center and in the vicinity of the lines connecting the midpoints of its sides. On these regions, there are portions where the thickness is at the maximum, i.e. $h = h_{max}$. The

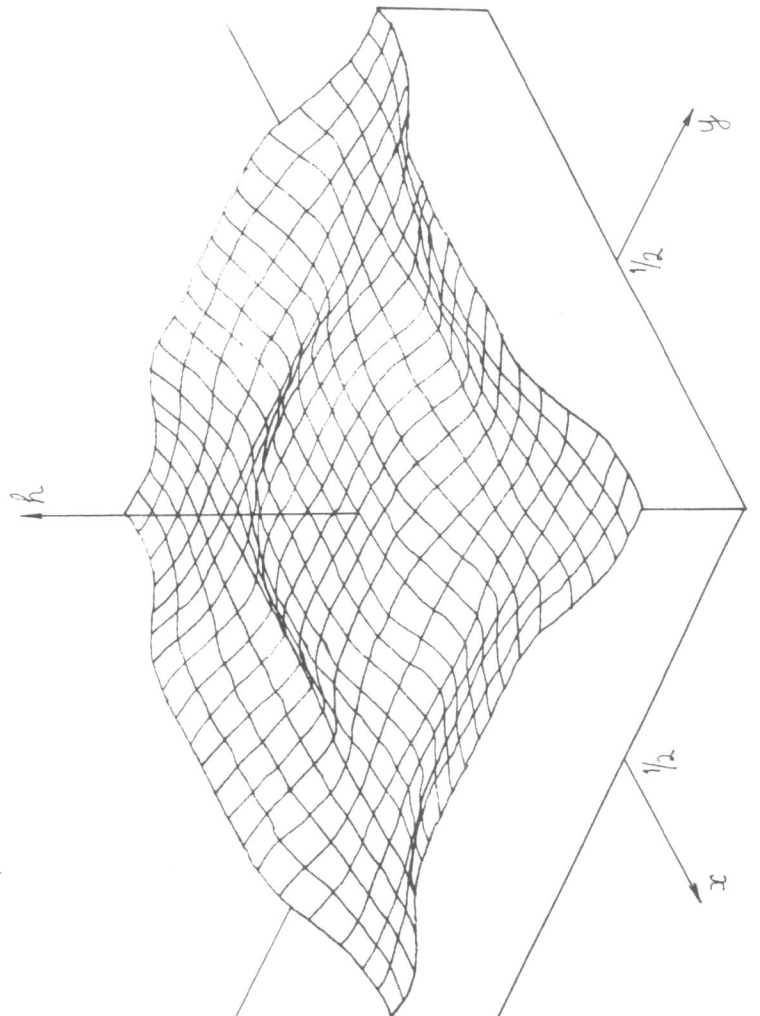

Figure 2.1.   Optimum Thickness Distribution For A Built-In (Clamped) Plate

350

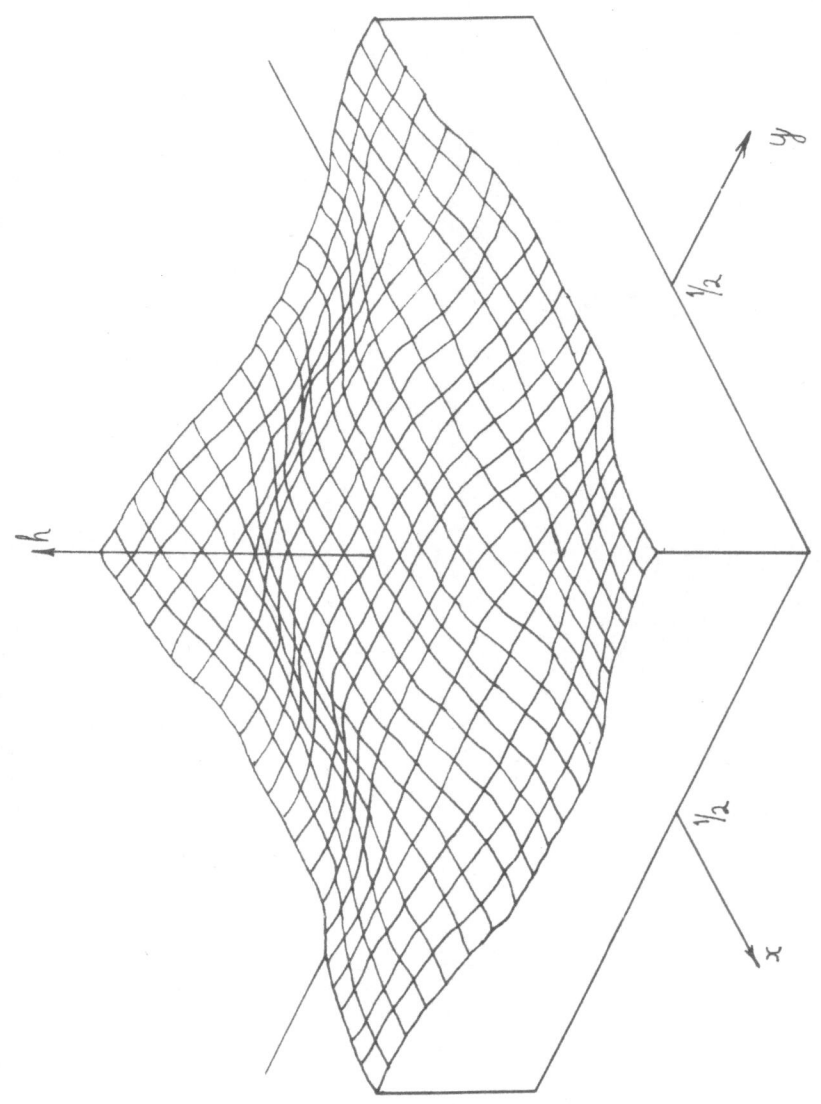

Figure 2.2.   Optimum Thickness Distribution For A Simply Supported Plate

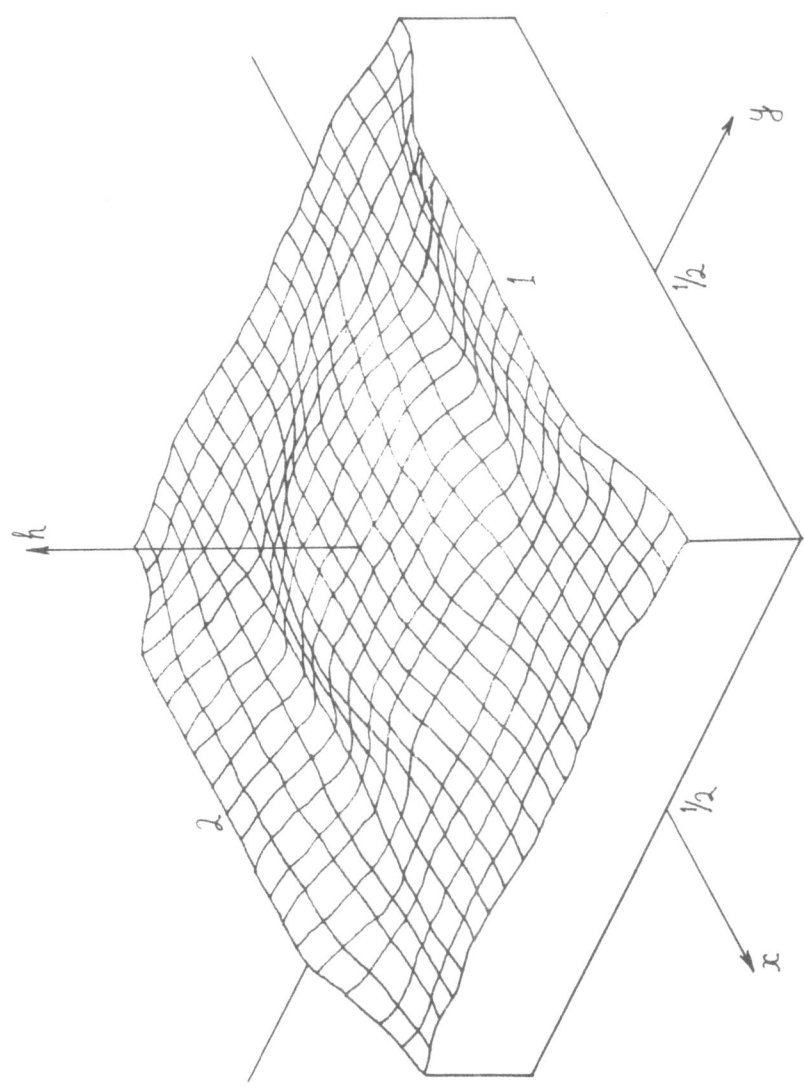

Figure 2.3.   Optimum Thickness Distribution For Built-In--Simply Supported Plate

qualitative behavior of h along these lines recalls the optimum distribution of material in a rigidly clamped beam. At the corners of the plate, as calculations indicate, the use of material is less effective and here the thicknesses are at the minimum, i.e. $h = h_{min}$. Between the regions in which $h(x,y) = h_{max}$ and $h(x,y) = h_{min}$, there are transition zones in which $h_{min} \leq h \leq h_{max}$.

It has been noted that as h_{max} increases the qualitative behavior of the solution is maintained, but the region in which plate material is concentrated becomes pronounced. The formation of an optimum beam system has been observed, with decreasing beam width and increasing beam height. For the case of a cubic dependence $D \sim h^3$, constraints $h_{min} \leq h \leq h_{max}$ are essential for the problem to be correctly posed. The deflection w_0 corresponding to the optimum plate is $w_{0*} = 0.00098$, and for a plate of constant thickness $w_0 = 0.00139$.

The thickness distribution of a simply supported plate is shown in Fig. 2.2. The solution shows that the bulk of the material is concentrated in the center and along the diagonals of the plate. In these directions, the thickness of the optimum plate is at the upper limit, i.e. $h = h_{max}$. For a simply supported plate it is more advantageous to place the material in the corners, where cohesion of the material and the supports is more pronounced. As Fig. 2.2 shows, $h = h_{min}$ in the regions adjacent to the midpoints of the plate edges, i.e. the material cannot function effectively in these regions. In this case, $w_{0*} = 0.00197$ for the optimum plate and, for a plate of constant thickness, $w_0 = 0.00274$.

Calculations were also made for the case in which two opposite edges ($x = \pm 1/2$, $-1/2 \leq y \leq 1/2$) are built-in, while two are simply supported. The thickness distribution for the optimum plate is shown in Fig. 2.3 (built-in edges are indicated by numbers 1 and 2 in Fig. 2.3). For a plate of constant thickness $h(x,y) = 1$, maximal deflection is $w_0 = 0.00169$, while for the optimum plate, $w_{0*} = 0.00091$.

2.6 Nonsymmetric Cases

The foregoing algorithm was used to calculate the optimum
thickness distribution of plates for cases in which the location
of the point of maximum deflection is not known beforehand. Calculation was made for square plates (Ω: $-1/2 \leq x < 1/2$, $-1/2 \leq y$
$< 1/2$) for various boundary conditions and values of the parameters h_{min} and h_{max}. The load q was taken to be constant,
$q(x,y) = 1$. Two cases were considered: (1) three edges were
simply supported (indicated by the numbers 1, 2, and 3 in Fig.
2.4) and one edge is built-in (indicated by the number 4 in
Fig. 2.4) and (2) three edges are built-in (indicated by numbers
1, 2, and 3 in Fig. 2.5) and one edge is simply supported
(indicated by the number 4 in Fig. 2.5).

Figure 2.4 shows the calculated distribution of thickness
for the optimal plate in the first case. The thickness distribution found corresponds to the following parameter values:
h_{max} = 1.2, h_{min} = 0.8, V = 1, and p = 100. The calculation
shows that plate material is concentrated at the center, at the
clamped (built-in) edge, and at the corners formed by the simply
supported edges. In these regions the plate thickness is
maximal, i.e. $h = h_{max}$. Utilization of the material is less
efficient at corners adjacent to the built-in edge and at the
centers of the simply supported edges, where $h = h_{min}$. The
values of the functionals J and J_p for plate with the thickness
distribution shown were $J_* = 0.00219$ and $J_{p*} = 0.00209$, while for
the plate of constant thickness J = 0.00295 and J_p = 0.0028.

Figure 2.5 shows the optimum thickness distribution corresponding to the second case. The values of the parameters h_{max},
h_{min}, V, and p are the same as for the plate shown in Fig. 2.4.
In this case, as seen from Fig. 2.5, the location of the material
at the center of the plate and at the center of the built-in
edges is efficient. Concentration of the material at the plate
corners is inefficient. The values of J and J_p for a plate
having the indicated thickness distribution equalled $J_* = 0.0012$
and $J_{p*} = 0.00114$, while for plates of constant thickness,
J = 0.0017 and J_p = 0.00161. The advantage gained with respect
to J_p by optimization amounts to 29.3% in this case.

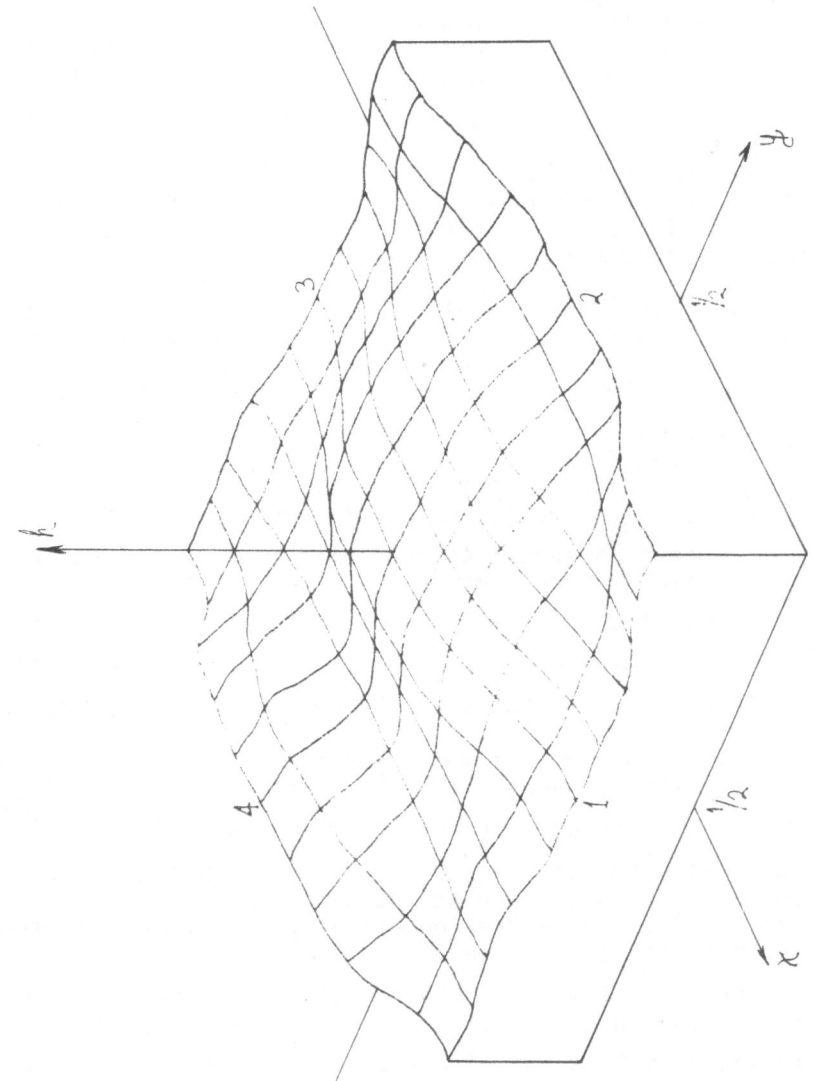

Figure 2.4.   Optimum Thickness Distribution For Simply Supported–Built-In Plate

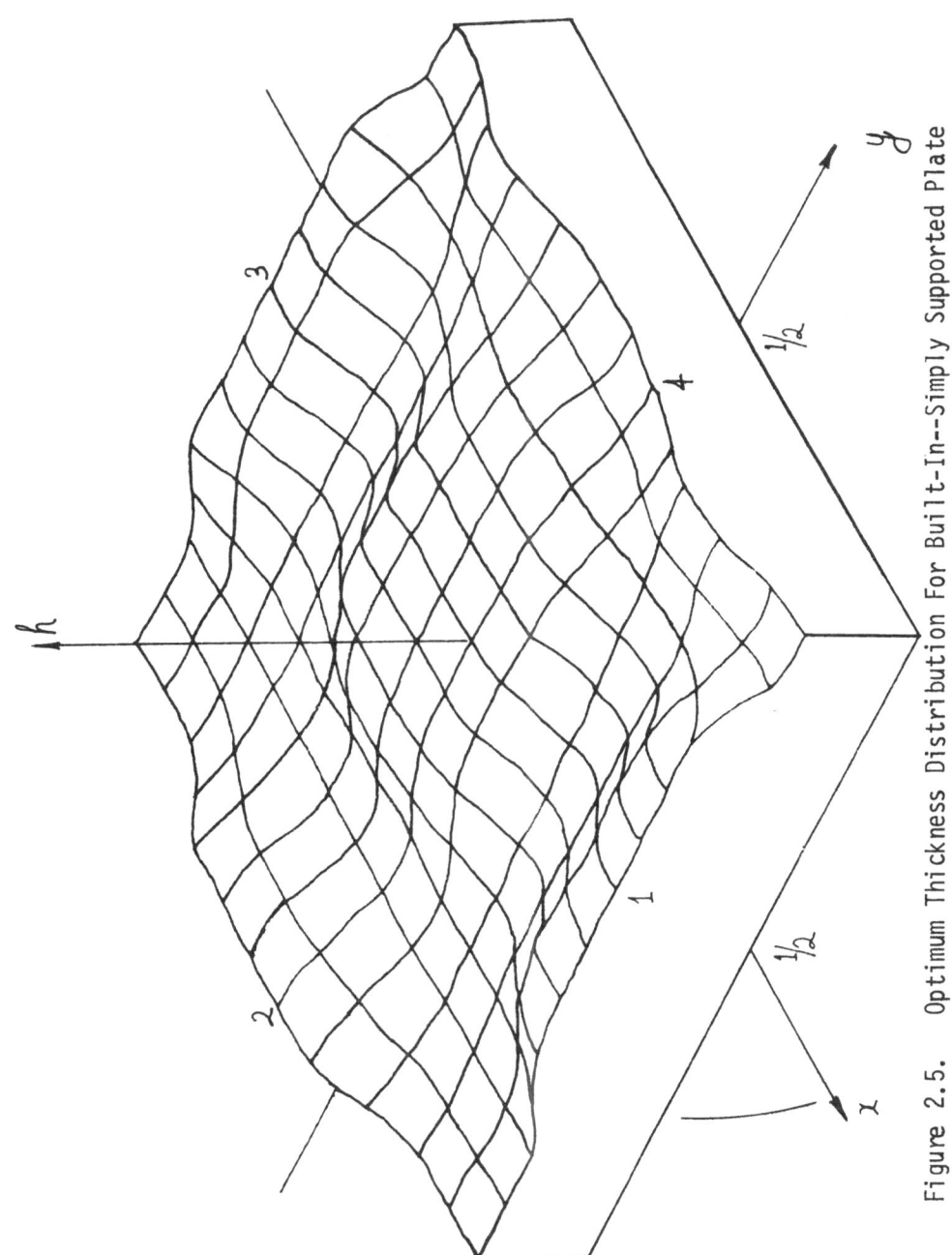

Figure 2.5.   Optimum Thickness Distribution For Built-In--Simply Supported Plate

3. DESIGN OF PLATES FOR MINIMUM STRESS

The problem of minimizing the maximal value of deflection in the region Ω was treated in Section 2. For applications it is also interesting to determine the optimum thickness distribution of elastic plates for which the minimum of the maximal (over the region occupied by the material) value of stress intensity is realized. The optimization problem for stress in a plate under bending is now treated.

One may characterize the stress state of the material at each point (x, y, ζ) by the magnitude g of the second invariant of the stress tensor deviator

$$g \equiv \frac{1}{3} [\sigma_x^2 + \sigma_y^2 - \sigma_x \sigma_y + 3(\tau_{xy}^2 + \tau_{xz}^2 + \tau_{yz}^2)] \tag{3.1}$$

where σ_x, σ_y, $\ldots$, and τ_{yz} the components of the stress tensor and $(x, y, \zeta) \in \Omega_h = \Omega \times [-h/2, h/2]$.

For convenience of solving the problem, one uses the dimensionless variables of Section 2 and introduces the dimensionless variables

$$\sigma_x' = \frac{\sigma_x}{E}, \ \sigma_y' = \frac{\sigma_y}{E}, \ \ldots, \ \tau_{yz}' = \frac{\tau_{yz}}{E}, \ \text{and} \ g' = \frac{12(1-\nu^2)S^{3/2}g}{VE^2}$$

Since the problem statement will be in terms of the dimensionless variables, the primes will be omitted. The bending equation and the boundary conditions are presented in Eq. 2.1. If the deflection function is found as the solution of the boundary-value problem of Eq. 2.1, then the nonzero components of the stress tensor are calculated according to the formulas

$$\sigma_x = -\frac{\zeta V}{(1 - \nu^2)S^{3/2}} (w_{xx} + \nu w_{yy})$$

$$\sigma_y = -\frac{\zeta V}{(1 - \nu^2)S^{3/2}} (w_{yy} + \nu w_{xx})$$

$$\tau_{xy} = -\frac{\zeta V}{(1 + \nu)S^{3/2}} w_{xy}$$

$$\tau_{xz} = -\frac{v^2(h^2-4\zeta^2)}{8S^3(1-v^2)h^3}\left[(1-v)(h^3w_{xy})_y+(h^3w_{xx})_y+v(h^3w_{yy})_x\right]$$

$$\tau_{yz} = -\frac{v^2(h^2-4\zeta^2)}{8S^3(1-v^2)h^3}\left[(1-v)(h^3w_{xy})_x+(h^3w_{yy})_x+v(h^3w_{xx})_y\right]$$

Using these formulas, one may represent the function g as the sum of two components

$$g = 4\zeta^2 g_1 + \frac{v^2(h^2-4\zeta^2)}{S^3h^6}\, g_2$$

where

$$g_1 = (w_{xx}+vw_{yy})^2 + (w_{yy}+vw_{xx})^2 - (w_{xx}+vw_{yy})(w_{yy}+vw_{xx})$$

$$+ 3(1-v^2)w_{xy}^2$$

$$g_2 = \frac{3}{16}\left\{\left[(1-v)(h^3w_{xy})_y + (h^3(w_{xx}+vw_{yy}))_x\right]^2\right.$$

$$\left.+\left[(1-v)(h^3w_{xy})_x + (h^3(w_{yy}+vw_{xx}))_y\right]^2\right\}$$

The optimization problem consists of finding a distribution $h(x,y)$ among the continuous thickness distributions that satisfy the conditions of Eq. 2.2, such that the maximum stress intensity

$$J(h) = \max_{xy\zeta} g((x,y,\zeta) \in \Omega_h)$$

is minimized. That is,

$$J_* = \min_h J(h) = \min_h \max_{xy\zeta} g \tag{3.2}$$

The general case is investigated in the present study. Here the position of the point at which the stress intensity is maximized is not known beforehand, but is found as a part of solving the problem, at the same time as the optimum thickness distribution is sought.

Note that g depends explicitly on ζ, so one should determine the internal maximum with respect to ζ. It is not difficult to show that

$$\max_{\zeta} g = \max\left\{ h^2 g_1, \ \frac{v^2}{s^3 h^2} \ g_2 \right\}$$

Consider how the case of thin plates, for which the coefficient v^2/s^3 is small and $h_{min} \neq 0$. Then, the first expression in curly brackets is maximal

$$J = \max_{xy} h^2 g_1$$

Using the method described in Section 2, one may reduce the original problem to minimization of the functional

$$J_p = \left[\ \iint_{\Omega} (h^2 g_1)^p dx dy \ \right]^{1/p}$$

The first variation of the functional J_p may be written, using the method of Section 2.3, as

$$\delta J_p = \iint_{\Omega} \Lambda \delta h dx dy \qquad\qquad (3.3)$$

where

$$\Lambda = 2h\chi[(1-v+v^2)(w_{xx}^2+w_{yy}^2) + (4v-1-v^2)w_{xx}w_{yy} + 3(1-v^2)w_{xy}^2]$$

$$- [w_{xx}v_{xx}+w_{yy}v_{yy}+v(w_{xx}v_{yy}+w_{yy}v_{xx}) + 2(1-v)w_{xy}v_{xy}]$$

with

$$\chi \equiv (h^2 g_1 / ||h^2 g_1||_{L_p})^{p-1}$$

Here, v is the conjugate function, determined as the solution of the boundary-value problem

$$L(h)v = \Phi_p$$

$$(v)_{\Gamma} = 0, \ \left(\frac{\partial v}{\partial n}\right)_{\Gamma_2} = 0, \ \left(h^3\left[\Delta v - \frac{1-v}{R} \frac{\partial v}{\partial n}\right]\right)_{\Gamma_1} = 0 \qquad (3.4)$$

where

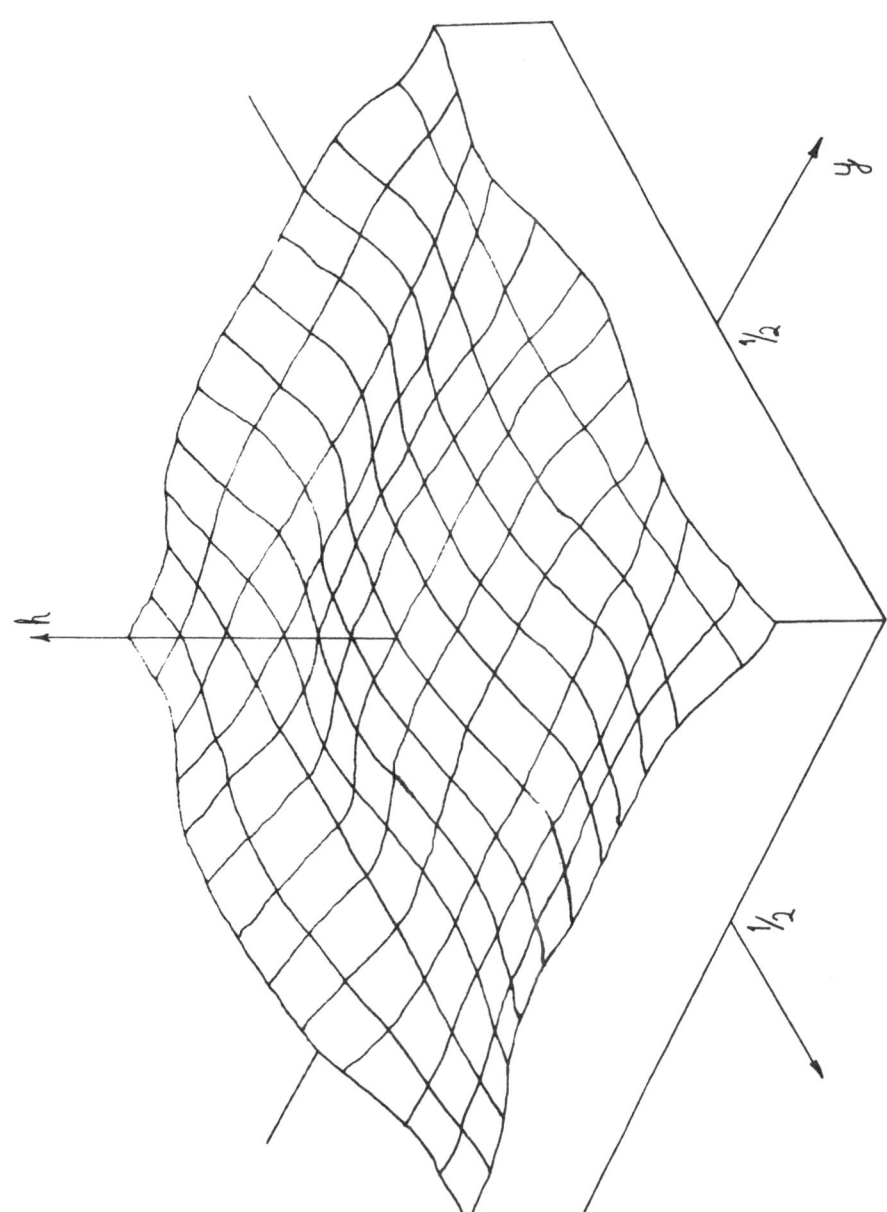

Figure 3.1. Optimum Thickness Distribution Of A Plate For Stress

$$\Phi_p \equiv 2(1-\nu+\nu^2)[(\chi h^2 w_{xx})_{xx} + (\chi h^2 w_{yy})_{yy}]$$

$$+ (4\nu-1-\nu^2)[(\chi h^2 w_{xx})_{yy} + (\chi h^2 w_{yy})_{xx}] + 6(1-\nu^2)(\chi h^2 w_{xy})_{xy}$$

The boundary value problems of Eqs. 3.4 and 2.7 are not the same, because of the different definition of the functions Φ_p.

Equations 3.3 and 3.4 give one the opportunity to apply to the problem of strength optimization the same algorithm as we used for the problem of minimization of maximum deflection. Calculations have been carried out for a square plate ($-1/2 \leq x \leq 1/2$, $-1/2 \leq y \leq 1/2$) that is simply supported on the edge $y = -1/2$, $-1/2 \leq x \leq 1/2$ and built-in on the other three edges. The load was taken to be constant $q(x,y) = 1$. The parameters of the plate were $h_{min} = 0.8$ and $h_{max} = 1.2$. Figure 3.1 shows the thickness distribution found via calculation. The plate of minimum deflection shown in Fig. 2.5 and the plate of minimum stress intensity shown in Fig. 3.1, are determined for the same boundary conditions, loadings, and parameters h_{min} and h_{max}. There are insignificant differences between the corresponding thickness distributions.

REFERENCES

1. Banichuk, N.V., "On Optimal Shapes Of Elastic Plates In Bending Problems," Izv. AN SSSR, MTT [Mechanics of Solids], No. 5, 1975.
2. Banichuk, N.V., Kartvelishvili, V.M., and Mironov, A.A., "Numerical Solution Of Two-Dimensional Optimization Problems For Elastic Plates," Izv. AN SSSR, MTT [Mechanics of Solids], No. 1, 1977.
3. Banichuk, N.V., Kartvelishvili, V.M., and Mironov, A.A., "Optimization Problems With Local Performance Criteria In The Theory Of Plate Bending," Izv. AN SSSR, MTT [Mechanics of Solids], No. 1, 1978.
4. Banichuk, N.V., Kartvelishvili, V.M., and Mironov, A.A., "The Numerical Method For Two Dimensional Optimization Problem In The Theory of Elasticity," Proc. Of The V USSR Conference On The Numerical Methods For Solving The Elasticity And Plasticity Problems, Part 2 (in Russian), Novosibirsk, VC SO AN SSSR, 1978.
5. Timoshenko, S.P., and Woinowsky-Krieger, S., Theory of Plates And Shells, 2nd edition, McGraw-Hill Book Company, New York, 1959.

6. Prager, W., and Taylor, J.E., "Problems Of Optimal Structural Design," Journ. of Appl. Mech., Vol. 35, No. 1, 1968.
7. Onat, E.T., Schumann, W., and Shield, R.T., "Design Of Circular Plates For Minimum Weight," Z. Ang. Math. Phys., Vol. 8, 1957.
8. Huang, N.C., "Optimal Design Of Elastic Beams For Minimum-Maximum Deflection," J. Appl. Mech., No. 4, 1971.
9. Seyranian, A.P., "Optimum Design Of Beams With Deflection Constraints," Izv. AN Arm. SSR, Mekhanika, No. 6, 1975.
10. Haug, E.J., Jr., and Kirmser, P.G., "Minimum-Weight Design Of Beams With Inequality Constraints On Stress And Deflection," Journ. of Appl. Mech., Vol. 34, No. 4, 1968.
11. Banichuk, N.V., "Game Problems In The Theory Of Optimal Design," Proc. IUTAM Symp. Optimization in Structural Design, Springer-Verlag, Berlin, 1975.
12. Banichuk, N.V., "Minimax Approach To The Structural Optimization Problems," Journ. of Optimizat. Theory and Appl., Vol. 20, No. 1, 1976.
13. Valentine, F.A., The Problem Of Lagrange With Differential Inequalities As Added Side Conditions, Contribution to the Calculus of Variations, 1933-1937, University of Chicago Press, 1937.
14. Savin, G.N., Stress Concentration Around Holes [in Russian], Gostekhizdat, Moscow - Leningrad, 1951.

VARIATIONAL FORMULATION AND NUMERICAL METHODS IN OPTIMAL DESIGN*

Carlo Cinquini
Istituto di Scienza e Tecnica delle Costruzioni,
Università degli Studi di Pavia, Pavia, Italy

ABSTRACT

Variational formulations of plastic and elastic optimization problems are dealt with, by means of the Lagrangian multiplier method. Methods of solution are discussed and some examples are presented, based on analytical and numerical approaches.

1. INTRODUCTION

In the past few years, optimal design problems have been studied by several authors by means of the variational formulations. In this way, both mathematical and physical features of the problem are pointed out, and numerical methods of solution can be deduced.

The author has dealt with structural optimization problems in several papers, using the Lagrange multiplier method. The optimal design for prescribed plastic collapse load was studied in Refs. 1 to 3, by a static method. A similar approach was presented in Ref. 4. Optimality conditions were found by the author from the variational formulation. Physical approaches have allowed several other authors to find analogous results (see e.g. Refs. 5 to 9).

* Research partially supported by Italian National Research Council (C.N.R.).

From the mathematical viewpoint, it is to be emphasized that in many cases of plastic optimal design, a suitable choice of the design variable and objective function allows one to formulate a convex problem, for which optimality conditions are both necessary and sufficient [1]. Starting from the optimality conditions, analytical solutions can be obtained, at least for rather simple problems [5,6,9,10].

From the viewpoint of the numerical techniques, linear programming is an effective numerical tool for treating plastic optimization problems, if the discretization of the structure leads to a minimum problem for a linear function that is subject to linear constraints (see e.g. Refs. 11 to 13). In the case of two dimensional structures, some difficulties can arise in relation to plastic yield linearization and to discretization of the structure. An iterative procedure, based on a variational formulation that makes use of the well-known finite element elastic techniques, is proposed in Ref. 14, starting from a kinematic approach.

Elastic optimization problems are dealt with in several papers, where bounds are placed on linear or rotational structural displacement (see e.g. Refs. 15 to 19). In Ref. 20, a variational formulation of elastic optimization problems with prescribed maximum deflection is proposed and an optimality condition is found. Using a unified Lagrange multiplier method, a variety of elastic optimization problems can be studied [21-22]. The set defined by the constraints is not generally convex for elastic optimal design problems, so optimality conditions are only necessary. Nevertheless, analytical solutions can be obtained for simple problems.

With regard to numerical procedures (formulated by discretizing the problem), many difficulties are encountered if mathematical programming algorithms are investigated, because of the non linear constraints on the problem. The variational formulation allows one to solve the problem, using a non linear algebraic system. Numerical solutions for beams and plates are found with iterative procedures, by using finite element techniques [23-26].

In the present paper, the variational formulation of plastic and elastic optimization problems is recalled, methods of solution are discussed, and some examples are presented for each problem. Multi-criteria optimal design is also treated.

2. GENERAL REMARKS

One or two-dimensional structures in bending are the subject of the present paper. The usual assumptions regarding smallness of the thickness and deflections are adopted. External loads and general layout are given. The design variable h represents a local geometric dimension. If Ω is the domain defined by the structure geometry, the cost function $\Gamma(h)$ that is to be minimized is obtained by integrating a specific cost function $\gamma(h)$ over Ω, to obtain

$$\Gamma(h) = \int_{\Omega} \gamma(h)d\Omega \qquad (2.1)$$

The mechanical behavior of the elastic structures is described in terms of generalized strains q and the associated stresses Q. For the plastic problem, generalized strain rates $\dot{q}$ are associated with the stresses Q. The function $s = s(h)$ indicates the specific stiffness of elastic structures and the specific plastic resistance for problems in plastic fields. External loads are denoted by L. Technological constraints are introduced to represent upper and lower bounds on the design variable in the form

$$h_{min} \leq h \leq h_{max} \qquad (2.2)$$

3. OPTIMAL DESIGN FOR PRESCRIBED COLLAPSE LOAD

If plastic optimization problems are approached in a static way, the equilibrium and the plasticity rule represent constraints. If a suitable linear differential operator is denoted by ϕ, equilibrium and plasticity conditions may be written as

$$L + \phi Q = 0 \qquad (3.1)$$

$$F(Q,s) \leq 0 \qquad (3.2)$$

The optimal plastic design problem can be formulated as the search for the minimum of the cost function $\Gamma(h)$ of Eq. 2.1, subject to the constraints of Eqs. 2.2, 3.1, and 3.2 and appropriate boundary conditions.

By means of Lagrangian multipliers, the constrained optimization problem is transformed into an unconstrained search for stationarity points of the following functional:

$$L = \lambda \int_{\Omega} \gamma(h)d\Omega + \int_{\Omega} \eta(L + \phi Q)d\Omega + \int_{\Omega} \mu F(Q,s)d\Omega$$

$$+ \int_{\Omega} \alpha(h - h_{max})d\Omega + \int_{\Omega} \beta(h_{min} - h)d\Omega \tag{3.3}$$

where λ, μ, α, $\beta \geq 0$. The stationarity conditions of L provide the relations of Eqs. 2.2, 3.1, and 3.2 and the conditions

$$\mu F(Q,s) = 0 \tag{3.4}$$

$$\alpha(h - h_{max}) = 0 \tag{3.5}$$

$$\beta(h_{min} - h) = 0 \tag{3.6}$$

$$\lambda \gamma_{,h} + \mu F_{,h} + \alpha - \beta = 0 \tag{3.7}$$

$$\phi^* \eta + F_{,Q} = 0 \tag{3.8}$$

where ϕ^* is the differential operator adjoint to ϕ. Suitable boundary conditions are to be taken into account in order to obtain Eq. 3.8.

A physical interpretation of the Lagrangian multipliers is suggested by the form of Eqs. 3.4 and 3.8 and by the boundary conditions [1-3]. If the material has an associated flow rule, the function η represents linear displacement rates. Thus, one has $\phi^* \eta = \dot{q}$ and the normality rule expressed by Eq. 3.8. An arbitrary numerical value can be assumed for the multiplier λ. Its physical dimensions depend on the significance on η and on the choice of the cost function. Equation 3.7 is the optimality condition, in which the term $\mu F_{,h}$ can be seen to represent the derivative of the specific dissipated power with respect to the design variable. This derivative is equal to the derivative of the cost function, to within the slack terms that are due to the technological constraints.

It should be noted that, in the case of optimal design for piecewise constant h, in the optimality condition the derivative of the dissipated power is integrated on each element having the same h.

A similar form of the optimality condition can also be obtained by a kinematic approach [14,21].

Note also that, if the material presents a non-associated flow rule, starting from the static approach, the function η does not represent the actual linear displacement rates. Analogously, by the kinematic approach, plasticity and equilibrium conditions can be found, which do not refer to the actual moments [21].

4. OPTIMAL ELASTIC DESIGN FOR PRESCRIBED MAXIMUM DEFLECTION

Optimal elastic design problems with bounds on the deflections of the structure are considered in this section. In particular, for the displacement function u an upper bound u^+ and a lower bound $-u^-$ are prescribed. The constraints of the problem can be written as:

$$-u^- \leq u \leq u^+ \tag{4.1}$$

$$L + \phi Q = 0 \tag{4.2}$$

$$Q + s\psi u = 0 \tag{4.3}$$

where ϕ and ψ are linear differential operators. Combining Eqs. 4.2 and 4.3, one has

$$L - \phi(s\psi u) = 0 \tag{4.4}$$

The optimal elastic design problem is now formulated as the search for the minimum of the cost function $\Gamma(h)$ of Eq. 2.1, subject to the constraints of Eqs. 2.2, 4.1, and 4.4 and appropriate boundary conditions.

The problem may be transformed into an unconstrained search for the stationarity conditions of the functional

$$L = \lambda \int_\Omega \gamma(h)d\Omega + \int_\Omega \eta(L - \phi(s\psi u))d\Omega + \int_\Omega \mu^+(u - u^+)d\Omega$$

$$+ \int_\Omega \mu^-(-u - u^-)d\Omega + \int_\Omega \alpha(h - h_{max})d\Omega + \int_\Omega \beta(h_{min} - h)d\Omega \tag{4.5}$$

where λ, μ^+, μ^-, α, $\beta > 0$. The stationarity conditions of L provide the relations of Eqs. 2.2, 4.1, 4.4, and, taking into account suitable boundary conditions, the equations

$$\mu^+(u - u^+) = 0 \tag{4.6}$$

$$\mu^-(-u - u^-) = 0 \tag{4.7}$$

$$\alpha(h - h_{max}) = 0 \tag{4.8}$$

$$\beta(h_{min} - h) = 0 \tag{4.9}$$

$$\lambda\gamma_{,h} + \phi^*\eta\ s_{,h}\Psi u + \alpha - \beta = 0 \tag{4.10}$$

$$\Psi^*(\phi^*\eta s) + \mu^+ - \mu^- = 0 \tag{4.11}$$

where ϕ^* and Ψ^* are the differential operators adjoint to ϕ and Ψ, respectively

A physical interpretation of the functions η and $(\mu^+ - \mu^-)$ is suggested by Eq. 4.11. Its structure is the same as that of Eq. 4.4, because of the linearity of the operators ϕ, Ψ, ϕ^*, and Ψ^*. Equation 4.11 then represents the elastic equilibrium equation for an elastic problem that is adjoint to the actual problem. The variables η and $(\mu^+ - \mu^-)$ represent, respectively, displacements and loads. Note that the functions μ^+ and μ^- vanish unless u is equal to u^+ or u^-, as can be seen from Eqs. 4.6 and 4.7. Under suitable hypotheses of smoothness, u reaches its extremum values at isolated points. Therefore, μ^+ and μ^- are Dirac functions (concentrated loads) applied at these points. An arbitrary numerical value can be assumed for the multiplier λ, the physical dimensions of which depend on the choice of the cost functions. The optimality condition of Eq. 4.10 can be seen to depend on the curvatures of both the actual and adjoint problems. The product of these two curvatures and the derivative of the stiffness, with respect to the design variable, represents the derivative of the specific mutual elastic energy. It is equal to the derivative of the objective function, to within the slack terms due to the technological constraints.

In the case of optimal design for piecewise constant h, in the optimality condition, the derivative of the specific mutual elastic energy is integrated on each element have the same h.

5. ANALYTICAL SOLUTIONS FOR PROBLEMS OF OPTIMAL PLASTIC DESIGN

Numerous examples have been proposed for the plastic optimization problems. Some classical absolute minimum cost solutions are given on the basis of the optimality condition, obtained by using limit analysis Theorems (see e.g. Refs. 5 and 8).

Starting from the variational formulation, integration of the system of partial differential equations (or inequalities), constituted by the static and kinematic relations and the optimality condition gives the solution of the optimization problem. If the material has an associated flow rule, the analytical minimum that is reached represents a correct solution. Otherwise, only a lower (static approaches) or upper (kinematic approaches) bound of the actual optimal design can be obtained [23-25].

For beams with prescribed piecewise constant design function, some examples are given in Ref. 10. The physical interpretation of the dual problem is very useful in investigating the solutions.

A general approach to the problem of absolute minimum cost design is presented here for the case of beams in bending, starting from the variational formulation given in the foregoing. For the present, let $\gamma(h) = h^m$ and $s(h) = K h^n$. Taking into account the well-known equilibrium and plasticity relations, in the absence of technological constraints, Eq. 3.7 yields, for $h \neq 0$,

$$\lambda m h^{m-1} - \mu K n h^{n-1} = 0 \tag{5.1}$$

and from Eq. 3.8,

$$\left. \begin{array}{ll} \eta'' + \mu = 0 , & \text{if } Q > 0 \\ \eta'' - \mu = 0 , & \text{if } Q < 0 \end{array} \right\} \tag{5.2}$$

solving Eq. 5.1 for μ and substituting into Eq. 5.2, one has

$$\left. \begin{array}{ll} \eta'' = - \dfrac{\lambda}{K} \dfrac{m}{n} h^{m-n} , & \text{if } Q > 0 \\[2mm] \eta'' = \dfrac{\lambda}{K} \dfrac{m}{n} h^{m-n} , & \text{if } Q < 0 \end{array} \right\} \tag{5.3}$$

The optimal solution must satisfy $|Q| = s$, so setting $\lambda = \dfrac{n}{m} K^{m/n}$, Eq. 5.3 becomes

$$\left. \begin{array}{ll} \eta'' = - Q^{\frac{m}{n} - 1} , & \text{if } Q > 0 \\[3mm] \eta'' = (-Q)^{\frac{m}{n} - 1} , & \text{if } Q < 0 \end{array} \right\} \tag{5.4}$$

In the sections $x = \bar{x}_i$, where Q vanishes, $h = 0$ is to be expected and Eqs. 5.1 to 5.4 are of no value.

For statically determinate structures, the moment function Q can be calculated by integrating the equilibrium equation of Eq. 3.1. The assumption $s = |Q|$ provides the design function h and the collapse mechanism, defined by Eq. 5.4. Therefore, a correct solution in the sense of limit analysis is achieved, which verifies the optimality condition.

The foregoing considerations suggest an approach for closed-form solution of statically indeterminate beam problems. The abscissae $\bar{x}_i$ can be treated as unknown parameters. Then, a statically determinate problem is to be solved. If the solution of this problem can be achieved, the minimum conditions of the cost function with respect to $\bar{x}_i$ provide an optimum solution for the statically indeterminate structure.

As an example, consider a fixed-end beam that is subjected to a uniformly distributed load p (Fig. 5.1). Here, $h = (\frac{p}{2K}|\bar{x}^2 - x^2|)^{1/n}$ can be found. For m = n one has, as known, $\bar{x} = \frac{1}{4}$. For m = 2n, $\bar{x} = \frac{1}{2\sqrt{3}}$ can be obtained [20].

6. ANALYTICAL SOLUTIONS FOR PROBLEMS OF OPTIMAL ELASTIC DESIGN

Closed-form solutions of problems of optimal elastic design with prescribed maximum deflection can be obtained by integration of the system of partial differential equations (or inequalities), consisting of behavioral constraints and optimality conditions. Except for some simple cases, this integration is complex. Furthermore, it is to be emphasized that the necessary optimality conditions are used as if they were also sufficient, in order to find this kind of solutions. Nevertheless, some analytical examples are proposed in Refs. 14 to 20.

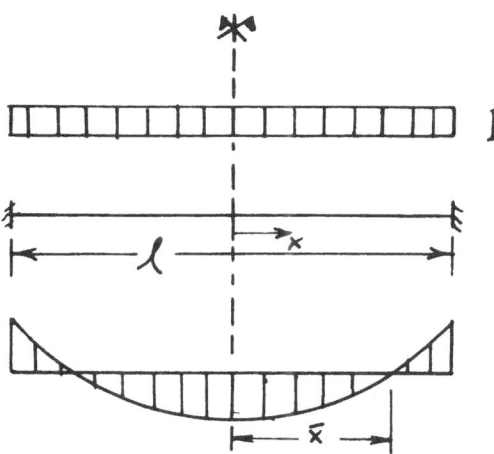

Figure 5.1   Fixed End Plastic Beam

The study of solutions for beams in bending is proposed here, in the absence of technological constraints. The case of statically determinate structures is first treated before studying a method that may be used for the statically indeterminate beams. Let $\gamma(h) = h^m$ and $s(h) = Kh^n$. Equations 4.4, 4.10, and 4.11 are then, respectively,

$$L - (K\,h^n u'')'' = 0 \tag{6.1}$$

$$\lambda m\,h^{m-1} - K\,n\,h^{n-1}\,u''\eta'' = 0 \quad , \text{ for } h \neq 0 \tag{6.2}$$

$$(K\,h^n \eta'')'' + \mu^+ - \mu^- = 0 \tag{6.3}$$

For statically determinate beams, the bending moment function

$$Q = -K\,h^n u'' \tag{6.4}$$

related to the actual problem can be calculated by twice integrating Eq. 6.1. Similarly, if an analytical form can be proposed for the adjoint load, the moment function

$$Q^* = -K\,h^n \eta'' \tag{6.5}$$

can be found by twice integrating Eq. 6.3. By substituting into Eq. 6.2, the function h can be calculated to within some multiplicative constants. The subsequent integration of Eqs. 6.1 and 6.3, taking into account the boundary conditions on u and η, fully solves the problem.

Consider now a beam of length ℓ, subject to a uniformly distributed load p. The structure is simply supported at the left end and fixed at the right end and an internal hinge is located at a section $x = \bar{x}$ (Fig. 6.1). Let x^* denote the unknown abscissa of the section where $u = u^+$, where $x^* < \bar{x}$ is assumed. The adjoint problem involves a concentrated load F, acting at $x = x^*$. The moment functions can be expressed as

$$Q = \frac{p}{2}\,x(\bar{x} - x) \tag{6.6}$$

$$Q^* = F\,\frac{\bar{x} - x^*}{\bar{x}} \quad , \text{ in } [0, x^*] \tag{6.7}$$

$$Q^* = F\,\frac{x^*}{\bar{x}}(\bar{x} - x) \quad , \text{ in } [x^*, \ell] \tag{6.8}$$

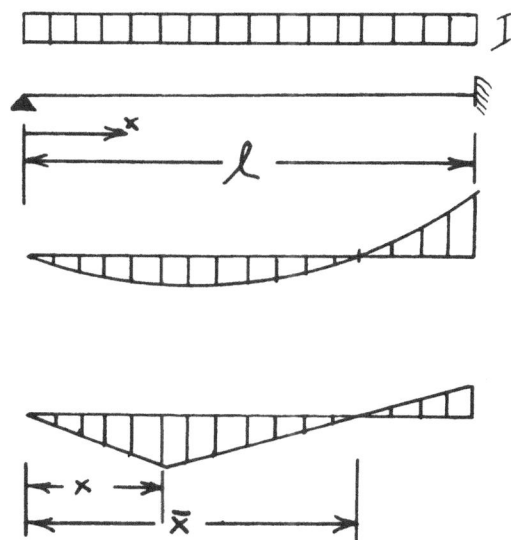

Figure 6.1   Simply Supported, Fixed Elastic Beam.

Then, from Eqs. 6.4, 6.5, 6.2,

$$h = \left[\frac{pF}{2K^2} \frac{x^* - \bar{x}}{\bar{x}} (x - \bar{x}) x^2 \right]^{\frac{1}{m+n}} \quad , \quad \text{in } [0, x^*] \qquad (6.9)$$

$$h = \left[\frac{pF}{2K^2} \frac{x^*}{\bar{x}} (x - \bar{x})^2 x \right]^{\frac{1}{m+n}} \quad , \quad \text{in } [x^*, \ell] \qquad (6.10)$$

Substituting into Eqs. 6.4 and 6.5, after integration, provides the solution.

As a specific example with $m = n$, one has

$$u'' = -K_1 \sqrt{\frac{\bar{x} - x}{x - x^*}} \, \bar{x} \quad , \quad \text{in } [0, x]$$

$$u'' = -K_1 \sqrt{\frac{\bar{x}}{x^*}} \, x \quad , \quad \text{in } [x, \bar{x}[$$

$$u'' = K_1 \sqrt{\frac{\bar{x}}{x^*}} \, x \qquad , \quad \text{in }]\bar{x}, \ell]$$

$$\eta'' = -K_2 \sqrt{\frac{\bar{x} - x^*}{\bar{x}(\bar{x} - x)}} \qquad , \quad \text{in } [0, x]$$

$$\eta'' = -K_2 \sqrt{\frac{x^*}{\bar{x} \, x}} \qquad , \quad \text{in } [x, \bar{x}[$$

$$\eta'' = K_2 \sqrt{\frac{x^*}{\bar{x} \, x}} \qquad , \quad \text{in }]\bar{x}, \ell]$$

with $K_1 = \sqrt{\frac{p}{2F}}$ and $K_2 = \sqrt{\frac{F}{2p}}$. The integration constants and the unknown parameters F and x^* must now be calculated. For this problem, the solution needs 14 boundary conditions, which are

$$\left. \begin{array}{l} u = 0 \\ \eta = 0 \end{array} \right\} \qquad , \quad \text{at } x = 0$$

$$\left. \begin{array}{ll} u = u^+ \, , & u' = 0 \\ \Delta u = 0 \, , & \Delta u' = 0 \\ \Delta \eta = 0 \, , & \Delta \eta' = 0 \end{array} \right\} \qquad , \quad \text{at } x = x^*$$

$$\left. \begin{array}{l} \Delta u = 0 \\ \Delta \eta = 0 \end{array} \right\} \qquad , \quad \text{at } x = \bar{x}$$

$$\left. \begin{array}{ll} u = 0 \, , & u' = 0 \\ \eta = 0 \, , & \eta' = 0 \end{array} \right\} \qquad , \quad \text{at } x = \ell$$

In order to solve statically indeterminate problems, Eq. 6.2 implies that $u'' \eta'' > 0$, to within singular points where h = 0. In these sections ($x = \bar{x}_i$) both bending moments Q and Q^* of the actual and adjoint problems must vanish. By treating the values $\bar{x}_i$ as unknown parameters, the optimization problem of a statically indeterminate beam can be solved as a search for the minimum cost solution, in a set of statically determinate solutions. For a beam that is simply supported at the left end and fixed at the right end, the foregoing solution of the statically determinate problem can be taken into account. The value of $\bar{x}$ must be calculated a posteriori, by minimizing the cost function with respect to the parameter $\bar{x}$. In particular, $x^* = 0.4213\ \ell$ and $\bar{x} = 0.6989\ \ell$ can be calculated.

The next example is a fixed-end beam of length ℓ, subject to a uniformly distributed load p (Fig. 6.2). The maximum linear displacement is at the middle-span. Accordingly, the adjoint problem involves a concentrated load F, acting in this section. The moment functions can now be expressed as

$$Q = \frac{P}{2}\,(\bar{x}^2 - x^2) \tag{6.11}$$

$$Q^* = \frac{F}{2}\,(\bar{x} - x) \tag{6.12}$$

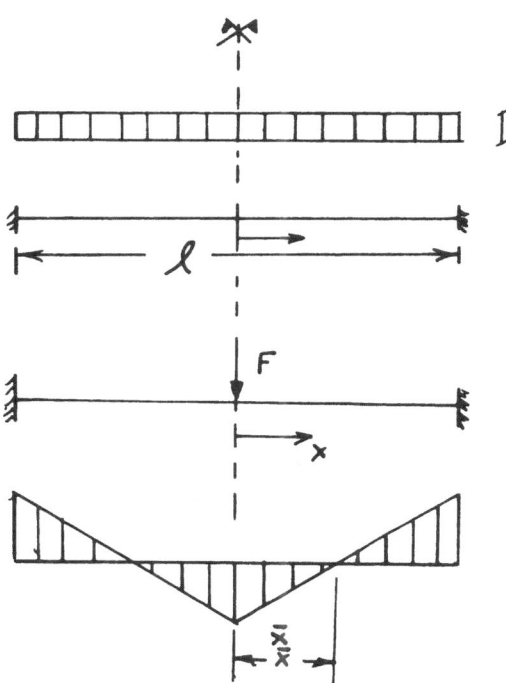

Figure 6.2   Fixed End Elastic Beam

From Eqs. 6.1, 6.2, and 6.3, if $\lambda = K\frac{n}{m}$ is assumed, one has

$$h = \left[\frac{pF}{4K^2} (x^2 - \bar{x}^2)(x - \bar{x}) \right]^{\frac{1}{m+n}} \tag{6.13}$$

$$u'' = \frac{p}{2K}(x^2 - \bar{x}^2) \left[\frac{pF}{4K^2}(x^2 - \bar{x}^2) \right]^{-\frac{n}{m+n}} \tag{6.14}$$

$$\eta'' = \frac{F}{2K}(x - \bar{x}) \left[\frac{pF}{4K^2}(x^2 - \bar{x}^2)(x - \bar{x}) \right]^{-\frac{n}{m+n}} \tag{6.15}$$

Integrating Eqs. 6.14 and 6.15 provides the solution. The integration constants and the parameter F are found by the boundary conditions. The optimal value of the abscissa $\bar{x}$ is then found by minimizing the cost function $\Gamma(h)$.

In particular, for m = n, one has

$$u'' = -\sqrt{\frac{p}{F}} \frac{1}{\sqrt{x+\bar{x}}} \qquad , \quad \text{in } [0,\bar{x}[$$

$$u'' = \sqrt{\frac{p}{F}} \frac{1}{\sqrt{x+\bar{x}}} \qquad , \quad \text{in }]\bar{x},\frac{\ell}{2}]$$

$$\eta'' = -\sqrt{\frac{F}{p}} \sqrt{x+\bar{x}} \qquad , \quad \text{in } [0,\bar{x}[$$

$$\eta'' = \sqrt{\frac{F}{p}} \sqrt{x+\bar{x}} \qquad , \quad \text{in }]\bar{x},\frac{\ell}{2}]$$

7. NUMERICAL SOLUTIONS FOR OPTIMAL ELASTIC DESIGN

In order to study a numerical solution procedure, a discretized form of the functional L defined by Eq. 4.5 can be found, by formulating the problem with a finite element technique. By the stationarity conditions L, a system of nonlinear algebraic equations is found [27-29].

Iterative methods are to be sought for an automated numerical solution. Starting from a given design, the optimal solution is determined by systematic modifications of the structure. At each step, a feasible non-optimal solution is achieved. The knowledge of the optimality condition becomes fundamental in seeking the variations to be made in depth distribution at each iteration. The procedure makes use of the only necessary conditions, as if they were also sufficient. Then, for some problems a local minimum (or maximum) solution, instead of the global optimum, can be found. This difficulty can be removed by means of suitable numerical expedients [27].

Figures 7.1 and 7.2 show two solutions of problems of absolute minimum cost, calculated by means of the numerical procedure dealt with in Ref. 22. The specific objective function and the stiffness are taken as $\gamma(h) = h$ and $s = K h^2$. The values of h are related to the uniform stiffness solution.

In Figs. 7.3 and 7.4, two solutions for annular plates in bending are shown. These examples and the discretized approach for annular or circular plates in bending are described in Ref. 26.

8.   MULTI-CRITERIA OPTIMAL DESIGN

By using the variational formulation, no difficulties are met in formulation of optimal design problems with alternative behavioral constraints. The functional L must simply include all the constraints, which consequently modifies the optimality conditions. In Ref. 21, the problem of optimal design is dealt with for beams with a prescribed maximum deflection and an assigned plastic collapse load.

Consider for example a cantilever beam loaded by a uniformly distributed load, as shown in Fig. 8.1. The specific elastic stiffness is taken as $K h^2$ and the specific plastic resistance as $S h$.

If the optimum design depends only on the elastic constraint, by assuming $\lambda = 1$ in the optimality condition, one has [21]

$$h = \left(\frac{p}{Ku^+}\right)^{1/2} \frac{\ell}{2} x$$

$$u = u^+ \frac{1}{\ell^2} (x^2 - 2\ell x + \ell^2)$$

$$\eta = \frac{1}{(p K u^+)^{1/2}} \frac{\ell}{2} (x \ln \frac{x}{\ell} + \ell - x)$$

376

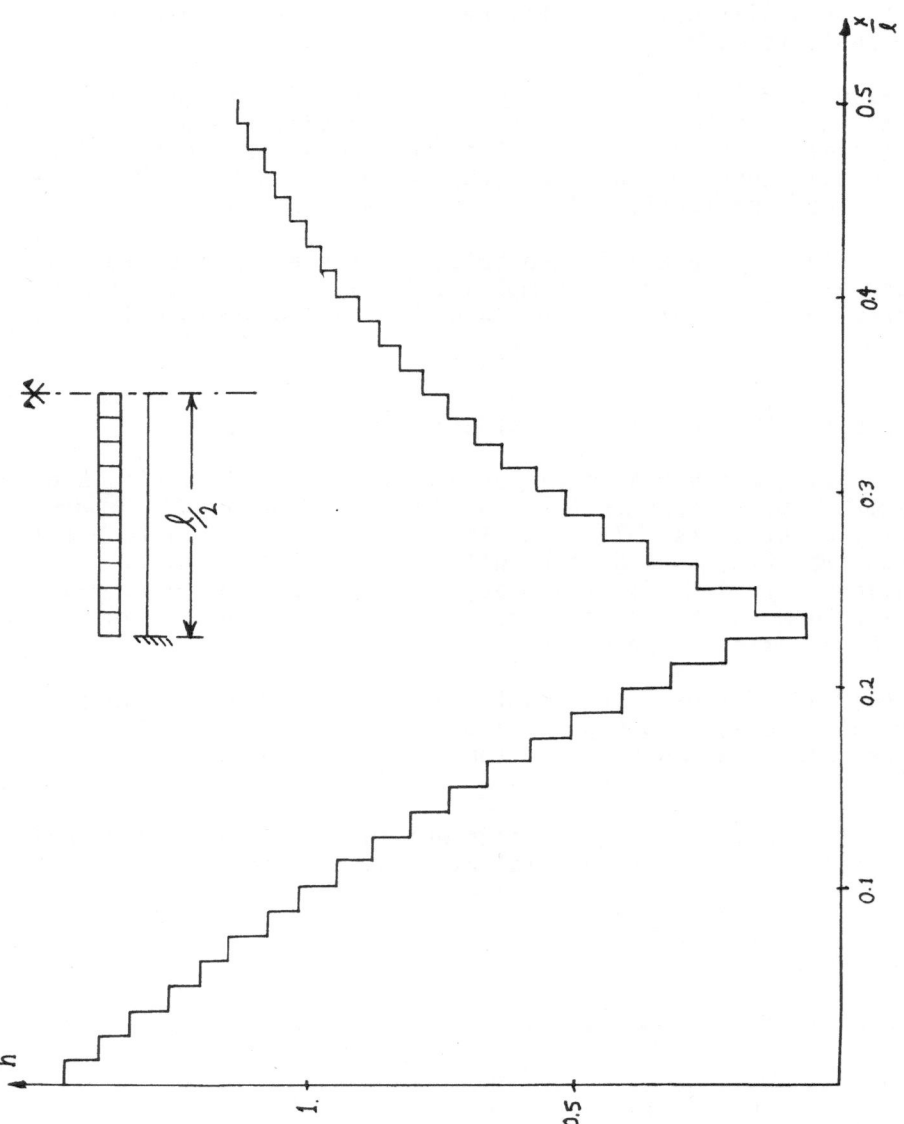

Figure 7.1 Discretized Optimum Clamped Beam

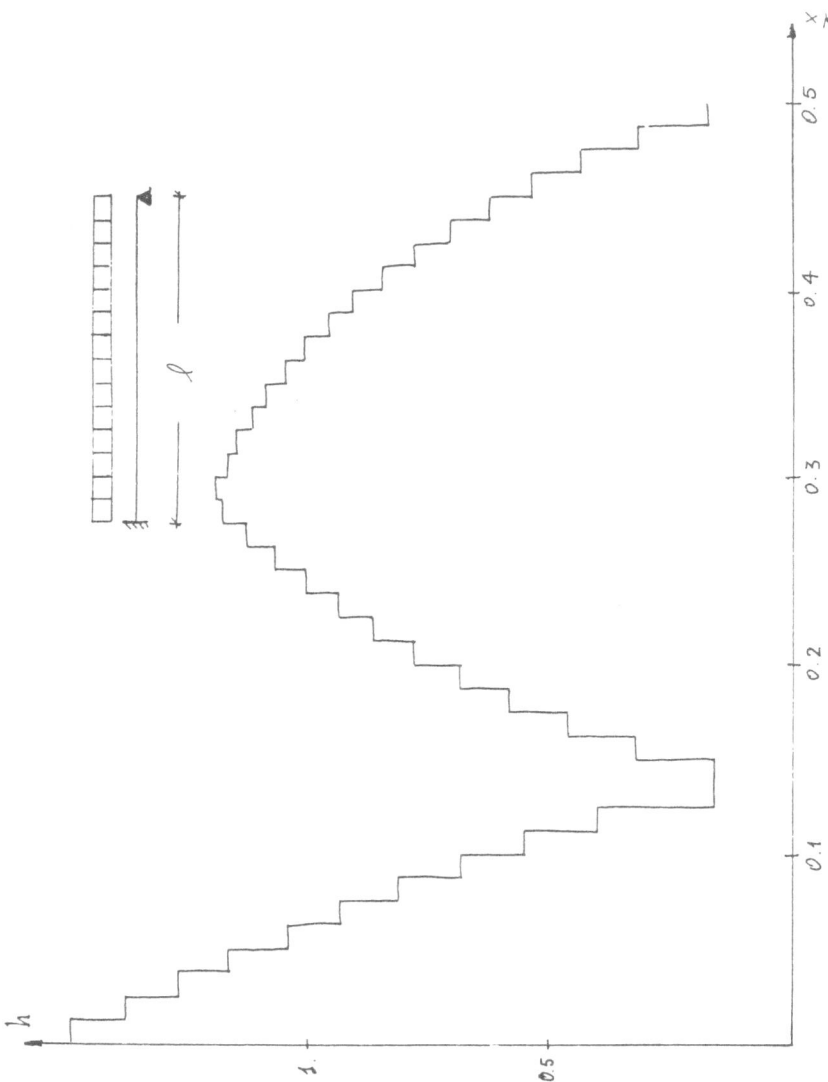

Figure 7.2   Discretized Optimum Clamped-Simply Supported Beam

R = 10 cm

P = 100 kg/cm

$\bar{w}(a) = 0.05$ cm

a/R = 0.25

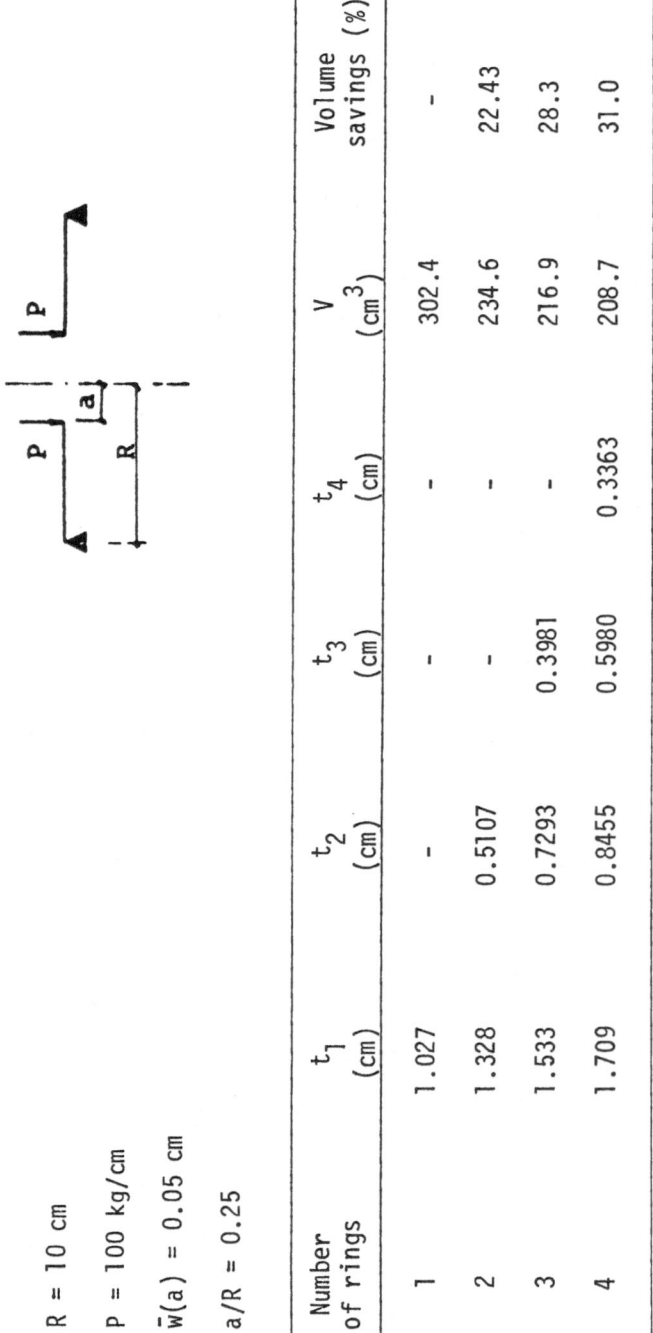

Number of rings	t_1 (cm)	t_2 (cm)	t_3 (cm)	t_4 (cm)	V (cm^3)	Volume savings (%)
1	1.027	-	-	-	302.4	-
2	1.328	0.5107	-	-	234.6	22.43
3	1.533	0.7293	0.3981	-	216.9	28.3
4	1.709	0.8455	0.5980	0.3363	208.7	31.0

Figure 7.3 Discretized Optimum Simply Supported Annular Plate

R = 10 cm

P = 100 kg/cm

$\bar{w}(a)$ = 0.05 cm

a/R = 0.25

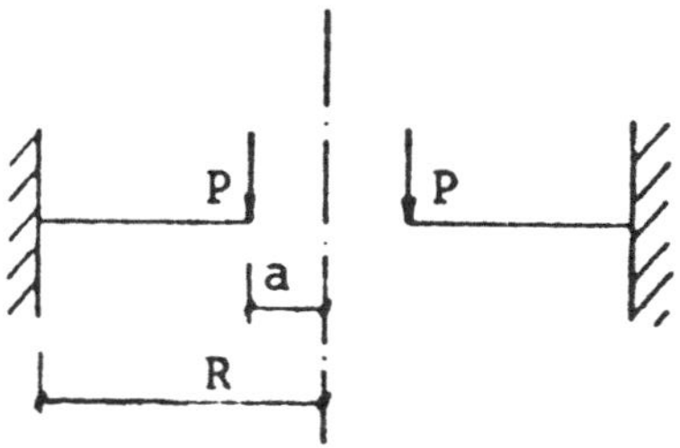

Number of rings	t_1 (cm)	t_2 (cm)	t_3 (cm)	t_4 (cm)	V (cm^3)	Volume savings (%)
1	0.6875	-	-	-	202.50	-
2	1.0411	0.3599	-	-	178.20	12.97
3	1.1191	0.3180	0.4760	-	162.53	19.73
4	1.3097	0.4994	0.2565	0.439	152.90	24.47

Figure 7.4 Discretized Optimum Clamped Annular Plate

380

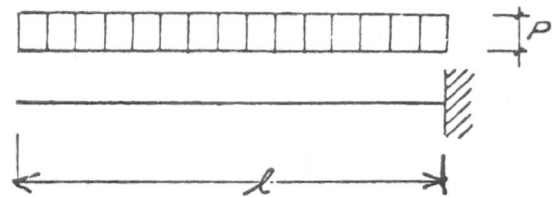

Figure 8.1   Cantilever Beam

If, on the contrary, the optimum design depends only on the plastic constraint, one finds

$$h = \frac{c}{S} \frac{p}{2} x^2$$

where c is the load factor.

Finally, a lower bound for h is introduced, because in both cases the design variable vanishes at x = 0.   The bound imposed is

$$h \geq h_{min}$$

Then, three different solutions can be conceived, depending on the numerical values of the parameters of the problem (see Fig. 8.2).

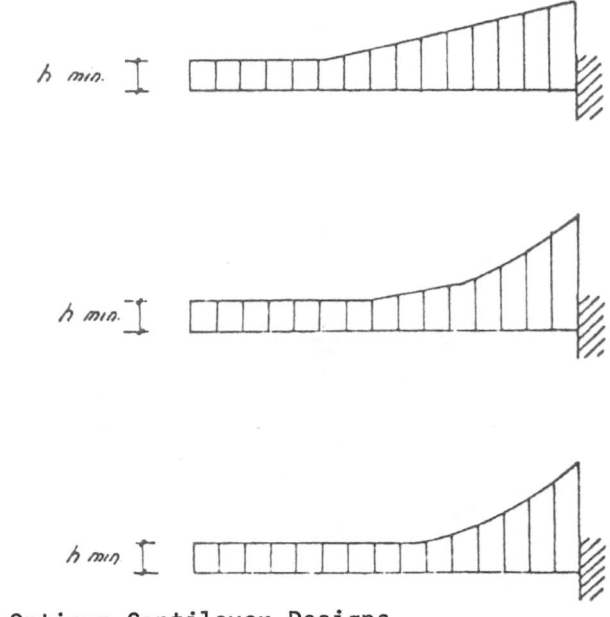

Figure 8.2   Optimum Cantilever Designs

REFERENCES

1. Cinquini, C. and Mercier, B., "Minimal Cost in Elastoplastic Structures," _Meccanica_, Vol. 11, no. 4, 1976, pp. 219-226.
2. Cinquini, C., Lamblin, D., and Guerlement G., "Variational Formulation of the Optimal Plastic Design of Circular Plates," _Computer Methods in Applied Mechanics and Engineering_, vol. 11, 1977, pp. 19-30.
3. Cinquini, C., "Formulazione Variazionale del Problema di Progettazione Ottimale di Piastre Inflesse," _Proc. III Congr. AIMETA_, Cagliari (I), 1976.
4. Sacchi, G., "A Variational Formulation of the Optimal Plastic Design Problem with Linear and Convex Cost Function," _IUTAM Symposium Opt. in Structural Design_, Springer-Verlag, 1975, pp. 104-110.
5. Drucker, D.C., and Shield, R.T., "Design for Minimum Weight," _Proc. 9th Int. Congr. Applied Mechanics_, Brussels, Book 5, 1956, pp. 212-222.
6. Gross, O. and Prager, W., "Minimum Weight Design for Moving Loads," _Proc. 4th U.S. Nat. Congr. Appl. Mech., ASME_, New York, Vol. 2, 1962, p. 1047.
7. Prager, W. and Shield, R.T., "A General Theory of Optimal Plastic Design," _J. Appl. Mech._, Vol. 34, No. 1, 1967.
8. Save, M., "An Unified Formulation of the Theory of Optimal Plastic Design with Convex Cost Functions," _J. Struct. Mech._, Vol. 1, No. 2, 1972, pp. 267-276.
9. Rozvany, G.I.N., _Optimal Design of Flexural Systems_, Pergamon Press, Sydney, 1976.
10. Guerlement, G., Lamblin, D., and Cinquini, C., "Dimension-nement Plastique de Coût Minimal Avec Constraintes Technologiques de Poutres Soumises à Plusieur Ensembles de Charges," _J. de Méc. Appl._, Vol. 1, no. 1, 1977, pp. 1-25.
11. Cohn, M.A., _Analysis and Design of Inelastic Structures_, University of Waterloo Press, Waterloo, 1972.
12. Maier, G., Srinivasan, R., and Save, M., "On Limit Design of Frames Using Linear Programming," _Proc. of the Int. Symposium on Computer-Aided Structural Design_, Univ. of Warwick, Coventry, Vol. A2.32-A2.59, 1972.
13. Sacchi, G., Maier, G., and Save, M., "Limit Design of Frames for Movable Loads by Linear Programming," _IUTAM Sympsium Opt. in Structural Design_, Springer-Verlag, 1975, pp. 415-432.
14. Cinquini, C., "Structural Optimization of Plates of General Shape by Finite Elements," _J. Struct. Mech._, to appear.
15. Chern, J.M. and Prager, W., "Optimum Design for Prescribed Compliance Under Alternative Loads," _J. Opt. Th. Appl._, Vol. 5, 1970, pp. 424-431.
16. Save, M., "A General Criterion for Optimal Structural Design," _J. Opt. Th. Appl._, Vol. 15, 1975, pp. 119-129.

382

17. Shield, R.T. and Prager, W., "Optimal Structural Design for Given Deflection," J. Appl. Math. Phys. ZAMP, Vol. 21, 1970, pp. 513-523.
18. Prager, W., "Optimal Design of Statically Determinate Beams for Given Deflection," J. Mech. Sc., Vol. 13, 1971, p. 893.
19. Huang, N.C., "Optimal Design of Elastic Beams for Minimum-Maximum Deflection," J. Appl. Mech., Vol. 38, 1971, pp. 1078-1081.
20. Cinquini, C., "Optimal Elastic Design for Prescribed Maximum Deflection," J. Struct. Mech., Vol. 7, no. 1, 1979, pp. 21-34.
21. Cinquini, C., and Sacchi, G., "Problems of Optimal Design for Elastic and Plastic Structures," Journal de Mécanique Appliquée, Vol. 4, no. 1, 1980 pp. 31-59.
22. Sacchi, G., "Optimal Structural Design Problems," Ist. Naz. Alta Matematica; seminari su "Problemi di Frontiera Libera," Pavia, 12/9/1979.
23. Radenkovic, D., "Théorie des Charges Limites, Extension à la Mécanique des Sols," Sém. de Plasticité, Ec. Polytechnique, Pub. Sc. et Tech. Min. Air., n. NT.116, 1961.
24. Palmer, A.C., "A Limit Theorems for Materials with Non-Associated Flow Laws," J. de Méc., Vol. 5, No. 2, 1966, pp. 217-222.
25. Sacchi, G. and Save, M., "A Note on the Limit Loads of Non Standard Materials," Meccanica, Vol. 3, no. 1, 1968, pp. 43-45.
26. Hegemier, G.A. and Tang, H.T., "A Variational Principle, the Finite Element Method, and Optimal Structural Design for Given Deflection," IUTAM Symposium Opt. in Struct. Design, Springer-Verlag, 1975, pp. 464-483.
27. Cantù, E. and Cinquini, C., "Iterative Solution for Problems of Optimal Elastic Design," Comp. Meth. Appl. Mech. Eng., Vol. 20, 1979, pp. 257-266.
28. Lamblin, D., Cinquini, C., and Guerlement, G., "Finite Element Iterative Methods for Optimal Elastic Design of Circular Plates, Comp. & Struct., to appear.
29. Cantù, E. and Cinquini, C., "Soluzione Ottimale di Telai Soggetti a Limitazioni di Spostamenti", Costr. Met. Vol. 1, pp. 9-14, 1980.

A NOTE ON THE OPTIMAL ELASTIC DESIGN FOR GIVEN DEFLECTION*

Carlo Cinquini and Giannantonio Sacchi

Istituto di Scienza e Tecnica delle Costruzioni,
Università degli Studi di Pavia, Pavia, Italy

Istituto di Scienza e Tecnica delle Costruzioni,
Politecnico di Milano, Milan, Italy

ABSTRACT

An approximate method of solution is presented for elastic optimization of beams in bending, with bounds set on displacement. By means of a piecewise linearization of the optimality condition, some important characteristics of the solution may be determined in a simple way. Zero points of the bending moments and, in some cases, the point of maximum deflection are explicitly found. In order to demonstrate feasibility of the method, some simple problems are dealt with. The influence of technological constraints is also discussed.

1. INTRODUCTION

This paper is concerned with optimal elastic design of beams in bending. The usual assumptions regarding the smallness of beam thickness and deflections are adopted. The structural layout and external loads $q = q(x)$ are given. The displacement $y_0 = y(x_0)$, at a given or unknown point, is constrained as follows:

$$- y^- \leq y_0 \leq y^+ \tag{1.1}$$

---

*Research partially supported by Italian National Research Council (C.N.R.).

where $y^- \geq 0$ and $y^+ \geq 0$ are given.

For the sake of simplicity, the structure treated here is a sandwich beam. The stiffness s and specific cost function C depend linearly on the design variable $h(x)$, i.e.

$$s = \alpha h \qquad (1.2)$$

$$C = ch \qquad (1.3)$$

More general assumptions concerning design dependence of section properties, such as $s = \alpha h^m$ and $C = ch^n$, do not involve very important difficulties. In special cases, the design variable may be subjected to technological constraints of the form

$$h_{min} \leq h \leq h_{max} \qquad (1.4)$$

In any case, a lower bound on h must be imposed, at least in the sense of $h \geq 0$. The curvature k of the beam, due to the bending moment M, is given by

$$k = - y'' = \frac{M}{s} \qquad (1.5)$$

where a prime denotes one derivative with respect to the abscissa.

The optimal design problem can be formulated [1 to 5] as the search for a minimum of the cost function

$$z = \int_b C h \, dx$$

under the constraints of Eqs. 1.1 and 1.4 and the elastic equilibrium equation

$$q - (sk)'' = 0 \qquad (1.6)$$

By means of a variational formulation, in the absence of technological constraints, the variables R, Q, and k* of the dual problem are introduced, with the following constraints:

$$R(y_0 - y^+) = 0 \qquad (1.7)$$

$$Q(-y_0 - y^-) = 0 \qquad (1.8)$$

$$C_{,h} - s_{,h}kk^* = 0 \qquad (1.9)$$

$$-(sk^*)_{,xx} + R - Q = 0 \qquad (1.10)$$

Equation 1.10 may be interpreted as the elastic equilibrium equation for the adjoint problem, where k* is the curvature function and (R - Q) is the load. The functions R and Q are different from zero only in sections of the beam where $y = y^+$ and $y = -y^-$, respectively (see Eqs. 1.7 and 1.8). The optimality condition is expressed by Eq. 1.9, which states that the product of the two curvatures and the derivative of the stiffness with respect to h is equal to the marginal specific cost [2,3]. An optimal design can be obtained by solving both the actual and adjoint problems, taking into account the optimality condition. It appears very difficult to apply theorems of existence and uniqueness to this set of equations, because the set defined by the constraints is non-convex. Only in the simple case of statically determinate structures can the existence of a solution be proved.

Generally speaking, except for some simple statically determinate problems, the analytical solutions of the system of necessary conditions is very complex and numerical iterative methods may involve a great deal of computation [3]. In this paper, a simple approximate method is presented which allows one to find some important characteristics of the solution by means of desk tools only.

2. APPROACH TO THE PROBLEM BY PIECEWISE LINEARIZING THE OPTIMALITY CONDITION

Taking into account Eqs. 1.2 and 1.3, the optimality condition of Eq. 1.9 can be written as

$$kk^* = C/\alpha \equiv A' \qquad (2.1)$$

A relationship between the actual and adjoint curvatures that may be represented by the hyperbola of Fig. 2.1 is obtained. A piecewise linearization is proposed, as follows:

$$|k + k^*| = A, \qquad A > 0 \qquad (2.2)$$

The form of such linearized curves are shown in Fig. 2.2. The error introduced strictly depends on the evaluation of A with respect to A', in particular on the ratio A^2/A'.

The well-known Mohr Theorem may now be used. In order to calculate the displacements of the actual beam, an analogical beam with suitable boundary conditions is considered. The

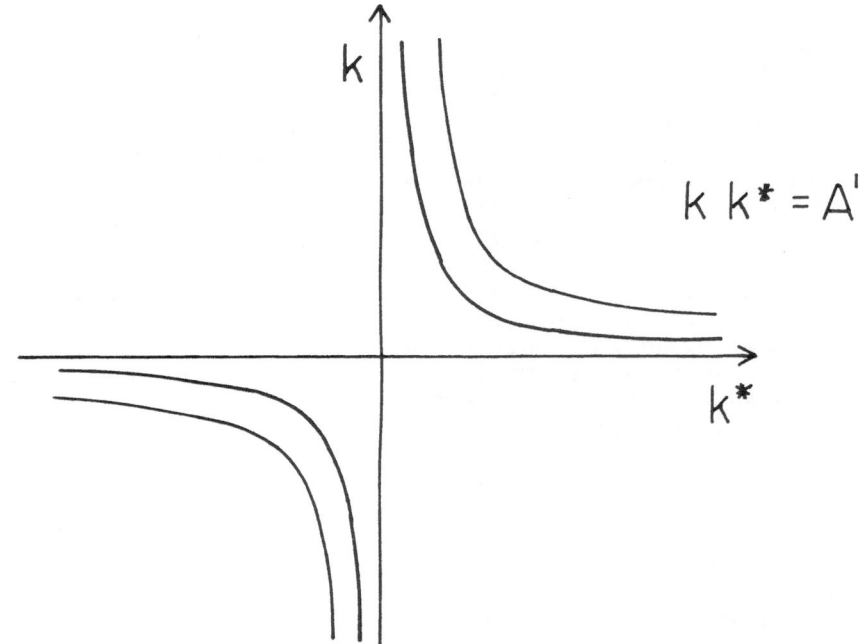

Figure 2.1   Curvature Hyperbola.

analogical loads are constituted by the actual curvature, and the analogical moments can be conceived as the actual displacements. Equilibrium of the analogical structure implies compatibility of the actual strains.

Taking into account the linearized form (Eq. 2.2) of the optimality condition and the elastic equilibrium equations (Eqs. 1.6 and 1.10), in an approximate approach to the optimal design problem, both the actual load 9 and adjoint load R or Q may be conceived as acting on the beam at the same time. The optimality condition of Eq. 2.1 requires that the zero points of the actual and adjoint problems are coincident in the optimal solution. On the other hand, the approximate form of the optimality condition implies that the analogical beam is subjected almost everywhere to a constant load A. By means of the analogical beam equilib- rium, the zero points of the analogical load, i.e. of the curva- tures $(k + k^*)$, can be found. The abscissa of the maximum analogical bending moment, if unknown, can also be calculated, so that an approximate evaluation of the abscissa of the maximum actual displacement can be obtained.

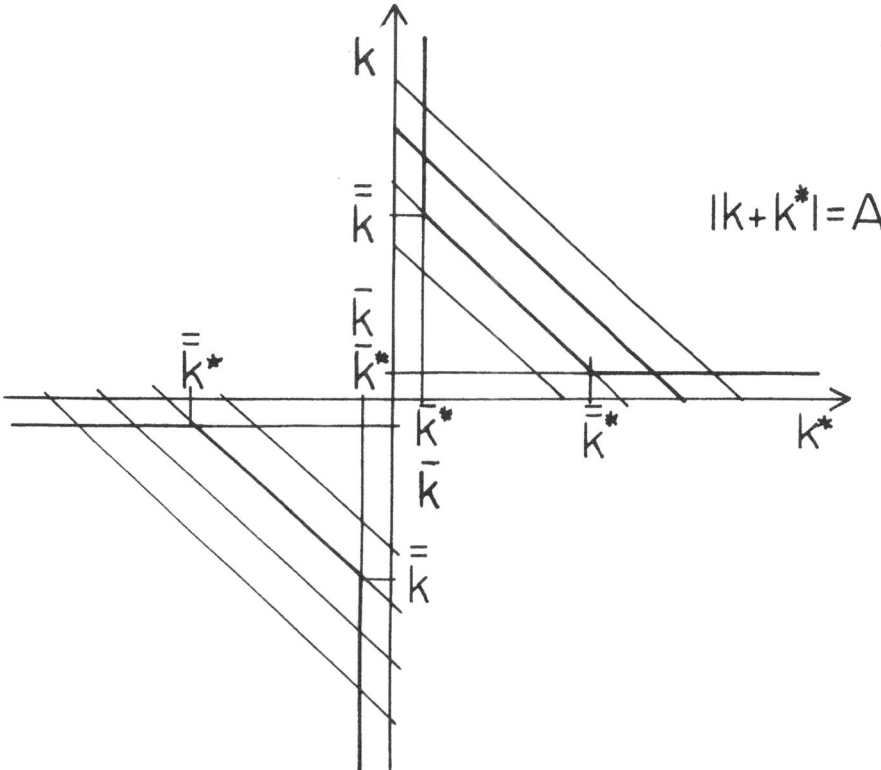

Figure 2.2   Linearized Curvature Relations.

In the absence of technological constraints, the design variable h vanishes with the actual bending moment. Then, the approximate method allows one to locate the zero points of the design variable and in some cases the maximum deflection abscissa, before actually calculating the optimum design. Starting from these characteristics of the solution, evaluated in an approximate way, the optimization problem can be solved without introducing any further approximations.

Using the virtual work principle, it is possible to write

$$Ry_0 = \int_0^\ell s\ k\ k^* dx \qquad (2.3)$$

Taking into account Eqs. 2.1 and 1.2, one has

$$R \, y_0 = \int_0^{\ell} s \, A' \, dx = A'\alpha \int_0^{\ell} h \, dx \qquad (2.4)$$

On the other hand, Eq. 2.2 allows one to write

$$|M + M^*| = A\alpha h \qquad (2.5)$$

and

$$\int_0^{\ell} |M + M^*| dx = A\alpha \int_0^{\ell} h \, dx \qquad (2.6)$$

Finally, having assumed $M^* = R \, M^{*\prime}$, from Eqs. 2.4 and 2.6, one has

$$\int_0^{\ell} |M + R \, M^{*\prime}| dx = A/A' \, R \, y_0 \qquad (2.7)$$

Equation 2.7 allows one to determine the value of the dual force R, once an admissible value for y_0 and an admissible error, given by A^2/A', have been chosen. Now, Eq. 2.1 provides

$$M \, R \, M^{*\prime} = c \, \alpha \, h^2 \qquad (2.8)$$

The function $h(x)$ can be calculated and an approximate solution of the optimal design problem is achieved.

3. EXAMPLE

Five examples are now considered, to illustrate the fore-going ideas.

3.1 Clamped, Simply Supported Beam

A beam fixed at the left end and simply supported at the right end is first considered (Fig. 3.1(a)). It is subjected to a uniformly distributed load q. Selfweight and technological constraints are neglected. The analogical beam is shown in Fig. 3.1(b). The analogical load A must be chosen in such a way to ensure the equilibrium of the structure. Thus, the zero point of the actual curvature is approximately calculated as $x = 0.701 \, \ell$, with an error of 1.1%, with respect to the exact value $x = 0.699 \, \ell$ [5].

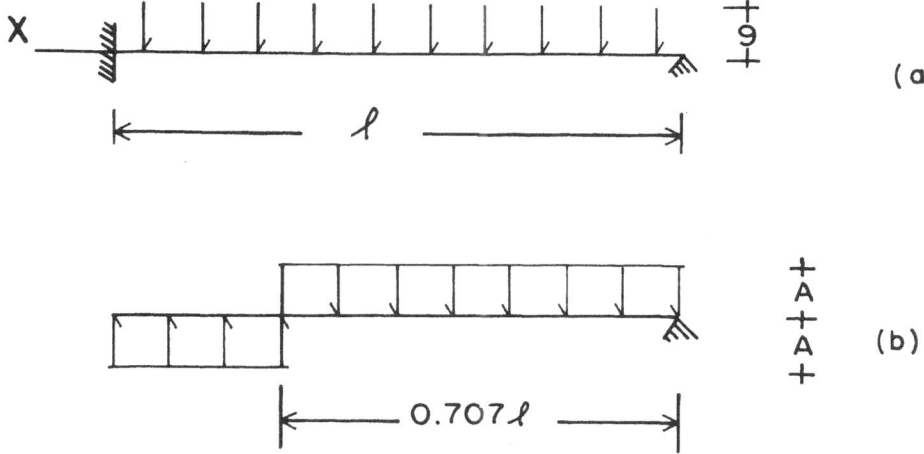

Figure 3.1   Clamped, Simply Supported Beam.

The maximum of the analogical bending moment function is achieved at $x_0 = 0.414$ ℓ. Then $y(x_0) = 0.110$ $A\ell^2$ represents a good evaluation of the maximum of the actual displacements. If the prescribed maximum deflection is assumed, e.g., as $y^+ = 1/500$ ℓ, $A = 0.018/\ell$ can be calculated. In order to have an error corresponding to the linearization depicted in Fig. 3.2, $A' = 0.56 \times 10^{-4}/\ell^2$ must be assumed, so that $A^2/A' = 5.79$ can be found.

By means of the procedure proposed in Section 2, the adjoint load is calculated as $R = 0.016$ $q\ell$ and for the design variable one has

$$q(0.353\ell x - 0.5x^2) \, 0.016 \, q\ell \, . \, 0.414 \, x = c\alpha h^2,$$

$$0 < x < 0.141\ell$$

$$q(0.353\ell x - 0.5x^2) \, 0.016 \, q\ell \, (0.414\ell - 0.586 \, x) = c\alpha h^2,$$

$$0.414\ell < x < \ell$$

(3.1)

The shape of optimal design solutions is shown in Fig. 3.3.

Note that the numerical results provide a very small value for the ratio $\dfrac{R}{q\ell}$, which justifies the approximation introduced by considering both loads as acting at the same time.

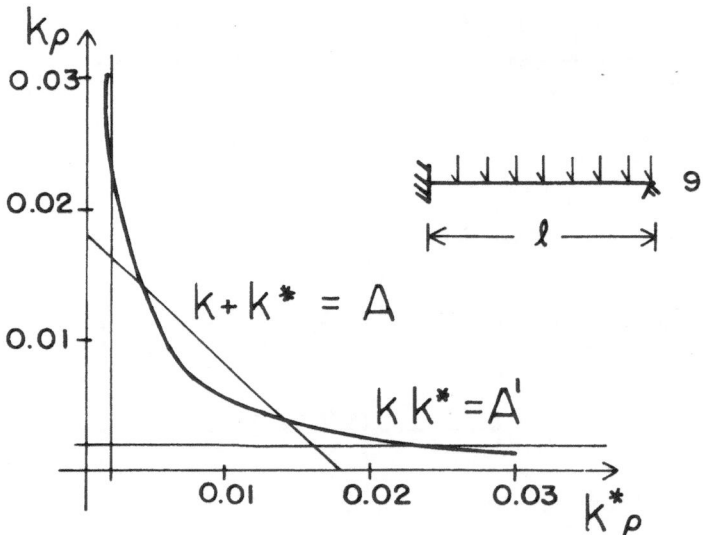

Figure 3.2 k vs k*.

3.2 Clamped Beam

The clamped beam of Fig. 3.4(a) is now considered. The load q(x) is uniformly distributed. Self weight influence and technological constraints are neglected. The analogical beam is depicted in Fig. 3.4(b) where the analogical load A is shown acting to ensure the equilibrium of the structure.

The zero points of the curvatures can be very easily calculated as $x = \ell/2(1\pm0.5)$, with an error of 0.4% with respect to the exact solution, $x = \ell/2(1\pm0.498)$ [6]. The maximum displacement, at the middle span, can be calculated as $y_0 = 0.065 \ A\ell^2$. With $y^+ = 1/500 \ \ell$, $A = 0.031/\ell$ can be found. Taking into account the numerical approximation shown in Fig. 3.2, the value of the adjoint load is found as $R = 0.086 \ q\ell$.

The design variable function h can be deduced from the relation

$$(q\ell. \ 0.5x - 0.05x^2 - q\ell^2 . \ 0.094) \ 0.086 \ q\ell. \ (0.5x - 0.125\ell)$$

$$= c\alpha h^2 \tag{3.2}$$

The form of the function h is shown in Fig. 3.5.

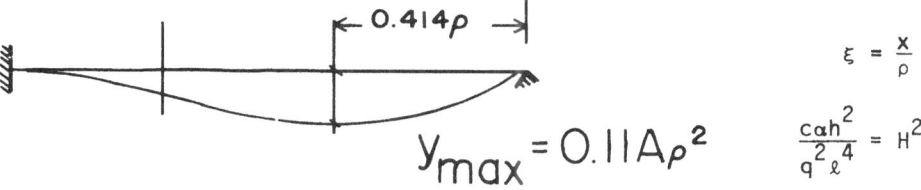

$$\xi = \frac{x}{\rho}$$

$$\frac{c\alpha h^2}{q^2 \ell^4} = H^2$$

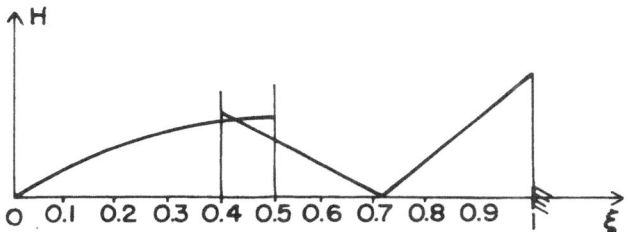

ξ	$H \times 10^3$
0	0
0.1	4.48
0.2	8.17
0.3	10.98
0.4	12.71
0.5	9.98
0.6	5.63
0.7	0.36
0.8	5.74
0.9	12.59
1.0	20.11

$$(0.353\xi - 0.5\xi^2) \, 0.016 \cdot 0.414\xi = H^2$$

$$0 \leq \xi \leq 0.414$$

$$(0.353\xi - 0.5\xi^2) \, 0.016(0.414 - 0.586\xi) = H^2$$

$$0.414 \leq \xi \leq 1$$

Figure 3.3   Solution of Example 3(1).

392

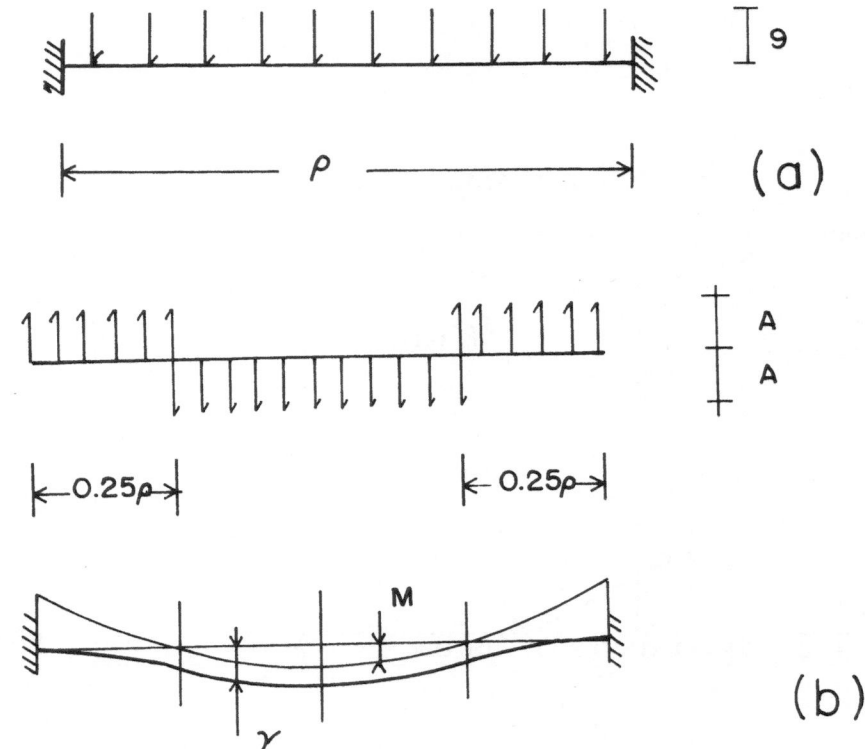

Figure 3.4 Solution of Example 3(2).

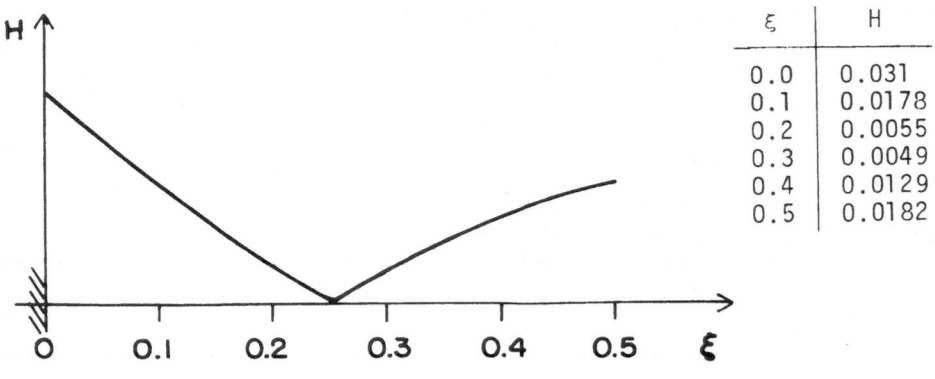

ξ	H
0.0	0.031
0.1	0.0178
0.2	0.0055
0.3	0.0049
0.4	0.0129
0.5	0.0182

$$\xi = \frac{x}{\rho} \quad (0.5\xi - 0.5\xi^2 - 0.094)0.086(0.5\xi - 0.125) = H^2$$

Figure 3.5 Optimum Solution of Clamped Beam.

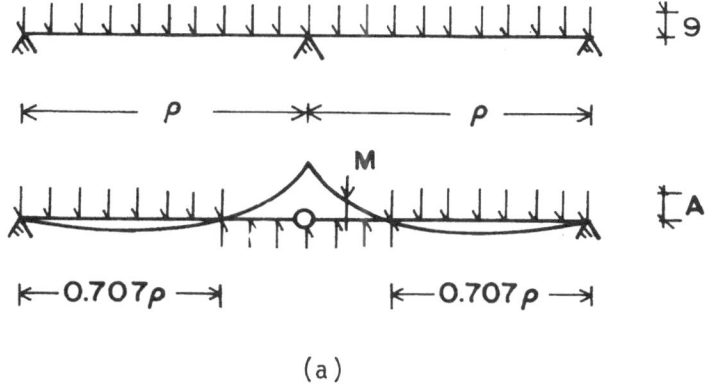

(a)

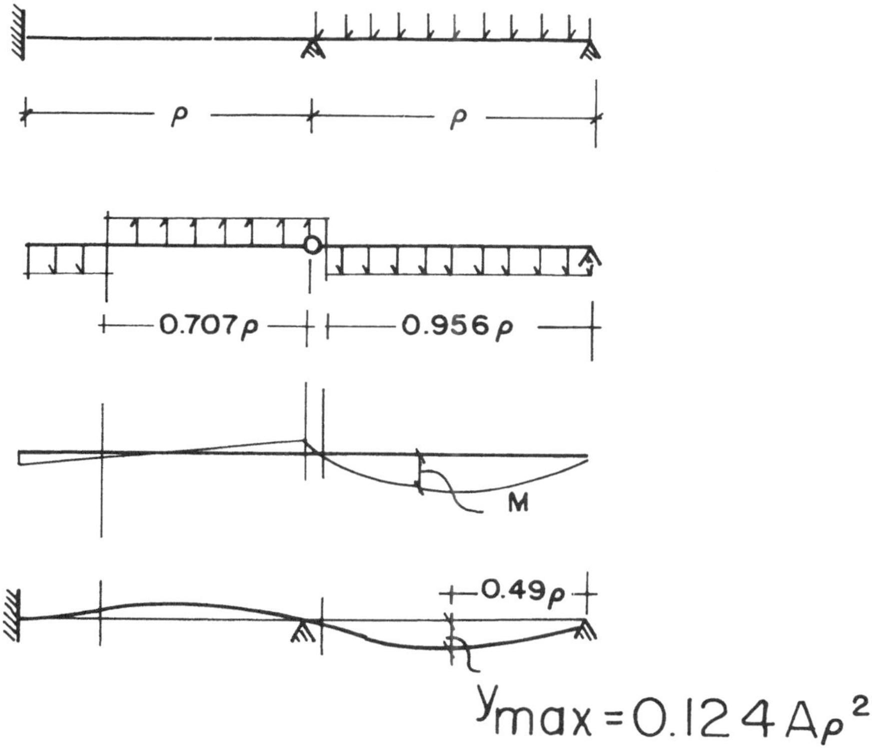

$$y_{max} = 0.124\,A\rho^2$$

(b)

Figure 3.6 Two Beams with Multiple Supports.

3.3 Beams With Multiple Supports

In Fig. 3.6, some simple examples of beams are drawn. For the beam of Fig. 3.6(a), the results shown in Section 3.1 can be used. The same results can be applied for the first span of the beam shown in Fig. 3.6(b); a zero point for the curvature is obtained in the second span, as an optimal design solution.

For the structure shown in Fig. 3.7, the approximate method does not allow one to find a complete optimal solution, with the same approximation as in the above studied examples. In fact, in the span of length ℓ, the compatibility of the deflections and the optimality (or equilibrium) condition cannot be fulfilled at the same time. By considering the structure to be constituted by two independent beams, this difficulty may be avoided.

3.4 Simply Supported, Two Span Beam

In Fig. 3.8, a simply supported, two-span beam is presented. Only the second span is subjected to a uniformly distributed load. In the absence of technological constraints, the design variable is required to vanish along the first span, in the optimal solution. A more realistic design can be obtained by imposing a lower bound on the design variable. In order to calculate the solution, the load acting on the first span of the analogical beam must be assumed to be linearly varying, because the stiffness is constant and the actual bending moment is a linear function. The solution is depicted in Fig. 3.8.

3.5 Portal Frames

A portal frame is considered (Fig. 3.9) with a uniformly distributed load. The maximum displacement is, of course, at the middle-span. Taking into account the symmetry of the problem, four zero points of the curvature function can be conceived.

For the columns, the results obtained in Section 3.1 can be used. In this way, the nodal rotation can be calculated as $0.414A\ell_2$. Then, a fixed-end beam that is subjected to a uniformly distributed load may be considered, with the prescribed rotations at its ends. Equilibrium of the analogical beam provides the zero points as

$$x = \frac{\ell_1}{2}\left[1 \pm (\frac{1}{2} + 0.414\,\frac{\ell_2}{\ell_1})\right]$$

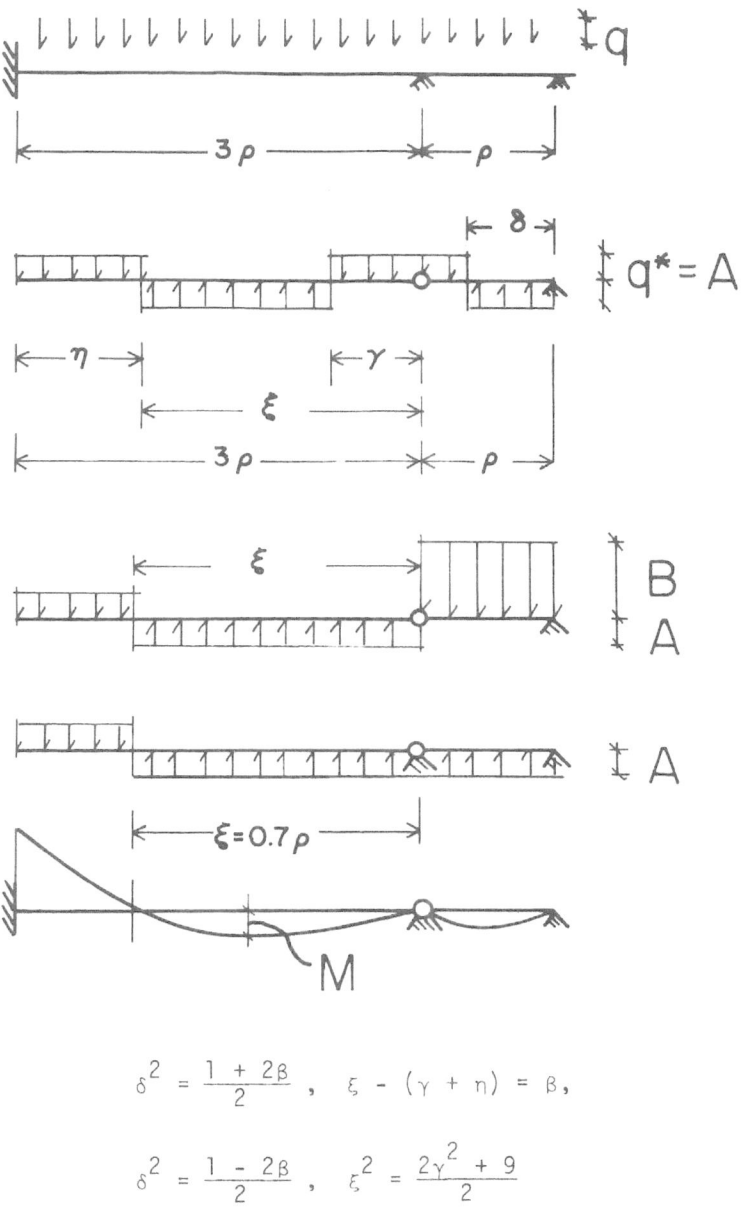

$$\delta^2 = \frac{1 + 2\beta}{2} \ , \quad \xi - (\gamma + \eta) = \beta,$$

$$\delta^2 = \frac{1 - 2\beta}{2} \ , \quad \xi^2 = \frac{2\gamma^2 + 9}{2}$$

Figure 3.7 Another Beam with Multiple Supports.

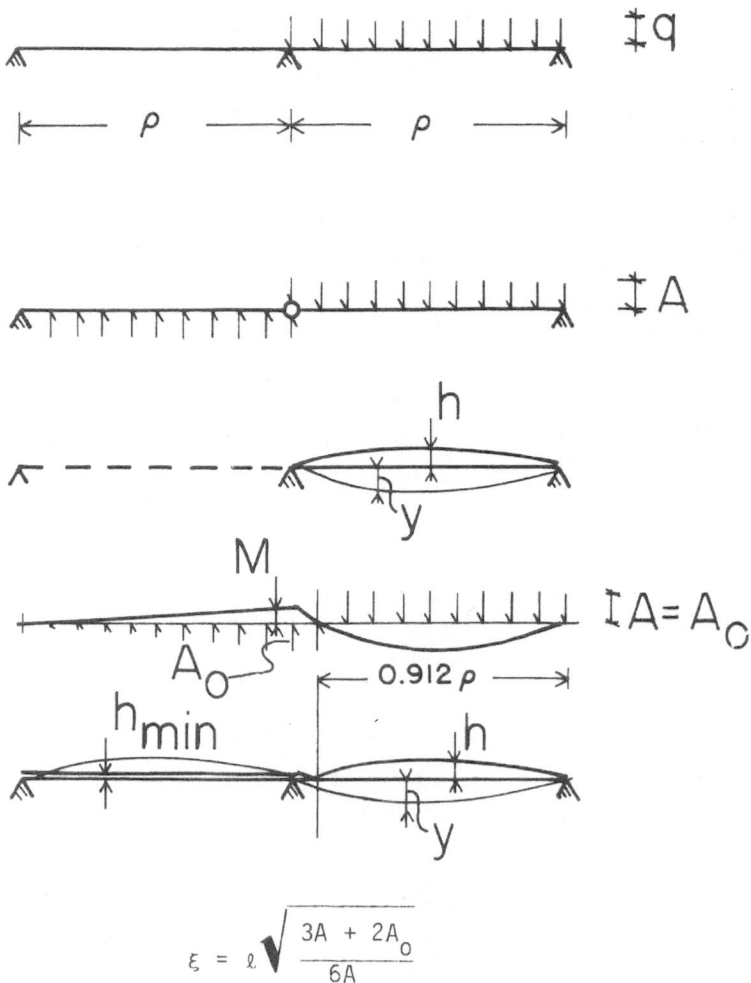

$$\xi = \ell \sqrt{\frac{3A + 2A_o}{6A}}$$

Figure 3.8 Simply Supported, Two Span Beam.

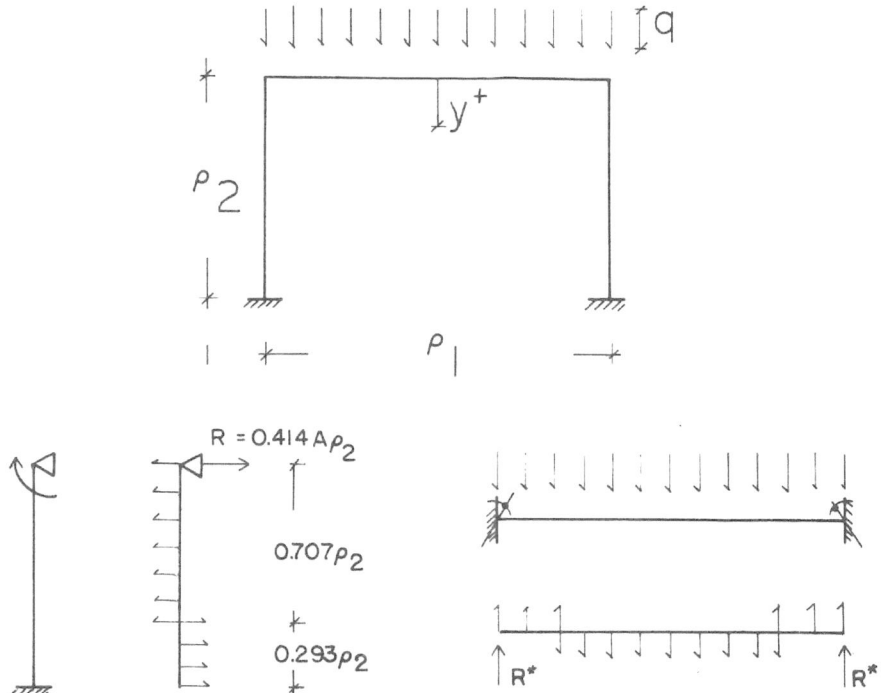

Figure 3.9   Portal Frame with Distributed Load.

This solution is acceptable only for $0 \le x \le \ell_1$, i.e. for $\ell_2 \le 1.208\ell_1$. Otherwise, in the optimal solution the bending stiffness can be seen to vanish along the columns. A realistic design is obtained by means of a minimum value prescribed to the columns.

The same portal frame, with different load conditions, may require a lower bound on the design variables for every value of the ratio ℓ_2/ℓ_1. In the case of the portal frame in Fig. 3.10, $h = h_{min}$ must be assumed for the design of the beam. In fact, the compatibility of the nodal rotations cannot be verified if the optimality condition is fulfilled along all the structure.

4.   CONCLUSIONS

The approximate method described in this paper can be employed for different cases of one dimensional structures. The

398

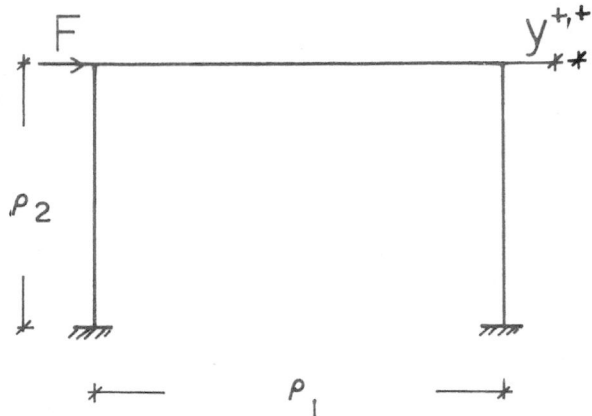

Figure 3.10   Portal Frame with Lateral Load.

calculations are very simple, with respect to the analytical procedures required by the exact closed-form solutions. A good
degree of approximation can be obtained. The results can also
be usefull for having a good starting solution for iterative
numerical procedures.

REFERENCES

1.   Rozvany, G.I.N. and Mroz, Z., "Analytical Methods in Struc-
     tural Optimization," Applied Mechanics Reviews, Vol. 30,
     N. 11, 1977, pp. 1461-1470.
2.   Cinquini, C. and Sacchi, G., "Problems of Optimal Design for
     Elastic and Plastic Structures," Journal de Mécanique
     Appliquée, Vol. 4, No. 1, 1980, pp. 31-59.
3.   Sacchi, G., "Optimal Structural Design Problems," Ist. Naz.
     Alta Matematica; Seminari su "Problemi di Frontiera Libera",
     Pavia, 1979.
4.   Niordson, F. and Olhoff, N., Variational Methods in Opti-
     mization of Structures, Report No. 161, The Danish Center
     for Applied Mathematics and Mechanics, 1979.
5.   Cinquini, C., "Variational Formulation and Numerical Methods
     in Optimal Design," Optimization of Distributed Parameter
     Structures (Ed. E.J. Haug and J. Cea), Sijthoff & Nordhoff,
     Alphen ann den Rijn, Netherland, 1980.
6.   Sander, G. and Fleury, C., "A Mixed Method in Structural
     Optimization," Int. J. for Num. Meth. in Engrg., Vol. 13,
     1978, pp. 385-404.

ON THE OPTIMUM SHAPE OF COLUMNS*

Vadim Komkov and Edward J. Haug.

Department of Mathematics and Math Reviews, University of Michigan, Ann Arbor, Michigan

Materials Division, College of Engineering, University of Iowa, Iowa City, Iowa 52242

ABSTRACT

Optimal design criteria for a buckled column are analyzed. An accurate model of column buckling, including effects of "usually" ignored compressive stresses and nonlinearity, results in a nonlinear eigenvalue problem that reduces in the limit to the Rayleigh eigenvalue problem. In general, it is shown to exhibit entirely different behavior. In particular, the analysis of the nonlinear eigenvalue problem reveals that some "optimal designs" based on the elementary beam theoretic model may in fact be far from optimum.

1. INTRODUCTION

Optimal design of columns with a constraint placed on Euler's buckling load has been considered by Keller [1], Tajbaksh and Keller [2], and by Keller and Niordson [3]. These papers were followed by extensive research of related problems of Olhoff, Masur, and others [4,5,6,7,8]. However, the optimal designs proposed by Keller and his co-workers have recently come under criticism, particularly the designs that exhibit singularities in the form of zero-cross-sectional areas. An alternate approach of Olhoff and Rasmussen [5] and of Olhoff and Taylor [6] addresses

---

*Work sponsored by NSF Grant No. Eng 77-19967.

difficulties associated with a minimal corss sectional area constraint, by implying that such a constraint is unnecessary for a sufficiently small value of the minimal cross-sectional area.

An attempt is made here at a fresh start in re-examining the problem and relating it closer to physical experience. The analysis of Keller, Olhoff, Taylor, etc., is modified to correct for an essential deficiency of the linear buckling model for a column whose cross-sectional area becomes very small, even at a single point along its length. Reasons for this difficulty are first pointed out, followed by development of a refined nonlinear model of column buckling.

The usual treatment of Euler's column theory goes somewhat along the lines of reasoning as seen in column one of Fig. 1.1.

Using a variety of assumptions (there exists a neutral surface, plane cross-sections remain plane, and the stress in each fiber is proportional to longitudinal strain so that stresses vary linearly with the distance from the neutral surface), one derives the relation

$$\rho^{-1} = \frac{M(x)}{EI(x)}$$

where ρ is the radius of curvature of the column, ρ^{-1} being given by the geometric formula

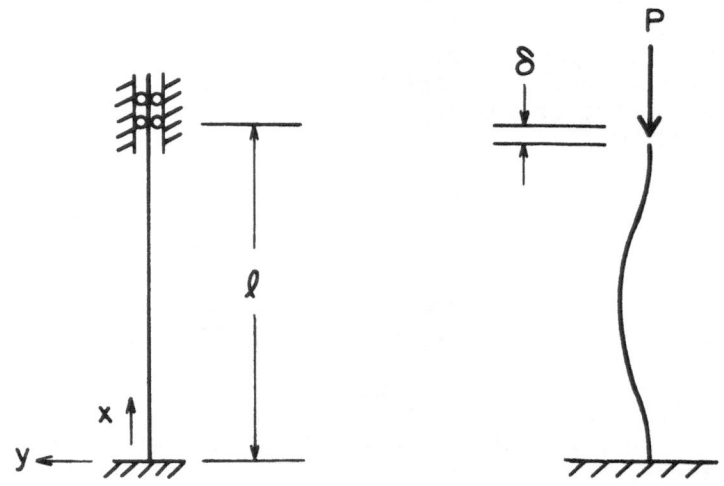

Figure 1.1   Column

$$\rho^{-1} = \frac{y''}{[1 + (y')^2]^{3/2}}$$

where $' = d/dx$.

At this point in the linear theory, one argues that y' is small and $[1 + (y')^2]^{3/2} \approx 1$, leading to the approximate formula $EI(x)y'' = M(x)$ for beam bending. The bending energy is then approximately

$$U = 1/2 \int_0^\ell \frac{M^2(x)}{EI(x)} dx = 1/2 \int_0^\ell EI(x) (y'')^2 dx \qquad (1.1)$$

The longitudinal displacement of the top of the column is calculated by assuming that the length of the beam does not change, since the effects of compression are negligible compared to the bending effects. Hence, one can write

$$\ell = \int_0^{\ell-\delta} [1 + (y')^2]^{1/2} dx$$

and after some manipulation obtain the change in length δ as

$$\delta = \int_0^\ell \left[\left(1 + \frac{(y')^2}{2} + \frac{(y')^4}{8} + - \ldots\right) - 1 \right] dx$$

$$\approx 1/2 \int_0^\ell (y')^2 dx$$

Equating the work done by the force P to the strain energy, one obtains the eigenvalue relation

$$1/2 \int_0^\ell [EI(y'')^2] dx - \frac{P}{2} \int_0^\ell (y')^2 dx = 0 \qquad (1.2)$$

for buckling load P. Let $y_n(x)$ be an eigenfunction associated with the corresponding eigenvalue P_n. Then,

$$P_n = \frac{\int_0^\ell EI(y_n'')^2 dx}{\int_0^\ell (y_n')^2 dx} \qquad (1.3)$$

The smallest buckling load P_1 can be approximated by means of Rayleigh's theorem, asserting that the buckling mode minimizes the quotient of Eq. 1.3. Higher buckling modes can be approximated by minimizing P_n of Eq. 1.3, subject to the condition of orthogonality with respect to the lower modes, presuming nonexistence of multiple eigenvalues.

Unfortunately, the assumption made at each step of this development are often forgotten and Eq. 1.2 and its differential operator counterpart

$$(EI(x)y")" + Py" = 0 \qquad\qquad (1.4)$$

are used to represent column buckling in most circumstances. This equation and the quotient of Eq. 1.3 are used as the starting point of numerous investigations and advanced mathematical discussions, which may lead to violations of the conditions under which Eqs. 1.2, 1.3 and 1.4 were derived. The effect of the design (i.e., the choice of $I(x)$) on the behavior of solutions of Eq. 1.4, subject to various boundary conditions at $x = 0$ and $x = \ell$, and related conditions of optimality can be studied as a mathematical project. However, the significance of such findings to actual beam or column design practice cannot be fully ascertained, until one has checked some crucial features of such designs. This is particularly important in optimization studies in which singularities are encountered.

In the following section, the "usual" optimal design analysis is modified by including one of the effects that is usually "safely" neglected. It is shown that this effect becomes significant in some considerations of the optimal shape and cannot be ignored if singular optimal shapes are considered.

In the entire analysis of sections 2 and 3 the deflection $y(x)$ is assumed to be an element of the Sobolev space $H_0^2(0,\ell)$ and all integrals introduced in these sections exist in the Riemann sense. The existence of certain terms is questioned only in the limiting transition to the classical Euler's theory as $(EA)^{-1} \to 0$, where A denotes the cross sectional area of the beam.

2.   FORMULATION OF A NONLINEAR BUCKLING PROBLEM

The validity of Hooke's law and assumptions stated in the preceeding part of this paper are retained, with two exceptions. Nonlinear terms in representing effects of deflection and compression on the buckling mode are retained.

The process of assuming the buckled shape is analyzed by considering virtual work performed by the axial buckling force P.

As P is increased to the critical value P_{cr}, the column is initially deflected axially and the length is decreased from its original value ℓ_1 to the value

$$\ell = \ell_1 - \int_0^\ell \frac{P}{A(x)E} \, dx \ .$$

As P reaches the value P_{cr}, two separate effects takes place, as indicated in Fig. 2.1. Ignoring any changes in the value of the compressive stress, one obtains the axial deflection due to a purely geometric change in the configuration, denoted δ_G, by assuming that the length of the column remains the same. The shape is assumed to be smooth; i.e., $y \in C^1(0,\ell)$.

The purely geometric effects of change from the straight shape $y \equiv 0$ to the buckled shape $y = \tilde{y}(x)$, involve the equality

$$\int_0^\ell \cos \theta ds = \ell - \delta_G$$

where $\theta(x) = \arctan (y'(x))$ is the angle between the axis of the column and the x-axis. Using a Taylor series expansion of $\cos \theta$, one has

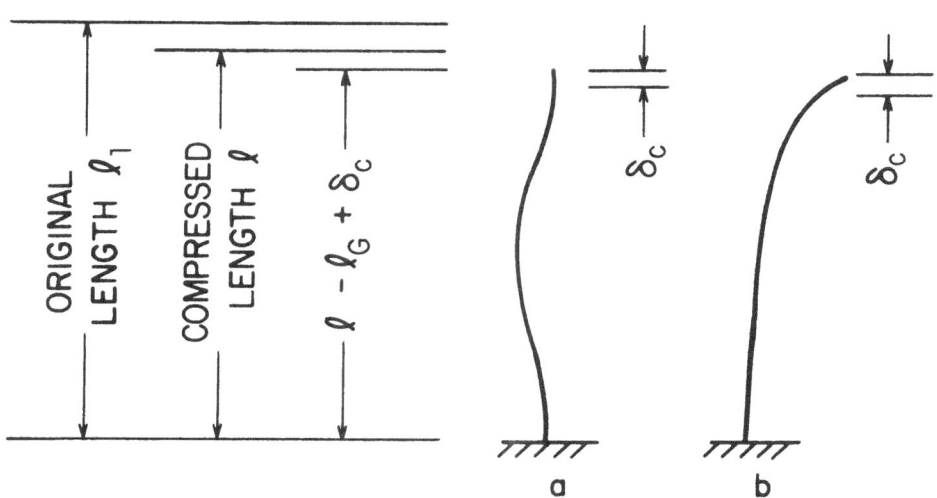

Figure 2.1   Deflection of Column.

$$\int_0^\ell \left(1 - \frac{\theta^2}{2!} + \frac{\theta^4}{4!} + \dots\right) ds = \ell - \delta_G$$

or

$$\delta_G = \int_0^\ell \left(\frac{\theta^2}{2!} - \frac{\theta^4}{4!} + \dots\right) ds + O(\theta^6) \tag{2.1}$$

The second effect to be accounted for concerns the change in length of the column that is caused by the change in the value of the compressive force from P to Pcosθ along the column. The corresponding change in the length of the column is therefore

$$\delta'_C = \int_0^\ell \frac{Pcos\theta(s) - P}{EA(s)} ds$$

The corresponding change in the dimension of the column along the x-axis is given by

$$\delta_C = \int_0^\ell \frac{Pcos^2\theta - Pcos\theta}{EA(s)} ds \tag{2.2}$$

This change is negative, corresponding to a "recovery" effect. That is, δ_G and δ_C have opposite signs

The total deflection along the x-axis is given by

$$\delta = \delta_G + \delta_C = \int_0^\ell \left(\frac{\theta^2}{2!} - \frac{\theta^4}{4!} + \dots O(\theta^6)\right) ds$$

$$- \int_0^\ell \frac{P}{EA(s)} \left(\frac{\theta^2}{2} + \frac{7}{24}\theta^4 + \dots O(\theta^6)\right) ds \tag{2.3}$$

The strain energy of the deflected beam is given by

$$SE = P^2 \int_0^\ell \frac{cos^2\theta(s)}{2EA(s)} ds + 1/2 \int_0^\ell EI(s)(\theta')^2 ds$$

$$= \frac{P^2}{2} \int_0^\ell \frac{\left(1 - \theta^2 + \frac{\theta^4}{3} - \dots\right)}{EA(s)} ds + 1/2 \int_0^\ell EI(s)(\theta')^2 ds$$

In this paper it is assumed that θ is an element of the Sobolev space $H_0^1[0,\ell]$, so that all integrals that appear, for example those appearing in Eqs. 2.4, 2.5, 2.6, and 2.7, are defined in the usual Riemann sense, possibly as improper integrals.

As the column buckles, the following change takes place in the value of the strain energy:

$$\Delta SE = 1/2 \int_0^\ell EI(s)(\theta')ds + \frac{P^2}{2} \int_0^\ell \frac{\cos^2\theta - 1}{EA(s)} \, ds$$

$$= 1/2 \int_0^\ell EI(s)(\theta')^2 ds - \frac{P^2}{2} \int_0^\ell \frac{\left(\theta^2 - \frac{\theta^4}{3} + \ldots\right)}{EA(s)} \, ds \qquad (2.4)$$

As the beam changes its shape, the work performed by the force P is simply

$$P\delta = P \int_0^\ell \left(\frac{\theta^2}{2} - \frac{\theta^4}{24} + \ldots\right) ds + \frac{P^2}{2} \int_0^\ell \left(\frac{\theta^2 + 7/12\theta^4 + \cdots}{EA(s)}\right) ds \qquad (2.5)$$

The net change in the total potential energy is thus given by

$$\Delta V = P\delta - \Delta SE = P \int_0^\ell \left(\frac{\theta^2}{2} - \frac{\theta^4}{24} + \ldots\right) ds$$

$$+ \frac{P^2}{2} \int_0^\ell \frac{(11/12 \, \theta^4 - \ldots)}{EA(s)} \, ds - 1/2 \int_0^\ell EI(s)(\theta')^2 ds$$

Buckling will occur if there exists a function $\theta(s) \in H_0^1[0,\ell]$, $\theta \neq 0$, such that $\Delta V \geq 0$. This is equivalent to the inequality

$$P^2 \int_0^\ell \frac{(11/24 \, \theta^4 - \ldots)}{EA(s)} \, ds + P \int_0^\ell \left(\frac{\theta^2}{2} - \frac{\theta^4}{24} + \ldots\right) ds$$

$$- 1/2 \int_0^\ell EI(s)(\theta')^2 ds \geq 0 \qquad (2.6)$$

for some admissible function $\theta(s)$. Considering only positive values of P, this inequality implies

$$P \geq \left\{ \frac{11}{12} \int_0^{\ell} \frac{\theta^4}{EA(s)} \, ds \right\}^{-1} \left\{ - \int_0^{\ell} \left(\frac{\theta^2}{2} - \frac{\theta^4}{24} \right) ds \right.$$

$$+ \left[\left(\int_0^{\ell} \left(\frac{\theta^2}{2} - \frac{\theta^4}{24} \right) ds \right)^2 + \left(\int_0^{\ell} \frac{11/12 \; \theta^4}{EA(s)} \, ds \right) \right.$$

$$\left. \left. \left(\int_0^{\ell} EI(s)(\theta')^2 ds \right) \right]^{1/2} \right\} \equiv F(\theta) \tag{2.7}$$

where powers of θ of order six and higher have been omitted.

The pair $(\tilde{\theta}(s), P_{cr})$ is called a critical pair if $\tilde{\theta}(s) \neq 0$, $F(\theta)$ attains an extremal value when $\theta = \tilde{\theta}$, and P_{cr} is equal to $F(\tilde{\theta})$. One must now show that such critical pairs exist and investigate the uniqueness of the critical vector $\tilde{\theta}(s)$.

Denote

$$A = \frac{11}{12} \int_0^{\ell} \frac{\theta^4}{EA(s)} \, ds \tag{2.8}$$

$$B = \int_0^{\ell} \left(\frac{\theta^2}{2} - \frac{\theta^4}{24} \right) ds \tag{2.9}$$

$$C = \int_0^{\ell} EI(s)(\theta')^2 ds \tag{2.10}$$

The Frechet derivative of $F(\theta)$ is given by

$$\frac{\partial F}{\partial \theta} = A^{-1} \left\{ \left[- \frac{\partial B}{\partial \theta} + \frac{1}{2} (B^2 + AC)^{-1/2} \left(2B \frac{\partial B}{\partial \theta} \right) + \frac{\partial A}{\partial \theta} C + A \frac{\partial C}{\partial \theta} \right] \right.$$

$$\left. - F(\theta) \frac{\partial A}{\partial \theta} \right\} \tag{2.11}$$

The necessary condition for an extremum of $F(\theta)$ is then

$$\left[- \frac{\partial B}{\partial \theta} + \frac{1}{2} (B^2 + AC)^{-1/2} \; 2B \frac{\partial B}{\partial \theta} + \frac{\partial A}{\partial \theta} C + A \frac{\partial C}{\partial \theta} \right] - F(\theta) \frac{\partial A}{\partial \theta} = 0 \tag{2.12}$$

where direct computation yields

$$\frac{\partial A}{\partial \theta} = \frac{11}{3} \left(\frac{\theta^3}{EA(s)} \right)$$

(2.13)

$$\frac{\partial B}{\partial \theta} = \theta - \frac{1}{6} \theta^2$$

(2.14)

$$\frac{\partial C}{\partial \theta} = 2(EI(s)\theta')'$$

(2.15)

Substituting from Eqs. 2.13 to 2.15 into Eq. 2.12, one has at $\theta = \tilde{\theta}$

$$\left(-\tilde{\theta} + \frac{1}{6} \tilde{\theta}^3 \right) + \frac{1}{2} \left[\left(\int_0^\ell \left(\frac{\tilde{\theta}^2}{2} - \frac{\tilde{\theta}^4}{24} \right) ds \right)^2 \right.$$

$$\left. + \left(\frac{11}{12} \int_0^\ell \frac{\tilde{\theta}^4}{EA(s)} ds \right) \left(\int_0^\ell EI(s)(\tilde{\theta}')^2 ds \right) \right]^{-1/2}$$

$$\times \left[\left(2 \int_0^\ell \left(\frac{\tilde{\theta}^2}{2} - \frac{\tilde{\theta}^4}{24} \right) ds \right) \left(\tilde{\theta} - \frac{1}{6} \tilde{\theta}^3 \right) + \frac{11}{3} \frac{\tilde{\theta}^3}{EA(s)} \left(\int_0^\ell EI(s)(\tilde{\theta}')^2 ds \right) \right.$$

$$\left. - \frac{11}{6} (EI(s)\tilde{\theta}')' \left(\int_0^\ell \frac{\tilde{\theta}^4}{EA(s)} ds \right) \right] - \frac{11}{3} \tilde{\theta}^3 (EA)^{-1} F(\tilde{\theta}) = 0$$

(2.16)

Since $F(\tilde{\theta}) = P_{cr}$, one obtains a necessary condition for extremality of the total potential energy in the form of an eigen-value equation 2.16. Also note that at any point $\tilde{x}$ where $\tilde{\theta}(\tilde{x}) = 0$, all terms of Eq. 2.16 vanish except

$$- \frac{11}{6} (EI(s)\tilde{\theta}')' \left(\int_0^\ell \frac{\tilde{\theta}^4}{EA(s)} ds \right)$$

Hence, where $\tilde{\theta}(\tilde{x}) = 0$, $(EI(\tilde{x})\tilde{\theta}')'(\tilde{x}) = 0$.

Theorem 1: For small values of $\theta^4(EA)^{-1}$ the inequality of Eq. 2.7 can be replaced by

$$P \geq \frac{\int_0^\ell EI(\theta')^2 ds}{\int_0^\ell \left(\theta^2 - \frac{\theta^4}{12} \right) ds}$$

(2.17)

If the term $\theta^4/12$ is neglected, this is exactly the Rayleigh quotient inequality.

Proof: A general form of l'Hospital's formula is used as $(EA)^{-1} \to 0$. Recalling that $\lim\limits_{a\to 0} \left[\dfrac{-b + \sqrt{b^2 + 4ac}}{2a} \right] = \dfrac{c}{b}$, one may substitute

$$a = \int_0^\ell \frac{11/24 \; \theta^4}{EA} \; ds$$

$$b = \int_0^\ell \left(\frac{\theta^2}{2} - \frac{\theta^4}{24} \right) ds$$

$$c = 1/2 \int_0^\ell EI(\theta')^2 \; ds$$

to complete the proof.

This theorem points out that the classical theory is applicable if $\theta^4 (EA)^{-1}$ is small compared to θ^2 and compared to $EI(\theta')^2$. The foregoing analysis shows, however, that such conclusions can not be supported if the cross-sectional area is very small and the term $(EA^{-1})\theta^4$ is comparable in magnitude to $EI(\theta')^2$. This observation becomes important in the subsequent analysis concerning optimization of design.

It may also be noted that some simplifications are possible if boundary conditions are assigned at $s = 0$ and $s = \ell$. The conditions $y(0) = y(\ell) = 0$ can be restated in the form

$$\int_0^\ell \sin\theta(s)ds = 0, \quad \text{or} \quad \int_0^\ell \theta(s) \; ds = \int_0^\ell \left(\frac{\theta^3}{3!} - \frac{\theta^5}{5!} + \ldots \right) ds$$

Observe also that the equation $F'(\theta)|_{\theta=\tilde{\theta}} = 0$ is a necessary condition for a local minimum of potential energy, corresponding to a local maximum of $F(\theta)$. This condition, written out in full, is given by Eq. 2.16. If one integrates Eq. 2.16 with respect to s and observes that

$$\int_0^\ell \left(\theta - \frac{\theta^3}{6} \right) ds = 0$$

he obtains the following "averaged" relation

$$
\frac{1}{2}\left[\frac{11}{12}\left(\int_0^\ell \frac{\tilde\theta^4}{EA(s)}\,ds\right)\left(\int_0^\ell EI(s)(\tilde\theta')^2ds\right)\right.
$$

$$
\left.+\left(\int_0^\ell\left(\frac{\tilde\theta^2}{2}-\frac{\tilde\theta^4}{24}\right)ds\right)^2\right]^{-1/2}
$$

$$
\times\left\{\frac{11}{3}\left(\int_0^\ell\frac{\tilde\theta^3}{EA(s)}\,ds\right)\left(\int_0^\ell(EI(s)(\tilde\theta')^2ds\right)\right.
$$

$$
\left.-\frac{11}{6}\left[EI(s)\tilde\theta'\right]_0^\ell\int_0^\ell\frac{\tilde\theta^4}{EA(s)}\,ds\right\}=\frac{11}{3}\left(\int_0^\ell\frac{\tilde\theta^3}{EA(s)}\,ds\right)F(\tilde\theta)
$$

If the end conditions are such that $\theta'(0) = \theta'(\ell)$, this equality becomes

$$
\frac{1}{2}\left[\frac{11}{21}\left(\int_0^\ell\frac{\tilde\theta^4}{EA(s)}\,ds\right)\left(\int_0^\ell EI(\tilde\theta')^2ds\right)+\left(\int_0^\ell\left(\frac{\tilde\theta^2}{2}-\frac{\tilde\theta^4}{24}\right)ds\right)^2\right]^{-1/2}
$$

$$
\times\left[\left(\int_0^\ell\frac{\tilde\theta^3}{EA(s)}\,ds\right)\left(\int_0^\ell EI(\tilde\theta')^2ds\right)\right]=\left(\int_0^\ell\frac{\tilde\theta^3}{EA(s)}\,ds\right)F(\tilde\theta)
$$

or

$$
F(\tilde\theta)=\left[\left(\int_0^\ell\left(\frac{\tilde\theta^2}{2}-\frac{\tilde\theta^4}{24}\right)ds\right)^2+\frac{11}{12}\left(\int_0^\ell EI(\tilde\theta')^2ds\right)\right.
$$

$$
\left.\times\left(\int_0^\ell\frac{\tilde\theta^4}{EA(s)}\,ds\right)\right]^{-1/2}\times\frac{1}{2}\int_0^\ell EI(\tilde\theta')^2ds \qquad (2.18)
$$

Equation 2.18 gives the numerical value of $F(\theta)$ at the critical point $\theta = \tilde\theta$.

From Eq. 2.18, it becomes apparent that for very small values of $\theta^4(EA)^{-1}(s)$, $0 \le s \le \ell$, the critical value of $F(\tilde\theta)$ is equal to the Rayleigh quotient, when θ is approximated by $y'(x)$ and terms involving θ^4 are ignored. Moreover, the value $F(\tilde\theta)$ is smaller than the corresponding value given by the Rayleigh quotient.

Equation 2.18 indicates that the numerators of the two quotients are roughly equal, while the denominator of Eq. 2.18 appears to be larger. An important corollary of this observation will be stated formally as a theorem.

Theorem 2: Within an approximation of the order $\int_0^\ell (\theta^6)ds$, the value P_{cr} given by Eq. 2.18 is always lower than the critical load predicted by the classical form of the Rayleigh quotient; i.e., by Eq. 1.3.

This theorem shows that the estimate provided by the classical form of Rayleigh quotient is too stringent. One might be tempted to conclude that in fact buckling occurs at a slightly lower value of the critical load if residual terms of order four (or even higher) are taken into account. Such conclusions appear to be refuted on physical grounds. Until the column buckles, the value of $y'(x)$ is very close to zero along the entire length. Hence, only the lowest order terms in $y'(x)$ need to be taken into account to compute the initiation of the buckling process. What Eq. 2.6 supplies, and what Theorem 2 states is that after buckling takes place, the buckling critical load slightly decreases. If a column carries a load higher than Euler's critical load and the load is slowly reduced, it will not return to a straight shape. When the load reaches Euler's value; it will remain in the buckled shape until the load is smaller than the corresponding value of P_{cr} given by Eq. 2.18.

In the terminology of the calculus of variations, $y \equiv 0$ is a local minimum for the functional of Eq. 2.18 if $P_{Euler} > P > P_{cr}$, but it is not a global minimum. Hence, it appears that one should not worry about the lower value of P_{cr} when dealing with an initially straight column, and in computing the buckling collapse of a structure, the higher value of the Rayleigh quotient given by Eq. 1.3 can be used in computations.

This approach would be correct for a perfectly static column. However, if one considers random loads (wind gusts), random deflections (mild earthquake), or dynamic effects of a very short duration, the main concern of the designer is the ability of the structure to "straighten itself out" when such loads of very short duration are removed. Particularly if in a deflection of short duration one can not assume that y' is small, Eq. 2.18 must be used instead of the classical formula of Eq. 1.3.

Proof of Theorem 2: The proof hinges on a simple algebraic inequality. Consider the positive zero of the quadratic function $y = ax^2 + bx - c$, where a, b, and c are positive real numbers, and the root of the corresponding linear equation $0 = bx - c$. It is claimed that the value $x = c/b$ is always greater than the larger of the two values of x corresponding to the zeros of $y(x)$. The case of equal roots is trivial since $-b/2a$ is negative.

Suppose that the claim is false and $-\frac{6}{2a} + \frac{b^2 + 4ac}{2a} \geq \frac{c}{b}$.

Then $\sqrt{b^2 + 4 ac} \geq \frac{2ac}{b} + b$ $(a > 0)$, so $b^2 + 4ac \geq b^2 + 4ac + \frac{4a^2c^2}{b^2}$,

which is impossible. The proof of the theorem follows if one identifies

$$a = \int_0^\ell \frac{11/24\ \theta^4}{EA(s)}\ ds$$

$$b = \int_0^\ell \left(\frac{\theta^2}{2} - \frac{\theta^4}{24}\right) ds$$

$$c = \frac{1}{2} \int_0^\ell EI(s)(\theta')^2 ds$$

and equates P_{cr} with the corresponding value of $F(\theta)$. Then, the algebraic inequality becomes

$$P_{cr} < \frac{c}{b} = \frac{\int_0^\ell EI(s)(\theta')^2 ds}{\int_0^\ell \left(\frac{\theta^2}{2} - \frac{\theta^4}{24}\right) ds}$$

The denominator is the value of δ_G, with terms smaller than $\int_0^\ell \left(\frac{\theta^6}{720}\right) ds$ omitted.

Using the identity

$$\int_0^\ell \cos\theta\, ds = \int_0^{\ell - \delta_G} [\cos(\text{arc tan } y')][1 + (y')^2]^{1/2} dx$$

the denominator can be replaced by $\frac{1}{2} \int_0^\ell (y')^2 dx$, while $(\theta'_\ell)^2 ds$ can be replaced by $(y'')^2 dx$, with the resulting error of the order of $\int_0^\ell (y')^6 dx$. This complete the proof of Theorem 2.

Care must be exercised in converting formulas involving $\theta(s)$ into formulas written in terms of $y(x)$ and its derivatives.

For example,

$$\int_0^\ell \cos\theta ds \neq \int_0^\ell [\cos(\arctan y')] \; [1 + (y')^2]^{1/2} dx$$

the error being of the order of δ_G. Thus y' and θ are not inter-changeable. A comparison of various formulas can be made by re-lating these results to a parallel development given in Appendix 2. An alternate form of Theorem 2 is given as follows:

Theorem 2a: In the buckling process described by the in-equalities of Eqs. 2.6 and 2.7, the load that initiates the buckling process is larger than the load necessary to maintain the buckled state.

This phenomenon has been known to experimentalists, but was explained by inertia of the material, which failed to follow exactly Hooke's law, and by other so-called "effects", such as the Bauschinger effect, which were supposedly related to consti-tutive properties of the materials tested.

Another important departure from classical theory is indicated by the theorem that follows; showing that, subject to an assumption that the angle θ is not too large, an equilibrium position can be reached for an arbitrary value of $P \geq P_{cr}$. This indicates the existence of a continuous spectrum of the buckling operator.

Theorem 3: Suppose that for a given $P > 0$ a function $\theta(x)$ found in $W_4^1 [0,\ell]$ such that $\theta \not\equiv 0$, $\theta < 1$, and $P > F(\theta)$. Then if θ is "small", there exists $\tilde{\theta}$ such that $F(\tilde{\theta}) = \max_{\theta \in W_4^1[0,\ell]} F(\theta)$, where $F(\theta)$ is the negative of the total potential energy.

Proof: It suffices to show that the operator representing the second derivative of $F(\theta)$ (in the Frèchet sense) is negative definite at the point $\tilde{\theta}$, where $F'(\tilde{\theta}) = 0$ for sufficiently small values of θ. Denote by u and v

$$u \equiv -B + [B^2 + AC]^{1/2}$$

$$= -\int_0^\ell \left(\frac{\theta^2}{2} - \frac{\theta^4}{24}\right) ds + \left[\left(\int_0^\ell \left(\frac{\theta^2}{2} - \frac{\theta^4}{24}\right) ds\right)^2\right.$$

(equation cont'd)

$$+ \left(\int_0^\ell \frac{11/12\ \theta^4}{EA(s)}\ ds \right) \times \left(\int_0^\ell EI(s)(\theta)^2 ds \right)^{1/2}$$

$$v \equiv A = \frac{11}{12} \int_0^\ell \left(\frac{\theta^4}{A(s)} \right)\ ds$$

Then, $F(\theta) = \frac{u}{v}$ and $F'(\theta) = \frac{u'}{v} - F(\theta) \frac{v'}{v} = 0$ is the necessary condition for the stationary behavior of $F(\theta)$. Now

$$F'' = \frac{(u'' - Fv'')\ v - (u' - Fv')\ v'}{v^2}$$

at the critical point of F. It is clear that $F'' < 0$ if

$$(u'' - Fv'')\ v < (u' - Fv')\ v'$$

Since $v(\theta) > 0$, this is equivalent to

$$(u'' - Fv'') < \left(\frac{u'}{v} - F \frac{v'}{v} \right) v' = F'(\theta)v' = 0\ .$$

Expanding the derivatives of u and v, this is

$$- \frac{\partial^2 B}{\partial\theta^2} + 1/2 \frac{\partial}{\partial\theta} \left\{ (B^2 + AC)^{-1/2} \left(2B \frac{\partial B}{\partial\theta} + \frac{\partial}{\partial\theta}\ (AC) \right) \right\} < F(\theta) \frac{\partial^2 A}{\partial\theta^2}$$

or

$$\frac{\partial}{\partial\theta} \left\{ (B^2 + AC)^{-1/2} \left(\frac{\partial}{\partial\theta}\ B^2 \right) + \frac{\partial}{\partial\theta}\ (AC) \right\} < F(\theta) \left[1 + \theta^2 \left(\frac{1}{3} + \frac{11}{EA} \right) \right]$$

Equivalently,

$$(B^2 + AC)^{-1/2} \left(\frac{\partial^2 B^2}{\partial\theta^2} + \frac{\partial^2 AC}{\partial\theta^2} \right) < F(\theta) \left[1 + \theta^2 \left(\frac{1}{3} + \frac{11}{EA} \right) \right]$$

$$+ \frac{1}{2}\ (B^2 + AC)^{-1/2} \left(\frac{\partial B^2}{\partial\theta} + \frac{\partial AC}{\partial\theta} \right)^2 (B^2 + AC)^{-1}$$

or

$$0 < F(\theta) \left[1 + \theta^2 \left(\frac{1}{3} + \frac{11}{EA}\right)\right] + \frac{1}{2} (B^2 + AC)^{-1/2}$$

$$\left\{\left[(B^2 + AC)^{-1} \left(\frac{\partial}{\partial\theta} (B^2 + AC)\right)^2\right] - 2 \frac{\partial^2}{\partial\theta^2} (B^2 + AC)\right\} \quad (2.19)$$

Writing out the expression inside the curly brackets of Eq. 2.19, in terms of powers of θ, one obtains a leading term $(1 - \theta^2)$; a positive quantity. Hence, for small values of θ, the desired inequality is correct and Theorem 3 is proved.

The buckling process is fully reversable if and only if the following equality is satisfied for the pair $(P_{cr}, \tilde{\theta})$:

$$\int_{\delta=0}^{\bar{\delta}=\delta_G+\delta_G'} P(\delta)d\delta = \frac{1}{2} \int_0^\ell EI(s)(\tilde{\theta}')^2 ds$$

$$- P_{cr} \int_0^\ell \frac{(\tilde{\theta}^2 - \tilde{\theta}^4/3 + \ldots)}{EA(s)} ds \quad (2.20)$$

with

$$P(0) = P_{Euler}$$

$$P(\bar{\delta}) = P_{cr}.$$

This could be offered as a formal theorem with an appropriate proof. Full reversibility of the buckling process implies that whatever kinetic energy is developed during the buckling phenomenon, it is fully converted to potential energy. Hence, the entire work performed by the outside load is converted into the increase in potential energy, which may be identified with the increase in total strain energy. But this is exactly the statement of Eq. 2.20.

The implications of this observation are quite far-reaching. One consequence is the following seemingly paradoxical observation. If one ignores kinetic energy effects, one can not apply a constant load $P > P_{cr}$ to the buckling process. This is of course a nonsensical result, which merely shows that purely static phenomena connot exist. For the sake of completeness this result is offered as a theorem.

Theorem 4: Buckling caused by a constant force P cannot be a fully reversible process.

Proof: Suppose to the contrary that the total work performed by the force P is entirely converted into strain energy in the deflected position. Hence, $P\delta = \Delta SE$. Ignoring terms $\dfrac{\tilde{\theta}^6}{720}$ and smaller, one arrives at the equality

$$P^2 \int_0^\ell \frac{11/12\ \tilde{\theta}^4}{EA(s)}\ ds + P \int_0^\ell \left(\tilde{\theta}^2 - \frac{\tilde{\theta}^4}{12}\right) ds - \int_0^\ell EI\left(\frac{d\tilde{\theta}}{ds}\right)^2 ds = 0$$

The positive root is given by the right hand side of the in-equality of Eq. 2.7, which is written as

$$P = A^{-1}\left\{-B + (B^2 + 4AC)^{1/2}\right\} = F(\tilde{\theta})$$

On the other hand, the shape $\tilde{\theta}$ is such that the potential energy of the column assumes a local minimum in H_0. Using the "averaged" minimal property of Eq. 2.18, one obtains

$$A^{-1}\left\{-B + (B + AC)^{1/2}\right\} = F(\theta) = \frac{1}{2}(B^2 + AC)^{-1/2}C$$

This can be specifically solved for B in terms of AC. After some manipulation, one can derive the relation $B^2 = (1/4)AC$. But this is impossible, since B contains no design parameters $A(x)$ or $I(x)$, while AC is design dependent. Since so far the design is arbitrary and the discussion is valid for any admissible designs, a contradiction is obtained, which completes the proof of Theorem 4.

This result is hardly surprising if one considers the implications of perfect elastic behavior and no dissipation. The inequality $P_{Euler} > P_{cr}$ indicates that some kinetic energy must be created and in fact one can easily estimate a lower bound on the maximum value of the kinetic energy. With no dissipation, the interchange between kinetic and potential energy will take place in a manner resembling perfectly elastic vibration of a mass-spring system. However the static model has no ability of reflecting such purely dynamic phenomena. As before, the fault lies not with the conclusions reached by a mathematical investigation of a given model, but with the model itself. However, the equilibrium formulas derived above could serve as a starting point for a dynamic investigation of the buckling process.

3. STATIC OPTIMALITY CRITERIA

It is desired to choose the cross-sectional area $A(x)$ to maximize the buckling load, subject to the constraint that the volume of material in the column is fixed. In normalized form, this is

$$\int_0^{\ell} Ads = 1 \tag{3.1}$$

This problem is equivalent to maximizing $F(\theta,A)$ with respect to A. A necessary condition is $\frac{dF(\theta,A)}{dA} = 0$, for a simple roof, subject to the condition of Eq. 3.1. This may be restated by requiring

$$\frac{d}{dA}\left[F(\theta,A) + \lambda \int_0^{\ell} Ads\right] = 0 ,$$

where λ is a Lagrangian multiplier. Note that

$$\frac{d}{dA} F(\theta,A) = \frac{\partial F}{\partial A} + \frac{\partial F}{\partial \theta} \frac{\partial \theta}{\partial A}$$

and since the condition for buckling is $\frac{\partial F}{\partial \theta} = 0$' the necessary condition for optimality becomes simply

$$\frac{\partial F(\theta,A)}{\partial A} = -\lambda = \text{constant} \tag{3.2}$$

Let A, B, and C, have the same meaning as in Eqs. 2.8 to 2.10. Then, Eq. 3.2 becomes

$$- \frac{11}{24} \theta^4 \left[A(B^2 + AC)^{-1/2} + \left\{-B + (B^2 + AC)^{1/2}\right\}\right]$$

$$+ AE^2A^2(\theta')^2 \frac{dI}{dA} = \text{constant} \tag{3.3}$$

If $I = kA^2$, which is common in practice, $\frac{dI}{dA} = 2kA$ and Eq. 3.3 assumes the form

$$(\theta)^2A^3 = C_1 \theta^4 \left[(B^2 + AC)^{-1/2} + F(\theta,A)\right] + C_2A^{-1} \tag{3.4}$$

where C_1 and C_2 are known constants.

To fully realize the implications of Eq. 3.4, compare it with the more classical linear buckling model that utilizes the Rayleigh quotient as given by Eq. 1.3. The problem of optimizing the Rayleigh quotient, subject to a constant volume constraint, is treated by looking for stationary behavior of the functional

$$\Phi(A,y) = \int_0^\ell EI(y'')^2 \, dx + \lambda_1 \int_0^\ell A(x) \, dx + \lambda_2 \int_0^\ell (y')^2 \, dx$$

the second multiplier involving a normalization condition

$$\int_0^\ell (y')^2 \, dx = 1$$

Suppose again that $I(x) = kA^2(x)$. Then, as before,

$$\frac{dI}{dy} = \frac{\partial \Phi}{\partial A} = (2kEA(y'')^2 + \gamma_1 = 0$$

or

$$\tilde{A}(y'')^2 = \text{constant} \tag{3.5}$$

is the necessary condition for optimality of $\tilde{A}(x)$.

An immediate consequence of Eq. 3.5 is that $y''(x)$ cannot be equal to zero. On the other hand $y''(x)$ must change signs on $[0,\ell]$. Otherwise $y'(x)$ is a monotone function. Hence, conditions $y'(0) = y'(\ell)$, or $y(0) = y'(0) = 0$, $y(\ell) = 0$ are impossible to satisfy. Hence y'' must be discontinuous. However, at a point x of discontinuity of y'', Eq. 3.5 may be multiplied by EkA^3 to obtain

$$EkA^3(\tilde{x}-) = EkA^4(\tilde{x}-)(Y''(\tilde{x}-))^2 = EkP^2y(\tilde{x}-) = EkP^2y(\tilde{x})$$

$$EkA^3(\tilde{x}+) = EkA^4(\tilde{x}+)(y''(\tilde{x}+))^2 = EkP^2y(\tilde{x}+) = EkP^2y(\tilde{x}),$$

so

$$Ek\left[A^3(\tilde{x}+) - A^3(\tilde{x}-)\right] = M(\tilde{x}+) - M(\tilde{x}-)$$

where $M(x)$ is bending moment. Since no couples act on the beam, the moment must be continuous, hence

$$A(\tilde{x}+) = A(\tilde{x}-)$$

However, since $y''(\tilde{x})$ is discontinuous and $A(\tilde{x})(y''(\tilde{x}))^2$ is continuous at $\tilde{x}$, it is required that $A(\tilde{x}) = 0$.

From Eq. 3.5, one obtains

$$y''(x) - IC/A(\tilde{x}) \tag{3.6}$$

where C is a constant. Thus, $|y''(x)|$ must approach ∞ as $x \to \tilde{x}$. If the limit from each side has the same algebraic sign, then $\frac{dM}{dx}$ has a different sign as $x \to \tilde{x}$ from the left and the right. However, this implies that the shear in the beam is discontinuous at $\tilde{x}$. But this is impossible, since there are no lateral loads acting on the beam at $x = \tilde{x}$. Thus y'' must approach limits of different sign from the left and right at $\tilde{x}$.

Therefore, in a design problem in which the boundary conditions imply the existence of inflection points (i.e., y'' changes its sign at some interior points of $[0,\ell]$), if no constraints are placed on the magnitude of stress or on a minimal size of a cross-section, optimality of design implies the existence of singular points. This fact may become disguised by numerical procedures used in obtaining the optimal shape. The choice of finite element or finite difference procedure may introduce an averaging effect into the numerical computation and instead of a clear singularity, only a necking effect may be displayed.

Consider now applicability of the results to optimum columns. The term

$$\int_0^\ell \frac{f(\theta,P)}{EA(s)} \, ds \; ,$$

which appears in our equations is of importance only if $[EA(s)]^{-1}$ is sufficiently large over a part of the length of the column. It becomes crucial, invalidating the classical theory, if the integral does not exist. Hence, we attempt to classify the results of optimization theory for buckled columns in terms of convergence of this integral.

In the clamped-clamped case it was shown by Olhoff and Rasmussen in [5] that the "simple-minded" miniminzation of the first eigenvalue leads to incorrect results, because of coalescence of the first and second eigenvalues. Raising further the value of the first eigenvalue by "improving" the design becomes counterproductive, because of the crossover effect.

From these arguments follows a conclusion that the truly optimal design can not have a zero cross-sectional area. Hence, in the clamped-clamped case a careful analysis will reveal that the linear theory is at least not completely invalidated.

This is not entirely the case in the optimal design of a clamped-free column (Figure 2.1(b) Here the optimal design leads to a zero cross-sectional area at the end of the column. The problem of convergence of the integral

$$\int_0^\ell \frac{f(P,\theta)}{AE} \, ds$$

becomes crucial. Analysis of the optimum shapes derived in papers of Olhoff reveal that this integral converges if $I = \alpha A^2$, but diverges if $I = \beta A$, which is the case in a sandwich column design. For such designs, the "optimal" designs obtained by linear theory must be disregarded and the non-linear term (including A^{-1}) introduced in this paper must be included in the analysis to "keep the theory honest".

In the theory of vibrating membranes, plates and beams some complications of the static theory of columns do not arise. V. I. Arnol'd offered a simple proof that in linear systems of this type, multiple eigenvalues cannot occur. The idealized shapes derived by Olhoff reveal zero cross-sectional areas. To duplicate our non-linear theory of columns, we can introduce a very small axial compressive load and study the effects of the corresponding term

$$\int_0^\ell \frac{f(P,\theta)}{EA} \, ds \ .$$

The shape of the "optimal" cross-sectional area reveals that this integral diverges for every value of the load P. Again it is this non-linear term which prevents the occurence of singularities in cross-sectional area design.

One could also ask if other neglected terms, which are usually "very small" will significantly affect the optimal design. For example, should one include the effect of shear in the design of buckled columns, or vibrating beams? The answer appears to be: yes. In the design of a column that is hinged at one end and clamped at the other, if one assumes that $I = \alpha A^2$, then the non-linear theory seems to contribute very little, if only the compressive terms

$$\int_0^\ell \frac{f(P,\theta)}{EA} \, ds$$

are included.

Using the classical theory, the optimal design appears to exhibit zero cross sections. Inclusion of our nonlinear compressive term seems to affect this design very little. However this design has absolutely no resistance to shear at the point of the interior hinge. Here, consideration of the usually neglected

420

and usually very small shear term may salvage the theory by preventing the occurence of singularities.

APPENDIX 1: FRÈCHET AND GATEAUX DIFFERENTIATION AND FORMAL RULES OF FRÈCHET CALCULUS

Let B_1 and B_2 be any normed spaces. A mapping $f : B_1 \to B_2$ is said to be Gateaux differentiable at $\bar{x} \in B_1$ if for any $h \in B_1$ and for any constant t such that $\bar{x} + th$ is in the domain of f, there exists a linear map $f'(x)$ defined in some neighborhood of $\bar{x}$, $f'(x) : B_1 \to B_2$ such that

$$f(\bar{x} + th) - f(\bar{x}) = tf'(\bar{x})h + r(\bar{x},h,t)$$

where $\lim_{t \to 0} \frac{||r||}{t} = 0$ for all $h \in B_1$. In the Hilbert space setting, continuity of f' at $\bar{x}$ implies validity of the Riesz representation theorem for the specific case when $B_2 = R$. Hence in the case $f : H \to R$, $f'(x)h$ is an inner product, and $f'(x)h = \langle z,h \rangle$. The operator z is called the Frechet derivative of f. Higher order derivatives are defined analogously. If f and y are vectors in a Hilbert space, then $\frac{\partial f}{\partial y}$ is a tensor product, and $\langle \frac{\partial x}{\partial y} x, x \rangle$ is a scalar. Thus $\frac{\partial f}{\partial y}$ can be regarded as an operator from H to H. More specifically, if $y(x)$ and x are n-dimensional vectors, then $\frac{\partial y}{\partial x}$ is an nxn matrix, which is in fact the Jacobian matrix.

The following rules of calculus are easily checked to be correct:

$$\frac{\partial}{\partial x} (\Phi_1 + \Phi_2) = \frac{\partial}{\partial x} \Phi_1 + \frac{\partial}{\partial x} \Phi_2$$

$$\frac{\partial}{\partial x} (c\Phi) = c \frac{\partial}{\partial x} \Phi$$

$$\frac{\partial}{\partial x} (uv) = u \frac{\partial v}{\partial u} + \frac{\partial u}{\partial x} v$$

$$\frac{\partial}{\partial x} \left(\frac{u}{v} \right) = v^{-2} \left(v \frac{\partial u}{\partial x} - u \frac{\partial v}{\partial x} \right)$$

$$\frac{\partial}{\partial x} \Phi(\psi(x)) = \frac{\partial \Phi}{\partial \psi} \frac{\partial \psi}{\partial x}$$

where Φ is a functional $H_2 \rightarrow R$, and ψ is a map : $H_1 \rightarrow H_2$, which itself is an element of a suitable Hilbert space. In particular, if A is a map A : $H \rightarrow H$ that has an adjoint A^* : $H \rightarrow H$, then

$$\frac{\partial}{\partial x} < Ax,x > = Ax + A^*x \ .$$

For a detailed exposition on the theory of Frèchet differentiation, see Refs. 17 and 18. For applications to continuum mechanics, see Ref. 19.

APPENDIX 2: PROBLEM FORMULATION IN y'

Because of the traditional development of the theory and for purposes of comparison with existing formulas, parallel results are derived in the x-y coordinate representation. The equation $\theta = (\arctan (y'))$ is approximated by $\theta = y' - \frac{1}{3} (y')^3 + \frac{1}{5} (y')^5 - \dots$ ($|y'| < 1$). Using the Taylor formula for cos and arctan, one has

$$\cos\theta = \cos(\arctan (y')) = 1 - \frac{(y')^2}{2} - \frac{(y')^4}{8} + \dots 0(y')^6$$

$$\cos^2\theta = \cos^2 (\arctan (y')) = 1 - (y')^2 \dots + 0(y')^6$$

Hence,

$$\delta_C \approx P \int_0^{\ell} \frac{-(y')^2/2 + (y')^4/8}{EA(x)} \, dx \qquad A(2.1)$$

and during the buckling process, the change in the distance along the x-axis between the ends of the column is given by

$$\delta_G + \delta_C = \int_0^{\ell} \left(1 - \frac{P}{EA(x)}\right) \left(\frac{(y')^2}{2} - \frac{(y')^4}{8}\right) dx$$

where terms of order $(y')^6$ and higher are ignored.

The work performed during the buckling process by the force P is thus

$$P \int_0^\ell \left[\left(1 - \frac{P}{EA(x)} \right) \left(\frac{(y')^2}{2} - \frac{(y')^4}{8} \right) \right] dx$$

The total work performed by the force P is therefore

$$\frac{P^2}{2E} \int_0^\ell \frac{dx}{A(x)} + P \int_0^\ell \left[\left(1 - \frac{P}{EA(x)} \right) \left(\frac{(y')^2}{2} - \frac{(y')^4}{8} \right) \right] dx \quad \text{(A2.2)}$$

The total strain energy is given by

$$V = \frac{P^2}{2E} \int_0^\ell \frac{\cos^2\theta(x)}{A(x)} \, dx + \frac{E}{2} \int_0^\ell \frac{I(x)(y'')^2}{[1 + (y')^2]^{3/2}} \, dx$$

$$= \frac{P^2}{2} \int_0^\ell \frac{(1 - (y')^2}{EA(x)} \, dx + \frac{1}{2} \int_0^\ell \frac{EI(x)(y'')^2}{[1 + (y')^2]^{3/2}} \quad \text{(A2.3)}$$

The total potential energy is

$$\text{TPE} = V - P\delta$$

and V and Pδ are given by Eqs. A2.3 and A2.2, respectively. The principle of minimum total potential energy requires that the variation of the total potential energy be zero; i.e.,

$$\delta\text{TPE} = 0 = \int_0^\ell \left\{ - \frac{y' \delta y' P^2}{EA} + \frac{EIy'' \delta y''}{[1 + (y')^2]^{3/2}} - \frac{3}{2} \frac{EI(y'')^2 y' \delta y'}{[1 + (y')^2]^{5/2}} \right.$$

$$\left. - P \left(1 - \frac{P}{EA} \right) \left(y' \delta y' - \frac{1}{2} (y')^3 \delta y' \right) \right\} dx$$

Integrating by parts and using boundary conditions, this reduces to

$$0 = \int_0^\ell \left\{ P^2 \frac{d}{dx} \left(\frac{y'}{EA} \right) + \frac{d^2}{dx^2} \left(\frac{EIy''}{[1 + (y')^2]^{3/2}} \right) + \frac{3}{2} \frac{d}{dx} \right.$$

$$\left. \left(\frac{EI(y'')^2 y'}{[1 + (y')^2]^{5/2}} \right) + P \frac{d}{dx} \left(1 - \frac{P}{EA} \right) \left(y' - \frac{(y')^3}{2} \right) \right\} \delta y \, dx$$

$$\text{(A2.4)}$$

for all δy satisfying boundary conditions; i.e., all $\delta y \in H_0^2(0,\ell)$ $\cap L_4$. Since $H_0^2(0,\ell)$ is dense in $L_2(0,\ell)$, one has

$$\frac{d^2}{dx^2}\left(\frac{EI(y'')}{[1+(y')^2]^{3/2}}\right) + \frac{d}{dx}\left[\frac{P^2 y'}{EA} + \frac{3EI(y'')^2 y'}{2[1+(y')^2]^{5/2}}\right.$$
$$\left. + P\left(1 - \frac{P}{EA}\right)\left(y' - \frac{(y')^3}{2}\right)\right] = 0$$

or

$$\frac{d^2}{dx^2}\left(\frac{EIy''}{[1+(y')^2]^{3/2}}\right) + \frac{d}{dx}\frac{3EI(y'')^2 y'}{2[1+(y')^2]^{5/2}} + Py'$$
$$- P\left(1 - \frac{P}{EA}\right)\frac{(y')^3}{2} = 0 \tag{A2.5}$$

Multiplying by y and integrating by parts, one has

$$\int_0^\ell \left\{\frac{EI(y'')^2}{[1+(y')^2]^{3/2}} - \frac{3EI(y'')^2(y')^2}{2[1+(y')^2]^{5/2}} - P(y')^2 \right.$$
$$\left. + P\left(1 - \frac{P}{EA}\right)\frac{(y')^4}{2}\right\} dx = 0$$

or

$$KP^2 + BP - S = 0 \tag{A2.6}$$

where

$$S = \int_0^\ell \frac{EI(y'')^2}{[1+(y')^2]^{3/2}}\left[\frac{2-(y')^2}{2(1+y')^2}\right]dx \tag{A2.7a}$$

$$B = \int_0^\ell \left[(y')^2 - \frac{(y')^4}{2}\right]dx \tag{A2.7b}$$

$$K = \int_0^\ell \frac{(y')^4}{2EA}dx \tag{A2.7c}$$

424

Thus, the critical load is

$$P_{cr} = \frac{1}{2} K^{-1} \left\{ -B + \sqrt{B^2 - 4KS} \right\}$$ (A2.8)

REFERENCES

1. Keller, J.B., "The Shape of the Strongest Column," Arch. Rat. Mech. Anal. 5, 1960, pp. 275-285.
2. Tadjbakhsh I. and Keller, J.B., "Strongest Column and Isoperimetric Inequalities for Eigenvalues," J. Appl. Mech. 9, 1962, pp. 159-164.
3. Keller, J.B. and Niordson, F.I., "The Tallest Column," J. Math. Mech. 16, 1966, pp. 433-446.
4. Olhoff, N., Optimal Design Against Structural Vibration and Instability, Ph.D. Thesis, Technical University of Denmark, Lyngby, Denmark, 1978.
5. Olhoff, N. and Rasmussen, S.H., "On Single and Bimodal Optimum Buckling Loads of Clamped Columns," Int. J. Solids Structures, 13, 1977, pp. 605-614.
6. Olhoff, N. and Taylor, J.E., "Designing Continuous Columns for Minimal Total Cost of Material and Interior Supports," J. Struct. Mech. 6, #4, 1978, pp. 367-382.
7. Freudenthal, J.C., "Constrained Optimal Design of Columns Against Buckling," J. Struct. Mech., 1, 1972, pp. 79-89.
8. Masur, E.F., "Optimality in the Presence of Discreteness and Discontinuity," Optimization in Structural Design, (Ed. A. Sawczuk and Z. Mróz), Springer-Verlag, Berlin, 1975, pp. 441-453.
9. Olhoff, N., "On Singularities, Eigenfrequencies and Formation of Stiffeners in the Optimum Design of Plates," Optimization in Structural Design, (Ed. A. Sawczuk and Z. Mróz), Springer-Verlag, Berlin, 1975, pp. 82-99.
10. Masur, E.F., A written discussion to reference 9 above, pp. 97-103.
11. Borri, M. and Mantegazza, P., "Efficient Solution of Quadratic Eigenvalue Problems Arising in Dynamic Analysis of Structures," Comput. Methods in Applied Mech. and Eng., 12, 1977, pp. 19-31.
12. Fawzy, I. and Bishop, R.E.D., "On the Dynamics of Nonconservative Systems," Proc. Roy. Soc., London, A, 352, 1976, pp. 25-40.
13. Hermann, G., Dynamic Stability of Structures, Pergamon Press, New York, 1967.
14. Wilkinson, J.D., The Algebraic Eigenvalue Problem, Oxford Press, Oxford, 1965.
15. Hsu, C.S., "On the Parametric Excitation of a Dynamic System Having Multiple Degrees of Freedom," J. Appl. Mech., 30, 1963, pp. 367-372.

16. Rebiere, J.P. and Sahraoui, S., "Parametric Resonance by Combined Frequencies of Cantilever Bars under Periodic Axial Load by the Elastic Joint Method," Mech. Res. Comm., 5, #1, 1978, pp. 39-44.

17. Vainberg, M.M., Variational Methods for the Investigation of Nonlinear Operators, Holden Day, San Francisco, 1963.

18. Nashed, M.Z., "Differentiability and Related Properties of Nonlinear Operators: Some Aspects of the Role of Differentials in Nonlinear Functional Analysis," Nonlinear Functional Analysis and Applications (Ed. L.B. Rall), Academic Press, New York, 1971, pp. 103-109.

19. Komkov, V., "On a Variational Formulation of Problems in the Classical Mechanics of Solids," Int. J. Eng. Sci., 6, 1968, pp. 695-720.

20. Olhoff, N. and Niordson, F.I., "Some Problems Concerning Singularities of Optimal Beams and Columns, ZAMM, 59, 1979, T16-T26.

Part 2

NUMERICAL OPTIMIZATION METHODS

OPTIMAL REMODELING THEORY AND APPLICATIONS*

John E. Taylor

The University of Michigan, Ann Arbor, Michigan
48109, U.S.A.

ABSTRACT

A theory of optimal remodeling is presented, in which a
basic structure is given, and the global extent of modification
is prescribed. Optimality criteria for solution of the general
problem are derived and applied to several examples. Successive
remodeling is shown to converge to the solution of an optimal
design problem without remodel constraints.

1. INTRODUCTION

In contrast to the objective in most treatments of optimal
structural design, the focus in the optimal remodeling problem is
on the best way to alter a given structure. A goal might be
simply to predict the optimum modification to an arbitrary
specified original structure. Various possible forms of altera-
tion correspond to optimal reinforcement, least-damaging reduction
in weight, or simply efficient redistribution of material.

The development for optimal remodeling applied to problems
in which the mode of structural response is governed by a variational principal. A variational formulation for the design
problem is presented here in a form that differs only slightly
from the material of Ref. 1. While the analytics of this

---

*The work reported in this paper received partial support from
the National Science Foundation.

formulation are quite similar to what is expected in more usual optimal structural design problem statements, the results lend themselves to a rather broader interpretation. For example, one might regard the conventional structural optimization problem to be imbedded in the optimal remodeling problem. A solution to the former results in the case that unlimited resource is applied in remodeling for optimum reinforcement.

A distinction between conventional and remodel optimal design methods might be made as well on the basis that a solution to the latter depends on the form of the prescribed original structure. Also, the remodel formulation provides a means for the prediction of evolutionary (sequential) changes in design, as a possibility. This idea, as well as other forms of application, are considered in detail in the development and the set of examples to follow.

2. VARIATIONAL FORMULATION

The formulation is demonstrated for linearly elastic continuum structures. In the problem of design for minimum compliance, which is used here to exemplify the development, structural response is associated with the extremum of the potential energy

$$\pi(D;\underline{u}) = \int_\beta [S(D) \, \eta(\underline{u}) - \underline{\rho} \, \underline{u}] \, dx \qquad (2.1)$$

Here $D(x)$ represents design, stiffness $S(D)$ is taken to depend with power $n > 1$ on the design, and u represents the response (state) vector. The function $\eta(u)$ identifies strain energy per unit stiffness per unit extent of the structure. In the case $n = 1$, for example, η measures specific strain energy, i.e., energy per unit volume of material. In general the quantities in Eq. 2.1 are to be interpreted in a form that is appropriate to the particular type of one or two dimensional structure being considered.

An initial design, say $D_0(x)$, is taken to be specified. Modifications $D_+(x)$ and $D_-(x)$ of this original structure represent, respectively, the addition or the removal of material. Thus the remodeled structure D is given by

$$D(x) = D_0(x) + D_+(x) - D_-(x)$$

The objective in the present formulation is to predict modifications D_+ and D_-, within the (isoperimetric) constraints

$$\int_{\beta} D_+ \, dx = R_+ \tag{2.2}$$

$$\int_{\beta} (D_+ - kD_-) dx = R_n \tag{2.3}$$

The value of R_+ indicates the total magnitude of resource for reinforcement, while $R_n \gtrless 0$ measures the amount out of this total that is added to or removed from the structure. The coefficient k measures the percentage of removed material that is available for reinforcement.

The optimal remodel design problem is governed by the functional

$$\Phi(D_+; D_-; \underset{\sim}{u}; \ldots) = \int_{\beta} [S(D)n(\underset{\sim}{u}) - \underset{\sim}{p} \underset{\sim}{u}] dx$$

$$- \Gamma[\int_{\beta} D_+ \, dx - R_+]$$

$$- \Lambda[\int_{\beta} (D_+ - kD_-) \, dx - R_n]$$

$$+ \int_{\beta} [\gamma D_+ + \lambda D_-] \, dx \tag{2.4}$$

comprised of potential energy Π, augmented by constraints of Eqs. 2.2 and 2.3. The role of the last term in Eq. 2.4 is simply to assure that $D_+ \geq 0$ and $D_- \geq 0$. The extremum of Φ is a saddle point, i.e.

$$\text{Arg} \left[\underset{D_+, D_-}{\max} \left(\underset{\underset{\sim}{u}}{\min} \Phi \right) \right] = (D_+^*; D_-^*; u^*) \tag{2.5}$$

identifies the optimum remodel design D_+^*; D_-^* and the associated response $\underset{\sim}{u}^*$ (for prescribed values R_+, R_-, and k).

The necessary conditions for this problem are

$$\delta_{\underset{\sim}{u}} \Phi = 0 \iff \delta_{\underset{\sim}{u}} \Pi = 0 \tag{2.6}$$

$$\frac{\partial S}{\partial D_+} \eta = \Gamma + \Lambda - \gamma \qquad (2.7)$$

$$\frac{\partial S}{\partial D_-} \eta = - k\Lambda - \lambda \qquad (2.8)$$

$$\gamma D_+ = 0 \qquad (2.9)$$

$$\gamma D_- = 0 \qquad (2.10)$$

$$D_+ \geq 0, \quad D_- \geq 0 \qquad (2.11)$$

$$\gamma \geq 0, \quad \lambda \geq 0 \qquad (2.12)$$

From (switching) Eqs. 2.9 and 2.10, a design modification $D_+ \neq 0$ implies $\gamma = 0$, while a modification $D_- \neq 0$ implies $\lambda = 0$. For clarity, the two forms of modification are taken to be mutually exclusive and the entire structure is covered by intervals β_+, β_-, and β_0 corresponding to positive, negative, or zero modification.

For convenience in writing, the symbol B is introduced for the quantity $(\partial S/\partial D)\eta$. From Eqs. 2.7 to 2.12,

$$\left.\begin{array}{ll} B_+ \triangleq \dfrac{\partial S}{\partial D_+} \eta = \Gamma + \Lambda, & x \in \beta_+ \\[2mm] B_- \triangleq - \dfrac{\partial S}{\partial D_-} \eta = k\Lambda, & x \in \beta_- \\[2mm] k\Lambda \leq (B_0 \triangleq \dfrac{\partial S}{\partial D_0} \eta) \leq \Gamma + \Lambda, & x \in \beta_c = \beta - (\beta_+ + \beta_-) \end{array}\right\} \quad (2.13)$$

The measure B of unit-energy-weighted stiffness-gradient is identified as design sensitivity. Equations 2.13 indicates that, for the optimally modified structure, the design sensitivity takes on constant, maximum and minimum values, respectively, over intervals where material has been added or removed. The value of k is restricted to be non-negative. Also, from the third of Eqs. 2.13, clearly $k < (1 + \Gamma/\Lambda)$.

In the special case $n = 1$ (e.g. trusses, or idealized sandwich beams and plates),

$$\frac{\partial S}{\partial D_+} = -\frac{\partial S}{\partial D_-} = \frac{\partial S}{\partial D_0} \overset{\Delta}{=} c$$

where c is a constant. Thus, the sensitivity is proportional to the specific energy η and Eqs. 2.13 reduce to

$$\left.\begin{array}{ll} B = c\eta = \Gamma + \Lambda \quad, & x \in \beta_+ \\[2mm] c\eta = k\Lambda \quad, & x \in \beta_- \\[2mm] k\Lambda \leq c\eta \leq \Gamma + \Lambda \quad, & x \in \beta_c \end{array}\right\} \qquad (2.14)$$

A schematic representation of these results is given in Figure 2.1. The dashed curve in Fig. 2.1 depicts the specific energy distribution in the structure, before modification.

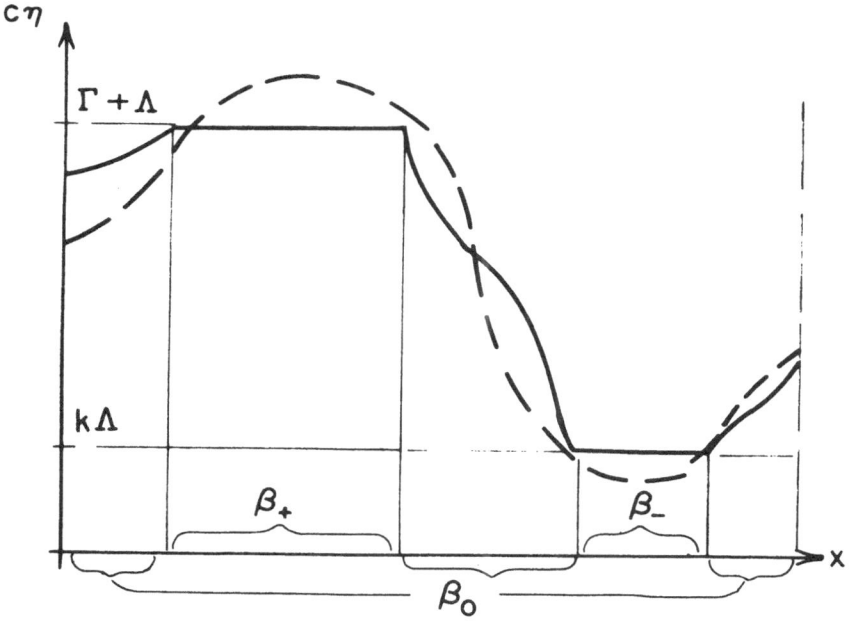

Domain of the Structure

Figure 2.1   A Schematic Representation of Eq. 2.14; Given R_+, R_n, and k

There exists a pair of limit values, say R_n^*; R_+^*, for the measures of resource for remodeling such that the solution corresponding to $R_n = R_n^*$; $R_+ = R_+^*$ is precisely the optimum design result for the conventional statement of the structural optimization problem. Consider a set of solutions in which each solution is associated with a pair of values of resource from an ordered set of such pairs. The corresponding (ordered) solutions represent an evolution from the original design toward the limit optimum design. Note that the (local) measures of change in design need not vary monotonically with the evolution.

The several distinct types of design modification problems accommodated via the formulation of this section are identified in Table 2.1

TABLE 2.1   TYPES OF DESIGN MODIFICATION

R_+	R_n	k	Type
$\emptyset$	$\emptyset$	0	Reinforcement only
$\emptyset$	0	$\emptyset$	Redistribution
0	$\emptyset$	$\emptyset$	Weight reduction only
$\emptyset$	$\emptyset$	$\emptyset$	Combined redistribution and net change

The non-monotonic change in design may occur in either the second or the fourth of these types. Examples are furnished in the next section.

It is possible to show [1] that Eqs. 2.6 to 2.12 are sufficient, as well as necessary, for the optimal remodel solution. In other words, if a solution to the system exists, it is unique. Accordingly, one might expect to find that computational procedures based on the variational formulation carry a reasonable assurance of convergence.

3.   EXAMPLE PROBLEMS

3.1   Minimum Compliance of an Axially Loaded Bar

The design for minimum compliance of an axially loaded bar is presented as a very simple example of reinforcement-only modification for a continuum structure [1]. The initial shape is prescribed as (see Fig. 3.1(a))

$$D_0 = A[2 + \cos(5\pi x/2L)]$$

The equilibrium equation of Eq. 2.6 and optimality equation of Eq. 2.7 are

$$(ED\, u')' + p = 0$$

$$\eta = E\, u'^2/2 = \Gamma - \gamma$$

The elastic modulus E is taken to be constant and D has the role of the design variable. Numerical results for an end-loaded bar (p = 2EA) are presented for each of two stages of remodel. The design modifications are shown in Figs. 3.1(b) and 3.1(c) and the original and modified specific energies are given in Fig. 3.2.

3.2 Minimum Compliance of an Elastic Plate

In the same design problem, but for an elastic plate, the optimality equation of Eq. 2.7 takes the form

$$B = ch^2[(w,_{xx} + w,_{yy})^2 + 2(1 - \nu)(w,_{xy}^2 - w,_{xx}\, w,_{yy}0] = \Gamma - \gamma$$

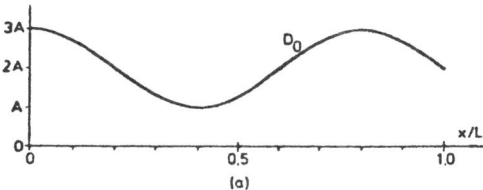

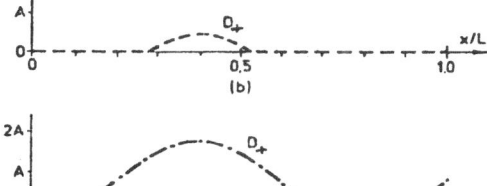

Figure 3.1   Original Design and Design Modifications for an Axially Loaded Bar.

436

Plate thickness h(x,y) is the design variable.

Results for a uniformly loaded, clamped, square plate at three stages of modification away from a flat initial design are given in Figs. 3.3, 3.4, and 3.5. These remodel solutions were obtained by K.-T. Cheng at the DTH Denmark. They are reported in part in Ref. 2. The results show graphically an evolution of a pattern of stiffener-like buildup on the plate.

3.3 Minimum Compliance of a Truss

Redistribution remodel design for minimum compliance of a truss is treated as a third example. The potential energy for the discrete structure is given by

$$\Pi = \sum_{\substack{\text{members} \\ m}} (A_{0m} + A_{+m} - A_{-m}) \, \ell_m \, E \, \epsilon_m^2 \, /2 - \sum_\alpha P_\alpha u_\alpha$$

in terms of design A_m for the mth member, strain ϵ_m, member length ℓ_m, loads P_α, and nodal displacement components u_α.

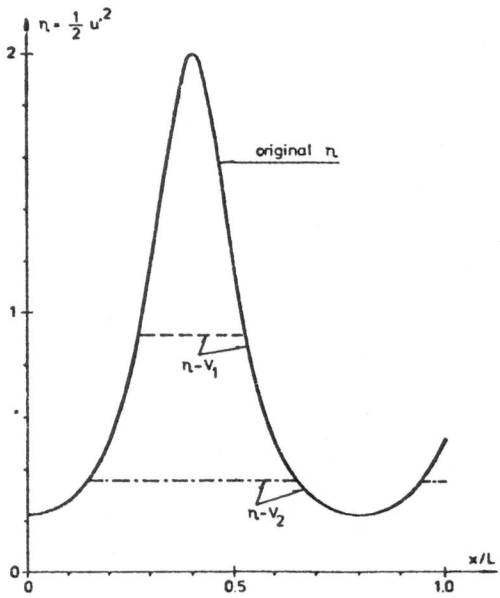

Figure 3.2 Specific Energies for the Axially Loaded Bar

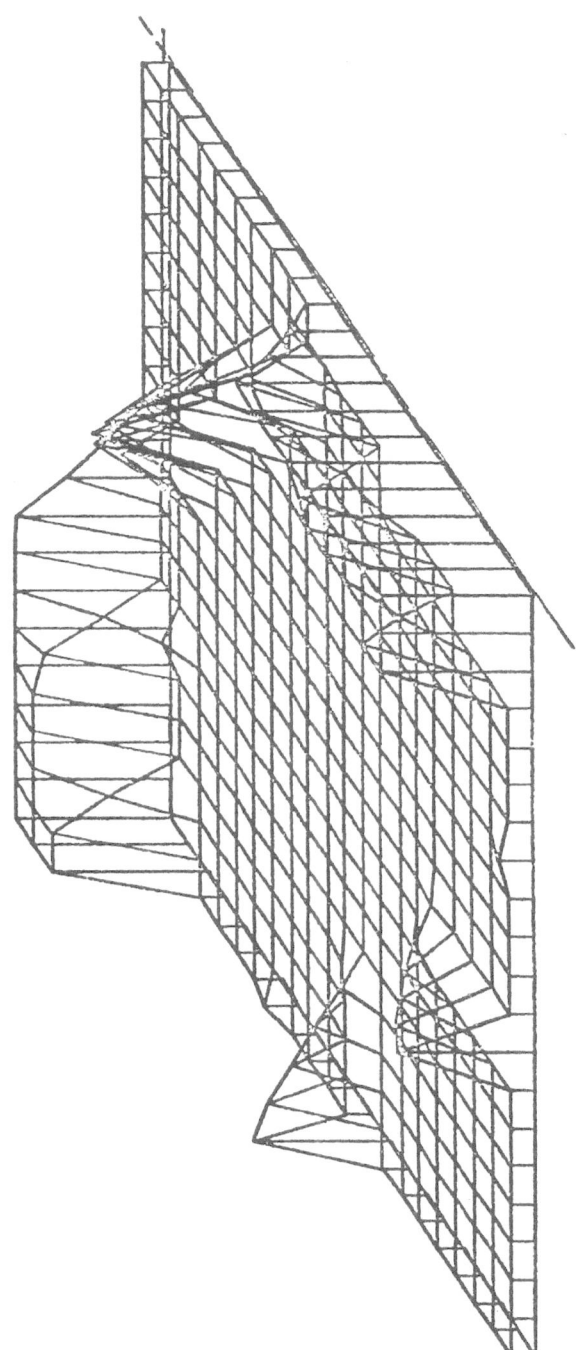

Figure 3.3   Uniformly Loaded, Clamped Square Plate; First Stage Remodel

438

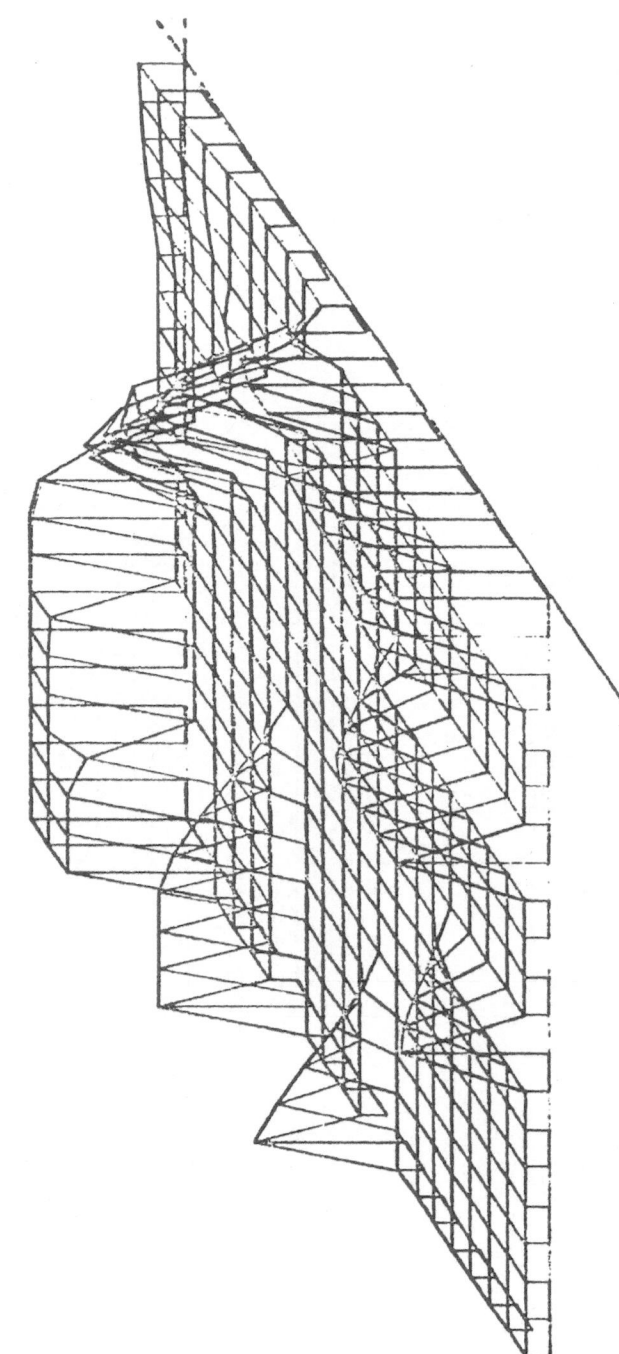

Figure 3.4   Uniformly Loaded, Clamped Square Plate; Second Stage Remodel

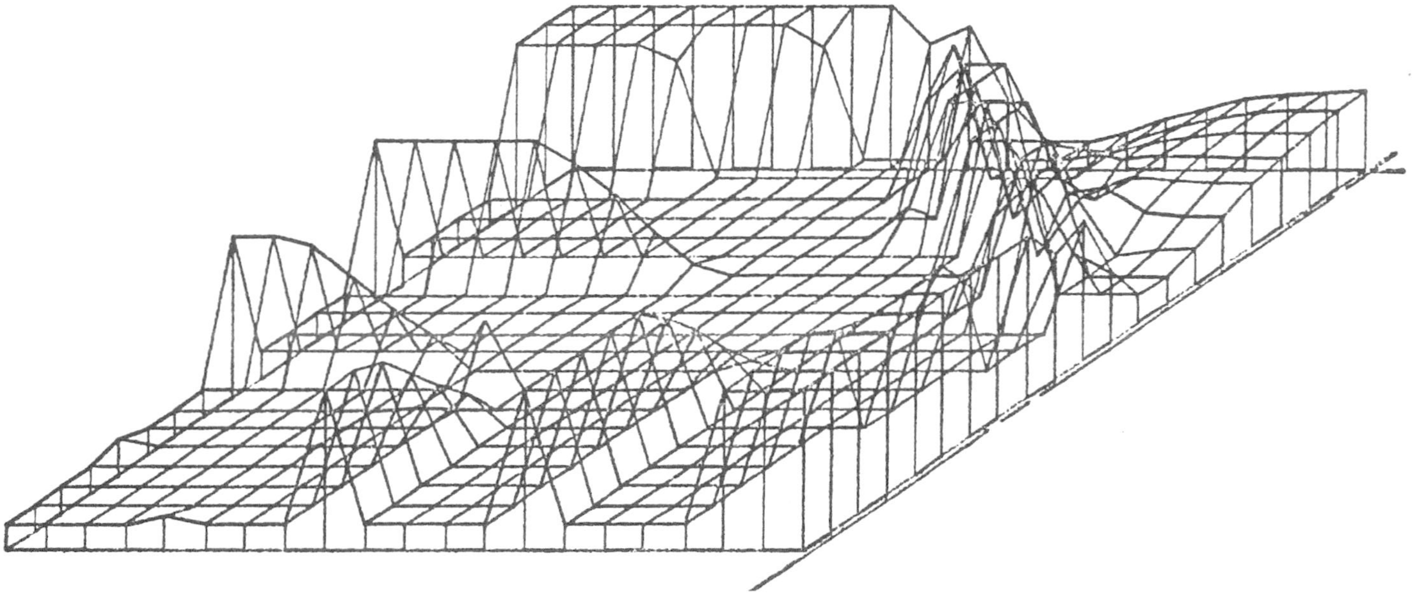

439

Figure 3.5 Uniformly Loaded, Clamped Square Plate; Third Stage Remodel

The governing function for the design problem, the counterpart to Eq. 2.4, is

$$\Phi = \Pi - \Gamma \left(\sum_m A_{+m} \ell_m - R_+ \right)$$

$$- \Lambda \left(\sum_m (A_{+m} - A_{-m}) \ell_m - R_n \right)$$

$$+ \sum_m (\gamma_m A_{+m} + \lambda_m A_{-m})$$

A solution procedure is based on a set of necessary conditions similar to Eqs. 2.6 to 2.12.

In this example problem, the symmetrically loaded and geometrically symmetric truss is given an unsymmetric initial design. The layout of the truss is given in Fig. 3.6, while member areas for the initial design and for several stages of remodel are shown in Fig. 3.7. The curves of Fig. 3.8 show the redistribution of specific energy and the reduction in compliance over successive intervals of design change. The design represented by the last design interval shown is the limit optimum design. According to the results in Fig. 3.7, member areas A_1 and A_5 are first increased and subsequently reduced in the evolution process.

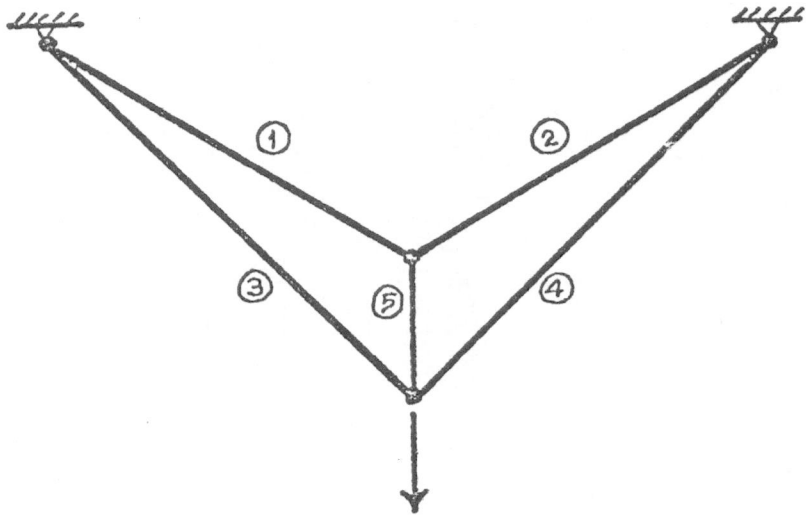

Figure 3.6   Truss Layout and Member Numbers

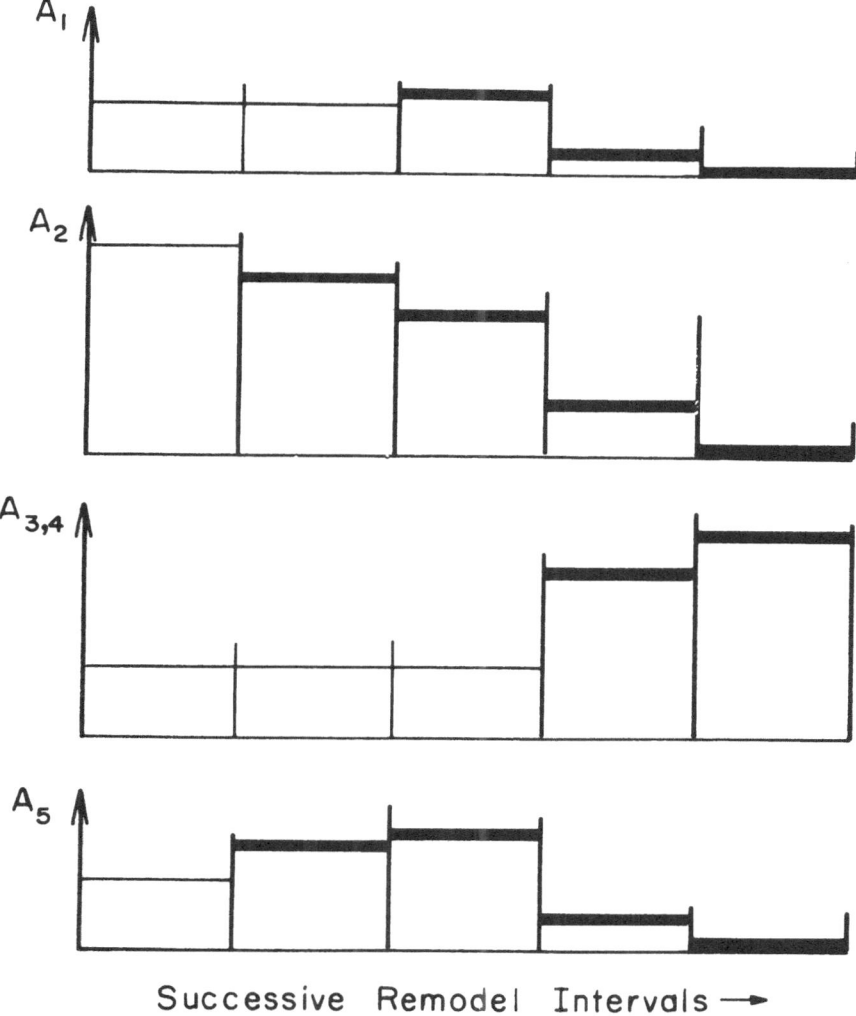

Figure 3.7   Truss Member Designs for the Initial Structure and
Four Successive Modifications

442

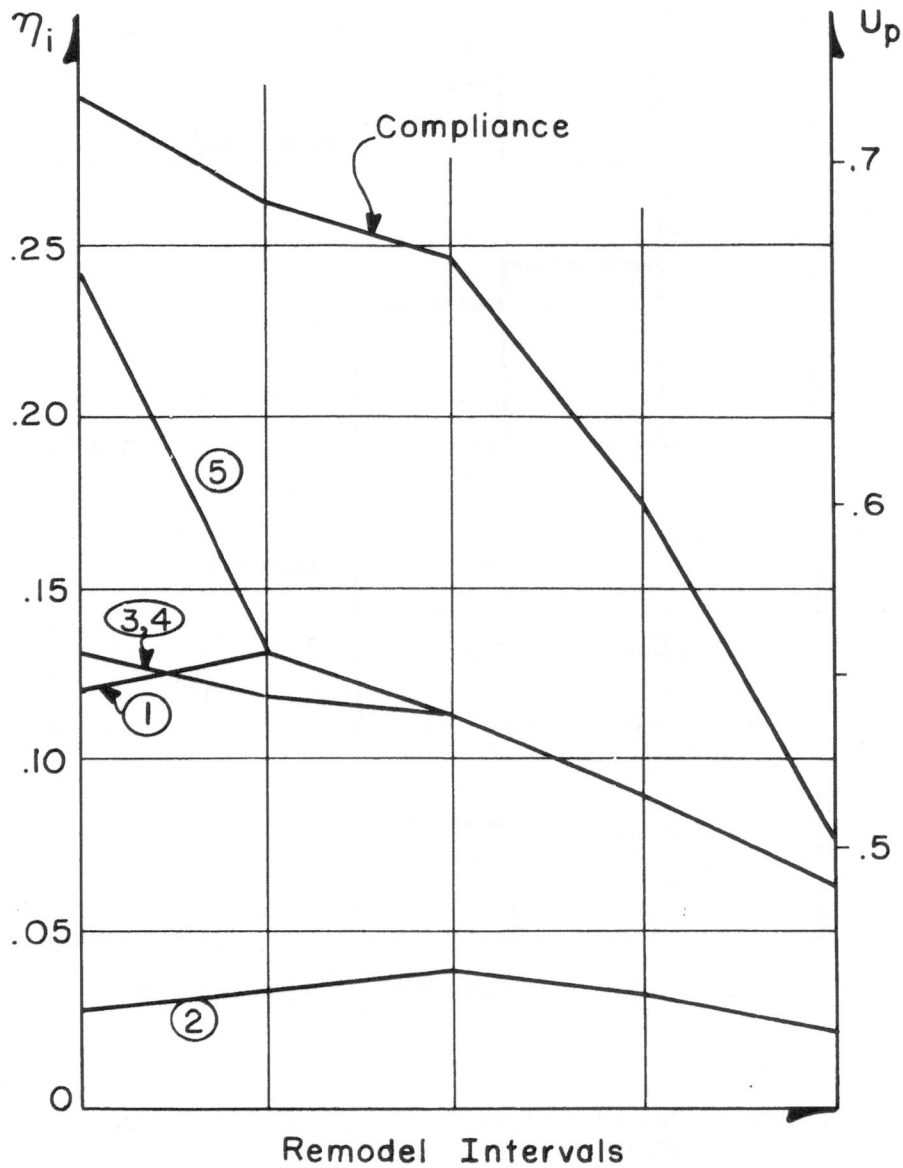

Figure 3.8 Truss Member Specific Energies and Compliance Over
the Design Evolution

443

3.4 Maximum Buckling Load of a Column-Like Mechanism

Modification to maximize the Euler load of a spring connected, rigid link, column-like mechanism is treated as a second
example of discrete-structure remodeling. The model is portrayed
in Fig. 3.9. The symbol R_k represents spring stiffness. Results
are presented for a three-link end supported system as an example.
Successive designs are shown in Fig. 3.10 and the corresponding
changes in Eigenvalues are indicated in Fig. 3.11. A notable
feature of the problem is that the two eigenvalues coalesce at
an intermediate point of the evolution in design. Figures 3.10
and 3.11 include optimal remodel results from two distinct
starting structures.

REFERENCES

1. Olhoff, N. and Taylor, J.E., On Optimal Structural Remodel-
    ing, J. Opt. Th. and Appl., Vol. 27, 1979, pp. 571-582.
2. Cheng, K.-T. and Olhoff, N., An Investigation Concerning
    Optimal Design of Elastic Plates, DTH-Denmark Manuscript,
    March, 1980.

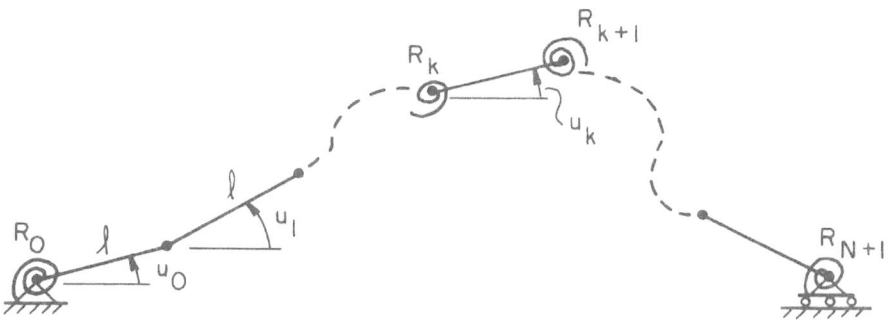

Figure 3.9 Schematic of a Column-Like Mechanism

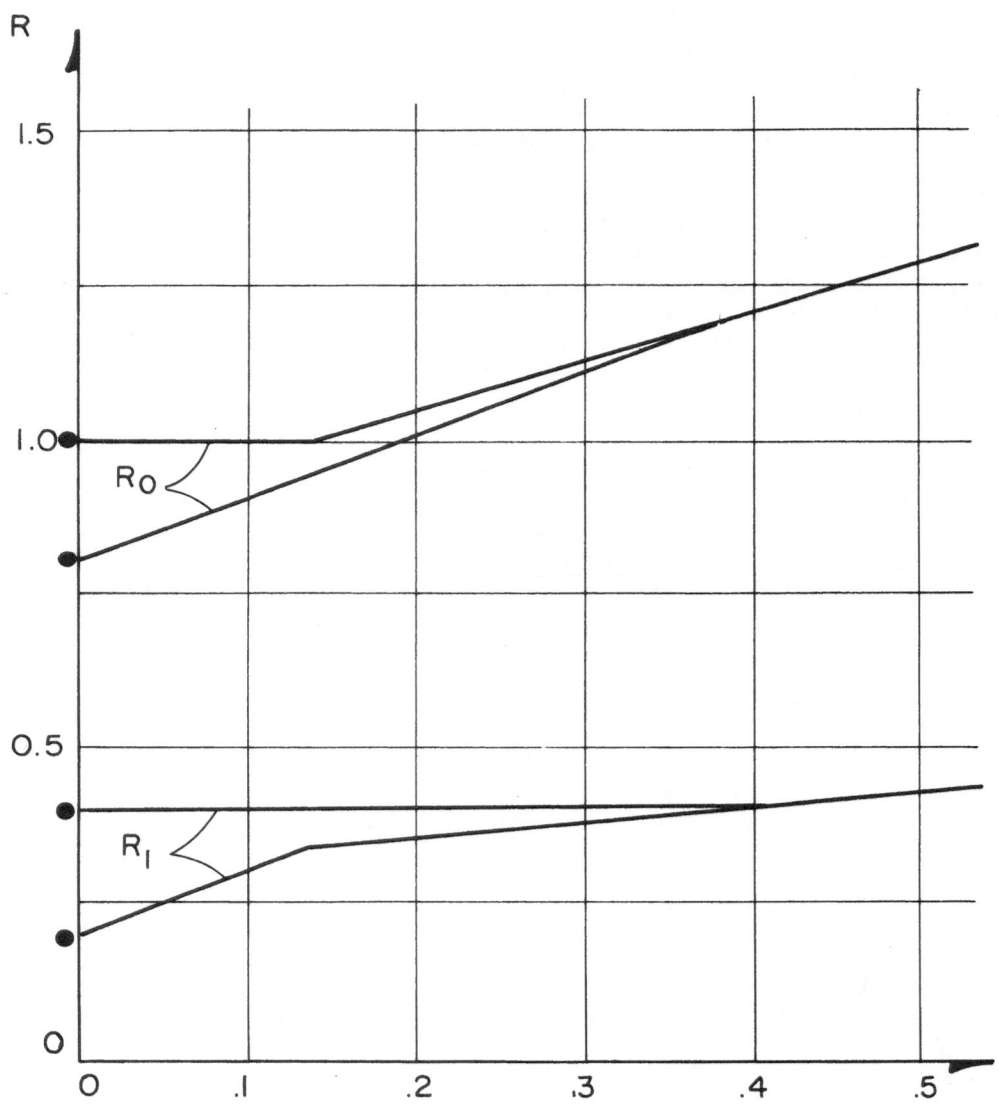

Figure 3.10   Spring Stiffnesses Over Design Evolution for Each
of Two Distinct Starting Designs of the Column
Mechanism

445

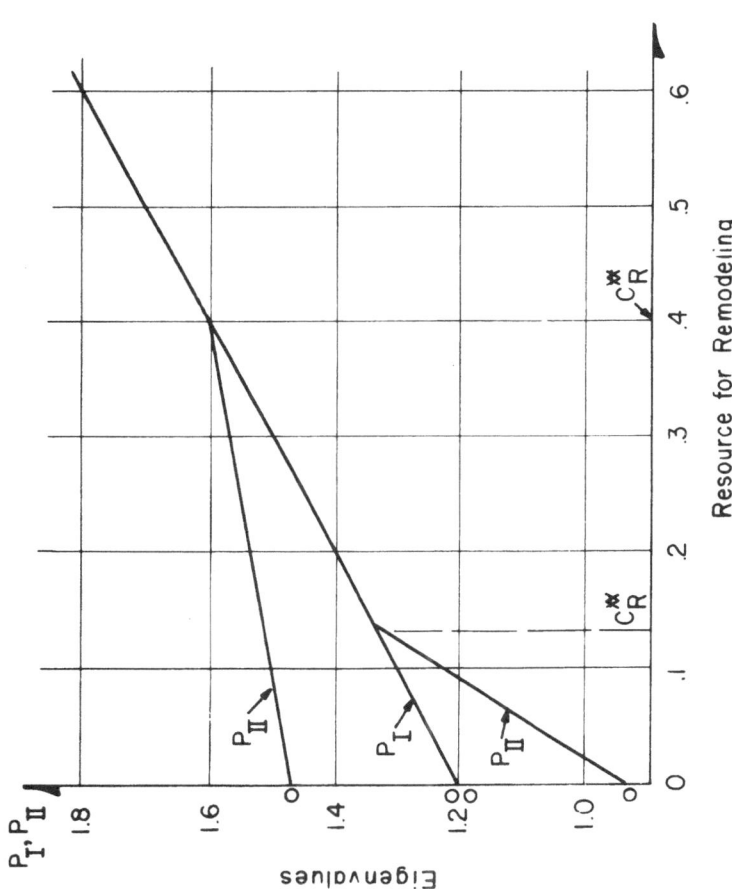

Figure 3.11   Eigenvalues Over Design Evolution for Each of Two Distinct Starting Designs of the Column Mechanism

A GRADIENT PROJECTION METHOD FOR STRUCTURAL OPTIMIZATION[*]

Edward J. Haug

Materials Division, College of Engineering,
The University of Iowa, Iowa City, Iowa   52242

ABSTRACT

A variational formulation of the equations of elasticity and a general optimal design formulation for distributed parameter structures are used to obtain explicit design derivatives needed for optimization. A function space gradient projection method is then developed to iteratively modify a design estimate to correct constraint errors and reduce cost. A computational algorithm is presented to implement the method, using finite element structural analysis methods. The method is illustrated for optimization of beams and plates with constraints on stress, displacement, and natural frequency.

1.   THE PROBLEM TREATED

A broad class of structural optimization problems can be formulated in a variational setting that lends itself to numerical solution methods. The class of problems discussed here is characterized by the equations of linear elasticity for structural response to load and for vibration. Design arises in a nonlinear way, however, and requires design sensitivity analysis methods of the kind presented in Refs. 1 and 2.

The design variable is denoted by the function $u(x)$ in $L^{\infty}(\Omega)$ and the structure is presumed to be linearly elastic, so small displacement theory holds. The state equation for the system is expressed as the differential operator equation

[*]Research supported by NSF Project No. ENG 77-19967

$$A(u)z = f(u) \quad , \qquad z \in D_A \tag{1.1}$$

where Ω is the domain of definition of the independent variable x and A, which depends on design, is a linear symmetric differential operator. The homogeneous boundary condition for the system are incorporated into the definition of the domain of the symmetric operator A,

$$D_A = \{z \in H^m(\Omega); \quad Cz = 0 \text{ in } \Gamma\} \tag{1.2}$$

where $H^m(\Omega)$ is a Sobolev space [1] and Γ is the boundary of the domain Ω.

When natural frequency or buckling load constraints are imposed on the structure, one has to solve an eigenvalue problem of the form

$$A(u)y = \zeta B(u)y \tag{1.3}$$

with

$$y \in D_A \subset D_B \tag{1.4}$$

where the symmetric operator B is associated with the mass of the structure for vibration and geometrical stiffness for buckling. The eigenvalue ζ is proportional to the square of the natural frequency of the structure for vibration and to buckling load for buckling. The function y is the corresponding eigenfunction. It is convenient to normalize the eigenfunction with respect to B, as follows:

$$\int_\Omega y^T B(u)y d\Omega = 1 \tag{1.5}$$

Typical performance constraints imposed on the structure involve bounds on stress, displacement, member size, and natural frequency. These constraints can be expressed analytically in the form

$$\phi_i(u(x)) \leq 0 \quad , \qquad i = 1,2,\ldots,k \qquad x \in \Omega \tag{1.6}$$

and

$$\psi_j = g_j(\zeta) + \int_\Omega G_j(u,z,x)d\Omega \left\{ \begin{array}{ll} \leq 0 \ , & j = 1,\ldots,r' \\[2ex] = 0 \ , & u = r'+1,\ldots,r \end{array} \right. \tag{1.7}$$

Stress and displacement constraints of the form $\eta(u(x),z(x),x) \leq 0$ for all x in Ω, are transformed to an equivalent functional constraint

$$\psi_j = \int_\Omega (\eta + |\eta|)d\Omega = 0 \qquad (1.8)$$

Finally, the cost function to be minimized is

$$\psi_0 = g_0(\zeta) + \int_\Omega G_0(z,u,x)d\Omega \qquad (1.9)$$

As noted, the cost function may depend on natural frequency, buckling load, and weight of the structure. The optimal design problem can now be defined as follows: find a design variable u that minimizes the cost function of Eq. 1.9 and satisfies Eqs. 1.1 to 1.7.

2. DESIGN SENSITIVITY ANALYSIS

As a first step in solution of the foregoing problem, derivatives of functionals are needed. As shown in Refs. 1 and 2, one solves an adjoint equation associated with Eq. 1.1 for an adjoint variable λ^j corresponding to each active functional constraint $\psi_j \geq 0$,

$$A\lambda^j = -\frac{\partial G_j^T}{\partial z} , \qquad \lambda^j \in D_A \qquad (2.1)$$

The differential of the constraints and cost functionals, for simple eigenvalues ζ is, then [1,2]

$$\delta\psi_j = \frac{\partial g_j}{\partial \zeta} \left[a_{u,\delta u}^{(1)}(y,y) - \zeta b_{u,\delta u}^{(1)}(y,y) \right] + \int_\Omega \frac{\partial G_j}{\partial u} \delta u d\Omega$$

$$+ a_{\alpha,\delta u}^{(1)}(z,\lambda^i) , \qquad j = 0,1,\ldots,r \qquad (2.2)$$

where

$$a_u(z,\lambda) = (A(u)z,\lambda) \qquad (2.3)$$

$$b_u(y,y) = (B(u)y,y) \qquad (2.4)$$

and $a_{u,\delta u}^{(1)}(z,\lambda)$ and $b_{u,\delta u}^{(1)}(y,y)$ are the Frechet differentials (first variations of $a_u(z,\lambda)$ and $b_u(y,y)$ with respect to u) of the forms of Eqs. 2.3 and 2.4, respectively. In calculating these differentials, the forms of Eqs. 2.3 and 2.4 are reduced to lowest order form z and y, through integration by parts. The differentials are then calculated as the first variation with respect only to u. Note that evaluation of these differentials requires that the state equations of Eqs. 1.1 and 1.3 be solved and that the adjoint equations of Eq. 2.1 be solved. If finite element methods are employed, the same stiffness matrix is used for all solutions, so computational cost need not be prohibitive.

To facilitate development of the gradient projection method to follow, the differentials of Eq. 2.2 are written in the form

$$\delta\psi_j = \int_\Omega \Lambda^{j^T} \delta u d\Omega \qquad (2.5)$$

This is possible since the differentials of Eq. 2.2 are linear in δu. Finally, direct differentiation of Eq. 1.6 gives the differential of the pointwise design constraint as

$$\delta\phi_i = \phi_u \delta u \qquad (2.6)$$

3. A GRADIENT PROJECTION METHOD

A simple method of numerical optimization is to iteratively modify a nominal design u by moving in a direction δu that decreases the cost most rapidly. This objective may be looked on as a form of structural remodeling presented in Ref. 3. Indeed it is a special case of remodeling with small design change and a linearized model. From Eq. 2.5

$$\delta\psi_0 = \int_\Omega \Lambda^{0^T} \delta u d\Omega \qquad (3.1)$$

By the Schwarz inequality,

$$|\delta\psi_0| \leq ||\Lambda^0|| \; ||\delta u|| \qquad (3.2)$$

with equality only if $\delta u = -\eta \Lambda^0$ for some real η, where $||\cdot||$ is the $L^2(\Omega)$ norm. Thus to maximize $|\delta\psi_0|$ over all δu with $||\delta u||$ fixed, one should select

$$\delta u = -\eta \Lambda^0 \qquad (3.3)$$

From Eq. 3.1,

$$\delta\psi_0 = -\eta \int_\Omega \Lambda^{0^T} \Lambda^0 \, d\Omega = -\eta \, ||\Lambda^0||^2 \tag{3.4}$$

so it is clear that η should be positive to decrease ψ_0. Thus the direction of steepest descent is simply given by Eq. 3.3, with $\eta > 0$.

In the presence of constraints, however, the direction of steepest descent may not be admissible. To determine the constrained direction of steepest descent, first define the set of ε-active constraints

$$\tilde{\psi} = \{\psi_j : \psi_j(u,z,\zeta) \geq -\varepsilon\} \tag{3.5}$$

$$\tilde{\phi}(x) = \{\phi_i : \phi_i(u(x)) \geq -\varepsilon\} \tag{3.6}$$

where $\varepsilon > 0$ is small. The objective now is to find δu to minimize $\delta\psi_0$ subject to the conditions

$$\delta\tilde{\psi} = \int_\Omega \tilde{\Lambda}^T \delta u d\Omega \leq 0 \tag{3.7}$$

$$\delta\tilde{\phi}(x) = \tilde{\phi}_u \delta u \leq 0 \quad , \qquad x \in \Omega \tag{3.8}$$

where δu is normalized by the condition

$$||\delta u||_W = \int_\Omega \delta u^T W \delta u d\Omega = 1 \tag{3.9}$$

and W is a weighting matrix.

This is a simple minimization problem without differential equation side conditions, so Kuhn-Tucker conditions may be applied. They guarantee existence of multipliers ν, $\tilde{\gamma} \geq 0$, and $\tilde{\mu}(x) \geq 0$ such that

$$\frac{\partial H}{\partial \delta u} = 0 \tag{3.10}$$

where

$$H = \Lambda^{0^T} \delta u + \tilde{\gamma}^T \tilde{\Lambda}^T \delta u + \tilde{\mu}(x)^T \tilde{\phi}_u \delta u + \nu \delta u^T W \delta u \tag{3.11}$$

and

$$\tilde{\gamma}_j \int_\Omega \Lambda^{j^T} \delta u d\Omega = 0 \tag{3.12}$$

$$\tilde{\mu}_i(x)\phi_{i,u} \; \delta u = 0 \; , \qquad x \in \Omega \tag{3.13}$$

From Eq. 3.10

$$\delta u = -\frac{1}{2\nu} W^{-1} \left[\Lambda^0 + \tilde{\Lambda}\tilde{\gamma} + \tilde{\phi}_u^T \, \tilde{\mu}(x) \right] \tag{3.14}$$

Presuming for the moment that the inequalities of Eqs. 3.7 and 3.8 are equalities, Eqs. 3.8 and 3.14 yield

$$\tilde{\phi}_u W^{-1} \; \tilde{\phi}_u^T \; \tilde{\mu} = -\tilde{\phi}_u \; W^{-1} \; \Lambda^0 - \tilde{\phi}_u W^{-1} \; \tilde{\Lambda}\tilde{\gamma} \tag{3.15}$$

If ε-active constraints are independent, the coefficient matrix of $\tilde{\mu}$ in Eq. 3.15 is nonsingular. Solving Eq. 3.15 for $\tilde{\mu}$ and substituting into Eq. 3.14 yields

$$\delta u = -\frac{1}{2\nu} W^{-1} \left\{ \left[I - \tilde{\phi}_u^T \, M_{\phi\phi}^{-1} \; \tilde{\phi}_u W^{-1} \right] \Lambda^0 + \left[I - \tilde{\phi}_u^T \, M_{\phi\phi}^{-1} \tilde{\phi}_u W^{-1} \right] \tilde{\Lambda}\tilde{\gamma} \right\}$$

where
$$\tag{3.16}$$

$$M_{\phi\phi} = \tilde{\phi}_u W^{-1} \; \tilde{\phi}_u^T \tag{3.17}$$

Substituting from Eq. 3.16 into Eq. 3.7, as an equality, yields an equation for $\tilde{\gamma}$,

$$M_{\psi\psi}\tilde{\gamma} + M_{\psi\psi_0} = 0 \tag{3.18}$$

where

$$M_{\psi\psi} = \int_\Omega \tilde{\Lambda}^T W^{-1} \left[I - \tilde{\phi}^T M_{\phi\phi}^{-1} \; \tilde{\phi}_u W^{-1} \right] \tilde{\Lambda} d\Omega \tag{3.19}$$

$$M_{\psi\psi} = \int_\Omega \tilde{\Lambda}^T W^{-1} \left[I - \tilde{\phi}^T \, M_{\phi\phi}^{-1} \; \tilde{\phi}_u W^{-1} \right] \Lambda^0 d\Omega \tag{3.20}$$

As shown in Ref. 4, the matrix $M_{\psi\psi}$ of Eq. 3.19 is positive semi-definite, hence likely nonsingular.

Defining the projected gradient as

$$\delta u^1 = W^{-1} \left[I - \tilde{\phi}_u^T \, M_{\phi\phi}^{-1} \; \tilde{\phi}_u W^{-1} \right] [\Lambda^0 + \tilde{\Lambda}\tilde{\gamma}] \tag{3.21}$$

where $\tilde{\gamma}$ is determined from Eq. 3.18, the desired design change is written as

$$\delta u = -\eta \; \delta u^1 \tag{3.22}$$

where $\eta > 0$ is a step size parameter.

Recall that in solving for the multipliers $\tilde{\mu}(x)$ and $\tilde{\gamma}$, it was "assumed" that the linearized constraints of Eqs. 3.7 and 3.8 remain active. To check this assumption, one should evaluate $\tilde{\mu}(x)$ and $\tilde{\gamma}$, which are to be nonnegative, from Eqs. 3.15 and 3.18. From sensitivity analysis it is expected that $\tilde{\mu}(x)$ and $\tilde{\gamma}$ are sensitivity coefficients of optimum cost with respect to variation in the constraints of Eqs. 3.12 and 3.13. Thus, at any point where a component of $\tilde{\mu}(x)$ is negative the corresponding constraint in $\tilde{\phi}(x)$ can be relaxed and a reduction in cost achieved. Similarly, if a component of $\tilde{\gamma}$ is negative, the corresponding constraint in ψ can be relaxed and a reduction in cost achieved. These modifications in $\tilde{\phi}$ and $\tilde{\psi}$ are thus made to obtain a refined direction of constrained steepest descent δu^1.

In case there are constraint violations in Eqs. 1.6 and 1.7, one would like to choose a change δu^2 in design to correct these errors. Let $\Delta\tilde{\psi}$ and $\Delta\tilde{\phi}(x)$ be the error corrections desired, normally $\Delta\tilde{\psi} = -\tilde{\psi}$ and $\Delta\tilde{\phi}(x) = -\tilde{\phi}(x)$. Then one seeks a design change δu^2 to satisfy

$$\delta\tilde{\phi}(x) = \tilde{\phi}_u(x)\delta u^2 = \Delta\tilde{\phi}(x) \quad , \qquad x \in \Omega \tag{3.23}$$

$$\delta\tilde{\psi} = \int_\Omega \tilde{\Lambda}^T \delta u^2 d\Omega = \Delta\tilde{\psi} \tag{3.24}$$

It is desirable to achieve the constraint error corrections of Eqs. 3.23 and 3.24 with the smallest norm of δu^2; i.e., choose δu^2 to minimize

$$J = \frac{1}{2} \, ||\delta u^2||_W^2 \equiv \frac{1}{2}\int_\Omega \delta u^2 W \delta u^2 d\Omega \tag{3.25}$$

The necessary condition for δu^2 is thus

$$W\delta u^2 + \tilde{\Lambda}\alpha + \tilde{\phi}_u^T \beta(x) = 0 \tag{3.26}$$

where α and $\beta(x)$ are multipliers. Substituting from Eq. 3.26 into Eq. 3.23,

$$M_{\phi\phi} \, \beta(x) = -\tilde{\phi}_u W^{-1} \, \tilde{\Lambda}\alpha - \Delta\tilde{\phi}(x) \tag{3.27}$$

Solving Eq. 3.27 for $\beta(x)$, substituting into Eq. 3.26 for δu^2, and substituting the result into Eq. 3.24 yields

$$M_{\psi\psi}\alpha = -(M_{\psi\phi} + \Delta\tilde{\psi}) \tag{3.28}$$

where

$$M_{\psi\phi} = \int_{\Omega} \tilde{\Lambda}^T W^{-1} \; \tilde{\phi}_u^T \; M_{\phi\phi}^{-1} \; \Delta\tilde{\phi}(x)d\Omega \tag{3.29}$$

One may now solve Eq. 3.28 for α and substituting into Eq. 3.27 for $\beta(x)$. These results may then be substituted into Eq. 3.26 for δu^2 to obtain

$$\delta u^2(x) = -W^{-1} \left[\tilde{\Lambda}\alpha + \tilde{\phi}_u^T \; \beta(x) \right] \tag{3.30}$$

As a computational algorithm, one can separately carry out constraint error correction with δu^2 and then descent with δu^1. This procedure is recommended if the constraint errors are large. If constraint errors are small however, separation of the correction and descent steps is computationally inefficient. A simultaneous descent-correction algorithm can be used by simultaneously calculating δu^1 and δu^2 and forming

$$\delta u = -\eta\delta u^1 + \delta u^2 \tag{3.31}$$

As a final note, selection of the step size parameter η may be initially based on a desired reduction $\Delta\psi_0 < 0$ in cost. Substituting $-\eta\delta u^1$ into Eq. 2.5 for $\delta\psi_0$, one may require.

$$\Delta\psi_0 = -\eta \int_{\Omega} \Lambda^{0^T} \delta u^1 \; d\Omega$$

Thus

$$\eta = -\Delta\psi_0 \Big/ \int_{\Omega} \Lambda^{0^T} \delta du^1 \Omega \tag{3.32}$$

The parameter η may be adjusted by the user to speed convergence of the algorithm. If descent is slow, η is increased. If oscillation occurs, η is decreased. More refined step size selection, such as one dimensional search, may be employed if the user desires.

4. OPTIMIZATION OF STRUCTURES WITH STRESS CONSTRAINTS

If stress constraints are imposed, which is required by most design situations, then a special form of state equation formulation may be considered. In the formulation presented in Refs. 1

454

and 2, fourth order operator equations with displacement as the state variable are employed for beam and plate problems. Since stress is given in terms of second derivatives of displacement, the functional constraint involving stress in beams is of the form

$$\psi = \int_0^\ell G(z,z'',u)ds \tag{4.1}$$

Then

$$\delta\psi = \int_0^\ell [G_z \, \delta z + G_{z''} \, \delta z'' + G_u \, \delta u]dx \tag{4.2}$$

To write this explicitly in terms of δz, two integrations by parts lead to

$$\delta\psi = \int_0^\ell \left\{ \left[G_z + \frac{d^2}{dx^2}(G_{z''}) \right] \delta z + G_u \delta u \right\} dx \tag{4.3}$$

The adjoint variable method of design sensitivity analysis of Section 2 can be employed to write $\delta\psi$ of Eq. 4.3 explicitly in terms of δu. However, the term $d^2(G_{z''})/dx^2$ must be evaluated. Since G depends on z'', fourth order derivatives of z will arise. This is a severe practical problem, since finite element methods used to solve the beam equations do not generally provide accurate estimates of the fourth derivative of displacement. Thus, one would have to resort to higher order elements to be able to carry out the calculations required for design sensitivity analysis.

An alternate approach that avoids the foregoing computational difficulty is to formulate the structural differential equations as a system of second order, with both displacement and moment as state variables. With this formulation, stresses are algebraic functions of the state variables and the foregoing complication does not arise. If one is careful to write the state equations in symmetric form, the design sensitivity analysis method of Section 2 can then be applied directly. This is the approach employed in the next two sections for optimization of beams and plates with stress constraints.

5. OPTIMAL DESIGN OF BEAMS WITH STRESS, DISPLACEMENT, AND NATURAL FREQUENCY CONSTRAINTS

5.1 Computational Formulation

As examples of the operator formulation of the design problem, with distribution of the design variables over one space dimension and complex constraints, two types of beam, each of length ℓ, are considered. One is simply supported and the other is fixed at both ends. Their cross sections are to be I sections and all dimensions of the cross sections vary in the same ratio, as a function of x. In this case, $I = \alpha u^2$, where u is the cross sectional area and α is a constant that depends on a shape selected for the section.

Denoting displacement by z_1 and moment by z_2, one may write the beam equation in symmetric ordinary differential operator form as

$$A(u)z \equiv \begin{bmatrix} -z_2'' \\ \\ -z_1'' - \dfrac{z_2}{E\alpha u^2} \end{bmatrix} = \begin{bmatrix} q(x) \\ \\ 0 \end{bmatrix} \tag{5.1}$$

where $z_i'' \equiv (d^2 z_i)/(dx^2)$ and $q(x)$ is applied load. When the beam is simply supported at both ends, the boundary conditions are

$$z_1(0) = z_1(\ell) = z_2(0) = z_2(\ell) = 0 \tag{5.2}$$

When the beam is clamped at its ends, the boundary conditions are given by

$$z_1'(0) = z_1'(\ell) = z_1(0) = z_1(\ell) = 0 \tag{5.3}$$

The operator equation for free vibration of the beam is given by

$$A(u)y \equiv \begin{bmatrix} -y_2'' \\ \\ -y_1'' - \dfrac{y_2}{E\alpha u^2} \end{bmatrix} = \zeta \begin{bmatrix} uy_1 \\ \\ 0 \end{bmatrix} \equiv \zeta \beta(u)y \tag{5.4}$$

where $\zeta = \rho\omega^2$, ρ is the mass density of the beam material, and ω is the natural frequency. The vector function y must also satisfy the boundary conditions of Eq. 5.2 for a simply supported beam and Eq. 5.3 for a clamped beam.

A minimum volume (equivalently minimum weight) design objective is treated here. Thus, the objective function is

$$\psi_0 = \int_0^\ell u(x)dx \qquad (5.5)$$

In order that the beam can support shear, a lower bound u_0 is placed on the cross section area. Analytically, this is

$$\phi = -u(x) + u_0 \leq 0 \qquad (5.6)$$

To preclude problems with resonance at an unwanted frequency, one imposes a lower bound ζ_0 on the lowest eigenvalue of Eq. 5.4, as

$$\psi_1 = \zeta_0 - \zeta \leq 0 \qquad (5.7)$$

Bending stress is required to be less than an allowable stress σ_0, so

$$\eta_1(x,z,u) = s|z_2|/u^{3/2} - \sigma_0 \leq 0 \quad , \qquad x \in [0,\ell] \qquad (5.8)$$

where s is a constant that is associated with the geometry of the cross section. Finally, one may wish to bound the extreme displacement under a given load. This may be expressed analytically as

$$\eta_2(x,z,u) = |z_1(x)| - \Delta \leq 0 \quad , \qquad x \in [0,\ell] \qquad (5.9)$$

The last two conditions, Eqs. 5.8 and 5.9, are replaced by equivalent functional constraints of the form

$$\psi_2 = \int_0^\ell \{\eta_1 + |\eta|\} dx$$

$$= \int_0^\ell \left\{ s|z_2|/u^{3/2} - \sigma_0 + \left| s|z_2|/u^{3/2} - \sigma_0 \right| \right\} dx = 0 \qquad (5.10)$$

and

$$\psi_3 = \int_0^\ell \{\eta_2 + |\eta_2|\}dx = \int_0^\ell \left\{ (|z_1| - \Delta) + \left| |z_1| - \Delta \right| \right\} dx = 0 \qquad (5.11)$$

To calculate design derivatives, first form the bilinear forms for each problem,

$$a_u(z,\lambda) = (A(u)z,\lambda) = \int_0^\ell \left[-z_2''\lambda_1 - z_1''\lambda_2 - \frac{z_2\lambda_2}{E\alpha u^2} \right] dx \qquad (5.12)$$

$$b_u(y,y) = (B(u)y,y) = \int_0^\ell uy_1^2 \, dx \qquad (5.13)$$

The differentials with respect to u are simply

$$a_{u,\delta u}^{(1)}(z,\lambda) = \int_0^\ell \frac{2z_2\lambda_2}{E\alpha u^3} \, \delta u \, dx \qquad (5.14)$$

$$b_{u,\delta u}^{(1)}(y,y) = \int_0^\ell y_1^2 \delta u \, dx \qquad (5.15)$$

Direct application of Eq. 2.2 to the functional ψ_1 of Eq. 5.7 yields

$$\delta\psi_1 = -\delta\zeta = -\int_0^\ell \left[\frac{2y_2^2}{E\alpha u^3} - \zeta y_1^2 \right] \delta u dx \qquad (5.16)$$

Thus, in Eq. 2.5

$$\Lambda^1 = -\frac{2y_2^2}{E\alpha u^3} + \zeta y_1^2 \qquad (5.17)$$

Substitution from Eq. 5.10 into Eq. 2.1 yields

$$A(u)\lambda^2 = - \begin{cases} 0 \, , & \text{if } \eta_1 > 0 \\[2ex] \{s \, \text{sgn}(z_2)/u^{3/2}\}\{1 + \text{sgn}(s|z_2|/u^{3/2} - \sigma_0)\} \, , \\ & \qquad\qquad\qquad \text{if } \eta_1 \geq 0 \quad (5.18) \end{cases}$$

where $\text{sgn}(z) = a/|a|$ if $a \neq 0$ and is 0 if $a = 0$. Since the operator A is symmetric, the boundary conditions on λ^2 are the same as those on z.

Similarly, through substitution from Eq. 5.11 into Eq. 2.1, one obtains

$$A(u)\lambda^3 = - \begin{bmatrix} \text{sgn}(z_1)\{1 + \text{sgn}(|z_1| - \Delta)\} \\ \\ 0 \end{bmatrix} \qquad (5.19)$$

with the same boundary conditions on λ^3 as on z.

Once λ^2 and λ^3 are computed, one may substitute into Eq. 2.2 to obtain

$$\delta\psi = \int_0^\ell \left[\left\{ -\frac{3}{2} s|z_2|/u^{5/2} \right\} \right.$$

$$\left. \left\{ 1 + \text{sgn}(s|z_2|/u^{3/2} - \sigma_0 \right\} + \frac{2\lambda_2^2 z_2}{E\alpha u^3} \right] \delta u\, dx \qquad (5.20)$$

so in Eq. 2.5

$$\Lambda^2 = \left\{ -\frac{3}{2} s|z_2|/u^{5/2} \right\} \left\{ 1 + \text{sgn}(s|z_2|/u^{3/2} - \sigma_0) \right\} + \frac{2\lambda_2^2 z_2}{E\alpha u^3} \qquad (5.21)$$

Further,

$$\delta\psi_3 = \int_0^\ell \frac{2\lambda_2^3 z_2}{E\alpha u^3} \delta u\, dx \qquad (5.22)$$

so

$$\Lambda^3 = \frac{2\lambda_2^3 z_2}{E\alpha u^3} \qquad (5.23)$$

Finally, since ψ_0 does not depend on z or ζ,

$$\Lambda^0 = 1.0 \qquad (5.24)$$

All quantities required to implement the steepest descent algorithm of Section 3 are now available. A finite element method is employed to solve the boundary-value problems. Numerical results from this efficient analysis method are then used to implement the steepest descent optimization algorithm.

5.2 Numerical Results for Simply Supported Beams

As a numerical example of simply supported beams of Fig. 5.1, the following data are employed: ℓ = 240 in., ρ = 7.34 × 10^{-4} lb -sec^2/in.^{4}, α = 2.26, s = 0.8681, σ_0 = 24 ksi, q(x) = 250 lb/in., and u_0 = 4.361 in.

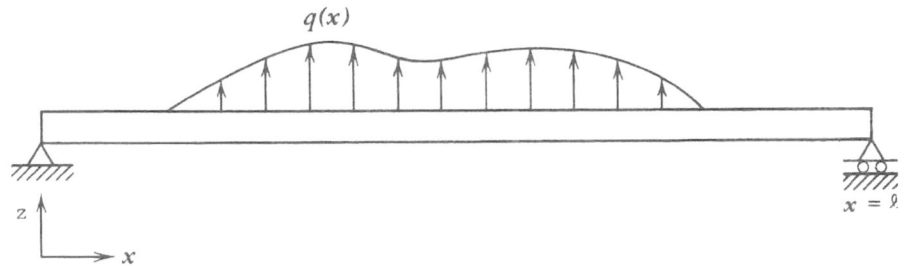

Figure 5.1   Simply Supported Beam with Static Load

Six sets of constraint conditions are employed to provide insights into the effect of differing design requirements. These cases are summarized in Table 5.1, where the optimum volume and an indication of which constraints are tight are given. The optimum design variable and associated stress distribution are given in Table 5.2. Profiles of half the symmetric optimum beams are given in Fig. 5.2.

TABLE 5.1   SUMMARY RESULTS FOR SIMPLY SUPPORTED BEAMS

	Stress (psi)	Constraints Displ (in.)	Freq (rad/sec)	Cost (in.^{3}) (Optimum Vol.)	Remarks
1	24000*			2900	max z_1 = 0.790 in. ω = 192 rad/sec
2	24000*	0.70*		3064	ω = 196.3 rad/sec
3	24000	0.65*		3176	ω = 199.5 rad/sec
4	24000*		195*	2979	max z_1 = 0.758 in.
5	24000*	0.70*	195	3064	ω = 196.3 rad/sec
6	24000*	0.70*	197*	3069	

*indicates a constraint is tight.

TABLE 5.2 OPTIMUM DESIGN VARIABLES (in.²) AND ASSOCIATED STRESS (psi) DISTRIBUTION FOR SIMPLY SUPPORTED BEAMS

i	Stress Constr		Str & Displ (0.7 in.) Constr		Str & Displ (0.65 in.) Constr		Str & Freq (195 rad/sec) Constr		Str, Displ (0.7 in.) & Freq (197 rad/sec) Constr	
	$u(x_i)$	$\sigma(x_i)$	$u(x_i)$	$\sigma(x_i)$	$u(x_i)$	$\sigma(x_i)$	$u(x_i)$	$\sigma(x_i)$	$u(x_i)$	$\sigma(x_i)$
1	16.170	24043.	18.292	19982.	19.062	18784.	16.185	24008.	17.756	20894.
2	16.159	24002.	18.169	20132.	18.978	18858.	16.156	24009.	17.694	20948.
3	16.065	24022.	17.804	20590.	18.737	19072.	16.061	24032.	17.509	21113.
4	15.941	23988.	17.672	20552.	18.366	19398.	15.924	24027.	17.451	20943.
5	15.751	23984.	17.233	20957.	17.902	19793.	15.744	23999.	16.774	21823.
6	15.506	23983.	16.772	21318.	17.383	20204.	15.510	23974.	16.679	21498.
7	15.129	24165.	16.256	21696.	16.904	20460.	15.207	23979.	16.257	21365.
8	14.850	23973.	15.720	22010.	16.337	20775.	14.855	23959.	15.738	21971.
9	14.408	24017.	15.088	22410.	15.740	21032.	14.544	23679.	15.238	22079.
10	13.901	24053.	14.481	22622.	15.182	21073.	14.180	23346.	14.686	22151.
11	13.355	24007.	13.813	22823.	14.463	21301.	13.754	22969.	14.063	22216.
12	12.717	24010.	13.070	23044.	13.663	21560.	13.254	22567.	13.358	22302.
13	12.014	23987.	12.249	23300.	12.783	21855.	12.663	22168.	12.562	22435.
14	11.206	24041.	11.224	23982.	11.513	23086.	11.966	21788.	11.675	22607.
15	10.342	23979.	10.340	23986.	10.652	22941.	11.143	21439.	10.693	22809.
16	9.3482	23972.	9.3274	24052.	9.4728	73500.	10.173	21117.	9.6069	23010.
17	8.2256	23908.	8.1956	24040.	8.2216	23926.	9.0216	20815.	8.3938	23193.
18	6.8730	24072.	6.8745	24064.	7.0098	23370.	7.6406	20537.	7.0116	23362.
19	5.3009	24162.	5.3232	24009.	5.4220	23356.	5.9523	20306.	5.3616	23752.
20	4.3610	16394.	4.3610	16394.	4.3610	16394.	4.3610	16394.	4.3610	16394.
21	4.3610	0.0	4.3610	0.0	4.3610	0.0	4.3610	0.0	4.3610	0.0

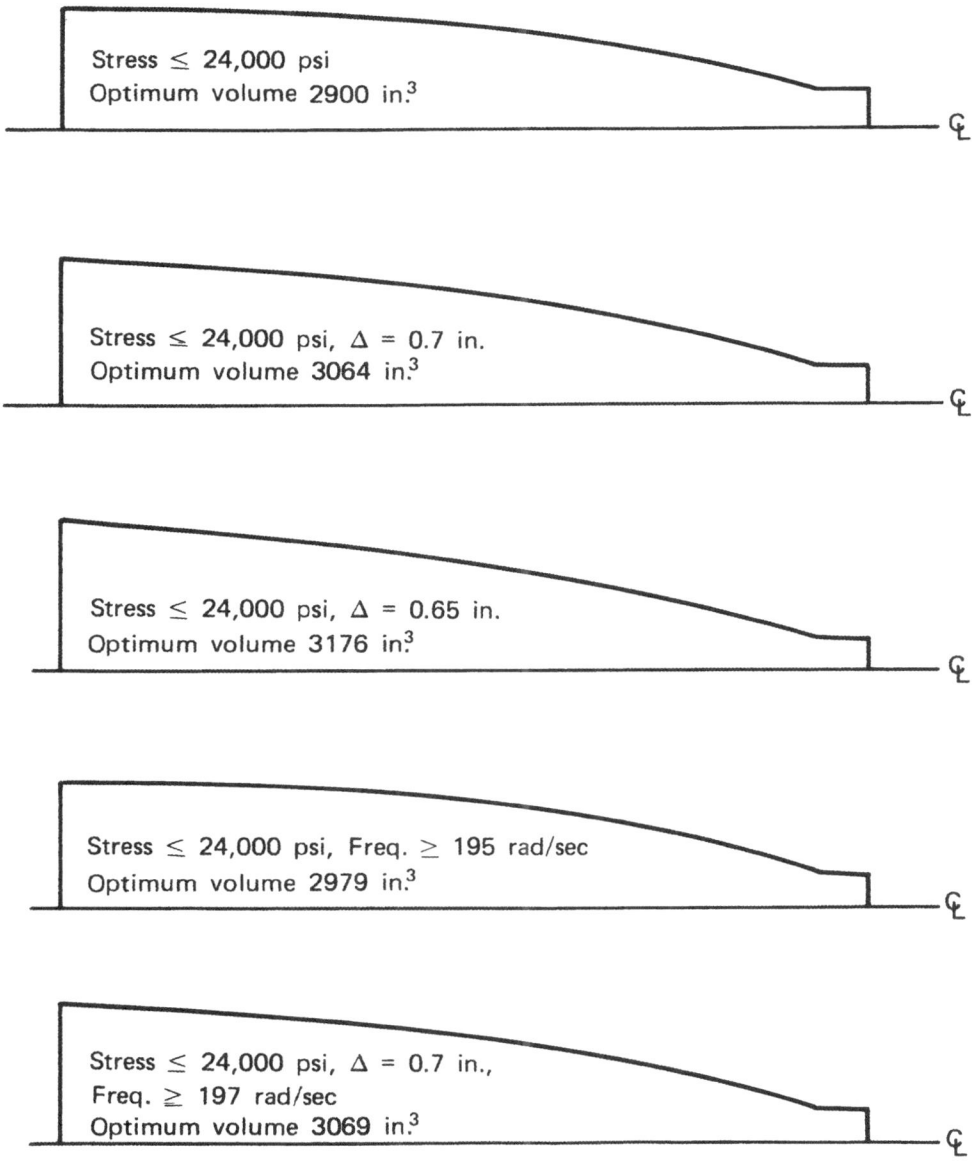

461

Stress ≤ 24,000 psi
Optimum volume 2900 in.^{3}

Stress ≤ 24,000 psi, Δ = 0.7 in.
Optimum volume 3064 in.^{3}

Stress ≤ 24,000 psi, Δ = 0.65 in.
Optimum volume 3176 in.^{3}

Stress ≤ 24,000 psi, Freq. ≥ 195 rad/sec
Optimum volume 2979 in.^{3}

Stress ≤ 24,000 psi, Δ = 0.7 in.,
Freq. ≥ 197 rad/sec
Optimum volume 3069 in.^{3}

Figure 5.2   Variation of Cross-Sectional Area of
             the Right-Half Span of Optimum Beams

462

5.3 Numerical Results for Clamped Beams

As a numerical example for clamped beams of Fig. 5.3, the
same data of Section 5.2 are employed. Five sets of constraint
conditions are employed to analyze the effect of varying types
and combinations of constraints. Table 5.3 provides a summary of
results for these various cases. Optimum design variables and
the associated stress distribution are given in Table 5.4. Profiles of half the symmetric optimum beams are given in Fig. 5.4.

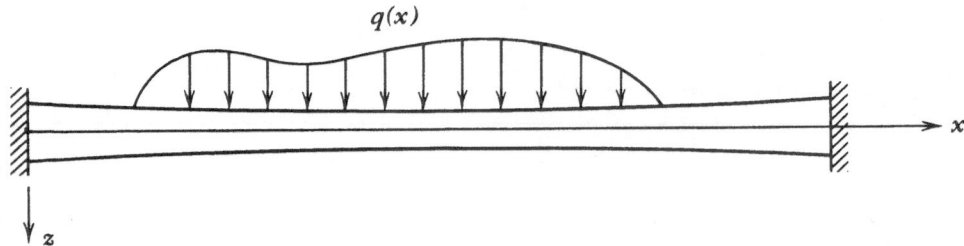

Figure 5.3 Clamped Beams

TABLE 5.3 SUMMARY RESULTS FOR CLAMPED BEAMS

	Stress (psi)	Constraints Displ (in.)	Freq (rad/sec)	Cost (in.^{3}) (Optimum Volume)
1	24000*			1569
2	24000*	0.45*		1668
3	24000*	0.40*		1786
4	24000*		440*	1659
5	24000*	0.45*	440*	1708

*indicates a constraint is tight.

6. OPTIMAL DESIGN OF PLATES WITH STRESS, DISPLACEMENT, AND NATURAL FREQUENCY CONSTRAINTS

6.1 Computational Formulation

As example problems with distribution of design variables
over two space dimensions, consider two rectangular plates; one
simply supported along all four edges and the other clamped along
all four edges. One wishes to determine plate thickness $u(x_1,x_2)$

TABLE 5.4 OPTIMUM DESIGN VARIABLES (in.^{2}) AND ASSOCIATED STRESS (psi) DISTRIBUTION FOR CLAMPLED BEAMS

i	Stress Constr		Str & Displ (0.45 in.) Constr		Str & Displ (0.40 in.) Constr		Str & Freq (440 rad/sec) Constr		Str, Displ (0.45 in.) & Freq (440 rad/sec) Constr	
	$u(x_j)$	$\sigma(x_j)$	$u(x_j)$	$\sigma(x_j)$	$u(x_j)$	$\sigma(x_j)$	$u(x_j)$	$\sigma(x_j)$	$u(x_j)$	$\sigma(x_j)$
1	5.8632	23930.	8.1338	17183.	8.8337	17436.	5.1011	24232.	5.9708	20491.
2	5.8183	23929.	8.0402	17312.	8.7913	17412.	5.0687	24123.	5.8923	20629.
3	5.6814	23933.	7.7642	17701.	8.6689	17323.	4.9405	24000.	5.6597	21042.
4	5.4448	23972.	7.3191	18353.	8.4848	17099.	4.6545	24300.	5.2572	21884.
5	5.0901	24139.	6.7251	19270.	8.2438	16699.	4.3610	23791.	4.8455	22167.
6	4.6710	23978.	6.0057	20446.	7.9671	16013.	4.3610	19931.	4.3650	22072.
7	4.3610	21862.	5.1800	21880.	7.6614	14955.	4.3610	15213.	4.3610	17385.
8	4.3610	16287.	4.4045	22413.	6.0798	17769.	4.3610	9638.7	4.3610	11810.
9	4.3610	9854.6	4.3610	16317.	5.2497	17275.	4.3610	3205.8	4.3610	5377.3
10	4.3610	2562.9	4.3610	9025.0	4.5144	14741.	4.3610	4085.9	4.3610	1914.4
11	4.3610	5588.0	4.3610	874.16	4.3610	7374.2	4.3610	12237.	4.3610	10065.
12	4.3610	14598.	4.3610	8135.6	4.3610	1635.6	5.3377	15691.	4.6049	17580.
13	4.4450	23775.	4.3733	17928.	4.3610	11503.	6.4845	17160.	5.9660	18088.
14	5.6155	24083.	5.3840	20942.	4.8421	18998.	7.4028	18917.	7.3477	18137.
15	6.7994	24022.	6.5580	21857.	5.9906	20997.	8.3188	20275.	8.5783	18575.
16	7.9536	24035.	7.6146	22857.	6.8581	23446.	9.1947	21509.	9.6843	19242.
17	9.1092	24011.	8.6375	23686.	8.7529	20993.	10.437	21375.	10.663	20131.
18	10.255	24026.	9.7654	23925.	10.114	20858.	11.011	23251.	11.587	21037.
19	11.403	24042.	10.972	23853.	11.006	22120.	12.257	22985.	12.578	21666.
20	12.538	24115.	12.112	24000.	11.927	23123.	13.166	23675.	13.230	23095.
21	13.655	0.0	13.434	23659.	13.248	0.0	14.193	0.0	15.401	0.0

464

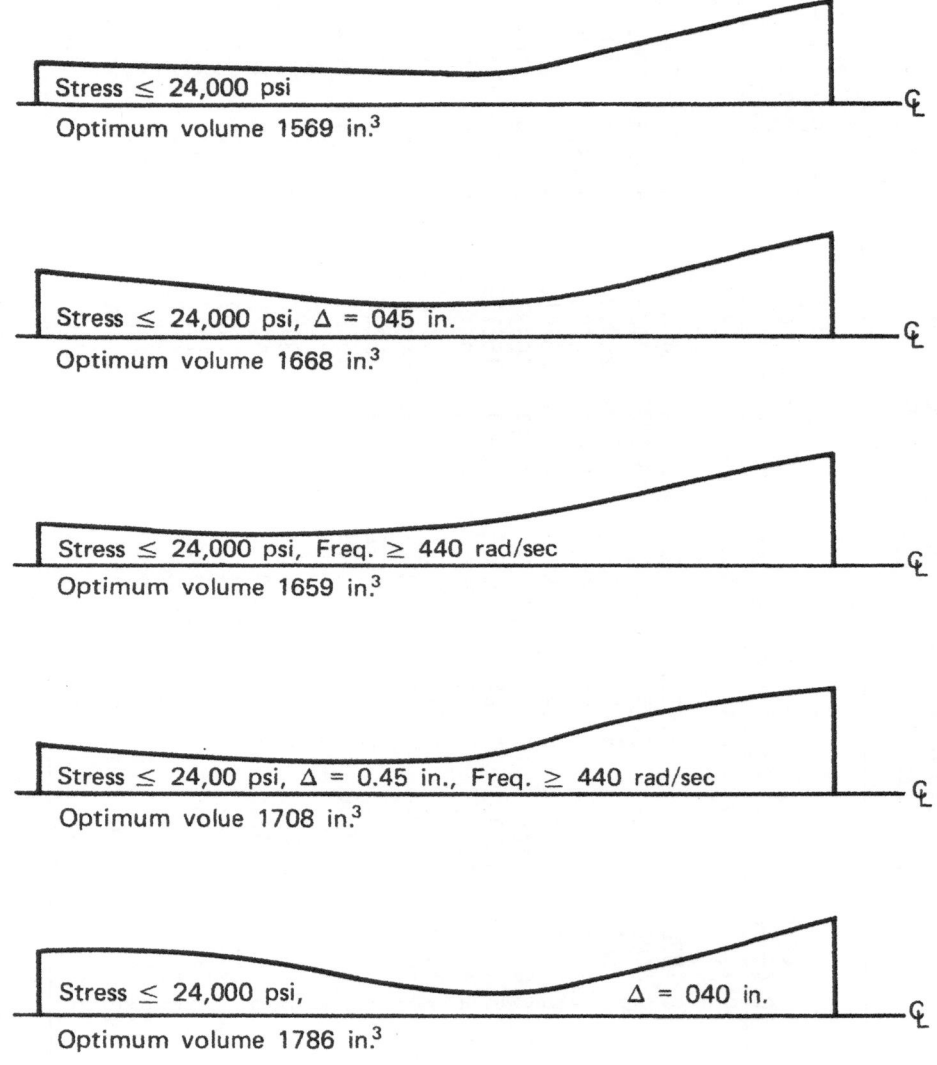

Figure 5.4   Variation of Cross-Sectional Area of the
Right Half Span of Optimum Beam (Clamped Support)

to minimize its weight, subject to constraints on natural frequency, stress, and deflection.

Denoting bending moments per unit length acting on the cross sections normal to x_1 and x_2 axes by z_1 and z_2, respectively, torsional moment by z_3, and displacement by z_4, one obtains the following operator equation for a plate, due to a lateral load $q(x_1,x_2)$:

$$A(u)z \equiv \begin{bmatrix} - z_{4,11} - \dfrac{12}{Eu^3}(z_1 - \mu z_2) \\[2mm] - z_{4,22} - \dfrac{12}{Eu^3}(z_2 - \mu z_1) \\[2mm] 2z_{4,12} - \dfrac{24(1+\mu)}{Eu^3} z_3 \\[2mm] - z_{1,11} - z_{2,22} + 2z_{3,12} \end{bmatrix} = \begin{bmatrix} 0 \\[2mm] 0 \\[2mm] 0 \\[2mm] q(x_1,x_2) \end{bmatrix} \tag{6.1}$$

When the plate is simply supported, the boundary conditions are

$$z_1\left(\pm \frac{a}{2},\, x_2\right) = z_2\left(x_1,\pm \frac{b}{2}\right) = z_4\left(\pm \frac{a}{2},x_2\right) = z_4\left(x_1,\pm \frac{b}{2}\right) = 0 \tag{6.2}$$

If the boundary is clamped, the boundary conditions are

$$z_{4,1}\left(\pm \frac{a}{2},x_2\right) = z_{4,2}\left(x_1,\pm \frac{b}{2}\right) = z_4\left(\pm \frac{a}{2},x_2\right) = z_4\left(x_1,\pm \frac{b}{2}\right) = 0 \tag{6.3}$$

Vibration of the plate is governed by the related eigenvalue problem

$$A(u)y = \zeta \begin{bmatrix} 0 \\ 0 \\ 0 \\ uy_4 \end{bmatrix} \equiv \zeta B(u)y \tag{6.4}$$

where $\zeta = \rho\omega^2$, ρ is mass density of the plate material, and ω is the natural frequency of the plate. The variable y in Eq. 6.4 satisfies the boundary condition of Eq. 6.2 or 6.3. It may be verified that the operators $A(u)$ and $B(u)$ are symmetric, under each of these sets of boundary conditions.

For minimum volume, the cost function is

$$0 = \int_{-b/2}^{b/2} \int_{-a/2}^{a/2} u(x_1,x_2)dx_1\, dx_2 \qquad (6.5)$$

As in the case of the beam, a lower bound u_0 is placed on the thickness of the plate, so

$$\phi = -u(x_1,x_2) + u_0 \leq 0 \qquad (6.6)$$

Also, a lower bound is placed on natural frequency,

$$\psi_1 = \zeta_0 - \zeta \leq 0 \qquad (6.7)$$

Stress in the plate should not be excessive; hence, at the surface of the plate, Von Mises' yielding criterion must not be violated, so it is required that

$$\eta_1 = \left[\frac{1}{u^2 \sigma_{yp}}\right]^2 (36z_1^2 - 36z_1z_2 + 36z_2^2 + 108z_3^2) - 1 \leq 0 \qquad (6.8)$$

where σ_{yp} is the material yield stress. Finally, the displacement constraint is

$$\eta_2 = |z_4| - \Delta \leq 0 \qquad (6.9)$$

Equations 6.8 and 6.9 must hold at each point in the plate, so they are replaced by the equivalent functional constraints

$$\psi_2 = \int_{-b/2}^{b/2} \int_{-a/2}^{a/2} (\eta_1 + |\eta_1|)\, dx_1\, dx_2 \qquad (6.10)$$

$$\psi_3 = \int_{-b/2}^{b/2} \int_{-a/2}^{a/2} \left\{(|z_4| - \Delta) + \Big||z_4| - \Delta\Big|\right\} dx_1\, dx_2 = 0 \qquad (6.11)$$

To calculate design derivatives, first form the bilinear forms

$$a_u(z,\lambda) = (A(u)z,\lambda) = \int_{-b/2}^{b/2} \int_{-a/2}^{a/2} \left[-z_{4,11}\lambda_1 - \frac{12}{Eu^3}(z_1 - \mu z_2)\lambda_1 \right.$$

$$\left. - z_{4,22}\lambda_2 - \frac{12}{Eu^3}(z_2 - \mu z_1)\lambda_2 + 2z_{4,12}\lambda_3 \right. \qquad \begin{array}{r}(6.12)\\ \text{Cont.}\end{array}$$

$$- \frac{24(1 + \mu)}{Eu^3} \, z_3\lambda_3 - z_{1,11}\lambda_4 - z_{2,22}\lambda_4$$

$$+ 2z_{3,12}\lambda_4 \Bigg] \, dx_1 \, dx_2 \tag{6.12}$$

$$b_u(y,y) = \int_{-b/2}^{b/2} \int_{-a/2}^{a/2} uy_4^2 \, dx_1 \, dx_2 \tag{6.13}$$

The differentials with respect to u are

$$a_{u,\delta u}^{(1)}(z,\lambda) = \int_{-b/2}^{b/2} \int_{-a/2}^{a/2} \left(\frac{36}{Eu^4}\right) \Bigg[(z_1 - \mu z_2)\lambda_1 + (z_2 - \mu z_1)\lambda_2$$

$$+ (1 + \mu)z_3\lambda_3 \Bigg] \delta u \, dx_1 \, dx_2 \tag{6.14}$$

$$b_{u,\delta u}^{(1)}(y,y) = \int_{-b/2}^{b/2} \int_{-a/2}^{a/2} y_4^2 \, \delta u \, dx_1 \, dx_2 \tag{6.15}$$

Direct application of Eq. 2.2 to the function ψ_1 of Eq. 6.7 yields

$$\delta\psi_1 = -\delta\zeta = \int_{-b/2}^{b/2} \int_{-a/2}^{a/2} \Bigg[\zeta y_4^2 - \frac{36}{Eu^4} \, y_1^2 - 2\mu y_1 y_2 + y_2^2$$

$$+ 2(1 + \mu)y_3^2 \Bigg] \delta u \, dx_1 \, dx_2 \tag{6.16}$$

so

$$\Lambda^1 = \zeta y_4^2 - \frac{36}{Eu^4} \, y_1^2 - 2\mu y_1 y_2 + y_2^2 + 2(1 + \mu)y_3^2 \tag{6.17}$$

Since Eqs. 6.10 and 6.11 depend on the state variables, the adjoint variables λ^2 and λ^3 must be found by solving

$$A(u)\lambda^2 = - \begin{bmatrix} \left(\dfrac{1}{u^2\sigma_{yp}}\right)^2 (72z_1 - 36z_2)(1 + \text{sgn } \eta_1) \\ \\ \left(\dfrac{1}{u^2\sigma_{yp}}\right)^2 (72z_2 - 36z_1)(1 + \text{sgn } \eta_1) \\ \\ \left(\dfrac{1}{u^2\sigma_{yp}}\right)^2 (216z_3)(1 + \text{sgn } \eta_1) \\ \\ 0 \end{bmatrix} \tag{6.18}$$

and

$$A(u)\lambda^3 = \begin{bmatrix} 0 \\ 0 \\ 0 \\ \text{sgn}(z_4)[1 + \text{sgn}(|z_4| - \Delta)] \end{bmatrix} \tag{6.19}$$

where λ^2 and λ^3 must satisfy the same boundary conditions as the state variable z. Application of Eq. 2.2 to Eqs. 6.10 and 6.11 yields

$$\delta\psi_1 = \int_{-b/2}^{b/2} \int_{-a/2}^{a/2} \left\{ \frac{36}{Eu^4} \left[(z_1 - \mu z_2)\lambda_1^2 + (z_2 - \mu z_1)\lambda_2^2 \right. \right.$$

$$\left. + 2(1 + \mu)z_3\lambda_3^2 \right] - \frac{4}{u^5\sigma_{yp}^2} (36z_1^2 - 36z_1z_2 + 36z_2^2 + 108z_3^2)$$

$$\left. \times (1 + \text{sgn } \eta_1) \right\} \delta u dx_1 \ dx_2 \tag{6.20}$$

so

$$\Lambda^2 = \frac{36}{Eu^4} \left[(z_1 - \mu z_2)\lambda_1^2 + (z_2 - \mu z_1)\lambda_2^2 + 2(1 + \mu)z_3\lambda_3^2 \right]$$

$$- \frac{4}{u^5 \sigma_{yp}^2} (36z_1^2 - 36z_1z_2 + 36z_2^2 + 108z_3^2)(1 + \text{sgn } \eta_1)$$

(6.21)

and

$$\delta\psi_3 = \int_{-b/2}^{b/2} \int_{-a/2}^{a/2} \frac{36}{Eu^4} \left[(z_1 - \mu z_2)\lambda_1^3 + (z_2 - \mu z_1)\lambda_2^3 \right.$$

$$\left. + 2(1 + \mu)z_3\lambda_3^3 \right] \delta u dx_1 \, dx_2 \qquad (6.22)$$

so

$$\Lambda^3 = \frac{36}{Eu^4} \left[(z_1 - \mu z_2)\lambda_1^3 + (z_2 - \mu z_1)\lambda_2^3 + 2(1 + \mu)z_3\lambda_3^3 \right] \quad (6.23)$$

Finally, since ψ_0 does not depend on z or ζ,

$$\delta\psi_0 = \int_{-b/2}^{b/2} \int_{-a/2}^{a/2} \delta u dx_1 \, dx_2 \qquad (6.24)$$

so

$$\Lambda^0 = 1 \qquad (6.25)$$

6.2 Numerical Results for Simply Supported Rectangular Plates

As a numerical example, a 72 in. square simply supported rectangular plate is considered. The material properties are E = 3.0×10^7 psi, μ = 0.3, and ρ = 7.43×10^{-4} lb-sec^2/in.^{4}. A uniform load of q = 1.39 psi acts over the plate. The design problem is to find the variable thickness of the plate, such that the plate is of minimum weight and satisfies certain constraints. Three typical design constraints are considered; the stress constraint of Eq. 6.8 with σ_{yp} = 36 ksi, the displacement constraint of 6.10 with Δ = 4 in. and 3 in., and the frequency constraint of Eq. 6.7 with ω_0 = 55 rad/sec (i.e., $\zeta_0 = \rho\omega_0^2$ = 2.248 lb/in.^{4}). The minimum allowable thickness of the plate is chosen as u_0 = 0.05 in., for all cases. Since the load and the kinematic conditions of the problem suggest symmetry of the stated and design variables, only a quarter of the square plate has been used in solving Eqs. 6.1, 6.4, 6.18 and 6.19. A finite element method is used for numerical analysis. Results are presented in Figs. 6.1 through 6.4 for three cases. Table 6.1 summarizes the output data.

The first case considers only stress constraints. Results in Fig. 6.1 show built-up regions near the center, at the corner, and in mid-region. The minimum thickness constraint was never tight. The maximum displacement and the natural frequency at the solution are shown in Table 6.1 along with results for the other cases.

A second case is treated, with stress and displacement constraints. For and allowable displacement of 4 in., 12 iterations were necessary to obtain a solution (Fig. 6.2). Nine more iterations were required to obtain the solution for a 3 in. allowable displacement, shown in Fig. 6.3. The shape of the optimum plate is similar to case one, but the center of the plate is thickened, while part of the edge is of minimum allowable thickness. The displacement at the center is 3.03 in., violating the displacement constraint by 1%. The average computer time for each case on an IBM 360-65 was 10 seconds per iteration.

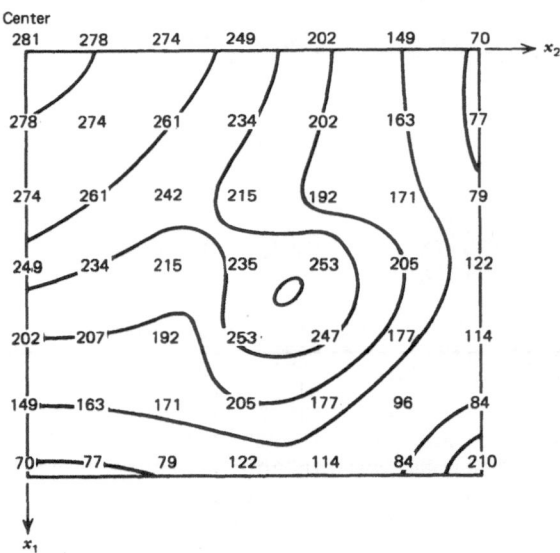

Figure 6.1   Thickness Profile for Stress Constraint (in 10^{-3} in.)

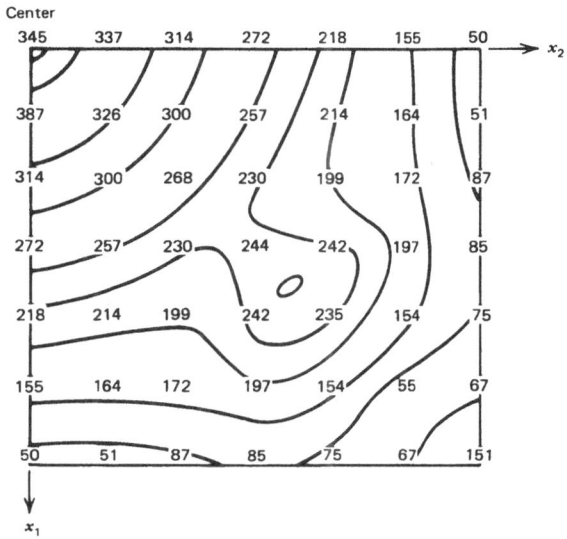

Figure 6.2   Thickness Profile for Stress Constraints
             Displacement $\leq$ 4 in. (in 10^{-3} in.)

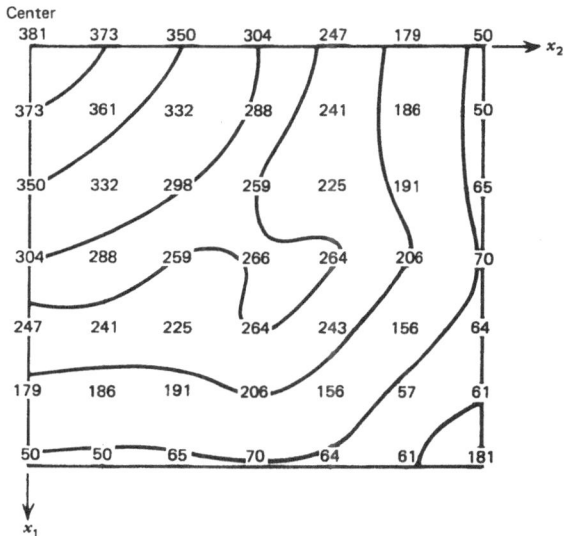

Figure 6.3   Thickness Profile for Stress Constraint,
             Displacement $\leq$ 3 in. (in 10^{-3} in.)

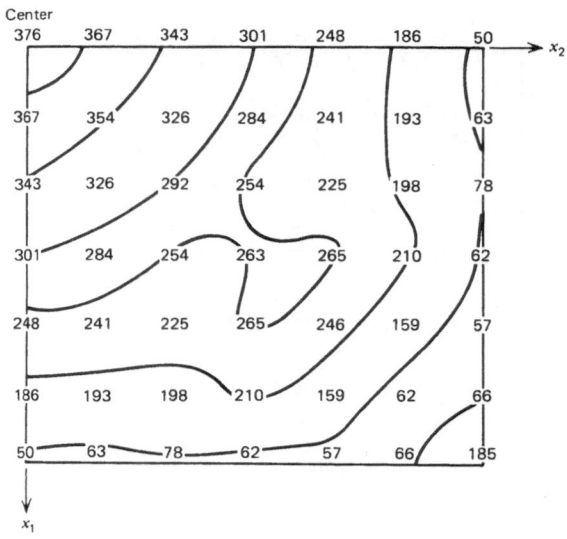

Figure 6.4   Thickness Profile for Stress Constraint Displacement
             ≤ 3 in., Frequency ≥ 55 rad/sec (in 10^{-3} in.)

A third case is treated, with stress, displacement, and
frequency constraints. Starting from the solution of the second
case with Δ = 3 in., the result shown in Fig. 6.4 is obtained
after 6 iterations. The average computer time per iterations was
11 seconds. The result shows that the frequency constraint tends
toward thickening the mid-region and the corner, with a general
shape similar to the previous cases. The optimum shapes for the
three cases treated here are strongly influenced by the stress
constraint.

TABLE 6.1   SUMMARY RESULTS FOR SIMPLY SUPPORTED PLATE

Case	Constraints Imposed	Displacement at Center (in.)	Frequency (rad/sec)	Volume (in.^{3})
1	Stress	4.933	48.3	1012
2	Stress Displacement (Δ = 4 in.)	4.046	49.9	1033
	Stress Displacement (Δ = 3 in.)	3.031	54.8	1130
3	Stress Displacement (Δ = 3 in.) Frequency	3.05	55.04	1132

REFERENCES

1. Haug, E.J. and Rousselet, B., "Design Sensitivity Analysis of Static Response Variations," _Optimization of Distributed Parameter Structures_ (Ed. E.J. Haug and J. Cea), Sijthoff & Noordhoff, Alphen aan den Rihn, Netherlands, 1980.
2. Haug, E.J. and Rousselet, B., "Design Sensitivity Analysis of Eigenvalue Variations," _Optimization of Distributed Parameter Structures_ (Eds. E.J. Haug and J. Cea), Sijthoff & Noordhoff, Alphen aan den Rihn, Netherlands,1980.
3. Taylor, J.E., "Optimal Remodeling Theory and Applications," _Optimization of Distributed Parameter Structures_ (Ed. E.J. Haug and J. Cea), Sijthoff & Noordhoff, Alphen aan den Rihm, Netherlands, 1980.
4. Haug, E.J. and Arora, J.S., _Applied Optimal Design_, Wiley-Interscience, New York, 1979.

474

DISTRIBUTED PARAMETER STRUCTURAL OPTIMIZATION FOR DYNAMIC RESPONSE[*]

Edward J. Haug and Jasbir S. Arora

Materials Division, College of Engineering
The University of Iowa, Iowa City, Iowa 52242

ABSTRACT

Problems of optimization of structures that are subjected to dynamic excitation are formulated in a distributed parameter setting. Partial differential equations of structural dynamics are used as state equations and adjoint variable design sensitivity analysis is carried out. Derivations of dynamic response measures are calculated using finite element solutions of state and adjoint differential equations. Beam and plate optimization problems with constraints on deflection, stress, natural frequency, and design variables are solved in the distributed parameter setting. The optimization algorithm is then specialized to a finite dimensional design space and a variety of examples are solved.

1.   PROBLEM FORMULATION

A major class of structural optimization problems involves spatial distribution of a design variable and transient dynamic performance constraints. Performance of these systems is described by a family of hyperbolic partial differential equations. The domain of the independent variable is a product space $\Omega \times T$, where $\Omega \subset R^K$ and $T = [0, \tau]$. The spatial boundary of Ω is denoted as Γ, so the lateral boundary of the set $\Omega \times T$ is $S = \Gamma \times T$, as shown in Fig. 1.1. Here, the scalar time dimension is denoted by t and the vector space dimension is denoted by a vector x.

---

[*]Research supported by NSF Project No.   ENG77-19967

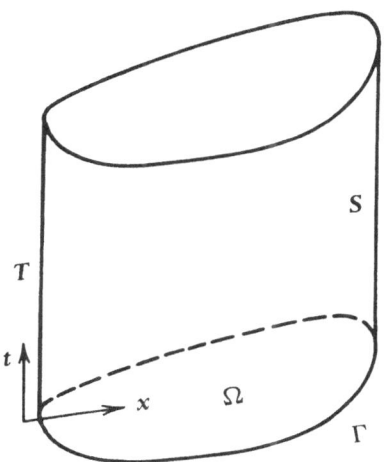

Figure 1.1   Domain of the Independent Variables

To distinguish the class of mechanical design problems from feedback control problems, the design variables considered here are only space dependent. That is, $u(x)$ is defined on Ω, for all t. The variable b is a vector of scalar design parameters. These design variables and parameters are to be chosen when the system is constructed and do not vary with the time variable, t.

The state of the system, generally a displacement field, is denoted by $z(x,t)$ and is both space and time dependent. The dynamic system problems treated here are described by initial-boundary-value problems that are linear in the state variable and can be written in differential operator notation as

$$A(u,b)z + N(u,b)z = Q(u,b,x,t) \qquad \text{in } \Omega \times T \qquad (1.1)$$

with boundary conditions

$$Bz = 0 \qquad \text{on } S = \Gamma \times T \qquad (1.2)$$

and initial conditions

$$Cz = r(x) \qquad \text{in } \Omega \text{ , with } t = 0 \qquad (1.3)$$

In Eq. 1.1, the linear differential operators A and N are spatial and temporal operators, respectively. The coefficients in these operators may depend on the design variables $u(x)$ and parameters b. The operator A will be presumed symmetric for z satisfying homogeneous boundary conditions in Eq. 1.2.

The cost functional ψ_0 and constraints are taken in the form

$$\psi_0 = \int_0^\tau \iint_\Omega f_0(z,u,b)d\Omega \; dt \tag{1.4}$$

$$\psi_j = \int_0^\tau \iint_\Omega f_j(z,u,b)d\Omega \; dt \quad \begin{cases} = 0 \; , & j=1,\ldots,r' \\[2mm] \leq 0 \; , & j=r'+1,\ldots,r \end{cases} \tag{1.5}$$

and

$$\phi_i(x,u(x)) \quad \begin{cases} = 0 \; , & i=1,\ldots,q' \\[2mm] \leq 0 \; , & i=q'+1,\ldots,q \end{cases} \quad x \in \Omega \tag{1.6}$$

The arguments of all functions appearing in Eqs. 1.1 to 1.6 may depend on x and t explicitly. These variables are suppressed here for notational convenience. It may also be noted that the state variable does not appear in the constraints of Eq. 1.6. It is presumed that constraints of the form

$$\eta(x,t,u,z,b) \leq 0 \; , \qquad \text{in } \Omega \times T \tag{1.7}$$

have been replaced by an equivalent functional constraint

$$\int_0^\tau \iint_\Omega [\eta + |\eta|]d\Omega \; dt = 0 \tag{1.8}$$

Finally, it may be noted that functionals involving integration only over Ω can be transformed to the form of one of the two multiple integrals of Eqs. 1.4 and 1.5 by multiplying by $1/\tau$ and integrating from zero to τ. No loss in computational efficiency results from this transformation.

2.   DESIGN SENSITIVITY ANALYSIS

The problem formulation presented here is especially well suited to dynamic system optimal design. Special features of the problem can be exploited to obtain an effective computational algorithm. Prior to developing a steepest descent programming algorithm for this dynamic system problem, it is necessary to determine the effect of a perturbation in design, $(\delta u, \delta b)$, on the functionals in Eqs. 1.2 and 1.5. Sensitivity analysis of the constraints of Eq. 1.6 is very simple and may be summarized in algebraic form as

$$\delta\phi_i = \frac{\partial\phi_i}{\partial u}\,\delta u \quad, \qquad x \in \Omega \tag{2.1}$$

The linearized form for perturbation of ψ_j is

$$\delta\psi_j = \int_0^\tau \iint_\Omega \left[\frac{\partial f_j}{\partial z}\,\delta z + \frac{\partial f_j}{\partial u}\,\delta u + \frac{\partial f_j}{\partial b}\,\delta b\right] d\Omega\, dt \tag{2.2}$$

The next step in developing design sensitivity data is elimination of δz from Eq. 2.1, through use of adjoint equations associated with Eqs. 1.1 to 1.3.

As a first step in design sensitivity analysis, form the scalar product of both sides of Eq. 1.1 with a function $\lambda(x,t)$, to obtain the identity

$$\int_0^\tau \left[\iint_\Omega \lambda^T A(u,b)z\,d\Omega\right] dt + \iint_\Omega \left[\int_0^\tau \lambda^T N(u,b)z\,dt\right] d\Omega$$

$$= \int_0^\tau \iint_\Omega \lambda^T Q(u,b,x,t)\,d\Omega\, dt \tag{2.3}$$

where iterated integrals have been used. Denote

$$a_{u,b}(\lambda,z) = \iint_\Omega \lambda^T A(u,b)z\, d\Omega \tag{2.4}$$

$$n_{u,b}(\lambda,z) = \int_0^\tau \lambda^T N(u,b)z\, dt \tag{2.5}$$

where integration by parts is used to reduce the order of derivatives of λ and z that appear in $a_{u,b}$.

While the bilinear form $a_{u,b}(\lambda,z)$ is symmetric, for all λ and z of sufficient regularity and satisfying the boundary conditions of Eqs. 1.2, the temporal bilinear form $n_{u,b}(\lambda,z)$ is not symmetric. Integration by parts in Eq. 2.5 allows one to write

$$n_{u,b}(\lambda,z) = \int_0^\tau \lambda^T N(u,b)z\, dt$$

$$= \int_0^\tau z^T N^*(u,b)\lambda\, dt + H(\lambda,z)\,\Big|_0^\tau \tag{2.6}$$

where the operator N* is the adjoint of N.

Using the bilinear forms of Eqs. 2.4 and 2.5, Eq. 2.3 may be written as

$$\int_0^\tau a_{u,b}(\lambda,z)dt + \iint_\Omega n_{u,b}(\lambda,z)d\Omega$$

$$= \int_0^\tau \iint_\Omega \lambda^T Q(u,b,x,t)d\Omega\ dt \tag{2.7}$$

Taking the first variation of both sides of Eq. 2.7, with λ held fixed yields the following identity among variations δu and δb in design and variation δz in state:

$$\int_0^\tau a_{u,b}(\lambda,\delta z)dt + \iint_\Omega n_{u,b}(\lambda,\delta z)d\Omega$$

$$= \int_0^\tau \iint_\Omega \lambda^T[Q_u\delta u + Q_b\delta b]d\Omega\ dt$$

$$- \int_0^\tau a_{\delta u,\delta b}^{(1)}(\lambda,z)dt - \iint_\Omega n_{\delta u,\delta b}^{(1)}(\lambda,z)d\Omega \tag{2.8}$$

where $a_{\delta u,\delta b}^{(1)}$ and $n_{\delta u,\delta b}^{(1)}$ are variations of $a_{u,b}$ and $n_{u,b}$ in the direction $(\delta u,\delta b)$.

Using symmetry of $a_{u,b}$ and the adjoint relation of Eq. 2.6 Eq. 2.8 can be written in the form

$$\int_0^\tau \iint_\Omega \delta z^T[A(u,b)\lambda + N*(u,b)\lambda]d\Omega\ dt$$

$$= - \iint_\Omega H(\lambda,\delta z)\Big|_0^\tau d\Omega + \int_0^\tau \iint_\Omega \lambda^T[Q_u\delta u + Q_b\delta b]d\Omega\ dt$$

$$- \int_0^\tau a_{\delta u,\delta b}^{(1)}(\lambda,z)dt - \iint_\Omega n_{\delta u,\delta b}^{(1)}(\lambda,z)d\Omega \tag{2.9}$$

where λ is required to satisfy the boundary conditions of Eq. 1.2. Since z satisfies the initial conditions of Eq. 1.3, $C\delta z(x,0) = 0$. This will generally cause the $H(\lambda,\delta z)|_0 = 0$. Since z is arbitrary at $t = \tau$, one can require that

$$H(\lambda(x,t), \delta z(x,\tau)) = 0 \quad , \qquad x \in \Omega \qquad (2.10)$$

for arbitrary $\delta z(x,\tau)$ to define terminal conditions on λ. Denote the resulting terminal conditions as

$$C*\lambda = 0 \quad \text{in } \Omega , \qquad \text{with } t = \tau \qquad (2.11)$$

The identity of Eq. 2.9 may now be used to write $\delta\psi_j$ of Eq. 2.2 explicitly in terms of δu and δb. One may now require λ^j to be the solution of the terminal-boundary-value problem

$$A(u,b)\lambda^j + N*(u,b)\lambda^j = \frac{\partial f_j^T}{\partial z} \qquad \text{in } \Omega \times T \qquad (2.12)$$

$$B\lambda^j = 0 \qquad \text{on } S = \Gamma \times T \qquad (2.13)$$

$$C*\lambda^j = 0 \qquad \text{in } \Omega , \qquad \text{with } t = T \qquad (2.14)$$

With this λ^j and Eq. 2.9, $\delta\psi_j$ of Eq. 2.2 may now be written as

$$\delta\psi_j = \iint_\Omega \left[\int_0^\tau \left(\frac{\partial f_i}{\partial u} + \lambda^{j^T} Q_u \right) dt - n_{1,0}^{(1)}(\lambda^j, z) \right] \delta u \, d\Omega$$

$$- \int_0^\tau a_{\delta u, \delta b}^{(1)}(\lambda^j, z) dt$$

$$+ \left\{ \iint_\Omega \left[\int_0^\tau \left(\frac{\partial f_i}{\partial b} + \lambda^{j^T} Q_b \right) dt - n_{0,1}^{(1)}(\lambda^j, z) \right] d\Omega \right\} \delta b$$

$$\equiv \ell^{j^T} b + \iint_\Omega \Lambda^j(x) \, \delta u \, d\Omega \qquad (2.15)$$

where $n_{1,0}^{(1)}$ and $n_{0,1}^{(1)}$ are $n_{\delta u, \delta b}^{(1)}$ evaluated at $\delta u = 1$, $\delta b = 0$ and $\delta u = 0$, $\delta b = 1$, respectively.

One now has sensitivity data that is independent of time. That is, the sensitivity data is given in the design variable space. One might view the process of eliminating this time dependence as collapsing the time variable out of the sensitivity analysis problem.

3. A GRADIENT PROJECTION COMPUTATIONAL ALGORITHM

A gradient projection method may now be readily developed, using the total design differentials or sensitivity coefficients that have been calculated in design space. One wishes to select a change in design $(\delta b, \delta u)$ that decreases $\delta \psi_0$ of Eq. 2.2 as much as possible, while satisfying constraints

$$\delta \tilde{\psi}_j = \ell^{j^T} \delta b + \iint_\Omega \tilde{\Lambda}^{j^T} \delta u \, d\Omega \begin{cases} = \Delta \tilde{\psi}_j, & \alpha=1,\ldots,r' \\ \\ \leq \Delta \tilde{\psi}_j, & \alpha=r'+1,\ldots,r \end{cases} \qquad (3.1)$$

for all j such that $\tilde{\psi}_j \geq -\varepsilon$, and

$$\delta \tilde{\phi}_i = \frac{\delta \tilde{\phi}_i}{\partial u} \delta u \leq \Delta \tilde{\phi}_i \, , \qquad x \in \Omega \qquad (3.2)$$

for all i and $x \in \Omega$ such that $\phi_i(x, u(x)) \geq -\varepsilon$, and

$$\delta b^T W_b \delta b + \int_\Omega \delta u^T W_u \delta u \, d\Omega \leq \xi^2 \qquad (3.3)$$

where $\varepsilon > 0$ is a small parameter. Here, a super $\sim$ denotes constraints that are active or violated and must be enforced. In Eq. 3.1, $\Delta \tilde{\psi}_j$ is the correction in constraint error desired, normally $\Delta \tilde{\psi}_j = -\tilde{\psi}_j$. Similarly, $\Delta \tilde{\phi}_i$ is a correction in pointwise constraint error, normally $\Delta \tilde{\phi}_i = -\tilde{\phi}_i$. The constraint of Eq. 3.3 is simply a form of step size restrictions to assure that $(\delta b, \delta u)$ is small enough so that the linear approximations used in computing design sensitivity coefficients are sufficiently accurate. Finally W_b and W_u are simply positive definite weighting martices matrices and ξ is a small parameter.

The conditions of Ref. 1 may now be directly applied to write a computational algorithm. Define

$$\tilde{\gamma} \equiv -M_{\psi\psi}^{-1} [2\nu (\Delta \tilde{\psi} - M_{\psi\phi}) + M_{\psi\psi_0}] \qquad (3.4)$$

and

$$\widetilde{\mu}(x) \equiv - \left(\frac{\partial \widetilde{\phi}}{\partial u} W_u^{-1} \frac{\partial \widetilde{\phi}^T}{\partial u}\right)^{-1} \left[2\nu\Delta\widetilde{\phi} + \frac{\partial \widetilde{\phi}}{\partial u} W_u^{-1} (\Lambda^0 + \widetilde{\Lambda}^T\widetilde{\gamma})\right] \quad (3.5)$$

where

$$M_{\psi\psi_0} \equiv \int_\Omega \widetilde{\Lambda}^T W_u^{-1} \left[I - \frac{\partial \widetilde{\phi}^T}{\partial u} \left(\frac{\partial \widetilde{\phi}}{\partial u} W_u^{-1} \frac{\partial \widetilde{\phi}^T}{\partial u}\right)^{-1} \frac{\partial \widetilde{\phi}}{\partial u} W_u^{-1}\right] \Lambda^0 \, d\Omega$$

$$+ \widetilde{\ell}^T W_u^{-1} \ell^0 \quad (3.6)$$

$$M_{\psi\psi} \equiv \int_\Omega \widetilde{\Lambda}^T W_u^{-1} \left[I - \frac{\partial \widetilde{\phi}^T}{\partial u} \left(\frac{\partial \widetilde{\phi}}{\partial u} W_u^{-1} \frac{\partial \widetilde{\phi}^T}{\partial u}\right)^{-1} \frac{\partial \widetilde{\phi}}{\partial u} W_u^{-1}\right] \widetilde{\Lambda} d\Omega$$

$$+ \ell^T W_b^{-1} P \quad (3.7)$$

and

$$M_{\psi\phi} \equiv \int_\Omega \widetilde{\Lambda}^T W_u^{-1} \frac{\partial \widetilde{\phi}^T}{\partial u} \left(\frac{\partial \widetilde{\phi}}{\partial u} W_u^{-1} \frac{\partial \widetilde{\phi}^T}{\partial u}\right)^{-1} \Delta\widetilde{\phi} \, d\Omega \quad (3.8)$$

Then,

$$\delta b = - \frac{1}{2\nu} \delta b^1 + \delta b^2 \quad (3.9)$$

and

$$\delta u(x) = - \frac{1}{2\nu} \delta u^1(x) + \delta u^2(x) \quad (3.10)$$

where

$$\delta b^1 = W_b^{-1}[\ell^0 - \widetilde{\ell} M_{\psi\psi}^{-1} M_{\psi\psi_0}] \quad (3.11)$$

$$\delta b^2 = W_b^{-1} \widetilde{\ell} M^{-1}[\Delta\widetilde{\psi} - M_{\psi\phi}] \quad (3.12)$$

$$\delta u^1(x) = W_u^{-1} \left[I - \frac{\partial \tilde{\phi}^T}{\partial u} \left(\frac{\partial \tilde{\phi}}{\partial u} W_u^{-1} \frac{\partial \tilde{\phi}^T}{\partial u} \right)^{-1} \frac{\partial \tilde{\phi}}{\partial u} W_u^{-1} \right]$$

$$\times \left[\Lambda^0 - \tilde{\Lambda} M_{\psi\psi}^{-1} M_{\psi\psi_0} \right] , \qquad x \in \Omega \qquad (3.13)$$

$$\delta u^2(x) = W_u^{-1} \left[I - \frac{\partial \tilde{\phi}^T}{\partial u} \left(\frac{\partial \tilde{\phi}}{\partial u} W_u^{-1} \frac{\partial \tilde{\phi}^T}{\partial u} \right)^{-1} \frac{\partial \tilde{\phi}}{\partial u} W_u^{-1} \right]$$

$$\times \left[\tilde{\Lambda} M_{\psi\psi}^{-1} (\Delta \tilde{\psi} - M_{\psi\phi}) \right] + W_u^{-1} \frac{\partial \tilde{\phi}^T}{\partial u} \left(\frac{\partial \tilde{\phi}}{\partial u} W_u^{-1} \frac{\partial \tilde{\phi}^T}{\partial u} \right)^{-1} \Delta \tilde{\phi} ,$$

$$x \in \Omega \qquad (3.14)$$

If the constraints are all satisfied, $\delta u^2(x)$ and δb^2 will all be zero. In this case,

$$\Delta \psi_0 = - \frac{1}{2\nu} \left[M_{\psi_0\psi_0} - M_{\psi\psi_0}^T M_{\psi\psi}^{-1} M_{\psi\psi_0} \right] \qquad (3.15)$$

where

$$M_{\psi_0\psi_0} = \ell^{0^T} W_b^{-1} \ell^0 + \int_\Omega \Lambda^{0^T} W_u^{-1}$$

$$\times \left[I - \frac{\partial \tilde{\phi}^T}{\partial u} \left(\frac{\partial \tilde{\phi}}{\partial u} W_u^{-1} \frac{\partial \tilde{\phi}^T}{\partial u} \right)^{-1} \frac{\partial \tilde{\phi}}{\partial u} W_u \right] \Lambda^0 \, d\Omega \qquad (3.16)$$

One can now specify a reasonable, desired reduction $\Delta \psi_0 < 0$ in ψ_0 and determine the parameter ν from Eq. 3.15 to yield the desired reduction in the cost function.

In summary form, one has the following <u>gradient projection</u> <u>algorithm</u>:

Step 1. Make an engineering estimate $u^{(0)}(x)$ and $b^{(0)}$ of the optimum design function and parameter.

Step 2. Solve Eqs. 1.1 to 1.3 for $z^{(0)}(x,t)$ corresponding to $u^{(0)}(x)$ and $b^{(0)}$.

Step 3. Check constraints and form vectors to tight and violated constraints $\tilde{\psi}$ and $\tilde{\phi}$, of Eqs. 1.5 and 1.6.

Step 4. Solve Eqs. 2.12 to 2.14 corresponding to both the funtionals ψ_0 and $\tilde{\psi}_j$; for λ^0 and λ^j, respectively.

Step 5. Compute $\Lambda^0(x)$, ℓ^0, $\tilde{\Lambda}(x)$, and $\tilde{\ell}$ in Eq. 2.15.

Step 6. Choose constraint corrections $\Delta\tilde{\psi}_j$ and $\Delta\tilde{\phi}_i$.

Step 7. Compute $M_{\psi\psi_0}$, $M_{\psi\psi}$, and $M_{\psi\phi}$ in Eqs. 3.6 through 3.8.

Step 8. Choose $\nu > 0$ and compute $\tilde{\gamma}$ and $\tilde{\mu}(x)$ in Eqs. 3.4 and 3.5. If any components of $\tilde{\gamma}$ with $j > r'$ or $\tilde{\mu}(x)$ with $i > q'$ are negative, redefine $\tilde{\psi}$ or $\tilde{\phi}(x)$ by deleting corresponding terms and return to Step 5.

Step 9. Compute $\delta u^1(x)$, $\delta u^2(x)$, δb^1, and δb^2 in Eqs. 3.11 through 3.14.

Step 10. Compute

$$u^{(1)}(x) = u^{(0)}(x) - \frac{1}{2\nu} \delta u^1(x) + \delta u^2(x)$$

and

$$b^{(1)} = b^{(0)} - \frac{1}{2\nu} \delta b^1 + \delta b^2$$

Step 11. If the constraints are satisfied and $||\delta u^1(x)||$ and $||\delta b^1||$ are sufficiently small, terminate. Otherwise, return to Step 2 with $u^{(0)}(x)$ and $b^{(0)}$ replaced by $u^{(1)}(x)$ and $b^{(1)}$, respectively, and continue.

Finally, it was found that when constraints $\eta(x,t,z,u) \leq 0$ are treated by the equivalent function formulation, $\psi = \int_0^\tau \iint_\Omega (\eta + |\eta|) d\Omega \, dt = 0$, an ε-active constraint treatment was helpful. The sensitivity analysis is performed with

$$\psi_\varepsilon = \int_0^\tau \iint_\Omega [(\eta + \varepsilon) + |\eta + \varepsilon|] d\Omega \, dt$$

i.e., the right side of the adjoint Eq. 2.12 is

$$\frac{\partial f^T}{\partial z} = [1 + \text{sgn}(\eta + \varepsilon)] \frac{\partial \eta^T}{\partial z}$$

In this way, even though η may be positive over a very small sub-set of $\Omega \times T$, numerical error in adjoint calculation is avoided

484

since $\eta + \varepsilon$ is positive over a larger domain. The desired change in ψ is retained as $\Delta\psi = -\int_0^\tau \iint_\Omega (\eta + |\eta|)d\Omega \, dt$.

It may be noted that if there is no spatial variable in the problem and the state variable reduces to a time dependent vector, as is the case when one treats a finite element structure, the optimization problem is considerably simplified. In this case, there is only a design parameter and spatial operator $A(b)$ is just a matrix. The above algorithm applies and considerable efficiencies can be effected by taking advantage of the simplified mathematical structure. An in-depth treatment of this finite dimensional dynamics problem is carried out in Section 6 of this paper. Attention is restricted in examples of Sections 4 and 5 to distributed design over one (beam) and two (plate) space dimensions.

4. DYNAMIC OPTIMIZATION OF BEAMS

As a first example, one wishes to distribute material along a beam (one space dimension) to minimize weight, subject to constraints on transient dynamic response. For simplicity, beams treated herein are loaded by uniform spatially distributed, time varying loads. For numerical implementation of the gradient projection algorithm, a finite element analysis method is employed (Ref. 2) to obtain approximate solutions of the equations of motion. The computer used was an IBM 360/65.

For the simply supported beam of Fig. 4.1, one wishes to minimize total weight, subject to the conditions that displacement is always within specified bounds and the cross section is bounded by predetermined strength requirements. The design variable is the cross sectional area $u(x)$ and geometrically similar cross sections are employed, so $I = \alpha u^2$. The system dynamic equations are

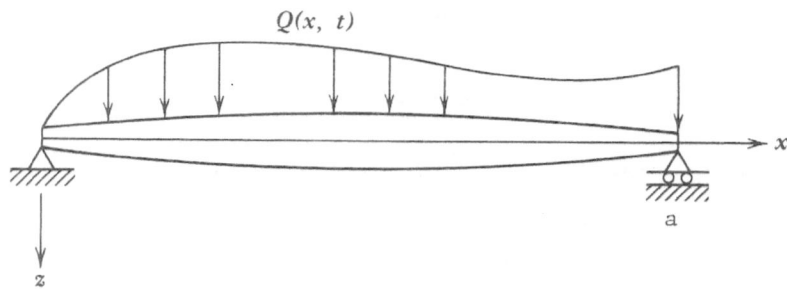

Figure 4.1   Simply Supported Beam with Dynamic Load

$$\left[E\alpha u^2 z''\right]'' + \rho u\ddot{z} = Q(x,t) \qquad \text{in } \Omega \times T$$

where $\dfrac{\partial z}{\partial x} \equiv z'$ and $\dfrac{\partial z}{\partial t} \equiv \dot{z}$, with boundary conditions

$$z(0,t) = z''(0,t) = 0 \qquad \text{in } T$$

$$z(a,t) = z''(a,t) = 0 \qquad \text{in } T$$

and initial conditions

$$z(x,0) = \dot{z}(x,0) = 0 \qquad \text{in } \Omega$$

$$(4.1)$$

In this problem, the space domain is $\Omega = [0,a]$ and Γ is just the two points $x = 0$ and a. One may readily verify that the operator $Az = \dfrac{\partial^2}{\partial x^2}(E\alpha u^2 z'') \equiv (E\alpha u^2 z'')''$ is symmetric and that $Nz = \rho u\ddot{z}$. For this operator A,

$$a_{u,b}(\lambda,z) = \int_0^a \lambda (E\alpha u^2 z'')'' dx$$

$$= \int_0^a E\alpha u^2 \lambda'' z'' dx$$

$$(4.2)$$

and

$$a^{(1)}_{\delta u,\delta b}(\lambda,z) = \int_0^a 2E\alpha u\delta u\lambda'' z'' dx \qquad (4.3)$$

Similarly, for the temporal operator N,

$$n_{u,b}(\lambda,z) = \int_0^\tau \lambda\rho u\ddot{z}\,dt$$

$$= \int_0^\tau z\rho u\ddot{\lambda}\,dt + (\lambda\rho u\dot{z} - \dot{\lambda}\rho uz)\Big|_0^\tau$$

$$(4.4)$$

and

$$n^{(1)}_{\delta u,\delta b}(\lambda,z) = \int_0^\tau z\rho\delta u\ddot{\lambda}\,dt + (\lambda\rho\delta u\dot{z} - \dot{\lambda}\rho\delta uz)\Big|_0^\tau \qquad (4.5)$$

Thus, in Eq. 2.6,

$$N*(u,b)\lambda = \rho u \ddot{\lambda}$$

$$H(\lambda,z) = \rho u(\lambda \dot{z} - \dot{\lambda} z)$$

$$(4.6)$$

Since δz and $\delta \dot{z}$ are arbitrary at $t = \tau$, Eqs. 2.10 and 4.6 yield terminal conditions on λ as

$$\lambda(\tau) = \dot{\lambda}(\tau) = 0 \tag{4.7}$$

The weight to be minimized is

$$\psi_0 = \int_0^a \gamma u dx = \int_0^\tau \int_0^a \frac{\gamma}{\tau} u(x) dx \, dt \tag{4.8}$$

where γ is material weight density. The displacement constraint

$$|z(x,t)| \leq d \qquad \text{in } \Omega \times T \tag{4.9}$$

is transformed to functional form as

$$\psi_1 = \int_0^\tau \int_0^a \left[|z| - d + \left| |z| - d \right| \right] dx \, dt = 0 \tag{4.10}$$

Finally, a lower bound is placed on the cross sectional area. This is

$$\phi_1 = -u(x) + u_0 \leq 0 \qquad \text{in } \Omega \tag{4.11}$$

where $u_0 > 0$ is given

The functional constraint of Eq. 4.10 requires solution of an adjoint problem. The adjoint equation of Eq. 2.12 is just the differential equation of Eq. 4.1, with the applied load replaced by

$$\bar{Q} = \frac{\partial f_1}{\partial z} = \text{sgn}(z) [1 + \text{sgn}(|z| - d)] \tag{4.12}$$

and with initial conditions replaced by terminal conditions

$$\lambda(x,\tau) = \dot{\lambda}(x,\tau) = 0 \qquad \text{in } \Omega \tag{4.13}$$

By inspection $\Lambda^0 = \frac{\gamma}{\tau}$ and $\frac{\partial \phi_1}{\partial u} = -1$. From Eqs. 2.15, 4.3, 4.5 (with an integration by parts), and 4.10,

$$\Lambda^1(x) = \int_0^\tau [-2E\alpha u\lambda''z'' + \rho\ddot{\lambda}\ddot{z}]dt \tag{4.14}$$

All information is now available to implement the optimization algorithm of Section 3.

As an example, a simply supported beam with a = 40 in., α = 0.3, E = 30 × 10^6 psi, γ = 0.28 lb/in.^{3} and $\dot{z}(x,0)$ = 0, is loaded uniformly with

$$Q(x,t) = \begin{cases} 9 \sin(80\pi t) \text{ lb/in.,} & 0 \le t \le 0.0125 \text{ sec} \\ 0 \text{ lb/in.,} & 0.0125 \le t \le 0.0150 \text{ sec} \end{cases}$$

Constraint parameters are d = 0.4 in. and u_0 = 0.2513 in.^{2}. Due to symmetry, only half of the beam is analyzed numerically. The half beam is divided into ten finite elements, with a total of 20 degrees of freedom. Five eigenvectors were used for modal analysis. The load is assumed concentrated at nodal points with

$$Q_1(t) = \begin{cases} 18 \sin(80\pi t) \text{ lb,} & 0 \le t \le 0.0125 \text{ sec} \\ 0 \text{ lb,} & 0.0125 \le t \le 0.0150 \text{ sec} \end{cases}$$

at each point for analysis. The initial design shown in Column 1 of Table 4.1 weighed 2.184 lb and $||\delta u^1|| \equiv \left[\int_0^a [\delta u^1]^2 dx\right]^{\frac{1}{2}}$ = 1.252. The displacement constraint was violated by 4.5% in the sixth iteration and remained active until the twenty-second iteration, where $||\delta u^1||$ = 0.0559. Results are shown in Table 4.1 and the optimum profile and the volume reduction history are plotted in Fig. 4.2 and 4.7, respectively.

Numerical results obtained using the finite dimensional formulation of Section 6 are also given in Table 4.1, for comparative purposes.

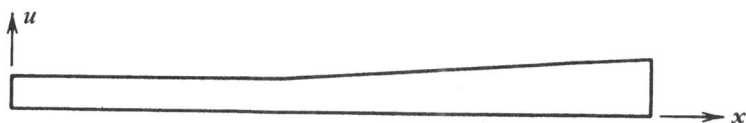

Figure 4.2   Optimum Area Profile of Half of
             Simply Supported Beam

488

To further illustrate the method, the fixed-fixed beam of Fig. 4.3 is treated. The only modification required is to change the boundary conditions in Eq. 4.1 to

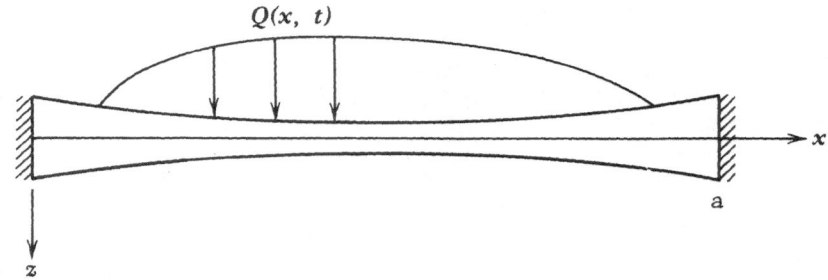

Figure 4.3   Clamped Beam with Dynamic Load.

$$\left.\begin{array}{l} z(0,t) = \dfrac{\partial z}{\partial x}(0,t) = 0 \\[2mm] z(a,t) = \dfrac{\partial z}{\partial x}(a,t) = 0 \end{array}\right\} \quad \text{in } T \qquad (4.15)$$

The remaining problem formulation and equations required for implementation of the gradient projection method remain the same. Implementation, thus requires only minor modifications in the computer program used to solve the simply supported beam problem. The load and beam properties are assumed to be the same as in the simply supported case.

Design constraint parameters are $d = 0.1$ in. and $u = 0.2413$ in.^{2}. Only half the beam is modeled, but with one end of the half beam fixed. A 19 degree of freedom finite element model was employed and five eigenvectors were used for modal analysis. The first design resulted in $||\delta u^1|| = 1.252$ without violating any constraints. The displacement constraint was first violated in the seventh iteration and remained active during the design process, except in iterations 12, 16, 25, and 26. The convergence measure was $||\delta u^1|| = 0.0235$ in the final design shown in Table 4.1. The optimum profile and the design volume history is shown in Figs. 4.4 and 4.7, respectively.

Figure 4.4   Optimum Area Profile of Half of Clamped Beam

For completeness, the cantilever beam of Fig. 4.5 is treated. Here, the boundary conditions of Eq. 4.1 are changed to

$$\left. \begin{array}{l} z(0,t) = \dfrac{\partial z}{\partial x}(0,t) = 0 \\[2em] \dfrac{\partial^2 z}{\partial x^2}(a,t) = \dfrac{\partial^3 z}{\partial x^3}(a,t) = 0 \end{array} \right\} \quad \text{in } T \tag{4.16}$$

All other equations in the formulation and solution remain the same. Only minor changes in the computer program are required to solve this problem.

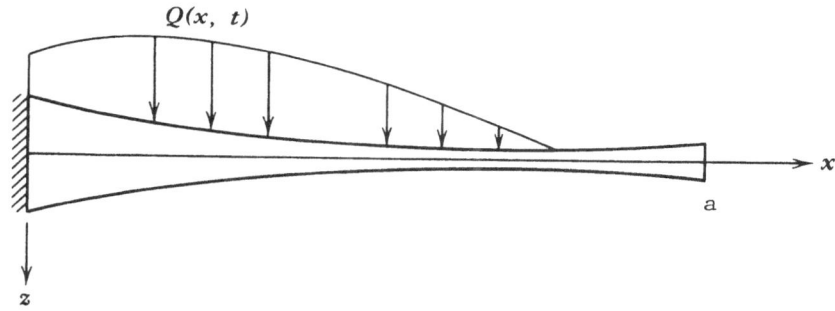

Figure 4.5   Cantilever Beam with Dynamic Load

The length of a cantilever beam is chosen to be half the length of the two beams previously, with the same load and other properties. Constraint parameters are as in the simply supported case. The cantilever beam is partitioned into ten elements for analysis, so it has 20 degrees of freedom. Five eigenvectors were used in modal analysis. At the initial design, $||\delta u^1|| = 1.208$. At the final iteration, $||\delta u^1|| = 0.147$. About half the beam, at the free end, reached the lower design variable bound before the displacement constraint was violated in the eighth design. This constraint violated was corrected to 0.58%. Design results are listed in Table 4.1 and the optimum profile and the volume history are shown in Figs. 4.6 and 4.7, respectively.

Figure 4.6   Optimum Area Profile of Cantilever Beam

490

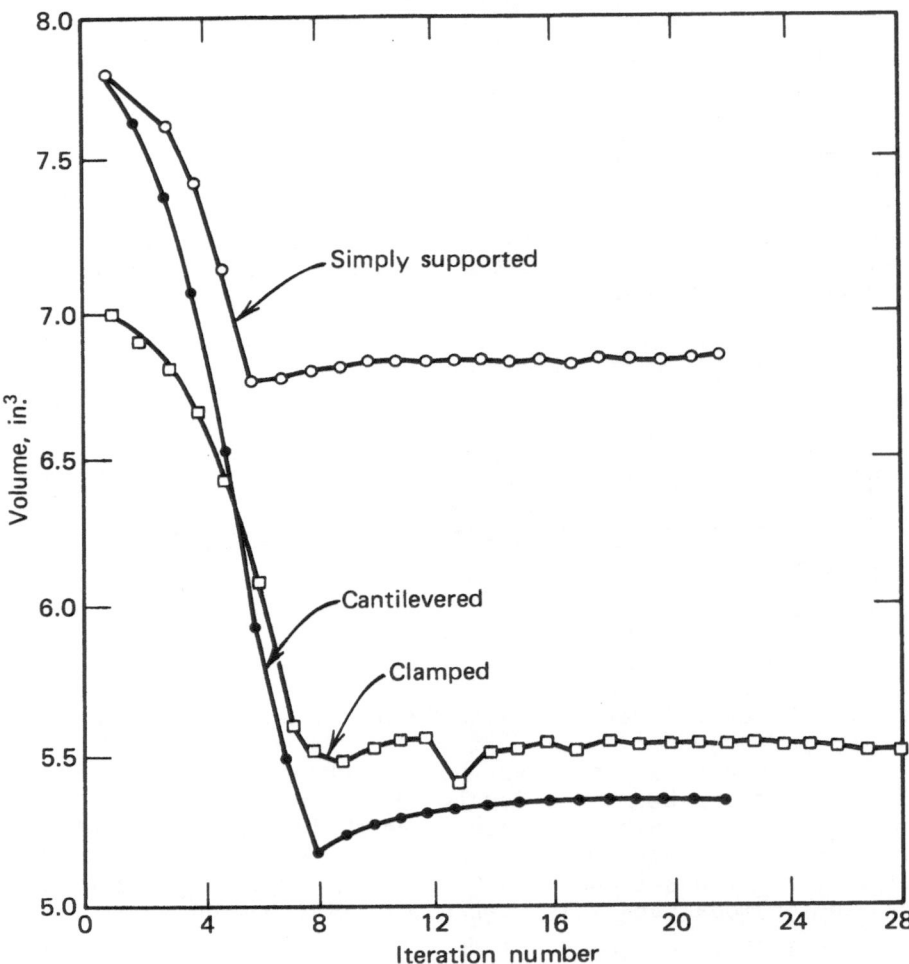

Figure 4.7   Volume-Iteration Curves of Examples

TABLE 4.1 INITIAL AND FINAL BEAM DESIGNS

Examples Designs	Simply Supported Beam			Clamped Beam			Cantilever Beam		
	Initial	Final	Discrete*	Initial	Final	Discrete*	Initial	Final	Discrete*
u_1 (in.^{2})	0.3000	0.2513	0.2513	0.4000	0.3511	0.3351	0.4800	0.3183	0.3143
u_2	0.3200	0.2513	0.2513	0.3800	0.3471	0.3331	0.4600	0.2975	0.2974
u_3	0.3400	0.2513	0.2513	0.3600	0.3045	0.3080	0.4400	0.2848	0.2812
u_4	0.3600	0.2808	0.2513	0.3400	0.2513	0.2513	0.4200	0.2683	0.2861
u_5	0.3800	0.3313	0.3702	0.3200	0.2513	0.2513	0.4000	0.2516	0.2513
u_6	0.4000	0.3794	0.3975	0.3000	0.2513	0.2513	0.3800	0.2513	0.2513
u_7	0.4200	0.4103	0.4093	0.3200	0.2513	0.2513	0.3600	0.2513	0.2513
u_8	0.4400	0.4223	0.4520	0.3400	0.2513	0.2513	0.3400	0.2513	0.2513
u_9	0.4600	0.4228	0.4540	0.3600	0.2513	0.2932	0.3200	0.2513	0.2513
u_{10}	0.4800	0.4231	0.3777	0.3800	0.2513	0.2650	0.3000	0.2513	0.2513
Wt (lb)	2.1840	1.9174	1.9409	1.9600	1.5466	1.5629	2.1840	1.4991	1.5046
Violation %					0.03			0.58	
Time (sec)		25.45	67.87		32.81	14.42		28.27	25.80
Time/Iter (sec)		1.16	2.42		1.17	2.06		1.28	1.98

*Results obtained with finite dimensional method, Section 6.

5. DYNAMIC OPTIMIZATION OF RECTANGULAR PLATES

To illustrate the distributed parameter method on a higher dimensional problem, optimal design of rectangular plates (two space dimensions) is considered. Simply supported plates are subjected to uniform spatially distributed, time varying loads and their transient dynamic response is constrained. One seeks to determine the thickness variation $u(x_1, x_2)$ over the plate to minimize weight, subject to dynamic response constraints.

For the simply supported rectangular plate of Fig. 5.1, one wishes to choose the thickness variation $u(x_1, x_2)$ to minimize weight, subject to the conditions that displacement is within prescribed bounds and the thickness is bounded uniformly away from zero. Using subscript notation for partial differentiation, the system dynamic equations in this case may be written as (Ref. 3)

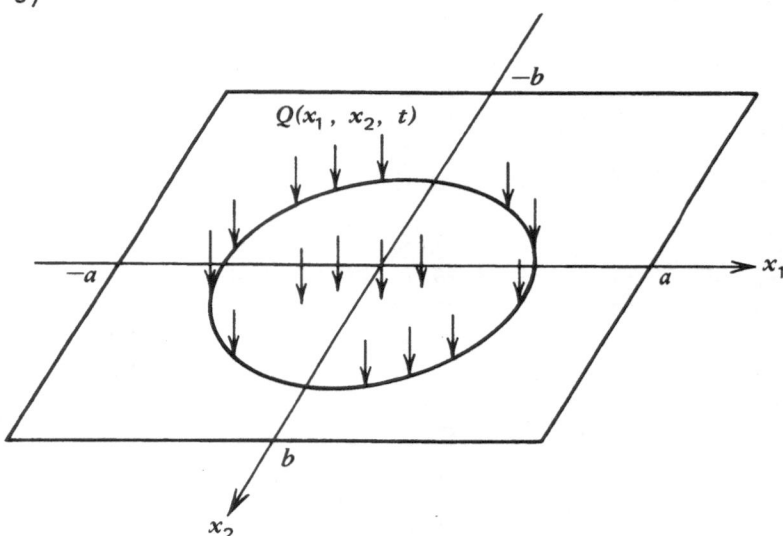

Figure 5.1   A Rectangular Plate with Dynamic Load

$$[D(u)(z_{11} + \mu z_{22})]_{11} + [D(u)(z_{22} + \mu z_{11})]_{22}$$

$$+ [2(1 - \mu)D(u)z_{12}]_{12} + \rho u \ddot{z} = Q(x_1, x_2, t) \quad \text{in } \Omega \times T$$

with boundary conditions

$$z(x_1,x_2,t) = 0$$

$$z_{11}(\pm a,x_2,t) = 0 \qquad \left. \right\} \quad \text{in } \Omega \times T$$

$$z_{22}(x_1,\pm b,t) = 0$$

$$\left. \right\} (5.1)$$

and with initial conditions

$$z(x_1,x_2,0) = 0$$

$$\dot{z}(x_1,x_2,0) = 0 \qquad \left. \right\} \quad \text{in } \Omega$$

Here, $D(u) = \dfrac{Eu^3}{12(1 - \mu^2)}$

One may verify that the operator

$$Az \equiv [D(u)(z_{11} + \mu z_{22})]_{11} + [D(u)(z_{22} + \mu z_{11})]_{22}$$

$$+ [2(1 - \mu)D(u)z_{12}]_{12} \tag{5.2}$$

is symmetric and $Nz \equiv \rho u \ddot{z}$ is the same as for the beam. Further,

$$a_{u,b}(z,\lambda) = (A(u)z, \lambda) = \iint_\Omega D(u)[z_{11}\lambda_{11} + \mu z_{22}\lambda_{11} + z_{22}\lambda_{22}$$

$$+ \mu z_{11}\lambda_{22} + 2(1 - \mu)z_{12}\lambda_{12}]d\Omega \tag{5.3}$$

and

$$a^{(1)}_{\delta u,\delta b}(z,\lambda) = \iint_\Omega D^1(u)\delta u[z_{11}\lambda_{11} + \mu z_{22}\lambda_{11} + z_{22}\lambda_{22}$$

$$+ \mu z_{11}\lambda_{22} + 2(1 - \mu)z_{12}\lambda_{12}]d\Omega \tag{5.4}$$

The temporal bilinear form $n_{u,b}(z,\lambda)$ and its design varia-
tion $n^{(1)}_{\delta u,\delta b}(z,\lambda)$ are the same as for the beam in Eqs. 4.4 and 4.5.
Similarly, the terminal conditions on λ are as in Eq. 4.7.

The weight to be minimized is

$$\psi_0 = \iint_\Omega \gamma u dx = \iint_\Omega \int_0^\tau \frac{\gamma}{\tau} u dx dt \tag{5.5}$$

where γ is the material weight density. The displacement constraint

$$|z(x_1,x_2,t)| \leq d \quad \text{in } \Omega \times T$$

is transformed to functional form as

$$\psi_1 = \iint_\Omega \int_0^\tau \Big[(|z| - d) + \big| |z| - d \big| \Big] dx dt = 0 \tag{5.6}$$

Finally, a lower bound is placed on the plate thickness

$$\phi_1 = -u(x_1,x_2) + u_0 \leq 0 \quad \text{in } \Omega \tag{5.7}$$

The functional constraint of Eq. 5.6 requires solution of an adjoint problem. The adjoint equation is the differential equation of Eq. 5.1, with the applied load replaced by

$$\bar{Q} = \frac{\partial f_1}{\partial z} = \text{sgn}(z) [1 + \text{sgn}(|z| - d)] \quad \text{in } \Omega \tag{5.8}$$

and with initial conditions replaced by terminal conditions

$$\lambda(x_1,x_2,\tau) = \dot{\lambda}(x_1,x_2,\tau) = 0 \quad \text{in } \Omega \tag{5.9}$$

By inspection, $\Lambda^0 = \frac{\gamma}{\tau}$ and $\frac{\partial \phi_1}{\partial u} = -1$. From Eqs. 2.15, 4.5, 5.4 and 5.6,

$$\Lambda^0(x_1,x_2) = -\int_0^\tau \Big\{ D'(u)[\lambda_{11}z_{11} + \mu\lambda_{11}z_{22} + \lambda_{22}z_{22}$$

$$+ \mu\lambda_{22}x_{11}] + 2(1 - \mu)D'(u)\lambda_{12}z_{12} - \rho\lambda\ddot{z} \Big\} dt \tag{5.10}$$

All information is now available to implement the optimization algorithm of Section 3.

As a numerical example, a square plate $a = b = 20$ in. is chosen, with material properties $E = 30 \times 10^6$ psi, $\mu = 0.3$, and $\gamma = 0.28$ lb/in.^{3}. The plate is loaded with a time varying load that is uniformly distributed over Ω,

$$Q(x_1,x_2,t) = \begin{cases} \sin(80\pi t) \text{ lb/in.}^2, & 0 \le t \le 0.0125 \text{ sec} \\ 0 \text{ lb/in.}^2, & 0.0125 \le t \le 0.0175 \text{ sec} \end{cases}$$

Constraint parameters are d = 0.4 in. and u_0 = 0.0625 in.

In Fig. 5.2, one quarter of the plate shown is divided into 25 finite elements for analysis. The finite element model has 75 degrees of freedom and six eigenfunctions were employed for modal analysis. The load was discretized in the same manner as in the beam problems. The initial design estimate weighed 20.61 lb. The maximum deflection was 0.428 in. at t = 0.0165 sec. In the twelfth design iteration, the constraint was completely satisfied. Numerical results are shown in Table 5.1. Note that the thickness of the plate is symmetric with respect to the bisector of the x_1' - x_2' axes in Fig. 5.2. A sketch of the resultant thickness contours is shown in Fig. 5.3.

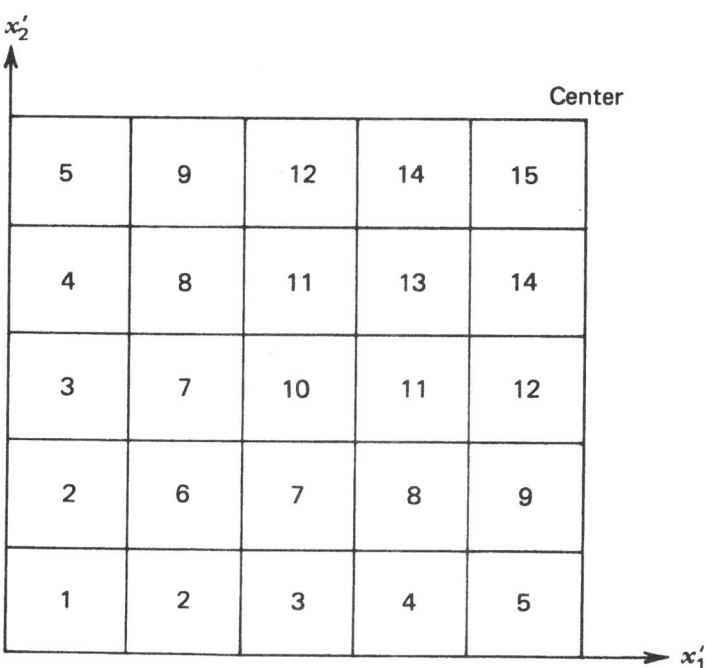

Figure 5.2   Finite Element Model of Quarter of Square Plate

496

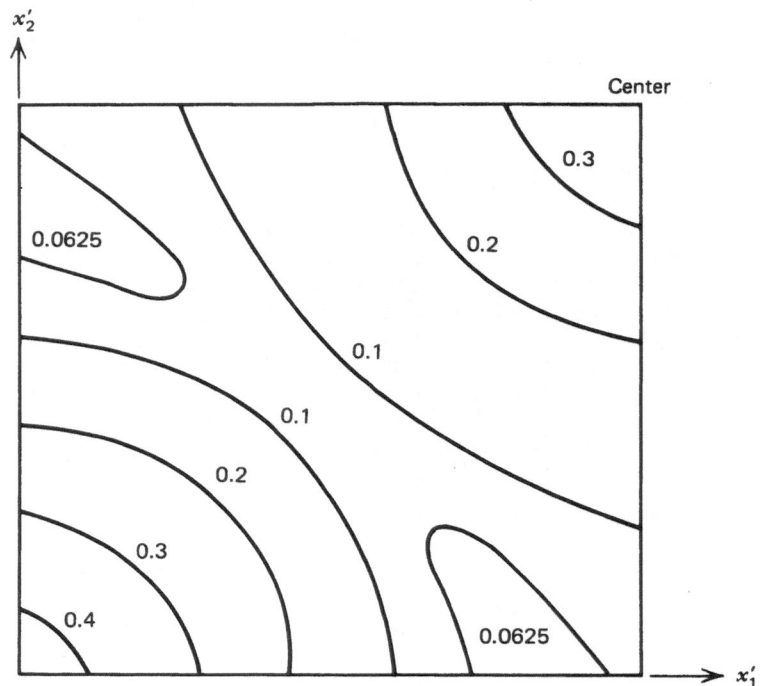

Figure 5.3   Optimum Thickness Contours of Simply Supported Plate

     While convergence properties for this problem were not as good as with the beam problems, convergence from several design estimates led to the same design weight given in Table 5.1.

     The plate optimization problem for a clamped plate follows directly from the preceding analysis. The only modification required is to change the boundary conditions in Eq. 5.1 to

$$z(x_1,x_2,t) = 0$$

and

$$\frac{\partial z}{\partial x_1}(\pm a,x_2,t) = 0$$

$$\frac{\partial z}{\partial x_2}(x_1,\pm b,t) = 0$$

$$\text{in } \Gamma \times T \qquad (5.11)$$

The remaining problem formulation and equations required for implementation of the gradient projection method are the same as in the simply supported plate.

As a numerical example, a plate with the same planar dimensions and material properties is subjected to the same load as the simply supported plate. Constraint parameters are d = 0.1 in. and u_0 = 0.0625 in. In Fig. 5.2 one quarter of the plate is shown divided into 25 elements for analysis. The resulting finite element model has 65 degrees of freedom and six eigenfunctions were used for analysis. The initial design estimate had a weight of 25.58 lb. Constraint violation was found to be about 4.6% in the second iteration. The optimum design was found to have a design weight of 21.30 lb, with a maximum displacement constraint violation of 1.5%. Numerical results are shown in Table 5.1 and the resultant thickness contours are shown in Fig. 5.4. Numerical calculations were more stable in this problem than the simply supported case. Several initial design estimates led to the same optimum.

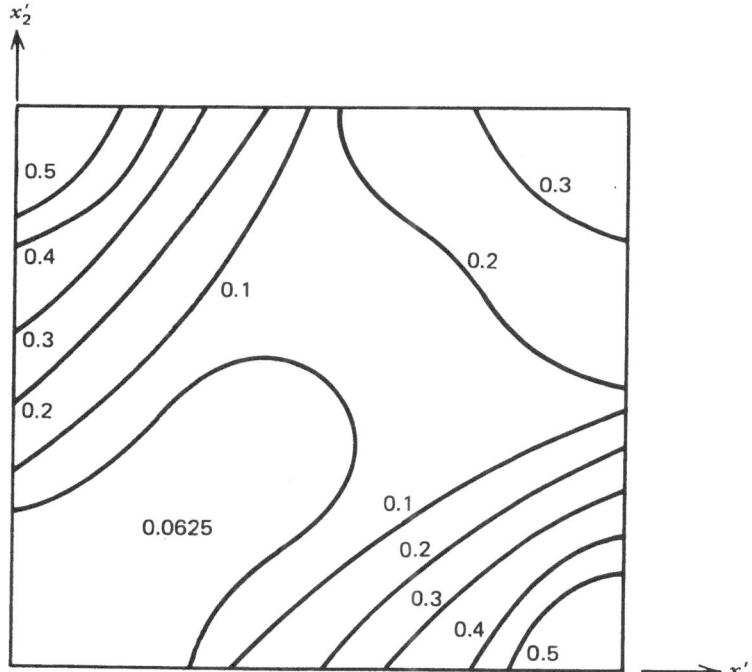

Figure 5.4 Optimum Thickness Controus of Clamped Plate

6.   FINITE DIMENSIONAL FORMULATION

The general method of design sensitivity analysis presented in Section 2 can be applied to systems described by a finite

TABLE 5.1   INITIAL AND FINAL PLATE DESIGNS

Example	Simply Supported Plate		Clamped Plate	
Design	Initial	Final	Initial	Final
u_1 (in.)	0.3600	0.3993	0.1100	0.0625
u_2	0.2600	0.2791	0.1100	0.0625
u_3	0.1600	0.1316	0.2200	0.1475
u_4	0.1100	0.0625	0.4200	0.3482
u_5	0.1100	0.0911	0.6200	0.5554
u_6	0.2600	0.2635	0.1100	0.0625
u_7	0.1600	0.1020	0.1100	0.0628
u_8	0.1100	0.0788	0.1200	0.0779
u_9	0.1100	0.1335	0.2200	0.2641
u_{10}	0.1600	0.1330	0.1100	0.0685
u_{11}	0.1600	0.1856	0.1100	0.0797
u_{12}	0.1600	0.1892	0.1200	0.1438
u_{13}	0.2600	0.2865	0.2200	0.1666
u_{14}	0.2600	0.2816	0.3200	0.2739
u_{15}	0.3600	0.3708	0.4200	0.3665
Wt. (lb)	20.61	20.12	25.58	21.31
Violation %	7.03	0	0	1.51
Time (sec, IBM 360-65)		121.81		133.57
Time/Iter (sec, IBM 360-65)		10.15		9.54

number of design parameters. Also the gradient projection method of Section 3 can be used to find optimal solutions of these systems subjected to dynamic loads. The purpose of this section is to present finite dimensional formulation of these methods.

6.1 Problem Formulation

Since most practical design problems use finite element models for stress analysis, these models will be used in the design formulation. Let $b \in R^k$ represent a design variable vector and $z(t) \in R^n$ represent a vector of generalized nodal coordinates. The dynamic equations of motion for the system are given in matrix form as

$$M(b)\ddot{z}(t) + K(b)z(t) = Q(b,t) \tag{6.1}$$

with initial conditions or

$$\dot{z}(0) = 0 \ , \qquad z(0) = 0 \tag{6.2}$$

Here $M(b)$ is a mass matrix for the system, $K(b)$ is a stiffness matrix, and $Q(b,t)$ is a vector of effective nodal loads. It is noted that Eq. 6.1 and the corresponding homogeneous boundary conditions have been obtained after certain transformations of the dependent variable [4,5]. These transformations also account for support motion of the structure, such as ground motion due to earthquakes.

The response $z(t)$ of the system is usually obtained by using the modal superposition method. This requires a solution of the eigenvalue problem

$$M(b)V\Omega - K(b)V = 0 \tag{6.3}$$

where V is an $n \times m$ matrix whose columns are m eigenmodes of the system and Ω is an $m \times m$ diagonal matrix of the corresponding eigenvalues. The number of eigenmodes necessary to obtain a solution of Eqs. 6.1 and 6.2 is normally much less than the dimensions of these equations, so $m \ll n$.

The design objective is to minimize weight or cost of the system, which may be written as

$$\psi_0 = \psi_0(b) \tag{6.4}$$

Other design objectives can be treated in the problem formulation without difficulty. It is required that displacements and stresses at critical points of the structure be within specified limits throughout the time interval under consideration. These constraints can be expressed as

$$|z_i(t)| - \bar{z}_i \leq 0 \qquad \text{for all t and some i} \qquad (6.5)$$

and

$$|\sigma_j(z,b)| - \bar{\sigma}_j \leq 0 \qquad \text{for all t and some j} \qquad (6.6)$$

where $\bar{z}_i$ and $\bar{\sigma}_j$ are specified bounds on displacements and stress. The constraint inequalities 6.5 and 6.6 can be combined and expressed in vector form as

$$\eta(z,b,t) \leq 0 \qquad \text{for all t} \qquad (6.7)$$

where inequality applies to each component of the vector function. Just as in Eq. 8, the state variable inequality constraint of Eq. 6.7 may be transformed to an equivalent integral constraint

$$\int_0^\tau (|\eta| + \eta_i)dt = 0 \qquad (6.8)$$

Since $\eta_i(z,b,t)$ is a continuous function of t, the integrand of Eq. 6.8 is continuous and nonnegative. If any constraint in expression 6.7 is violated, the corresponding functional constraint in Eq. 6.8 is violated and vice versa. Finally, natural frequency and design parameter constraints are expressed as

$$\omega_j^L \leq \omega_j \leq \omega_j^U \qquad \text{for some j} \qquad (6.9)$$

and

$$b_r^L \leq b_r \leq b_r^U \qquad \text{for some r} \qquad (6.10)$$

where ω_j^L and ω_j^U are lower and upper bounds on the jth eigenvalue ω_j and b_r^L and B_r^U are lower and upper bounds on the rth design parameter.

6.2 Design Sensitivity Analysis

Before the gradient project method of Section 3 can be applied to the problem formulated in the preceding, gradients of the constraints must be calculated. It will be shown in the following that the general design sensitivity analysis procedure of Section 2 can be directly used to obtain gradients of the functional constraints of Eq. 6.8. For this purpose, consider a general functional constraint of the form

$$\psi_j \equiv \int_0^\tau f_j(z,b,t)dt = 0 \tag{6.11}$$

One could go through the formal perturbation analysis of the constraint of Eq. 6.11 and develop an expression for its gradient [4-6]. However, it will be shown that the general method presented in the operator notation in Section 2 can be directly used to calculate design sensitivity coefficients.

Comparing Eqs. 1.1 and 6.1, one observes that operator $A(u,b)$ is the matrix operator $K(b)$, $N(u,b)z = M(b)\ddot{z}$, and $Q(u,b,x,t)$ is the vector $Q(b,t)$. In the notation of Eqs. 2.4 and 2.5 one has simply

$$a_b(\lambda,z) = \lambda^T K(b)z \tag{6.12}$$

$$n_b(\lambda,z) = \int_0^\tau \lambda^T M(b)\ddot{z}dt \tag{6.13}$$

where $\lambda = \lambda(t)$ is an n-vector of adjoint variables. Integrating Eq. 6.13 by parts twice and noting that $M(b)$ is symmetric, one obtains

$$n_b(\lambda,z) = \int_0^\tau z^T M(b)\ddot{\lambda}dt + (\lambda^T M(b)\dot{z} - \dot{\lambda}^T M(b)z)_0^\tau \tag{6.14}$$

and

$$n_{\delta b}(\lambda,z) = \int_0^\tau \frac{\partial}{\partial b}(z^T M(b)\ddot{\lambda})\delta b dt + \frac{\partial}{\partial b}(\lambda^T M(b)\dot{z} - \dot{\lambda}^T M(b)z)_0^\tau \delta b \tag{6.15}$$

Thus in Eq. 2.6

$$N^*(b)\lambda = M(b)\ddot{\lambda}$$

$$H(\lambda,z) = \lambda^T M(b)\dot{z} - \dot{\lambda}^T M(b)z \left.\right\} \tag{6.16}$$

Since δz and $\delta \dot{z}$ are zero at $t = 0$ and δz and $\delta \dot{z}$ are arbitrary at $t = \tau$, Eqs. 2.10 and 6.16 yield terminal condition on λ as

$$\lambda(\tau) = \dot{\lambda}(\tau) = 0 \tag{6.17}$$

The governing differential equation for the adjoint variable λ is given from Eq. 2.12 as

$$K(b)\lambda + M(b)\ddot{\lambda} = \frac{\partial f_j^T}{\partial z} \tag{6.18}$$

Now substituting various quantities in Eq. 2.15, the design sensitivity vector ℓ^j for the constraint of Eq. 6.11 can be identified as

$$\ell^j = \left[\int_0^\tau \frac{\partial f_j}{\partial b} + \lambda^{j^T} \frac{\partial Q}{\partial b} - \lambda^{j^T} \frac{\partial}{\partial b}(K(b)z) - \lambda^{j^T} \frac{\partial}{\partial b}(M(b)\ddot{z}) \right]^T \tag{6.19}$$

It is observed that Eq. 6.19 is identical to the one obtained in Refs. 5 and 6 by going through the formal perturbation analysis (Eq. 46 of Ref. 5 and Eq. 31 of Ref. 6).

Gradient calculations for eigenvalue constraints of Eq. 6.9 are well known [4-6] and are not repeated here. The gradient of and eigenvalue ω_j is given as [4-6]

$$\frac{\partial \omega_j}{\partial b} = \frac{\partial}{\partial b} \left\{ \left[v^{j^T} K(b) v^j \right] - \omega_j \frac{\partial}{\partial b} \left[v^{j^T} M(b) v^j \right] \right\} / M_j^* \tag{6.20}$$

where

$$M_j^* = v^{j^T} M(b) v^j$$

7. FINITE DIMENSIONAL EXAMPLES

In this section, optimal design of several systems that may be modeled by beam finite elements are considered. In all examples, weight or volume of the system is to be minimized, given as

$$\psi_0(b) = \sum_{i=1}^{NE} \rho_i \ell_i A_i(b_i) \tag{7.1}$$

where ρ_i is the density, ℓ_i is the length, and $A_i(b_i)$ is the cross-sectional area of the ith element and NE is the total

number of elements.   All designs are obtained using the program
of Ref. 7.

7.1   Examples of Beam Design

Beam design examples presented in Section 4 are solved again
using the finite dimensional formulation.   Each beam was divided
into 10 finite elements and the modal superposition method was
used to solve the state and adjoint equations (Eqs. 6.1 and 6.18).
Final results for these examples with the finite dimensional
method are given in Table 4.1.   In comparison with the results
obtained using continuous formulation, the finite dimensional
results are almost identical.   However, computational time for
each design iteration with the finite dimensional formulation is
substantially higher.   The primary reason for this is that there
is only one displacement constraint in the continuous formulation.
However, in the finite dimensional formulation, the displacement
constraint is imposed at each nodal point.   Thus more design sensitivity calculations are needed in the finite dimensional case
to impose the displacement constraint.   The total computing time
for all the problems varies and it is difficult to draw any conclusions based on the total computing effort.   The reason being
that the total computing time depends on the number of iterations
required for convergence, which is highly dependent on the step
size parameter selected by the designer.

As another example, optimal design of a tubular tipped cantilever beam subjected to angular acceleration at its base is considered.   The beam is shown in Fig. 7.1.   This example is an

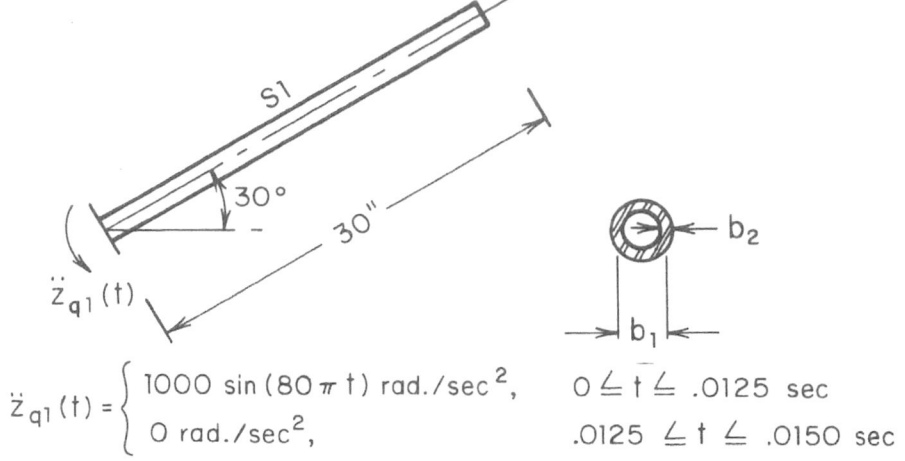

$$\ddot{Z}_{q1}(t) = \begin{cases} 1000 \sin(80\pi t) \text{ rad./sec}^2, & 0 \leq t \leq .0125 \text{ sec} \\ 0 \text{ rad./sec}^2, & .0125 \leq t \leq .0150 \text{ sec} \end{cases}$$

Figure 7.1   Tipped Tubular Cantilever Beam

idealization of a gun tube subjected to angular acceleration loads at the breech. The angle of tip for the beam is 30°, $E = 30 \times 10^6$ psi, and $\rho = 0.28$ lb/in.^{3}. The fixed end is forced to rotate in a prescribed manner, as shown in Fig. 7.1. The beam is divided into four elements of equal length for analysis. Axial deformation is considered, so the system has 12 degrees of freedom. Design constraints for the problem are listed in Table 7.1.

TABLE 7.1   DESIGN CONSTRAINTS FOR TIPPED CANTILEVER BEAM

The vertical and angular deflections of the free end must be less than or equal to 0.6 in. and 0.04 rad., respectively $\|\sigma_i(t)\| \leq 45,000$ psi,   for all t and i Natural Frequency Constraint: $30 \leq f_1 \leq 60 \leq f_2 \leq 400$ Hz $0.8 \leq b_1 \leq 2.0$ in., $0.1 \leq b_2 \leq 0.15$ in.

There are two design parameters for the beam, the mean radius b_1, and the wall thickness b_2.

The first five of nine eigenvectors, from subspace iteration, were used in the analysis. The starting design is given in Table 4. The frequencies at this design were 67.20, 419.8, 1,175, 1,706, and 2,294 Hz. This design resulted in $\|\delta b^1\| = 6.225$ in the first iteration. The stopping criteria was satisfied in the nineteenth iteration, where $\|\delta b^1\| = 8.182 \times 10^{-6}$ and no constraint violation occurred. The final design weight was 2.723 lb with frequencies of 46.36, 290.1, 814.9, 1,600, and 1,706 Hz. The fifth frequency of 1,706 Hz corresponds to axial deformation. At the optimum, the maximum vertical and angular displacements of the free end were 0.582 in. and 0.0308 rad., respectively. The maximum stress at the fixed end was 44,700 psi. The total computing time was 22.50 sec for 20 iterations. Numerical results are shown in Table 7.2 and Fig. 7.2 (Solution 1).

This problem was also solved by starting from a different design point with an initial design weight of 3.958 lb and initial frequencies of 45.25, 283.0, 795.1, and 1.706 Hz. This design resulted in $\|\delta b^1\| = 4.082$ in the first iteration. At the fifth iteration, $\|\delta b^1\| = 2.413 \times 10^{-6}$. The final design weight was 2.671 lb with frequencies of 45.48, 284.6, 799.5, 1,570, and 1,706 Hz. At the optimum, maximum stress at the fixed end was 45,200 psi. The total computing time was 9.65 sec for six

iterations. Numerical results are shown in Table 7.2 and
Figure 7.2 (Solution 2).

TABLE 7.2   INITIAL AND FINAL DESIGNS FOR TIPPED CANTILEVER BEAM

Design		Solution 1		Solution 2	
		Initial	Final	Initial	Final
S1	b_1	1.500	1.032	1.000	1.012
	b_2	0.110	0.100	0.150	0.100
Wt. lb.		4.354	2.723	3.958	2.671
Time/Iteration (sec IBM 360-65)		1.13		1.61	

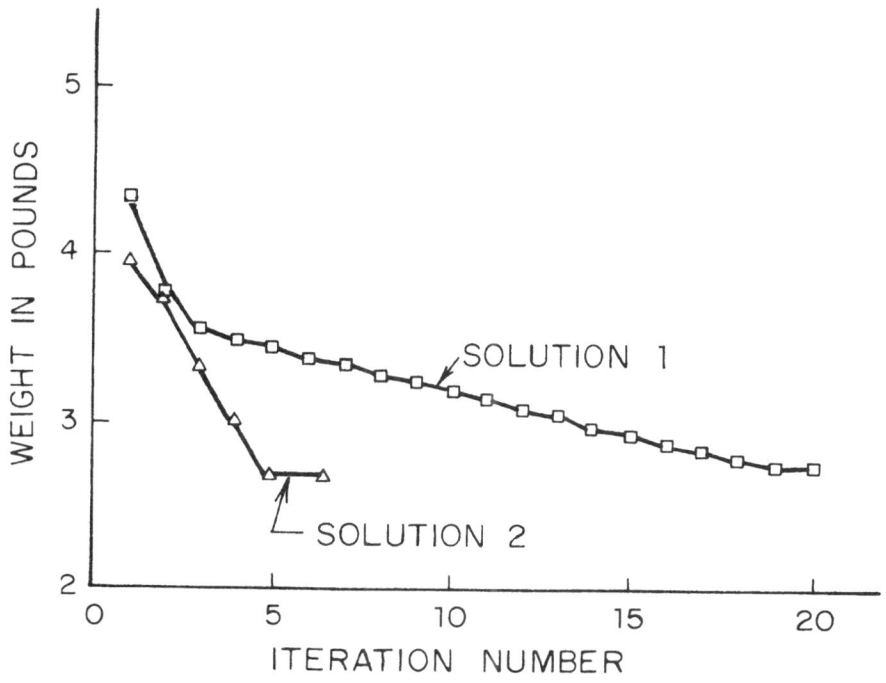

Figure 7.2   Weight-Iteration Curves for Tipped Cantilever Beam

7.2   A Seven Member Truss-Frame

Seven tubular members are pinned together and two concentrated masses are attached at the nodes shown in Fig. 7.3.   The
fixed end of the structure is forced to move in a prescribed manner.   This problem is taken from Ref. 8, for comparison purposes.
All members are divided into two elements of equal length for
analysis, so the system has 41 degrees of freedom.   The mean radius and the wall thickness of each member are taken as design
parameters.   Therefore there are a total of 14 design parameters.
The frame is designed for two constraint cases listed in Table 7.3.

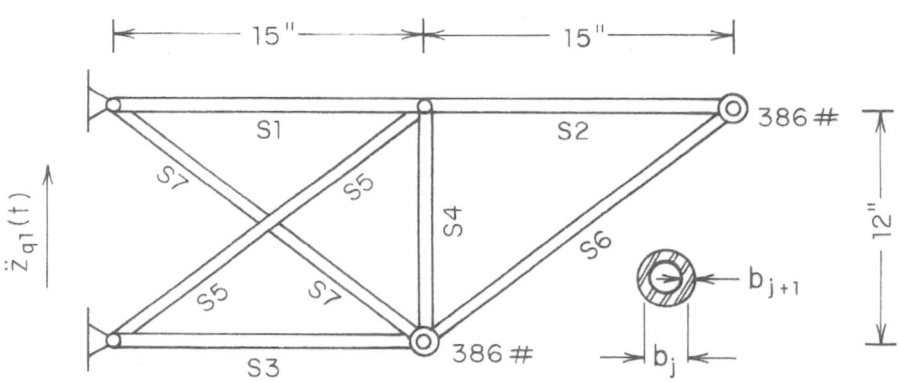

$$\ddot{z}_{q1}(t) = 3125 \sin(250t) \text{ in/sec}^2, \quad 0 \le t \le .0125664 \text{ sec}$$

Figure 7.3   Structure and Load for Seven Member Truss Frame

The first 11 of 16 eigenvectors, from subspace iteration,
were used in the solution for Case A.   The starting design weight
was 10.083 lb., with a mean diameter of 1.0 in. and wall thickness
of 0.1 in. for all members.   The frequencies were 31.43, 97.73,
135.0, 191.3, 308.5, 308.7, 501.7, 505.2, 505.3, and 784.5 Hz.
This initial design resulted in $||\delta b^1|| = 3.86$.

The first nine cycles were executed by ignoring the dynamic
response constraints.   The remaining design cycles were executed
with these constraints included and the dynamic response was
checked, out to 0.0188496 sec, with $\ddot{z}_{q1}(t) = 0$ in./sec^2 in the
interval 0.0125664 < t < 0.0188496 sec.   No displacement or stress
constraints were violated, throughout the design process.   At the
ninth iteration, $||\delta b^1|| = 9.670 \times 10^{-3}$.   The final design weight
was 6.212 lb, with all member sizes at their lowest bounds except

members S1 and S3. The frequencies were 25.00, 76.30, 106.2, 151.2, 186.7, 186.9, 186.9, 305.9, 328.4, 337.6, and 477.6 Hz. The maximum vertical deflection at the upper-right node was 0.165 in. The maximum stress occurred in the middle of member S1, at a level of 48,600 psi. The total computing time was 83.28 sec for 13 iterations. Optimum results are shown in Table 7.4 and Fig. 7.4.

TABLE 7.3   DESIGN CONSTRAINTS FOR SEVEN MEMBER TRUSS FRAME

Case A
Rectilinear displacement components of $z(t)$ of all nodal points must be less than or equal to 0.2 in.
$\|\sigma_i(t)\| \leq 50{,}000$ psi, for all t and i
$25 \leq f_1 \leq 90$ Hz
$0.6 \leq b_j \leq 4.0$ in., $\quad 0.1 \leq b_{j+1} \leq 0.12$ in., $\qquad j = 1,3,\ldots,13$

Case B
Rectilinear displacement components of $z(t)$ of all nodal points must be less than or equal to 0.1 in.
$\|\sigma_i(t)\| \leq 36{,}000$ psi, for all t and i
$25 \leq f_1 \leq 90$ Hz
$0.6 \leq b_j \leq 4.0$ in., $\quad 0.1 \leq b_{j+1} \leq 0.12$ in., $\qquad j = 1,3,\ldots,13$

In Ref. 8, the wall thickness was fixed at 0.1 in., and the starting mean diameter was 2.5 in. for all members. The starting design weight was 25.21 lb and the least design weight was found as 10.60 lb with $b_1 = 1.18$ in., $b_3 = 1.46$ in., $b_5 = 1.28$, $b_7 = b_9 = 0.63$ in., $b_{11} = 1.06$ in., and $b_{13} = 1.14$ in. The stress bounds used in Ref. 8 in members S1, S3, S6, and S7 reached 50 ksi and the stress bound in member S2 was close to this value.

TABLE 7.4 INITIAL AND FINAL DESIGNS OF SEVEN
MEMBER TRUSS-FRAME

Design		Case A		Case B		
		Initial	Final	Initial	Final 1	Final 2
S1	b_1	1.000	0.646	0.700	1.128	1.141
	b_2	0.100	0.101	0.100	0.107	0.108
S2	b_3	1.000	0.600	0.700	0.756	0.785
	b_4	0.100	0.100	0.100	0.101	0.102
S3	b_5	1.000	0.663	0.700	1.798	1.810
	b_6	0.100	0.101	0.100	0.120	0.120
S4	b_7	1.000	0.600	0.700	0.600	0.600
	b_8	0.100	0.100	0.100	0.100	0.100
S5	b_9	1.000	0.600	0.700	0.600	0.600
	b_{10}	0.100	0.100	0.100	0.100	0.100
S6	b_{11}	1.000	0.600	0.700	1.140	1.155
	b_{12}	0.100	0.100	0.100	0.111	0.112
S7	b_{13}	1.000	0.600	0.700	1.272	1.271
	b_{14}	0.100	0.100	0.100	0.120	0.120
Wt lb		10.083	6.212	7.058	11.812	11.951
Time/Iter (sec, IBM 360-65)			6.41		20.54	21.54

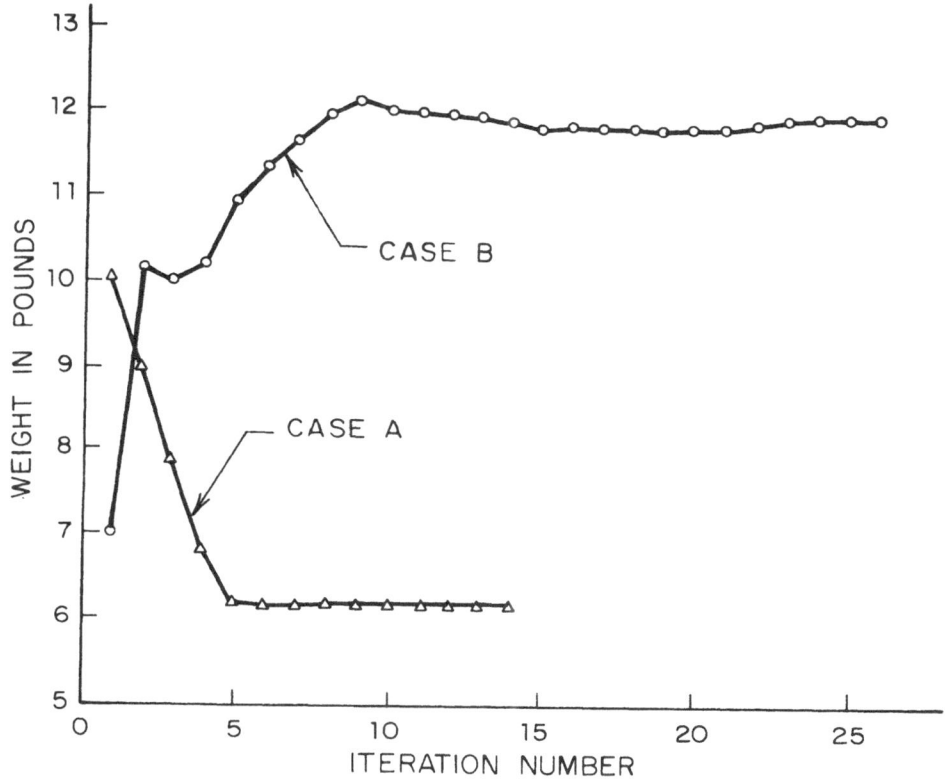

Figure 7.4   Weight-Iteration Curves for Seven Member Truss Frame

A comparison of results shows that the design weight reported herein is about 40 per cent lighter than the result reported in Ref. 8.   The actual maximum stresses in members S1, S3, S6, S7, and S2 were 48,600, 40,200, 37,000, 23,100, and 43,800 psi, respectively, in the present work.   Results of the present study indicate that members need not be fully stressed or at their lower bounds at the optimum point, contrary to what was implied by the results of Ref. 8.   The fact that no constraint is tight for member S3 is a result of the frequency constraint of this design.

Case A is redesigned with the same load as in Fig. 7.3, but with the constraints of Case B shown in Table 7.3.   Design iterations were started with a weight of 7.058 lb, a mean diameter of 0.7 in., and a wall thickness of 0.1 in. for all members.   The frequencies were 26.31, 81.84, 113.1, 160.2, 217.0, 217.2, 217.3, 354.7, 355.9, 356.0, and 554.5 Hz.   The measure of convergence $||\delta b^1||$ was 2.296.   The maximum deflection of the free end was

was 0.132 in. and the maximum stress occurred in members S1 and S3. The iterative process was terminated at the twenty-first iteration, where $||\delta b^1|| = 0.0073$, with no constraint violated. The design weight was 11.812 lb, with frequencies of 35.84, 92.48, 159.0, 186.8, 242.9, 351.17, 383.0, 391.8, 477.8, 563.8, and 822.2 Hz. The computing time for this solution was 431.38 sec for 21 iterations. This solution is shown in Final Design 1 in Table 7.4. Starting from this design point, dynamic response was checked out to 0.015708 sec, with $\ddot{z}_{q1}(t) = 0$ in./sec^2 in the interval

0.0125664 < t < 0.015708 sec. It was found that member S2 was overstressed by 6 per cent. Taking another 141.17 sec for five iterations, the computation showed $||\delta b^1|| = 2.395 \times 10^{-2}$ and the stress violation was brought down to 1 per cent. The final design weight was 11.951 lb, with frequencies of 36.17, 93.41, 160.1, 186.8, 244.1, 356.4, 391.7, 397.2, 477.8, 570.2, 822.2 Hz. This solution is shown as Final Design 2 in Table 7.4.

7.3 Two Story Frame

As a final example of finite dimensional applications, a two story-two bay frame shown in Fig. 7.5 is considered. This example is taken from Ref. 9. The frame is made of six I-columns and four I-beams. The moment of inertial I has been chosen to be the design variable for each element of the structure. A uniformly distributed weight of 10 lb/in. has been applied to the beam members of the frame and the structure is subjected to a ground shock load, simulating earthquake disturbances. Dimensions and loading of the structure, as well as generalized displacements coordinates, are shown in Fig. 7.5. Two cases for this problem are considered. In the Case I axial deformation is neglected, so if each member of the structure is modeled as one element, the finite-element equations have eight degrees of freedom. In the Case II, axial deformation of members is included. Thus, each node has three degrees of freedom and the total number of generalized coordinates is 18. The constraints for this design problem are on displacements, stress, fundamental frequency, and design variables. They are given in Table 7.5.

TABLE 7.5   DESIGN CONSTRAINTS FOR TWO-STORY, TWO-BAY FRAME

$$|z_4(t)| \leq 3 \text{ in.}, \quad |z_8(t)| \leq 3 \text{ in.}, \quad \text{for all } t$$

$$|\sigma_i(t)| \leq 30{,}000 \text{ psi}, \quad \text{for all } t \text{ and } i = 1,2,\ldots,20$$

$$f_1 \geq 4.775 \text{ Hz}, \quad 290 \leq b_n \leq 20{,}300 \text{ in.}^4, \quad n = 1,2,\ldots,10$$

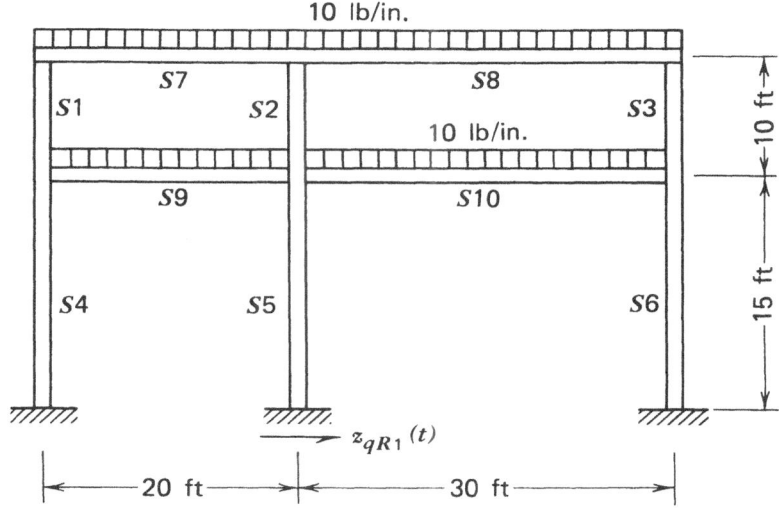

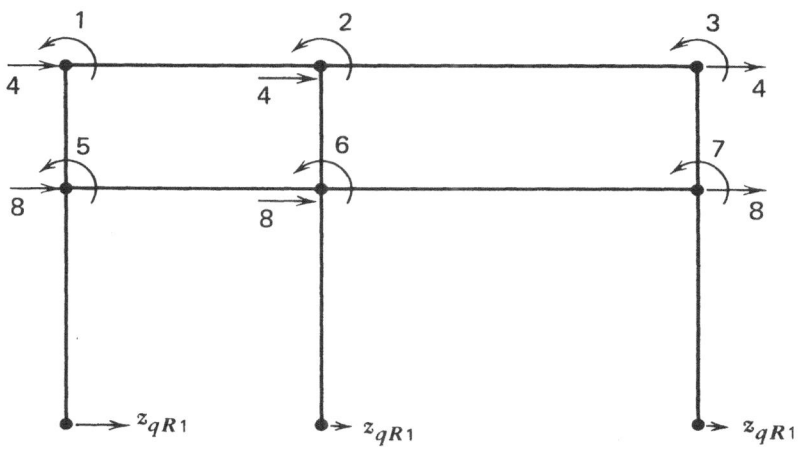

Figure 7.5   Two-Story Planar Frame

Material constants are modulus of elasticity $E = 3 \times 10^7$ psi and specific weight, $\rho = 0.28$ lb/in.[3]. Formulas for sectional properties are given as in Ref. 9:

$$S = (60.6I + 84,000)^{1/2} - 290 \quad , \qquad 0 \le I \le 9000$$

$$S = \frac{I - 8056.3}{1.876} \quad , \qquad 9000 \le I \le 20,300$$

$$A = 0.465(I)^{1/2} \quad , \qquad 0 \le I \le 9000$$

$$A = \frac{I + 2300}{256} \quad , \qquad 9000 \le I \le 20,300 \quad ,$$

where the moment of inertial I is given in in.[4], the bending S in cubic inches, and the area A in square inches. At time t = 0, the structure is at rest and the base undergoines a transient ground motion.

$$z_{q1}(t) = \begin{cases} \sin 30\ t\ \text{in.,} & 0 \le t \le \pi/30 \\ \\ 0 & \pi/30 < t \quad . \end{cases}$$

The total time interval used for integration was 0.16 sec and the time grid interval was 0.002 sec. The initial design is given in Table 7.6. In Case I, the first three of eight eigenvectors calculated by the subspace iteration were used in dynamic analysis. In the II, the first four of eighteen calculated eigenvectors were used. After considerable numerical experimentation, feasible designs for both cases were obtained and are given in Table 7.6. Cassis and Schmit [9] have treated the problem of Case I and their solution is also given in Table 7.6. The present solutions represent 26.9 percnet and 19.8 percent improvement on the design presented by Cassis and Schmit [9]. At the final design, the natural frequencies for Cases I and II were 9.96, 23.55, and 31.33 and 10.36, 34.74, 37.40 and 58.28 Hz respectively.

Cassis and Schmit [9] have presented a method for solving comparable problems. Their method is now briefly compared with the foregoing method and numerical results analyzed. There are three major steps in each design cycle of both optimization algorithms: (1) analysis of the structure under dynamic loads, (2) design sensitivity analysis, and (3) calculation for optimum design change. The efficiency of either method can be improved by increasing the efficiency of calculations in any of the three steps. In the first step, Cassis and Schmit have used the model matrix transformation approach to compute dynamic displacements. However, they use all eigenvectors in their dynamic and sensitivity analyses. This is unnecessary in most structural problems, because higher modes do not make a significant contribution to the dynamic response.

Design sensitivity analysis is the next major point of comparison, since once design derivatives have been calculated, any of the well-known nonlinear programming techniques may be used to compute improvements in design. Cassis and Schmit compute derivatives of constraint functions by calculating the sensitivity matrix $\partial z(t)/\partial b$. For the dynamic problem at hand, the displacement vector $z(t)$ is available only through the modal transformation. Thus, in calculating $\partial z(t)/\partial b$, they compute $\partial V/\partial b$, derivatives of all eigenvectors with respect to all design variables. The number of such derivatives is quite large and the computational effort is extensive.

TABLE 7.6   INITIAL AND FINAL DESIGNS OF THE
TWO-STORY, TWO-BAY FRAME

Member	Design Variables	Initial	Final Design (in.^{4}) Axial Def. Neglected	Axial Def. Included	Cassis [9] Design
S1	b_1	3,000	1,289	1,442	1,868
S2	b_2	3,000	1,402	1,903	3,126
S3	b_3	3,000	290	1,938	3,757
S4	b_4	3,000	3,806	3,654	4,523
S5	b_5	3,000	1,855	4,452	5,940
S6	b_6	3,000	5,916	6,644	6,784
S7	b_7	3,000	1,373	810	1,636
S8	b_8	3,000	290	290	327
S9	b_9	3,000	2,206	2,338	2,233
S10	b_{10}	3,000	290	290	2,055
Wt. (lb)		14,976	9,982	10,850	13,532
Time/Iter (sec,IBM 360-65)			2.81	10.77	2.24

In contrast, the state space method of Section 6 requires only calculation of a number of adjoint vectors from Eq. 6.18 that is equal to the number of tight constraints that involve the state variable. Further, this calculation is numerically efficient, since Eq. 6.18 is exactly the same as the original equation of motion Eq. 6.1. Thus, the previously calculated eigenvectors are used to construct the adjoint vectors. Therefore,

514

the design sensitivity analysis for the problem in the state space setting should be more efficient than the chain rule differentiation used in Ref. 9.

The third step in the optimal design algorithm is to compute an improved design. In the version of SUMT used in Ref. 9, and many other nonlinear programming techniques, a one-dimensional search must be performed to compute a design change. This search will be computationally quite expensive, since an augmented cost functional must be evaluated several times.

Therefore, the method with state space design sensitivity analysis is efficient by a factor of $(2.24 \times 8)/2.81 = 6.4$ for computations in each design iteration. Here, a factor of 8 is estimated as the ratio of computing speeds of the two computing machines (IBM 360-91/IBM 360-65).

REFERENCES

1. Haug, E.J., "A Gradient Projection Method for Structural Optimization," Optimization of Distributed Parameter Structures (Eds. E.J. Haug and J. Cea), Sijthoff & Noordhoff, Alphen aan den Rihn, Netherlands, 1980.
2. Zienkiewicz, O.C., The Finite Element Method in Engineering Science, McGraw-Hill, 1971.
3. Wang, C.T., Applied Elasticity, McGraw-Hill, New York, 1953.
4. Feng, T.T., Arora, J.S., and Haug, E.J., Optimal Design of Elastic Structures Under Dynamic Loads, Technical Report No. 18 (also Ph.D. dissertation of the first author), Division of Materials Engineering, The University of Iowa, Iowa City, Iowa, May 1975.
5. Feng, T.T., Arora, J.S., and Haug, E.J., "Optimal Structural Design Under Dynamic Loads," Int. J. for Num. Methods in Engrg., Vol. 11, No. 1, 1977, pp. 39-52.
6. Haug, E.J., Arora, J.S., and Feng, T.T., "Sensitivity Analysis and Optimization of Structures for Dynamic Response," J. of Mechanical Design, ASME, Vol. 100, No. 2, April 1978, pp. 311-318.
7. Feng, T.T., Arora, J.S., and Haug, E.J., Optimal Design of Elastic Structures under Dynamic Loads: A User's Manual for Program DYSTROP-I, Technical Report No. 33, Division of Material Engineering, The University of Iowa, Iowa City, Iowa, May 1977.
8. Fox, R.L., and Kapoor, M.P., "Structural Optimization in the Dynamic Response Regime: A Computational Approach," AIAA J., Vol. 8, No. 10, 1970, pp. 1798-1804.
9. Cassis, J.H., and Schmit, L.A., "Optimal Structural Design with Dynamic Constraints," J. of the Structural Division, ASCE, Vol. 102, No. ST10, 1976, pp. 2053-2071.

10. Arora, J.S., and Haug, E.J., "A Discussion on the Paper, Optimal Structural Design with Dynamic Construction," J. of the Structural Division, ASCE, Vol. 103, No. ST10, October 1977, pp. 2071-2074.

REMARKS ON THE OPTIMAL SHAPE OF THE FIXED-FIXED COLUMN

K. K. Hu and Philip G. Kirmser

College of Engineering, Kansas State University
Manhattan, Kansas 66506

ABSTRACT

The fixed-fixed column is an unusually difficult problem, all
published solutions of which the authors are aware of are incorrect, yet remarkably close to what appears to be a solution of
the problem. Although plausible, it is not clear that statements
of the problem based on multiplicity of eigenvalues are equivalent
to simple statements of optimality. In this paper, some new
theory and several numerical solutions are presented. Although
these show some progress, at present this problem remains incompletely solved.

1.   INTRODUCTION

The problem of finding the optimum shapes of columns is old,
difficult, and, for some boundary conditions, incompletely solved.
Lagrange [1] tried, but failed to find the best shape for a pin-
ended column. Clausen's solution [2] was forgotton, and the
problem was solved again, independently, by Keller [3]. Tadjbakhsh
and Keller [4] extended this solution to find optimum shapes for
fixed-fixed and fixed-pinned columns. Their optimum shapes include zero cross-sections, which amount to hinges. Barnes [5],
in order to avoid these internal hinges, re-solved the problem
using upper and lower bound constraints on cross-sections. All
these papers establish conditions for optimum shapes, using
calculus of variations. Several isoperimetric inequalities are
established.

Taylor [6], and Taylor and Liu [7] showed that energy methods could be used to solve some of the same problems, and established bounds using them. Proceeding in a different direction, Haug [8] outlined methods for solving structural optimization problems using mathematical programming and exhibited a series of optimal solutions for certain pin-ended columns with additional constraints, such as on maximum allowable stress.

Olhoff and Rasmussen [9] found that Tadjbakhsh and Keller's solution [4] for the fixed-fixed column was incorrect, because symmetrical buckling was assumed to take place, and the shape presented, with its zero cross-sections, actually buckled in an asymmetrical mode, i.e. the symmetrical eigenfunction for the shape found does not have the lowest eigenvalue. Stated another way, the eigenfunction having the lowest eigenvalue does not belong to the allowable set that was used in deriving the optimum shape.

Although Olhoff and Rasmussen's solution is an imaginative improvement that considers the possibility that repeated eigenvalues occur in the solution for the optimal fixed-fixed column, several open questions remain. An easy calculation shows that the second variation of the functional used for optimization is $\int_0^1 (\delta\alpha_1)(\delta\alpha_2)[(1-\gamma)(y_1'')^2 + \gamma(y_2'')^2] \, dx$. This is indefinite for arbitrary variations $\delta\alpha_1$ and $\delta\alpha_2$. Thus, they have found only a stationary value for their functional. There is no proof that this value is a maximum.

The method used for calculation involves only the two lowest eigenfunctions of the differential operator for the fixed-fixed column. The authors believe that this limits the admissible class of functions used to evaluate the functional to too restrictive a set. Olhoff and Rasmussen's solution, although far superior to Tadjbakhsh and Keller's, in fact has a much smaller error of the same type.

It is not clear that Tadjbakhsh and Keller's problem, which is that of finding the shape of a fixed-fixed column of unit volume so that its buckling load is the greatest possible, is the same as Olhoff and Rasmussen's, which is that of finding the shape of a fixed-fixed column of unit volume such that its two lowest eigenvalues are a maximum. The authors know of no proof that shows that these problems are equivalent.

A proof is presented, in this paper that, for sufficiently smooth shapes, the multiplicity of eigenvalues for the fixed-fixed column cannot exceed two. A new, partial solution for the fixed-fixed column that yields the highest buckling load yet found is

518

then presented. Finally, some reasons to try to prove or disprove
the equivalence of Tadjbakhsh and Keller's and Olhoff and
Rassumssen's optimization problems are discussed.

2.   ON THE MULTIPLICITY OF EIGENVALUES FOR THE FIXED-FIXED
      COLUMN

      It seems to be intuitively obvious that the shape of the
strongest column should yield a lowest eigenvalue of multiplicity
greater than one. If the lowest eigenvalue were simple, the shape
of the column could be altered to make this eigenvalue greater, at
least until the lowest two eigenvalues were equal. But then,
couldn't the lowest two be increased until at least the lowest
three were equal, then the lowest, four, etc? After all, the
fixed-fixed column equation has an infinite number of eigenvalues
and eigenfunctions. The situation is clarified by the following
theorem.

      Theorem 2.1: For stiffness and deflection functions that
are representable by convergent power series, the multiplicity of
eigenvalues for the fixed-fixed column problem is less than or
equal to two.

      Proof: The fundamental differential equation is

$$[\beta(x) \ y''(x)]'' + P \ y''(x) = 0 \tag{2.1}$$

with the boundary conditions

$$y(0) = y'(0) = y(1) = y'(1) = 0 \tag{2.2}$$

Under the hypotheses stated,

$$y(x) = a_2 x^2 + a_3 x^3 + \ldots + a_j x^j + \ldots \tag{2.3}$$

and

$$\beta(x) = b_0 + b_1 x + b_2 x^2 + \ldots + b_j x^j + \ldots \tag{2.4}$$

Substitution of these series into the fundamental differential
equation of Eq. 2.1, equating the coefficients of like powers
of x, and rearranging, leads to

$$
\overset{D}{
\begin{pmatrix}
(b_2 + \frac{P}{2\times1}) & b_1 & b_0 & 0 & 0 & \cdots & 0 \\
b_3 & (b_2 + \frac{P}{3\times2}) & b_1 & b_0 & 0 & \cdots & 0 \\
b_4 & b_3 & (b_2 + \frac{P}{4\times3}) & b_2 & b_0 & \cdots & 0 \\
\vdots\; b_{N-1} & \vdots & \vdots & \vdots & \vdots & \cdots & b_0 \\
\frac{1}{2\times1} & \frac{1}{3\times2} & \frac{1}{4\times3} & \cdots & \cdots & b_1 & b_1 \\
 & & & & (b_2 + \frac{P}{(N-1)(N-2)}) & \frac{1}{N(N-1)} & \frac{1}{(N+1)N} \\
1 & \frac{1}{2} & \frac{1}{3} & \cdots & \cdots & \frac{1}{N-1} & \frac{1}{N}
\end{pmatrix}
}
\begin{pmatrix}
a_2 \\ a_3 \\ a_4 \\ \vdots \\ a_{N+2}
\end{pmatrix}
=
\begin{pmatrix}
0 \\ 0 \\ 0 \\ \vdots \\ 0
\end{pmatrix}
\qquad (2.5)
$$

$$(N \times N) \qquad\qquad (N \times 1) \qquad (N \times 1)$$

In this equation, N is taken large enough so that the errors caused by using finite series are negligible and D is a non-singular diagonal matrix. Now, the first (N-2) rows of the b-matrix are independent, because the coefficient b_0 cannot be zero (it it were, there would be a hinge at the fixed end of the column). This means however that the remaining coefficients b_j and buckling load P are determined, so the deficiency of the coefficient matrix cannot exceed two, which in turn guarantees that no more than two linearly independent solutions for y can exist for any P. Thus, there is no need to consider multiplicity of eigenvalues greater than two in finding the optimum fixed-fixed column of unit volume. It should be remembered, however, that this theorem says nothing about the equivalence of the simply stated optimization problem and the maximum of the minimum repeated eigenvalue optimization problem.

3. THE TADJBAKHSH-KELLER OPTIMAL FORM IS NEARLY CORRECT

The shape of the fixed-fixed column that was found by Tadjbakhsh and Keller is shown in Fig. 3.1. As Olhoff and Rasmussen discovered, the normalized buckling load for this column is 30.51, which is that of the lowest asymmetrical mode, instead of 52.638 found by Tadjbakhsh and Keller, which is that of the lowest symmetrical mode.

From engineering intuition, the column shown in Fig. 3.1 could be strengthened by the addition of very little material at the places of zero cross-sectional area, which are at the quarter points. This should strengthen the asymmetrical mode of buckling markedly, while scarcely altering the buckling load of the symmetric mode.

Figure 3.2 shows the buckling loads for the lowest symmetric and asymmetric modes of the filled in Tadjbakhsh-Keller optimal shape, i.e., a column whose cross-section is defined by the maximum of the Tadjbakhsh-Keller area or a given constant value. The abscissa shown in Fig. 3.2 is the rato of the minimum to maximum cross-sectional areas. At 0, the filled-in shape is Tadjbakhsh and Keller's, at 1, the column is of uniform cross-section. One see immediately that the maximum of the minimum buckling loads, 52.3112 by quadratic interpolation, is close to 52.3563, the value found by Olhoff and Rasmussen. In the calculations, the volume of the filled-in column was reduced to one by appropriate scaling.

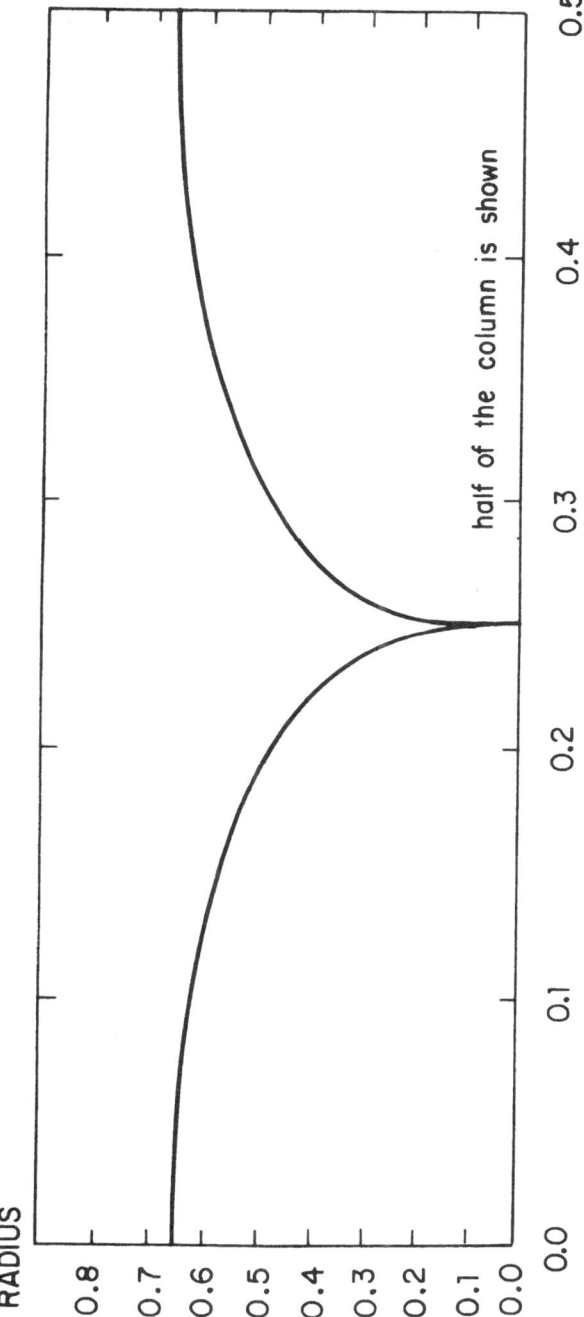

Figure 3.1   The Tadjbakhsh-Keller Optimal Solution

522

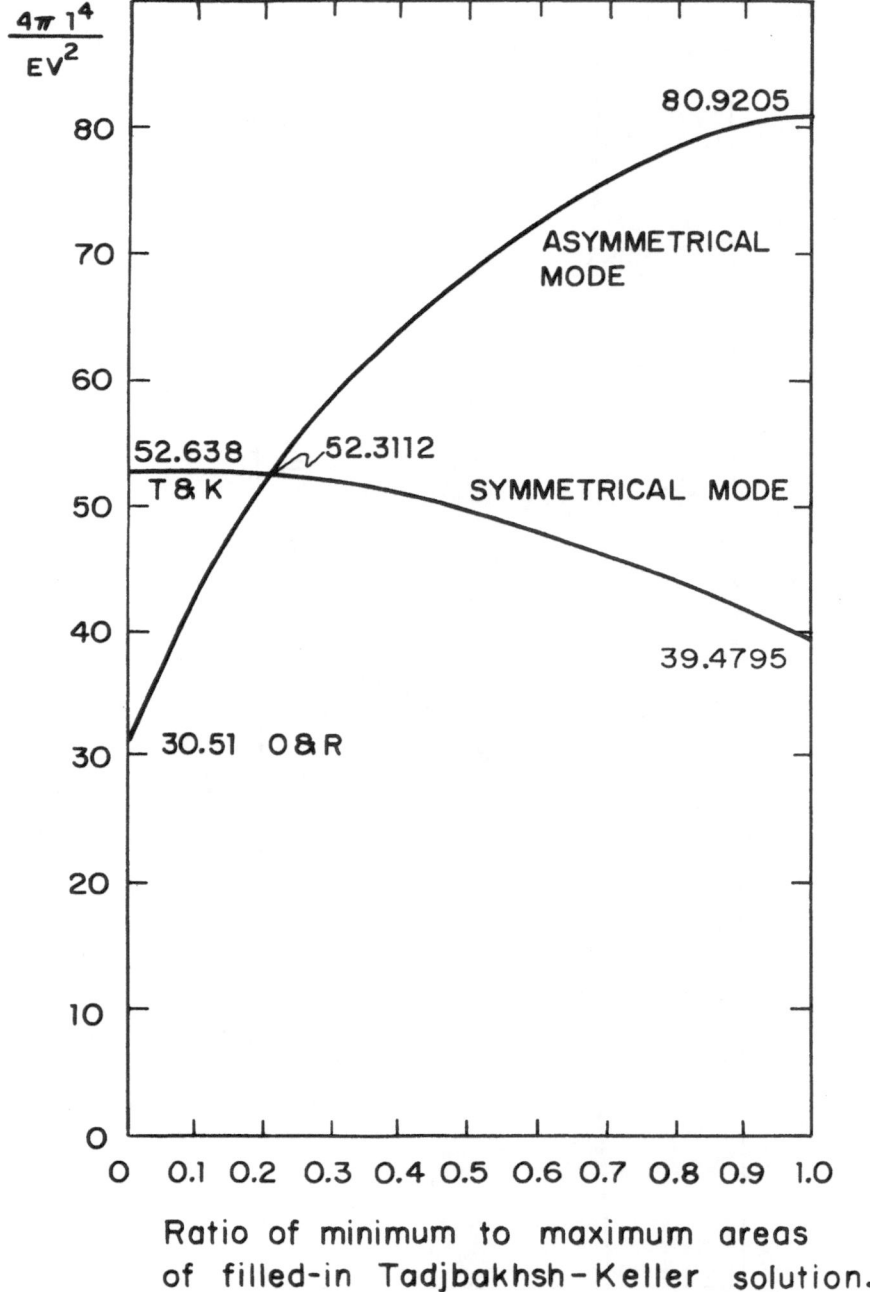

Figure 3.2   Eigenvalues for the Lowest Symmetric and Asymmetric
Modes for the Filled-In Tadjbakhsh-Keller Solution

EIGENVALUE

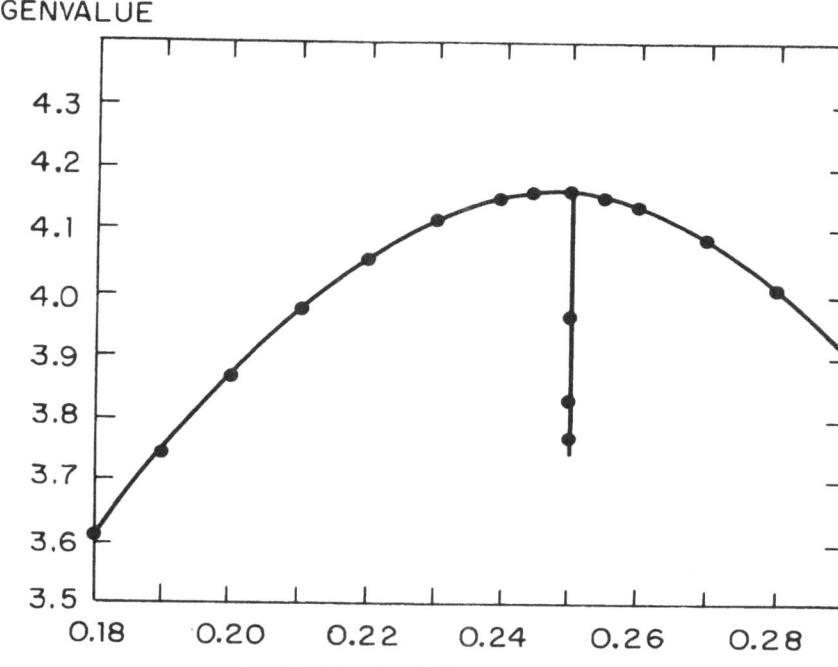

LOCATION OF THE THIN SPOT

Figure 3.3   The Lowest Eigenvalue of the Filled-In Shape as a
Function of the Location of the Thin Spot

Figure 3.3 shows the maximum of the minimum of the lowest
two eigenvalues of the best filled-in Tadjbakhsh-Keller cross-
sections for fixed-fixed columns, calculated relatively crudely,
in order to show that the best locations for the cross-sections
of zero area are at the quarter points. The best filled-in eigenvalue found was 4.155 (this value must be multiplied by 4π for
comparison to the normalized eigenvalues shown in Fig. 3.2).
Several lower double eigenvalues, which are lss than the optimum,
were also found. These are shown in Fig. 3.3.

4.   A NEW ALGORITHM

Where should material be added to strengthen a column? This
question is answered in this section.

The fundamental equation for the buckling of a fixed-fixed
column is

$$\beta(x) \ y''(x) + P \ y(x) = M - Hx \tag{4.1}$$

where

$\beta(x)$ = stiffness of the column

$y(x)$ = deflection

P = buckling load

M = moment at the fixed support

H = shearing force at the fixed support

Taking variations gives

$$(\beta+\delta\beta)(y''+(\delta y)'') + (P+\delta P)(y+\delta y) = (M+\delta M) - (H+\delta H)x \tag{4.2}$$

where δ is the variation. Multiplying and saving only the linear first order terms yields

$$\delta\beta(y'') + (\delta P)y = -\beta(\delta y)'' - P(\delta y) + (\delta M-(\delta H)x). \tag{4.3}$$

The right hand side of Eq. 4.3 is nearly zero, because the first two terms represent the negative changes in $M - Hx$, brought about by admissible variations in solution of Eq. 4.1, and the last two terms are their positive counterparts. From this, it follows that

$$(\delta P)y = - (\delta\beta)y'' \tag{4.4}$$

Multiplying both sides of Eq. 4.4 by y'', integrating from 0 to 1, and rearranging yields

$$\delta P = \frac{\int_0^1 (\delta\beta)(y'')^2 dx}{\int_0^1 (y')^2 dx} \tag{4.5}$$

This equation relates changes in buckling load δP to variations in stiffness $\delta\beta$. It shows that an optimum way to add stiffness to a column is to choose

$$\delta\beta = k(y'')^2 \tag{4.6}$$

where k is a suitably chosen constant. Such a choice reaffirms the engineering principle that in order to keep something from bending, find out where it bends, and add material there.

A new algorithm for finding optimal shapes for a fixed-fixed column of unit volume is the following:

(1) Starting from a given shape, find the eigenfunctions $y(x)$ that corresponds to the lowest eigenvalue.

(2) Add a small amount of material according to $\delta\beta = k(y'')^2$, where the arbitrary, small number k is the step size.

(3) Rescale the new $\beta + \delta\beta$ to keep the volume constant.

(4) Recalculate the eigenfunction that corresponds to the lowest eigenvalue and continue until the lowest two eigenvalues are equal. This algorithm is outlined in more detail in the Appendix.

Once the lowest two eigenvalues are equal, the algorithm fails, because the eigenfunction that corresponds to the lowest, now double eigenvalue, is not uniquely determined, i.e. it is not known where to add the material to keep the column from buckling.

5. RESULTS

In an attempt to speed convergence, the algorithm of adding stiffness where the column is found to bend most, as previously discussed, was modified according to the following formulas:

$$\delta\beta = C\left(\frac{W_1 \lambda_a^2}{\lambda_a^2 + \lambda_s^2} (y_s'')^2 - \frac{W_2 \lambda_s^2}{\lambda_a^2 + \lambda_s^2} (y_a'')^2 \right), \text{ if } \lambda_s < \lambda_a \quad (5.1)$$

$$\delta\beta = C\left(\frac{-W_2 \lambda_a^2}{\lambda_a^2 + \lambda_s^2} (y_s'')^2 + \frac{W_1 \lambda_s^2}{\lambda_a^2 + \lambda_s^2} (y_a'')^2 \right), \text{ if } \lambda_a < \lambda_s \quad (5.2)$$

where

λ_a and λ_s = the lowest asymmetric and symmetric buckling mode, respectively

y'' = the second derivative of the corresponding eigenfunction designated by the subscript

W_1 and W_2 = weighting constants

C = an arbitrary constant that controls the step size used.

These formulas add stiffness to the lower mode and subtract it from the higher mode, for either the symmetric or asymmetric buckling modes, whichever order they occur in.

Results obtained are shown in Table 5.1.

TABLE 5.1   RESULTS OF CALCULATIONS USING
THE NEW ALGORITHM

W_1	W_2	Estimated Double Eigenvalue*	Best Values λ_s	Obtained** λ_a
1.0	1	4.152742	4.149023	4.152743
1.25	1	4.165444	4.166973	4.167283
1.5	1	4.167471	4.171677	4.181495
1.75	1	4.168562	4.171883	4.181128
2.0	1	4.169241	4.170437	4.212756
2.25	1	4.169730	4.170955	4.186852
2.5	1	4.169881	4.167879	4.260274
3.0	1	4.170268	4.168704	4.249844
4.0	1	4.170746	4.170805	4.169593
5.0	1	4.171071	4.170060	4.196496
1.0	0	4.172025	4.172029	4.172069

*From simple, linear interpolation of successive values at the cross-over of eigenvalues, i.e. where successive computations of eigenvalues changed from $\lambda_s < \lambda_a$ to $\lambda_a < \lambda_s$, or vice-versa.

**These should be multiplied by 4π before comparison to the values given by Tadjbakhsh-Keller or by Olhoff-Rasmussen.

The best shape obtained, with weights $W_1 = 1$ and $W_2 = 0$ (which corresponds to the first and simplest algorithm previously discussed), is shown in Fig. 5.1, with an enlargement of the boxed outline shown in Fig. 5.2. The buckling load obtained is 52.4273, as compared to 52.3563 calculated by Olhoff and Rasmussen, and 52.6378 the erroneous value found by Tadjbakhsh and Keller. The shape obtained is not symmetrical about the quarter points. It has its thinnest spot at about 0.246, with an area of 0.2134, which is somewhat smaller than that found by Olhoff and Rasmussen.

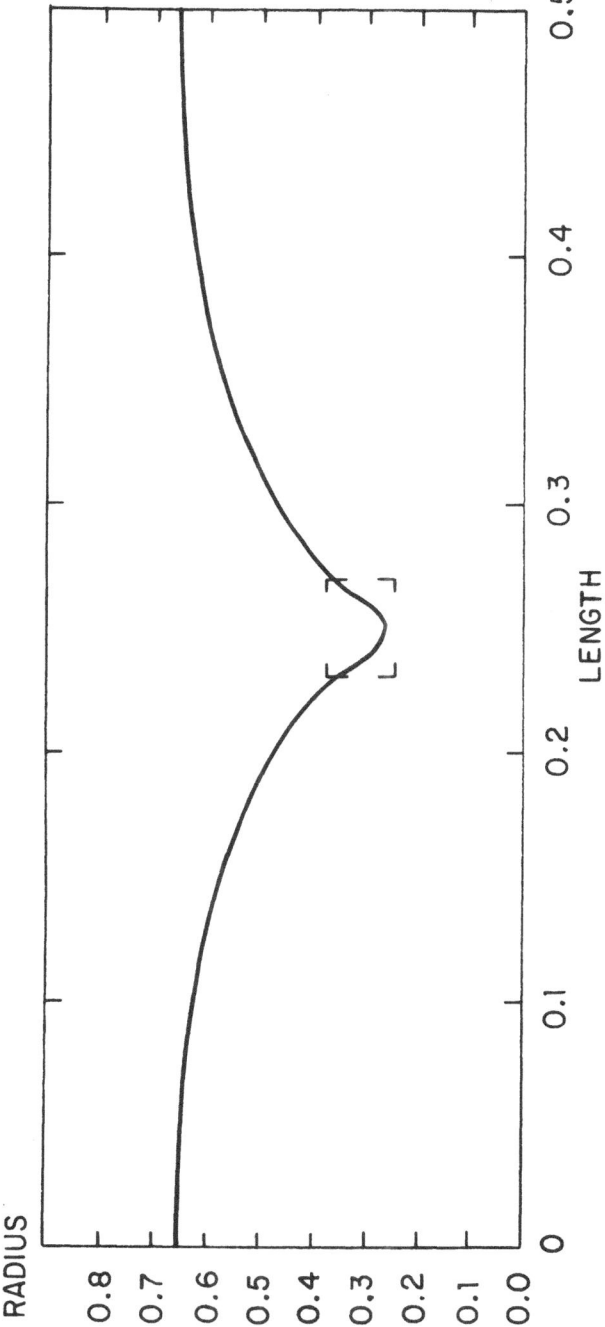

Figure 5.1 The Optimal Form for the Fixed-Fixed Column

528

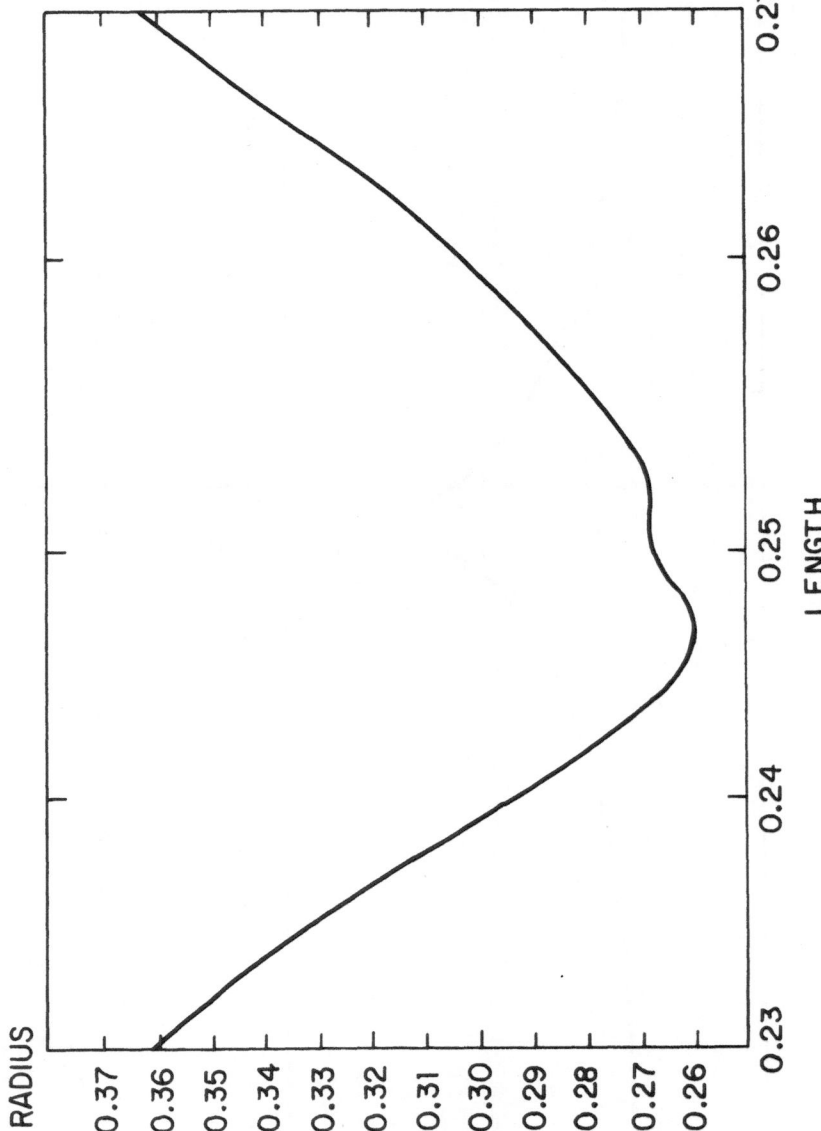

Figure 5.2 Detail of the Optimal Form for the Fixed-Fixed Column

6. CONCLUDING REMARKS

The buckling load found, 52.4273, is only slightly larger than that found by Olhoff and Rasmussen. Is this difference real, or only an indication of the accuracy of calculation? The authors believe it is real, since the difference is in the third digit and calculations are accurate to four or five digits, as indicated by a comparison of the buckling load found for the filled-in Tadjbakhsh-Keller design with that found by Olhoff and Rasmussen. The value found is only a lower bound. It has not been demonstrated to be the maximum value possible.

Although the new algorithm presented here has yielded the strongest fixed-fixed column yet discovered, it has defects. It fails once the lowest two eigenvalues have merged, whether or not the merged values are the maximum possible, because the eigenfunctions corresponding to repeated eigenvalues are not uniquely determined. It is sensitive to the step size used. If too large a step size is used, poor results are obtained. If too small a step size is used, convergence is slow. It is also sensitive to starting shapes. Starting from the Tadjbakhsh-Keller shape leads to the strongest fixed-fixed column yet found, but starting from a uniform column, it converges to a repeated eigenvalue about 5% lower than the maximum.

Use of this algorithm implies a maximum of the smallest double eigenvalue formulation of the optimization problem. Whether or not this formulation is equivalent to finding the shape of the strongest column of unit volume is not known. The equivalence of the two problems depends on the shapes of the lowest eigenvalue curves in the vicinity of their intersection, as shown in Fig. 6.1. These curves are simplified, since the dependence on shape is a function of infinitely many parameters, not as the simple one parameter families shown. An analytical proof of the equivalence of the two formulations is desirable, for it is unlikely that equivalence can be either proved or disproved by numerical examples.

REFERENCES

1. Lagrange, J.L., _Oevres_, Vol. 2, pp. 125-170, Cauthier-Villars, 1868. (summarized in I. Todhunter and K. Pearson, A History of the Theory of Elasticity and of the Strength of Materials. Vol. 1, pp. 66-67, Cambridge, England, 1886).
2. Clausen, T., "Uber die Form architektonischer Saulen," _Bulletin Physico-Mathematiques et Astronomiques_, Tome 1, 1849-1853, pp. 279-294. (summarized in I. Todhunter and K. Pearson), Vol. 2, pp. 325-329, Cambridge, England, 1893.

530

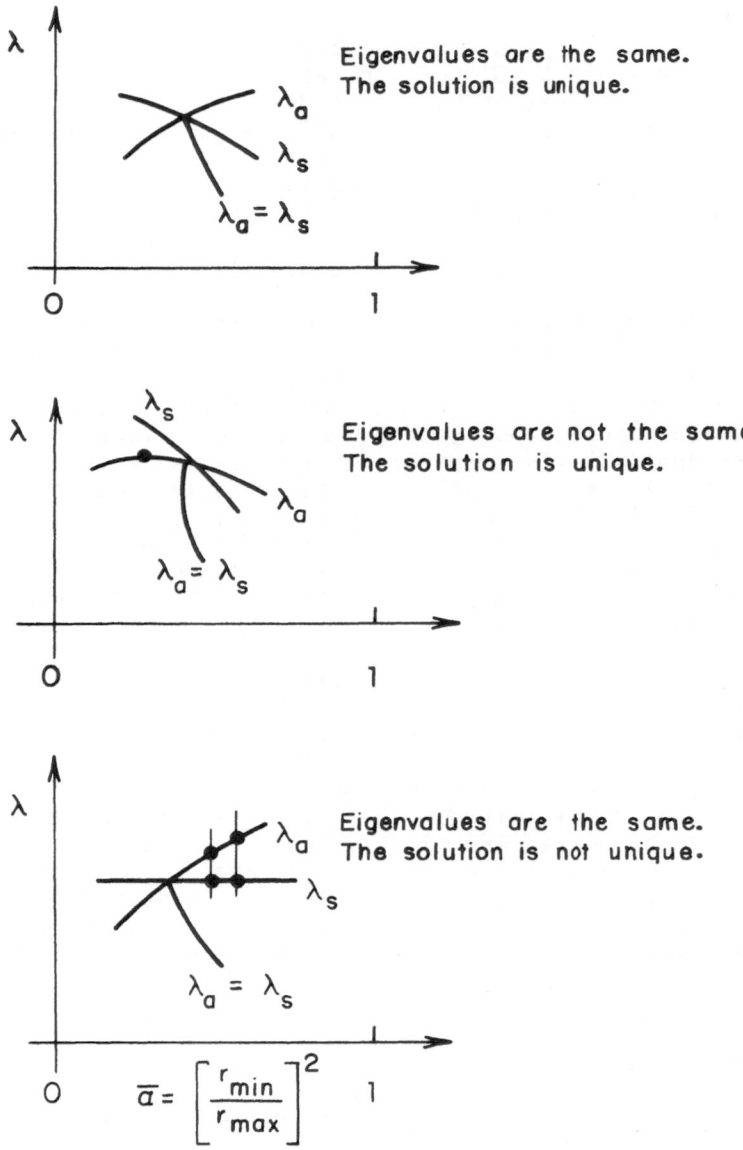

Eigenvalues are the same.
The solution is unique.

Eigenvalues are not the same.
The solution is unique.

Eigenvalues are the same.
The solution is not unique.

There is no proof that maximizing the lowest two eigenvalues is equivalent to optimizing the fixed-fixed column.

Figure 6.1   Possible Forms for the Equivalence of the Two Problems on Optimization

3. Keller, J.B., "The Shape of the Strongest Columns," Archives for Rational Mechanics and Analysis, Vol. 5, 1960, p. 275.
4. Tadjbakhsh, I. and Keller, J.B., "Strongest Columns and Isoperimetric Inequalities for Eigenvalues," J. Appl. Mech. Vol. 28, No. 1, 1962, pp. 159-164.
5. Barnes, E.R., Quarterly J. Appl. Math., Vol. 35, 1977, p. 393.
6. Taylor, J.E., "The Strongest Column: An Energy Approach," J. Appl. Mech., Vol. 34, No. 2, 1967, p. 486.
7. Taylor, J.E. and Liu, C.Y., "Optimal Design of Columns," AIAA J., Vol. 6, No. 8, 1968, p. 1497.
8. Haug, E.J., "Two Methods of Optimal Structural Design," Proc. 11th Midwestern Mech. Conf., Iowa State University Press, Ames, Iowa 1969, pp. 847-859.
9. Olhoff, N and Rasmussen, S.H., "On Single and Bimodal Optimum Buckling Loads of Clamped Columns", International Journal of Solid Structures, Vol. 13, 1977, pp. 605-614.

APPENDIX: AN OUTLINE OF THE CALCULATIONS USED

The steps used in the calculations made to add material where the column bends most are as follows:

(1) Select an admissible function y_0'', symmetric or asymmetric with respect to the center of the fixed-fixed column.

(2) From ther iterated form of the differential equation

$$(\beta y_{n+1}'')'' = -P y_n''$$

determine by integration

(3) $(\beta y_{n+1}'')' = -\lambda \int_0^x y_n'' \, d\xi + C_1$

and

(4) $(\beta y_{n+1}'') = -\lambda \int_0^x \int_0^\eta y_n'' \, d\xi d\eta + C_1 x + C_2$

from which, since $\beta(x)$ is never zero

(5) $y_{n+1}'' = - \dfrac{\lambda}{\beta(x)} \int_0^x \int_0^\eta y_n'' \, d\xi d\eta + C_1 \dfrac{x}{\beta(x)} + \dfrac{C_2}{\beta(x)}$

(6) Integrate twice more to get

$$y'_{n+1} = -\lambda \int_0^x \frac{\int_0^S \int_0^\eta y''_n d\xi d\eta}{(s)} ds + C_1 \int_0^x \frac{S}{\beta(s)} ds$$

$$+ C_2 \int_0^x \frac{ds}{\beta(s)} + C_3 \qquad (C_3 = 0)$$

$$y_{n+1} = -\lambda \int_0^x \int_0^t \frac{\int_0^S \int_0^\eta y''_n d\xi d\eta}{\beta(s)} ds \, dt + C_1 \int_0^x \int_0^t \frac{sds}{\beta(s)} dt$$

$$+ C_2 \int_0^x \int_0^t \frac{ds}{\beta(s)} dt + C_4 \qquad (C_4 = 0)$$

(7) y_{n+1}, y'_{n+1}, and y''_{n+1} have the forms

$$y''_{n+1} = -\lambda G_1(x) + C_1 f_1(x) + C_2 g_1(x)$$

$$y'_{n+1} = -\lambda G_2(x) + C_1 f_2(x) + C_2 g_2(x)$$

$$y_{n+1} = -\lambda G_3(x) + C_1 f_3(x) + C_2 g_3(x)$$

(8) Determine the constants C_1 and C_2 as follows:

(a) For the symmetric case,

$$(\beta y''_{n+1})' \big|_{x=0} = 0$$

Therefore $C_1 = 0$.

$$y'_{n+1} \big|_{x=\frac{\ell}{2}} = 0$$

Therefore, $C_2 = \dfrac{\lambda G_2(\frac{\ell}{2})}{g_2(\frac{\ell}{2})}$.

(b) For the asymmetric case, the moment condition

$$\beta y''_{n+1} \big|_{x = \frac{\ell}{2}} = 0$$

and the deflection condition

$$y_{n+1}\Big|_{x=\frac{\ell}{2}} = 0$$

leads to the matrix equation

$$\lambda \left\{ \begin{array}{c} G_1\left(\frac{\ell}{2}\right) \\ G_3\left(\frac{\ell}{2}\right) \end{array} \right\} = \left(\begin{array}{cc} f_1\left(\frac{\ell}{2}\right) & g_1\left(\frac{\ell}{2}\right) \\ f_3\left(\frac{\ell}{2}\right) & g_3\left(\frac{\ell}{2}\right) \end{array} \right) \left\{ \begin{array}{c} C_1 \\ C_2 \end{array} \right\}$$

from which

$$\left\{ \begin{array}{c} C_1 \\ C_2 \end{array} \right\} = \frac{\left(\begin{array}{cc} g_3 & -g_1 \\ -f_3 & f_1 \end{array} \right) \left\{ \begin{array}{c} G_1\left(\frac{\ell}{2}\right) \\ G_3\left(\frac{\ell}{2}\right) \end{array} \right\}}{f_1 g_3 - g_1 f_3}$$

(9) Determine the eigenvalues from the Rayleigh quotient

$$\lambda_{n+1} = \frac{\int_0^1 \beta (y''_{n+1})^2 \, dx}{\int_0^1 (y'_{n+1})^2 \, dx}$$

(10) Terminate the iteration when

$$\left| \frac{\lambda_{n+1} - \lambda_n}{\lambda_n} \right| < \varepsilon$$

where ε is a pre-assigned, small, positive number.

NOTE: The half Simpson's rule is used in integration, i.e.,

$$\int_0^h f dx = \frac{h}{12} (5f(0) + 8f(h) - f(2h))$$

$$\int_0^{2h} f dx = \frac{h}{12} (5f(2h) + 8f(h) - f(0))$$

A NUMERICAL METHOD FOR OPTIMIZATION OF STRUCTURES WITH REPEATED EIGENVALUES*

Kyung K. Choi and Edward J. Haug

Department of Materials Enigneering, University of Iowa, Iowa City, Iowa, U.S.

ABSTRACT

A numerical method for solving structural optimization problems in which double eigenvalues may occur is formulated. Recent results concerning the derivative of eigenvalues of structures [1] is used in design sensitivity analysis of the eigenvalues and the gradient projection method of Ref. 2 is applied for optimization. Results are compared with the conventional engineering perturbation analysis method of treating the eigenvalue problem.

1. INTRODUCTION

The structure to be optimized may be approximated by finite element methods. The eigenvalue problem that arises in vibration or buckling is then of the form

$$A(u)y = \zeta B(u)y \qquad (1.1)$$

where $u \in R^n$, $y \in R^k$ and $A(u)$ and $B(u)$ are $k \times k$ matrices. In the problems considered here, $A(u)$ and $B(u)$ are symmetric and positive definite.

The statement of the problem is as follows: find $u \in R^n$ that minimizes the cost function $\psi_0(u)$, subject to Eq. 1.1 and the constraints

*Research supported by National Science Foundation Project No. ENG 77-19967.

$$\psi_i(u) = \zeta_0 - \zeta_i \leq 0 \quad , i = 1,2 \tag{1.2}$$

$$\phi_j(u) = c_j - u_j \leq 0 \quad , j = 1,2,\ldots,n \tag{1.3}$$

where ζ_0 and c_i are given constants. Here it is assumed that the cost function $\psi_0(u)$ is differentiable.

2. DESIGN SENSITIVITY ANALYSIS

The basic idea used in developing an iterative optimization method in Ref. 2 is to first construct an estimate u^0 of the solution and then find a small change δu such that $u^0 + \delta u$ is an improved design. Thus, it is necessary to determine how changes δu in design change the cost function $\psi_0(u)$ and constraints ψ_i and ϕ_j. In this analysis, δu is required to be small so that the linearization of the functions of the problem is valid around point u^0. By linearizing the cost function $\psi_0(u)$ and Eqs. 1.2 and 1.3 around u^0, one has

$$\delta\psi_0 = \frac{\partial\psi_0}{\partial u}(u^0)\,\delta u \tag{2.1}$$

$$\delta\psi_i = -\zeta^{(1)}_{u^0,\delta u,i} \quad , \qquad i = 1,2 \tag{2.2}$$

$$\delta\phi_j = -\delta u_j \quad , \quad j = 1,2,\ldots,n \tag{2.3}$$

It will be convenient to obtain an explicit expression in Eq. 2.2 for $\zeta^{(1)}_{u^0,\delta u,i}$. If ζ_1 and ζ_2 are simple eigenvalues at u^0, then one can use engineering perturbation analysis of Eq. 1.1 to obtain

$$\zeta^{(1)}_{u^0,\delta u,i} = \sum_{k=1}^{n}\left(\left(\frac{\partial A(u^0)}{\partial u_k} - \zeta(u^0)\frac{\partial B(u^0)}{\partial u_k}\right)y^i, y^i\right)\delta u_k \tag{2.4}$$

where y^i are $B(u^0)$ - orthonormal eigenvectors of Eq. 1.1 corresponding to $\zeta_i(u^0)$. However if $\zeta_1(u^0) = \zeta_2(u^0)$, then Eq. 2.4 is not valid in general [3] and in Ref. 1, it is shown that eigenvalues have only directional derivatives $\zeta^{(1)}_{u^0,\delta u,i}$, which are the eigenvalues of the matrix

$$M(u^0, \delta u) = \left[\sum_{k=1}^{n} \left(\left(\frac{\partial A(u^0)}{\partial u_k} - \zeta(u^0) \frac{\partial B(u^0)}{\partial u_k} \right) y^\ell, y^m \right) \delta u_k \right] \tag{2.5}$$

where y^ℓ and y^m are any $B(u^0)$ - orthonormal eigenvectors. Note that unlike the simple eigenvalue case, if $\zeta_1(u^0) = \zeta_2(u^0)$, then $\zeta_{u^0, \delta u, i}^{(1)}$ are not linear in δu. This nonlinearity motivates the following quasilinearization procedure.

For any pair of $B(u^0)$ - orthonormal eigenvectors y^1 and y^2 corresponding $\zeta(u^0)$, let

$$\begin{aligned}
\bar{y}^1 &= y^1 \cos a + y^2 \sin a \\
\bar{y}^2 &= -y^1 \sin a + y^2 \cos a
\end{aligned} \tag{2.6}$$

then $\bar{y}^i$ are $B(u^0)$ - orthonormal for any angle a. Now define the matrix

$$\bar{M}(u^0, \delta u) = \left[\sum_{k=1}^{n} \left(\left(\frac{\partial A(u^0)}{\partial u_k} - \zeta(u^0) \frac{\partial B(u^0)}{\partial u_k} \right) \bar{y}^\ell, \bar{y}^m \right) \delta u_k \right] \tag{2.7}$$

If one chooses $a(\delta u)$, depending on δu, such that

$$\begin{aligned}
\bar{M}_{12}(u^0, \delta u) &= \tfrac{1}{2} \sin 2a \left(-M_{11}(u^0, \delta u) + M_{22}(u^0, \delta u) \right) \\
&+ \cos 2a \, M_{12}(u^0, \delta u) = 0
\end{aligned} \tag{2.8}$$

Then the directional derivatives are

$$\left. \begin{aligned}
\zeta_{u^0, \delta u, 1}^{(1)} &= \cos^2 a \, M_{11}(u^0, \delta u) + \sin 2a \, M_{12}(u^0, \delta u) \\
&+ \sin^2 a \, M_{22}(u^0, \delta u) \\[2em]
\zeta_{u^0, \delta u, 2}^{(1)} &= \sin^2 a \, M_{11}(u^0, \delta u) - \sin 2a \, M_{12}(u^0, \delta u) \\
&+ \cos^2 a \, M_{22}(u^0, \delta u)
\end{aligned} \right\} \tag{2.9}$$

Note that, since $M_{\ell m}(u^0, \delta u)$ are linear in δu, one can write $\zeta^{(1)}{}_{u^0, \delta u, i}$ as functions of δu. However, they are not linear in δu, as mentioned before. By treating a as an undetermined parameter and incorporating Eq. 2.8 as an equation of constraint, Eq. 2.9 gives an expression for $\zeta^{(1)}{}_{u^0, \delta u, i}$ needed in Eq. 2.2. Also note that if $\zeta_1(u^0)$ and $\zeta_2(u^0)$ are simple eigenvalues, then by letting a = 0 in Eq. 2.9 one has Eq. 2.4. Now the quasi-linearization of the optimization problem becomes

$$\delta \psi_0 = \frac{\partial \psi 0}{\partial u}(u^0)\delta u$$

$$\delta \psi_i = \ell^{\psi i}(u^0, a)\delta u \quad , \quad i=1,2 \tag{2.10}$$

$$\delta \phi_i = -\delta u_j \quad , \quad j=1,2,\ldots,n$$

with the requirement that a is determined by

$$\delta \psi_3 \equiv \bar{M}_{12}(u^0, \delta u) = \ell^{\psi 3}(u^0, a)\delta u = 0 \tag{2.11}$$

where $\ell^{\psi i}(u^0, a)$, i=1,2, can be found from Eq. 2.9 and $\ell^{\psi 3}(u^0, a)$ can be found from Eq. 2.8.

3.   A GRADIENT PROJECTION OPTIMIZATION METHOD

The problem of determining δu to minimize (maximize the negative of) $\delta \psi_0$ must deal with the inequality constraints of Eqs. 1.2 and 1.3. The iterative method to be used here is simply to ignore any inequality constraint that is considerably negative before an iteration begins. If, on the other hand, a constraint function is ε-active, it is enforced. That is, it is required that $\delta \psi_i \leq - \psi_i$ for each i=1,2 with $\psi_i \geq - \varepsilon$, $\delta \psi_3 = 0$, and $\delta \phi_j \leq - \phi_j$ for each j=1,2,\ldots,n with $\phi_j \geq - \varepsilon$. Define

$$A = \{i: \psi_i(u^0) = \zeta_0 - \zeta_i(u^0) \geq - \varepsilon, \ i=1,2\}$$

$$B = \{j: \phi_j(u^0) = c_j - u_j^0 \geq - \varepsilon, \quad i=1,2,\ldots,n\} \tag{3.1}$$

and column vectors

$$\tilde{\psi} = \begin{bmatrix} \psi_i(u^0); \ i \in A \\ 0 \end{bmatrix}$$

$$\tilde{\phi} = [\phi_j(u^0); \ j \in B]$$

$\left.\rule{0pt}{40pt}\right\}$ (3.2)

Also define matrices

$$\Lambda^{\tilde{\psi}}(a) = \left[\ell^{\psi_i T}(u^0,a), \ \ell^{\psi_3 T}(u^0,a); \ i \in A \right]$$

$$\Lambda^{\tilde{\phi}} = \left[\frac{\partial \phi_j(u^0)}{\partial u}; \ j \in B \right]$$

(3.3)

Then the quasi-linearized problem is to find δu and a to minimize $\delta\psi_0$ of Eq. 2.10 subject to

$$\Lambda^{\tilde{\psi} T}(a)\delta u + \tilde{\psi} \leq 0$$

(3.4)

$$\Lambda^{\tilde{\phi}}\delta u + \tilde{\phi} \leq 0$$

(3.5)

and

$$\delta u^T W \delta u - \xi^2 \leq 0$$

(3.6)

Equation 3.6 is introduced to ensure that δu is sufficiently small, where W is chosen as a positive definite weighting matrix and ξ is a small number.

The Kuhn-Tucker necessary conditions [2] can be applied to this quasi-linearized problem, which states that there exist multipliers $\gamma = \begin{bmatrix} \gamma_i; \ i \in A \\ \gamma_3 \end{bmatrix}$, $\gamma_i \geq 0$ for $i \in A$, $\mu = [\mu_j; \ j \in B] \geq 0$, and $\gamma_0 \geq 0$ that define

$$H = \left[\frac{\partial\psi_0}{\partial u}(u^0) + \gamma^T\Lambda^{\tilde{\psi} T}(a) + \mu^T\Lambda^{\tilde{\phi}} \right]\delta u + \gamma^T\tilde{\psi} + \mu^T\tilde{\phi}$$

$$+ \ \gamma_0(\delta u^T W \delta u - \xi^2)$$

(3.7)

such that

$$\frac{\partial H}{\partial \delta u} = \frac{\partial\psi_0}{\partial u}(u^0) + \gamma^T\Lambda^{\tilde{\psi} T}(a) + \mu^T\Lambda^{\tilde{\phi}} + 2\gamma_0\delta u^T W = 0$$

(3.8)

and

$$\frac{\partial H}{\partial a} = \left[\gamma_1 \ell_a^{\psi_1}(u^0,a) + \gamma_2 \ell_a^{\psi_2}(u^0,a) + \gamma_3 \ell_a^{\psi_3}(u^0,a) \right] \delta u = 0 \qquad (3.9)$$

where subscript a denotes differentiation with respect to a.

The idea is that if ζ_1 and ζ_2 are simple eigenvalues at u^0, then, by letting a = 0 and $\ell^{\psi_3}(u^0,a) = 0$, one has correct sensitivity coefficients from Eq. 2.9 and one can ignore the necessary condition of Eq. 3.9. On the other hand, if the eigenvalues become close enough so that the eigenvalue analysis package program (e.g., Eispack [4] or subspace iteration method [5]) treats them as same eigenvalues, one has to treat a as an undetermined parameter and use the necessary condition of Eq. 3.9 to find a by an inner loop iteration procedure.

With this information, one obtains an extension of the gradient projection method of Ref. 2. The resulting algorithm is summarized as follows:

Step 1. Make an engineering estimate of the optimum design u^0.

Step 2. Solve Eq. 1.1 for $\zeta_1(u^0)$ and $\zeta_2(u^0)$ and the corresponding y^1 and y^2.

Step 3. Check the constraint $\phi_j(u^0)$ and form $\tilde{\phi}$ of Eq. 3.2 . compute $\Lambda^{\tilde{\phi}}$ in Eq. 3.3 and $\frac{\partial \psi_0}{\partial u}(u^0)$ in Eq. 2.10.

Step 4. (A) If $\zeta_1(u^0) = \zeta_2(u^0)$ and $\psi_1(u^0) = \psi_2(u^0)$ is ε - active, let

$$\tilde{\psi} = \begin{bmatrix} \zeta_0 - \dfrac{\zeta_1(u^0) + \zeta_2(u^0)}{2} \\[2ex] \zeta_0 - \dfrac{\zeta_1(u^0) + \zeta_2(u^0)}{2} \\[2ex] 0 \end{bmatrix}$$

and go to step 5(A). Otherwise, go to step 4(B).

(B) Let $a = 0$ and $\ell^{\psi}3(u^0, a) = 0$. Form $\tilde{\psi}$ of Eq. 3.2 and compute $\Lambda^{\tilde{\psi}}(0)$ in Eq. 3.3. Go to Step 5(B).

Step 5,(A) Choose step size $\gamma_0 > 0$. Use the necessary condition of Eq. 3.9 to find a. To do this, for each a, one has to compute $\Lambda^{\tilde{\psi}}(a)$ in Eq. 3.3 and $\Lambda^{\psi}_a(a)$. Also, for each a, compute $M_{\psi\psi_0}$, $M_{\psi\psi}$, $M_{\psi\phi}$, γ, and δu from the following equations [2]

$$M_{\psi\psi_0} = \Lambda^{\tilde{\psi}T}(a)W^{-1}(I - \Lambda^{\tilde{\phi}T}\Lambda^{\phi^{-1}}\Lambda^{\tilde{\phi}}W^{-1})\frac{\partial\psi_0}{\partial u}^T(u^0) \tag{3.10}$$

$$M_{\psi\psi} = \Lambda^{\tilde{\psi}T}(a)W^{-1}(I - \Lambda^{\tilde{\phi}T}\Lambda^{\phi^{-1}}\Lambda^{\tilde{\phi}}W^{-1})\Lambda^{\tilde{\psi}}(a) \tag{3.11}$$

$$M_{\psi\phi} = \Lambda^{\tilde{\psi}T}(a)W^{-1}\Lambda^{\tilde{\phi}T}\Lambda^{\phi^{-1}}\Lambda^{\tilde{\phi}}_{\tilde{\phi}} \tag{3.12}$$

where

$$\Lambda^{\phi} = \Lambda^{\tilde{\phi}}W^{-1}\Lambda^{\tilde{\phi}T} \tag{3.13}$$

and

$$\gamma = M_{\psi\psi}^{-1}\left[2\gamma_0(\tilde{\psi} - M_{\psi\phi}) - M_{\psi\psi_0}\right] \tag{3.14}$$

$$\delta u = -\frac{1}{2\gamma_0}\delta u^1 + \delta u^2 \tag{3.15}$$

where

$$\delta u^1 = W^{-1}\left(I - \Lambda^{\tilde{\phi}T}\Lambda^{\phi^{-1}}\Lambda^{\tilde{\phi}}W^{-1}\right)\left[\frac{\partial\psi_0}{\partial u}^T(u^0) - \Lambda^{\tilde{\psi}}(a)M_{\psi\psi}^{-1}M_{\psi\psi_0}\right] \tag{3.16}$$

$$\delta u^2 = W^{-1}\left(I - \Lambda^{\tilde{\phi}T}\Lambda^{\phi^{-1}}\Lambda^{\tilde{\phi}}W^{-1}\right)\Lambda^{\tilde{\psi}}(a)M_{\psi\psi}^{-1}(M_{\psi\phi} - \tilde{\psi})$$
$$- W^{-1}\Lambda^{\tilde{\phi}T}\Lambda^{\phi^{-1}}\tilde{\phi} \tag{3.17}$$

compute

$$\mu = -\Lambda^{\phi^{-1}}\left[\Lambda^{\tilde{\phi}}W^{-1}\left(\frac{\partial\psi_0}{\partial u}(u^0) + \Lambda^{\tilde{\psi}}(a)\gamma\right) - 2\gamma_0\tilde{\phi}\right] \tag{3.18}$$

go to step 6(A).

(B) Compute $M_{\psi\psi_0}$, $M_{\psi\psi}$, $M_{\psi\phi}$ in Eqs. 3.10 - 3.12. Choose step size $\gamma_0 > 0$ and compute γ and μ in Eqs. 3.14 and 3.18 respectively. Go to step 6(B).

Step 6.(A) If any components of γ (except γ_3) are negative, delete the corresponding components of ψ and $\Lambda^\Psi(a)$. If any components of μ are negative, delete the corresponding components of ϕ and $\Lambda^{\tilde{\Phi}}$ and return to Step 5(A). Otherwise, go to Step 7.

(B)If any components of γ are negative, delete the corresponding components of ψ and $\Lambda^\Psi(0)$. If any components of μ are negative, delete the corresponding components of ϕ and $\Lambda^{\tilde{\Phi}}$ and return to Step 5(B). Otherwise, go to Step 7.

Step 7. Compute δu^1 and δu^2 of Eqs. 3.16 and 3.17.

Step 8. Compute

$$u^{j+1} = u^j - \frac{1}{2\gamma_0}\delta u^1 + \delta u^2$$

Step 9 If the constraints are satisfied and $||\delta u^1||$ is sufficiently small, terminate. Otherwise, return to Step 2 with u^j replaced by u^{j+1}.

4.   OPTIMAL DESIGN OF A COLUMN WITH REPEATED EIGENVALUES

In this section, the computational algorithm of Section 3 is applied to the column problem presented in Section 2.2 of Ref. 3.

The governing eigenvalue equation for the column was given as

$$K(b)y = \zeta Dy \tag{4.1}$$

where

$$K(b) = \begin{bmatrix} b_0^2 + 4b_1^2 + b_2^2 & -2b_1^2 - 2b_2^2 & b_2^2 & 0 \\ -2b_1^2 - 2b_2^2 & b_1^2 + 4b_2^2 + b_3^2 & -2b_2^2 - 2b_3^2 & b_3^2 \\ b_2^2 & -2b_2^2 - 2b_3^2 & b_2^2 + 4b_3^2 + b_4^2 & -2b_3^2 - 2b_4^2 \\ 0 & b_3^2 & -2b_3^2 - 2b_4^2 & b_0^2 + b_3^2 + 4b_4^2 \end{bmatrix} \qquad (4.2)$$

and

$$D = \begin{bmatrix} 2 & -1 & 0 & 0 \\ -1 & 2 & -1 & 0 \\ 0 & -1 & 2 & -1 \\ 0 & 0 & -1 & 2 \end{bmatrix} \qquad (4.3)$$

The design problem was formulated as follows: find $b \in R^4$ that minimizes

$$\psi_0 = \sum_{j=1}^{4} b_j \qquad (4.4)$$

subject to eigenvalue constraints

$$\psi_i = \zeta_0 - \zeta_i < 0 , \qquad i = 1,2 \qquad (4.5)$$

and design variable constraints

$$\phi_j = c_j - b_j < 0 , \qquad j = 1,2,3,4 \qquad (4.6)$$

where ζ_i are the smallest two eigenvalues of Eq. 3.1 and c_i are given constants.

In Table 4.1, the numerical result for the known optimum point A of Fig. 2.5 of Ref. 3 is given. The method converges precisely to the known solution, which is a symmetric column. Table 4.2 provides numerical result for the same point A using the engineering perturbation analysis Eq. 2.4. For iterations

1 to 11, the result is the same as in Table 4.1. However, as the eigenvalues coalese, the eigenvalue analysis package program Eispack [4] gives two eigenvectors that are neither symmetric nor antisymmetric (see Section 3.4 of Ref. 3) and the method does not converge. Note that the column becomes asymmetric as iteration continues in Table 4.2. Another result for the known optimum point B of Fig. 2.5 of Ref. 3 is given in Table 4.3. Table 4.4 is the numerical result for the same point using the engineering perturbation analysis Eq. 2.4. Note that the same kind of difficulties happen in this case. In Table 4.5, the numerical result of the problem with asymmetric design variable lower bounds, $c_1 = c_2 = 0.4$ and $c_3 = c_4 = 0.001$. The optimum column in this case is asymmetric. Table 4.6 is the result of the same problem using engineering perturbation analysis of Eq. 2.4. The result of Table 4.5 converges, whereas the result of Table 4.6 does not converge.

5.   CONCLUSION

The eigenvalue perturbation formula of Eq. 2.9 for repeated eigenvalues has been used to extend the gradient projection method of Ref. 2. The column example in Section 4 shows that the new method converges to the correct solution, whereas the engineering perturbation analysis method fails to converge. There is one point yet to be resolved in the new method. That is, "When can one regard the two numerically determined eigen-values as the same?" In the examples of Tables 4.1 and 4.3, one knows that the optimum column is symmetric [3], so one can set the initial design estimate to be symmetric. Then as long as two eigenvalues are different, the eigenvectors y^1 and y^2 will be symmetric and antisymmetric, which give symmetric $\delta u = - \frac{1}{2\gamma_0} \delta u^1 + \delta u^2$. Hence, one can regard the eigenvalues to be equal when y^1 and y^2 are neither symmetric nor antisymmetric. However, if the optimum column is asymmetric, there is no simple method to identify that the two eigenvalues are numerically same. In the example of Table 4.5, two eigenvalues are regarded to be equal if $|\zeta_1 - \zeta_2|/\zeta_1 \leq 2 \times 10^{-6}$. The number 2×10^{-6} is chosen by the experience based on the two previous symmetric examples. But to treat the general problem, it is necessary to investigate the eigenvalue analysis package that is used in the algorithm. Even though only discrete system optimization has been considered in this paper, the result can be extended to distributed systems.

544

REFERENCES

1. Haug, E.J. and Rousselet, B, "Design Sensitivity Analysis in Structural Mechanics II: Eigenvalue Variations," J. Str. Mech., Vol. 8, No. 2, 1980.
2. Haug, E.J. and Arora, J.S., Applied Optimal Design, Wiley, New York, 1979.
3. Choi, K.K. and Haug, E.J., "Optimization of Structures with Repeated Eigenvalues," Optimization of Distributed Parameter Structures (Eds. E.J. Haug and J. Cea), Sijthoff & Noordhoff, Alphen aan den Rijn, Netherlands, 1980.
4. Garbow, B.S., Boyle, J.M., Dongarra, J.J., and Moler, C.B., Matrix Eigensystem Routines--Eispack Guide Extension, Springer-Verlag, New York, 1977.
5. Bathe, K.J. and Wilson, E.L., Numerical Methods in Finite Element Analysis, Prentice-Hall, New Jersey, 1976.

TABLE 4.1 NUMERICAL RESULTS FOR THE POINT A

$b_0 = .7$
$\zeta_0 = .467870503298$

$b_1 = .32956309281$
$b_2 = .67043769071\,9$

$c_1 = c_2 = .01$

ITER	a	ζ_1	ζ_2	$\|\delta b^1\|$	ψ_0	b_1	b_2
1		.4594	.8274		2.6	.65	.65
2		.4680	.7877	1.406	2.5383	.6004	.6688
3		.4679	.7278	1.410	2.4385	.5506	.6687
4		.4679	.6660	1.409	2.3394	.5507	.6690
5		.4678	.6045	1.407	2.2404	.4510	.6692
6		.4678	.5455	1.405	2.1418	.4013	.6696
7		.467811	.490751	1.402	2.0437	.3518	.6701
8		.441440	.467779	1.397	1.9462	.3024	.6707
9		.46783150	.46885895	2.33×10^{-14}	2.00184145	.33054604	.67037468
10		.46787049	.46787162	5.0×10^{-14}	2.00000217	.32956342	.67043766
11		.46787050503298	.46787050503300	2.56×10^{-11}	2.00000000000004	.32956309282	.67043769070720
12	.0244	.46787050503297	.46787050503299	1.46×10^{-12}	2.00000000000000	.32956309281	.67043769070719
13	.5883	–	–	9.40×10^{-13}	–	–	–
14	.0428	–	–	5.94×10^{-13}	–	–	–
15	.1472	–	–	3.84×10^{-13}	–	–	–
16	.7189	–	–	2.48×10^{-13}	–	–	–

TABLE 4.2   NUMERICAL RESULTS FOR THE POINT A

(USING ENGINEERING PERTURBATION EQ. 2.4)

ITER	ζ_1	ζ_2	$\|\delta b^1\|$	ψ_0	b_1	b_2	b_3	b_4
1-11			SAME AS TABLE 4.1					
12	.46776806	.46797187	.015	1.99998897	.32987882	.66947438	.67084270	.32929307
13	.46494673	.47072637	.170	1.99861560	.33049804	.66967300	.66403914	.33440543
14	.46766381	.46804646	.058	1.99995734	.32904922	.67135059	.66958068	.32997685
15	.46430963	.47140081	.174	1.99837340	.33143128	.66887362	.66318049	.33488800
16	.46765381	.46805754	.053	1.99995567	.32911510	.67145320	.66945372	.32993364

TABLE 4.3 NUMERICAL RESULTS FOR THE POINT B

$b_0 = 3.0$ $b_1 = .63485377173$ $c_1 = c_2 = .01$

$\zeta_0 = .51808086472$ $b_2 = .36514622827$

ITER	a	ζ_1	ζ_2	$\|\delta b^1\|$	ψ_0	b_1	b_2
1		.8434	.9223		2.6	.80	.50
2		.7292	.7635	2.0	2.4	.75	.45
3		.6195	.6229	2.0	2.2	.70	.40
4		.4906	.5249	2.0	2.0	.65	.35
5		.51849228	.51883999	1.44×10^{-14}	2.00093554	.63500133	.36546644
6		.51808920	.51808940	3.99×10^{-11}	2.00000031	.63485539	.36514476
7		.51808909073965	.51808909098978	2.52×10^{-6}	2.000000000000	.63485381071	.365144630843
8	.0068	-	-	9.82×10^{-9}	-	.634855381040	.365144630859
9	.7853	-	-	9.88×10^{-9}	-	.634855381009	.365144630874
10	.3924	-	-	9.94×10^{-9}	-	.634855380977	.365144630889
11	1.4946	-	-	9.99×10^{-9}	-	-445	-905
12	.7807	-	-	1.00×10^{-8}	-	-913	-920

TABLE 4.4   NUMERICAL RESULTS FOR THE POINT B

(USING ENGINEERING PERTURBATION EQ. 2.4)

ITER	ζ_1	ζ_2	$\|\delta b^1\|$	ψ_0	b_1	b_2	b_3	b_4
1-7				SAME AS TABLE 4.3				
8	.51656062	.51962485	.619	1.99807881	.63290171	.36613769	.36613769	.63290172
9	.51809189	.51866395	.437	2.00021863	.63467469	.36543466	.36543462	.63467466
10	.51808844	.51809009	1.52×10^{-4}	2.00000027	.63485514	.36514421	.36514526	.63485565
11	.51655742	.51962805	.619	1.99808273	.63294900	.36622468	.36604664	.63286241
12	.51735602	.51925686	.468	2.00006053	.63433444	.36483685	.36599438	.63489486

TABLE 4.5 NUMERICAL RESULTS FOR ASYMMETRIC DESIGN

$b_0 = .7, \quad \zeta_0 = .5, \quad c_1 = c_2 = .4, \quad c_3 = c_4 = .001$

ITER	a	ζ_1	ζ_2	$\|\delta b^1\|$	ψ_0	b_1	b_2	b_3	b_4
1		.4900	.9229		2.8	.7	.7	.7	.7
2		.5000	.8085	1.414	2.6286	.6000	.7143	.7143	.6000
3		.5000	.6738	1.413	2.4290	.5001	.7144	.7144	.5001
4		.5000	.5472	1.413	2.2295	.4002	.7146	.7146	.4002
5		.4804	.5003	.998	2.1299	.4000	.7147	.7147	.3004
6		.4971	.5027	.104	2.1558	.4000	.7215	.7058	.3285
7		.4797	.5179	.530	2.1317	.4000	.6693	.7056	.3568
8		.4533	.4986	1.001	2.0853	.4000	.7164	.7082	.2607
9		.4973	.5040	.055	2.1584	.4000	.7247	.7037	.3301
10		.4881	.5093	.453	2.1433	.4000	.6777	.7223	.3433
11		.4969	.5015	.300	2.1517	.4000	.7135	.7088	.3294
12		.499626	.499980	.149	2.158265	.40000	.704466	.725177	.328621
13		.499791	.500440	1.082	2.159292	.400218	.704351	.725323	.329400
14		.499954	.500360	.831	2.159683	.400000	.704709	.725720	.329253
15		.499999	.500411	.349	2.159871	-	.704681	.725833	.329357
16		.49999997	.50039256	.291	2.15985748	-	.70463463	.72588520	.32933766
17		.50000000	.50037450	.289	2.15983670	-	.70458241	.72593259	.32932171
18		.50000000	.50035664	.289	2.15981588	-	.70452990	.72597966	.32930633
19		.50000000	.50033882	.288	2.15979514	-	.70447748	.72602665	.32929101
20		.50000000	.50032102	.288	2.15977446	-	.70442516	.72607359	.32927571
21		.50000000	.50030326	.287	2.15975385	-	.70437295	.72612046	.32926044
22		.50000000	.50028553	.287	2.15973331	-	.70432084	.72616728	.32924519
23		.50000000	.50026783	.286	2.15971284	-	.70426885	.72621403	.32922997
24		.50000000	.50025016	.286	2.15969244	-	.70421695	.72626072	.32921477

n									
25	.32919959	.72630736	.70416516	–	2.15967211	.285	.50023252	.50000000	
26	.32918444	.72635393	.70411348	–	2.15965185	.285	.50021491	.50000000	
27	.32916932	.72640044	.70406190	–	2.15963166	.284	.50019732	.50000000	
28	.32915422	.72644689	.70401043	–	2.15961153	.283	.50017977	.50000000	
29	.32913914	.72649328	.70395906	–	2.15959148	.283	.50016225	.50000000	
30	.32912408	.72653961	.70390779	–	2.15957149	.283	.50014476	.50000000	
31	.32910906	.72658588	.70385663	–	2.15955157	.282	.50012730	.50000000	
32	.32904405	.72663209	.70380557	–	2.15953172	.282	.50010986	.50000000	
33	.32907907	.72667824	.70375462	–	2.15951193	.281	.50009246	.50000000	
34	.32906411	.72672433	.70370377	–	2.15949221	.281	.50007509	.50000000	
35	.32104918	.72677036	.70365302	–	2.15947256	.280	.50005774	.50000000	
36	.32903427	.72681633	.70360238	–	2.15945298	.280	.50004043	.50000000	
37	.32901939	.72686224	.70355183	–	2.15943346	.280	.50002315	.50000000	
38	.32900453	.72690809	.70350139	–	2.15941401	.280	.50000589	.50000000	
39	.329003790684	.726910383123	.703498881270	–	2.159413055076	.279	.500005033892	.49999999991	
40	.329003050515	.726912671379	.703496363290	–	2.159412085183	.279	.500004173200	.49999999991	
41	.329002303871	.726914961896	.703493349685	–	2.159411115453	.279	.500003310438	.50000000000	
42	.329001584011	.726917242365	.703491319466	–	2.159410145842	.279	.500002456545	.49999999950	
43	.329000706916	.726919579667	.703488887928	–	2.159409174510	.279	.500001552847	.49999997497	
44	.329001304783	.726921242645	.703485436126	–	2.159407983554	.310	.500001375343	.49999976663	
45	.328999104107	.726920684905	.703487032788	–	2.159406821800	.134	.500000470539	.49999529458	
46	.328999104109	.726920684909	.703487032789	–	2.159406821805	3.59×10^{-15}	.500000470540	.499999529460	.3892
47	–	–	–	–	–	2.82×10^{-16}	–	–	.3832
48	–	–	–	–	–	4.50×10^{-16}	–	–	.9372
49	–	–	–	–	–	6.28×10^{-16}	–	–	.2877
50	–	–	–	–	–	1.27×10^{-15}	–	–	.9755

TABLE 4.6 NUMERICAL RESULTS FOR ASYMMETRIC DESIGN

(USING ENGINEERING PERTURBATION EQ. 2.4)

ITER	ζ_1	ζ_2	$\|\delta b^1\|$	ψ_0	b_1	b_2	b_3	b_4
1-45				SAME AS TABLE 4.5				
46	.49998182	.50000184	.435	2.15940352	.40000000	.70348621	.72691546	.32900185
47	.49999901	.50000099	.134	2.15940619	-	.70349022	.72691726	.32899871
48	.49999740	.50000259	.435	2.15940185	-	.70348709	.72691187	.32900289
49	.49999866	.50000134	.134	2.15940577	-	.70349234	.72691498	.32899845
50	.49999688	.50000312	.435	2.15940074	-	.70348767	.72690949	.32900358

MULTIPLE EIGENVALUES AND SUPREMUM NORM CONSTRAINTS

Bernard Rousselet

Université de Nice, Département de Mathematique
06034 Nice Cédex, France

ABSTRACT

This paper presents an algorithm based on design sensitivity [1] and necessary condition of optimality [2] that have very recently been obtained. The necessary condition appears at first glance to be useless for numerical purposes. Unfortunately, the derivation of this algorithm is so new that there is no proof of convergece (which, at any rate, seems a difficult task) and no numerical results are yet obtained. An algorithm is first sketched for the case of an optimum eigenvalue that is simple. It is then shown how to modify the algorithm for the case of repeated eigenvalues. Finally, a supremum norm constraint is treated and it is shown that a similar algorithm may be implemented to solve it.

1. OPTIMUM AT A SIMPLE EIGENVALUE

1.1 The Optimization Problem

The vibration of buckling state of a typical structural system is a solution of

$$a_u(y,v) = \zeta b_u(y,v) \qquad (1.1)$$

where $y \neq 0$ is an eigenfunction corresponding to the eigenvalue ζ. Equation 1.1 must hold for every v in some space of functions V that depends on the specific problem. The design variable u is a function that is usually the variable thickness of the structure, to be chosen from a space u of admissible designs.

Many structural mechanics examples may be found in Ref. 1. Specific examples for which the optimum is at a repeated eigen-value are discussed in Ref. 2. In all these examples, it may be mathematically shown that a_u is a coercive bilinear form on V and b_u is, in most cases, equivalent to the scalar product of $L^2(\Omega)$. Both bilinear forms are self-adjoint and depend smoothly on u. The solutions ζ of Eq. 1.1 constitute a discrete and denumerable set of real positive numbers, which are all eigenvalues of finite multiplicity.

The optimization problem to be addressed here is to find a design u to minimize

$$\psi(u) = \int_\Omega u \, dx \qquad (1.2)$$

where Ω is the domain over which the structure is distributed, subject to the constraint $\zeta(u) \geq \zeta_0 > 0$ on the smallest eigenvalue $\zeta(u)$ of Eq. 1.1 and the requirement $u(x) \geq u_1 > 0$ almost every-where in Ω if u is not smooth. The lower bound u_1 is usually a constant, referred to as the design constraint. The set $A = \{x : u(x) = u_1\}$ is called the set of active design constraint. This set is defined up to a set of measure zero if $u(x)$ is not smooth.

In case $\zeta(u)$ is simple at the optimum u_0, it depends smoothly on u in the neighborhood of u_0 and the usual Lagrange multiplier rule applies in its classical way. When $u(x) \geq u_1 > 0$ is not active, there exist real members $\lambda_0 \geq 0$ and $\lambda_1 \geq 0$, not both zero such that

$$\lambda_0 \int_\Omega \bar{u} \, dx - \lambda_1 \, \zeta'(u_0)\bar{u} = 0 \qquad (1.3)$$

for every $\bar{u} \in U$. When the constraint $u(x) \geq u_1 > 0$ is active, Eq. 1.3 is valid only for functions $\bar{u}$ with support in the set Ω-A. The necessary condition is [4,5]

$$\lambda_0 \int_\Omega (u - u_0)dx - \lambda_1 \, \zeta'(u_0)(u - u_0) \geq 0$$

for every $u \geq u_1$.

Notice that if $\lambda_1 = 0$, then λ_0 is also zero, contradicting the hypothesis that λ_0 and λ_1 are not both zero. Hence one may set $\Lambda = \lambda_0/\lambda_1$, and rewrite Eq. 1.3 as

554

$$\Lambda \int_{\Omega} \bar{u} \, dx - \zeta'(u_0)\bar{u} = 0 \qquad\qquad (1.4)$$

1.2   Toward A Computational Algorithm

Note that this is the necessary condition (as long as Λ is known) of minimization of

$$L(\Lambda,u) = \Lambda \int_{\Omega} u \, dx - \zeta(u) \qquad\qquad (1.5)$$

Hence, for known Λ, an appropriate algorithm is

$$u_{n+1} = u_n - \rho_n[\Lambda - \zeta'(u_n)] \qquad\qquad (1.6)$$

where ρ_n is an appropriate normalizing factor. This is a gradient type method for $u \rightarrow L(\Lambda,u)$ [6].

To be more specific, as shown in Ref. 1,

$$\zeta'(u) \, \bar{u} = a_{a,\bar{u}}^{(1)} (y_u,y_u) - \zeta(u) \, b_{u,\bar{u}}^{(1)} (y_u,y_u) \qquad\qquad (1.7)$$

where y_u is an eigenfunction of Eq. 1.1, associated with $\zeta(u)$, normalized by $b_u(y_u,y_u) = 1$, and $a_{u,\bar{u}}^{(1)}$ and $b_{u,\bar{u}}^{(1)}$ denote the derivative of the bilinear forms a_u and b_u with respect to the design variable u, evaluated in the direction $\bar{u}$, y being held fixed.

Denote now

$$M_{u,\bar{u}} = a_{u,\bar{u}}^{(1)} (y_u,y_u) - \zeta(u) \, b_{u,\bar{u}}^{(1)} (y_u,u) \qquad\qquad (1.8)$$

and M_u^+ the integrand of $M_{u,\bar{u}}$, i.e.

$$M_{u,\bar{u}} = \int_{\Omega} M_u^+ \, \bar{u} \, dx \qquad\qquad (1.9)$$

Then the necessary condition of Eq. 1.4, with Eq. 1.7, implies that

$$\Lambda = M_u^+ \qquad\qquad (1.10)$$

at least where the design constraint is not active.

In an iterative procedure M_u^+ will generally not be constant, even where the design variable is not active. However, one may consider computing an approximation of Λ with the equation

$$\Lambda^{(p+1)} = \frac{1}{\text{mes}(\Omega-A^{(p)})} \int_{\Omega-A}(p) \; M_u^+(p) \; dx \tag{1.11}$$

where $A^{(p)}$ is the set $\{x : u^{(p)}(x) = u_1\}$, defined up to a set of measure zero if $u^{(p)}$ is not smooth.

1.3 The Complete Algorithm

The algorithm proposed here is as follows:

(1) Start from an estimated design $u^{(0)}$, with $p = 0$.

(2) Solve Eq. 1.1 for $u = u^{(p)}$.

(3) Compute:

$$\Lambda^{(p+1)} = \frac{1}{\text{mes}(\Omega-A^{(p)})} \int_{\Omega-A} \; M_u^+(p) \; dx \tag{1.12}$$

(4) Start an iterative procedure by setting
$v_0^{(p+1)} = u_0^{(p+1)} = u^{(p)}$ (set $n = 0$). Solve Eq. 1.1
for $\zeta_u^{(p+1)}$, with $u = v_n^{(p+1)}$

$$v_{n+1}^{(p+1)} = v_n^{(p+1)} - \rho_n[\Lambda^{(p+1)} - M_u^+(p+1)] \tag{1.13}$$

where $u_{n+1}^{(p+1)}(x)$ is set equal to $v_{n+1}^{(p+1)}(x)$ if
$v_{n+1}^{(p+1)}(x) \geq u_1$ and is set to u_1 if not. After a
terminal check, restart the procedure at step (4) or
set

$$u^{(p+1)} = u_{n+1}^{(p+1)} \tag{1.14}$$

after last use of Eq. 1.13.

(5) If $(\Lambda^{(p+1)}, u^{(p+1)}, \zeta_u^{(p+1)})$ satisfies some terminal

check, write the computed solution. If not, set p
equal to p+1 and go back to step (3).

Before switching to the case of repeated case eigenvalues,
one should note that the choice of $\Lambda^{(p+1)}$ given by Eq. 12 may
also be defined as the value of $\mu \in R$ that minimizes

$$\int_{\Omega-A}(p) |M_u^\dagger(p) - \mu|^2 dx \qquad (1.15)$$

2. OPTIMUM AT A REPEATED EIGENVALUE

2.1 Toward A Computational Algorithm

The optimization problem is as in Section 1.1, but with
$\zeta(u_0)$ not necessarily simple. Instead of Eq. 1.3, a necessary
optimality condition is

$$\lambda_0 \int_\Omega \bar{u} \, dx - \lambda_1 \int_\Omega \int_{S_{m-1}} (M_u^\dagger(u_0) \, \bar{u} \, X, X) d\mu(X) \, dx = 0 \qquad (2.1)$$

where λ_0 and λ_1 are nonnegative numbers, not both zero; S_{m-1} is
the unit sphere in R^m, where m is the multiplicity of $\zeta(u_0)$;
is a positive measure on S_{m-1} of the total mass unity; i.e.

$$\int_{S_{m-1}} d\mu(X) = 1$$

X denotes the variable in R^m; and $M_{(u,\bar{u})}$ is an m x m matrix, the
entries of which are

$$M_{ij}(u,\bar{u}) = a_{u,u}^{(1)} (y_i,y_j) - \zeta(u_0) \, b_{u,u}^{(1)} (y_i,y_j)$$

where y_i (i=1,...,m) is a basis of the eigenspace of $\zeta(u_0)$ that
is orthonormalized by the condition

$$b_u(y_i,y_j) = \delta_{ij}$$

where δ_{ij} is the Kronecker delta. Also, $M_{ij}^\dagger(u_0)\bar{u}$ is the integrand
of $M_{ij}(u,\bar{u})$; i.e.

$$M_{ij}(u,\bar{u}) = \int_{\Omega} M_{ij}^{+}(u)\bar{u}\ dx$$

Equation 2.1 holds only for $\bar{u}$ with support in Ω-A. This necessary condition is derived in Ref. 2. A short presentation is provided in Ref. 3.

As in Section 1, $\lambda_1 \neq 0$ and one may set $\Lambda = \lambda_0/\lambda_1$, which reduces Eq. 2.1 to

$$\Lambda \int_{\Omega} \bar{u}\ dx - \int_{\Omega} \int_{S_{m-1}} (M_u^{+}(u_0)\ \bar{u}X,\ X)\ d\mu(X)\ dx = 0 \tag{2.2}$$

As in Section 1, one gets from Eq. 2.2 that, for $\bar{u}$ with support in Ω-A,

$$\Lambda = \int_{S_{m-1}} (M_u^{+}(u_0)\ X,\ X)d\mu(X) \tag{2.3}$$

Following the approach of Section 1.3, if u_0 is known an effective way of computing Λ and μ is to minimize

$$\int_{\Omega-A} \left| \Lambda - \int_{S_{m-1}} (M_u^{+}(u_0)\ X,\ X)d\mu(X) \right|^2 dx \tag{2.4}$$

Naturally, in numerical computations one approximates the measure μ by a convex combination of Dirac measures

$$\sum_{k=1}^{K} \mu_k\ \delta_{X_k} \tag{2.5}$$

with $\mu_k \geq 0$ and $\Sigma\ \mu_k = 1$ (see Ref. 7). The integer K will have to be chosen large enough to match the approximation of Eq. 1.1 of the structure, X_k will be distributed on a regular mesh on S_{m-1}, and a thorough analysis of the choice of K and X_k is deserved. With this approximation, Λ and μ_k (k=1,...,K) will be computed by minimizing

$$\int_{\Omega-A} \left| \Lambda - \sum_{k=1}^{K} (M_u^{+}(u_0)X_k,X_k)\mu_k \right|^2 dx$$

Naturally, the integral over Ω-A is in turn approximated with a standard quadratic equation.

Now, in fact, one has to compute also u. To find an algorithm, one rewrites Eq. 2.2 in the following way:

$$\int_{\Omega} \left[\Lambda - \int_{S_{m-1}} (M_u^+(u_0)X, X)d\mu(X) \right] \bar{u} \, dx = 0$$

Comparing with Eq. 1.17, one sees that

$$\int_{S_{m-1}} (M_u^+(u_0)X,X)d\mu(x)$$

plays the role of $M_{u_0}^+$, so

$$u_{n+1} = u_n - \rho_n \left[\Lambda - \int_{S_{m-1}} (M_u^+(u_n) \, X,X)d\mu(X) \right]$$

2.2 The Algorithm

The algorithm sketched so far is not generally needed to start the optimization procedure from an initial estimate, since repeated eigenvalues seem to appear only in the neighborhood of the optimum in some special cases. Thus, one may start with a usual gradient projection method such as that of Ref. 8. If one reaches a point u that satisfies the necessary condition for simple eigenvalues, up to the computing errors, that is fine. In some instances, however, the algorithm will lead to a design u with repeated eigenvalues, up to computing accuracy. The usual steepest descent method for simple eigenvalue constraint then works very poorly, as discussed in Ref. 9.

From the number of clustering eigenvalues one then has an estimate of the multiplicity m to be expected. One may now switch to the following algorithm:

(1) For the starting design $u^{(0)}$ first choose K, the number of Dirac measures used to approximate μ, and the points X_k to discretize S_{m-1}. Then define $A^{(p-1)}$, the set of active design constraints for the design $u^{(0)}$.

(2) Solve Eq. 1.1 for $u^{(p)}$ and define the set $A^{(p)}$, where p begins at 0.

(3) Compute $\Lambda^{(p+1)}$ and $\mu_k^{(p+1)}$ (k=1,...,K) by minimizing

$$\int_{\Omega-A^{(p)}} \left| \Lambda - \sum_{k=1}^{K} (M_u^+(u^{(p)})x_k, x_k)\mu_k \right|^2 dx$$

(4) Carry out the following iterative procedure for $u^{(p+1)}$:

   (1) set $v_0^{(p+1)} = u_0^{(p+1)} = u^{(p)}$, i.e. start with n = 0

   (2) solve Eq.1.1 for $u = u_n^{(p+1)}$

   (3) put $v_{n+1}^{(p+1)} = v_n^{(p+1)} - \rho_n \left[\Lambda^{(p+1)} - \sum_{k=1}^{K} (M_u^+(u_n^{(p+1)}) \times x_k, x_k)\mu_k^{(p+1)} \right]$

   (4) if $v_{n+1}^{(p+1)}(x) \geq u_1$ set $u_{n+1}^{(p+1)}(x) = v_{n+1}^{(p+1)}(x)$, otherwise $u_{n+1}^{(p+1)}(x) = u_1$

   (5) after some terminal check, go back to step (2) (with n = n+1), or put $u^{(p+1)} = u_{n+1}^{(p+1)}$

(5) Determine if $(\Lambda^{(p+1)}, \mu^{(p+1)}, \ldots, \mu_K^{(p+1)}, u^{(p+1)})$ satisfies some terminal check. If so, write the computed solution. If not, set p equal to p+1 and go back to step (2).

In step (2) of the algorithm, the integral over $\Omega-A^{(p)}$ is approximated by a standard quadrature formula.

In step (4) the inequalities and equalities are, in practice, only considered at the points of a mesh on Ω, which may be the same as for solving Eq. 1.1. As was explained in the introduction, neither proof nor numerical results are yet obtained. An alternate algorithm is provided in Ref. 9.

3. NECESSARY OPTIMALITY CONDITION WITH A SUPREMUM NORM CONSTRAINT

Necessary conditions of optimality for problems with a supremum norm constraint are now derived without going into all the technical details. The problem is stated for a simple model structure, a clamped beam under static load. The state equation

560

for this problem is

$$(E\alpha u^2 z_u'')'' = f, \quad \text{in }]0,\ell[$$
$$z(0) = 0 = z'(0), z(\ell) = 0 = z'(\ell) \tag{3.1}$$

where E is Young's modulus, u is the variable cross-sectional area*, α is a positive constant that depends on the shape of the cross-section (αu^2 is the moment of inertia of cross-section area), f is a distributed load, and z_u is the displacement. The subscript u emphasizes the dependence of the solution of Eq. 3.1 on u.

The functional to be minimized is

$$J(u) = \rho \int_0^\ell u \, dx \tag{3.2}$$

which is the weight of the structure, where ρ is the density of the material.

The constraints to be satisfied are

$$u(x) \geq \mu_1 > 0, \quad \text{a.e. in }]0,\ell[\tag{3.3}$$

and

$$H(u) \leq C_2 \tag{3.4}$$

is a displacement constraint, where

$$H(u) = \text{Sup}\{|z_u(x)| : x \in [0,\ell]\} \tag{3.5}$$

The only trouble is that H(u) is not even Gateaux differentiable and is not convex. However, it may be shown that it has directional (one-sided) derivatives and a subdifferential.

Indeed, set

$$\phi(z) = \text{Sup}\{|z(x)| : x \in [0,\ell]\} \tag{3.6}$$

so that $H(u) = \phi(z_u)$. The mapping $\mu \to z_u$ is Frechet derivable from $L^\infty(0,\ell)$ into $H_0^2(0,\ell)$ (see Ref. 10) and hence into $C([0,\ell])$. On the other hand, ϕ is convex and continuous in $C([0,\ell])$, hence

---

*Belonging to some Banach space U, e.g., $U = L^\infty(0,\ell)$.

it has one-sided directional derivatives and its sub-differential $\partial\phi(z)$ consists of all regular measures μ that satisfy

$$\int_{[0,\ell]} d\mu^+ + \int_{[0,\ell]} d\mu^- = 1 \qquad (3.7)$$

where μ^+ and μ^- are the positive and negative parts of $\mu(\mu = \mu^+ - \mu^-)$. Moreover, these measures are concentrated, respectively, on the sets

$$S_+(z) = \{x_0 \in [0,\ell] : z(x_0) = \mathrm{Sup}\{|z(x)| : x \in [0,\ell]\}\}$$
$$\qquad (3.8)$$
$$S_-(z) = \{x_0 \in [0,\ell] : z(x_0) = -\mathrm{Sup}\{|z(x)| : x \in [0,\ell]\}\}$$

(see Ref. 5, Section 4.5.2)

Hence, one can apply Theorem 4.4.2 of Ref. 5 to obtain

$$\partial H(u) = [z'_{\mu,\cdot}]^T \, \partial\phi(z_u) \qquad (3.9)$$

where T denotes the transpose of the linear mapping $\bar{u} \to z'_{u,\bar{u}}$ from $L^\infty(0,\ell)$ into $C([0,\ell])$. It is thus a linear mapping from the measures on $[0,\ell]$ into $(L^\infty(0,\ell))'$. One now deduces that

$$\partial H(u) = \{u^* \in (L^\infty(\Omega))' : <u^*,\bar{u}> = \int_{[0,\ell]} z'_{u,\bar{u}}(x) \, d\mu(x)\} \quad (3.10)$$

where μ satisfies Eq. 3.7, with the support properties of Eq. 3.8.

If the design constraint is not active, the necessary optimality condition may thus be written as

$$\lambda_0 \int_0^\ell \bar{u} \, dx + \lambda_1 \int_{[0,\ell]} z'_{u_0,\bar{u}} \, d\mu = 0 \qquad (3.11)$$

for any $\bar{u} \in L^\infty(0,\ell)$, where $\lambda_0 \geq 0$, $\lambda_0 \geq 0$, and $\lambda_0\lambda_1 \neq 0$.

If the design constraint was active, it would be

$$\lambda_0 \int_0^\ell (u - u_0) dx + \lambda_1 \int_{[0,\ell]} z'_{u_0,u-u_0} \, d\mu \geq 0 \qquad (3.12)$$

which should hold for any u that satisfies the design constraint $u \geq u_1$.

   To get really useful necessary conditions, the dependence on $\bar{u}$ in the second integral of Eq. 3.11 should be made more explicit. This will be carried out with an adjoint equation.

   Recall (see Ref. 10) that $z'_{u_0,\bar{u}}$ is the solution of the equation obtained by formally taking the derivative of the variational equation for z. For the beam, this is

$$E\alpha \int_0^\ell u^2 \frac{d^2}{dx^2} z'_{u_0,\bar{u}} \frac{d^2v}{dx^2} dx = -2E\alpha \int_0^\ell u_0\bar{u} \frac{d^2z_{u_0}}{dx^2} \frac{d^2v}{dx^2} dx \quad (3.13)$$

Which should hold for every smooth v that satisfies the boundary conditions of Eq. 3.1. Consider the solution $p_{u_0;u}$, called the adjoint state, of the following equation

$$E\alpha \int_0^\ell u_0^2 \frac{d^2p}{dx^2} \frac{d^2\phi}{dx^2} dx = \int_{[0,\ell]} \phi d\mu \quad (3.14)$$

which should hold for every ϕ, that satisfies the boundary conditions of Eq. 3.1. Here, μ is the measure used in the necessary condition of Eq. 3.11.

   The existence of the solution of this equation falls into the classical variational method. One must only note that since $H_0^2(0,\ell)$ is topologically included in $C([0,\ell])$, $\mu \in (H_0^2(0,\ell))'$. Setting $\phi = z'_{u_0,\bar{u}}$ in Eq. 3.14 and $v = p_{u_0;\mu}$ in Eq. 3.13, one gets

$$\int_{[0,\ell]} z'_{u_0,\bar{u}} d\mu = -2E\alpha \int_0^\ell u_0 \frac{d^2p_{u_0;\mu}}{dx^2} \bar{u} dx \quad (3.15)$$

Equation 3.15 enables one to write Eq. 3.11 in the following way:

$$\int_0^\ell \left(\lambda_0 - 2\lambda_1 E\alpha u_0 \frac{d^2z_{u_0}}{dx^2} \frac{d^2p_{u_0;\mu}}{dx^2} \right) \bar{u} dx = 0$$

The multiplier λ_1 cannot be zero, otherwise $\lambda_0 = 0$, which is excluded. Set $\Lambda = \lambda_0/\lambda_1$ to get

$$\int_0^\ell \left(\Lambda - 2E\alpha u_0 \frac{d^2z_{u_0}}{dx^2} \frac{d^2p_{u_0;\mu}}{dx^2} \right) \bar{u} dx = 0 \quad (3.16)$$

If the design constraint is active, one has in the same way

$$\int_0^\ell \left(\Lambda - 2E\alpha \; u_0 \; \frac{d^2 z_{u_0}}{dx^2} \; \frac{d^2 p_{u_0;\mu}}{dx^2} \right) (u - u_0) \; dx \geq 0 \qquad (3.17)$$

for any $u \geq u_1$, a.e. in $]0,\ell[$.

4. ALGORITHM FOR SUPREMUM NORM CONSTRAINT

A numerical algorithm for supremum norm constraints will be based on Eq. 3.16. In this equation, one must be aware that $p_{u_0;\mu}$ depends on μ, which is unknown. However, the dependence is linear. As in Section 2.1, one may approximate this measure by a linear combination of Dirac measures

$$\mu_{[x_i,x_j]} \equiv \Sigma \; \mu_i^+ \; \delta_{x_i} - \Sigma \; \mu_j^- \; \delta_{x_j}$$

where μ_i^+ and μ_j^- are positive numbers and x_i and x_j are mesh points in $S_+(z)$ and $S_-(z)$, respectively. Actually, in an iterative method they will be mesh points in $S_+(z_{u_n})$ and $S_-(z_{u_n})$.

If one denotes by $p_{u_{0_j} \delta_{x_i}}$ the solution of Eq. 3.16 with $\mu = \delta_{x_i}$ then

$$p_{u_0;\mu_{[x_i,x_j]}} = \Sigma \mu_i^+ \; p_{u_0;\delta_{x_i}} - \Sigma \mu_i^- \; p_{u_0;\delta_{x_j}}$$

Hence, one may rewrite Eq. 3.16 for $\mu_{[x_i,x_j]}$ as

$$\int_0^\ell \left[\Lambda - 2E \; \alpha u_0 \; \frac{d^2 z_{u_0}}{dx^2} \left(\Sigma \mu_i^+ \; p''_{u_0;\delta_{x_i}} - \Sigma \mu_j^- \; p''_{u_0;\delta_{x_i}} \right) \right] \bar{u} \; dx = 0$$

Before describing the algorithm, its is suggested that one selects the points x_i and x_j just within the mesh used to solve Eq. 3.1. This is just a suggestion. The actual computation or the theory may suggest that in fact more nodes be used.

The computational algorithm proposed is as follows:

(1) Start from an estimated design $u^{(0)}$; set $p = 0$.

(2) Solve Eq. 3.1 for $u = u^{(p)}$; define $A^{(p)} = \{x \in [0,\ell]:$
$u^{(p)}(x) = u_1\}$; and estimate $S_+(z_u(p))$ and $S_-(z_u(p))$,
and grid these sets of points x_i and x_j.

(3) Compute $\Lambda^{(p+1)}$, $\mu_i^{(p+1)+}$ and $\mu_j^{(p+1)-}$ by minimizing

$$\int_{\Omega-A(p)} \left| \Lambda -2E\alpha\, u_p\, \frac{d^2 z_{u_0}}{dx^2} \left(\Sigma\, \mu_i^+\, P''_{u_i};\delta_{x_i} - \Sigma\, \mu_j^-\, P''_{u_{pj}}\delta_{x_j}\right) \right|^2 dx$$

The integral is approximated with some standard
quadratic formula.

(4) Find $u^{(p+1)}$ using the following iterative procedure:

1) set $v_0^{(p+1)} = u_0^{(p+1)} = u^{(p)}$ (i.e. set $n = 0$).

2) solve Eq. 3.1 with $u = u_n^{(p+1)}$. Estimate $S_+(z_{u_n} (p+1))$
and $S_-(z_{u_n} (p+1))$ and grid these sets with point x_i
and x_j.

3) put $v_{n+1}^{(p+1)} = v_n^{(p+1)} - \rho_n \left(\Lambda - 2E\alpha\, u_0\, \frac{d^2 z_{u_n} (p+1)}{dx^2} \right.$

$$\left. \cdot \left(\Sigma\, \mu_j^-\, P''_{u_n} (p+1);\delta_{x_i} - \Sigma\, \mu_j^-\, P''_{u_n} (p+1);\delta_{x_j}\right)\right)$$

4) if $v_{n+1}^{(p+1)} (x) \geq u_1$, set $u_{n+1}^{(p+1)} (x) = v_{n+1}^{(p+1)} (x)$,

otherwise $u_{n+1}^{(p+1)} (x) = u_1$

5) after some terminal check, go back to step (2)
(with $n = n+1$) or put $u^{(p+1)} = u_{n+1}^{(p+1)}$

(5) If $(\Lambda^{p+1}, \mu_i^{(p+1)+}, \mu_j^{(p+1)-}, u^{(p+1)})$ satisfies some
terminal check, write the computed solution. If not,
set p equal to $p+1$ and go back to step (3).

At the moment, there is neither proof of convergence, nor any numerical results obtained. However, the algorithm is quite feasible and, due to its close connection with the necessary condition, should be very promising.

The algorithm has been described for a very simple equation. However, it is clear that it may be extended to more complex structures (plates, planar elasticity, and multi-membered structures) with thickness as design variable. It may also be extended to shape optimal design.

REFERENCES

1. Haug, E.J. and Rousselet, B., "Design Sensitivity Analysis of Eigenvalue Variations", Optimization of Distributed Parameter Structures (Ed. E.J. Haug and J. Cea), Sijthoff & Nordhoff, Alphen aan den Rijn, Netherlands, 1980.
2. Choi, K.K., Haug, E.J. and Rousselet, B., Problems of Structural Optimization Involving Multiple Eigenvalues, Technical Report, Materials Division, College of Engineering, The University of Iowa, November, 1979.
3. Rousselet, B., Condition Nécessaire d'Optimalite en Presence de Valeurs Propres Multiples, Note C.R.A.S., Paris, to appear, 1980.
4. Pshenichnyi, B.N., Necessary Conditions for an Extremum, M. Dekker, New York, 1971.
5. Ioffe, A.D. and Tihomirov, V.M., Theory of Extremal Problems, North-Holland, Amsterdam, 1979.
6. Cea, J., Lectures on Optimization - Theory and Algorithms, Springer-Verlag, 1978.
7. Diendonné, J., Treatise on Analysis, Vol. 2, Academic Press, New York, 1970.
8. Haug, E.J., "A Gradient Projection Method for Structural Optimization", Optimization of Distributed Parameter Structures (Ed. E.J. Haug and J. Cea), Sijthoff & Nordhoff, Alphen aan den Rijn, Netherlands, 1980.
9. Choi, K.K., "A Numerical Method for Optimization of Structures with Repeated Eigenvalues", Optimization of Distributed Parameter Structures (Ed. E.J. Haug and J. Cea), Sijthoff & Nordhoff, Alphen aan den Rijn, Netherlands, 1980.
10. Haug, E.J. and Rousselet, B., "Design Sensitivity Analysis of Static Response Variations", Optimization of Distributed Parameter Structures (Ed. E.J. Haug and J. Cea), Sijthoff & Nordhoff, Alphen aan den Rijn, Netherlands, 1980.

Part 3

OPTIMIZATION OF STRUCTURES UNDER EARTHQUAKE LOADS

OPTIMAL DESIGN OF STRUCTURES UNDER DYNAMIC LOADING[*]

Karl S. Pister

Department of Civil Engineering, University of
California, Berkeley, California 94720

ABSTRACT

This paper briefly describes major components of the dynamic design problem for structures exposed to earthquake loading. The following aspects are reviewed: structural modeling, characterization of ground motion, simulation of structural response, and design objectives. A concluding section places the problem in the format of a nonlinear program. Companion papers enlarge on themes introduced in this paper.

1. INTRODUCTION

In the broadest sense, design of structures is not likely to be a decision process that will ever be susceptible to rational analysis. Indeed, one might even hope that conceptual design forever escapes domination by digital computation, a circumstance in which the human expression of artistry in design would have been removed. The present paper has a much more modest objective — to sketch briefly the phenomenology associated with a design process for structures exposed to dynamic loads after conceptual design has been accomplished. Structures whose geometric configuration has been assigned are dealt with, searching for optimal proportioning of size or stiffness of structural components. Although what is discussed here has applicability to a wide class of dynamically loaded structures and machines, attention here is

---

[*]This work was supported by the National Science Foundation under Grant ENV 76-04264.

focused on framed structures that are located on the surface of
the earth, in regions where the crust is periodically disturbed
by earthquakes.

A significant feature of design of structures of this class
is that their primary function is not to withstand earthquakes,
per se, i.e. multistory buildings provide enclosures for working
space and functionally deliver vertical gravity-induced floor
loads to the building skeleton. Functional dynamic loads may
arise from operation of equipment, but dynamic loads produced by
winds and earthquakes are clearly not associated with the building's function. Contrast this kind of design problem with that
of machines and aircraft, for example. Typically, this dichotomy
in structural design is resolved by requiring that the structure
satisfy serviceability requirements to avoid functional failure
and safety requirements to prevent major structural damage and
collapse in the event of frequently occurring an infrequently
occurring earthquakes, respectively [1]. These terms must be defined and given operational meaning in the design process, which
in turn depends directly upon principles of structural mechanics.
A survey of the interaction between mechanics and the process of
mechanical analysis and design (in nuclear reactor technology)
can be found in Ref. 2. The present paper gives a brief description of major components of the dynamic design problem; structural
modeling, characterization of ground motion, simulation of structural response, and design objectives. Design algorithm construction, computer-aided interactive implementation, and applications
may be found in Refs. 3, 4 and 5.

Before proceeding to this task, the author is obliged to
state the following caveat: complexity of the optimal design problem for real world structures ought to be self-evident. We are
constrained by lack of insight and experience, time, and money to
deal with imperfect models of the design process. Limitations not
withstanding, the philosophy to optimal design forces advance
quantification of what is important in the design itself and requires that some attention to overall balance is given. For
example, the state-of-the-art of contributing areas to the design
process requires careful evaluation and the establishment of
heirarchies of approximation.

2. STRUCTURAL MODELS

Structures such as buildings, bridges, etc., are obviously
distributed systems in terms of their mass, stiffness, and damping
properties. Information defining configuration, loading, and
material behavior of structural components can be incorporated
into the format of a mechanical field problem in mathematical
physics. Using the principle of momentum balance, one can

interconnect configuration, input data, and material properties through the device of an initial-boundary value problem, whose solution gives the mechanical state of the structure over space and time. The field problem is the simulation device that replaces the real-world prototype. The complexity of the resulting problem is in large measure dependent on the corresponding structural configuration, material constitutive model, and input data required to define the simulation model. In the world of steel and reinforced concrete structures, perfection in modeling is yet a distant goal. Therefore, every simulation model ought to be subjected to careful scrutiny to ensure its faithfulness to the prototype, insofar as critical features of performance are concerned. Models commonly employed in current practice can be found in Refs. 6 and 7.

The simulation problem requires that nonlinear partial differential equations be solved, a task that must be approached head-on by numerical analysis and digital computation. Accordingly, the structural modeling problem is intimately tied to the simulation of structural response. Currently, typical computer programs are based upon the finite element method, providing solutions of global forms of the balance of momentum. The end result of the modeling process is a system of coupled, ordinary differential equations with the structure

$$M\ddot{u} + F(u,\dot{u}) = v \tag{2.1}$$

where

   u = an N-dimensional generalized nodal point displacement
       vector

   M = discretized mass matrix

   F = discretized nodal constitutive vector reflecting
       material properties and kinematic relations

   v = discretized load vector

and the superposed dots denote derivatives with respect to time. Dependence of u and v on time t is assumed, even if not explicitly shown.

A great deal of experience and judgment is required to replace a structure by Eq. 2.1. For example, buildings are comprised of columns, girders, floors, walls, and architectural materials, inter alia; not the least of which are connections joining the various elements noted. Both the spatial discretization of the structure into a finite set of elements, as well as characterization of constitutive behavior of the elements remain

challenging problems. The latter problem, with particular reference to hysteresis and degradation of stiffness, is summarized in
Ref. 6. In many structural models, usually with little physical
justification, (2.1) is written

$$M\ddot{u} + C\dot{u} + K(u,\dot{u}) = v \qquad (2.2)$$

where C is an equivalent linear damping matrix and K is a nonlinear
stiffness vector that is dependent on the structural state.

The modeling problem is further aggravated by the phenomenon
of soil-structure interaction, which couples dynamic structural
response to the deformable, exciting medium, i.e. the motion of
the earth in the neighborhood of the structure. In short, free-
field records of earthquake ground motion, as recorded by seismo-
graphs, constitute only a first approximation of the ground motion
at a building site under certain conditions (such as massive
structures). In these instances, the simulation problem is vastly
more complicated [7,8]. The definition of the structure must be
enlarged to include the supporting medium on which the actual
structure is founded, resulting in a substantial increase in
dimension of the vector u in Eq. 2.1. In such cases, the forcing
function v in Eq. 2.1 is often associated with the motion of bed
rock in the earth's crust. In conclusion, it may be well to note
that structural modeling is replete with a number of optimization
problems, associated with identification and parameter estimation
for models of distributed systems [9].

3. GROUND MOTION

Excitation of a structure located on the earth's surface
during an earthquake is delivered via horizontal and vertical
ground accelerations that are associated with the release of
energy transmitted by waves emanating from the hypocenter of the
quake [7]. Although the location of most major faults is now a
matter of record, prediction of the area of occurrence, magnitude
of energy release, and date of the event is yet to be accomplished.
In addition, the precise mechanisms of energy release and transmission to the earth's surface is presently only imperfectly
understood and broadly described as a nonstationary random process.
This has the effect in Eqs. 2.1 and 2.2 of making v a random variable and u a function of a random variable. The ability to formulate models of the nonstationary random process, from which sta-
tistics of ground motion such as peak ground acceleration or other
measures thought to be suitable for design can be extracted, is
severely limited by the paucity of recorded data applicable to a
proposed building site. In other words, it is difficult to draw
inferences about the structure of the model simulating ground
motion, as well as the nature of the underlying random process

driving the model, i.e. the earthquake generation process itself. In a recent study, progress has been achieved in this area through the use of autoregressive, moving average (ARMA) models employing Box-Jenkins identification and estimation techniques on digitized ground acceleration time-series records [10]. It has been found that nonstationarity arises primarily from strong time-dependence of the variance of the underlying noise process. The variance envelope has been identified by a Kalman filter and the resulting model has been employed to synthesize artificial accelerograms, after once identifying the ARMA model and variance envelope from actual records [11].

Although this work shows promise for providing an appropriate characterization of ground motion for design purposes, the state-of-the-art relies heavily on the notion of ground motion response spectra (shock spectra) derived from smoothing a set of normalized records of actual earthquakes that are deemed appropriate for the building site. Ordinates of such spectral plots denote the maximum value of the response of a single degree of freedom, viscously damped, linear oscillator that is subjected to the ground motion. Spectral plots commonly include absolute acceleration of the mass of the oscillator or its relative velocity with respect to the ground (pseudovelocity), plotted as functions of period of the oscillator for a fixed percentage of critical damping. An informative treatment of the recording, analysis, and interpretation of accelerogram records can be found in Ref. 12. Utilization of response spectra in the dynamic analysis of structures is treated in Ref. 7 and 13.

So far, it has been tacitly assumed that ground motion characterization is to be utilized in connection with linear structural models for simulation of response. In the event that nonlinear structural models are needed for the design process (a certainty if strong-motion earthquakes are included), spectral methods are unsuitable, although modified response spectra for certain non-linear models have been proposed for design purposes [14,15]. In general, it is necessary to utilize a set of strong-motion records (real or artificially generated) that are deemed appropriate to the site as input and carry out whatever statistical analyses of structural response that are desired by Monte Carlo simulation. This leads one to consider next the topic of structural response simulation.

4. SIMULATION OF STRUCTURAL RESPONSE

In the previous sections, the general form of model typically employed in dynamic analysis of structures, along with a description of the exciting force supplied by ground motion, has been sketched. The design process calls for histories of structural response variables (needed for evaluation of structural performance) over the range of design loading conditions for trial values of design parameters. This set of operations produces points in an abstract design space, so that comparisons and decisions about the goodness of a trial design can be made. The effort required to produce reliable simulations of real structural response to dynamic loads is staggering, even if it is only one step in the design process. Comprehensive reviews of this area can be found in Refs. 8 and 16. Here, one should note the main features of response simulation.

As noted in Section 2, for the purpose of analysis and design for dynamic loads, it is customary to idealize complex structures as spatially discrete dynamic systems whose degrees of freedom are associated with motion of a finite set of nodal points, at which mass, damping, and internal restoring force properties of the components of the structure are represented. In general, spatial discretization is most easily carried out by employing the finite element method, although in cases such as rigid frames, discretization into beam and column elements follows in an obvious manner. For a wide class of problems, the resulting equations governing dynamic behavior of the now-idealized structure can be put in the form

$$D[z,u(z,\tau),t,v] = 0, \quad \tau \in [0,t], \quad t \in [0,T] \tag{4.1}$$

with initial conditions

$$u(z,0) = 0$$

In Eq. 4.1, $u(z,t)$ denotes the N-dimensional state vector of the structural system at time t, associated with Z, a P-dimensional, time-invariant design (parameter) vector that characterizes the structure. In structural dynamics problems, u is composed of ordered sets of generalized displacements and velocities of nodal points. The operator D, defining system dynamics, is a differential or integro-differential operator. As noted previously, v is the forcing function. Returning to Eq. 4.1, one notes that the present value of the state vector may depend on the past history (path of evolution) of the dynamic process (such as may occur in inelastic systems) and that T denotes the extent of the time period of interest, over which the system is observed. For earthquakes, this period may be of the order of 30 to 40 seconds. Note that Eqs. 2.1 and 2.2 are special cases of Eq. 4.1.

One may now briefly examine the major determinants of the simulation problem, for the spatially discretized structure. Consideration is limited to deterministic forcing functions (equivalently, one considers one realization of a random process) and excludes soil-structure interaction. In other terms,

structural response that is conditional to the ground motion is
discussed. As noted in Ref. 8, soil-structure interaction depends upon relative stiffness properties of the structure and
its underlying foundation, becoming particularly important for
massive, short-period structures such as gravity dams and nuclear
reactor structures. Furthermore, a common motion for all support
points of the structure may be assumed in most cases — long
earth dams and bridges being possible exceptions. A major simplification is usually introduced to decouple orthogonal translational
degrees of freedom, so that three independent equations of motion
for components of the state vector emerge. The quality of this
approximation evidently must be tested in each application by
giving proper attention to the configuration of the structure, as
reflected in its spatial distribution of mass and stiffness.
Furthermore, for linear systems, one may employ three statistically
independent translational components of ground motion to synthesize
the total response of the structure, under reasonable assumptions
for the statistical properties of the ground motion [17]. Bearing
in mind these caveats, the next hurdle faced is selection of
appropriate algorithms to solve the differential equations governing the system. Typically, different paths are followed for linear
and nonlinear response analysis.

4.1 Linear Structural Response

When structural response occurs in the linear range of the
dynamic operator D in Eq. 4.1, considerable simpliciation results
and Eq. 2.2 is used in the form

$$M\ddot{u} + C\dot{u} + Ku = -Mr\ddot{u}_g \tag{4.2}$$

where u = u(z,t) is the nodal displacement vector, M and C are
previously defined mass and damping matrices, and K is now a state-
independent stiffness matrix. The forcing function v is composed
of the ground acceleration time history $\ddot{u}_g$, appropriate to the
direction of the displacement u, and a unit influence coefficient
vector r, whose components take on values corresponding to each
translational degree of freedom, or are otherwise zero. Assuming
that mode shapes of the structure are orthogonal with respect to
the damping matrix, Eq. 4.2 can be solved by familiar techniques
of modal analysis [7,8,13]. Maximum values of displacement (or
internal member force and stress quantities derived therefrom) at
nodal degrees of freedom are usually estimated by the square root
of the sum of squares (SRSS) of modal displacements. The quality
of the estimate tends to be good for structures with well-sepa-
rated modal frequencies, but it may deteriorate if the frequencies
of lower modes are closely spaced. In many instances, in first
few modes of a structure capture of the main features of response
to an earthquake.

Use of this procedure requires specification of an earthquake response spectrum characterizing the ground acceleration in Eq. 4.2. Typically, the shape of such a spectrum is a smoothed envelope of a set of actual earthquake spectra that is scaled to the mean value of the maximum absolute acceleration (or some other statistic) defining the design loading for the structure. This procedure is not without difficulties.

4.2 Nonlinear Structural Response

In the nonlinear range of the operator D in Eq. 4.1, the form of the equation of motion of Eq. 2.2 is usually of the form

$$M\ddot{u} + C\dot{u} + F = -Mr\ddot{u}_g \tag{4.3}$$

The terms in Eq. 4.3, apart from F, correspond to those already defined in Eq. 4.2. However, the nodal internal force vector $F = F(t)$, as noted in Section 2, is for hysteretic structures a complex function of the deformation history of the structure, over the time interval $[0,t]$. The structure of F depends mainly on the constitutive assumptions employed to model material behavior. Equation 4.3 is solved numerically by approximating $u(t)$ and its derivatives at discrete points in time. Employing a step-by-step integration procedure, from prescribed initial conditions, a time-marching process carries forward the solution at discrete time points.

Two distinct phases appear in this process. First, the equation is linearized at the current state and estimates of the solution at the next state are obtained. Secondly, the internal forces at the new state are calculated and an error force is calculated. If the error is within a prescribed tolerance, the process is repeated for the next step. Otherwise, a Newton-Raphson type of iteration is used to reduce the error before proceeding to the next step. Integration operators of the Newmark class [7,8], which are implicit, single-step, two-parameter operators are wisely used. Extensive software is available for this purpose. Note that the discretized version of Eq. 4.3 requires digitized ground acceleration records. Such records are available for a very large number of recorded earthquakes [12], digitized at intervals of 0.02 seconds.

4.3 Sensitivity Analysis

An integral part of design algorithms that employ gradient techniques is sensitivity analysis, i.e. computation of the rate of change of response quantities u (or functions of u) with respect to design parameters Z. Such sensitivity matrices $\partial u/\partial Z$

for a linear system can be determined by solving the differential
equation obtained by differentiating Eq. 4.2 with respect to Z.
Expressions for the rate of change of maximum nodal displacements,
with respect to design parameters, can also be obtained explicitly;
see, for example Ref. 18.

For nonlinear systems, such as described by Eq. 4.3, one can
also obtain equations of motion for sensitivity matrices and
obtain values at discrete points in time by numerical analysis
[19]. Alternatively, the same information can be obtained by
finite difference approximations of the sensitivity matrices
obtained by solving Eq. 4.3 for a set of design parameters in a
linear neighborhood of some reference set [18].

5.   EARTHQUAKE-RESISTANT DESIGN PHILOSOPHY

If structural modeling, ground motion characterization, and
response simulation appear to be problem areas in which much
research still needs to be done, one could reasonably be dismayed
by the problems to be formulated, much less solved, in developing
a rational methodology for earthquake-resistant design. In the
Introduction to their book [7], Newmark and Rosenblueth state:

> In this text on earthquake engineering, we take for
> granted that the purpose of design in engineering is
> optimization, and that we deal with random variables.
> In the past, the orthodox viewpoint maintained that
> the objective of design was to prevent failure; it
> idealized variables as deterministic. This simple
> approach is still fruitful when applied to design
> under only mild uncertainty, and in situations in
> which the possibility of failure may be contemplated
> at such a distant future as to be almost irrelevant;
> but when confronted with the effects of earthquakes,
> this orthodox view seems so naive as to be sterile.
> In dealing with earthquakes, we must contend with
> appreciable probabilities that failure will occur
> in the near future. Otherwise, all the wealth of
> this world would prove insufficient to fill our needs:
> the most modest structures would be fortresses.

Although progress has been made in the nine years since
publication of this quotation, current design philosophy, regret-
tably, is more past-oriented and orthodox than forward-looking,
in the sense used by the authors. In a later chapter devoted to
design objectives, the authors go on to say:

Two principal aspects of earthquake resistant design
differentiate this discipline from other branches of
engineering. One concerns the enormous spread of uncertainty in the disturbance. The other concerns the
nature of the disturbance itself. Even if the detailed
characteristics of future ground motions were known
accurately, we could not be certain about the survival
of given structures; at present ignorance about structural characteristics is great. These circumstances
make it advisable to apply the theory of probability
and optimization techniques openly in design to a
more significant degree than in other engineering
disciplines. The traditional deterministic disguise
will do less well in earthquake engineering.

The explicit process of optimization requires
assessing a number of nearly "imponderable" losses
and benefits, whose value or associated utility
depends on the subject for whom the optimization is
done. For example, the cost of collapse to the owner
usually involves that of the building, minus its
insurance and salvage values, plus the expected cost
of lawsuits and of indemnities. To society, it involves
the cost of the building, minus its salvage value, plus
the expected cost of damage to, or destruction of, its
contents, and that of lives lost and injuries. And to
the engineer it signifies the expected cost of lawsuits
and that of his loss of prestige, plus the moral harm.

The moral, ethical, and technical aspects of the design problem are clearly intertwined and inescapable. Because of the scale
of potential destruction caused by severe earthquakes, it is
incumbent upon engineering to review and continuously upgrade the
quality of the design process to provide a basis for minimizing,
or at least controlling, losses to society. The task is complicated by non-congruence of the utility scales of the owner,
occupant, and designer, as well as that of society at large. In
no small measure is this responsible for the current state-of-the-
art of earthquake-resistant design. Let us, however, move from
this philosophical background to examine a possible format for
earthquake-resistant design.

5.1 Design Objective

In the profession of structural engineering, there is now
general agreement that a proposed structural design should meet
the following objectives [20]:

(1) Resistant minor earthquakes without damage.

(2)   Resist moderate earthquakes without structural damage, but possibly with some non-structural damage.

(3)   Resist major earthquakes, of the strongest experienced in California, without collapse, but with some structural, as well as non-structural, damage.

The problem here is to translate these broad objectives into quantifiable measures, i.e. first of all minor, moderate, and major earthquakes must be defined. Secondly, structural response variables have to be mapped into damage variables and collapse mechanisms must be defined. One may then introduce the notion of lifetime cost (LC) of a structure. This cost includes the initial cost of construction and the present value of the cost of future earthquake damage, to which the structure may be subjected. Note that such a cost function requires additional information — a mapping function from damage to monetary units and the acquisition of an expected earthquake exposure profile over the design life of the structure, so that total future damage can be calculated.

Here, in order to develop the notion of lifetime cost, one must accept two approximations. First, a model for the probability of occurrence of earthquakes of prescribed intensity is required — including a definition of intensity (Richter magnitude, peak ground acceleration, etc.). Secondly, for a realization of an event of given intensity, one needs a model relating intensity to response to damage. The modeling problem clearly calls for multivariate distributions of random variables. For the present purposes, assume that lifetime cost, or damage, is interchangeable with mean values of these quantities.

Implementation of the objectives cited suggests the need for at least a two-tier design philosophy, in terms of permissible load-response relations. For example, one might require linear elastic structural behavior under minor earthquakes, but permit increasing amounts of inealstic response, short of collapse, under earthquakes of increasing severity.

In formulating a model to reflect sources of cost, it is useful to distinguish between costs that are weakly, as opposed to strongly dependent on the vector of design variables. In making comparisons of alternative designs, the former add an approximately constant increment to the cost and may be disregarded.

In attempting to find a design that meets the stated objectives, at minimum lifetime cost, one must clearly constrain the search with additional restrictions on structural response. These will be discussed next.

5.2 Design Constraints

In the stated design objectives, the following constraints
have been noted: Minor earthquakes must be resisted without damage, and major earthquakes must not produce collapse. In the
former case, one can impose restrictions on structural response,
such that for earthquakes under some prescribed intensity (however measured) the structural response state is elastic. Similarly, various measures of collapse have been proposed in the
literature. One such philosophy for buildings is to require that
inelastic deformations occur in the girders, with columns remaining elastic to prevent collapse. This strong column, weak girder
constraint is satisfactory, provided that proper attention is
given to column-girder joint details, to permit adequate inelastic
rotation.

It must be recognized that very significant constraints are
placed on the design process by building codes. Such codes, in a
broad sense, serve as a protection for society to provide some
assurance that a design will perform satisfactorily. To ignore
such constraints, even where legally possible, usually invites
trouble for the designer. Code constraints can normally be incorporated into the constrained optimization process with no difficulty. An application of this philosophy to design of multistory
steel-frame buildings can be found in a report by Walker [21].

5.3 Further Remarks

A number of important, unresolved questions that deserve
further comment were raised in the previous sub-sections. First,
the characterization of damage in structures is in a primitive
state. In steel-framed buildings, damage to the structural skeleton is usually slight, except in the most severe cases. However,
secondary damage to glass, partition walls, and plumbing, heating,
and electrical systems is often severe. Such damage has in the
past been related to the maximum relative motion occurring between
successive floor levels (story drift), [21].

In reinforced concrete structures, under moderate to severe
earthquakes, the situation is quite different. Cracking and
spalling of structural members and walls often occurs, leading to
cost of rehabilitation and repair of structural damage, in addition to that described for steel-framed structures. In both
instances, it must be pointed out that structures, once damaged,
incur changes in damping and stiffness characteristics that influence their response to future earthquake excitation.

It is also appropriate to add a further caution, while considering design philosophy. The global success of a proposed

design is directly dependent on a great deal of local detailed design that lies outside the scope of optimal design. Attention to proper structural detailing and quality construction and inspection are absolute requisites for success. This perspective of the problem can be found in Ref. 22.

In this section, only one of many approaches to optimal design of structures has been described. An alternative is to place the problem in a reliability-risk format [20]. Although there is merit and attractiveness in this approach, it is not pursued here. However, satisfactory resolution of the design problem ultimately depends on recognition and resolution of a stochastic optimization problem. This introductory exposition is concluded with a description of the earthquake-resistant design problem, in a nonlinear programming format.

6. A FORMAT FOR OPTIMAL EARTHQUAKE-RESISTANT DESIGN

A class of optimal design problems of the type considered in the previous section can be expressed as

$$\min_{z} \ \max_{t \in T} \ [F\{R^{\ell}(z,t), R^{S}(z,t)\}]$$

such that

$$\max_{t \in T} \ G^{\ell}(R^{\ell}(z,t)) \leq \delta_1^{\ell}$$

(6.1)

$$\max_{t \in T} \ G^{S}(R^{S}\{z,t)) \leq \delta_1^{S}$$

$$H(z) \leq \delta_g \quad ,$$

where

$R^{\ell}, R^{S}: \mathbb{R}^{P} \times \mathbb{R} \to \mathbb{R}^{Q}$ is some function of structural response. The superscripts l and s refer to response when subjected to a large and a small earthquake respectively.

$T = [t_0, t_f]$ is the interval in which significant earthquake ground motion occurs

$z \in \mathbb{R}^{P}$ is the design parameter vector

P is the total number of design parameters

$F : \mathbb{R}^Q \to \mathbb{R}$ is some function of structural response, which is to be minimized

Q is the number of structural response functions

$G : \mathbb{R}^Q \to \mathbb{R}^M$ are time-dependent inequality constraints (functional constraints)

M is the number of functional inequality constraints

$H : \mathbb{R}^P \to \mathbb{R}^L$ are conventional inequality constraints

L is the number of conventional inequality constraints

$\delta_1 \in \mathbb{R}^M, \delta_2, \in \mathbb{R}^L$ are prescribed constraint bounds.

The optimal design problem formulated in Eq. 6.1 is not directly suitable for application of nonlinear programming techniques. An appropriate canonical form of the nonlinear programming problem can be expressed as

$$\min_{z} \{f^0(z)\}$$

such that

$$\max_{t \in T} \phi^j(z,t) \leqq 0 \quad , \qquad j = 1, \cdots, M$$

$$g^j(z) \leqq 0 \quad , \qquad j = 1, \cdots, L$$

where

$\phi^j : \mathbb{R} \times \mathbb{R} \to \mathbb{R}$ = *functional inequality constraints*

$z \in \mathbb{R}^P$ = *design parameter vector*

$f^0 : \mathbb{R}^P \to \mathbb{R}$ = *objective function*

$g^j : \mathbb{R}^P \to \mathbb{R}$ = *conventional inequality constraints*

The optimal design problem of Eq. 6.1 can be transcribed to the canonical form of Eq. 6.2 by augmenting the parameter vector z by a dummy cost parameter z_{P+1}. The dummy cost parameter is an upper bound on the objective function to be minimized, i.e.

$$z_{P+1} \geq \max_{t \in T} [F\{R^1(z,t), R^2(z,t)\}] \quad .$$

Thus, the minimization of z_{P+1} will imply the minimization of the actual objective function. The optimal design problem can then be written as

$$\min_{z} z_{P+1}$$

such that

$$\left.\begin{array}{l} \max_{t \in T} \ [F\{R^{\ell}(z,t), R^{S}(z,t)\}] - z_{P+1} \leqq 0 \\[2mm] \max_{t \in T} \ G^{\ell}\{R^{\ell}(z,t)\} - \delta_1^{\ell} \leqq 0 \\[2mm] \max_{t \in T} \ G^{S}\{R^{S}(z,t)\} - \delta_1^{S} \leqq 0 \\[2mm] H(z) - \delta_2 \leqq 0 \ . \end{array}\right\} \qquad (6.3)$$

Equation 6.3 is in canonical form with:

$$\left.\begin{array}{ll} f^0(z) = z_{P+1} & \\[2mm] \phi^1(z,t) = F\{R^{\ell}(z,t), R^{S}(z,t)\} - z_{P+1} & \\[2mm] \phi^j(z,t) = G_j^{\ell}\{R^{\ell}(z,t)\} - \delta_{1j}^{\ell}, & j = 2, \cdots, J_1 \\[2mm] \phi^j(z,t) = G_j^{S}\{R^{S}(z,t)\} - \delta_{1j}^{S}, & j = J_1, \cdots, M \\[2mm] g^j(z,t) = H_j(z) - \delta_{2j}, & j = 1, 2, \cdots, L \ . \end{array}\right\} \qquad (6.4)$$

A discussion of algorithms for solving this class of problems, together with necessary software for interactive, computer-aided implementation, and with applications, will be found in Refs. 3, 4 and 5.

REFERENCES

1. Clough, R.W., "Deficiencies in Current Seismic Design Procedures," Civil Engineering Frontiers in Environmental Technology, Department of Civil Engineering, University of California, Berkeley, 1971, pp. 115-127.

584

2. Pister, K.S., "On the Role of Mechanics in Reactor Technology", Proceedings of the Conference on Structural Analysis, Design and Construction in Nuclear Power Plants, Porto Alegre, Brazil, 1978, pp. 339-373.

3. Polak, E., "Algorithms for Optimal Design," Optimization of Distributed Parameter Structures (Ed. E.J. Haug and J. Cea), Sijthoff & Noordhoff, Alphen aan den Rihn, Netherlands, 1980.

4. Bhatti, M.A., Essebo, T., Nye, W., Pister, K.S., Polak, E., Sangiovanni, Vincentelli, A., and Tits, A., "A Software System for Optimization Based Interactive Computer Aided Design," Optimization of Distributed Parameter Structures, (Ed. E.J. Haug and J. Cea), Sitjhoff & Noordhoff, Alphen aan den Rihn, Netherlands, 1980.

5. Bhatti, M.A. and Pister, K.S., "Application of Optimal Design to Structures Subject to Earthquake Loading," Optimization of Distributed Parameter Structures (Ed. E.J. Haug and J. Cea), Sijthoff & Noordhoff, Alphen aan den Rihn, Netherlands, 1980.

6. Sozen, M.A., "Hysteresis in Structural Elements," Applied Mechanics in Earthquake Engineering (Ed. W.D. Iwan), American Society of Mechanical Engineers, New York, AMD-Vol. 8, 1974, pp. 63-98.

7. Newmark, N.M. and Rosenblueth, E., Fundamentals of Earthquake Engineering, Prentice-Hall, Englewood Cliffs, New Jersey, 1971.

8. Chopra, A.K., "Earthquake Analysis of Complex Structures," Applied Mechanics in Earthquake Engineering (Ed. W.D. Iwan) American Society of Mechanical Engineers, New York, AMD-Vol. 8, 1974, pp. 163-203.

9. Pister, K.S., "Constitutive Modeling and Numerical Solution of Field Problems," Nuclear Engineering and Design, Vol. 28, 1974, pp. 137-146.

10. Chang, M.K., et al., ARMA Models for Earthquake Ground Motions, Report No. UCB/EERC-79/19, Earthquake Engineering Research Center, University of California, Berkeley, California, 1979.

11. Nau, R.F., Oliver, R.M. and Pister, K.S., Simulating and Analyzing Artifical Non-stationary Earthquake Ground Motions, Operations Research Center Report, College of Engineering, University of California, Berkeley, California, in press.

12. Hudson, D.E., Reading and Interpreting Strong Motion Accelero-grams, Earthquake Engineering Research Institute Monograph, Berkeley, California, 1979.

13. Clough, R.W. and Penzien, J., Dynamics of Structures, McGraw-Hill, New York, 1975.

14. Newmark, N.M. and Hall, W.J., Procedures and Criteria for Earthquake-Resistant Design, Building Practices for Disaster Mitigation, Building Science Series 46, National Bureau of Standards, February 1973.

15. Murakami, M. and Penzien, J., Nonlinear Response Spectra for Probabilistic Seismic Design and Damage Assessment of Rein-forced Concrete Structures, Report No. EERC 75-38, Earthquake Engineering Research Center, University of California, Berkeley, California, 1975.

16. Vanmarcke, E.H., "Structural Response to Earthquakes," in Siesmic Risk and Engineering Decisions (Ed. C. Lomnitz and E. Rosenblueth) Elsevier, New York, 1976, pp. 287-337.
17. Penzien, J. and Watabe, M., "Characteristics of 3-Dimensional Earthquake Ground Motions," Earthquake Engineering and Structural Dynamics, 3, 1975, pp. 365-373.
18. Bhatti, M.A., "Optimal Design of Localized Nonlinear Systems with Dual Performance Criteria Under Earthquake Excitations," Report No. UCB/EERC-79/15, University of California, Berkeley, California, July 1979.
19. Ray, D., Pister, K.S. and Polak, E., "Sensitivity Analysis for Hysteretic Dynamic Systems," Computer Methods in Applied Mechanics and Engineering, 14, 1978, pp. 179-208.
20. Whitman, R.V. and Cornell, C.A., "Design," in Seismic Risk and Engineering Decisions (Ed. C. Lomnitz and E. Rosenblueth) Elsevier, New York, 1976, pp. 339-380.
21. Walker, N.D. Jr., "Automated Design of Earthquake-Resistant Multistory Steel Building Frames," Report No. UCB/EERC 77-12, Earthquake Engineering Research Center, University of California, Berkeley, California, 1977.
22. Dowrick D.J., Earthquake-Resistant Design, John Wiley & Sons, New York, 1977.

ALGORITHMS FOR OPTIMAL DESIGN*

Ed Polak

Department of Electrical Engineering and Computer
Sciences, University of California, Berkeley, California

ABSTRACT

This paper reviews a class of recently developed algorithms
for the solution of functional inequalities and of optimization
problems with functional inequality constraints, that arise in
the context of optimal design. These algorithms fall into the
categories of phase I - phase II methods of feasible directions,
recursive quadratic programming methods, and outer approximations
methods.

1. INTRODUCTION

This paper surveys algorithms for optimal design of structures subject to dynamic loads or disturbances. In particular,
algorithms that are applicable to the design of earthquake resistant structures are examined. In the case of such designs, one
has to cope with constraints of the form

$$\phi(z,y) \leq 0, \quad \text{for all } y \in Y \tag{1.1}$$

In Eq. 1.1, $z \in R^n$ is the design vector, to be computed, and
$y \in Y \in R^p$ may be a scalar, as when it represents time or frequency, or it may be a vector, as when it represents a construction tolerance or an element of a parametrized family of disturbances. Of course, y can also be a vector that combines a

---

*This research was supported by the National Science Foundation
ECS-79-13148 and ENV76-04264.

number of these quantities into one. When the function $\phi(z,y)$ represents a time response or a frequency response, it is normally found to be differentiable in (z,y). However, when it represents eigenvalues of an operator, it is generally not differentialbe, which leads to additional difficulties. Attention here is mostly concentrated on the case in which $\phi(z,y)$ is continuously differentiable in both variables.

In designing a structure, it may be more important to satisfy the given specifications than to optimize a particular criterion. Consider therefore algorithms for two types of problems. The first is that of finding a feasible vector:

$$\text{FEAS:} \quad \text{find a } z \in F \tag{1.2}$$

where

$$F = [z: g^j(z) \le 0, \quad j = 1,\ldots,p \quad \phi^j(z,y_j) \le 0$$
$$\text{for all } y_j \in Y^j, \, j = 1,\ldots,q] \tag{1.3}$$

The g^j are differentiable functions representing simple design constraints and the ϕ^j model the distributed, i.e. functional, constraints. The second problem considered is of the form

$$\text{OPT:} \quad \min[f(z) : z \in F] \tag{1.4}$$

One may categorize algorithms on the basis of whether the Y are intervals or multidimensional sets and of whether the ϕ^j are differentiable or not. Since the algorithms described here have a modular structure, it is convenient to present them in the form of Master Algorithms, which call any of a set of Subalgorithms. Furthermore, since the essential features of Master Algorithms are preserved when one ignores conventional constraints and assumes that there is only one functional constraint. Since this results in considerable notational simplification, consider here only the simplest form of F, viz.,

$$F = [z : \phi(z,y) \le 0, \quad \text{for all } y \in Y] \tag{1.5}$$

2. ALGORITHMS FOR COMPUTING FEASIBLE POINTS

2.1 Feasible Directions Type Algorithms

Feasible directions type algorithms can be used for solving the problem FEAS of Eq. 1.2, when the sets Y are compact intervals. They differ from each other by the choice of the search direction subalgorithm. Two such subalgorithms are now described for the simplified problem of Eq. 1.5. For every $z \in R^n$ let

$$m(z) = max [0, max\{\phi(z,y) : y \in Y\}] \qquad (2.1)$$

and for any $\varepsilon > 0$, let

$$
\begin{aligned}
Y_\varepsilon(z) = [y \in Y : &\phi(z,y) \geq m(z) - \varepsilon \text{ and } y \text{ is a} \\
&\text{local maximizer of } (z,.)] \qquad (2.2)
\end{aligned}
$$

The nature of the physical problems usually considered justify the following assumption:

(A1) For every $z \in R^n$ and for every $\varepsilon > 0$, the set $Y_\varepsilon(z)$ contains only a finite number of points.

The simplest search direction subalgorithm computes the search direction vector $h_\varepsilon(z)$ as a solution of

$$d_\varepsilon(z) = min [max\{<\nabla_x\phi(z,y),h > : y \in Y_\varepsilon(z)\} : h \in S] \qquad (2.3)$$

with S the unit hypercube centered at the origin. Thus, in view of assumption A1, Eq. 2.3 is seen to be a finite linear program. A better direction search subalgorithm computes the search direction $h_\varepsilon(z)$ as a solution of

$$d_\varepsilon(z) = min [|h| : h \in co\{\nabla_z\phi(z,y),y \in Y_\varepsilon(z)\}] \qquad (2.4)$$

which one recognizes as a finite quadratic program. A detailed discussion of such subalgorithms is found in Ref. 1, which were first introduced in Refs. 2 and 3, for optimization problems with functional constraints. For the case of ordinary constraints, the reader is referred to Ref. 1.

Master Algorithm 2.1 [4].

Parameters: $\varepsilon_0 > 0$, a, b, c $\in (0,1)$

Data: $z_0 \in R^n$

Step 0: Set i = 0.

Step 1: Set $\varepsilon = \varepsilon_0$.

Step 2: If $z_i \in F$ stop. Else, compute $Y_\varepsilon(z_i)$, and use a search direction finding subalgorithm to compute $d_\varepsilon(z_i)$ and $h_\varepsilon(z_i)$.

Step 3: If $d_\epsilon(z_i) > -\epsilon$, set $\quad = \alpha \times \epsilon$ and go to step 2.

Step 4: Compute the largest step size $s_i \in \{1, b, b^2, \ldots\}$ such that

$$m(z_i + s_i h (z_i)) < -s_i c \epsilon \qquad (2.5)$$

Step 5: Set $z_i + 1 = z_i + s_i h (z_i)$, set $i = i+1$ and go to

step 1 (or step 2).

Master Algorithm 2.1 has the following property (see Ref. 3): If the problem FEAS of Eq. 1.2 is such that $d_0(z) \neq 0$ for all $z \notin F$, and the sequence $\{z_i\}$ constructed by Master Algorithm 2.1 is bounded, then Master Algorithm 2.1 finds a $z \in F$ in a finite number of iterations.

In the form stated, Master Algorithm 2.1 cannot be programmed, since it lacks instructions for approximating $m(z)$. Implementable versions of this algorithm have also been presented in Refs. 2 and 3, which improve adaptively, on the basis of tests, the precision with which $m(z)$ is evaluated.

2.2 Newton's Method for Infinite Systems of Inequalities

Newton's method can also only be used when the sets Y_i are compact intervals and assumption A1 holds. In addition, one must require that all functions be three times continuously differentiable. In this case, the problem of Eq. 1.5 can be replaced by a problem specified by the finite set of inequalities:

$$\text{find a } z \in F = [z : \phi(z,y) \leq 0, \quad \text{for all } y \in Y_\epsilon(z)] \qquad (2.6)$$

where $\epsilon > 0$ is arbitrary. It is shown in Ref. 5 that, under some additional, rather technical assumptions on $\phi_y(z,y)$ and $\phi_{yy}(z,y)$, with $y \in Y_\epsilon(z)$, one can construct a locally, quadratically convergent version of Newton's method, for solving Eq. 2.6, as follows: Let $v(z) \quad R^n$ be defined by

$$v(z) = \arg \min [\|v\| : \phi(z,y) + \langle \nabla_z \phi(z,y), v \rangle \leq 0,$$

$$\text{for all } y \in Y] \qquad (2.7)$$

Then, given $z \in F$, set

$$z_{i+1} = A_1(z_i) = z_i + v(z_i) \qquad (2.8)$$

Note that because $Y_\epsilon(z)$ is finite, by assumption, the quadratic program of Eq. 2.7 does not present any exceptional difficulties and can be solved by standard QP codes. It is also possible to use the L_∞ norm in Eq. 2.7, which then makes this problem an LP. Since the radius of convergence of the process defined by Eq. 2.8 may be quite limited, it was proposed in Ref. 5 to stabilize it by switching over to Master Algorithm 2.1 when a certain test fails, as follows: First note that when the return from Step 5 to Step 1 is used, Master Algorithm 2.1 becomes one step and the ϵ accepted in Step 4 is a function of z_i only, so one can write it as $\epsilon(z_i)$. Hence, let $A_2(.)$ be the map from R^n into all subsets of R^n, defined by Master Algorithm 2.1, with $h_\epsilon(z)$ computed in Step 1 and s computed in Step 3 of Master Algorithm 2.1, so that $A_2(.)$ has the form

$$A_2(z) = z + sh_\epsilon(z) \qquad (2.9)$$

Master Algorithm 2.2.

Parameters: $\epsilon > 0$, $d \in (0,1)$, $K > 1$

Data: $z \in R^n$

Step 0: Set $i = 0$, $j = 0$.

Step 1: If $m(z_i) < 0$, stop.

Step 2: If a solution $v(z_i)$ of Eq. 2.7 exists and

$$|v(z_i)| < Kd^j \qquad (2.10)$$

set

$$z_{i+1} = A_1(z_i) \qquad (2.11)$$

Otherwise, set

$$z_{i+1} = A_2(z_i) \qquad (2.12)$$

Step 3: Set $i = i + 1$ and go to step 1.

Note that as long as the Newton process of Eq. 2.8 is well defined and shows signs of at least linear convergence, according to Eq. 2.10, one uses it to define z_{i+1}, otherwise, one switches to Master Algorithm 2.1. It is shown in Ref. 5 that (under suitable assumptions) Master Algorithm 2.2 has the following property: If the sequence $\{z_i\}$ constructed by Master Algorithm 2.2 is infinite and bounded, then there exists an i_0 such that Eq. 2.11 holds for all $i > i_0$, and $z_i \to z \in F$, quadratically (for some $z \in F$).

In the form stated, Master Algorithm 2.2 is conceptual, since it does not specify approximation rules form computing sets $Y_\epsilon(z_i)$. The reader will find an implementable form discussed in detail in [M4].

2.3 Outer Approximations Algorithms

The requirement that Y^j be intervals is now dropped. They may be any compact subsets of a Euclidean space. Outer approximations algorithms decompose the problem FEAS, which contains infinitely many inequalities, into an infinite sequence of problems of the form

> FEAS(k): find a $z \in F_k$

where

$$F_k = [z : g^j(z) \le 0, \; j = 1,\ldots,p, \; \phi^j(z,y_j) \le 0,$$

$$\text{for all } y_j \in Y_k^j, \; j = 1,\ldots,q] \tag{2.13}$$

or, in the simplified framework of Eq. 1.5,

$$F_k = [\; z: \phi(z,y) \le 0, \text{ for all } y \in Y_k] \tag{2.14}$$

One sees that the sets F_k are described by finite sets of inequalities, which can be solved very rapidly by algorithms, such as those described in Refs. 6 and 7. When certain simple assumptions are satisfied, these algorithms need only a finite number iterations to solve FEAS(k) and are called, as subalgorithms, by outer approximations master algorithms.

The simplest outer approximations algorithms are described by Eaves and Zangwill in Ref. 8. In these algoriths, given a problem FEAS(k), with a finite set Y_k, one first computes a $z_k \in F_k$ and then one computes

$$y_k \in \arg \max [\phi(z_k,y) : y \in Y_k] \tag{2.15}$$

The next problem, FEAS(k+1) is then constructed by setting

$$Y_{k+1} = Y_k \cup \{y_k\} \tag{2.16}$$

It is easy to show that any accumulation point z of an infinite sequence $\{z_k\}$, constructed in this manner, must be in F. However, since the cardinality of Y_k grows relentlessly, one may not be able to compute too many elements of such a sequence. Because of this, Eaves and Zangwill have proposed a scheme for constructing the Y_k, which periodically flushes out a number of elements from Y_k. Since their scheme had a number of shortcoming, including the fact that it was nonimplementable and high scaling dependent, Mayne, Polak, and Trahan [9] have proposed a more sophisticated method, which was implementable and which included features to minimize scaling dependence. However, it still retained at least one bad feature in the method of Ref. 8, viz. that convergence could be claimed not for accumulation points of the whole sequence $\{z_k\}$, but only for those of the subsequence at which the Y_k were flushed out. This difficulty, as well as a few other ones, were overcome by the scheme proposed by Gonzaga and Polak [10], which is now stated for the simplified problem of Eq. 1.5.

Master Algorithm 2.3 [4,10].

Parameters: K,L.

Data: $Y_0 \subset Y$.

Step 0: Set k = 0.

Step 1: Compute a $z_k \in F_k$.

Step 2: Compute

$$y \in \arg \max [\phi(z_k,y) : y \in Y] \tag{2.17}$$

Step 3: If $m(z_k) = 0$, stop. Otherwise, set

$$Y_{k+1} = [y_i : \phi(z_i,y_i) \geq K \{1/(1+i)^{i/L} - 1/(1+k)^{1/L}\},$$

$$i = 0,1,\ldots,k] \tag{2.18}$$

Step 4: Set k = k + 1 and go to step 1.

An examination of Eq. 2.18 shows that a point y_i will always be included in Y_i and, possibly, in a few subsequent Y_k. However, as k increases, the right hand side of the inequality in Eq. 2.18 also increases and, in most likelihood, y_i will be dropped from the Y_k after a few retentions. It is shown in Ref. 10 that any accumulation point z of a sequence {z}, constructed by Master Algorithm 2.3, is in F.

Again, as stated, Master Algorithm 2.3 is only conceptual since no rule is given for approximating $m(z_k)$. An implementable version can very easily be deduced from the schemes given in Ref. 10, where it is shown that one merely needs to evaluate $m(z_k)$ with progressively greater precision as k increases.

2.4 Nondifferentiable Constraints

When the functions ϕ^j are nondifferentiable, none of the above algorithms apply. However, when they are at least locally Lipschitz continuous, it becomes possible to make use of some of the known results in nondifferentiable optimization [11 to 14]. In particular, when the ϕ^j are eigenvalues of an operator and the Y are intervals, the problem becomes reasonably tractable, by an algorithm that is essentially of the form of Master Algorithm 2.1, which uses vectors belonging to an ε-bundle of generalized gradients [11], instead of gradients in Eq. 2.4. The construction of these vectors is quite complicated, so the interested reader is referred to Ref. 14, where such an algorithm is described in detail.

This concludes the discussion of algorithms for solving inequalities that arise in the context of engineering design and attention is now turned to optimization.

3. ALGORITHMS FOR COMPUTING OPTIMAL POINTS

3.1 Phase I - Phase II Feasible Directions Algorithms

Phase I - Phase II feasible direction algorithms can be used for solving the problem OPT of Eq. 1.4, when the Y are compact intervals and assumption A1 is satisfied. Just as the algorithms described in the preceding section, they differ from one another only by the choice of the search direction subalgorithm. There is a direct relationship between the direction subalgorithms used in Section 2.1 and the ones that are described here. Thus,

Eq. 2.3 becomes

$$d_\varepsilon(z) = \min[\max\{<\nabla f(z),h> \; : \; <\nabla_z \phi(z,y),h> \;,y \in Y_\varepsilon(z)\}: h \in S]$$

$$(3.1)$$

with $Y_\varepsilon(z)$ and S as in Eq. 2.3. Similarly, Eq. 2.4 becomes

$$d_\varepsilon(z) = \min[|h|| \; : \; h \in co \; \{\nabla f(z), \; \nabla_z \phi(z,y), \; y \in Y_\varepsilon(z)\}] \quad (3.2)$$

It is shown in Refs.2 and 3 that $d_0(z) = 0$ is an optimality condition that is equivalent to the F. John optimality condition. Hence, the functions $d_\varepsilon(z)$ are called optimality functions. Although the zeros of the two optimality functions defined by Eqs. 3.1 and 3.2 are identical, they do lead to algorithms with different computational properties. The one defined by Eq. 3.2 has certain self scaling properties that result in a superior algorithm.

Master Algorithm 3.1 [3].

Parameters: $\varepsilon > 0$, a, b,c, $\in (0,1)$.

Data: $z_0 \in R^n$.

Step 0: Set i = 0.

Step 1: Set $\varepsilon = \varepsilon_0$.

Step 2: If $d_0(z_i) = 0$, stop. Otherwide, use a search direction finding subalgorithm to compute $Y_\varepsilon(z_i)$, $d_\varepsilon(z_i)$, and $h_\varepsilon(z_i)$.

Step 3: If $d_\varepsilon(z_i) > -\varepsilon$, set $\varepsilon = a \times \varepsilon$ and go to step 2.

Step 4: Compute the largest step size $s_i \in \{1,b,b^2,....,\}$, such that if $z_i \in F$, then

$$z_i + s_i h_\varepsilon(z_i) \in F \tag{3.3}$$

and

$$f(z_i + s_i h_\varepsilon(z_i)) - f(z_i) < -s_i c\varepsilon \tag{3.4}$$

If $z \notin F$, then

$$m(z_i + s_i h_\varepsilon(z_i)) - m(z_i) < -s_i c \varepsilon \tag{3.5}$$

Step 5: Set $z_{i+1} = z_i + s_i h_\varepsilon(z_i)$, set $i = i + 1$, and go to step 1 (or Step 2).

Note that Master Algorithm 3.1 concentrates at first on finding a feasible point. Once such a point is found, the remainder of the sequence is feasible. Referring to Ref. 3, one finds that Master Algorithm 3.1 has the following property: If the problem OPT is such that $d_\varepsilon(z) < 0$ for all $z \notin F$, then any accumulation point z of a sequence $\{z_i\}$ constructed by the Master Algorithm 3.1 is feasible, i.e $z_i \in F$, and satisfies the F. John condition of optimality, i.e., $d_0(z) = 0$.

Again, since no rule for specifying approximations to $m(z)$ ahd $Y_\varepsilon(z)$ are given, Master Algorithm 3.1 is only conceptual. An implementable version, with the same convergence properties may be found in Refs. 3 and 4.

3.2 Recursive Quadratic Programming Algorithms

At present, there is strong sentiment in the mathematical programming community that recursive quadratic programming algorithms, such as those described in Ref. 15 to 19, which have evolved from Wilson's method [15], offer substantial advantages over other constrained optimization algorithms. Recall that Wilson's method is an appropriate form of Newton's method for solving the equations and inequalities in the Kuhn-Tucker conditions. At present, there is no recursive quadratic programming algorithm for solving OPT in the literature. However, based on what has been seen in Section 2.2, it is felt that for the case in which the sets Y^j are compact intervals and assumption Al holds, one can construct an RQP type algorithm for solving OPT, as follows: First, observe that when one computes $d_\varepsilon(z)$ by Eq. 3.2, the quadratic program returns a set of positive weights: $u(z,f), u(z,y), y \in Y_\varepsilon(z)$, summing to unity, such that

$$h_\varepsilon(z) = u(z,f) \nabla f(z) + \sum_{y \in Y_\phi(z)} u(z,y) \nabla_z \phi(z,y) \tag{3.6}$$

Assuming that $u(z,f) \neq 0$, one can rescale these weights, which are pontential F. John multipliers, by dividing them all by $u(z,f)$ to obtain a set of potential Kuhn-Tucker multipliers, which will be denoted by $u(z,y,e)$, to make their dependence on ε

more obvious. These multipliers can now be used to compute the corresponding second derivative matrix

$$H(z,\varepsilon) = \partial^2 f(z)/\partial z^2 + \sum_{y \in Y\phi(z)} u(z,y,\varepsilon)\partial^2 \phi(z,y)/\partial z^2 \qquad (3.7)$$

In this interpretation, given any $\varepsilon > 0$ and z_i, Wilson's method, applied to the problem

$$\min \, [f(z_i) : \phi(z_i,y) < 0, \, y \in Y_\varepsilon(z_i)] \qquad (3.8)$$

requires one to compute $v(z_i)$ as a solution of

$$\min[<\nabla f(z_i)v> + (1/2)< v,H(z_i,\varepsilon)v> :\phi(z_i y) +<\phi(z_i y),v> \leq 0,$$

$$\text{for all } y \in Y_\varepsilon(z_i)] \qquad (3.9)$$

and to set

$$z_{i+1} = A_1(z_i) = z_i + v(z_i) \qquad (3.10)$$

On the basis of the analysis referred to in Section 2.2, one can anticipate that, under suitable assumptions, the algorithm defined by Eq. 3.9 will be locally, quadratically, convergent. Let us denote the construction defined by the steps of Master Algorithm 3.1, with a return from Step 5 to Step 1, by

$$z_{i+1} = A_2(z_i) \qquad (3.11)$$

Next, denote by $\varepsilon(z_i)$ the value of ε that is accepted by Master Algorithm 3.1 at z_i, i.e. which permits passage from Step 3 to Step 4.

One can now follow the scheme in Master Algorithm 2.1 to obtain a globally convergent version of an RQP type method for Eq. 3.7. Note that this method requires one to solve two quadratic programs per iteration.

Master Algorithm 3.2.

Parameters: $d \in (0,1)$, $K > 1$.

Data: $z \in R^n$.

Step 0: Set $i = 0$, $j = 0$.

Step 1: Compute $\varepsilon(z_i)$, $Y_\varepsilon(z_i)$, and $H(z_i, \varepsilon(z_i))$ by means of Eq. 3.2.

Step 2: If a solution $v(z_i)$ of Eq. 3.9 exists for $\varepsilon = \varepsilon(z_i)$, and

$$|v(z_i)|| \leq Kd^j \tag{3.12}$$

set

$$z_{i+1} = A_1(z_i) \tag{3.13}$$

Otherwise, set

$$z_{i+1} = A_2(z_i) \tag{3.14}$$

Step 3: Set $i = i + 1$ and go to Step 1.

Thus, just as in the case of Master Algorithm 2.2, one accepts Wilson's method if it shows signs of converging, at least linearly, according to Eq. 3.12, and one reverts to Master Algorithm 3.1 otherwise. On the basis of the results in Refs. 5 and 18, one may conjecture as follows:

If (1) all functions are three times continuously differentiable, (2) all Kuhn-Tucker points satisfy second order sufficiency conditions and, (3) the conditions on $\phi_y(z,y)$ required for Master Algorithm 2.2 are satisfied,

Then, (1) all accumulation points of a sequence $\{z_i\}$, constructed by Master Algorithm 3.2, are Kuhn-Tucker points, and (2) if the sequence $\{z_i\}$ constructed by Master Algorithm 3.2 is bounded, then it converges quadratically to a Kuhn-Tucker point.

Master Algorithm 3.2 can be made implementable by using the same rules as those for Master Algorithms 2.2 and 3.1, and hence implementability is not a source of difficulty. What may prove to be a source of serious difficulty is the need to evaluate second derivatives. Fortunately, as can be seen from Refs. 17 to 19, there are a number of first order schemes available for approximating $H(z, \varepsilon(z))$. These schemes make use of either secant or quasi-Newton formulas.

3.3 Outer Approximations Algorithms

The requirement that the Y^j be intervals (see Section 2.3) is now dropped and they are allowed to be any compact subsets of a Euclidean space. The scheme presented in Ref. 3 is now described, where the decomposition of the problem OPT into a sequence of problems

$$OPT(k) : min[f(z) : z \in F_k] \tag{3.15}$$

is carried out essentially in the same manner as in solving inequalities, with one important exception. One can solve a problem FEAS(k) in a finite number of iterations, but one cannot solve OPT(k) in a finite number of iterations (or find a stationary point, for that matter). Hence, one must specify in what sense the solutions to OPT(k) should be approximated. Such a specification is incorporated in the master algorithm, below, which calls an ordinary constrained optimization subalgorithm to solve OPT(k). Without essential loss of generality, one again states the algorithm in terms of the simplified problem OPT, with F given by Eq. 1.5. First, given Y_k, a finite subset of Y, for any $z \in R^n$, define

$$m(z,Y_k) = max[0,max\{\phi(z,y) : y \in Y_k\}] \tag{3.16}$$

Next, for any $z \in R^n$ and $\varepsilon > 0$, define

$$Y_{k,\varepsilon}(z) = [y \in Y_k : \phi(z,y) > m(z,Y_{k-\varepsilon}] \tag{3.17}$$

Finally, define

$$d(z,Y_k) = min[|h|| : h \in co\{\nabla f(z),\nabla_z\phi(z,y), y \in Y_k(z)\}] \tag{3.18}$$

Note that $d(z,Y_k) = 0$ is a necessary condition of optimality for OPT(k), which, as already mentioned, is equivalent to the F. John condition.

Master Algorithm 3.3.

Parameters: ε_0, K, L > 0.

Data: $Y_0 \subset Y$

Step 0: Set k = 0, set $\varepsilon = \varepsilon_0$.

Step 1: Solve OPT(k), tothe extent of computing a z_k such that

$$d(z_k, Y_k) > - \varepsilon \tag{3.19}$$

and

$$m(z_k, Y_k) < \varepsilon \tag{3.20}$$

Step 2:   Compute a

$$y \in \arg\max[\phi(z_k, y) : y \in Y] \tag{3.21}$$

Step 3:   If $m(z_k, Y) = 0$, step.   Otherwise, set

$$Y_{k+1} = [y_j : \phi(z_j, y_j) > K\{1/(1+j)^{1/L} - 1/(1+k)^{1/L}\},$$

$$j = 1, \ldots k] \tag{3.22}$$

Step 4:   Set $\varepsilon = \varepsilon/2$, set $k = k + 1$, and go to Step 1.

The only part of Master Algorithm 3.3 that is implementable is the computation of the y_k and $m(z_k, Y)$. A simple rule is given in Ref. 10, which states that all that is needed is that these computations be carried out with progressively greater and greater precision, as the computation progresses (in the same manner as for OPT(k)). As far as convergence is concerned, the following result is given in Ref. 10:

(1) If the sequence $\{z_k\}$ constructed by Master Algorithm 3.3 is finite, then its last element is in F and satisfies the F. John condition of optimality for OPT i

(2) If the sequence $\{z_k\}$ constructed by Master Algorithm 3.3 is infinite, then all its accumulation points are in F and satisfy the F. John optimality condition for OPT.

Finally, it should be pointed out that the rule for constructing Y_{k+1}, as given in Eq. 3.22 and also in Eq. 2.18 can be substantially modified by using double subscripted sequences $\{\varepsilon_{j,k}\}$ in the right hand side of the inequalities in Eqs. 2.18 3.22, as explained in Ref. 3, with the one used in this paper being only one example of such a sequence.

600

3.4 Nondifferentiable Optimization Algorithms

At present, it appears that the entire arsenal of nondifferentiable optimization algorithms that apply to OPT, consists of those described in Refs. 14 and 20. These algorithms make use of outer approximation master algorithms and of either ordinary or nondifferentiable optimization subalgorithms. The lack of space precludes their being described here. The interested reader is referred to Res. 14 and 20.

4. CONCLUSION

We have given in this paper a brief summary of a class of algorithms which are applicable to optimal design. Some of these algorithms, such as feasible directions, have been used by us for a number of years and have been found to be quite reliable, though somewhat expensive. The newer algorithms, such as PQP are in the process of being programmed up for testing. Given their excellent behavior on ordinary mathematical programming problems, we expect that they will eventually prove to be a most valuable tool in optimal design, as well.

REFERENCES

1. Polak, E., Trahan, R. and Mayne, D.Q., "Combined Phase I - Phase II Methods of Feasible Directions," Mathematical programming, Vol. 17, No. 1, 1979, pp. 61-73.
2. Polak, E. and Mayne, D.Q., "An Optimization with Functional Inequality Constraints," IEEE Trans., Vol. AC-21, No. 2, 1976, pp. 184-193.
3. Gonzaga, G., Polak, E. and Trahan, R., "An Improved Algorithm for Optimization Problems with Functional Inequality Constraints," IEEE Trans., Vol. AC-25, No. 1, 1980, pp. 49-54.
4. Polak, E. and Mayne, D.Q., "Algorithms for Computer Aided Design of Control Systems by the Method of Inequalities," Proc. 1979 IEEE Conference on Decision and Control, Fort Lauderdale, Fl. Dec. 1979.
5. Mayne, D.Q. and Polak, E., A Quadratically Convergent Algorithm for Solving Infinite Dimensional Systems of Inequalities, University of Calfornia, Electronics Research Lab. Memo No. UCB/ERL M80/11, 1980.
6. Mayne, D.Q., Polak, E. and Heunis, A.J., Solving Nonlinear Inequalities in a Finite Number of Iterations, Publication 70/3, Department of Computing and Control, Imperial College, London, 1979. (To appear in JOTA).
7. Polak, E. and Mayne, D.Q., "On the Finite Solution of Inequalities," IEEE Trans. Vol. AC-24, No. 3, 1979, pp. 443-445.

8. Eaves, B.C. and Zangwill, W.I., "Generalized Cutting Plane Algorithms," SIAM J. Control, Vol. 9, No. 4, 1971, pp. 529-542.

9. Mayne, D.Q., Polak, E., and Trahan, R., "An Outer Approximations Algorithm for Computer Aided Design Problems," JOTA, Vol. 28, No. 3, 1979, pp. 331-352.

10. Gonzaga, C. and Polak, E., "On Constraint Dropping Schemes and Optimality Functions for a Class of Outer Approximations Algorithms," SIAM, J. Control and Optimization, Vol. 17, No. 4, 1979, pp. 477-493.

11. Clarke, F.M., "Generalized Gradients and Applications," Trans. Amer. Math. Soc., Vol. 205, 1975, pp. 247-262.

12. Mifflin, R., "Semismooth and Semiconvex Functions in Constrained Optimization," SIAM J. Control and Optimization, Vol. 15, 1979, pp. 959-972.

13. Lemarechal, C., "An Extension of Davidon Methods to Non-differentiable Optimization Problems," Nondifferentiable Optimization, (M.L. Balinski and P. Wolfe eds.) Mathematical Programming study 3, Amsterdam, North Holland, 1965, pp. 95-109.

14. Polak, E. and Sangiovanni Vincentelli, A., "Theoretical and Computational Aspects of the Optimal Design Center, Tolerancing and Tuning Problem," IEEE Trans. Vol. CAS-26, No. 9, 1979, pp. 795-813.

15. Wilson, R.B., A Simplified Algorithm for Concave Programming, Ph.D. dissertation, Graduate School of Business Administration, Harvard University, Cambridge, Mass. 1963.

16. Robinson, S.M., "Perturbed Kuhn-Tucker Points and Rates of Convergence for a Class of Nonlinear Programming Algorithms," Mathematical programming, Vol. 7, 1976, pp. 1-16.

17. Han, S.P., "Dual Variable Metric Algorithms for Constrained Optimization," SIAM J. Control and Optimization, Vol. 11, 1977, pp. 546-566.

18. Polak, E. and Mayne, D.Q., A Robust Secant Method for Optimization Problems with Inequality Constraints, University of California, Electronics Research Lab. Memo No. UCB/ERL M79/2, 1979. (To appear in JOTA.)

19. Powell, M.J.D., "A Fast Algorithm for Nonlinearly Constrained Optimization Calculations," Presented at 1977 Dundee Conference on Numerical Analysis.

20. Polak, E., An Implementable Algorithm for the Optimal Design Centering, Tolerancing and Tuning Problem, University of California, Electronics Research Lab Report No. UCB/ERL M79/33, 1979. (To appear in JOTA.)

A SOFTWARE SYSTEM FOR OPTIMIZATION BASED INTERACTIVE COMPUTER-AIDED DESIGN*

M. Asghar Bhatti,[1] Tommy Essebo,[2] William Nye,[3]
Karl S. Pister,[1] E. Lucein Polak,[3] Alberto Sangiovanni-
Vincentelli,[3] and Andre Tits[3]

ABSTRACT

This paper describes an interactive·software system, called INTEROPTDYN, for optimization based computer-aided design of engineering systems. Criteria for interaction in general purpose optimization systems is developed. Some of the important commands and main features of the present system are presented.

1. INTRODUCTION

The term computer-aided design is used to describe a great variety of activities. In structures, computer-aided design often amounts to no more than simulation of structural systems, coupled with a trial-and-error design procedure. The designer first chooses an initial design configuration. The configuration is then analyzed by means of a computer program, which simulates

[1] Department of Civil Engineering, University of California, Berkeley.

[2] Department of Automatic Control, Institute of Technology, Lund, Sweden.

[3] Department of Electrical Engineering and Computer Sciences, University of California, Berkeley.

*This research was supported by Joint Services Electronics Program Grant F49620-79C-0178 and National Science Foundation Grants ENG77-16062 and ENV76-04264.

behavior of the physical system. By looking at the results of the computer simulation, the designer adjusts parameter values in an attempt to satisfy a set of given specifications, which are not met by the initial configuration and/or to obtain a better design, in terms of its performance. After the adjustment, a new simulation is performed and the overall procedure is iterated until a satisfactory design is obtained.

Over the last decade, research in computer simulation of structural systems has made considerable progress, resulting in a number of excellent simulation programs (e.g. see Refs. 1-4). Since the late 1960's, it has been felt that the trial-and-error design mode could and should be improved considerably [5,6]. In a cut-and-try design mode, the designer essentially assumes the role of a heuristic master optimization and inequality solving algorithm. Unfortunately, pure heuristics are generally inefficient in searching a multidimensional parameter space, hence they cannot be relied on to produce a feasible design, let alone an optimal one. Consequently, an efficient design procedure must make substantial use of optimization algorithms, to relieve the designer from the drudgery of search in the parameter space and to allow him to concentrate on the conceptual aspects of the design.

Despite considerable research activity in optimal design of structural systems, optimization techniques are not used as widely as might be expected. There are several reasons for this. Perhaps the most important one is that the optimization algorithms used until now have been too primitive for the task at hand. For example, they are not capable of solving non-convex problems and problems with dynamic constraints. Even in the simple cases, the cost benefit ratio has frequently been unfavorable, because algorithms failed to converge to a solution in a reasonable amount of computer time. This situation may be caused by ill-conditioning of the mathematical programming problem into which the design problem was translated, by the weak convergence properties of the algorithms used (e.g. penalty function with conjugate gradients as subroutine for unconstrained optimization), or by a poor choice of initial design and/or algorithm parameters. Since any algorithm for optimization based computer-aided design requires a number of simulations per iteration, and since the cost of simulation ranges from 15 secs for a simple problem to minutes for a realistic design problem, it is clear that slow convergence or no convergence at all may be considered as a very expensive accident!

Recently, new algorithms have been developed for general, non-convex design problems involving dynamic constraints [7-9]. At the same time, methods for early detection of ill-conditioning in the mathematical programming problem, into which the design problem was translated, are emerging. Also, heuristics are currently being developed that help avoid translation ill-conditioning. Since, in general, the transcription of a design problem into a mathematical programming problem is not unique, these heuristics suggest ways for changing the transcription of the design problem to eliminate ill-conditioning. However, these algorithms are still very sensitive to the choice of internal parameters and initial values of design parameters.

It appears that a new design methodology, based on interactive graphic computing, is indispensable. Interactive computing permits one to abort, stop and restart, or otherwise modify a computation as it progresses, resulting in very substantial savings not only in computing time, but also in the overall time needed to carry out a design. As an example, suppose that an initial design proposed by a designer fails to meet specifications. When using an interactive CAD system, he could identify this fact by observing the computations. He could stop the computation and either modify the structure of his design or experiment with relaxation of certain specifications. Next, in the case of ill-conditioning, he could change the description of the design problem into a different mathematical programming problem by observing the heuristic information displayed on the screen. Finally, he would be in an ideal situation to trade-off one desirable goal to obtain an improvement in another. Since it is obviously impossible either to compute or to display an entire multidimensional trade-off surface, interactive computing techniques are being developed that will enable the designer to find a satisfactory compromise solution, on the basis of a sequence of computations that he must guide interactively.

One of the goals of the research presented here is to develop a software system for optimization based, interactive, computer-aided design. A prototype interactive software system INTEROPTDYN (INTERactive OPTimization of DYNamical systems), built around the language INTRAC [10], is described in the following sections. Methods for interaction of the designer and an optimization package are discussed in section 2. In section 3, some of the features of the computer program INTEROPTDYN are described.

2. INTERACTIVE GRAPHICS IN OPTIMIZATION BASED CAD

When expressed as an optimization problem, most engineering design problems become a nonstandard mathematical programming problem of the form

minimize $f(x)$

subject to constraints

$$g^j(x) \leq 0, \ j=1,\ldots,q$$

$$\max_{p^i \in P^i} \phi^i(x,p^i) \leq 0, \ i=1,\ldots,m \qquad (2.1)$$

where x is the vector of design parameters. The components of the p^i's are parameters such as temperature, time, or frequency. The sets P^i, $i=1,\ldots,m$ are generally intervals or n dimensional boxes or other convex sets in R. It is assumed that the cost and constraint functions f and g^j, $j=1,\ldots,q$ are continuously differentiable. It is also assumed that the constraint functions ϕ^i, $i=1,\ldots,m$ of Eq. 2.1 are continuously differentiable with respect to x and Lipschitz continuous with respect to p^i. Of course the constraints of the form $\max_{p^i \in P^i} \phi^i(x,p^i) \leq 0$ are not continuously differentiable with respect to x. For the sake of simplicity, design problems in which these constraints are not present are considered in this section.

To illustrate the use of interactive graphics in solving such a problem, one may use a feasible directions method [19]. Let $F = \{x : g^j(x) \leq 0, \ j=1,\ldots,q\}$ be the feasible region for the design problem. Given $\varepsilon > 0$, the set $J_\varepsilon(x) = \{j \in \{1,\ldots,q\} : g^j(x) \geq -\varepsilon\}$ is called the set of indices of the ε-active constraints. One may now introduce the following <u>feasible direction algorithm</u> to discuss the use of graphic interaction in optimization:

Data: $\qquad x_0 \in F$

Parameters: $\quad \alpha, \ \beta \in (0,1); \quad \varepsilon_0 > 0; \quad \overline{\varepsilon} \geq 0$

Step 0: $\qquad$ Set $i=0$

Step 1: $\qquad$ Set $\varepsilon = \varepsilon_0$

Step 2 (Direction finding subprocedure): compute $h_\varepsilon(x_i)$ and $v_\varepsilon(x_i)$, where

$$h_\varepsilon(x_i) = -\text{argmin} \{||h||^2 : h = \lambda_0 \nabla f(x_i) + \sum_{j \in J_\varepsilon(x_i)} \lambda_j \nabla g^j(x_i);$$

$$\sum_{j \in J_\varepsilon(x_i)} \lambda_j = 1; \ \lambda_j \geq 0, \ j \in J_\varepsilon(x_i) \}$$

and $v_\varepsilon(x_i) = ||h_\varepsilon(x_i)||^2$

Step 3 (Termination criterion): If $\varepsilon \leq \bar{\varepsilon}$, compute $v_\varepsilon(x_i)$ and stop if $v_\varepsilon(x_i) \leq \bar{\varepsilon}$. Otherwise, proceed.

Step 4 (ε-reduction): If $v_\varepsilon(x_i) \leq \varepsilon$, set $\varepsilon = \frac{\varepsilon}{2}$ and go to Step 2. Otherwise, proceed.

Step 5 (Stepsize computation by Armijo rule [19]): Compute the smallest integer $k_i \geq 0$ such that

$$f(x_i + \beta^{k_i} h_\varepsilon(x_i)) - f(x_i) \leq -\alpha\beta^{k_i} ||h_\varepsilon(x_i)||^2 \qquad (2.2)$$

$$g^j(x_i + \beta^{k_i} h_\varepsilon(x_i)) \leq 0, \ j=1,\ldots,q \qquad (2.3)$$

Step 6:   Set $x_{i+1} = x_i + \beta^{k_i} h_\varepsilon(x_i)$, set $i = i+1$, and go to Step 1.

Under reasonably weak conditions on the feasible region F, the above algorithm is guaranteed to produce a sequence of design parameters whose accumulation points satisfy a first order necessary optimality condition [9]. Its computational efficiency depends critically upon the values of the parameters. Each of the parameters controls a particular phase of the optimization algorithm. For example, ε and $\bar{\varepsilon}$ control the termination of the algorithm, ε_0 controls which constraints have to be taken into account at the beginning of each iteration of the algorithm, in computing the descent direction, α and β control respectively the slope of the line in Figure 2.1 and the rate of reduction of the step size in the Armijo rule. Unfortunately, the optimum values of the parameters are problem dependent. A designer with knowledge of the mechanisms of the algorithm may see from the results of the early iterations that the computation is not progressing satisfactorily. In order to avoid continued inefficient use of the algorithm, the designer must be able to interrupt the computing process, analyze the causes of the unsatisfactory situation, change the values of the data or of the parameters, and restart the computing process. These actions can be accomplished efficiently only by interaction. However, in order to make interaction as effective as possible, the designer must be provided with indicators that can guide him in the detection of poor computational behavior of the algorithm. Moreover, these indicators should provide information that helps him decide how the parameters of the algorithm should be changed so as to improve its performance.

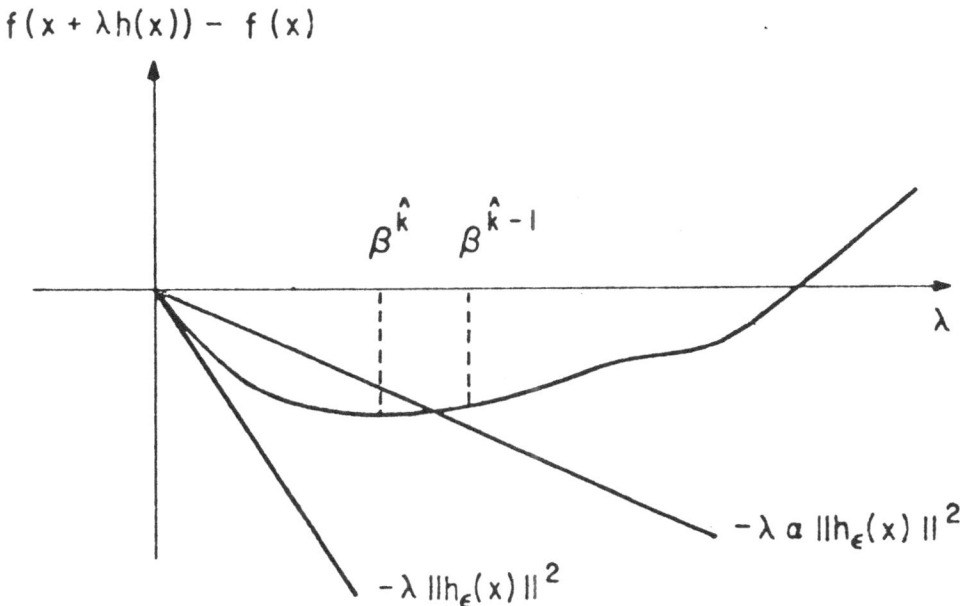

Figure 2.1   The Armijo Step-Size Calculation Procedure

For example, suppose that after computing the search direction $h_\epsilon(x_i)$ the algorithm gets hung up in the Armijo step loop because the test of Eq. 2.3 is not satisfied. This is certainly extremely undesirable, since in CAD problems each function evaluation requires expensive simulation. A designer may try to correct this situation be adjusting some of the parameters of the algorithm or by rescaling the problem. In order to do this, he has to detect which constraint is forcing the decrease of the step size in the Armijo step. Then he must be able to check if this constraint is in the set of the ε-active constraints. If this constraint is not in the set of ε-active constraints, then the search direction computation does not "see" this constraint and, as a consequence, the step size may have to be reduced considerably before the test of Eq. 2.3 is satisfied. In this case, the designer could increase ε so that the neglected constraint is in the set of ε-active constraints. When increasing ε, the designer must be careful not to make ε so large that too many other constraints become active, otherwise $h_ε(x_i)$ may become too small and ε would then be reduced in step 3, wasting several cycles in step 2. Furthermore, gradient computations are costly and should be kept to a minimum. If the designer finds that the impeding constraint is in the ε-active constraint set, then he may try a different strategy to improve performance of the algorithm. In this case, the poor computational behavior may be caused by bad scaling. Consider first the geometrical interpretation of the search direction. According to step 2, the search direction calculation problem turns out to be the negative of the nearest vector to the origin in the convex hull of the gradients of the cost and of the ε-active constraints. From the geometry of the direction calculation problem, one deduces that if the $L_∞$ norm of the gradient of the impeding constraint is very large, compared to the norms of the gradients of the other ε-active constraints and/or of the cost, then the search direction computed by solving the quadratic programming problem in step 2 may not take into account the gradient of the constraint, that is causing difficulty (see Fig. 2.3). This situation may be detected by looking at the angle between the search direction and the gradient of the constraint. If this angle is close to 90° then, $h_ε(x_i)$ does not adequately take into account the impeding constraint. To correct this situation, the designer could multiply the gradient of the limiting constraint by a "pushfactor" to make the $L_∞$ norm of this vector comparable to those of the other vectors considered in step 2.

It turns out that in almost all the critical phases of the optimization algorithm similar information should be made available to the designer. Of course, this information can be given numerically at each iteration of the algorithm and at each iteration of the subalgorithms (loops) inside the steps of the algorithm. However, it is easy to see that a massive quantity of numerical data, presented on a screen, may overwhelm a designer and jeopardize the efficiency of the design procedure. A much more efficient method for carrying out interaction is through computer graphics.

Consider the computational problem described above. To help the designer detect reasons for poor performance of the algorithm,

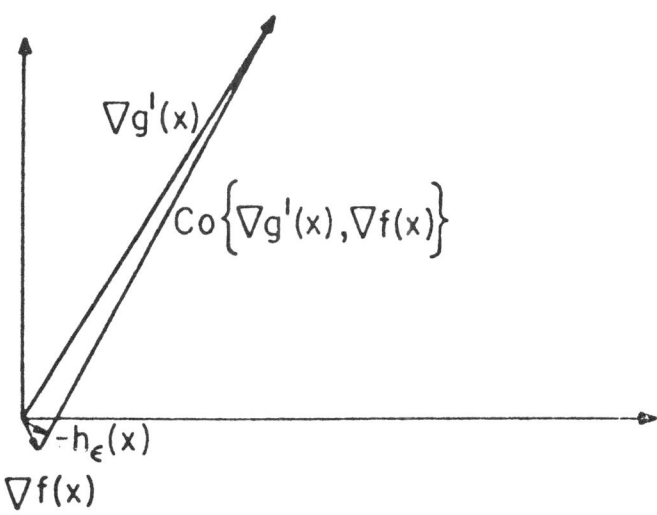

Figure 2.2   An Example of the Influence of Bad Scaling on the Search Direction Calculation

one could display a bar chart plotting the values of the constraints before entering the Armijo step and at each iteration within the Armijo step, side by side. By displaying also a line corresponding to zero and a line corresponding to ε, the designer could immediately detect if an impeding constraint is not in the set of the ε-active constraints by looking at the bar corresponding to the value of the constraint before the Armijo step is entered (see Fig. 2.3). The use of color graphics could further improve man-machine interaction. In the example described above, if the bars corresponding to constraints that are not satisfied are plotted in red, the bars corresponding to the constraints that are satisfied are plotted in green, and the ε-line is plotted in yellow, the designer could grasp all the information he needs at a glance. The need for graphical display of information is even stronger, if one considers a design problem with dynamic constraints. In this case, the designer needs to have a feeling for the change in the function $\phi^i(x,p^i)$ when x changes. While the plot of the function $\phi^i(x,p^i)$ with respect to p^i for a fixed x gives information that can be easily grasped by a designer, the numerical data printed on a screen cannot be absorbed without a lengthy analysis.

610

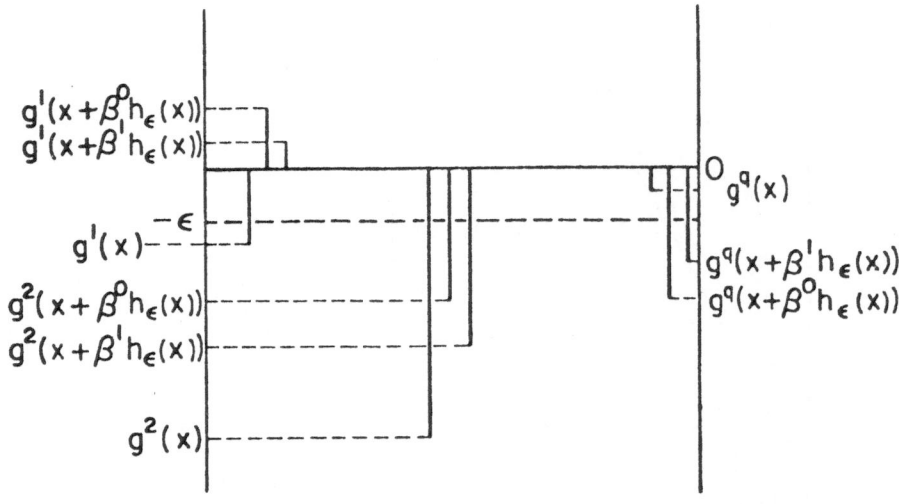

Figure 2.3   The Bar Chart of the Constraints

As noted before, the designer may need to perform additional
computations in order to gather relevant information to rescale
the problem.   It seems undesirable to load the main algorithm
with many side computations to cover all the possible needs of the
designer, partly because it is not possible to anticipate all
such needs and partly because the designer may come up with un-
foreseen tests that are particularly efficient for his problem.
Thus, an optimization based computer-aided design system should
permit improvised side computations on variables, vectors, and
matrices used in the optimization algorithm.   Consequently, the
system should incorporate a powerful scratchpad, capable of matrix
operations such as inversions, transpositions, and calculation of
condition numbers.

As a result, a prototype system has been designed with the
following criteria in mind:

(1) Ease of interaction should be emphasized.   A designer
should be able to interrupt the computing process, change the
parameter values, and restart the process.   Moreover he
should be able to control the flow of the algorithm by single
stepping through its loops.   (This feature is most useful in
diagnosing where the computation jammed up and what is the
probable cause of the jamming of the algorithm).
(2) Graphical display of quantities computed by the
optimization and simulation algorithms should be possible.

Color graphics should be used to enhance man-machine interaction.
(3) A powerful, high level, scratchpad for side computations on variables, vectors, and matrices used in the optimization algorithm should be available.

3. THE INTEROPTDYN SYSTEM

The INTEROPTDYN system is an experimental interactive software package for optimization based computer-aided design of dynamical systems, developed by the authors of this paper. The system is running on a DEC VAX 11/780 computer, granted by NSF for research on interactive computer-aided design of engineering systems. The operating system is a virtual memory version of UNIX (a Bell System trade mark) developed at the University of California, Berkeley. The system can be used to solve design problems of the form of Eq. 2.1, where the P^i's are intervals.
At present, the system consists of:

(1) A main program (OPTDYN) that is written in FORTRAN, implementing the Gonzaga-Polak-Trahan phase I-phase II method of feasible directions [9,11].
(2) The interpreter of an interactive language, INTRAC-C, that evolved from INTRAC, which is an elementary interactive language originally developed at the Department of Automatic Control, Lund Institute of Technology, Sweden [10]. The language INTRAC-C is written in FORTRAN and produces FORTRAN as intermediate code. The interactive language interpreted by INTRAC-C is referred to as the INTRAC-C language.
(3) A set of procedures (macros) that are written in the INTRAC-C language.

The heart of the system is INTRAC-C. It is an application specific extension of INTRAC, which is an elementary interaction language conceived in such a way that applications specific extensions are easy to construct. The INTRAC-C language has the following four sets of problem independent commands:

(1) The original INTRAC commands for assignment of variables, conditional and unconditional branching, looping, input, and output.
(2) Commands for interacting with the optimization package.
(3) Graphics commands.
(4) Scratchpad commands, i.e. powerful commands for algebraic manipulations of scalars, vectors, and matrices.

The INTRAC-C language allows the use of macros (procedures).
A macro is implemented in INTRAC-C as a text file. When a macro

is called by the user, INTRAC-C reads the file corresponding to the macro from mass storage and takes the appropriate action. The INTEROPTDYN system has a simple text editor to modify macros.

Variables in the INTRAC-C language can be local or global. Local variables are local to the macro level and are defined when they are first given a value in a read statement or in an assignment statement. Global variables are always accessible and may pass information between macros.

A feature that makes INTRAC-C particularly useful in interactive CAD is the possibility of suspending the execution of a macro by using the command SUSPEND. The execution of the macro can be resumed by using the command RESUME. When a macro is suspended, commands can be inputted from the terminal. A typical use of this feature would be the following: When executing a macro, the program may need some information from the user to perform effectively its task. Then the macro is suspended and a question-answering phase begins. In this phase, all the variables of the macro are accessible and the user can change values of variables that are local to the macro. When the interaction is ended, the macro is resumed and the computation progresses. In INTEROPTDYN, it is also possible to interrupt the execution of the macro externally, forcing it in suspended mode. This feature allows the user to abort an unsatisfactory run and to access parameters and variables in the program outside a fixed frame. Parameters and variables of the optimization algorithm that need to be changed interactively, or need to be accessible to INTRAC-C, must be deposited in the symbol table of INTEROPTDYN.

Among the application commands available in INTRAC-C, one finds the commands that handle the interaction with OPTDYN. The first step in developing these commands was to decide where interaction should take place. According to the considerations of Section 2, interaction should be implemented at each step of the main loop of the algorithm, as well as at each step of every internal loop. Thus, breakpoints have been inserted after the corresponding statement of OPTDYN. At each breakpoint, a subroutine INTCAL is called. This subroutine checks the condition associated with the break point. The condition may assume the following values:

(1) NEVER: In this case no action is taken and the control is returned to the main program.
(2) ALWAYS: In this case INTRAC-C is called and an interaction phase takes place.

The condition of a breakpoint can be changed by the HALT command of INTRAC-C. An INTRAC-C command has the general form:

The following notation is used in describing the syntax of commands:

(1) < > denotes that the enclosed term is not used literally, but is replaced by its appropriate value.
(2) { } groups terms together.
(3) [ ] groups terms together and denotes that the group is optional.

The HALT command has the following structure: HALT <breakpoint> <condition>, where <breakpoint> is the name of the breakpoint where the condition is to be set, and <condition> can be ALWAYS, NEVER, or an IF-clause followed by ALWAYS or NEVER.   The IF-clause is used to change the condition dynamically.   For example, the command HALT ARMIJO IF INTER>3 ALWAYS, sets the condition of the breakpoint ARMIJO to ALWAYS if the number of iterations in the Armijo step is larger than 3.

A number of other commands are available for the control of flow of INTEROPTDYN.   For the sake of brevity, they are not described here.

The variables in the symbol table of INTEROPTDYN can be changed by using the following commands:

(1) SET: The SET command has the following structure: SET<variable>=<argument>, where <argument> can be either <variable> or <number>.
(2) SETDIM: The SETDIM command changes the dimension of a variable in the symbol table.   Its syntax is SETDIM {ncol|nrow}(<variable>)=<argument>, where <ncol> and <nrow> are respectively the column and the row dimension of the variable.

The graphics commands of INTRAC-C can be grouped in the following two parts:

(1) Low level primitives for vector generation, initialization, terminal control, text output, positioning, and windowing.
(2) High level display functions.

All these commands can be executed on the following graphics interactive terminals: Tektronix 4027, Ramtek Micrographics, and HP 2648.   The first two terminals are color graphics terminals. About 15 low level primitives are available.   For the sake of brevity, only one of these is examined here; the VECTOR command.

It is used to draw a vector between two points.   Its syntax is:

VECTOR<x1><y1><x2><y2>

where, <x1> and <x2> are the x-coordinates of the two points and <y1> and <y2> are the y-coordinates of the two points.

Two high level display commands are available; a CURVE and a BAR command.   Both commands have the same syntax, which is

<command><array><ymin><ymax>[<topcolor>]

[<botcolor><threshold>]

where <array> contains the name of the array carrying the information to be displayed, <ymin> and <ymax> are used for the y-axis scaling, and the optional <topcolor> specifies the color to be used in the output if the second option is not used.   If the optional <botcolor> is given, then a numeric <threshold> must follow.   All entries in the <array> above the <threshold> will appear in the <topcolor>, while all entries below the <threshold> will appear in the <botcolor>.   The CURVE command plots all the entries in the <array>, while the BAR command produces a barchart. These commands implement, among others, the ideas discussed in Section 2 for using color graphics to plot indicators for the behavior of the optimization algorithm.

In addition to the main INTEROPTDYN symbol table, there is a second, or scratchpad symbol table, which is separate from the main symbol table.   This is used for the results of side computations.   In order to protect the main symbol table, scratchpad commands can access both symbol tables, but can only alter values in the scratchpad symbol table.   The most interesting commands of the scratchpad are:

(1) PDIM: This command creates arrays in the symbol table. Its syntax is: PDIM<array>[(<nrow>[:<ncol>])]<type>, where <array> is the name of the variable that is being created and <nrow> and <ncol> are the row dimension and the column dimension of the array, which can be given optionally.   If <nrow> and <ncol> are not given, then the variable being created is a scalar.   If <ncol> is not given then the variable being created is a column vector, and so on.   The symbol <type> indicates which type is to be attached to the variable.   The scratchpad set of commands accepts four types, namely, integer, real, double precision read, and complex.
(2) PMAT: This command is used to perform mathematical operations on arrays.   It takes the following two forms:

     (a) PMAT<array>={<array><number>}<op><array>, where
the first <array> is the name of the array in which the
result of the operation is stored, the second is the
first operand, the third is the second operand, and
<op> is one of the following matrix operations:
*(multiplication), +(addition), or -(subtraction).
     (b) PMAT<array>=<func><array>, where <func> can be:
INV(inversion), TRANS(transposition), TRACE(trace),
DET(determinant, or COND(condition number).
(3) PSCAL: This command is used to perform mathematical
operations on the scalars. Its syntax is similar to PMAT,
except for the <array> that is now replaced by <scalar>.
The operations available are the four basic operations and the
functions available are the functions allowed in FORTRAN.

This set of commands meets the specifications indicated in
section 2. For example, in step 2 of the algorithm, a quadratic
programming problem must be solved to find the search direction.
The Wolfe algorithm is used to solve this mathematical programming
problem. It is very important that the matrix of the linear
constraints be well conditioned for the algorithm to produce a
meaningful solution. If the search direction is not satisfactory,
the designer could check the conditioning of this matrix. To do
so, he would form a square matrix by multiplying the matrix by its
transpose, using the functions and the matrix operations provided
in the scratchpad set of commands (a transposition followed by
the multiplication of two matrices). Then he would compute the
condition number of this square matrix, using the COND function
provided in the PMAT command.

Finally, the last component of the INTEROPTDYN system is a
set of macros written in the INTRAC-C language, which have been
found to be sufficiently useful to warrant depositing them in the
library. These macros can be divided into the following three
groups:

(1) Macros that manage the execution of the optimization
algorithm.
(2) Macros that implement high level display functions.
(3) Macros that make the use of the scratchpad feature
easier.

The main macro of the first set is called RUN. This macro
enables the execution of a specified number of overall iterations
of the optimization algorithm. Its syntax is RUN<nitn>[<display>],
where <nitn> is an integer indicating the number of iterations one
wants to perform and <display> is the name of a macro that can be
coupled to RUN. This macro will be executed at each iteration of
the overall algorithm. The program will stop after the number of

iterations specified has been reached (of course, the program may stop before this number of iterations is completed if the optimal design is reached). When the program stops, the macro displays on the screen a set of questions indicating to the user possible changes of algorithm parameters before running more iterations. It is very useful to combine RUN with a display macro that prints or graphically displays the values of the cost and the constraints, while the computation is progressing. In fact, on the basis of the information displayed on the screen, the designer may decide to suspend execution of the macro RUN and to perform side computations or change the values of a few parameters, via the SET command.

Several macros are available that implement high level display functions. For example, the macro GRAPH is used to plot the values of an array. Its syntax is GRAPH<array><color><mark> [<indexl><index2>], where <array> is the name of the column vector whose entries are to be plotted, <color> is the name of the color to be used when plotting the array, and <mark> can assume either the value yes or no. In the first case, the points of the graph corresponding to entries of the array are marked with a small asterisk. The optional <indexl> and <index2> are used to plot only a subset of entries of the array, namely the ones between the elements with index equal to indexl and with index equal to index2. The macro computes the scale factors to fit the curve into a given window on the screen and it clips out the subarray defined by the indices. Thus, the option can be used to zoom in on a part of the graph that looks particularly interesting to the designer. Graphic macros are built hierarchically, so that macros at higher levels call macros at lower levels. GRAPH is a macro at an intermediate level. For example, PLTROW is a macro at high level that plots a specific row of an array and that calls GRAPH. Its syntax is

PLTROW<array(I:)><color>[<mark>]

where <array(I:)> is the name of a row to be plotted, <color> is the name of the color to be used to plot the array, and the optional <mark> indicates if the coordinate points forming the graph are to be marked. To give an example of how a macro is written in INTEROPTDYN, PLTROW is listed, as follows:

MACRO PLTROW H(I:) C; YESNO

The array is H(I:), a row vector extracted from a matrix H by picking up the I-th row, C is a local variable indicating the color of the plot, the ; separates the compulsory arguments from the optional ones, and YESNO is the argument that indicates if the plot has to be marked or not.

```
DEFAULT YESNO = N
```

This line of code indicates the default value of YESNO, which is no.

```
ROW oH = H(I:)
```

This line uses the macro ROW to create a new variable oH, which is a row vector.

```
TRS oHT = TRANS(oH)
```

This line uses the macro TRS to create a column vector that is obtained by transposing the row vector of the previous statement.

```
PREM oH
```

This line of code is needed to remove the variable oH, which has been created in the scratchpad symbol table by the previous command.

```
GRAPH oHT C YESNO
```

This line of code calls the macro GRAPH whose arguments are the column vector created in the previous statements, the color indicated in the arguments of the macro PLTROW, and the variable needed to determine if the plot has to be marked.

```
PREM oHT
```

Once oHT has been used, it should be removed from the symbol table of the scratchpad, so as not to waste memory.

```
END
```

This command indicates the end of the macro.

Macros are relatively easy to write, but they are inefficient, since the commands of a macro are interpreted every time a command is read, it is parsed and then executed. Therefore, macros involving loops take a long time to run. On the other hand, the commands are implemented by FORTRAN routines, are compiled, and are therefore much more efficient. However, they are more difficult to write than the macros.

The last set of macros available in INTEROPTDYN make the scratchpad commands easier to use. For example, the macro MM makes use of the matrix multiplication command much easier. When using the command PMAT A = B*C, one needs to create the variable A first, declaring its proper dimensions. The macro MM creates the variable with the right dimensions automatically. Its syntax is

MM <array> = <array> *<array>

where the first <array> indicates the name of the array in which the product will be stored and the second and third <array> indicate the name of the arrays to be multiplied. One need not declare the first <array> nor its dimensions.

4.   CONCLUSIONS

INTEROPTDYN is a prototype software package for optimization based, computer-aided design of engineering systems. Its main features are:

(1) Ease of interaction with the optimization algorithm that is implemented in the package.
(2) Extensive use of color graphics.
(3) A scratchpad subsystem used to perform side computations that are needed to monitor the behavior of the optimization algorithm and to improve it when unsatisfactory.

A number of powerful simulation packages are currently being incorporated, which can be called by INTEROPTDYN for function and derivative computation. As a result, a number of very powerful optimization based CAD packages are being obtained for use in different engineering fields. In parallel with this activity, a more advanced version of INTEROPTDYN is being developed, based on a new interactive language that should be more powerful than the INTRAC-C language.

REFERENCES

1.   Marcel, P.V., "Survey of General Purpose Programs", Proceedings, Second U.S.-Japan Seminar on Matrix Methods of Structural Analysis and Design, Berkeley, California, August 1972.
2.   ANSYS, Engineering Analysis Systems, User's Manual, Swanson Analysis Systems, Inc., Pennsylvania, U.S.A.
3.   Bathe, K.J., Wilson, E.L., and Peterson, F.E., SAP IV - A Structural Analysis Program for Static and Dynamic Response of Linear Systems, Report No. EERC 73-11, Earthquake Engineering Research Center, University of California, Berkeley, June 1973.

4. Mondkar, D.P. and Powell, G.H., ANSR-1, A general Purpose Program For Analysis of Nonlinear Structural Response, Report No. EERC 75-37, Earthquake Engineering Research Center, University of California, Berkeley, December 1975.

5. Schmidt, L.A. and Miura, H., Approximation Concepts for Efficient Structural Synthesis, NASA CR-2552, National Aeronautics and Space Administration, Washington, D.C., March 1976.

6. Haug, E.J. and Arora, J.S., Applied Optimal Design: Mechanical and Structural Systems, John Wiley & Sons, New York, 1979.

7. Gonzaga, C. and Polak, E., "On Constraint Dropping Schemes and Optimality Functions for a Class of Outer Approximations Algorithms", SIAM J. Control and Optimization, Vol. 17, pp. 477-493, July 1979.

8. Polak, E., "Algorithms for a Class of Computer-Aided Design Problems: A Review", Automatica, Vol. 15, pp. 531-538, 1979.

9. Gonzaga, C., Polak, E., and Trahan, R., An Improved Algorithm for a Class of Optimization Problems with Functional Inequality Constraints, Electronics Research Laboratory Memo No. UCB/ERL-M78/56, University of California, Berkeley, 1978.

10. Wieslander, J. and Elmqvist, H., INTRAC, A Communication Module for Interactive Programs, Department of Automatic Control Memo LUTFD2/(TFRT-3149)/1-060/1978, Lund Institute of Technology, Lund, Sweden, Aug. 1978.

11. Bhatti, M.A., Polak, E., and Pister, K.S., OPTDYN- A General Purpose Optimization Program for Problems with or without Dynamic Constraints, Report No. UCB/EERC-79/16, Earthquake Engineering Research Center, University of California, Berkeley, July 1979.

APPLICATIONS OF OPTIMAL DESIGN TO STRUCTURES SUBJECTED TO EARTHQUAKE LOADING*

M. Asghar Batti and Karl S. Pister

Department of Civil Engineering, University of California, Berkeley, California, U.S.A.

ABSTRACT

This paper presents three typical applications of optimization techniques to the design of structures that are subjected to earthquake loading: (i) minimum weight design, (ii) minimum lifetime cost design, and (iii) design for minimum (or maximum) structural response. Formulation of these problems is briefly described and some numerical results are presented.

1. INTRODUCTION

There has been considerable interest in the application of optimization techniques to the design of structures that are subjected to dynamic loading, in particular to earthquake loading. Pierson [1] and Rao [2] have presented literature surveys in the area. The purpose of this paper is to give some representative applications of mathematical programming techniques in the area of earthquake resistant design.

Most problems treated in the literature fall into one of the following categories:

---

*This research was conducted under National Science Foundation Grant ENV76-04264.

(1) <u>Minimum Weight Design</u>: The lightweight objective is to design the structure so that its weight is minimized. This seems to be the direct extension of ideas used in the aerospace industry, where weight is a crucial factor. Typical constraints include bounds on stresses, displacements and frequencies. Applications presented by Ray, Pister, and Chopra [3], Walker and Pister [4], Bertero and Kamil [5] and Zagajeski and Bertero [6] fall into this category.

(2) <u>Minimum Cost Design</u>: For civil engineering structures, weight in most cases is not an important consideration. A more important criterion is the cost itself, including the cost of fabrication, maintenance, repair, etc. Walker [7] has formulated and solved the problem of minimum cost design of multistory steel frames that are subjected to earthquakes.

(3) <u>Design for Minimum (or Maximum) Structural Response</u>: In certain cases, it might make more sense to minimize some structural response quantity directly. As an example, consider design of an earthquake isolation system for multistory buildings. Since the purpose of an isolation system is to minimize forces in the building, it will be logical to take the story shears as the objective function. Other examples include designing structures for maximum energy absorption, minimum accelerations in the equipment in buildings, etc. Bhatti [8] has designed an earthquake isolation system in which story shears are minimized, while Haug and Arora [10] have presented an example of design of a nonlinear impact isolation system in which the objective is to minimize the acceleration of the mass.

In this paper, three examples representing each of the above class of design problems are presented. Only a short description of the problems is given here, but original references are cited and the reader is referred to these for more details.

2. MINIMUM WEIGHT DESIGN OF A MULTISTORY STEEL SHEAR FRAME

A minimum weight multistory steel shear frame was presented by Ray, Pister, and Chopra [3]. The structural model is shown in Fig. 2.1. The floor diaphragms are rigid and axial deformations are neglected. Thus, the system has only one degree of freedom (in the lateral direction) at each floor. The design variables are the column moment of inertias. The objective function is the weight of the structure, which is expressed as

$$W = 2\gamma \sum_{i=1}^{N} A_i L_i \qquad (2.1)$$

where

622

γ = Density of steel.

N = Number of stories.

A_i = Area of column at the ith story.

L_i = Height of the ith story.

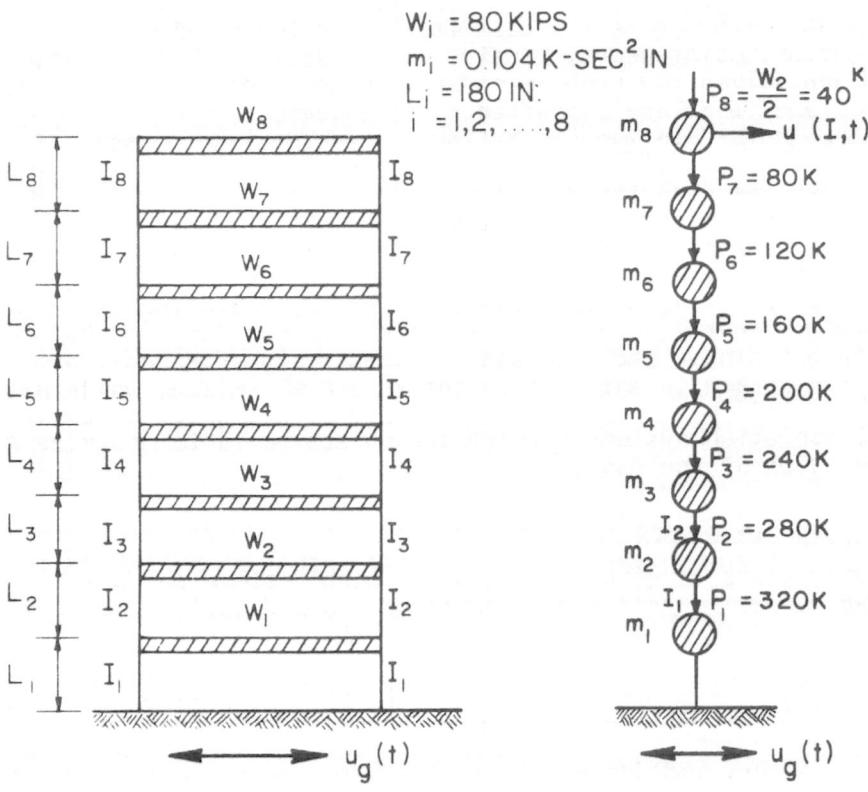

Figure 2.1   An Eight Story, Single Bay Shear Frame

For AISC standard wide flange sections, section properties can be expressed in terms of moment of inertias as

$$A = 0.8I^{1/2} \tag{2.2}$$

$$S = 0.78I^{3/4} \tag{2.3}$$

where, S is the Section modulus.   Using Eq. 2.2 in Eq. 2.1 and assuming equal story heights, one has

$$W(I) = 1.6\gamma L \sum_{i=1}^{N} I_i^{1/2} \tag{2.4}$$

Constraints are imposed on displacements and stresses.   To prevent damage to non-structural elements, constraints are imposed on relative story drift.   Relative story drift, denoted as $\delta_i(I,t)$ for the ith story, is given by

$$\delta_i(I,t) = u_i(I,t) - u_{i-1}(I,t)$$

$$\delta_1(I,t) = u_1(I,t) \tag{2.5}$$

where $u_i(I,t)$ is the lateral displacement at the ith story at time t.   Now, let $\delta_{maxi}(I)$ be the maximum over time of the relative story drift at the ith story.   Then, the constraints can be expressed as

$$[\delta_{maxi}(I)]^2 \leqq [\delta_{ai}]^2 \tag{2.6}$$

where, δ_{ai} is the set of prescribed allowable story drifts.   The stresses $\sigma_i(I,t)$, in shear frames, can be evaluated from

$$\sigma_i(I,t) = \frac{M_i(I,t)}{S_i(I)} + \frac{P_i}{A_i(I)}$$

where $\sigma_i(I,t)$ is the stress in the column at the ith floor at time t, $M_i(I,t)$ is the bending moment at the end of corresponding column, and P_i is corresponding axial force.   The maximum bending moment at the end of a column at i^{th} story, neglecting second order effects of P_i, can be expressed as

$$M_{maxi}(I) = 1/4[a_i L_i]\delta_{maxi} I_i$$

where $a_i = 24E/L_i^3$ and E is Young's modulus.   Thus, the maximum stress, $\sigma_{maxi}(I)$, can be expressed as

$$\sigma_{maxi}(I) = 1/4[a_i L_i / S_i(I)]\delta_{maxi} I_i + P_i / A_i(I) \qquad (2.7)$$

Then, the stress constraints can be written as

$$[\sigma_{maxi}(I)]^2 \leqq [\sigma_{ai}]^2 \qquad (2.8)$$

where $\{\sigma_{ai}\}$ is the set of prescribed allowable stresses.

In order to evaluate constraint equations 2.6 and 2.8, maximum relative story drifts are needed, which in turn require story displacements. The story displacements are computed by solving the equations of motion for the system, using a response spectrum technique. The earthquake input is expressed as the piecewise continuously differentiable, pseudo-velocity spectrum

$$V(T) = 14.903T - 11.705T^2 + 3.235T^3 \quad T < 1.5$$

$$= 6.938 + 0.875(T-11.5) \qquad T \geq 1.5$$

where $V(T)$ is spectrum ordinate in inch/sec and T is the time period of vibration in seconds. The design pseudo-velocity spectrum used is $(1/4)(2.7)V(T)$. References 3 and 9 may be consulted for further details of the response spectrum technique used.

The design problem can now be summarized as

$$\min_I W(I)$$

such that

$$[\sigma_{maxi}(I)]^2 \leqq [\sigma_{ai}]^2$$

$$[\delta_{maxi}(I)]^2 \leqq [\delta_{ai}]^2$$

$$-I_i \leqq 0$$

The problem is solved using a method of feasible directions, see Ref. 3 for details.

For each story i, the allowable maximum relative story drift δ_{ai} is chosen according to the following empirical rule:

$$\delta_{ai} = 0.0025L_i$$

Thus, for the example, δ_{ai} = 0.45 inches. The allowable maximum stress σ_{ai} in the column at floor i is taken so as to provide a factor of safety against yielding of 1.5. Thus the prescribed σ_{ai} is 24 ksi.

The initial and optimum design characteristics are compared in Table 2.1. Reduction in the cost function with the number of iterations is plotted in Figure 2.2. As expected, earthquake loading requires that the optimum elastic structure must be the most flexible structure in the feasible domain, when the objective function is directly proportional to the column stiffnesses. Figure 2.2 brings out one of the major advantages of the feasible directions method used here, which is the fact that since each iteration is associated with non-increasing cost, termination of the method at any iteration will always ensure an improvement over the design based on the preceding one.

3. MINIMUM LIFE-TIME COST DESIGN OF A MULTISTORY STEEL FRAME

The problem addressed here is selection of member sizes for a single-bay, multistory, unbraced steel frame with fully rigid connections. Uniformly distributed beam loads and earthquake-generated horizontal ground motion are considered. A typical frame is shown in Figure 3.1. The frame is symmetric about its vertical mid-plane, and members are to be selected from the set of A-36 rolled steel wide flange economy sections. Performance constraints for operating loads are introduced through typical code requirements, while for earthquake loads a dual criteria based on selection of moderate and strong design earthquakes is adopted. Minimization of lifetime cost of the structure, which is composed of initial (construction) cost and the cost of earthquake induced damage over its lifetime, is taken as the design objective. The following sections briefly sketch a methodology for formulating the problem and illustrate the method with an example. More detailed information can be found in Ref. 7.

3.1 Design Objective

To compare alternative choices of a given structure, a design objective must be chosen. Here, minimum lifetime cost (LC) is chosen as the objective. Only costs that are strongly related to the design variables are calculated. Costs that are relatively independent of the design vector merely add a constant to

TABLE 2.1   COMPARISON OF INITIAL AND OPTIMUM DESIGN CHARACTETISTICS FOR 8-STORY FRAME

Story	Col. Moment of Inertia (in^4)		Rel. Story Drift (in)		Col. Max. Stress (ksi)	
	Initial	Optimum	Initial	Optimum	Initial	Optimum
1	1600	673	0.14	.23	16	24*
2	1600	569	0.14	.26	15	24*
3	1000	463	0.20	.30	18	24*
4	1000	364	0.18	.35	15	24*
5	700	274	0.23	.41	16	24*
6	700	219	0.20	0.45*	13	22
7	400	174	0.27	0.45*	14	19
8	400	115	0.17	0.45*	8	15

*Active Constraints

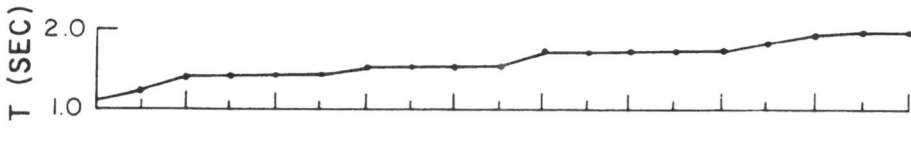

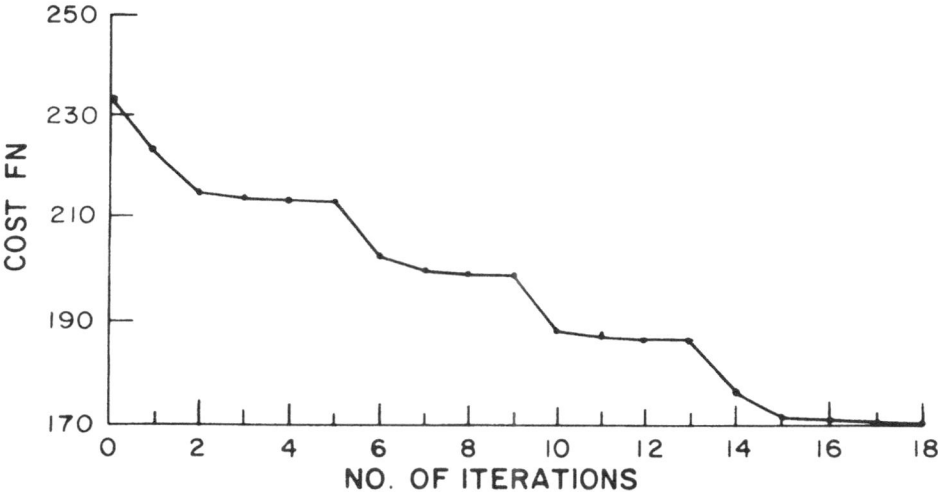

Figure 2.2 Cost Function and Natural Time Period Versus Number
of Iterations

the cost, producing no effect on the outcome of the design process.
Obviously, care must be exercised in selecting design variables
that are compatible with the design objective (cost). Here,
moment of inertia of the member cross-section is selected for
this purpose.

The LC associated with multistory framed buildings separates
into two categories: (1) the cost of construction and (2) the
cost of damage associated with structural overload, assumed here
to result from earthquake exposure.

Construction Costs. Design vector dependent construction
costs include costs of members, beam-column connections (including
welding), transportation, size extra charges, painting, etc. The
form of these cost functions will be indicated here; for details
see Ref. 7. If C_s denotes the unit cost of steel, the total

frame cost can be written as

628

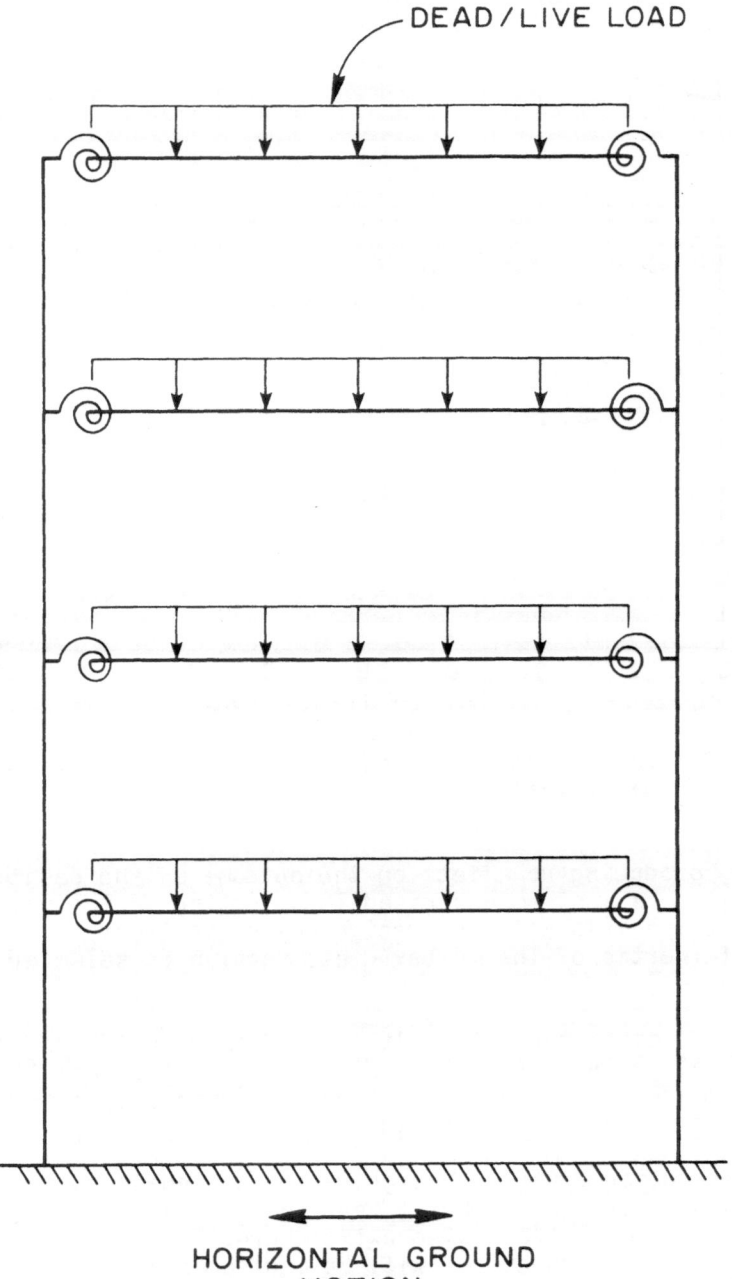

Figure 3.1   Typical, Single Bay Frame for Minimum Cost Design

$$\text{Total Cost} = C_s \; \gamma \sum_{m_i} A_i \; L_i \tag{3.1}$$

where A_i and L_i are the cross-sectional area and length of each member, γ is the unit weight of steel, and summation is taken over all frame members. To account for cost of connections, welding, and other member-related charges, it is possible to develop empirical equations that relate costs to section properties. Thus, the total construction cost C_c can be expressed in the form

$$C_c = \gamma \sum_{m_i} [C_s A_i + C_s' f_s(A_i)] \; + \; \gamma \sum_{g_i} [C_c' f_c(I_i) + C_w f_w(I_i)] \tag{3.2}$$

In Eq. 3.2, the following definitions have been introduced:

C_s' = unit cost of additional charge for members (transportation, etc.)

C_c' = unit cost of connection steel

C_w = unit cost of welding steel.

The functions f_s, f_c, and f_w can be determined by curve-fitting [7]. The symbols m_i and g_i in the summations denote "all members" and "all girders", respectively.

Damage Costs. To develop a model for cost of damage resulting from earthquake-induced overload, it is necessary to relate damage to structural response parameters and to identify an expected earthquake exposure hazard for the building lifetime.

The damage costs can be divided into three categories: (1) structural damage, (2) non-structural damage, and (3) down-time costs. The definition of structural damage is elusive. Fortunately, for steel framed buildings structural, as opposed to non-structural, damage is relatively unimportant, assuming the design prevents collapse of the structure. A suggested model, based on restoration of member ductility, can be found in Ref. 7.

Included in the category of non-structural damage are items such as interior and exterior walls, partitions, glazing, plumbing, electrical fixtures, etc. Taken collectively, the cost of damage for these items is much more significant than structural damage in steel framed buildings. From the above list, the principal contributions are from interior drywalls, glazing and masonry, if present. There is evidence to support the choice of story drift as an appropriate measure of non-structural damage. Utilizing available damage data, the damage ratio D_n, defined as the cost of damage repair divided by the cost of construction of the damaged items, can be expressed as [7]

$$D_n = 8.52 \, \delta \qquad (3.3)$$

where δ is the story drift in feet. Using Eq. 3.3 to compute the non-structural damage ratio D_n, the cost of damage per story can be developed. The total cost of non-structural damage is then obtained by summing over all the floors.

Repair of non-structural damage frequently requires temporary shutdown or relocation of activities, with resulting costs and revenue losses that affect the LC. In order to estimate this type of cost, some assessment of the susceptibility of the function of a building to such inconvenience cost must be made. As a first order approximation, it is assumed here that these costs are directly proportional to damage costs.

Life-Time Cost. The damage cost models developed apply to individual earthquakes. To obtain lifetime cost, it is necessary to make assumptions about the intensity and frequency distributions of earthquakes for the particular site and sum the damage costs over all expected earthquakes to obtain a lifetime exposure profile. This is accomplished by developing a model for the annual frequency of earthquakes that have a given peak acceleration at a site, utilizing a linear relation between log frequency and magnitude in connection with Housner "affected area" curves [7] for a fixed fault direction. A least squares fit of the resulting simulation gives

$$n = 3.44e^{-15.25a} \qquad (3.4)$$

where a is the acceleration normalized by gravity. The constants reflect seismicity appropriate to a Southern California site. The lifetime cost of non-structural damage per story can then be written as

$$C_{ns} = \int_0^{a_{max}} NnD_n \, da = \int_0^{a_{max}} d_t \, da \qquad (3.5)$$

where n is determined by Eq. 3.4, N is the structural life in years, D_n is obtained from Eq. 3.3, and $d_t = NnD_n$ is called the "lifetime damage profile". For multistory frames, story drifts at each floor level can be found from appropriate dynamic analysis, e.g., employing modal analysis and maximum modal response estimates for the assumed response spectra. Equation 3.5 is then evaluated at each story and the total damage cost is obtained by summation. This result, together with Eq. 3.2, provides the design objective in terms of a lifetime cost.

3.2 Performance Constraints

Design limitations are typically imposed via building codes. Here, building code constraints under operating loads are treated in the usual manner. However, criteria for earthquake loading follows a different course. Only the general outline of the constraint formulation scheme will be given. Details may be found in Ref. 7.

Constraints Under Operating Loads. Maximum moments in members are required to satisfy the condition

$$|M| \leq CM_p \tag{3.6}$$

where M is the moment under operating loads, M_p is the plastic moment, and C is a reduction coefficient, typically approximately equal to 0.6. For columns, M_p must be modified to reflect axial loading. Limitations on maximum beam deflection are also incorporated.

Because of lateral strength requirements of earthquake resistant frames, it is assumed here that sidesway stability requirements will not play a prominent role in the design process. This requirement, along with any other limitations thought to be necessary, can be easily incorporated.

Constraints Under Dynamic Loading. For response to earthquake loads, the following dual design criteria is adopted:

(i) The structure should respond elastically to a moderate earthquake of an intensity reasonably anticipated within its lifetime.
(ii) During a maximum credible (strong) earthquake, the structure may yield significantly but must avoid collapse.

Design earthquakes that are representative of the above conditions are typically selected on the basis of their probability of occurrence. A sample probability of occurrence curve is shown in Fig. 3.2, which is generated on the basis of a 50-year life expectancy for a Southern California site. Moderate earthquakes are chosen with a 50-80 percent probability of occurrence in mind, whereas strong earthquakes are picked in the 5-10 percent probability of occurrence range. Both are selected on the basis of a 50-70 year building life expectancy. Thus, two peak ground acceleration values, referred to as design earthquakes, are chosen to represent a moderate and strong earthquake. This is consistent with analysis procedures that employ response spectra. In the specification of dynamic constraints, dead/live load effects on the beams are accommodated, in addition to those resulting from the earthquake. No reduction of the live load from that specified for the static operating constrains is introduced. For a moderate earthquake, the structure is to respond elastically, hence, the maximum member moments must be less than each corresponding member yield moment, M_y.

In general, the same form of constraints on maximum beam and column moments carries over to the case of moderate earthquake, i.e., constraints have the form of Eq. 3.6, where M is now obtained by combining the separate effects of operating loads and earthquake loading.

The strong earthquake design criterion requires avoidance of structural collapse. The strong column-weak girder design is adopted, which in terms of ductility ratio, (defined as maximum total end rotation of a member divided by its elastic limit end rotation) means that the ductility demands of each member must be less than some specified allowable limit, which for columns is close to unity. Let M_T represent the total maximum moment,

i.e., the sum of the static and dynamic moments, in a particular member. Then strong earthquake constraints can be written

$$M_T \leqq \mu M_p \qquad\qquad (3.7)$$

where μ is the allowable ductility. As can be seen, this equation is identical in form to Eq. 3.6, with $c = \mu$. Hence all of the constraint developments for the moderate earthquake apply to the strong earthquake, with c equal to the allowable dultility in each member.

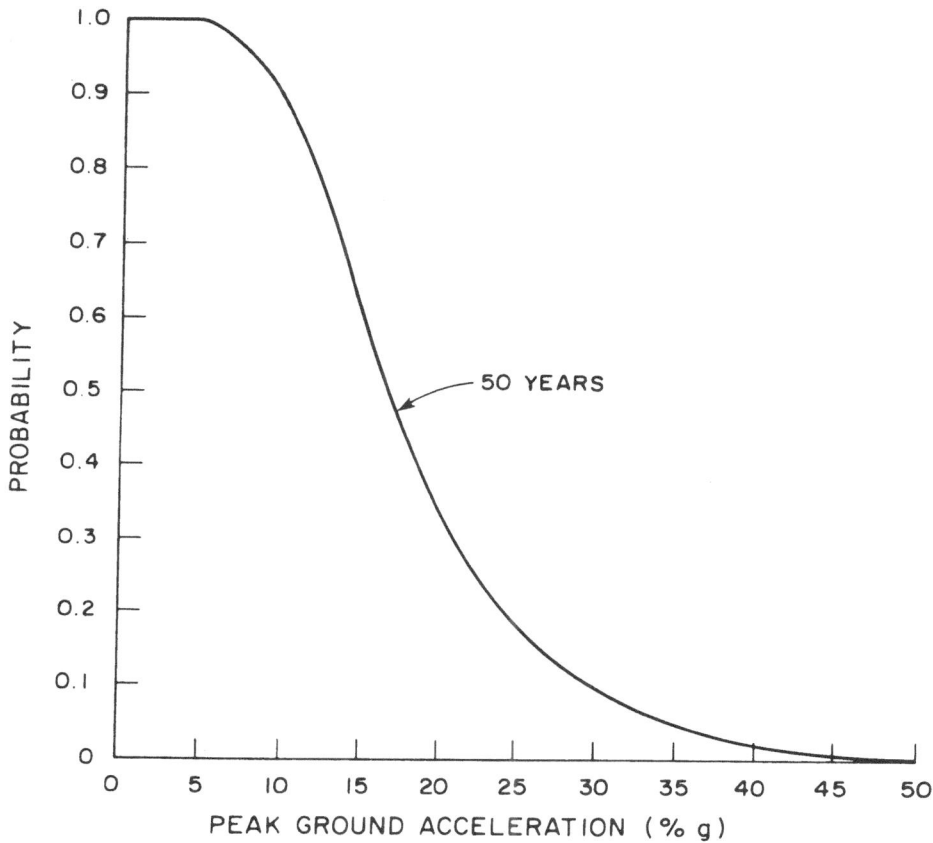

Figure 3.2   Probability of Occurrence of Peak Acceleration

3.3   Numerical Results

The methodology is now applied to the design of the four
story frame shown in Fig. 3.3.   The distributed loading is
45 kip for the roof beam and 72 kip for the floor beams.
Moderate and strong earthquakes are taken to have 0.12g and 0.35g
peak ground accelerations (corresponding to 80 percent and 5
percent probabilities of occurrence (Fig. 3.2), respectively.
In Eq. 3.6, reduction coefficients C are given values of 0.60 for
operating loading and 0.85 for dynamic loading associated with the
moderate earthquake.   A deflection of one inch is permitted at
the center of the beam span, under operating loads.   For the
strong design earthquake, ductility factors μ are selected as 1

for second and third story columns, 2 for first and fourth story columns, and 6 for beams.   Typical construction cost rates for California are assumed, along with allowances of 10 percent of construction cost for overhead and profit and 10 percent of total damage cost for down-time costs.   Structural damage is not accounted for, on the basis of earlier computational experience [7].   The lifetime damage profile is computed by following the procedure outlined previously and is shown in Fig. 3.4.

A three phase design algorithm is used to solve the problem. In phase 1, an unconstrained optimum design is obtained, starting from an arbitrary design point.   A coordinate descent algorithm is used, in which an unconstrained optimum is sought by searching, in turn, in each coordinate direction.   The task of phase 2 is to adjust the initial design to satisfy the violated constraints. A modified Newton method is used to solve the nonlinear constraint equations and a design that is close to the one obtained in phase 1 and also satisfies all the constraints is chosen. Phase 3 is the constrained optimal design phase.   A gradient projection algorithm is used for this purpose.   See Ref. 7 for details.   The performance of this design process is shown in Fig. 3.5.   The following optimal design is obtained:

$$X = (670, 660, 2377, 641, 1894, 710, 2003, 1039) \qquad (3.8)$$

where compact sections and fully rigid connections are assumed. The ductility demands on the optimum structure for the strong earthquake are (top story down):

Beams   = 0.69 0.66 0.99 1.12

Columns = 0.58 0.88 1.00 0.98

Note that the first story column ductility demand is only 0.98, even though the allowable is 2.   This is a good indication of the strong influence of the cost function on the columns.   The computed lifetime cost for the optimum frame is 4465.75, with a construction cost of 2158.33.

A comparison between the above minimum LC frame and the minimum construction cost frame could prove to be of interest. Hence, the minimum construction cost frame is computed and found to be

$$X = (318, 245, 818, 480, 1079, 587, 1204, 516) \qquad (3.9)$$

This structure is substantially more flexible than that of Eq. 3.8.   This increased flexibility is clearly reflected in the strong earthquake ductility demands, which are now

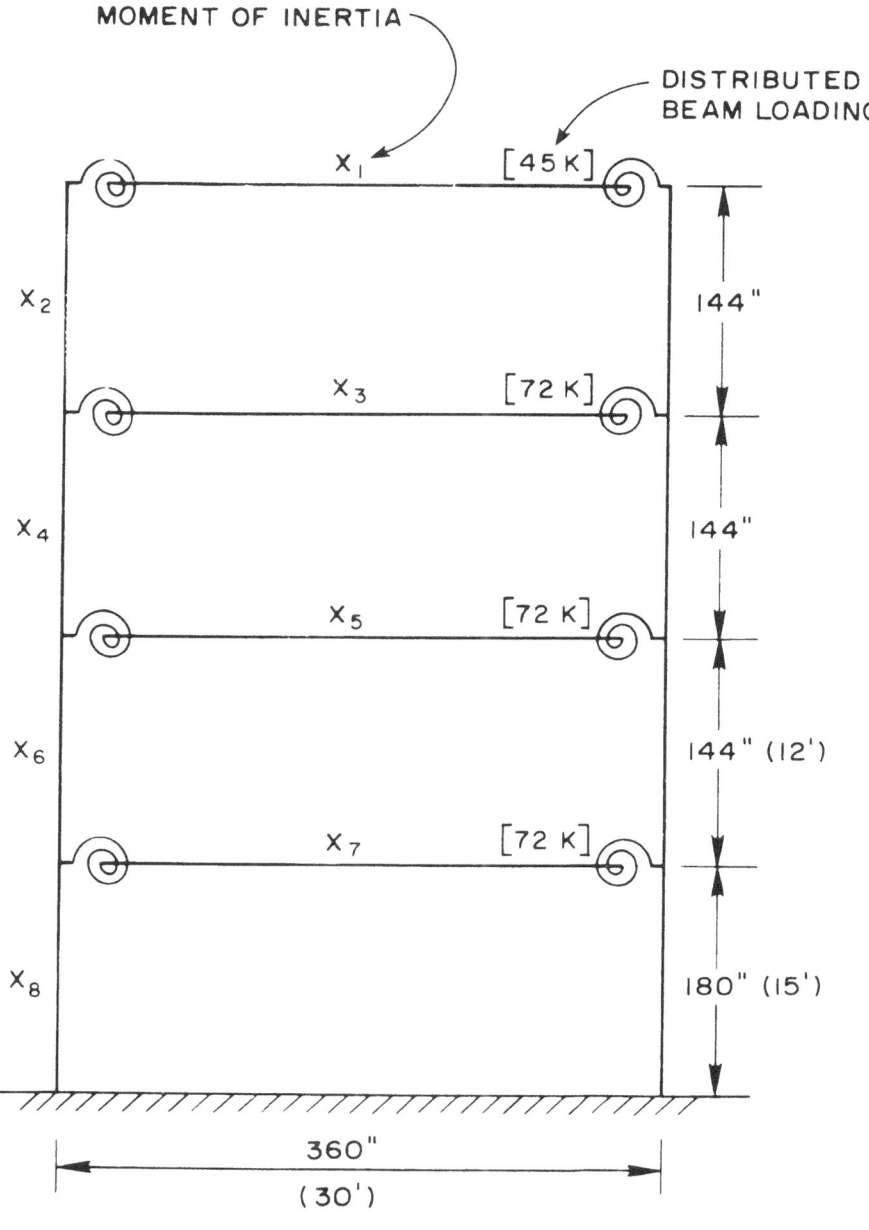

FRAME SPACING: 288" (24')

Figure 3.3   Four Story Frame For Minimum Cost Design Problem

636

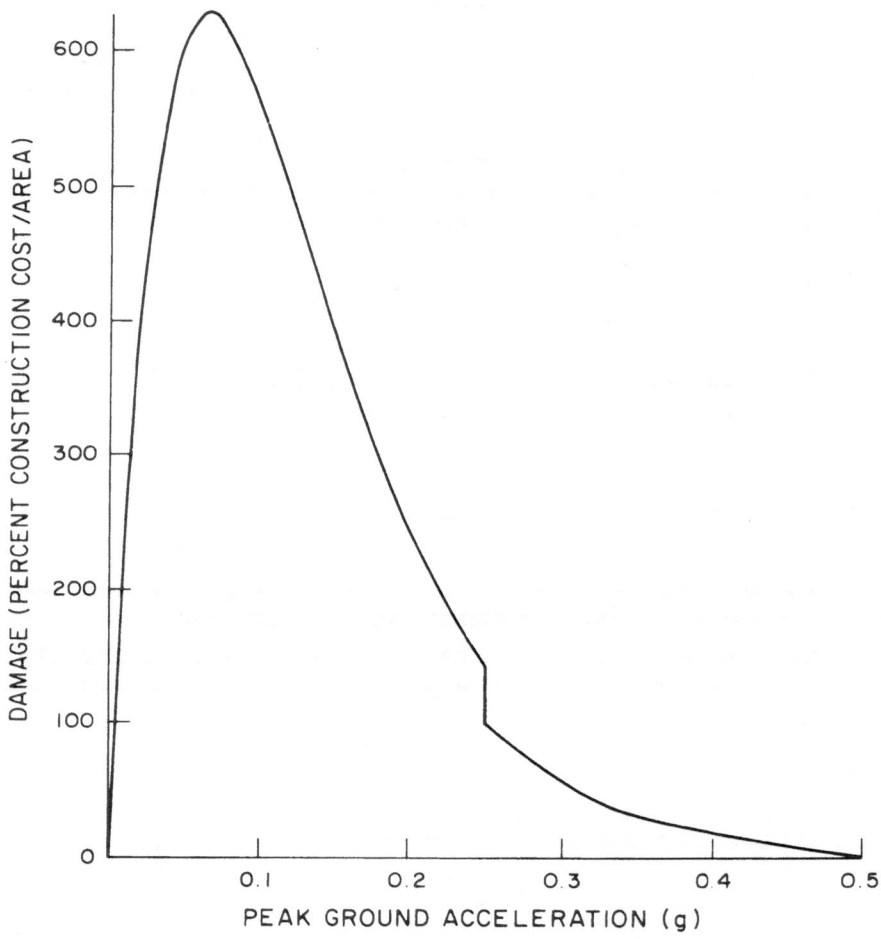

Figure 3.4   Life-Time Damage Profile for Four Story Example

Beams   = 1.10 1.20 1.27 1.33

Columns = 1.10 1.00 1.00 1.59

and show a significant increase over the minimum LC frame demands.
The construction cost for this frame is 1554.71, which represents
a 28 percent reduction from the minimum LC frame construction
cost.   The LC is now 4905.46, however, which represents a 10 percent increase in LC over that of Eq. 3.8.

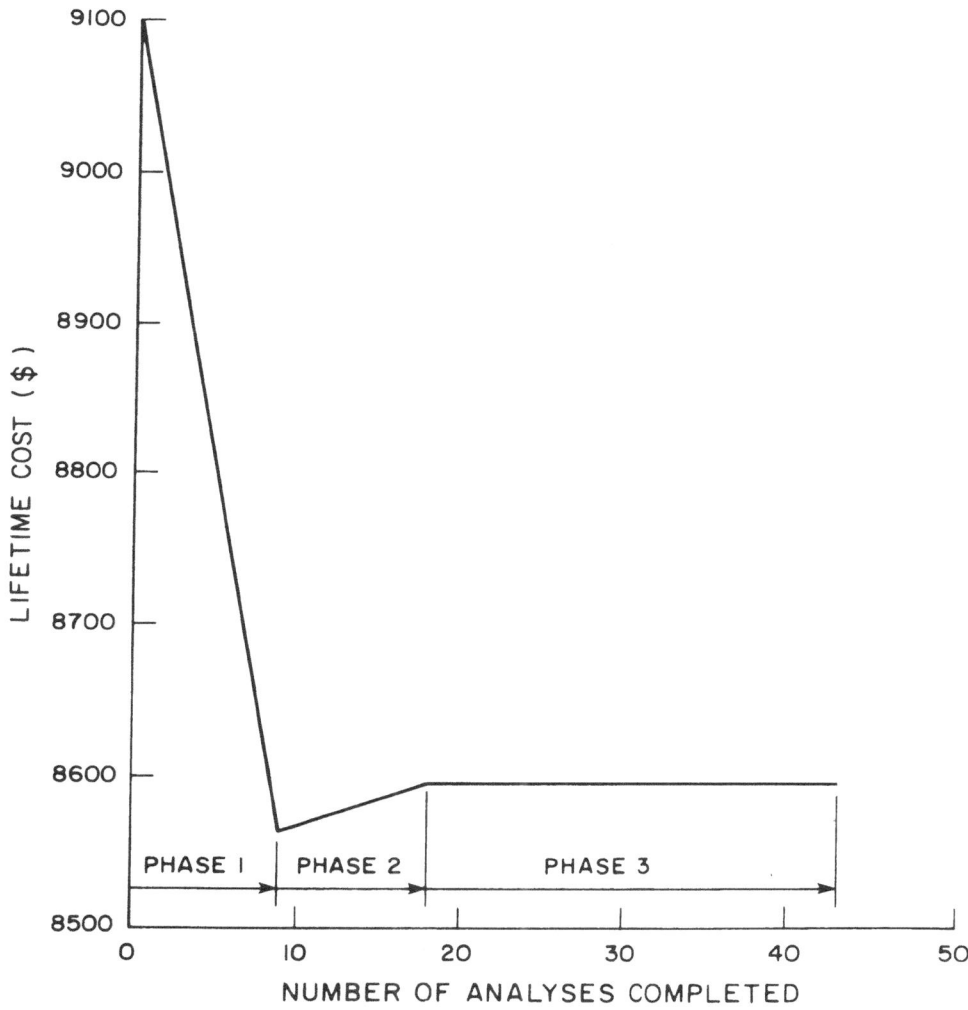

Figure 3.5   Algorithm Performance

The designs of Eqs. 3.8 and 3.9 both represent acceptable design practice, they are strikingly different, however, and reflect a very real difference in design philosophy.   This choice of design philosophy is clearly one that should be addressed by every prospective building sponsor, prior to design formulation.

4. DESIGN OF AN EARTHQUAKE ISOLATION SYSTEM

Optimal design of an earthquake isolation system for a three story steel frame, tested on the Earthquake Simulator at the Earthquake Engineering Research Center, University of California, Berkeley, is considered here.   The isolation system consists of rubber bearings and a mild steel energy absorbing device.   The details of the isolation system and a description of the test configuration, along with the test results, can be found in Ref. 11.   The tests show that for small earthquakes the structure behaves as if attached to a rigid foundation, while for strong earthquakes the foundation isolation system yields and absorbs amounts of energy equivalent to as much as 35 percent of critical viscous damping.   The question naturally arises, what is the "best" choice of the isolation system for a particular structure and site earthquake hazard?   This section formulates the design of energy-absorbing device as a constrained optimization problem, whose solution is obtained by utilizing a method of feasible directions [8].

4.1 The Test Structure

The test structure is a three story steel frame with added mass at each floor level, as shown in Fig. 4.1.   The structure is supported vertically by specially designed rubber bearings, whose properties are specified.   The bearings also provide nominal shear resistance.   An energy-absorbing device is linked to the base of the structure as shown.   This device acts as an hysteretic passive controller, supplying a time-dependent horizontal force to the base.   Selection of the mechanical properties of the controller constitutes the design problem.

4.2 Hysteretic Behavior of the Energy-Absorbing Device

As described in Refs. 8 and 11, the energy-absorbing device dissipates energy through hysteresis associated with cyclic inelastic torsion of a mild steel bar.   Based on work reported in Ref. 12, it has been found that an appropriate model for the hysteretic behavior of the device can be expressed by the equations

$$\dot{F}(t) = K_0\left[\dot{\delta}(t) - |\dot{\delta}(t)|\left[\frac{F(t)}{F_0} - S\right]^n\right] \tag{4.1}$$

$$S(t) = \alpha\left[\frac{\delta(t)}{\delta_0} - \frac{F(t)}{F_0}\right] \tag{4.2}$$

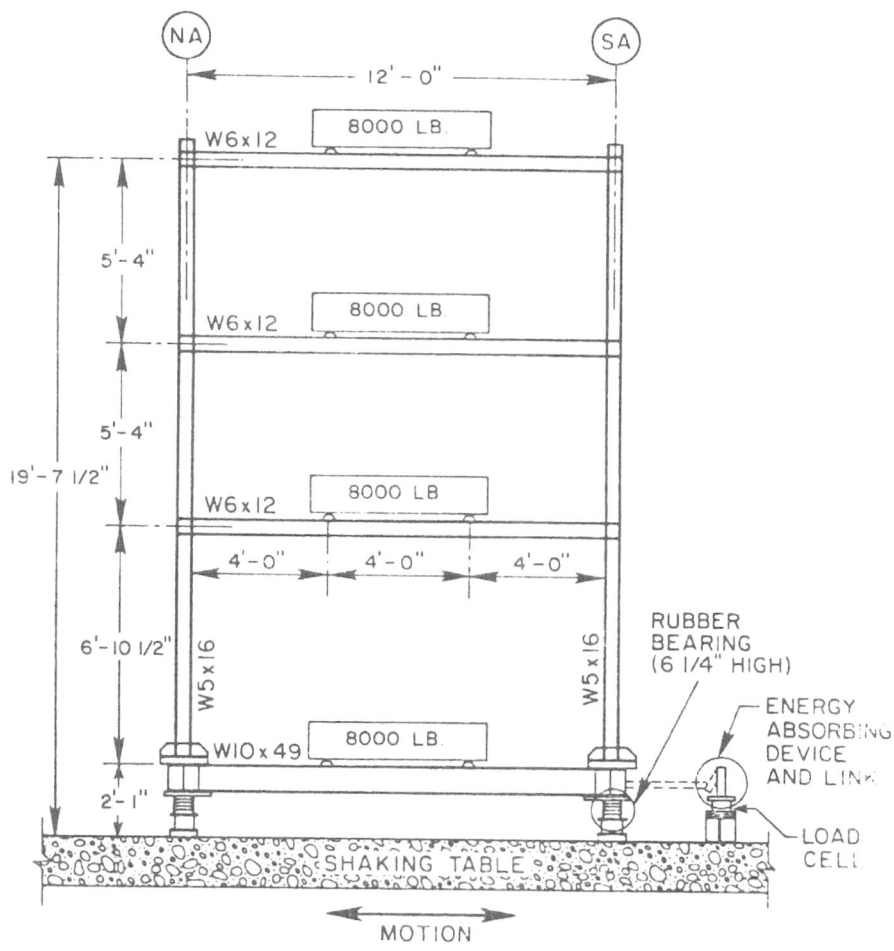

Figure 4.1   Test Structure with the Isolation System

where

F(t) = force in the energy-absorbing device

Ḟ(t) = time derivative of force in the energy-absorbing device

δ(t) = deformation of the energy-absorbing device

δ̇(t) = deformation rate of the energy-absorbing device

$$K_0 = F_0/\delta_0$$

F_0 = yield force

δ_0 = yield deformation

α = a constant that controls the slope after yielding,
$$K_y \approx K_0 \frac{\alpha}{1+\alpha}$$

n = a material parameter, taken as an odd integer that controls the sharpness of transition from the elastic to the inelastic region. As $n \to \infty$ the model approaches a bilinear model.

The parameters F_0, δ_0, α, and n are chosen so that predicted response from the model closely matches experimental response. Typical loops that are generated by this model, under sinusoidally varying deformation in time, are shown in Fig. 4.2.

4.3 Design Parameters

It is assumed here that the characteristics of the rubber bearings are fixed. Therefore, only the parameters of the energy-absorbing device are adjusted to obtain the optimum design. The two basic variables in design of energy absorbers are the elastic stiffness and the post-yield stiffness. These variables are controlled by the parameters F_0, δ_0, and α in the hysteretic model of the energy absorbers. The elastic stiffness is approximately equal to F_0/δ_0 and the post-yield stiffness is approximately equal to $\frac{F_0}{\delta_0}(\frac{\alpha}{1+\alpha})$. Thus, the design variables are F_0, δ_0, and α. Another parameter that may influence the design is the exponent n in the hysteretic model, but it is not considered as a variable in the present study.

4.4 Design Considerations

The purpose of an earthquake isolation system is to minimize some measure of response of the structure. There is a number of response quantities that could be minimized, e.g., the maximum acceleration in the structure, maximum base shear, maximum story shear, maximum interstory drift, etc. In order to get

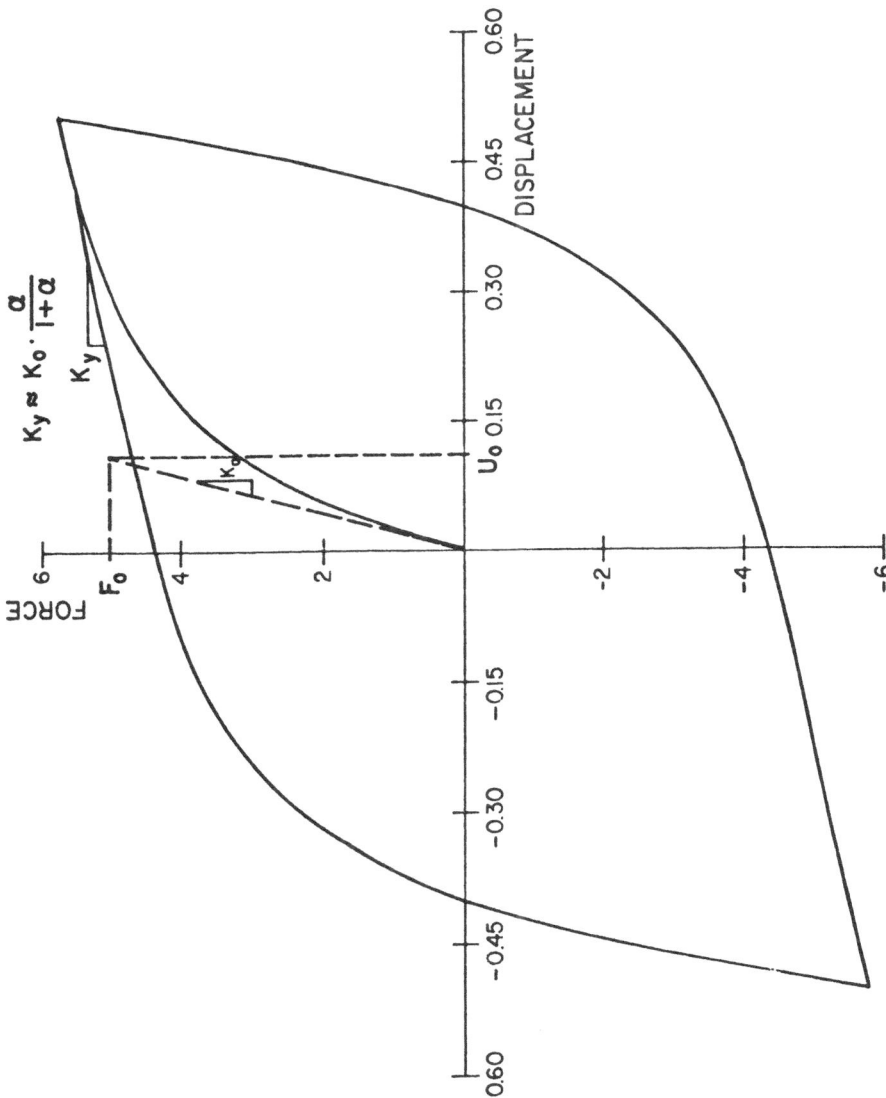

Figure 4.2   Hysteresis Loops Generated Under Sinusoidal Excitation

meaningful results, response constraints are also needed. Some of the constraints are dictated by the problem itself, e.g., the design parameters F_0, δ_0, and α are not allowed to attain

negative values. Such constraints constitute conventional inequality constraints. Constraints on the response quantities are also needed; e.g., when accelerations in the frame are minimized, the displacement at the base must not be arbitrarily large. These restrictions give rise to functional inequality constraints. The dual design criteria for earthquake resistant systems, mentioned in section 3.2, becomes even more critical in case of structures with an isolation system. This is so because the isolation system designed only for a large earthquake is very flexible and undergoes large displacements when the system is subjected to moderate earthquake and wind excitations. Thus, performance constraints on the system, under both a moderate and a strong earthquake, are needed.

In the following sections, an optimal design problem is formulated, based upon the above considerations, and numerical results are presented.

4.5 Optimal Design Problem

It has been observed that inter-story drift is one of the important parameters that controls damage to the structural and non-structural components. Since story shears are proportional to the inter-story drift, the objective function for this problem is chosen as the sum of squares of story shears, due to a large earthquake. Constraints are placed on the bottom floor displacement, due to both a large and a small earthquake, and on the force in the energy-absorber when it is subjected to a small earthquake. Mathematically, the problem can be expressed as

$$\min_{z} \max_{t \in T} \left[\sum_{j=1}^{3} \left\{ K_j [u_j^{\ell}(z,t) - u_{j+1}^{\ell}(z,t)] \right\}^2 \right]$$

subject to

$$\max_{t \in T} [u_4^{\ell}(t)]^3 \leq \delta^2$$

$$[u_{4,\max}^s]^2 \leq \delta_1^2 \qquad\qquad (4.3)$$

$$[F_{\max}^s]^2 \leq F_0^2$$

(Equation continued on next page)

$$F_0, \delta_0, \alpha > 0$$

where

u_j, $j = 1,4$, = story displacements (top down). Superscripts ℓ and s refer to a large and a small earthquake, respectively

K_j, $j = 1,4$, = story stiffnesses, (top down)

F_{max}^s = maximum force in the energy-absorbing device when it is subjected to a small earthquake

$z = [z_1, z_2, z_3]^T = [F_0, \delta_0, \alpha]^T$ is the design parameter vector

$T = [t_0, t_f]$, given time interval

δ = prescribed limit on u_4^ℓ (taken as 4 inches)

δ_1 = prescribed limit on u_4^s (taken as 0.25 inches)

By introducing a dummy cost variable z_4, the corresponding nonlinear programming problem can be written as [8]

$$\min_z z_4$$

subject to

$$\max_{t \in T} \left[\sum_{j=1}^{3} \left\{ K_j [u_j^\ell(z,t) - u_{j+1}^\ell(z,t)] \right\}^2 \right] \leq z_4$$

$$\max_{t \in T} [u_4^\ell(z,t)]^2 \leq \delta^2 \tag{4.4}$$

$$[u_{4,max}^s(z)]^2 \leq \delta_1^2$$

$$[F_{max}^s(z)]^2 \leq F_0^2$$

$$F_0, \delta_0, \alpha > 0$$

4.6 Numerical Results

Initial values of the design parameters for this problem are

F_0 = 5.0, δ_0 = 0.11, α = 0.064, and

dummy cost parameter, z_4 = 26.0

The Elcentro 1940 NS earthquake record, modified for the experiment (Fig. 4.3), is used to represent a strong earthquake. A design response spectrum with a maximum of 0.15g peak acceleration is used for a small earthquake. A method of feasible directions is used to solve the problem. See Ref. 8 for details. The optimal values obtained for these parameters are

F_0 = 4.337126 δ_0 = 0.25028 α = 0.0583057

z_4 = 16.89978.

Figure 4.4 shows the decrease in the cost parameter, which is the same as the decrease in the sum of squares of story shears, versus the number of design iterations. Story shears and bottom floor displacement time histories for both optimum and initial parameters, are plotted in Figs. 4.5 through 4.8. The optimal design is softer than the initial design, the stiffness being reduced from 45.5 to 17.33.

It is interesting to compare the results of this problem with the one solved in Ref. 13, which was essentially the same except that there were no constraints under small earthquake excitation. In that case the stiffness of the optimum energy absorber was 2.4, which is considerably lower than the value obtained here. Thus, the present design shows that the previous design was not satisfactory, since it will have excessive deformations under small excitations.

5.   CONCLUDING REMARKS

Three applications of mathematical programming techniques to the optimal design of earthquake resistant systems have been presented. These applications are based on different design philosophies and show a broad spectrum of problems that can be formulated and solved, using these techniques. Solution of these problems shows that significant improvements in the behavior and/ or reductions in the cost can be achieved by properly formulating the problem, in terms of a mathematical programming problem.

645

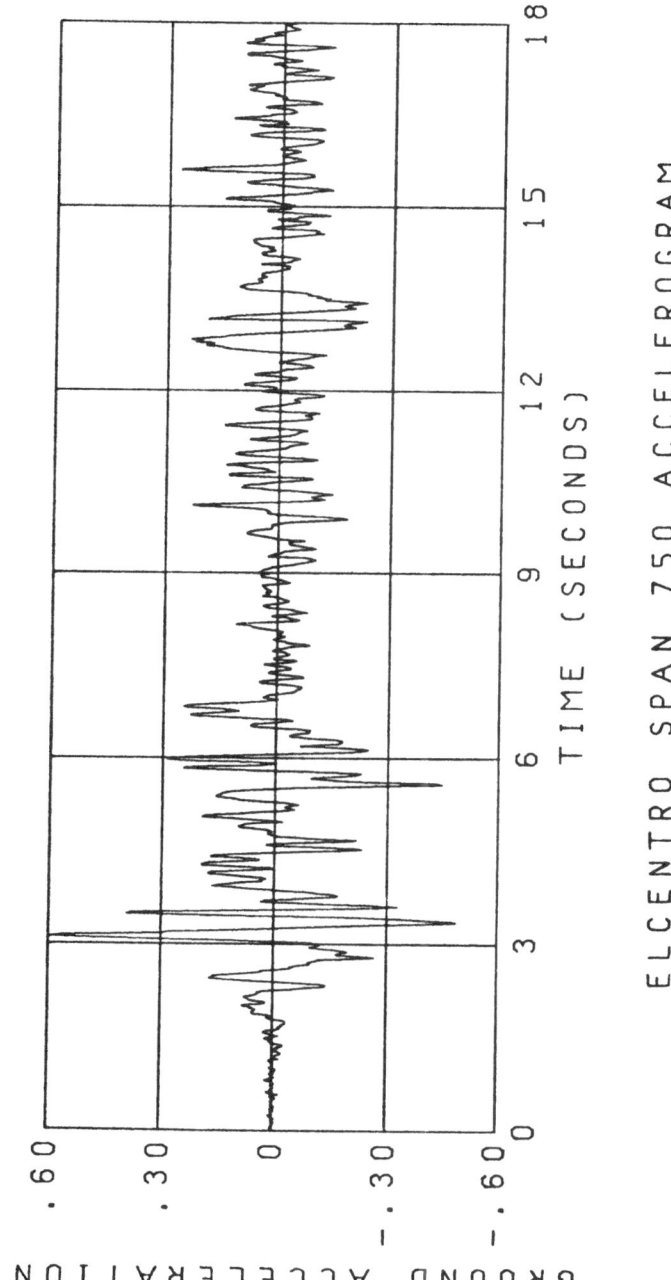

ELCENTRO SPAN 750 ACCELEROGRAM

Figure 4.3 Elcentro Span 750 Accelerogram

646

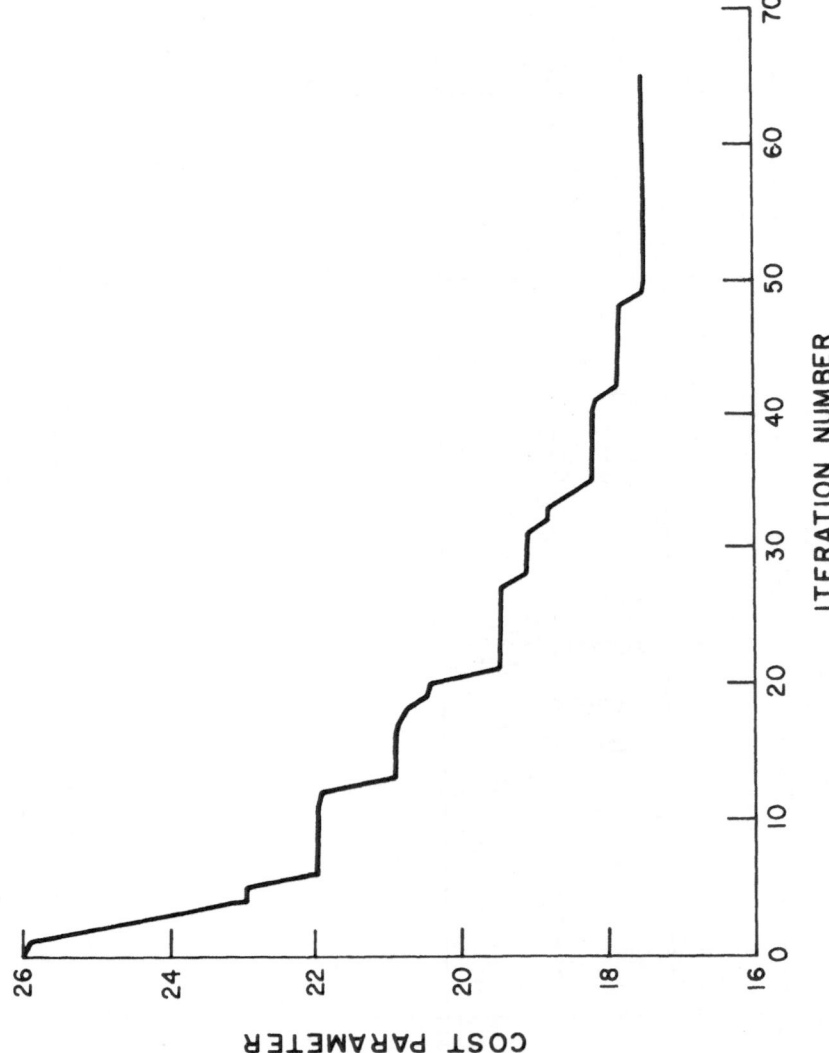

Figure 4.4   Cost Parameter Versus Iteration Number

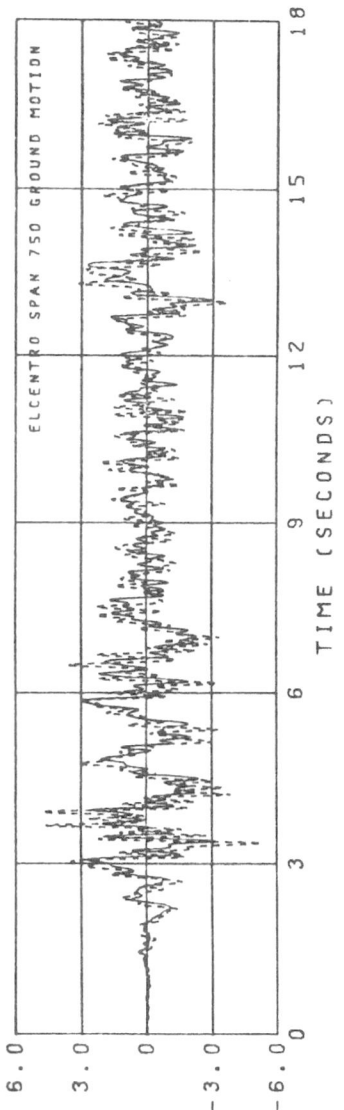

Figure 4.5   First Story Shear Time History

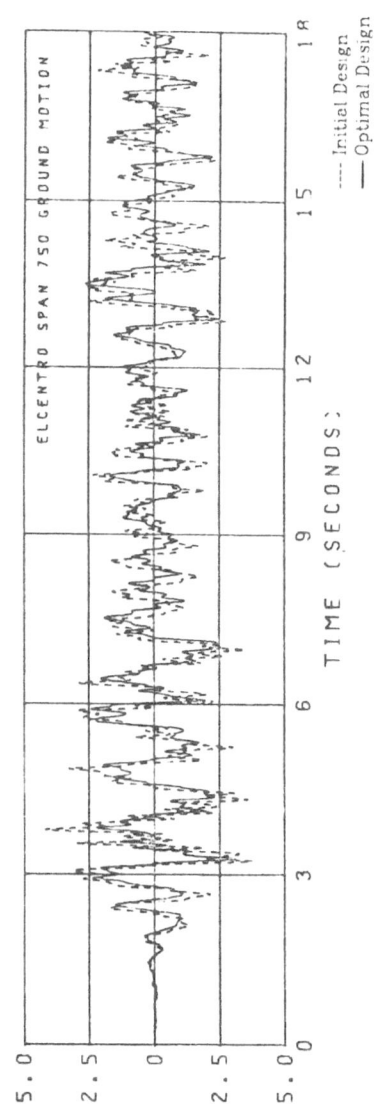

Figure 4.6   Second Story Shear Time History

648

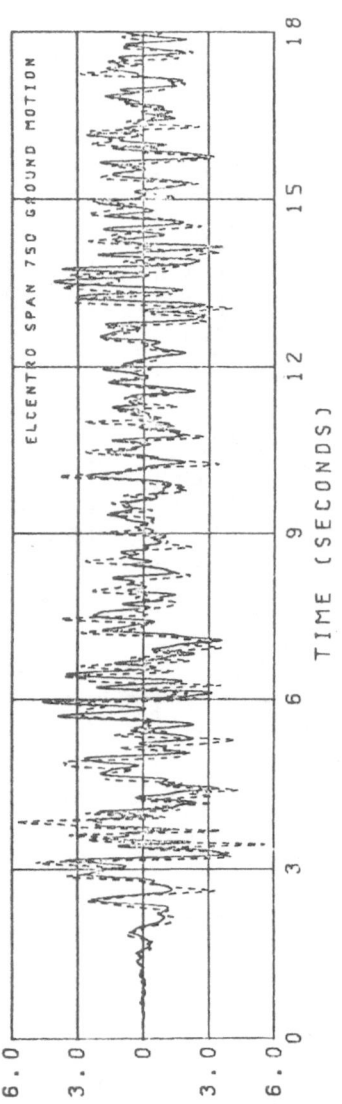

Figure 4.7 Third Story Shear Time History

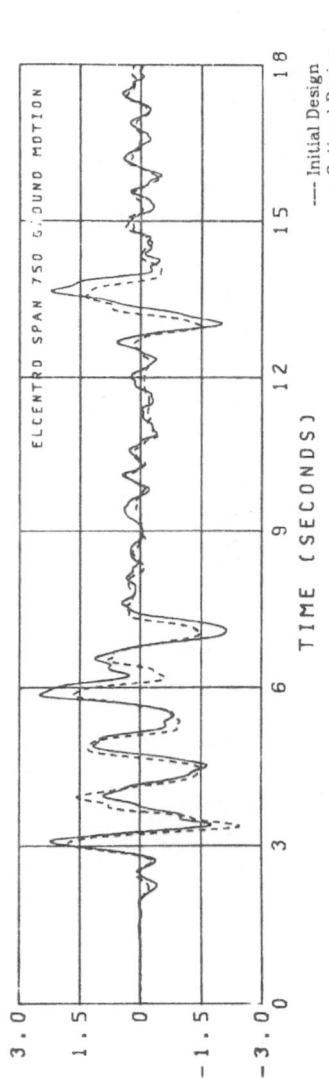

Figure 4.8 Base Displacement Time History

REFERENCES

1. Pierson, B.L., "A Survey of Optimal Structural Design Under Dynamic Constraints", International Journal of Numerical Methods in Engineering, Vol. 4, 1972, 491-499.

2. Rao, S.S., "Structural Optimization under Shock and Vibration Environment", The Shock and Vibration Digest, Vol. 11, No. 2, February 1979.

3. Ray, D., Pister, K.S., and Chopra, A.K., Optimum Design of Earthquake resistant Shear Buildings, Report No. EERC 74-3, Earthquake Engineering Research Center, University of California, Berkeley, January 1974.

4. Walker, N.D. and Pister, K.S., Study of a Method of Feasible Directions for Optimal Elastic Design of Framed Structures Subjected to Earthquake Loading, Report No. EERC 75-39, Earthquake Engineering Research Center, University of California, Berkeley, December 1975.

5. Bertero, V.V. and Kamil, H., "Nonlinear Seismic Design of Multistory Frames", Canadian Journal of Civil Engineering, Vol. 2, No. 4, December 1975.

6. Zagajesky, S.W. and Bertero, V.V., Computer-Aided Optimum Seismic Design of Ductile Reinforced Concrete Moment Resisting Frames, Report No. UCB/EERC-77/16, Earthquake Engineering Research Center, University of California, Berkeley, December 1977.

7. Walker, N.D., Automated Design of Earthquake resistant Multi-story Steel Building Frames, Report No. UCB/EERC-77/12, Earthquake Engineering Research Center, University of California, Berkeley, May 1977.

8. Bhatti, M.A., Optimal Design of Localized Nonlinear Systems with Dual Performance Criteria Under Earthquake Excitations, Report No. UCB/EERC-79/15, Earthquake Engineering Research Center, University of California, Berkeley, July 1979.

9. Clough, R.W. and Penzien, J., Dynamics Of Structures, McGraw-Hill, New York, 1975.

10. Haug, E.J. and Arora, J.S., Applied Optimal Design: Mechanical and Structural Systems, John Wiley & Sons, New York, 1979, p. 339.

11. Kelly, J.M., Eidinger, J.M., and Derham, C.J., A Practical Soft Story Earthquake Isolation System, Report No. UCB/EERC-77/27, Earthquake Engineering Research Center, University of California, Berkeley, November 1977.

12. Ozdemir, H., Nonlinear Transient Dynamic Analysis of Yielding Structures, Ph.D. Dissertation, Division of Structural Engineering and Structural Mechanics, Department of Civil Engineering, University of California, Berkeley, 1976.

13. Bhatti, M.A., Pister, K.S., and Polak, E., Optimal Design of an Earthquake Isolation System, Report No. UCB/EERC-78/22, Earthquake Engineering Research Center, University of California, Berkeley, October 1978.

650

EVALUATION OF FRAME SYSTEMS BASED ON OPTIMALITY CRITERIA WITH MULTICOMPONENT SEISMIC INPUTS, PERFORMANCE CONSTRAINTS, AND P-Δ EFFECT*

Franklin Y. Cheng

Civil Engineering, University of Missouri-Rolla, Rolla, MO 65401

ABSTRACT

This paper presents an optimum design technique that incorporates a variety of dynamic behavior constraints. Several 15-story frameworks are designed, using the computer program ODSEWS (Optimum Design of Static, Earthquake, and Wind Structures) that has been developed with consideration of two components of actual Seismic accelerograms, two structural models, and various parametric and side constraints. Results show the versatility of the design technique and the effect of interacting ground motions on structural systems.

1.   INTRODUCTION

Recent progress in computer technology has led to the development of sophisticated computer programs for the analysis of complex structures. When these programs are used for design, the relative stiffness of the constituent members must be assumed. If the preliminary stiffnesses are misjudged, repeated analyses will usually not yield an improved design, regardless the program's sophistication. Thus, use of an analysis program does not guarantee an efficient design and the conventional design method is in reality an art, rather than a science.

----
*This paper is based on partial results obtained under NSF Grant No. PFR78-5694.   Appreciation is extended to Mr. L.H. Sheng for his assistance in numerical solution of examples.

This paper presents a general optimum design technique for various types of structural systems. The application, however, is shown for plane steel structures that are subjected to interacting ground motions. The consistent mass method is employed in the structural analysis formulation, for which the second-order P-Δ effect [1], resulting from structural and nonstructural weight, and vertical ground motion are included. Constraints considered herein are imposed on stresses (axial, bending, and shear), displacements, natural frequencies, and cross sections. The design objective is minimum weight of a structural system.

2. GENERAL OPTIMIZATION FORMULATION

2.1 Kuhn-Tucker Conditions and Recursion Relations

The optimization technique is based on optimality criteria that are derived on the basis of Kuhn-Tucker conditions. Recursion relations are used as an iterative procedure to satisfy the optimality requirement for a structural design. A general problem in optimal structural design can be stated as follows: Minimize

$$W(x)$$

subject to constraints

$$g_j(x) \leq 0, \qquad j = 1, 2, \ldots, n \qquad (2.1)$$

$$x \geq 0$$

in which $W(x)$ is the objective function, x is the vector of design variables, $g_j(x)$ is the jth constraint function, and n is the total number of constraints. In a general optimal structural design problem, $W(x)$ can represent the total cost of the structure. The total weight of the structure has been widely used in most publications and is used as the objective function in this work.

The constraint functions $g_j(x)$ may represent bounds on frequencies or behavior constraints of stress and displacement limitations that are imposed on a structure. These behavior constraints may be required for both static loads and seismic excitations. The problem defined in Eq. 2.1 is an inequality form of constrained minimization. To develop an algorithm for solving this problem, necessary conditions of optimization known as the Kuhn-Tucker conditions must be considered.

The optimization technique considered in this paper can be applied to trusses and frameworks with, or without bracing members. The trusses are composed of two-force members. The members of frameworks that do not have bracing are subjected to combined bending and axial forces, acting as beam-columns, and framework having bracing include beam-columns and two-force members. Thus, the formulation for braced frames can be viewed as a generalized presentation for all three types of structures. Based on the Kuhn-Tucker necessary conditions, the problem of optimal design of a braced plane frame can be expressed as follows: Minimize

$$W(x) = \sum_{i=1}^{m} \rho_i n_i x_i \ell_i \tag{2.2}$$

subject to constraints

$$g_j(x) = y_j(x) - b_j \leq 0, \qquad j = 1, \ldots, n \tag{2.3}$$

$$x_i \geq x_i^o, \qquad i = 1, 2, \ldots, m \tag{2.4}$$

In this formulation, $W(x)$ is total weight of a structure that is composed of columns, girders, and bracing members; ρ_i is mass density of the ith member, n_i is the ratio of the cross-sectional area A_i to the moment of inertia x_i for the ith member; x_i is the primary design variable of the ith member; ℓ_i is the length of the ith member; $y_j(x)$ is an expression for the jth behavior constraint; b_j is the limitation imposed on the jth behavior constraint; x_i^o is the lower bound on the ith design variable; m is the total number of members; and n is the total number of behavior constraints.

The primary design variables x_i are the moments of inertia for the columns and girders and represent the cross-sectional areas for bracings and truss members. The ratio of the cross-sectional area to the moment of inertia for the ith bracing member is unity. The functions, $y_j(x)$, represent the stress, displacement, and natural frequency of a structure. By using the Kuhn-Tucker necessary conditions to characterize any local minimum, Eqs. 2.2 through 2.4 can be written for each design variable as

$$\frac{\partial W(x)}{\partial x_i} + \sum_{j=1}^{n} \lambda_j \frac{\partial y_j(x)}{\partial x_i} - s_i = 0, \qquad i = 1, \ldots, m \qquad (2.5)$$

with

$$\lambda_j [y_j(x) - b_j] = 0, \qquad j = 1, 2, \ldots, n \qquad (2.6)$$

$$s_i [x_i^0 - x_i] = 0, \qquad i = 1, 2, \ldots, m \qquad (2.7)$$

$$\lambda_j \geq 0, \qquad j = 1, 2, \ldots, n \qquad (2.8)$$

$$s_i \geq 0, \qquad i = 1, 2, \ldots, m \qquad (2.9)$$

The variables λ_j and s_i are the Lagrange multipliers associated with the constraints in Eqs. 2.3 and 2.4, respectively. By examining Eqs. 2.6 and 2.8, one can observe that when the jth constraint is active, $\lambda_j > 0$ and when the ith constraint is not active, the corresponding $\lambda_j = 0$. Similar observations can be applied to the conditions expressed in Eqs. 2.7 and 2.9.

For simplicity, the active side constraints can be separated from the active behavior constraints. Let N denote the number of active behavior constraints, then the separated form of the Kuhn-Tucker necessary conditions from Eq. 2.5 can be simplified to

$$\frac{\partial W(x)}{\partial x_i} + \sum_{j=1}^{N} \lambda_j \frac{\partial y_j(x)}{\partial x_i} = 0, \qquad i \in J \qquad (2.10)$$

$$\frac{\partial W(x)}{\partial x_i} + \sum_{j=1}^{N} \lambda_j \frac{\partial y_j(x)}{\partial x_i} \geq 0, \qquad i \in J_0 \qquad (2.11)$$

where J is the set of design variables for which $x_i > x_i^0$ and J_0 is the set of design variables with $x_i = x_i^0$. The separated form of the Kuhn-Tucker necessary conditions in Eqs. 2.10 and 2.11 can be combined into a single equation as follows:

$$-\frac{\sum_{j=1}^{N} \lambda_j \mu_{ji}}{\tau_i} \begin{cases} = 1, \ i \ \epsilon \ J, \\ \le 1, \ i \ \epsilon \ J_o, \end{cases} \quad i = 1, 2, \ldots, m \quad (2.12)$$

in which

$$\mu_{ji} = \frac{\partial y_j(x)}{\partial x_i} \quad (2.13)$$

$$\tau_i = \frac{\partial W(x)}{\partial x_i} \quad (2.14)$$

Equation 2.12 is the optimality criteria. Any design that satisfies these criteria is a relative minimum. However, these criteria are necessary but not sufficient to guarantee a global optimum.

The recursion relations can now be derived by using the optimality criteria of Eq. 2.12, with all $x_i > x_i^o$. Therefore, Eq. 2.12 becomes

$$-\frac{\sum_{j=1}^{N} \lambda_j \mu_{ji}}{\tau_i} = 1, \quad i = 1, 2, \ldots, m \quad (2.15)$$

For the convenience of computer programming, let

$$x_i = \Lambda \alpha_i, \quad i = 1, \ldots, m \quad (2.16)$$

where α_i are the relative design variables corresponding to the design variables x_i and Λ is a scaling factor. Multiplying both sides of Eq. 2.15 by $(\Lambda \alpha_i)^2$, rearranging terms, and taking the square root of both sides, one obtains

$$\Lambda \alpha_i = \alpha_i \left[-\frac{\sum_{j=1}^{N} \lambda_j \mu'_{ji}}{\tau'_i} \right]^{1/2}, \quad i = 1, \ldots, m \quad (2.17)$$

where

$$\mu'_{ji} = \Lambda\mu_{ji} \tag{2.18}$$

$$\tau'_i = \frac{\tau_i}{\Lambda} \tag{2.19}$$

and $\Lambda\alpha_i$ is the ith design variable, which is expressed as a function of α_i. The form of Eq. 2.17 suggests the following recursion relation for determining the design variable in each cycle:

$$(\Lambda\alpha_i)_{v+1} = (\alpha_i)_v \left[-\frac{\sum\limits_{j=1}^{N} \lambda_j\mu'_{ji}}{\tau_i} \right]_v^{1/2}, \qquad i = 1, \ldots, m \tag{2.20}$$

where the subscripts v and $v+1$ denote the cycles of iteration. The use of the recursion relation expressed in Eq. 2.20 is the same as forcing the final design to satisfy the Kuhn-Tucker necessary conditions of Eqs. 2.12 and 2.13.

2.2 Lagrange Multipliers for Multiple Constraints

At any stage in the design process in which Eq. 2.20 is used, it is possible that there is more than one active constraint. Therefore, it is necessary to find the Lagrange multipliers that correspond to the active constraints of the current design variables. Calculation of the multipliers can be accomplished from Eq. 2.15 by forming an auxiliary function $L(\lambda)$, as follows:

$$L(\lambda) = \sum_{i=1}^{J} \left[1 + \frac{\sum\limits_{j=1}^{N} \lambda_j\mu_{ji}}{\tau_i} \right]^2 \tag{2.21}$$

This is minimized by solving the set of linear equations

$$\frac{\partial L(\lambda)}{\partial \lambda_j} = 0, \qquad j = 1, \ldots, N \tag{2.22}$$

for the Lagrange multipliers λ_j. Equation 2.22 can then be written as

$$\underline{H}\vec{\lambda} = \vec{G} \tag{2.23}$$

where

$$H_{jk} = \sum_{i=1}^{J} E_{ji}E_{ki} , \quad E_{ji} = \frac{\mu_{ji}}{\tau_i} \left.\begin{array}{c} \\ \\ \\ \\ \\ \end{array}\right\}$$

$$G_k = - \sum_{i=1}^{J} E_{ki}$$

$$(2.24)$$

The matrix $\underline{H}$ has the property of being symmetrical. By solving Eq. 2.23, the Lagrange multipliers corresponding to the active constraints can be determined from the current design variables. These multipliers are then used in Eq. 2.20 for determining the design variables at the next cycle in the design process.

3. DYNAMIC BEHAVIOR CONSTRAINTS

The behavior constraint functions $y_j(x)$, considered in the optimal structural design problem, are limitations of displacements (flexibility) and stresses (stiffness) that are imposed on both static and seismic structures [2,3,4]. However, for seismic structures, an additional behavior constraint on natural frequencies is taken into account. The side constraints are limitations on the size and moment of inertia or cross-sectional area of the members. This paper presents only dynamic behavior constraints.

3.1 Dynamic Flexibility Constraints

The dynamic displacement constraint function can be expressed in terms of virtual work as

$$u_j(x,t) = \vec{Q}_j^T \vec{r}(x,t),$$

$$(3.1)$$

where $\vec{Q}_j$ is a load vector with unit force in the jth direction, the time function is also a unit value, and $\vec{r}(x,t)$ is a vector of generalized displacements attributable to the dynamic load $\vec{R}(t)$. Differentiating Eq. 3.1 with respect to the design variables x_i yields

$$\frac{\partial u_j(x,t)}{\partial x_i} = \vec{Q}_j^T \frac{\partial \vec{r}(x,t)}{\partial x_i},$$

$$(3.2)$$

where $\partial \vec{r}(x,t)/\partial x_i$ can be found by using the following motion equation:

$$[\underline{M_n} + \underline{M_s}]\ddot{\vec{r}}(x,t) + [\underline{K_s} - \underline{K_g}]\vec{r}(x,t) = \vec{R}(t) \qquad (3.3)$$

where $[\underline{M_n}]$ and $[\underline{M_s}]$ are nonstructural mass and structural mass matrices, respectively, $[\underline{K_s}]$ is the elastic stiffness matrix, and $[\underline{K_g}]$ is the geometric stiffness or second-order matrix. Damping is included in the response analysis and is therefore not considered in the above equation. In order to find $\partial \vec{r}(x,t)/\partial x_i$ in Eq. 3.2 one takes the first variation of Eq. 3.3 and then uses modal analysis and the approximate Rayleigh quotient equation (see Eq. 3.5). The final form of Eq. 3.2 becomes

$$\frac{\partial u_j(x,t)}{\partial x_i} = - \frac{1}{x_i}[\vec{q}_j^T(x,t)\underline{a_i^T} \underline{K_{si}} \underline{a_i} \vec{r}(x,t)$$

$$- p_i' \vec{q}_j^T(x,t)\underline{a_i^T} \underline{K_{gi}} \underline{a_i} \vec{r}(x,t)$$

$$- p^2\vec{q}_j^T(x,t)\underline{a_i^T} \underline{M_{si}} \underline{a_i}(x,t)], \qquad (3.4)$$

where $\vec{q}_j(x,t)$ is the displacement vector resulting from application of the force $\vec{Q}_j f(t)$, $\underline{a_i}$ is the compatibility matrix connecting the generalized coordinates of a structure and those of member i, P_i' consists of half the structural mass $(n_i \rho_i \ell_i/2)$ of member i that must have influence on the geometric stiffness matrix established on the basis of column members, and the natural frequency p is obtained by using the following approximate Rayleigh quotient:

$$p^2 = \frac{\vec{r}^T(x,t)\underline{K}\vec{r}(x,t)}{\vec{r}^T(x,t)\underline{M}\vec{r}(x,t)} \qquad (3.5)$$

where $\underline{K} = \underline{K_s} - \underline{K_g}$ and $\underline{M} = \underline{M_n} + \underline{M_s}$. The term μ_{ji}' in the recursion equation of Eq. 2.20 can be based on Eq. 3.4 as follows

$$\mu_{ji}' = \frac{\Lambda \partial u_j(x,t)}{\partial x_i} \qquad (3.6)$$

3.2 Dynamic Stiffness Constraints

A measure of dynamic stiffness may be described by the work done by the magnitude of the dynamic load vector $\vec{R}_o$ and the generalized displacement $\vec{r}(x,t)$, in the form

$$z(x,t) = \frac{1}{2} \vec{R}_o^T \vec{r}(x,t) \tag{3.7}$$

because the product $\vec{R}_o^T \vec{r}(x,t)$ is an inverse measure of the stiffness. Differentiating Eq. 3.7 with respect to the design variable x_i yields

$$\frac{\partial z(x,t)}{\partial x_i} = \frac{\vec{R}_o^T}{2} \frac{\partial \vec{r}(x,t)}{\partial x_i} \tag{3.8}$$

Similar to the derivation of Eq. 3.4, the final form of Eq. 3.8 is

$$\frac{\partial z(x,t)}{\partial x_i} = - \frac{1}{2x_i} [\vec{r}^T(x,t) a_i^T \underline{K_{si}} \; a_i \; \vec{r}(x,t)$$

$$+ \vec{r}^T(x,t) a_i^T \underline{K_{gi}} \; a_i \; \vec{r}(x,t)$$

$$- p^2 \vec{r}^T(x,t) a_i^T \underline{M_{si}} \; a_i \; \vec{r}(x,t)] \tag{3.9}$$

Consequently, the term μ'_{ji} in Eq. 2.20 can be expressed as

$$\mu'_{ji} = \frac{\Lambda \partial z(x,t)}{\partial x_i} \tag{3.10}$$

3.3 Frequency Constraints

The natural frequency ω_j of any mode of a structure can be obtained by using the Rayleigh quotient

$$\omega_j^2 = \frac{\vec{\phi}_j^T \; \underline{K} \; \vec{\phi}_j}{\vec{\phi}_j^T \; \underline{M} \; \vec{\phi}_j} \tag{3.11}$$

where $\vec{\phi}_j$ is the normal mode vector and ω_j represents not only the natural frequency but also a mathematical function of design. Differentiation of Eq. 3.11 with respect to the design variable x_i yields

$$\frac{\partial \omega_j^2}{\partial x_i} = \left(\vec{\phi}_j^T \, a_i^T \, K_{si} \, a_i \, \vec{\phi}_j - p_i' \, \vec{\phi}_j^T \, a_i^T \, K_{gi} \, a_i \, \vec{\phi}_j \right.$$
$$\left. - \omega_j^2 \, \vec{\phi}_j^T \, a_i^T \, M_{si} \, a_i \, \vec{\phi}_j \right) \bigg/ x_i \, \vec{\phi}_j^T \, \underline{M} \, \vec{\phi}_j \qquad (3.12)$$

Thus, the term μ'_{ji} in Eq. 2.20 becomes

$$\mu'_{ji} = \frac{\Lambda \partial \omega_j^2}{\partial x_i} \qquad (3.13)$$

4. NUMERICAL SCHEME AND DESIGN VARIABLES

The numerical scheme is sketched in the flow chart of Fig. 4.1, which shows the analysis and design procedures. The design variables may be studied from the typical built-up section of steel shown in Fig. 4.2, for which the moment of inertia I_o is considered as a primary design variable and the depth d, flange thickness t_f, and web thickness t_w are the secondary design variables. Expressions for the cross-sectional area A_o, the moment of inertia I_o, the section modulus S_o, and the shear flow v_o, of the section are

$$A_o = d^2 \left[\frac{t_w}{d} + 2 \frac{t_f}{d} \left(\frac{b}{d} - \frac{t_w}{d} \right) \right] \qquad (4.1)$$

$$I_o = d^4 \left[\frac{b}{2d} \left(\frac{t_f}{d} \right) \left(1 - \frac{t_f}{d} \right)^2 + \frac{1}{12} \frac{t_w}{d} \left(1 - \frac{2t_f}{d} \right)^3 \right] \qquad (4.2)$$

$$S_o = d^3 \left[\frac{b}{d} \left(\frac{t_f}{d} \right) \left(1 - \frac{t_f}{d} \right)^2 + \frac{1}{6} \frac{t_w}{d} \left(1 - \frac{2t_f}{d} \right)^3 \right] \qquad (4.3)$$

$$v_o = \frac{d^2\left[\frac{b}{2d}(\frac{t_f}{d})(1 - \frac{t_f}{d})^2 + \frac{1}{12}\frac{t_w}{d}(1 - 2\frac{t_f}{d})^3\right]\frac{t_w}{d}}{\left[\frac{b}{2d}(\frac{t_f}{d})(1 - \frac{t_f}{d}) + \frac{1}{8}\frac{t_w}{d}(1 - \frac{2t_f}{d})^2\right]} \qquad (4.4)$$

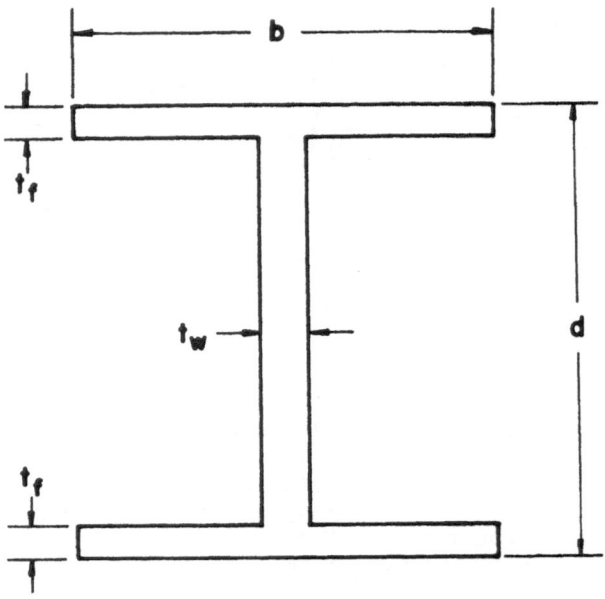

Figure 4.1   Built-up Steel Section

Upper and lower bounds on the design variables are imposed on the secondary design variables, instead of the primary design variables, as

$$d_{min} \le d \le d_{max}$$

$$(\frac{t_f}{d})_{min} \le \frac{t_f}{d} \le (\frac{t_f}{d})_{max}$$

$$(4.5)$$

However, the ratios of the minimum moment of inertia to the maximum moment of inertia of both girders and columns can be specified.   The width b and the ratio t_w/d of web thickness to depth of the built-up section are kept constant for each design variable.

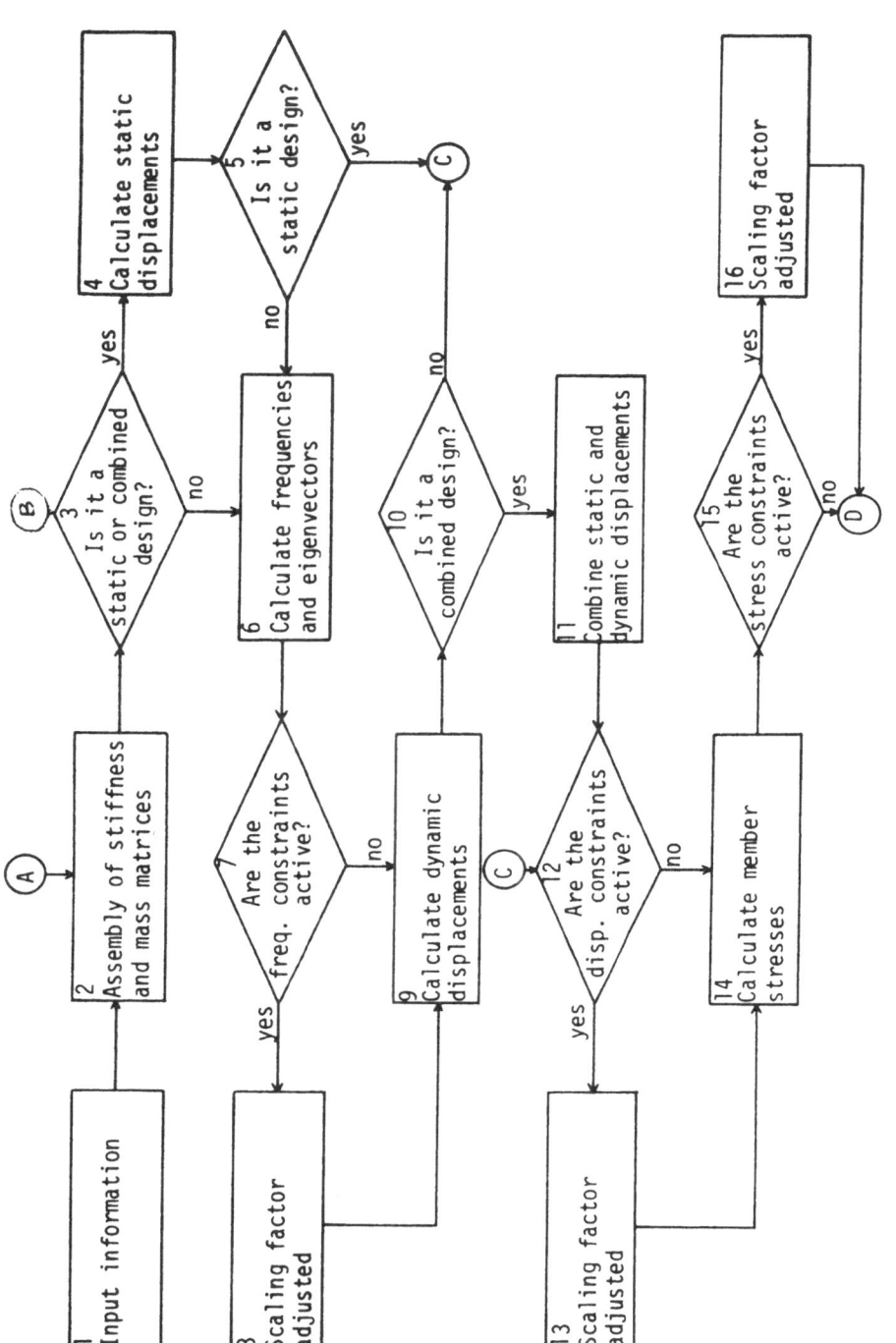

Figure 4.2   Flow Chart of Numerical Procedures (to be continued)

662

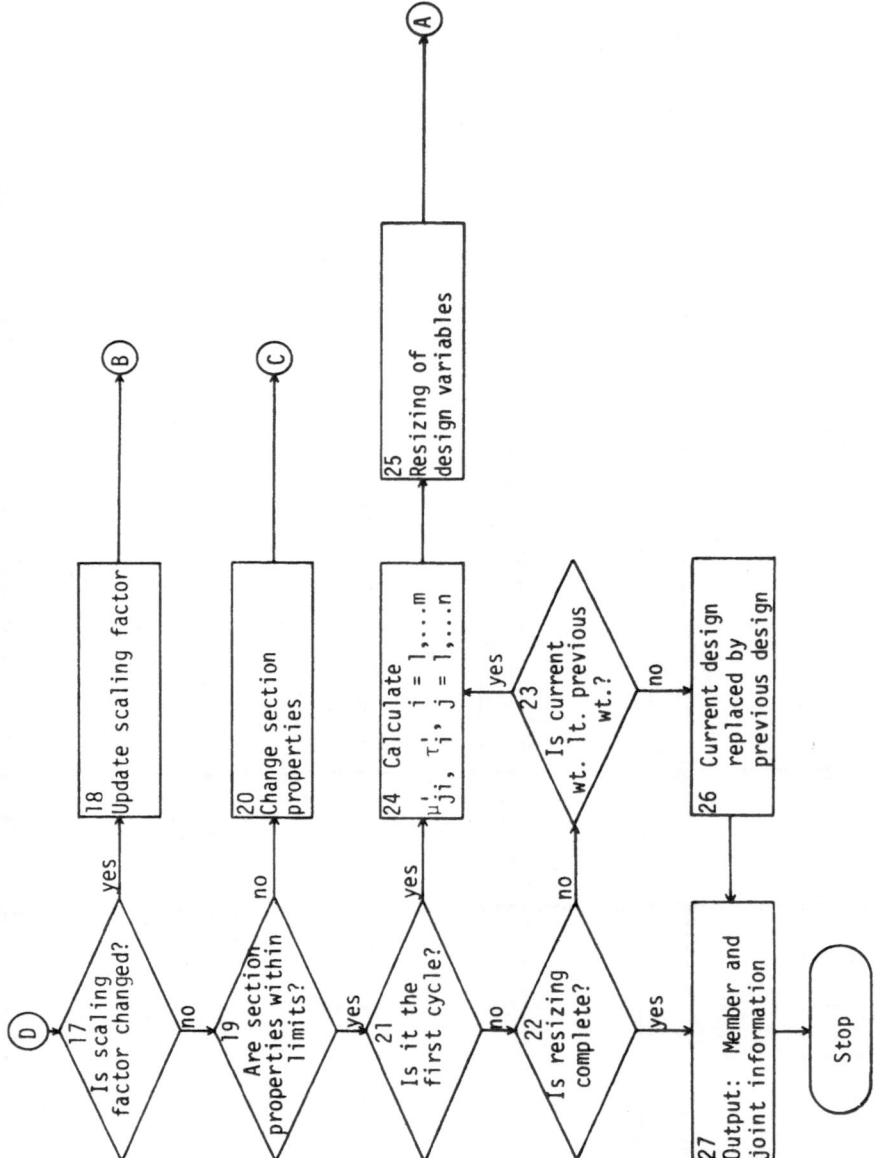

Figure 4.2 (continued) Flow Chart of Numerical Procedures

5. NUMERICAL RESULTS AND CONCLUSIONS

5.1 Numerical Results

The 15-story frames shown in Figs. 5.1 through 5.4 are designed for Models I and II of the accompanying figures. The span length is 21 ft (6.40 m), the floor height is 12 ft (3.66m), the dead load w on each floor (nonstructural mass) is 180 lbs/in (178.74 N/m), the modulus of elasticity E is 29,000 ksi (200.1 GN/m^2), and the mass density ρ of the construction material is 0.283 lbs/in^3 (783.34 kg/m^3). The dynamic excitation is due

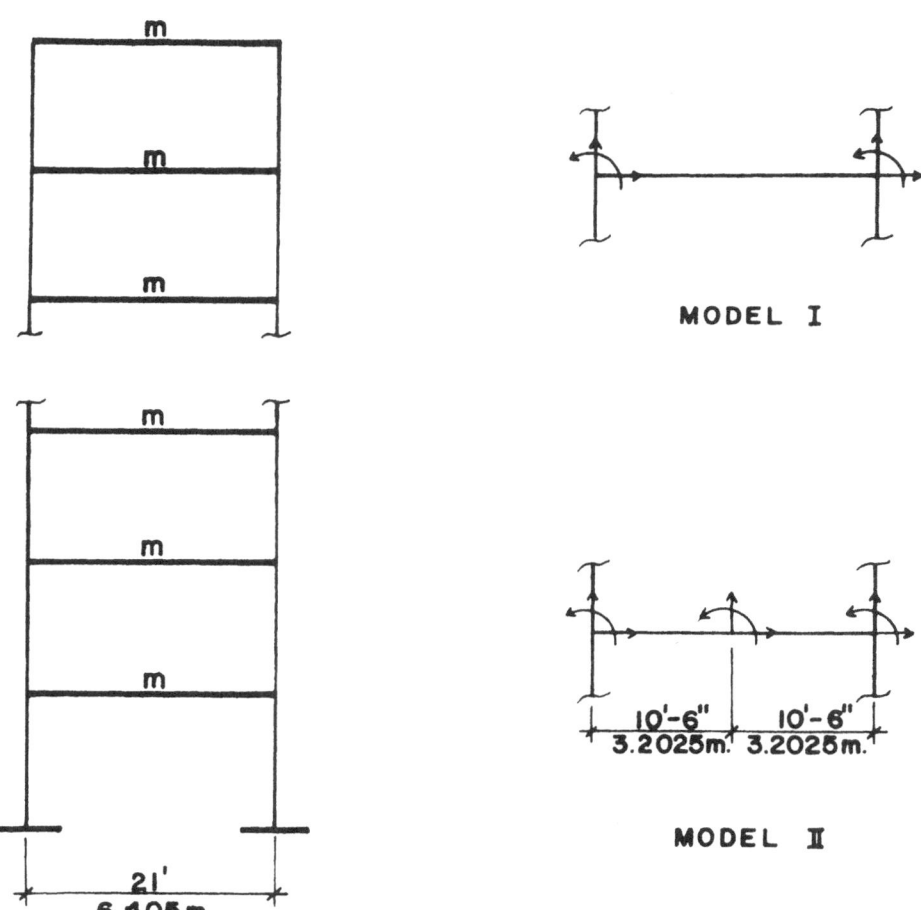

MODEL I

MODEL II

Figure 5.1 Unbraced Frame

to horizontal and vertical earthquake accelograms of El Centro, 1940, for which the acceleration spectra with 5% damping are given in Fig. 5.5. The allowable stress for combined bending and axial force is $\sigma < 29$ ksi (200.1 MN/m^2) and the allowable shear stress σ_v should be less than or equal to 0.65 σ. Although different allowable deflections may be imposed at any particular nodes, the allowable deflection considered herein is based on the general code provision, i.e., the relative displacement between floors is limited to 0.005 times the story height. Other constants are b = 12 in. (30.48 cm), d_{max} = 75 in (190.50 cm), d_{min} = 8 in (20.32 cm), $(t_f/d)_{min}$ = 0.023, and t_w/d = 0.02

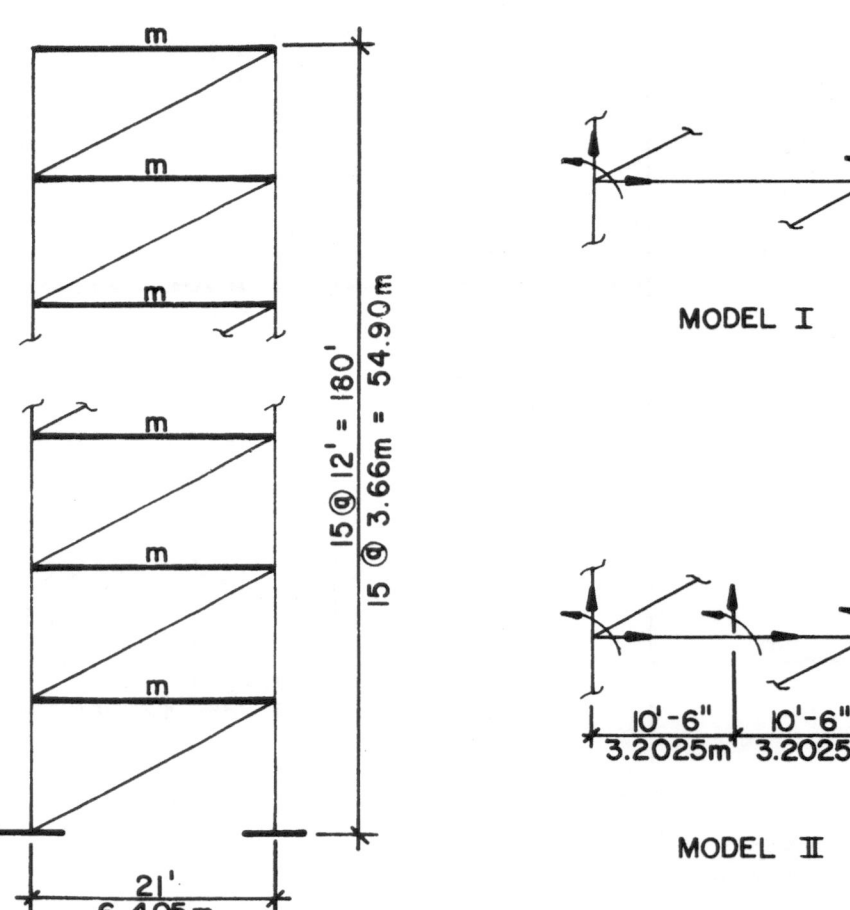

Figure 5.2   Single-Braced Frame

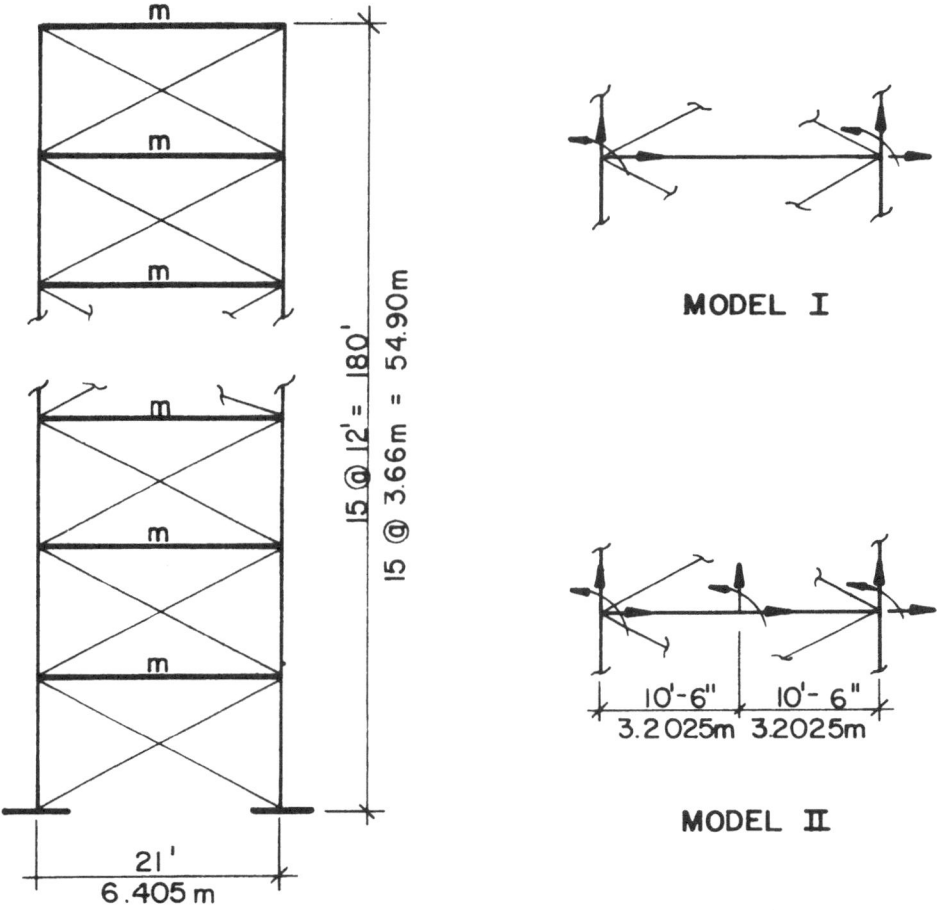

Figure 5.3   Double-Braced Frame

Four cases of the final design results for the unbraced frame of Fig. 5.1 are shown in Figs. 5.6 and 5.11, in which Case (a), designated H, signifies the design resulting from the horizontal ground motion only; Case (b), designated H+P-Δ(DL), is due to the horizontal ground motion plus the P-Δ effect of static load of nonstructural mass acting on girders; Case (c), designated H+V, indicates the horizontal and vertical earthquake components; and Case (d), designated H+V+P-Δ(DL+V), corresponds to the design obtained by considering horizontal and vertical earthquake components and the P-Δ effect of the vertical inertia forces that are associated with structural and nonstructural masses.   Figures 5.6 and 5.7 reveal the stiffness distribution for columns and

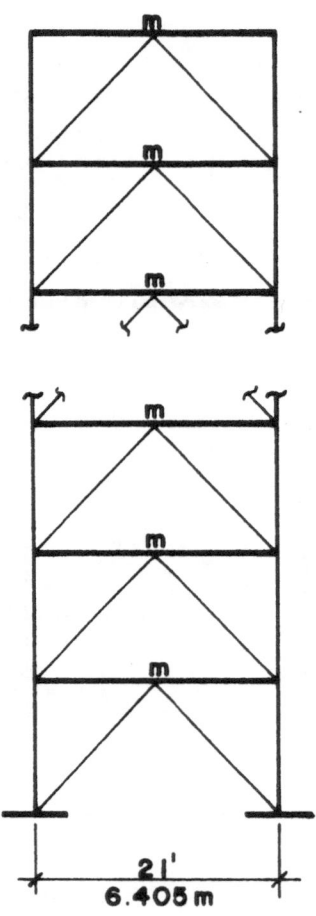

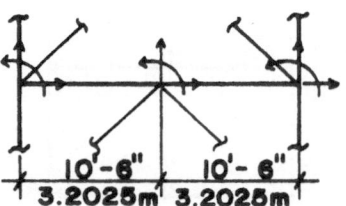

Structural Model

Figure 5.4   K-Braced Frame

girders of the system, for which the normalized combined stresses
(the ratio of the actual bending and axial stresses to the allowable stress) are shown in Figs. 5.8 and 5.9.   The structural
weight and the associated number of design cycles are given in
Fig. 5.10.   The ratios of the combined strain and kinematic energies of the individual modes to those of all the modes used in the
design are plotted in Fig. 5.11.

Because the response parameters of the unbraced-frame design
are also used for the braced frames, a careful examination of
Figs. 5.6 through 5.11 seems essential for studying other frame
systems and may reveal the following interesting observations:

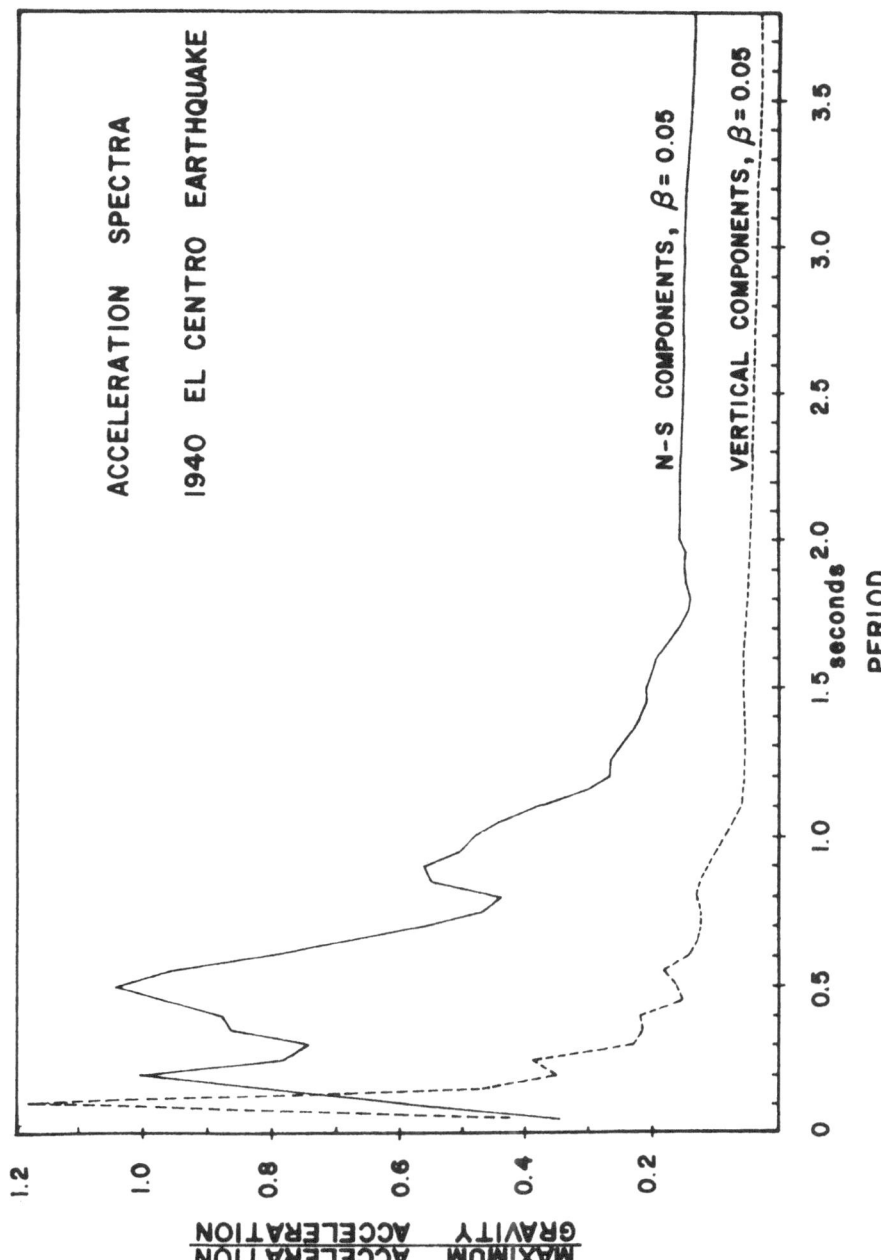

Figure 5.5   Horizontal and Vertical Earthquake Acceleration Spectra

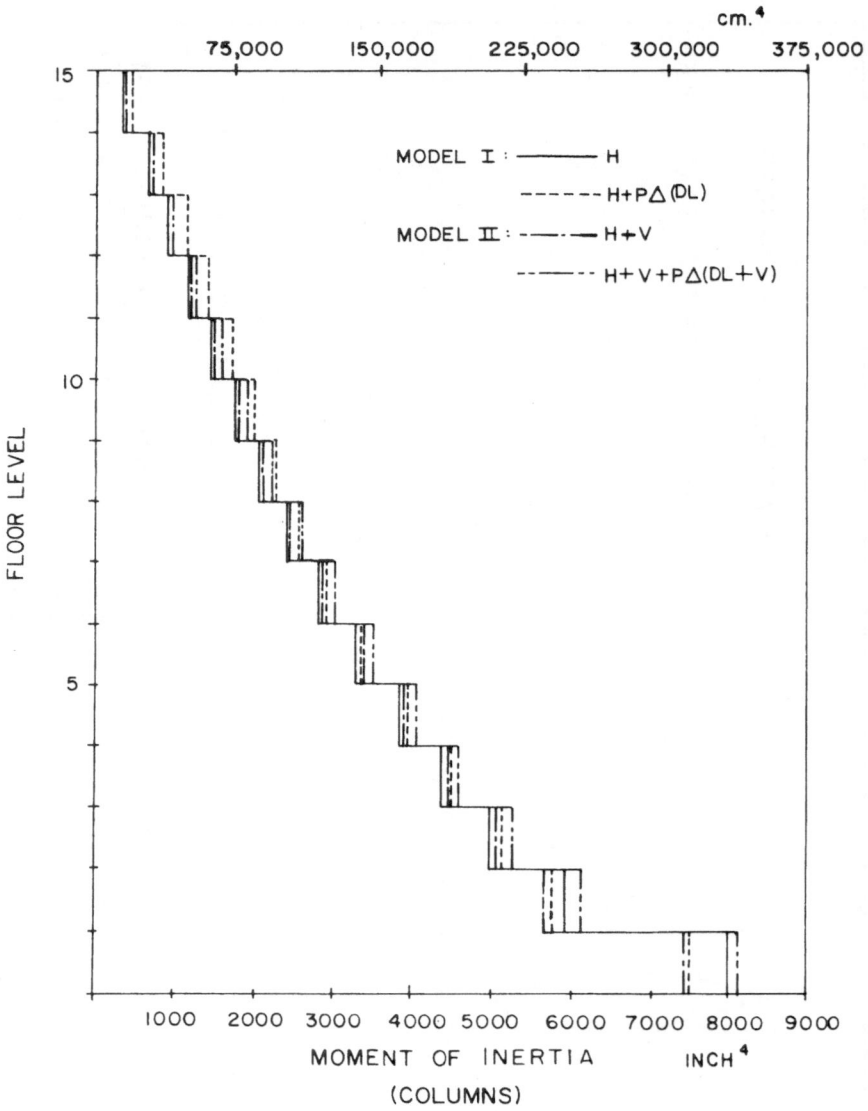

Figure 5.6   Moment of Inertia of Unbraced-Frame Columns

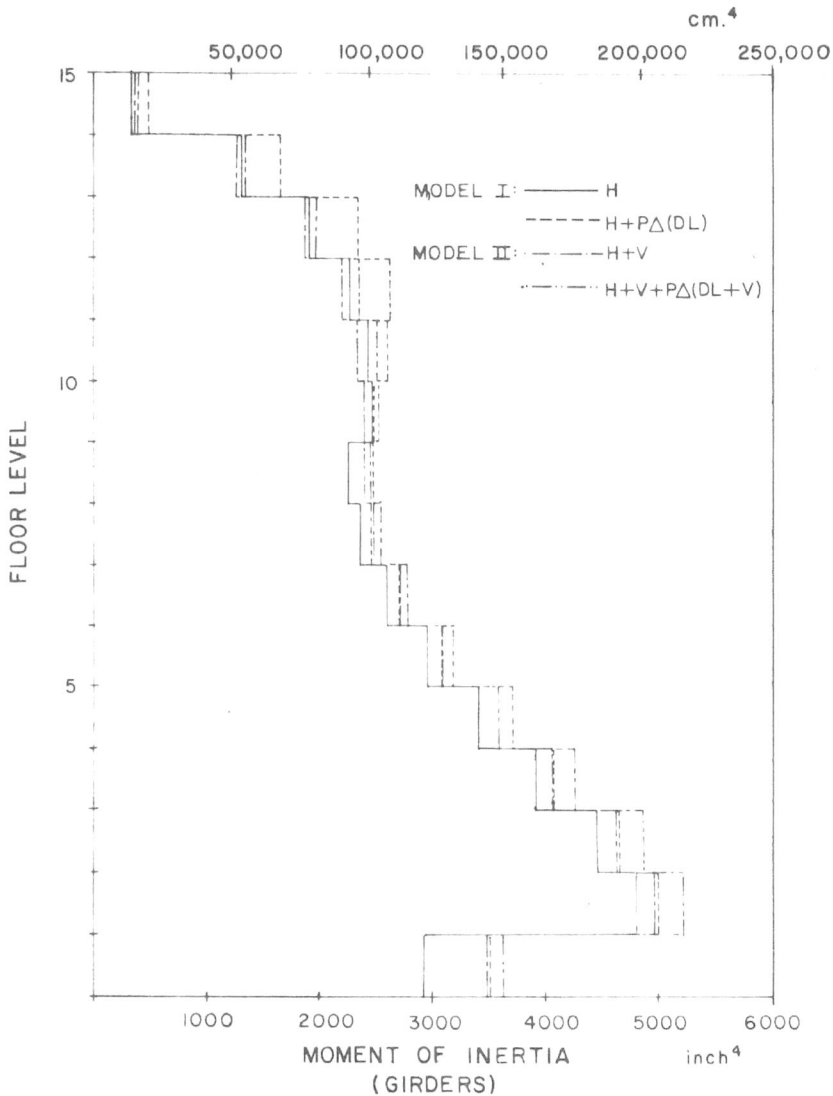

Figure 5.7   Moment of Inertia of Unbraced-Frame Girders

(1) The stiffness distribution of columns is greatest at the
support and gradually decreases from the support to the top floor.
To avoid too much difference between the moment of inertia of
columns or girders, one may use a control data card in the computer
program to specify how to vary the moment of inertia of the

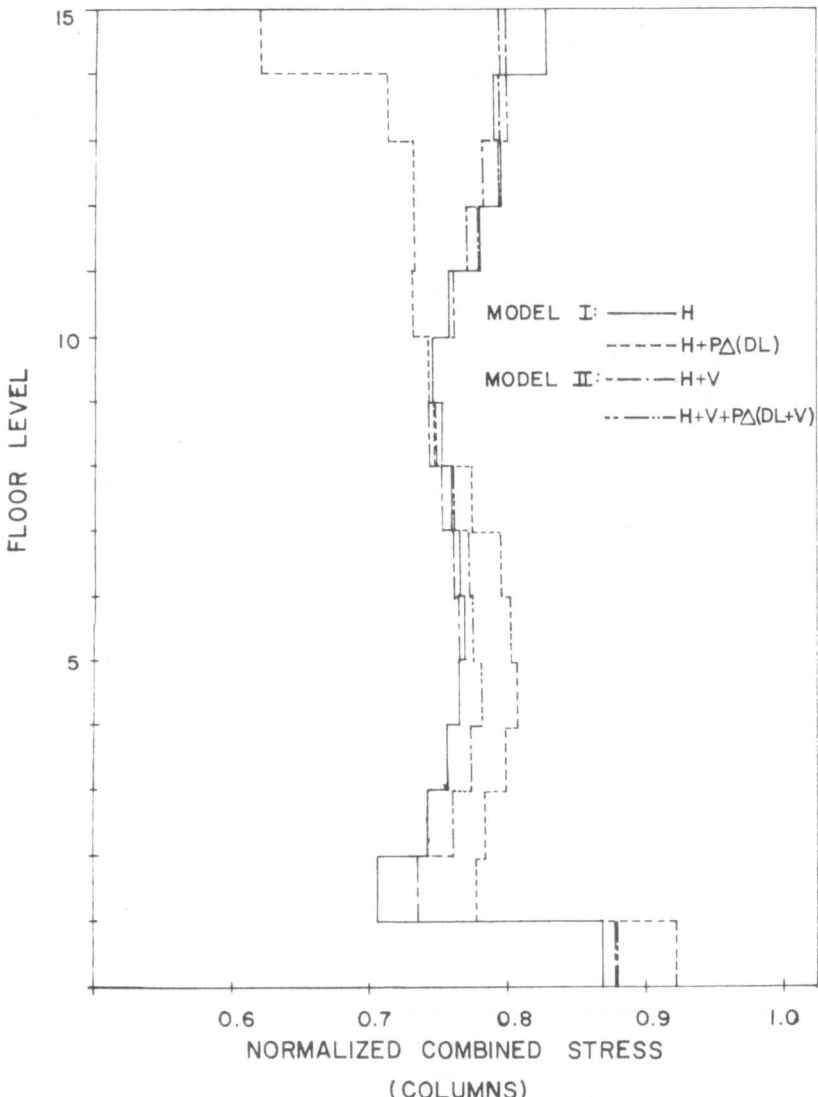

Figure 5.8   Normalized Combined Stresses of Unbraced-Frame
Columns

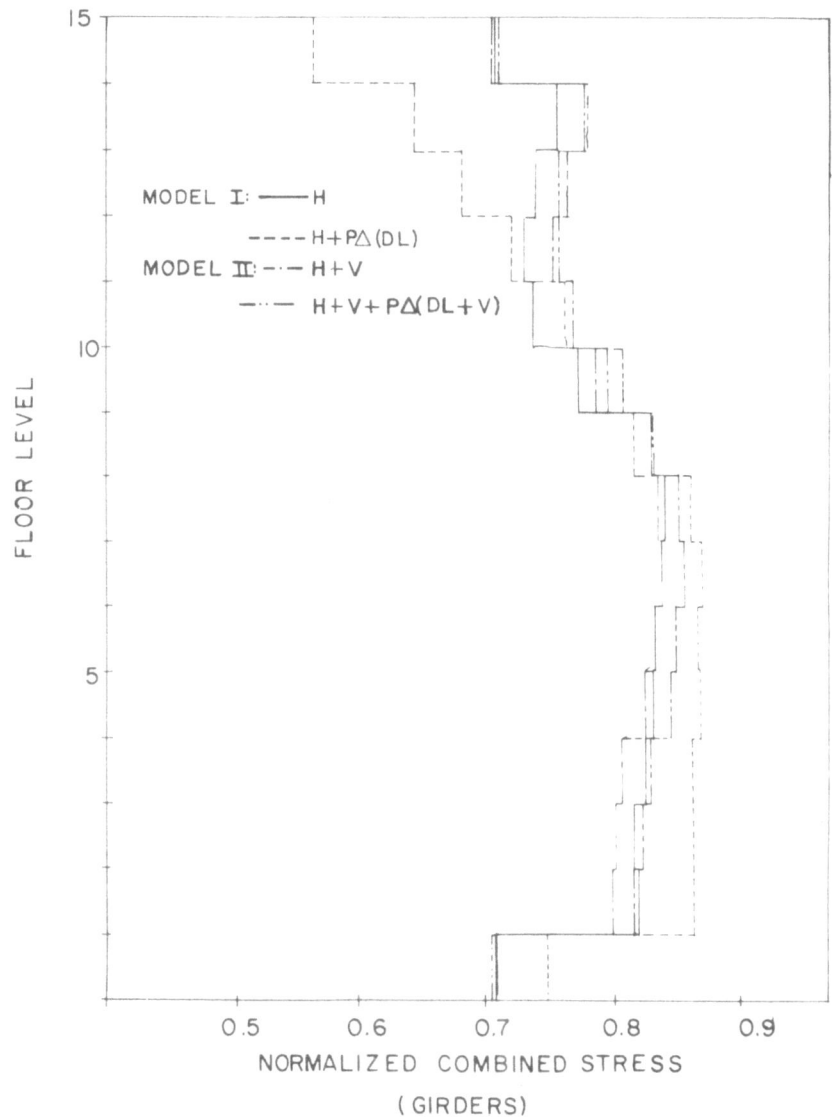

Figure 5.9   Normalized Combined Stresses of Unbraced-Frame
             Girders

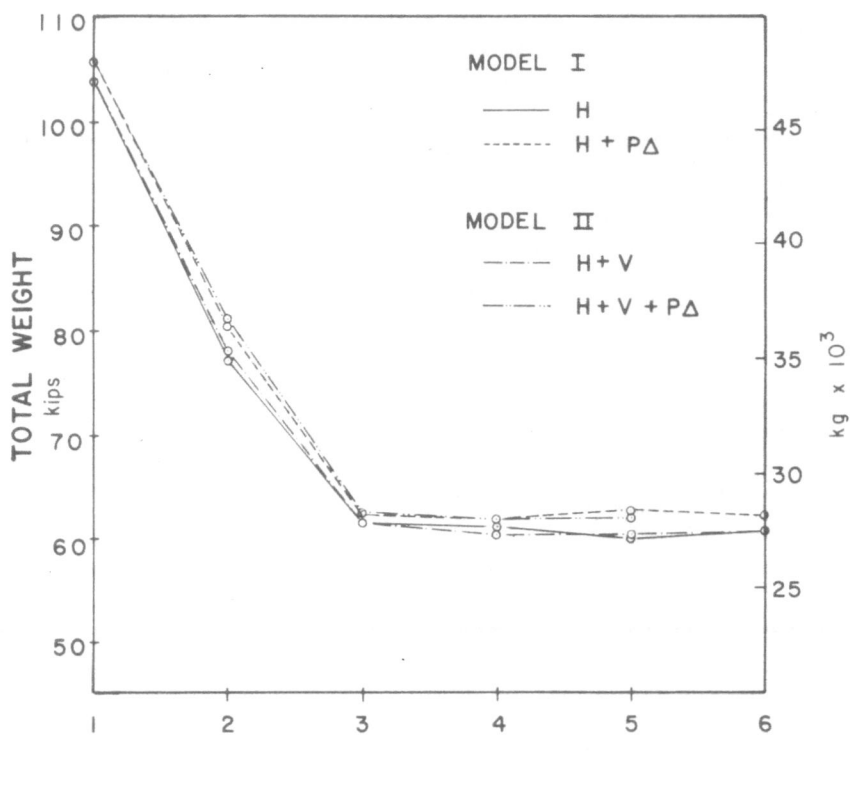

NUMBER OF ITERATIONS

Figure 5.10   Structural Weight Vs. Design Cycles of Unbraced
               Frame

is not as great as for the second through the fourth floors.   The
stiffness requirement is almost the same for the fifth through
eleventh floors.   (3) The actual combined stresses of both
columns and girders do not reach the allowable stress.   The
design is actually controlled by the allowable displacements.
(4) The design converges rapidly in the first three cycles and
does not require more than six cycles.   (5) The first three modes
and the first five modes are adequately accurate for the design
for Models I and II, respectively.

     The final design results for single-braced systems, in terms
of the moments of inertia and the cross-sectional areas for Models
I and II, are shown in Figs. 5.12 through 5.14, for all the four
loading cases.   The design-result comparisons corresponding to
Case (d) of Model II, for the single-, double-, and K-braced

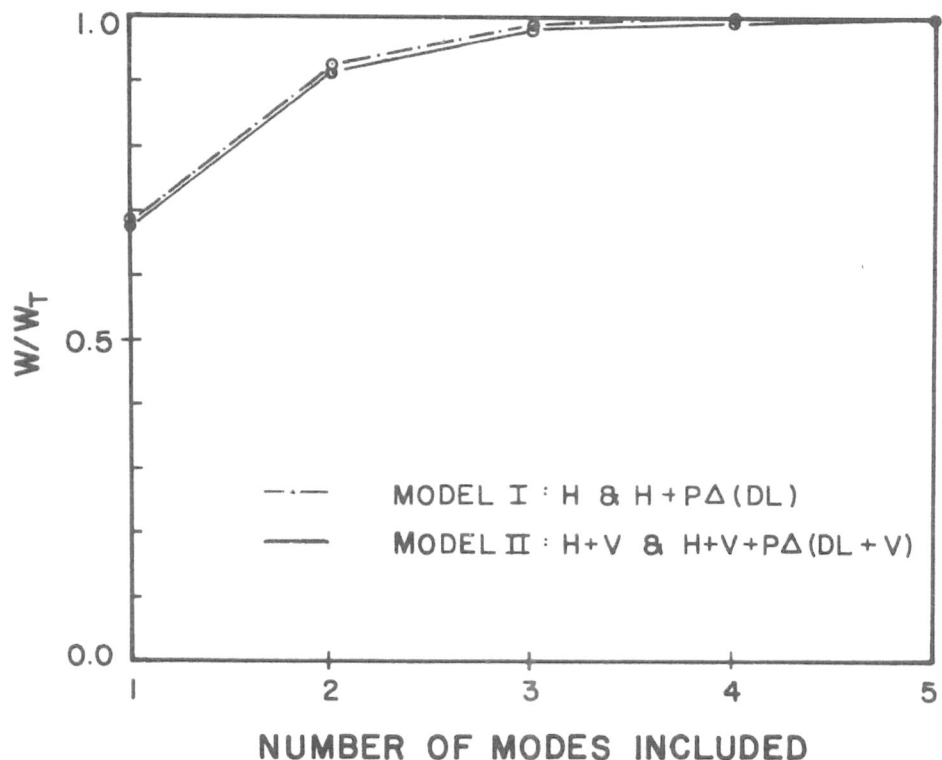

Figure 5.11   Contribution of Energy from Individual Modes in
                Unbraced-Frame Design

systems, are shown in Figs. 5.15 through 5.20.   The final results
for structural weight, natural periods, and displacements at the
top floor are given in Table 5.1.

5.2 Conclusions

        General observations of the results are as follows: (1) The
optimality criteria method is presented for five versatile design
conditions of braced and unbraced frames, with multicomponent inputs of static loads and seismic excitations.   (2) The inclusion
of the vertical seismic component and the P-Δ effect in Model II
yields the heaviest design, among all four cases.   (3) The moments
of inertia of columns of all the structures are the largest at the
base and then become gradually smaller from the bottom to the top
floor.   However, the moment of inertia of the first floor girder

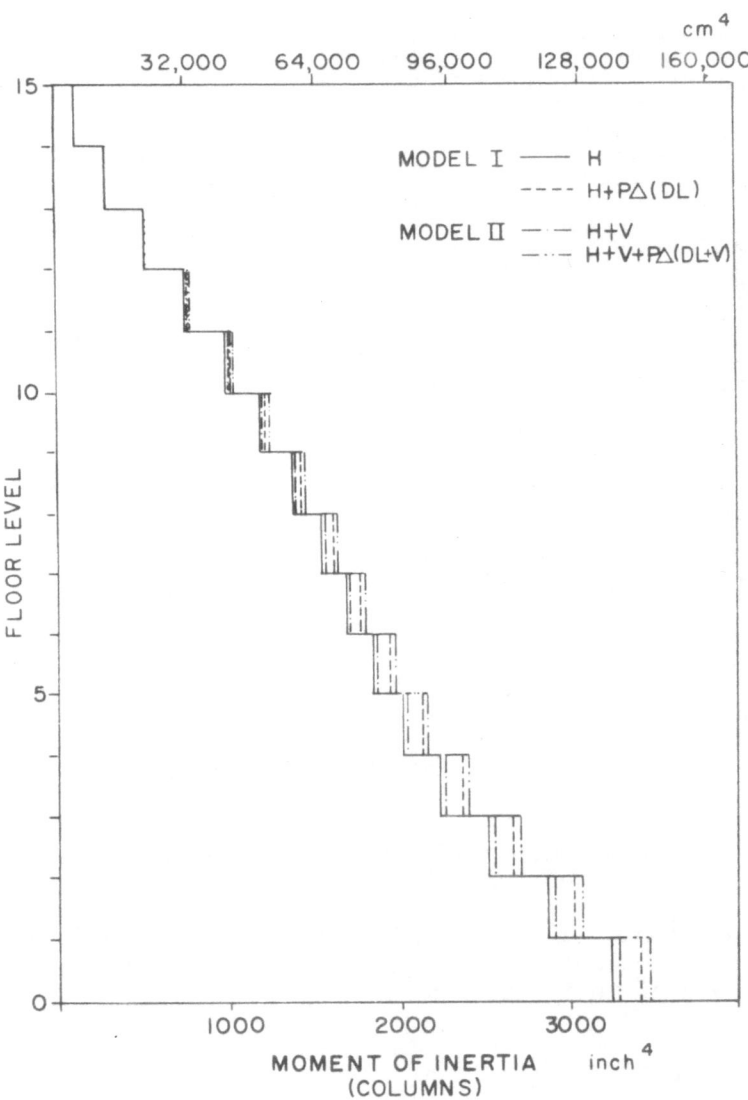

Figure 5.12   Moment of Inertia of Single-Braced-Frame Columns

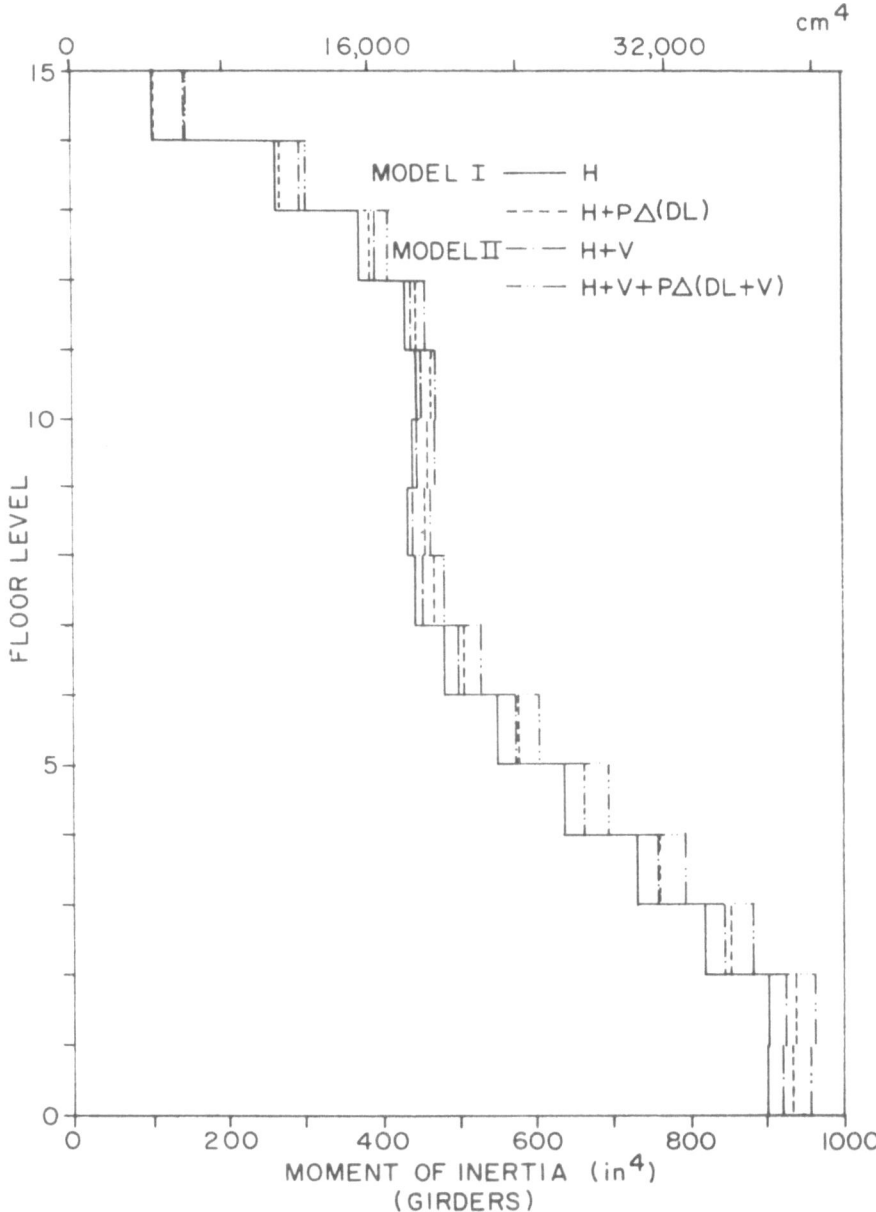

Figure 5.13   Moment of Inertia of Single-Braced Frame Girders

676

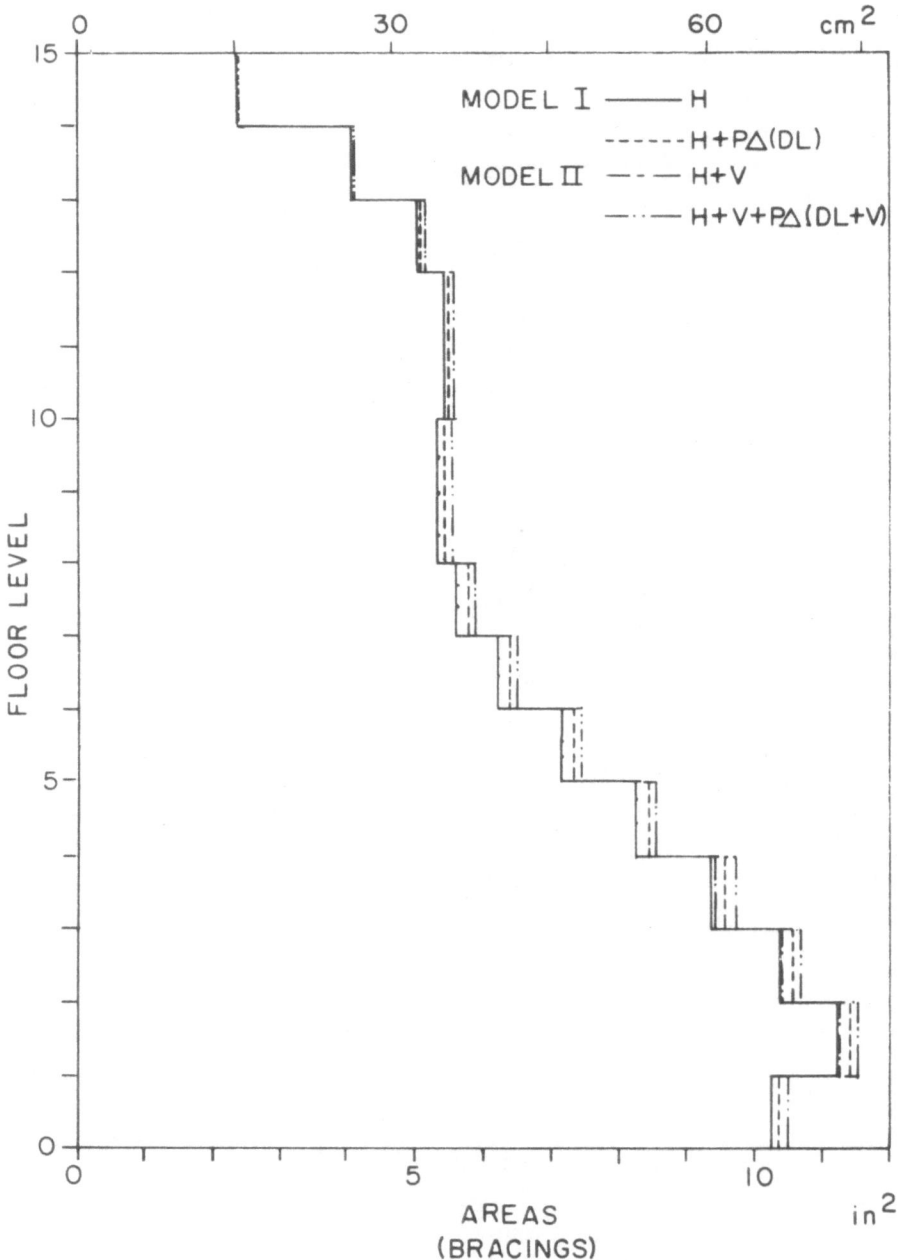

Figure 5.14   Bracing Areas of Single-Braced Frames

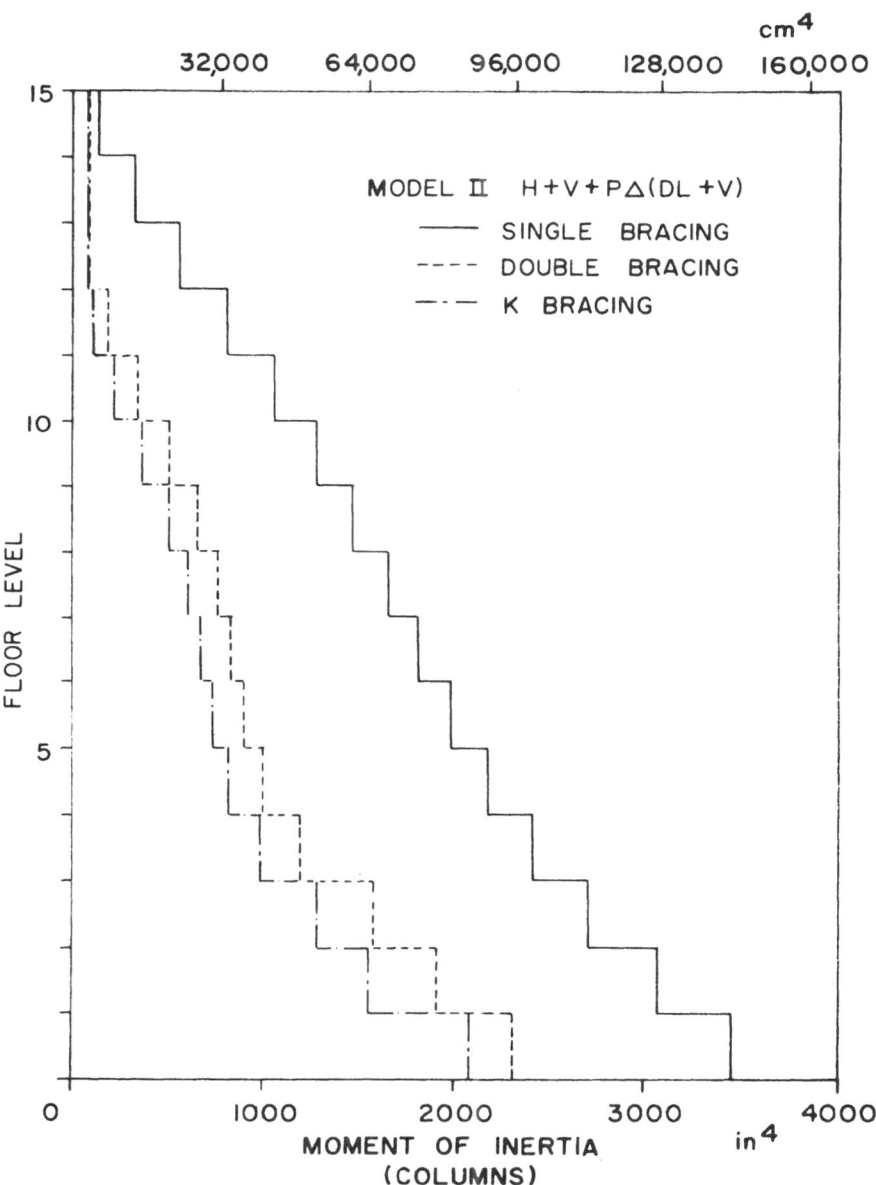

Figure 5.15   Comparison of Moment of Inertia of Columns of
Single-, Double-, and K-Braced Frames

678

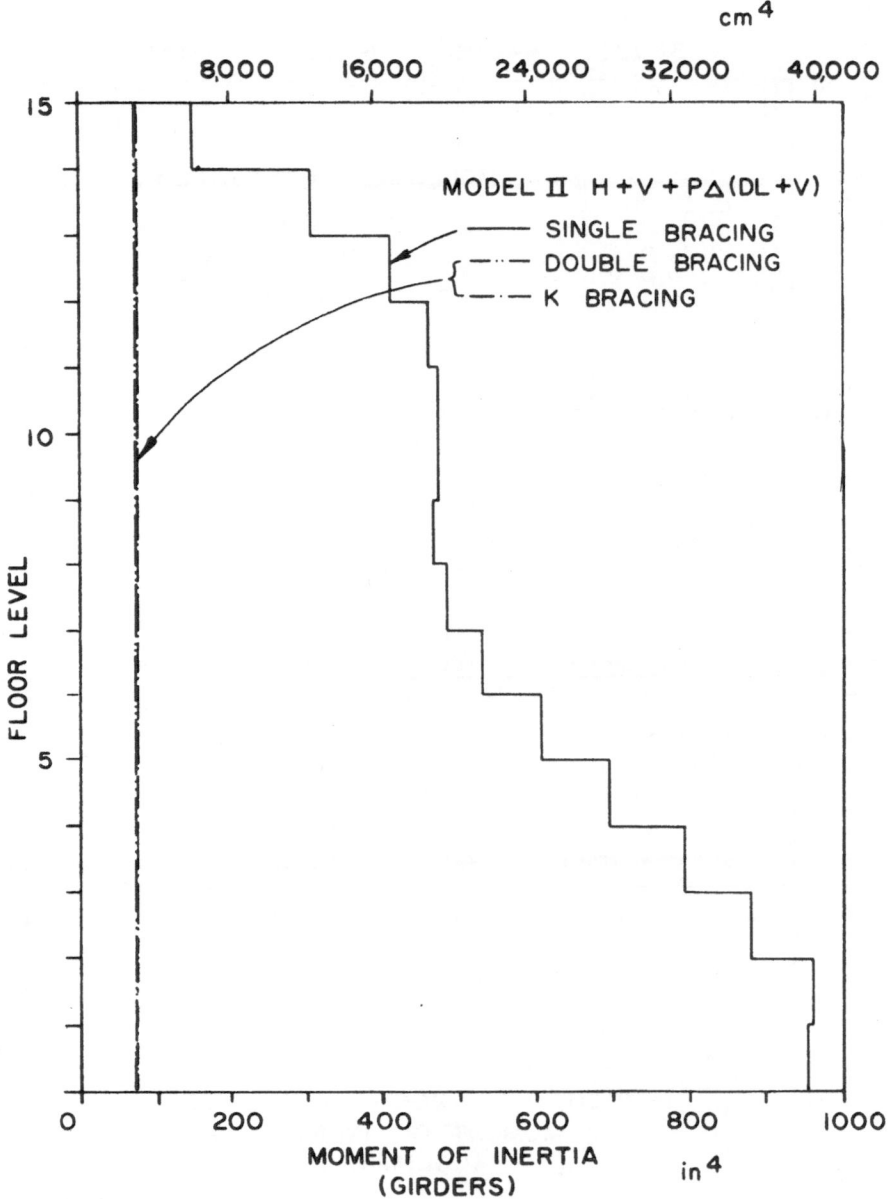

Figure 5.16   Comparison of Moment of Inertia of Girders of
              Single-, Double-, and K-Braced Frames

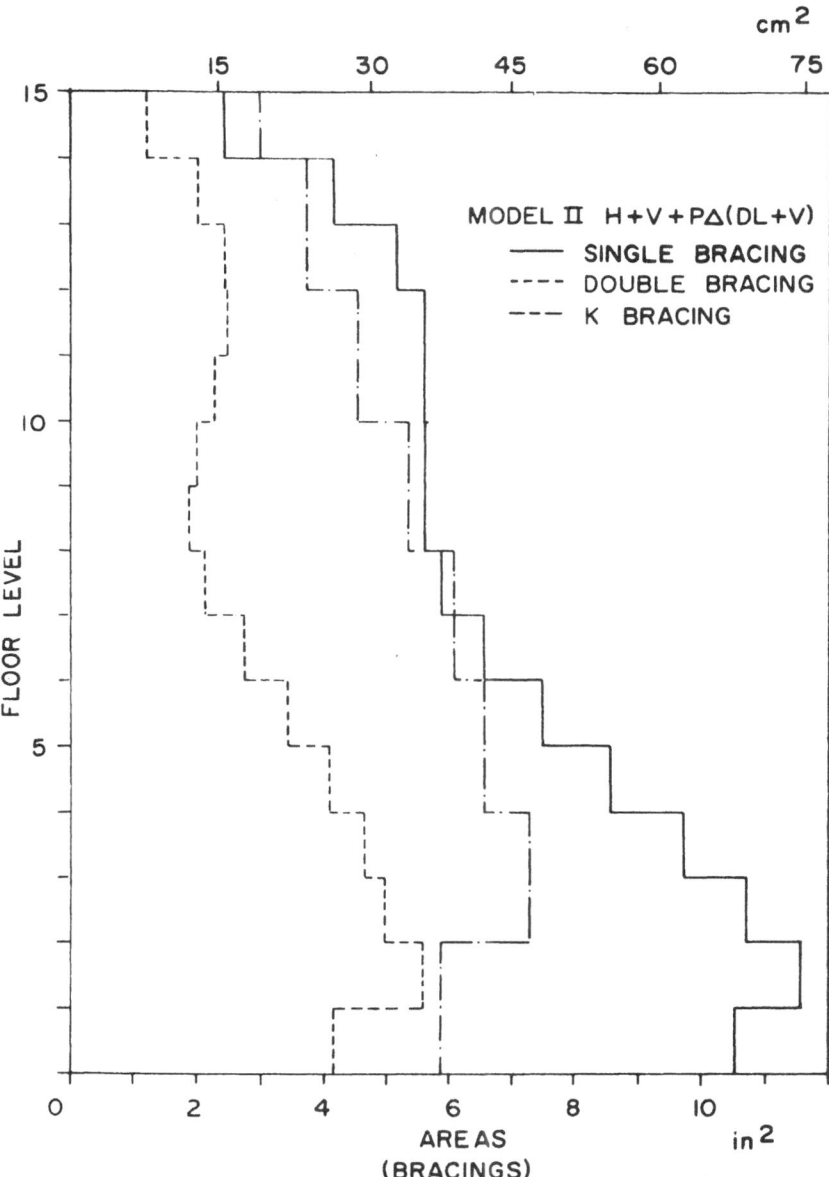

Figure 5.17   Comparison of Bracing Areas of Single-, Double-
and K-Braced Frames

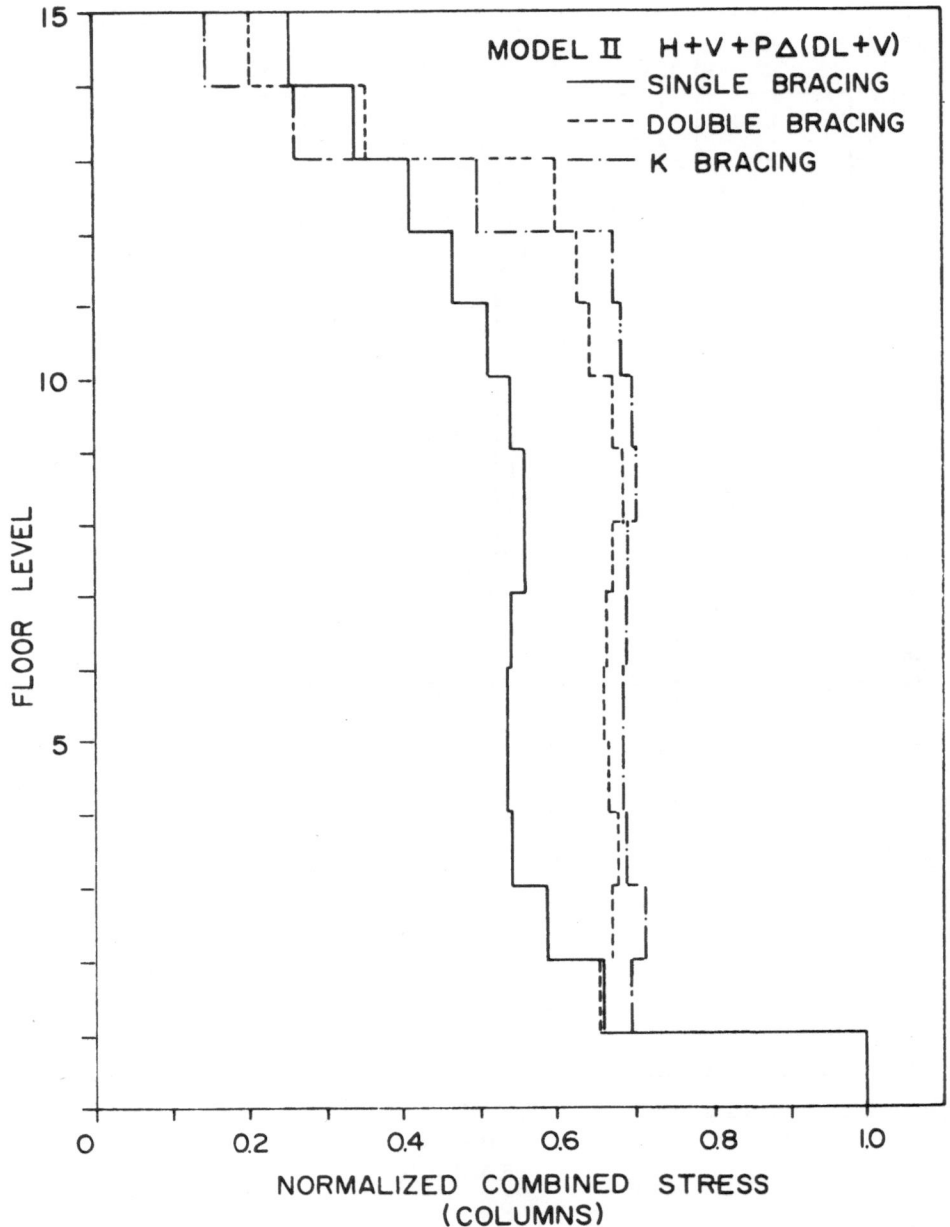

Figure 5.18   Comparison of Normalized Combined Stresses of
Columns of Single-, Double- and K-Braced Frames

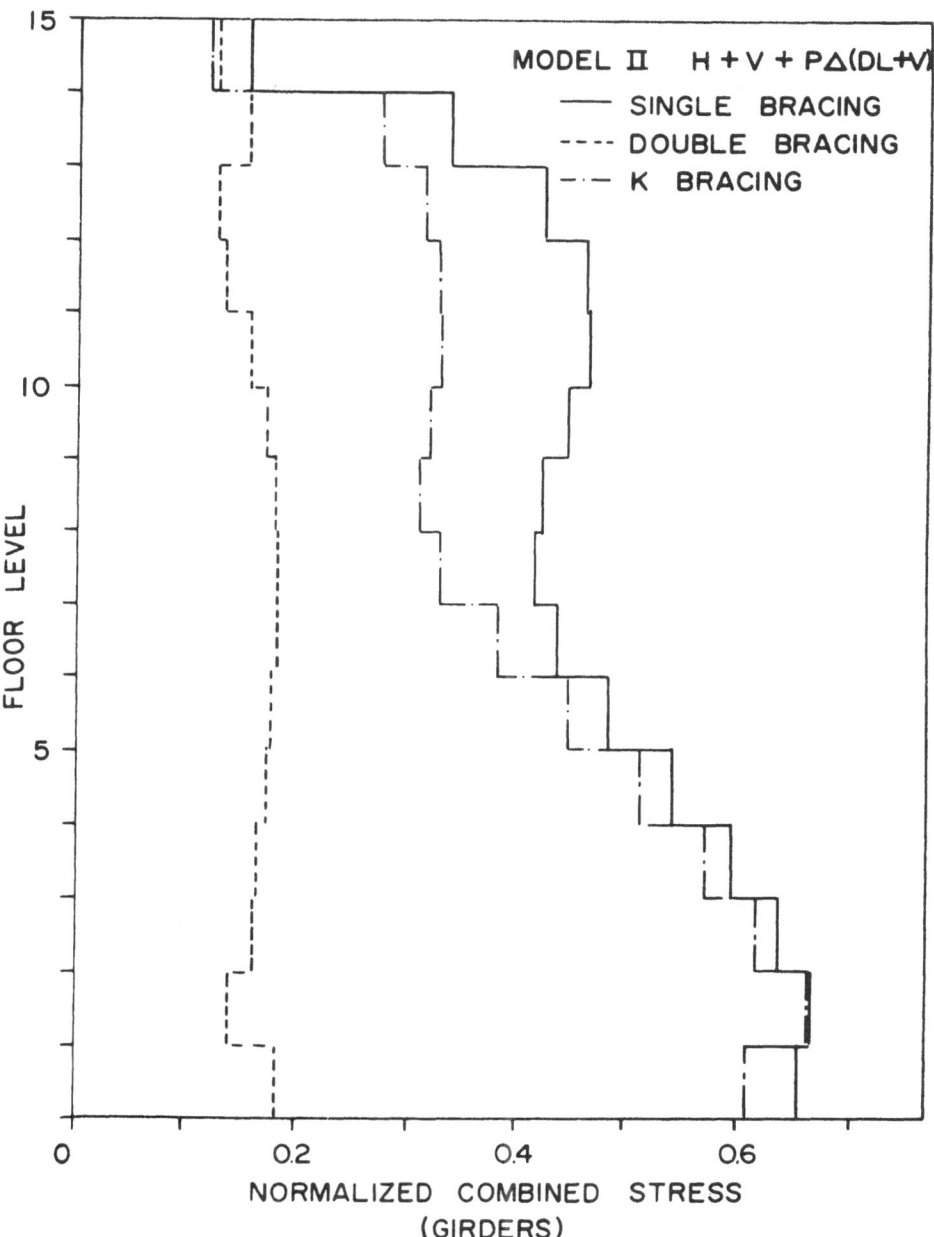

Figure 5.19   Comparison of Normalized Combined Stresses of
Girders of Single-, Double-, and K-Braced Frames

682

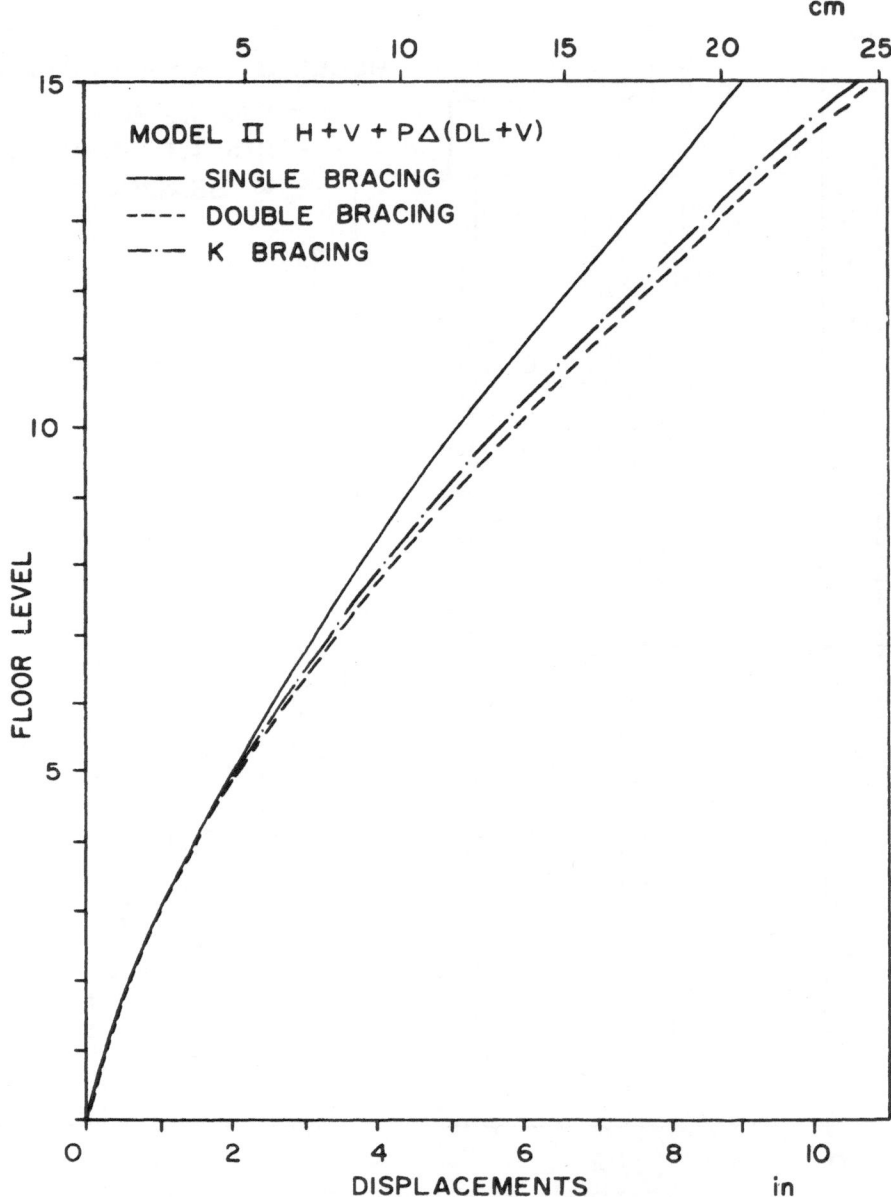

Figure 5.20   Comparison of Displacements of Single-, Double-, and K-Braced Frames

of the unbraced frame is smaller than those of the next three upper floors and those of the fifth through the eleventh floors are almost the same. The moments of inertia of girders of double- and K-braced systems are governed by the lower bound constraints. (4) The K-braced system demands the lightest structural design and seems to be the most favorable system, among these four. (5) The designs of all the braced frames are controlled by the combined stress of axial load and bending of the columns at the support, all other members are not fully stressed. (6) The displacements of the K- and double-braced systems are similar, but are larger than those of the single-braced frame.

TABLE 5.1   FINAL WEIGHTS, NATURAL PERIODS, AND DISPLACEMENTS AT TOP FLOOR (A = Unbraced, B=Single-Braced, C=Double-Braced, D=K-Braced) (1 kip-453 kg, 1 in.=2.54 cm)

Group	Case	Final Weight (kips)	Natural Period (sec.) 1	2	3	4	5	Disp. at Top Floor (in.)
A	a	60.44	2.152	0.694	0.404	0.282	0.212	10.80
	b	62.21	2.179	0.671	0.386	0.268	0.202	10.80
	c	60.77	2.158	0.692	0.402	0.281	0.217	10.80
	d	62.21	2.152	0.677	0.395	0.276	0.212	10.80
B	a	48.32	1.879	0.490	0.256	0.247	0.187	9.00
	b	47.69	1.876	0.479	0.274	0.252	0.209	9.05
	c	48.76	1.874	0.488	0.291	0.255	0.224	9.02
	d	50.01	1.874	0.485	0.288	0.253	0.221	9.02
C	d	31.55	2.134	0.509	0.398	0.361	0.335	10.80
D	d	28.61	2.111	0.519	0.274	0.269	0.178	10.55

REFERENCES

1.   Cheng, F.Y. and Botkin, M.E. (1966), "Nonlinear Optimum Design of Dynamic Damped Frames," Journal of the Structural Division, ASCE, Vol. 102, No. ST3, 609-628.
2.   Cheng, F.Y. and D. Srifuengfung (1978), "Optimum Structural Design for Simultaneous Multicomponent Static and Dynamic Inputs," International Journal for Numerical Methods in Engineering, Vol. 13, No. 2, 353-372.
3.   Venkayya, V.B. and Cheng, F.Y. (1966), "Resizing of Frames Subjected to Ground Motion," Proc. Intl. Symp. on Earthquake Structural Engineering, University of Missouri-Rolla, 1, 597-612.
4.   Venkayya, V.B. (1978), "Structural Optimization: A Review and Some Recommendations," Intl. J. Num. Methods Eng., 13(2), V.B. Venkayya, Ed.).

Part 4

FINITE DIMENSIONAL STRUCTURAL OPTIMIZATION

STRUCTURAL AND MECHANICAL DESIGN VIA OPTIMALITY CRITERION METHODS

Kenneth D. Willmert

Mechanical and Industrial Engineering Department,
Clarkson College, Potsdam, New York 13676

ABSTRACT

A summary of structural and mechanical design work done using
optimality criterion techniques is presented. Numerous examples
are treated, including truss and frame design under stress, displacement, and frequency constraints. Also discussed are problems involving plane elements and ones in which the members
possess several design variables each. The optimization of high
speed mechanical mechanisms, accounting for vibrational effects,
is another area presented using optimality criterion methods.

1. INTRODUCTION

Optimality criterion techniques have become of great interest
in the design of structural and mechanical systems because of
their efficiency. A number of techniques have been generated for
the design of large scale trusses and frames under static loading,
which have recently been extended to structures involving plane
elements (constant strain triangular and symmetric rectangular
shear panels). Constraints have included stress, displacement,
buckling, and minimum gauge limitations. Beams and frames under
natural frequency constraints have also been considered. An
optimality criterion method has been developed for structural
design in which each member has several design variables associated with it. The research reported here represents an extension
of optimality criterion methods to the design of high speed
mechanical mechanisms under vibrational effects. Current efforts
involve the extension of these techniques to transient dynamic
structural design, and to problems involving changes in topology.

The techniques presented here have been derived from the Kuhn-Tucker conditions, based for the most part on a basic assumption that at any stage of the optimization only one constraint is most active or most violated. This leads to a simple optimization technique involving very little calculation at each stage of the method. For example only one Lagrange multiplier need be calculated and only one constraint derivative is required. Methods based on this idea have been shown to be very efficient and able to locate a near optimal design in a small number of iterations (and analyses). The details of the techniques have been previously reported by Khan, Thornton, Syed and Willmert in a number of research papers [1-9]. As a result they will not be repeated here. Presented in this paper are the results of several examples solved by these methods and comparisons with results obtained with other methods, when available.

2.   TRUSS AND FRAME STRUCTURES SUBJECTED TO STRESS AND DISPLACEMENT CONSTRAINTS

In truss and frame design the problem is to select the cross sectional sizes (characterized by a single variable such as the area) of the members subjected to static loads. The design objective is minimum weight, subject to limitations on the maximum stress in the members and displacements at the nodes.

2.1 Three Member Frame

The first example considered is the three-member frame shown in Fig. 2.1. A single load condition is applied, with values as indicated in the diagram. The areas of the three members are taken as the design variables, with minimum volume (equivalent to weight) as the objective. The stress in the members (combined axial and bending) is limited to 24,000 psi and no displacement limits are imposed.

The problem was solved by an optimality criterion method and by two other techniques. These were a standard SUMT nonlinear mathematical programming approach (this is a general interior penalty function technique, using Powell's unconstrained minimization, which has seen wide application on a variety of design problems), and a special sequential linear programming method developed by Briggs and Willmert [10] and modified by Calafell and Willmert [11]. This linear technique was originally developed to provide a very efficient method of solving structural optimization problems. Tables 2.1 shows the resulting optimal designs from the three techniques. The problem was solved first assuming the ratio between the moment of inertia of the cross-section and the area to be 75 and the ratio between the section modulus and

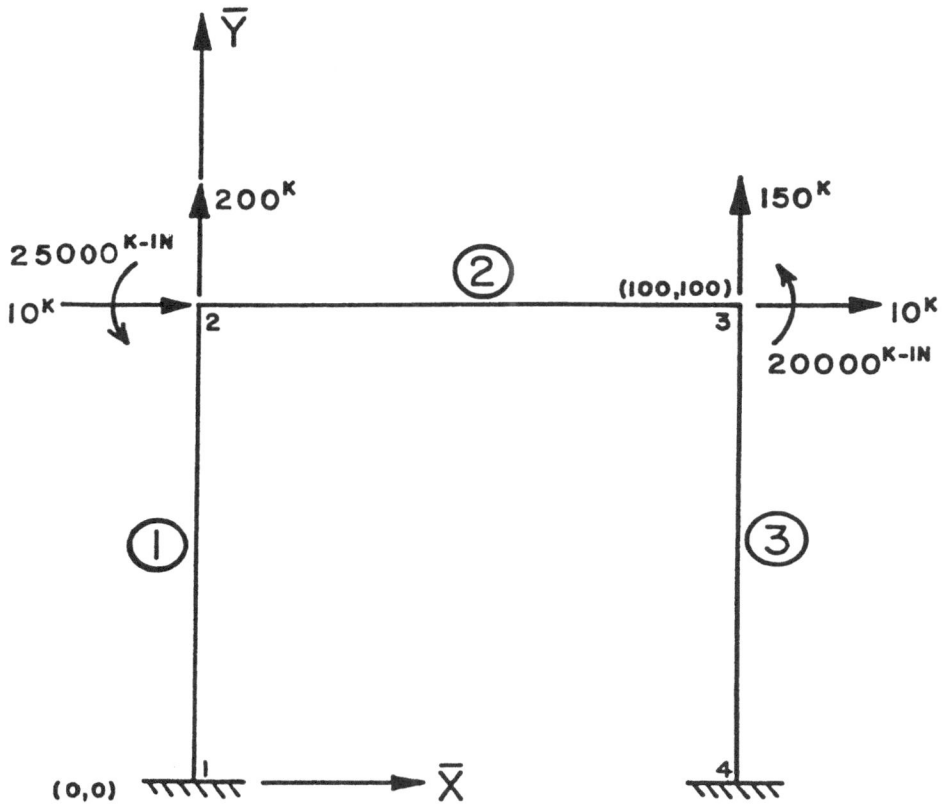

Figure 2.1 Three-Member Frame

the area to be 9, which are reasonable approximations for wide flage steel shapes. This assumption was used by Briggs and Willmert [10]. The results are shown in the first three columns of Table 2.1. The problem was then solved again using a complex relationship between moment of inertia, section modulus, and area as developed by Calafell and Willmert [11], which provides better correlation for economy wide flange sections. The results in this case are shown in the last two columns of Table 2.1.

As can be seen from the first three columns of Table 2.1, the linear, nonlinear, and optimality criterion techniques converged to approximately the same design. Comparison of the computational times (on an IBM 360/65) showed the optimality criterion technique to be far superior to both the linear and non-linear methods, even though the special linear method was better than the nonlinear technique. The starting point for both the optimality criterion and SUMT techniques was 90 sq. in. for all areas. The last two columns of Table 2.1 show that the technique can be used in the case of varying moment of inertia and section modulus relative to the areas. The computational advantage appears here as well.

TABLE 2.1.   RESULTS FOR THREE MEMBER FRAME

	Fixed Property			Variable Property	
	Briggs Ref. 10	This Paper	SUMT	This Paper	Calafell Ref. 11
No. of Cycles	6	3	7	5	7
CPU (sec)	10.19	1.32	42	1.22	11.85
A_1	19.74	20.15	19.68	18.527	18.44
A_2	105.38	104.95	105.43	78.258	78.38
A_3	30.13	30.32	30.12	35.042	32.37
Volume (in^3)	15525	15542	15526	13183	12919

2.2   Variable Cross-Sectional Size Three-Member Frame

As a second example, the cross-sectional size of the members in the three-member frame of the first example (same loading and geometry) is allowed to vary along their lengths, as shown in Fig. 2.2, in a piecewise constant manner.

Analysis of the structure was done using a finite-element technique. To save computational time in the analysis (although not required for a small problem of this type), a new super-element/subelement concept was introduced. Since the area in each subelement (see Fig. 2.2) is allowed to vary independently, a standard finite-element analysis of the frame using one element per member cannot be used. Each subelement would have to be taken as an element. Large numbers of subelements should be taken so that they can approximate closely a continuously vary-ing cross-sectional size. This would result in a large number of variables (deflections or internal forces) in the analysis, thus increasing computational time and storage. Generally, however, it is unnecessary to use a large number of finite elements per member for problems of this type to obtain accurate values for

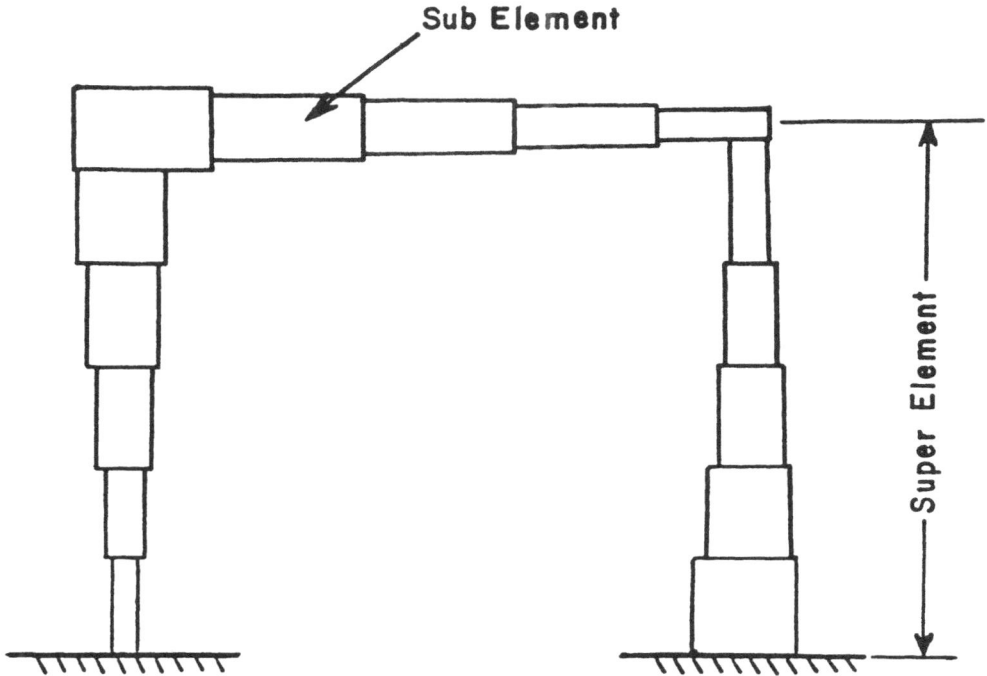

Figure 2.2 Superelement Representation of Three-Member Frame

stresses, which are required in the optimization. Thus the
superelement concept was introduced. Here the entire member is
taken as one element, but the integrals required in the finite-
element analysis are divided into pieces corresponding to the
subelements. The components of the element stiffness matrix for
each member are then composed of a sum of terms corresponding to
the subelements, each involving a separate cross-sectional area
as a variable.

Table 2.2 shows results obtained by applying the optimality
criterion technique to this problem using 1, 5, and 10 subelements per member. As expected, results for one subelement
were the same as those obtained in the previous example. Also,
the optimum volume of the structure decreased as the number of
subelements increased. This is due to the increased capability
of the structure to distribute the mass more effectively to
support the loads. The summary of Table 2.2 gives the computational times for each of the three cases and the number of cycles.
Note that the number of cycles (analyses) remained about the same
as the number of subelements (variables) increased, and the
computational time increased only slightly. The ten subelement

TABLE 2.2.   SUPER ELEMENT MODELS OF THREE MEMBER FRAME

(a)   One Sub element per member

Members	A
1	20.15
2	104.95
3	30.32

(b)   Five Sub elements per member

Members	A_1	A_2	A_3	A_4	A_5
1	12.18	13.15	14.13	15.10	16.08
2	109.43	105.35	101.27	97.18	93.10
3	26.24	25.04	25.53	26.73	27.94

(c)   Ten Sub elements per member

Members	A_1	A_2	A_3	A_4	A_5	A_6	A_7	A_8	A_9	A_{10}
1	11.73	12.20	12.67	13.13	13.60	14.07	14.54	15.0	15.47	15.94
2	109.58	107.73	105.48	103.43	101.38	99.33	97.28	95.23	93.23	91.14
3	26.07	25.49	24.92	24.41	24.98	25.56	26.13	26.71	27.28	27.86

(d)   Summary

No. of Sub Elements	1	5	10
No. of Cycles	3	4	4
CPU (sec)	1.32	1.59	1.99
Volume (in^3)	15542	14168	14013

per member case involved 30 variables, but the computational time was still very short. This is contrary to the usual effect on standard nonlinear mathematical programming methods as the number of variables increases, which is generally a drastic increase in computational time.

2.3 Ten Bar Truss

A cantilever truss that has been studied by many researchers [12,13,14,16,17] is shown in Fig. 2.3. Displacement limits of ±2.0 inches are imposed on all nodes, in both the horizontal and vertical directions. The limiting value of stress in each member is ± 25,000 psi. A single load condition is considered, with P_1 = 150 kips, P_2 = 50 kips. A lower limit of member size of 0.1 in^2 is enforced.

The optimality criterion technique was started with each cross-sectional area equal to 100 in^2. The final design is given in Table 2.3. The technique required only 9 function evaluations and compares very well with previously published optimal designs.

2.4 Seventy-two Member Space Truss

The structure shown in Fig. 2.4, has been studied previously in Refs. 12, 13, 14, 15, and 18. Stress limits of ± 25,000 psi

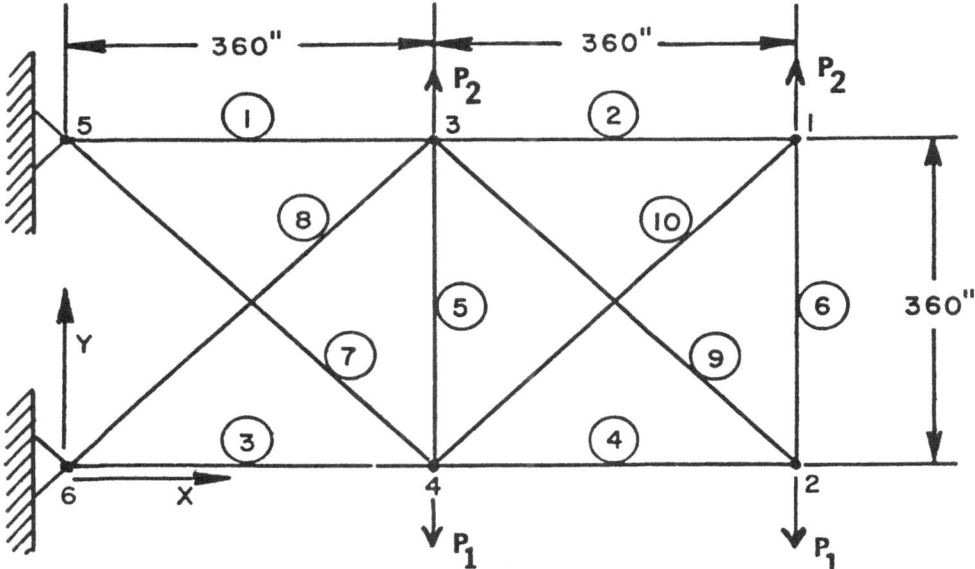

Figure 2.3   Ten Bar Truss

TABLE 2.3. COMPARISON OF FINAL DESIGNS FOR TEN BAR TRUSS

Final Cross-Sectional Areas (in^2)

Member No.	Schmit & Miura NEWSUMT Ref. 13	Schmit & Miura COMMIN Ref. 13	Schmit & Farshi Ref. 12	Venkayya Ref. 14	Gellatly & Berke Ref. 15	Dobbs & Nelson Ref. 15	Rizzi Ref. 17	This Paper
1	23.550	23.55	24.289	25.190	–	25.813	23.533	24.716
2	0.100	0.176	0.100	0.363	–	0.100	0.100	0.100
3	25.290	25.20	23.346	25.419	–	27.233	25.291	26.541
4	14.360	14.39	13.654	14.327	–	16.653	14.374	13.219
5	0.100	0.100	0.100	0.417	–	0.100	0.100	0.108
6	1.970	1.967	1.969	3.144	–	2.024	1.9697	4.835
7	12.390	12.400	12.670	12.083	–	12.776	12.389	12.664
8	12.810	12.860	12.544	14.612	–	14.218	12.825	13.775
9	20.340	20.410	21.971	20.261	–	22.137	20.328	18.438
10	0.100	0.100	0.100	0.513	–	0.100	0.100	0.10
Wt (lbs)	4676.96	4684.11	4691.84	4895.60	–	5059.7	4676.92	4792.52
Analyses	11	10	23	13	–	12	12	9

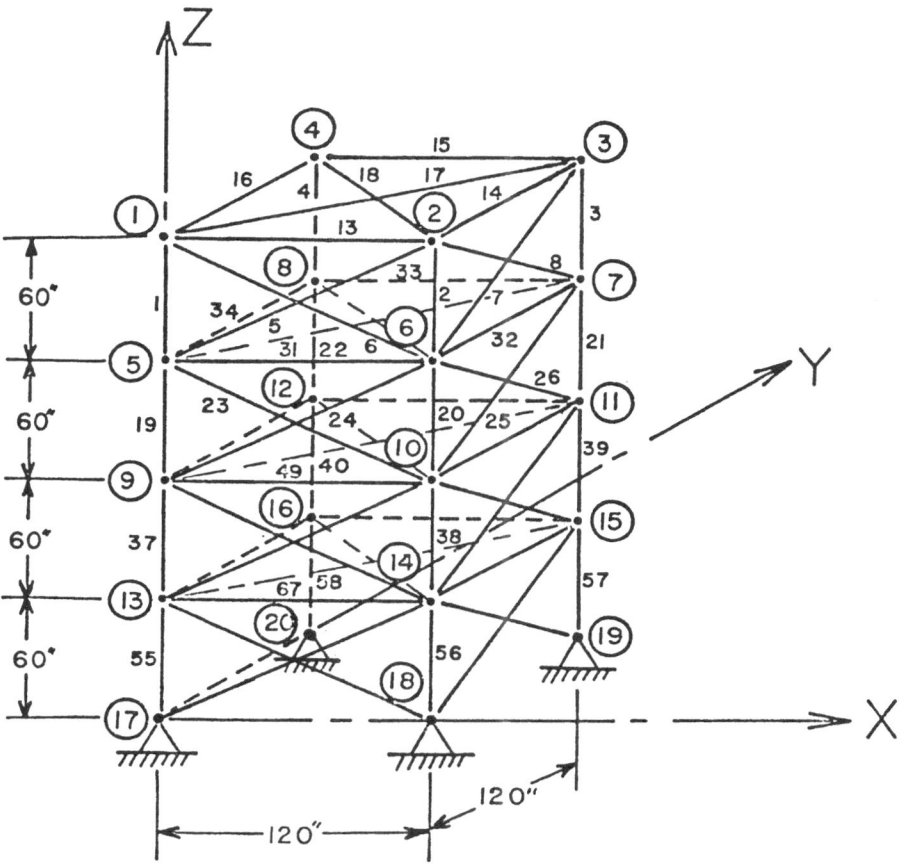

Figure 2.4 Seventy-two Member Špace Truss

are applied on all members. Displacement limits of ±0.25 inch in the x and y directions are imposed on the 4 top nodes. A lower limit of 0.1 in² is imposed on all members. Design variable linking is used. Members are placed in 16 groups as shown in Table 2.4. Thus, there are 16 independent design variables. Two load conditions are considered, as given in Table 2.5.

Table 2.6 gives the final design obtained in this research, and compares this with previous results. The design procedure was started with all members equal to 100 in². At the optimum, in the second load condition the first four members had stresses equal to their limiting values, while the displacements of node 1 in the x and y directions were near their specified limits.

TABLE 2.4. MEMBER LINKING GROUPS, SEVENTY-TWO MEMBER TRUSS

Design Variable Group Number	Members in Group
1	1, 2, 3, 4
2	5, 6, 7, 8, 9, 10, 11, 12
3	13, 14, 15, 16
4	17, 18
5	19, 20, 21, 22
6	23, 24, 25, 26, 27, 28, 29, 30
7	31, 32, 33, 34
8	35, 36
9	37, 38, 39, 40
10	41, 42, 43, 44, 45, 46, 47, 48
11	49, 50, 51, 52
12	53, 54
13	55, 56, 57, 58
14	59, 60, 61, 62, 63, 64, 65, 66
15	67, 68, 69, 70
16	71, 72

TABLE 2.5. LOAD CONDITIONS, SEVENTY-TWO MEMBER TRUSS

Load Condition	Node	Direction		
		x	y	z
1	1	5 K	5 K	-5 K
2	1	0	0	-5 K
	2	0	0	-5 K
	3	0	0	-5 K
	4	0	0	-5 K

2.5 Twenty-five Member Frame

The structure shown in Fig. 2.5 has one load condition, as indicated on the diagram. All members are 100 inches in length, except the diagonal elements. Stress limits are ±24,000 psi for all members. Two cases are considered. Case 1 has the above stress limits and displacements limites of ±3.0 inches at joints 1 through 6, in both x and y directions. Case 2 has the above stress limits and displacement limits of ± 0.05 inches at the same same joints. The minimum member size is 5 in^2.

TABLE 2.6. FINAL DESIGNS, SEVENTY-TWO MEMBER TRUSS

Final Cross-Sectional Areas (in^2)

Members of Group	Schmit & Miura		Schmit & Farshi	Venkayya	Gellatly & Berke	Berke & Khot	This Paper
	NEWSUMT Ref. 13	CONMIN Ref. 13	Ref. 12	Ref. 14	Ref. 15	Ref. 18	
1	0.1565	0.1558	0.1585	0.161	0.1492	0.1571	0.1494
2	0.5458	0.5484	0.5936	0.557	0.7733	0.5385	0.5698
3	0.4105	0.4105	0.3414	0.377	0.4534	0.4156	0.4434
4	0.5699	0.5614	0.6076	0.506	0.3417	0.5510	0.5192
5	0.5233	0.5228	0.2643	0.611	0.5521	0.5082	0.6234
6	0.5173	0.5161	0.5480	0.532	0.6084	0.5196	0.5231
7	0.1000	0.1000	0.1000	0.100	0.1000	0.1000	0.100
8	0.1000	0.1133	0.1509	0.100	0.1000	0.1000	0.1963
9	1.267	1.268	1.1067	1.246	1.0235	1.2793	1.2076
10	0.5118	0.5111	0.5792	0.524	0.5421	0.5149	0.5208
11	0.1000	0.1000	0.1000	0.100	0.1000	0.1000	0.100
12	0.1000	0.1000	0.1000	0.100	0.1000	0.1000	0.100
13	1.885	1.885	2.0784	1.818	1.4636	1.8931	1.7927
14	0.5125	0.5118	0.5034	0.524	0.5207	0.5171	0.5223
15	0.1000	0.1000	0.1000	0.100	0.1000	0.1000	0.100
16	0.1000	0.1000	0.1000	0.100	0.1000	0.1000	0.100
Final Wt (lbs)	379.640	379.792	388.63	381.2	395.97	379.67	386.718
Analyses	9	8	22	12	9	5	13

698

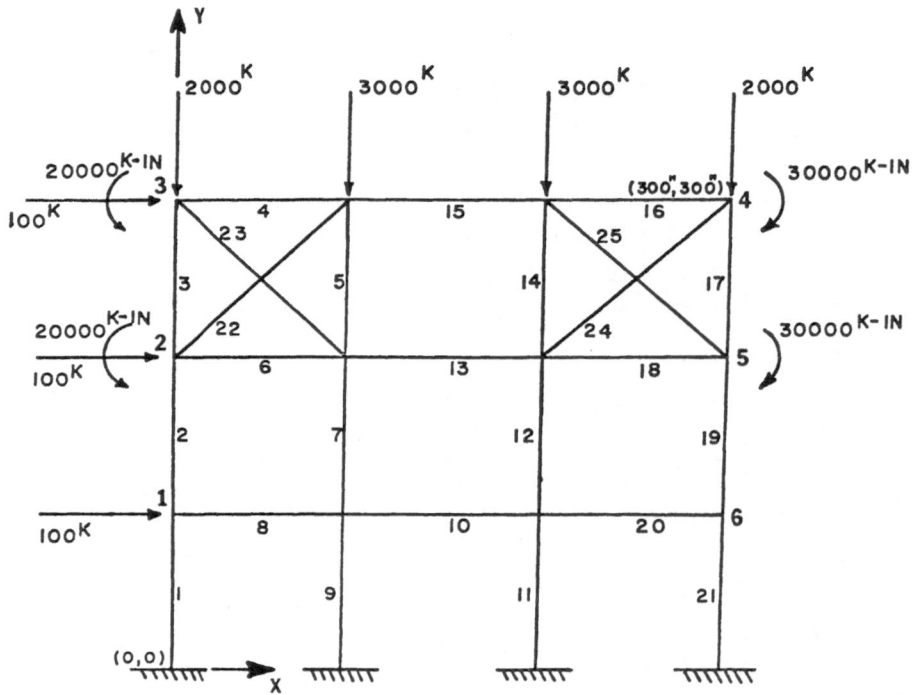

Figure 2.5   Twenty-five Member Frame

The resulting optimal designs are shown in Table 2.7.   Case 1 is compared with results from Ref. 10, with excellent agreement in the designs.   The method of this paper can be seen to produce the optimal design, with a drastic reduction in CPU time required for the method of Ref. 10.   The design for case 1 is fully stressed at the optimum and the displacement limits are inactive. The case 2 design is displacement constrained, with no active stress constraints.   No previous results were available for comparison in this case.

3.   STRUCTURES CONTAINING PLANE ELEMENTS

Structural design problems have also been solved involving different types of elements, particularly ones composed of truss elements, plane stress triangular elements, and symmetric rectangular shear panels.   The design variables for the plane elements are taken as the thickness of members.

The wing structure shown in Fig. 3.1 is assumed to be symmetric with respect to the middle surface.   As a result, only

TABLE 2.7.   FINAL DESIGN COMPARISON FOR
TWENTY-FIVE MEMBER FRAME

| | Member Cross-Sectional Areas (in^2) | | |
| | Case 1 | | Case 2 |
Member Number	Briggs Ref. 10	This Paper	This Paper
1	138.00	129.55	337.79
2	148.58	153.26	293.07
3	154.08	151.29	162.15
4	28.34	31.63	69.68
5	128.93	133.64	170.02
6	5.00	5.00	52.19
7	130.10	131.58	217.43
8	15.72	23.37	108.73
9	162.97	170.80	233.83
10	5.00	5.00	110.55
11	120.97	119.91	170.29
12	111.06	110.78	181.92
13	5.00	5.00	87.35
14	122.00	123.06	109.72
15	5.00	5.00	105.77
16	52.96	54.00	147.31
17	191.76	190.80	233.56
18	5.00	5.00	191.63
19	119.70	123.13	336.97
20	5.00	5.00	199.56
21	123.74	119.32	465.20
22	8.61	5.00	191.92
23	5.00	5.00	88.26
24	5.00	5.00	84.14
25	48.78	48.67	95.59
Vol. (in^3)	187421	188215	463523
Analyses	(a)	15	10
CPU (b) (sec)	1849.00	69.02	38.79

(a) Not Applicable

(b) All times on IBM 360/65

700

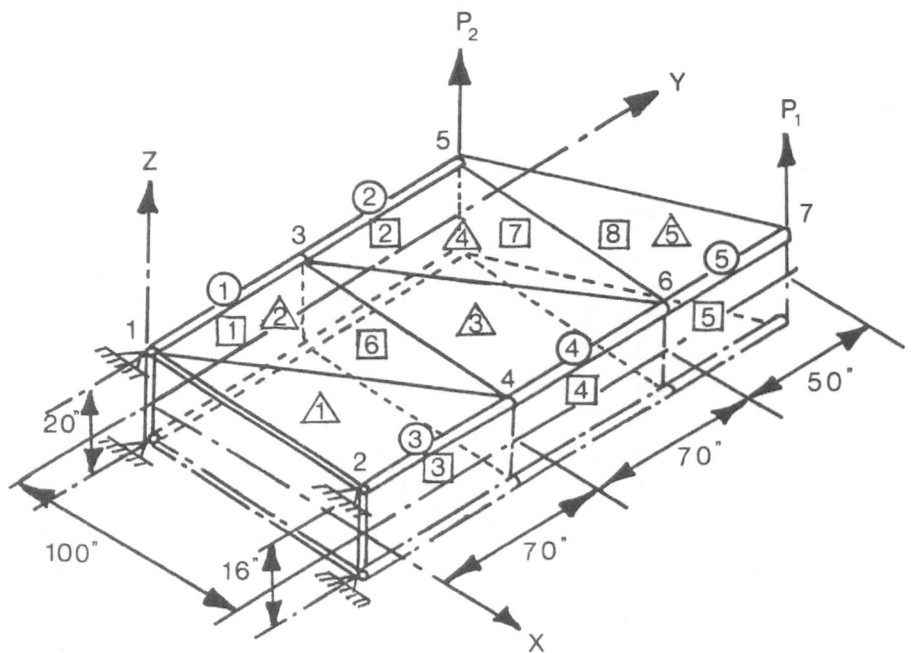

○ TRUSS ELEMENT

△ CST ELEMENT

□ SSP ELEMENT

Figure 3.1   Finite Element Model for Eighteen Element Wing
             Box Structure

the upper half of the structure is considered in the optimization.
In this half, there are 5 truss elements, 5 constant strain triangular elements, and 8 rectangular shear panel elements.   The
stress limit used for all elements is 10,000 psi, the lower bound
on truss areas is 0.1 in^2 and on plane stress element thicknesses
is 0.02 in.   Two different load conditions are considered; 5000
lbs. at node 7 (as shown in Figure 3.1), and 10,000 lbs. at node
5, both in the z direction.   Displacement limits of 2.0 inches
are imposed at all nodes, in all directions.

This problem was solved by several other researchers [13,19,
20], with results shown in Table 3.1.   It is difficult to compare
these designs directly, since different finite element models of
the structure were used in each case.   Thus some variation in the
optimum designs are observed.   However the structures are similar.

TABLE 3.1. FINAL DESIGNS FOR 18 ELEMENT WING BOX,
ALLOWABLE DISPLACEMENT 2.0 INCHES

Member No.	Schmit and Miura			Gallatly & Berke	Gallatly*	This Research
	CST Model 1 SSP	CST Model 2 SSP	CST Model 1 Shear Web			
TRUSS	$A_i(in^2)$	$A_i(in^2)$	$A_i(in^2)$	$A_i(in^2)$	$A_i(in^2)$	$A_i(in^2)$
1	4.045	3.151	2.229	0.6505	1.0431	0.9735
2	0.1001	0.1000	0.1001	0.1001	0.1036	0.1301
3	0.1001	0.1000	0.1000	0.2366	0.3508	0.2339
4	0.1330	0.2324	0.3202	0.2352	0.3315	0.3652
5	0.1002	0.1000	0.1001	0.1001	0.1035	0.1479
CST	$T_j(in)$	$T_j(in)$	$T_j(in)$	$T_j(in)$	$T_j(in)$	$T_j(in)$
1,2	0.08286	0.08641	0.1093	0.1328	**0.1441	0.11131
3,4	0.05363	0.05733	0.05911	0.0702	**0.0599	0.05139
5	0.03786	0.03932	0.04098	0.0449	0.0435	0.03555
SSP	$T_j(in)$	$T_j(in)$	$T_j(in)$	$T_j(in)$	$T_j(in)$	$T_j(in)$
1	0.3636	0.3851	0.09345	0.0876	0.0876	0.53075
2	0.2236	0.2152	0.09437	0.0889	0.0895	0.22626
3	0.1310	0.1361	0.07687	0.0808	0.0664	0.11120
4	0.1156	0.1004	0.07293	0.0768	0.0553	0.16274
5	0.09166	0.09113	0.07570	0.0815	0.0537	0.10408
6	0.02000	0.02000	0.02001	0.0200	0.0219	0.02070
7	0.02000	0.02000	0.02001	0.0200	0.0215	0.02070
8	0.03096	0.03090	0.02804	0.0337	0.0256	0.02994
Final Wt. (lbs)	402.97	403.35	357.82	387.67	389.8	429.38
Analyses Needed	9	11	9	4***	193	13

*The original design obtained by Gallatly was scaled up so that the triangular idealization of the cover plates satisfies stress constraints.

**Each portion was modeled by a quadrilateral element in the original work by Gallatly.

***Subsequent iterations gave heavier designs.

4.   STRUCTURES SUBJECTED TO FREQUENCY CONSTRAINTS

Some limited work has been done on structures subject to frequency constraints. Again the problem is to determine the cross-sectional sizes of the members of a structure so as to minimize the weight, but in this case the only constraint is on the natural frequencies. The lowest natural frequency is required to be a specified value.

4.1   Cantilever Beam With Concentrated Mass

A beam with concentrated mass at the end as shown in Fig. 4.1. In this example four finite elements are used along the length of the beam, with the areas of each as the design variables. The concentrated mass is 1.0 lb-sec^2/in. and the lowest natural frequency is specified to be 18.8 rad/sec.

This problem has also been solved by Turner [21], with results shown in Table 4.1. As can be seen, excellent agreement with his results were obtained using the optimality criterion technique.

TABLE 4.1.   RESULTS OF FREQUENCY CONSTRAINED CANTILEVER BEAM WITH CONCENTRATED MASS

Element	This Paper Mass ($\frac{lb\text{-}sec^2}{in}$)	Turner Mass ($\frac{lb\text{-}sec^2}{in}$)
1	2.114	2.126
2	1.871	1.845
3	1.337	1.299
4	0.558	0.535
Total Mass	5.880	5.805

4.2   Three Member Frame

A three-member frame with varying cross-sectional size along the length of each member is shown in Fig. 4.2. The members are divided into several finite elements, with each having its own independent cross-sectional size. The specified frequency is 2000 rad/sec.

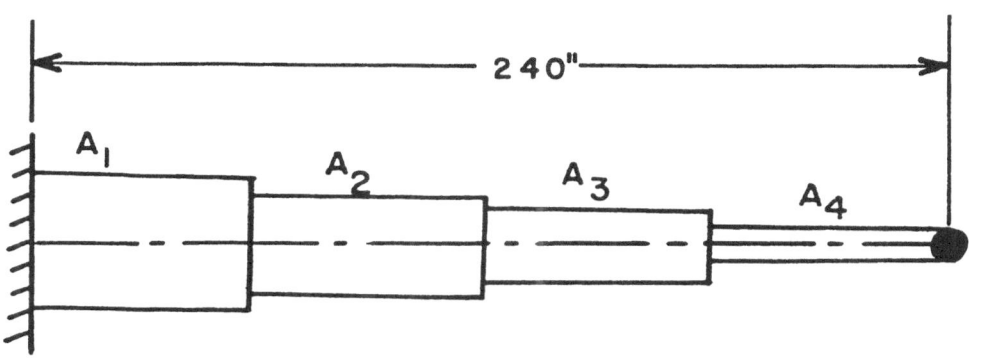

Figure 4.1   Cantilever Beam With Concentrated Mass

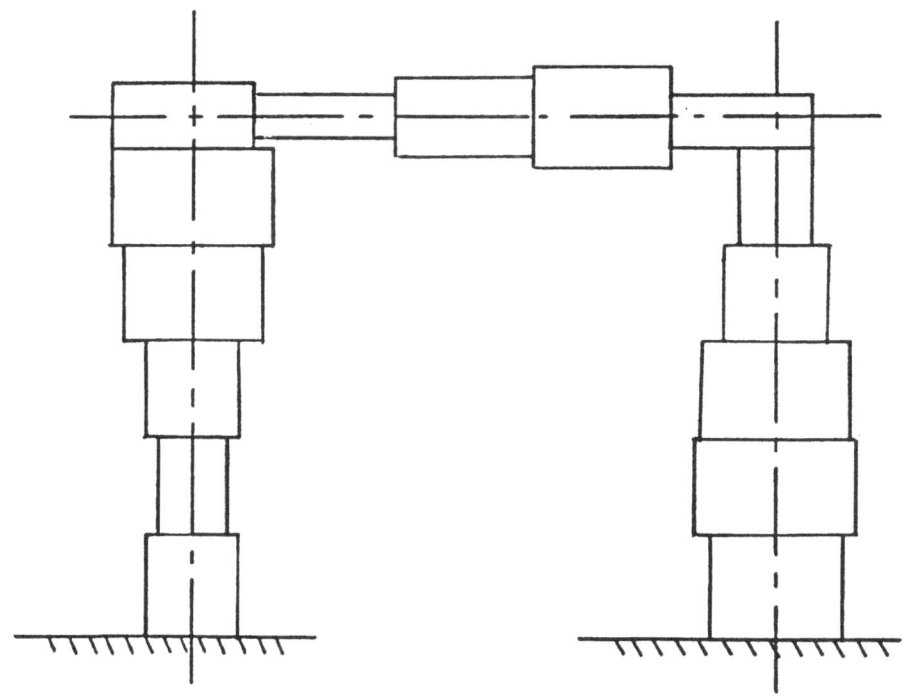

Figure 4.2   Variable Cross-Sectional Size Three Member
Frame Subjected to Frequency Constraint

The optimal design for 5 elements per member is shown in Table 4.2 for two values of η (a parameter in the optimization). No previously published results are available for comparison.

TABLE 4.2.   RESULTS OF FREQUENCY CONSTRAINED FRAME

		Final Design η = 0.1	Final Design η = 0.2
No. of Analyses		8	5
CPU (sec)		167.55	111.55
	A_1	0.55111	0.56137
	A_2	0.37666	0.38179
	A_3	0.16693	0.16953
	A_4	0.26627	0.24780
	A_5	0.33851	0.34349
	A_6	0.30497	0.29425
(Areas, in^2)	A_7	0.20394	0.19326
	A_8	0.09236	0.08431
	A_9	0.20394	0.19326
	A_{10}	0.30497	0.29425
	A_{11}	0.33851	0.34349
	A_{12}	0.26627	0.24780
	A_{13}	0.16693	0.16953
	A_{14}	0.37666	0.56137
	A_{15}	0.55111	0.56137
Wt (lbs)		0.9059	0.8975

5   STRUCTURES WITH MULTIPLE DESIGN VARIABLES PER MEMBER

Problems have also been solved in which there is more than one design variable per member. The particular case considered is structures consisting of I-beam members with cross-sections shown in Fig. 5.1. The height, width, and web and flange thicknesses are taken as design variables. Thus for each member

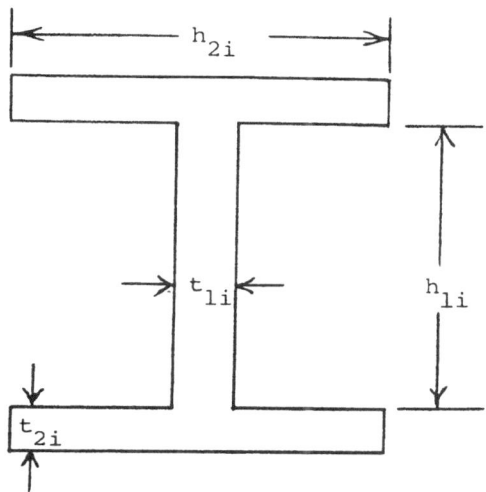

Figure 5.1   I-Beam Member

t_1, t_2, t_1/h_1, and t_2/h_2 are the design variables (the ratios t_1/h_1 and t_2/h_2 are used for convenience). The constraints imposed are limits on bending stress, yielding and local buckling in the flange and web. No displacement limits are imposed.

5.1   Beam With Applied Bending Moment

A beam subjected to end moments of 10^8 in-lbs, with a stress limit of 10^6 psi is first considered. This problem was also solved by Spunt [23], using a closed form technique, with results shown in Table 5.1. As can be seen, the values of the design variables obtained were identical. It is interesting to note that the optimality criterion method required only one analysis (iteration) to obtain this design.

TABLE 5.1.   RESULTS FOR DETERMINATE BEAM

Method	h_1 Web Height	h_2 Flange Height	t_1 Web Thickness	t_2 Flange Thickness	No. of Analyses
SUMT	15.5870	5.7282	0.6112	0.8428	1578
SPUNT	15.6420	5.6311	0.6130	0.8337	Closed Form
Optimality	15.6420	5.6311	0.6130	0.8337	1

5.2   Seventy-Member Frame

The seventy-member structure shown in Fig. 5.2, with the loading indicated, has 280 independent design variables and 280 constraints. The optimal design (although no previous results are available for comparison) was obtained in only 15 iterations (15 analyses) and is shown in Table 5.2.

6.   MECHANISM OPTIMIZATION

In the design of mechanical mechanisms, the link lengths are normally determined first so that the mechanism will perform a given task. Once this is accomplished the cross-sectional sizes of the members must then be selected. Generally this is done so as to produce the minimum weight mechanism, with limits on the stresses in the links. For slow moving mechanisms this can easily be accomplished by hand calculation. However, for high speed mechanisms, which may undergo dynamic vibration in addition to the normal rigid body motion, the problem becomes more diffi-cult, requiring some form of optimization method. Additional

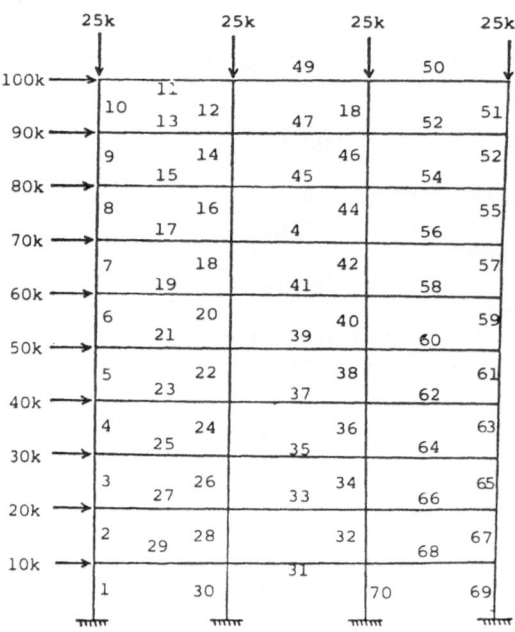

(Each member length is 100 in)

Figure 5.2   70 Member Frame

TABLE 5.2. RESULTS FOR SEVENTY MEMBER FRAME

Technique	This Paper 280 Variables			
Iterations	15			
Member No.	Width of Web (in)	Thickness of Web (in)	Width of Flange (in)	Thickness of Flange (in)
1	55.438	0.307728	20.233	0.42097
2	47.638	0.26404	17.386	0.36174
3	44.161	0.24478	16.117	0.33534
4	41.639	0.23080	15.197	0.31619
5	39.415	0.21846	14.385	0.29930
6	37.077	0.20551	13.532	0.28155
7	34.453	0.19096	12.574	0.26162
8	31.480	0.17448	11.489	0.23904
9	28.159	0.15608	10.277	0.21382
10	25.153	0.13942	9.180	0.19100
11	24.458	0.13557	8.926	0.18573
12	31.327	0.17364	11.433	0.23789
13	32.526	0.18028	11.871	0.24698
14	38.074	0.21103	13.896	0.28912
15	36.077	0.19997	13.167	0.27395
16	42.739	0.23689	15.598	0.32454
17	39.283	0.21774	14.337	0.29830
18	46.677	0.25872	17.035	0.35444
19	41.853	0.23198	15.275	0.31781
20	50.047	0.27740	18.265	0.38004
21	43.833	0.24295	15.997	0.33285
22	52.980	0.29365	19.336	0.40231
23	45.389	0.25158	16.565	0.34466
24	55.526	0.30777	20.265	0.42164
25	46.884	0.25987	17.111	0.35602
26	57.570	0.31854	20.975	0.43640
27	48.749	0.27020	17.792	0.37018
28	58.247	0.32285	21.258	0.44230
29	45.943	0.25465	16.767	0.34887
30	62.876	0.34851	22.948	0.47745
31	46.737	0.25905	17.057	0.35490
32	57.656	0.31957	21.042	0.43781
33	51.748	0.28682	18.886	0.39295
34	57.176	0.31691	20.867	0.43417
35	52.212	0.28940	19.056	0.39647
36	55.436	0.30727	20.232	0.42095
37	51.483	0.28536	18.789	0.39094
38	52.987	0.29369	19.338	0.40236
39	50.039	0.27736	18.263	0.37998
40	50.122	0.27781	18.293	0.38060
41	48.018	0.26615	17.525	0.36463
42	46.808	0.25944	17.083	0.35544
43	45.325	0.25123	16.542	0.34418
44	42.811	0.23729	15.624	0.32508
45	41.719	0.23124	15.226	0.31679
46	37.504	0.20788	13.688	0.28479
47	36.006	0.19957	13.141	0.27341
48	29.005	0.16077	10.586	0.22025
49	24.960	0.13834	9.109	0.18953
50	19.394	0.10749	7.078	0.14727
51	19.879	0.11018	7.255	0.15095
52	29.784	0.16509	10.870	0.22617
53	26.988	0.14959	9.850	0.20494
54	36.083	0.20000	13.169	0.27400
55	31.416	0.17413	11.466	0.23856
56	39.811	0.22066	14.529	0.30230
57	34.542	0.19146	12.607	0.26229
58	42.240	0.23413	15.416	0.32075
59	37.087	0.20556	13.535	0.28162
60	44.003	0.24390	16.059	0.33414
61	39.316	0.21792	14.349	0.29855
62	45.344	0.25133	16.549	0.34432
63	41.373	0.22932	15.100	0.31417
64	46.557	0.25806	16.992	0.35354
65	43.569	0.24149	15.901	0.33084
66	48.002	0.26606	17.519	0.36450
67	46.427	0.25733	16.944	0.35255
68	44.830	0.24848	16.361	0.34042
69	53.276	0.29529	19.444	0.40455
70	62.152	0.34449	22.683	0.47195
Volume (in³)	149739			

constraints may also be required, such as limits on the maximum deformations that exist in the links.

6.1 Inverted Slider Crank Mechanism

The mechanism shown in Fig. 6.1 has cross-sectional areas of the two links as the design variables. The weight of the slider is 0.772 lb and the stress limit is 5000 psi. No displacement limits are imposed.

The design problem was solved using two different types of analyses. The first was a complete vibrational analysis to determine the deformations and thus the internal stresses in the links. The second method was a quasi-static analysis, which is obtained by ignoring the mass matrix and accelerations in calculating the deformations, i.e.

$$\vec{X} = K^{-1} \vec{F}$$

In this case the deformations are still time varying, since the stiffness matrix K and dynamic force vector $\vec{F}$ are functions of time, due to the rigid body motion of the linkage.

The results of optimization are shown in Table 6.1 and are compared with the SUMT design. As can be seen the optimal areas are almost identical. However, the SUMT technique required considerably more computational time.

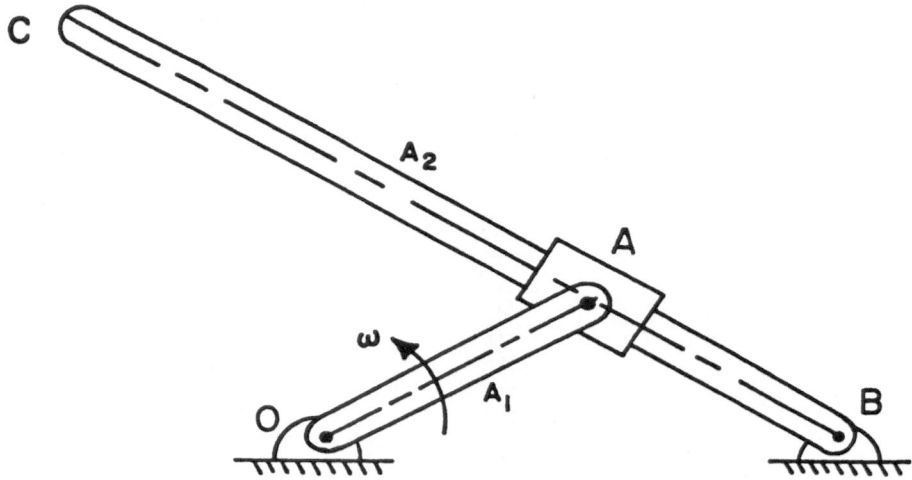

Figure 6.1   Inverted Slider Crank Mechanism

TABLE 6.1. INVERTED SLIDER CRANK RESULTS

Technique	SUMT	This Paper	This Paper
Analysis	Quasi-Static	Quasi-Static	Vibrational
Member			
1	2.5862	2.5839	2.7329
2	1.8578	1.8606	1.9721
CPU Time (sec)	836	6.25	73

6.2 General Eight-Link Mechanism

Another mechanism considered is the eight link mechanism
shown in Fig. 6.2. In this problem, stress limits of 4000 psi
are imposed on all members and the x and y deflections at nodes
2, 3, 7, and 9, as well as the deflection in the sliding direction at node 8, are limited to ±0.01 inches.

Optimal designs are shown in Table 6.2 using a quasi-static
and complete vibrational analyses. These were obtained in only
26 and 13 analyses respectively. No comparisons for these designs were available.

7. CONCLUSIONS

The optimality criterion technique developed in this research have been shown to be very efficient and adapatable to a
variety of structural and mechanical design problems. Numerous
examples have been solved, many of which could not be presented
here, which have verified the methods.

REFERENCES

1. Khan, M.R., Willmert, K.D., and Thorton, W.A., "A New
    Optimality Criterion Method for Large Scale Structures,"
    Proceedings of the AIAA/ASME 19th Structures, Structural
    Dynamics and Material Conference, April 1978.

710

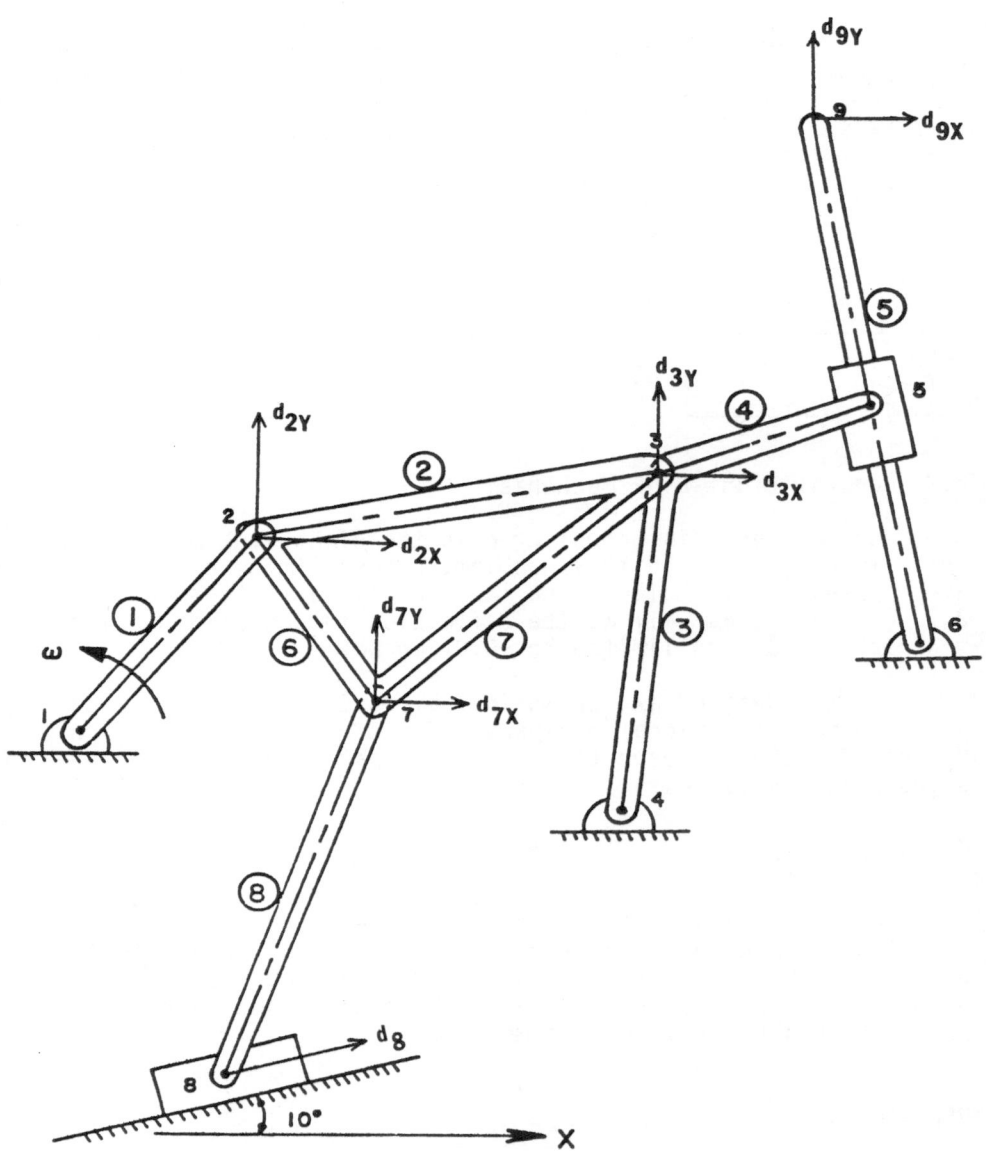

Figure 6.2   General Eight-Link Mechanism

TABLE 6.2. OPTIMAL GENERAL 8 LINK MECHANISM

Technique	Optimality Criteria	
Analysis CPU (sec)	Quasi-Static 355.84	Dynamic 2133.79
No. of Analyses	26	13
A_1	3.875	4.326
A_2	1.143	1.590
A_3	3.386	3.820
A_4	4.021	4.635
A_5	2.159	2.596
A_6	0.253	0.099
A_7	0.575	0.428
A_8	1.198	1.205
Vol. (in^3)	166.9	185.5
η	0.03	0.08

2. Khan, M.R., Thorton, W.A., and Willmert, K.D., "Optimality Criterion Techniques for Structures with Multiple Design Variables per Member," Proceedings of the AIAA/ASME/ASCE/AHS 20th Structures, Structural Dynamics and Materials Conferences, 1979.
3. Khan, M.R., Willmert, K.D., and Thornton, W.A., "An Optimality Criterion Method for Large Scale Structures," AIAA Journal, Vol. 17, No. 7, 1979.
4. Syed, M.E., Willmert, K.D., and Khan, M.R., "Optimality Criterion Technique Applied to Structures Composed of Different Element Types," Paper to be presented at AIAA/ASME/ASCE/AHS 21st Structures, Structural Dynamics and Materials Conference, Seattle, Washington, May 1980.

5. Khan, M.R., Willmert, K.D. and Thornton, W.A., "Automated Analysis/Design of High Speed Planar Mechanisms," Proceedings of the 5th OSU Applied Mechanisms Conference, Oklahoma, 1977.

6. Khan, M.R., Thornton, W.A., and Willmert, K.D., "Optimality Criterion Techniques Applied to Mechanical Design," Trans. ASME, J. of Mechanical Design, Vol. 100, No. 1, 1978.

7. Willmert, K.D., Thornton, W.A., and Khan, M.R., "A Hierarchy of Methods for Analysis of Elastic Mechanisms with Design Applications," ASME Paper No. 78-DET-56, 1978.

8. Thornton, W.A., Willmert, K.D., and Kahn, M.R., "Mechanism Optimization Via Optimality Criterion Techniques," Trans. ASME, J. of Mechanical Design, Vol. 101, No. 3, 1979.

9. Khan, M.R., Willmert, K.D., and Thornton, W.A., "A Computer Program Package for Large Scale Structural and High Speed Mechanism Design," Paper in the proceedings of the engineering software conference, Southhampton, England, 1979.

10. Briggs, W.J. and Willmert, K.D., "Optimum Design of Frames Using Linear Programming Techniques," Department of Mechanical and Industrial Engineering, Clarkson College, Office of Naval Research Contract No. N00014-76-0064, Report No. MIE-026, 1978.

11. Calafell, D.O. and Willmert, K.D., "Automated Resizing Optimization of Generally Loaded Planer Frames Via Linear Programming Techniques," Proceedings of the Symposium on Applications of Computer Methods in Engineering, USC, California, 1977.

12. Schmit, L.A. and Farshi, B., "Some Approximation Concepts for Structural Synthesis," AIAA Journal, Vol. 12, No. 5, 1974, pp. 692-699.

13. Schmit, L.A. and Miura, H., Approximation Concepts for Efficient Structural Synthesis, NASA CR-2552, 1976.

14. Venkayya, V.B., "Design of Optimum Structures," Journal of Computers and Structures, Vol. 1, No. 1-2, 1971, pp. 265-309.

15. Gellatly, R.A. and Berke, L., Optimal Structural Design, AFFDL-TR-70-165, 1971.

16. Dobbs, M.W. and Nelson, R.B., "Application of Optimality Criteria to Automated Structural Design," AIAA Journal, Vol. 14, No. 10, 1976, pp. 1436-1443.

17. Rizzi, P., "Optimization of Multiconstrained Structures Based on Optimality Criteria," Paper presented at the AIAA/ASME/SAE 17th Structures, Structural Dynamics, and Materials Conference, King of Prussia, PA., May 1976.

18. Berke, L. and Khot, N.S., "Use of Optimality Criteria Methods for Large Scale Systems," AGARD Lecture Series No. 70 on Structural Optimization, AGARD-LS-70, 1974, pp. 1-29.

19. Gallatly, R.A. and Berke, L., "Optimal Structural Design," AFFDL-TR-70-165, Air Force Flight Dynamics Laboratory, Wright-Patterson A.F.B., Ohio, 1975.

20. Gallatly, R.A., "Development of Procedures for Large Scale Automated Minimum Weight Structural Design," AFFDL-TR-66-180, 1966.
21. Turner, M.J., "Design of Minimum Mass Structures with Specified Natural Frequencies," AIAA Journal, Vol. 5, No. 3, 1967, pp. 406-412.
22. Spunt, L., Optimum Structural Design, Prentice-Hall, Inc., Englewood Cliffs, New Jersey, 1971.

ANALYSIS OF "ALLOWABLE STRESS" TYPE ALGORITHMS*

William R. Spillers

Civil Engineering, Rensselaer Polytechnic Institute
Troy, New York 12181

ABSTRACT

This paper informally explores convergence questions associated with allowable stress type algorithms now in use.

1.   INTRODUCTION

Before commenting upon some aspects of structural optimization it is helpful to dwell briefly on the ambience of the computing world in general and the world of mathematical programming in particular. For those who have worked with and watched commercial computers as they have evolved, starting from the 1950's, the most impressive aspect of computing remains hardware. During this period an evolution has occurred from early machines such as the 2000 word IBM 650 to present large scale computers whose immediate access memories can contain a million words or more. Furthermore, there seems to be no end to these developments in sight. There now exist serious discussions [1] of placing 10^7 to 10^8 transitors on a single chip, in which case a single chip could function as one of today's large computers. The question is then how to combine these chips or modules into a high performance system and what the software should be to communicate with it.

If the hardware picture is fantastic, the state-of-the-art of software is quite the opposite. A measure of this largely

---

*Research supported by the National Science Foundation.

dreary state is indicated by the recent work of Ragsdell [2] who attempted to rate the performance of several nonlinear algorithms having general capacilities. Of his 30 test problems, the average number of variables was 8. If the discussion of algorithms is extended to include some more simple cases such as linear and quadratic programming the picture certainly improves. There is of course the scenario in which hardware developments combine with increasing efficiency in linear programming, to make some variant of sequential linear programming attractive as a general mathematical programming algorithm. But that remains to be seen.

In a somewhat perverse way, then, the computing environment remains exciting and challenging. At the moment, there are unsolved questions of owning a computer versus networking. While declining hardware costs tend to make owning a computer attractive, declining communication costs tend to make networking and the concomitant advantages of a large system attractive. If hardware costs decline sufficiently, an obvious solution to the problem of choosing a mathematical programming algorithm will be to run them all in parallel. This in turn brings up the problem of the design community's general inexperience with parallel programming logic and questions of how a large computing system should be organized and a proper way to communicate with it.

2. STRUCTURAL OPTIMIZATION

Structural optimization [3,4] is at least as many-sided as the mathematical programming literature. There will be no attempt made here to deal with the entirety of structural optimization. Comments will be limited to the part that might be regarded as an extension of existing general purpose structural analysis computer programs. That is, the area to be addressed here is the one naturally associated with general purpose computer programs if one were to add optimization features to them. This is an area in which it is not uncommon to deal with thousands of parameters. It is an area in which it appears that structural engineers are far ahead of the mathematical programming literature. This is in no small part due to the fact that in structural design one is not dealing with general mathematical programming situations but one has recourse to the particular features of the equations of structures. There is an interesting parallel in optimization theory with the sparse matrix technology of linear analysis. Structural engineers pioneered in the use of sparse matrix technology, which was subsequently embraced by numerical analysts. It appears that there are overtones of the same process [5] now going on in optimization.

The theme to be pursued here in some detail was recently stated by Fleury [6]. To paraphrase him, he argues that there are two canonical forms of structural optimization, the mathematical programming version and the optimality criteria version, and that they are both mathematical programming. Certainly, a mathematical programming algorithm does not have to produce an exact result. It appears that the argument is even stronger. The work he refers to as optimality criteria methods involved pioneering efforts which may well eventually be adopted by the mathematical programming literature as developing important, if special purpose, algorithms.

In this paper the term "allowable stress algorithm" is used rather than "optimality criteria method", since it seems to be more descriptive of the mechanisms invoked in this process. While one may, in fact, argue about some of the more sophisticated versions of these methods, there appears to be no question that their basic motivation comes directly from early work relating to allowable stress design for trusses. In any case, the remainder of this paper will be used to discuss some basic features of these algorithms and point out some of the problems that persist when they are used.

3. ALLOWABLE STRESS TYPE ALGORITHMS

The most simple case of allowable stress design is surely the truss problem with a single loading condition, where it is common to iterate

$$A_i^{(n+1)} = |F_i^{(n)}|/\sigma^{allow} \tag{3.1}$$

Here

A_i = area of bar i

F_i = force in bar i

σ^{allow} = allowable stress

and the superscripts in parenthesis denote iterations. Note the curious side of Eq. 3.1, which deals simultaneously with safety and optimization. In 1973 Venkayya and his co-workers (7) demonstrated impressively the range of problems to which algorithms of this type can be applied. Motivation has remained another matter.

There are many points of view from which to attempt an analysis of the convergence of allowable stress algorithms. One may begin with the idea of embedding, which has been used extensively by Richard Bellman. In this case some of the conditions of the problem are neglected, which results in "embedding" the original problem in a larger space that may contain solutions that did not exist in the original or real formulation. One way to do this for structures is to neglect the constitutive equations and write the problem formulation in terms of forces. For discrete systems or discrete approximations, such as the finite element method, this appears as minimizing the objective function, or weight $t(F)$ written in terms of the member forces, while satisfying the force equations of node equilibrium

$$\tilde{N}F = P \tag{3.2}$$

where

F = member force matrix

P = node load matrix

If the Lagrangian is then formed as

$$L = t(F) + \tilde{\lambda}(P - \tilde{N}F) \tag{3.3}$$

using the matrix of Lagrange multipliers λ, the Kuhn-Tucker (optimality conditions) result as

$$\left. \begin{array}{c} \tilde{N}F = P \\ \\ \nabla t = N\lambda \end{array} \right\} \tag{3.4}$$

in which the Lagrange multipliers clearly play the role of node displacements. With some simple manipulation [9], it can be shown that Eqs. 3.4 have the appearance of the node method of structural analysis and that the associated member stiffness matrix can be derived from the Hessian matrix of the function t^2.

It has been shown [9] that monotone behavior can be expected for a wide class of problems, including of course Eq. 3.1, and that the embedding technique can be extended to more complex situations, such as design for optimal shape [10].

4. PROBLEMS WITH ALLOWABLE STRESS TYPE ALGORITHMS

It is a relatively easy matter to set up an example in which allowable stress design - at least the most obvious version of it - simply does not work. Lefcochelos [12] has done so for a simple case of truss design, with constant allowable stresses for two loading conditions. In his example each bar is not fully stressed under at least one loading condition, but there is a statically determinate substructure that is fully stressed under one loading condition or the other. Any iterative scheme of the type

$$A_i^{(n+1)} = \max \left[|F_{i1}^{(n)}|, |F_{i2}^{(n)}| \right] / \sigma^{\text{allow}} \tag{4.1}$$

is therefore doomed to failure. In Eq. 4.1, 1 and 2 in the subscripts refer to the two loading conditions.

In some cases, algorithms like that of Eq. 4.1 may provide approximate solutions that are quite useful, but that would be another approach entirely to the problem. It may be added here that failure of algorithms, such as that of Eq. 4.1, gave rise to the study [11] of constraints that constitutive equations place upon solutions that satisfy equilibrium.

Frequency constraints provide a case in which allowable stress type algorithms work, but in a rather curious manner. The most simple optimization problem with frequency constraints is one in which the mass matrix M is fixed and in which it is desired to find the stiffness matrix K to minimize the structural mass of a truss, given the frequency. Let

k_i = stiffness of member i ($k_i = A_i E / L_i$)

L_i = length of member i

A_i = area of member i

Δ_{ai} = "allowable" length change of member i, i.e.,

$$\Delta_{ai} = \sigma^{\text{allow}} L_i / E$$

ω^2 = square of the fundamental frequency of the structure

The optimization problem can be written

$$\left. \begin{array}{l} \text{minimize } \sum_i (\Delta_{ai})^2 k_i \\[2mm] \text{subject to } \omega^2(k_i) = c \end{array} \right\} \tag{4.2}$$

The Lagrange multiplier method can be applied to the problem of Eq. 4.2 to give

$$L = \sum_i (\Delta_{ai})^2 k_i + \lambda(c - \omega^2) \tag{4.3}$$

from which the Kuhn-Tucker conditions follow as

$$\left.\begin{array}{l} \partial L/\partial k_i = 0 \rightarrow (\Delta_{ai})^2 = \lambda \nabla \omega^2 \\[2ex] \partial L/\partial \lambda = 0 \rightarrow \omega^2 = c \end{array}\right\} \tag{4.4}$$

Now let δ be the eigenvector associated with the fundamental frequency of the structure, let Δ be the member displacement matrix associated with this eigenvector, and normalize δ with respect to the mass matrix. It can be shown by direct calculation that

$$\partial \omega^2/\partial k_i = (\Delta_i)^2 \tag{4.5}$$

Eq. 4.5 implies, using the first of Eqs. 4.4, that in the optimum design the member length changes are the allowable length changes or that the design is fully stressed.

It is curious that in a problem of this type, which is totally different from the problem of design for a given load, the concept of a fully-stressed design should emerge. In fact, matters are much worse since it can be shown that an allowable stress algorithm will behave monotonically.

As problem complexity increases, it becomes increasingly difficult to motivate an allowable stress algorithm. In addition to the two cases mentioned above, there is no obvious allowable stress algorithm for problems of optimal geometry, nor does one appear likely.

5. ANALYSIS OF ALLOWABLE STRESS TYPE ALGORITHMS

By analysis of an algorithm, of course, is meant some formal motivation or explanation of why an algorithm works. Surely the most basic approach is Newtonian. That is, if the optimization problem is reduced to solving a nonlinear system of equations and if the method of solution can be shown to involve formally linearizing the system at each point, then recourse can be had to various theorems concerning the convergence of Newton's method. In simple terms, convergence is usually guaranteed when the starting point is sufficiently close to the solution. If Newton's method sounds simple, it can be just the opposite since it may be necessary - or at least advantageous - to transform the system prior to linearizing it. There are sophisticated variations of Newton's method [13], which may be important to the user and make the difference between computational feasibility and infeasibility.

Allowable stress type algorithms do not immediately fall into any particular mathematical programming category. They may be defined to have two phases, analysis and re-design. In the analysis phase the stiffnesses are given and stresses and displacements are computed. This phase generally makes use of sparse matrix technology and can be done quite efficiently. In the re-design phase the computed stresses and displacements are used to compute new member stiffnesses.

One of the aspects of allowable stress iteration, which separates it from the common mathematical programming literature, is the absence of the computation of derivatives. Whether Newtonian methods are used for nonlinear systems or descent methods are used for direct optimization, the computation of derivatives is always prominent. It can, of course, be shown that structural analysis can be equivalent to computing derivatives in the following manner. Let the node method of structural analysis be written as

$$\tilde{N}KN\ \delta = P \tag{5.1}$$

If this expression is differentiated directly with respect to the ith member stiffness it follows that for fixed loads P

$$(\tilde{N}N)_i\ \delta\ +\ \tilde{N}KN\ \delta_{,i} = 0 \tag{5.2}$$

or that

$$\delta_{,i} = -(\tilde{N}KN)^{-1}\ N\ \Delta_i \tag{5.3}$$

Equation 5.3 implies that the derivative of the node displacements with respect to a particular member stiffness can be computed by solving the node method for the case in which the loads are defined from the given member displacements. Solving the node method is, at this point, directly related to computing derivatives.

The frequency constraint problem cited above is quite different. In that case, the nonhomogenous system of Eq. 5.1 is replaced by the homogeneous eigenvalue problem

6. CONCLUDING REMARKS

This paper has attempted to informally explore some of the characteristics of allowable stress type algorithms, as they are applied to a wide range of structural problems. The attempt has been to show that these algorithms have a generality beyond existing applications and should be the source of further study.

REFERENCES

1. Schwartz, J.T., Ultra-computers, N.Y. University Report, 1978.
2. Ragsdell, K.M., "On Some Experiments Which Delimit the Utility of Nonlinear Programming Algorithms", paper presented at the ORSA/TIMS meeting in Los Angeles, November, 1978.
3. Mathematical Programming Applications in Civil Engineering Design: Current Status and Future Trends (a collection of 3 articles compiled by Harry L. Jones), preprint 2897, ASCE, Spring Convention and Exhibit, Dallas Texas, April 25-29, 1977.
4. Venkayya, V., "Structural Optimization: A Review and Some Recommendations", Int. J. for Numerical Methods in Engineering, 13, 1978, pp. 203-228.
5. Nocedal, Jorge, "Updating Quasi-Newton Matrices with Limited Storage", IIMAS, Universidad Nacional Autonomade de Mexico, January, 1979.
6. Fleury, C., "A Unified Approach to Structural Weight Optimization", Computer Methods in Applied Mechanics and Engineering, Vol. 20, 1979, pp. 17-38.
7. Venkayya, V.B., Khot, N.S., and Berke, L., "Application of Optimality Criteria Approaches to Automated Design of Large Practical Structures", Second Symposium on Structural Optimization, AGARD-CP-123, Milan, Italy, April 1973.
8. Bazarra, M.S. and Shetty, C.M., Foundations of Optimization, Lecture Notes in Economics and Mathematical Systems #122, Springer-Verlag Berlin, 1976.
9. Spillers, William R., Iterative Structural Design, N. Holland, Amsterdam, 1975.
10. Spillers, William R. and Kountouris, George E., "Geometric Optimization Using a Simple Code Representation", Journal of the Structural Div. ASCE, Vol. 106, No. ST5, 1980, pp. 959-971.
11. Spillers, W.R. and Lefcochilos, E., "Elastic Realizability of Force and Displacement Systems", Quart. Appl. Math., 1980, pp. 411-421.

722

12. Lefcochilos, E., _Theory of Realizability for Linearly Elastic Truss-Structures and Related Applications in Optimum Structural Design_, Ph.D. Thesis, Dept. of Civil Engineering, R.P.I., 1979.
13. Dennis, J.E., Jr. and More, Jorge J., "Quasi-Newton Methods, Motivation and Theory", _SIAM Review_, 1976, pp. 46-89.

AN INTRODUCTION TO THE SOLUTION OF OPTIMAL STRUCTURAL DESIGN PROBLEMS USING THE FINITE ELEMENT METHOD

Alan J. Morris

Structures Department, RAE, Farnborough, Hants

ABSTRACT

A correspondence between distributed parameter and finite element structural optimization methods is first established. Duality based optimization methods are then presented, with computational aspects of finite element structural analysis and optimization methods leading to definition of an automated structural design system. Computational experience with the method is treated in a companion paper by Fleury.

1. INTRODUCTION

The purpose of the present paper is to illustrate the correspondence between the approach adopted in the solution of distributed parameter systems and that used for the type of finite dimensional problems that occur when structural optimization problems, associated with finite element models, are considered. The need for an automated design or structural optimization program, in real design, has evolved from the developments that have taken place in computerized structural analysis where a range of finite element computer programs are now available and routinely used in the design of major structural assemblies. Although the progress made in this area is satisfactory in many ways it represents a limited application of the power of the computer - particularly with regard to the class 6 or 'supercomputers'. A more effective use of this power, which has great potential in rapidly creating efficient structures, is to computerize the complete design cycle. At the present time the analysis part of the cycle is performed using a large system

of the NASTRAN, ASAS, PAFEC type, with the design engineer
intervening to make what he believes are appropriate changes to
the structure in order to achieve a satisfactory design. However,
this is a difficult task and for realistic design cases can only
be achieved by directly linking the analysis with some type of
optimization algorithm and thereby completely automating the
design process.

As is shown in this paper, selection of the specific
structural optimization algorithm is particularly important. In
order to illuminate the position and importance of the various
algorithms available in modern systems, such as the RAE STARS
package, the basic optimality conditions are re-examined in
Section 2, from a finite element viewpoint. By doing so one
observes how the uniform mutual strain energy density criterion
of Prager, Taylor, and Schield [1,2], devised from purely
structural arguments, emerge from the classical Kuhn-Tucker
conditions. In passing, it should be mentioned that this correspondence was noted much earlier by Hemp [3] for the special
case of frameworks. On the basis of the Kuhn-Tucker conditions,
section 3 also examines the dual problem and it is observed that
by fully exploiting the mathematical nature of the structures
problem, a useful dual problem can be generated.

The Kuhn-Tucker conditions provide the basis for examination
of several of the more useful structural optimization algorithms.
In particular emphasis is placed on how these algorithms are
effectively obtained by insisting that either various parts of
the Kuhn-Tucker conditions hold or, in one specific case, by
demanding the complete satisfaction of these optimality conditions.
As a by-product of this explanation it is seen how the history of
the subject of structural analysis has developed, by initially
employing algorithms that satisfied only the simplest part of
the Kuhn-Tucker condition, to give rise to the familiar stress-
ratioing algorithm. By pursuing this line of development, the
position in the structural optimization firmament of algorithms
of the optimality criterion and subsequently the pseudo-Newton
type is established. This section also indicates the important
role played by duality theory in the control of algorithms,
particularly with respect to active set strategy and convergence.

Finally, it is shown how these algorithmic aspects may be
drawn together in the solution of a specific problem, though
more emphasis on the solution of design problems is given by other
contributors. To make this solution machinery effective in the
design of realistic problems, evolution of advanced software and
algorithm systems exemplified by the RAE STARS package is required.
An outline description of this package is given in Ref. 4, but a
more extensive explanation is found in the RAE STARS programming
manuals.

2. OPTIMALITY CRITERIA AND DUALITY

2.1 Optimality Criteria

In this section, optimality conditions are developed for a structure that is modelled by finite elements. A minimum weight design is sought, subject initially to constraints on stresses and displacements, under the influence of static loads. It is assumed that the design variables are element thicknesses or cross-sectional areas and that the element stiffness matrices are linear in these variables.

Under these assumptions, the problem can be defined as one of finding a vector $x* = \{x_1^*, x_2^*, \ldots, x_n^*\}^t$ that minimizes

$$W(x) = \sum_{i=1}^{n} \omega_i x_i$$

subject to constraints

$$\bar{u}_j \geq u_j(x) , \qquad\qquad j = 1,\ldots,m$$

$$\bar{\phi}_i^2 \geq \phi_i^2(x) , \qquad\qquad i = 1,\ldots,n$$

$$\bar{x}_i \leq x_i , \qquad\qquad i = 1,\ldots,n$$

where ω_i represents the terms within the definition of element weight that remain constant, i.e., specific weight. The terms u_j represent the m nodal displacements, which are limited to the range $\bar{u}_j$ and may consist of all the nodal displacements of the entire finite element model or simply a subset of the total. The term $\phi_i^2(x)$ represents a yield criterion, such as that of Von Mises or similar, which is described in terms of stresses within the ith element. The limits $\bar{x}_i$ are gauge constraints, which represent practical lower limits to the design variables.

Guided by the lessons learned in the analysis of determinate structures [5], one redefines the problem using the inverse of the physical design variables. Thus the optimum design problem, called MWP, becomes one of finding a vector $z_i^* = 1/x_i^*$ ($i = 1,2,\ldots,n$) that minimizes

$$W(z) = \sum_{i=1}^{n} \frac{\omega_i}{z_i}$$

subject to constraints

$$\bar{u}_j \geq u_j(z) , \qquad\qquad j = 1,\ldots,m$$

$$\bar{\phi}_i^2 \geq \phi_i^2(z) , \qquad\qquad i = 1,\ldots,n \qquad\qquad \left.\rule{0pt}{55pt}\right\} \quad \text{MWP}$$

$$\frac{1}{\bar{x}_i} \geq z_i$$

The Lagrangian associated with this problem is given by

$$L(z,\lambda,\mu,\nu) = W(z) + \sum_{j=1}^{m} \lambda_j(u_j(z) - \bar{u}_j)$$

$$+ \sum_{i=1}^{n} \mu_i(\phi_i^2(z) - \bar{\phi}_i^2) + \sum_{i=1}^{n} \nu_i(z_i - \frac{1}{\bar{x}_i})$$

with the constraints

$$\lambda_j \geq 0 \qquad\qquad j = 1,\ldots,m$$

$$\mu_i \geq 0 , \quad \nu_i^* \geq 0 , \qquad\qquad i = 1,\ldots,n$$

The optimality conditions are obtained by applying the Kunh-Tucker conditions denoted KTC, to obtain

$$\nabla_z L(z^*,\lambda^*,\mu^*,\nu^*) = 0$$

$$\lambda_j^*(u_j(z^*) - \bar{u}_j) = 0 , \qquad\qquad j = 1,\ldots,m$$

$$u_i^*(\phi_i^2(z^*) - \bar{\phi}_i^2) = 0 , \qquad\qquad i = 1,\ldots,n$$

$$\nu_i^*(z_i^* - \frac{1}{\bar{x}_i}) = 0 , \qquad\qquad i = 1,\ldots,n \qquad\qquad \left.\rule{0pt}{90pt}\right\} \quad \text{KTC}$$

$$\lambda_j^* \geq 0, \qquad \mu_i^* \geq 0 , \qquad \nu_i^* \geq 0 , \quad \left.\rule{0pt}{12pt}\right\} j = 1,\ldots,m$$

$$\bar{u}_j \geq u_j(z^*) \quad \bar{\phi}_i^2 \geq \phi_i^2(z^*) , \quad \frac{1}{\bar{x}_i} \geq z^*_i , \left.\rule{0pt}{12pt}\right\} i = 1,\ldots,n$$

The first part of these optimality criterion can be written in the expanded form,

$$\frac{\partial L}{\partial z_k}(z^*,\lambda^*,\mu^*,\nu^*) = -\frac{\omega_k}{z_k^{*2}} + \sum_{j=1}^{m}\lambda_j^*\frac{\partial u_j}{\partial z_k}$$

$$+ \sum_{i=1}^{n}2\mu_i^*\phi_i\frac{\partial\phi_i}{\partial z_k} + \nu_k = 0$$

which requires the derivatives of stresses and displacements. To obtain these derivatives one turns to the finite element formulation, in which the applied loads {P} are related to the nodal displacements {u} through the global stiffness matrix [K] thus,

$$\{P\} = [K]\{u\}$$

Taking the derivatives with respect to the design variables one has

$$\{0\} = \left\{\frac{\partial K}{\partial z_k}\right\}\{u\} + \{K\}\left\{\frac{\partial u}{\partial z_k}\right\}$$

This, however, gives the rate of change of all displacements with respect to the design variable z_k. To extract the derivative of a specific displacement u_j, one pre-multiplies by a vector $\{e_h\}^t$ that contains zero values, except for a value of unity in the jth position. Thus,

$$\frac{\partial u_j}{\partial z_k} = \{e_j\}^t\left\{\frac{\partial u}{\partial z_k}\right\} = -\{e_j\}^t\{K\}^{-1}\left\{\frac{\partial K}{\partial z_k}\right\}\{u\}$$

where the pre-multiplication by $\{e_j\}^t$ is equivalent to the application of a unit load along u_j. Recalling that the global stiffness matrix is assembled from individual element stiffness matrices and that these are linear in terms of the original design variables, one notes that

$$\frac{\partial u_j}{\partial z_k} = \left\{\tilde{u}_j^k\right\}^t\left\{\frac{k^k}{z_k}\right\}\{u^k\}$$

Here $\{k^k\}$ represents the stiffness matrix of the kth element, $\{u^k\}$ is the vector of nodal displacements for this element and $\{\tilde{u}_j^k\}$ is the vector of nodal displacements associated with the kth element due to the application of a unit load acting at the point of application and along the constrained displacement u_j.

Recalling that stresses in a specific element are given by

$$\{\sigma^i\} = [D]\ \{u^i\}$$

where $\{\sigma^i\}$ is a vector of stresses in the element, due to the nodal displacements. Usually the matrix [D] is a function of position. In such cases, it is normal to choose a specified point, possibly the center in the case of flat membrane elements, and evaluate the stresses at that point. The Von Mises criterion can be written in the form

$$\phi_i^2 = \{\sigma^i\}^t\ [V]\{\sigma^i\}$$

where, for a flat membrane element defined in terms of the orthogonal coordinates ξ and η,

$$\{\sigma^i\}^t = \left\{\sigma_\xi^i,\ \sigma_\eta^i,\ \tau_{\xi\eta}^i\right\}$$

$$[V] = \begin{bmatrix} 1 & -1/2 & 0 \\ -1/2 & 1 & 0 \\ 0 & 0 & 3 \end{bmatrix}$$

Thus,

$$\frac{\partial(\phi_i^2)}{\partial z_k} = 2\left\{\frac{\partial\sigma^i}{\partial z_k}\right\}^t\ [V]\ \{\sigma^i\}$$

and

$$\left\{\frac{\partial\sigma^i}{\partial z_k}\right\} = [D]\ \{\tilde{u}^{ik}\}^t\left\{\frac{k^k}{z_k}\right\}\{u^k\}$$

The pseudo-displacements are now contained in the matrix $\tilde{u}^{ik}$ and correspond to displacements measured at the nodes of element k, due to the application of unit loads at the nodes of element i. Alternatively, one can write

$$\left\{ \frac{\partial \sigma^i}{\partial z_k} \right\} = \frac{1}{z_k} \{\tilde{\sigma}^{ik}\}$$

where $\{\tilde{\sigma}^{ik}\}$ represents the stresses induced in the ith element, due to the nodal forces on element k generated by the applied loads, being considered themselves as loads.

Using these expressions the Kuhn-Tucker conditions KTC for the problem MWP are now given by

$$\sum_{j=1}^{m} \lambda_j^* \left\{ \tilde{u}_j^k \right\}^t \{k^k\}\{u^k\} + \sum_{i=1}^{n} 2\mu_i^* \{\tilde{\sigma}^{ik}\}^t \{V\}\{\sigma^i\}$$

$$+ \nu_k^* z_k^* = \frac{\omega_k}{z_k^*} , \quad k = 1,\ldots,n$$

$$\lambda_j^*(u_j(z^*) - \bar{u}_j) = 0$$

$$\mu_i^*(\phi_i^2(z^*) - \bar{\phi}_i^2) = 0$$

$$\nu_i^* \left(z_i^* - \frac{1}{\bar{x}_i} \right) = 0 \qquad\qquad j = 1,\ldots,m$$

$$\qquad\qquad\qquad\qquad\qquad\qquad i = 1,\ldots,n$$

$$\lambda_j^* \geq 0 , \qquad \mu_i^* \geq 0 , \qquad \nu_i^* \geq 0 ,$$

$$\bar{u}_j \geq u_j(z^*) , \quad \bar{\phi}_i^2 \geq \phi_i^2(z^*) , \quad \frac{1}{\bar{x}_i} \geq z_i^*$$

Thus, in the case of a finite element formulation, the optimality conditions are obtained in terms of element stresses and displacements or pseudo-stresses or pseudo-displacement. These latter can be obtained in practice by applying multi-unit load conditions when a structural analysis is being performed.

Having derived a general form of the optimality criterion, one can now specialize these conditions to the particular case in which only one displacement constraint is present. For this problem, the Kuhn-Tucker conditions become

$$-\frac{\omega_k}{z_k^{*2}} + \lambda^* \{\tilde{u}^k\} \left\{\frac{k^k}{z_k^*}\right\} \{u^k\} = 0 \ , \quad k = 1,\ldots,n$$

$$\lambda^*(\bar{u} - u(z^*)) = 0$$

$$\lambda \geq 0, \quad \bar{u} \geq u(z^*)$$

where $\{\tilde{u}^k\}$ are the displacements at the nodes of the kth element, due to a unit load applied at the point where the single displacement constraint $\bar{u}$ is specified. If the Lagrange multiplier λ^* is interpreted as a pseudo-load, giving

$$\{\hat{u}^k\} \left\{\frac{k^k}{z_k^*}\right\} \{u^k\} = \omega_k, \qquad k = 1,\ldots,n$$

with $\{\hat{u}^k\}$ now the nodal displacements associated with a load of magnitude λ^*. Interpreting the left-hand side as a virtual strain energy one observes that the optimality condition requires that the virtual strain energy density of each element attains a previously specified constant value. If the specific weight of each element has the same value, then the optimality condition demands that the strain energy density is constant throughout the structure, a fact originally discovered by Prager and Taylor [1].

Although, for convenience the optimality conditions for statically loaded structures subject to simple constraints have been developed, many of the arguments apply in more complex situations. For example, in the case of a design subjected to forced vibrations, the role played by the stiffness matrix in the above equations is substituted by the dynamic stiffness matrix. In a similar manner, the inclusion of overall buckling constraints can also be readily incorporated.

2.2 Duality

The definition of optimality conditions obviously has a central position in the theory of optimal structures. Of almost equal importance, when solution algorithms are considered, is the role of duality theory. The dual problem associated with MWP is defined by demanding a set of vector that minimize

$$L(z,\lambda,\mu,\nu) = W(z) + \sum_{j=1}^{m} \lambda_j(u_j(z) - \bar{u}_j)$$

$$+ \sum_{i=1}^{n} \mu_i(\phi_i^2(z) - \bar{\phi}_i^2) + \sum_{i=1}^{n} \nu_i (z_i - \frac{1}{\bar{x}_i})$$

subject to the dual constraints

$$\frac{\partial L}{\partial z_k} = - \frac{\omega_k}{z_k^2} + \sum_{j=1}^{m} \lambda_j \frac{\partial u_j}{\partial z_k} + 2 \sum_{i=1}^{n} u_i \phi_i \frac{\partial \phi_i}{\partial z_k} + \nu_k = 0$$

$$k = 1,\ldots,n$$

By using Eulers theorem,

$$\sum_{k=1}^{n} z_k \frac{\partial L}{\partial z_k} = - W(z) + \sum_{j=1}^{m} \lambda_j u_j(z) + 2 \sum_{i=1}^{n} \mu_i \phi_i^2(z)$$

$$+ \sum_{k=1}^{n} \nu_k z_k = 0$$

which leads to a new form of the dual objective function:

$$2W(z) - \sum_{j=1}^{m} \lambda_j \bar{u}_j - \sum_{i=1}^{n} \mu_i(\phi_i^2(z) + \bar{\phi}_i^2) - \sum_{i=1}^{n} \frac{\nu_i}{\bar{x}_i}$$

which is again subject to the dual constraints. If feasible values are selected for the design variables z_i, $i = 1,\ldots,n$, then the dual problem is linear in terms of the multipliers λ, μ, and ν, which may now be regarded as dual variables of a linear programming problem.

Following Bartholomew [6] the function F is introduced, defined by

$$F = \sum_{j=1}^{m} \lambda_j \bar{u}_j + \sum_{i=1}^{n} \mu_i(\phi_i^2(z) + \bar{\phi}_i^2) + \sum_{i=1}^{n} \frac{\nu_i}{\bar{x}_i}$$

and the dual objective function becomes $2W - F$. With this abbreviated objective function, one can investigate the effect of the extra freedom provided by the structural scale factor. Thus, considering a structure having cross-sectional areas

sx_i, $i = 1,\ldots,n$, the dual objective function becomes $2Ws - Fs^2$. Maximizing this with respect to s gives

$$2W - 2Fs = 0$$

Hence, a new objective function can be derived, which yields a dual problem requiring the maximization of

$$\frac{W^2}{F}$$

subject to the constraints

$$\left.\begin{array}{l} \displaystyle\sum_{j=1}^{m} \lambda_j \frac{\partial u_j}{\partial z_k} + 2 \sum_{i=1}^{n} \mu_i \phi_i \frac{\partial \phi_i}{\partial z_k} + \nu_k = \frac{\omega_k}{z_k^2} , \quad k = 1,\ldots,n \\[12pt] \hspace{6cm} j = 1,\ldots,m \\[6pt] \lambda_j \geq 0 , \quad \mu_i \geq 0 , \quad \nu_i \geq 0 , \hspace{2cm} i = 1,\ldots,n \end{array}\right\} \quad \text{DP}$$

Solutions to this modified dual formulation give a closer bound on the optimum weight than the straightforward formulation delineated earlier.

As is described in the next section the Kuhn-Tucker conditions outlined above provide the basis of the most successful structural optimization algorithms. They have, therefore, a more important role than simply defining the conditions that must prevail when the optimum solution to the design problem is achieved. The role of the dual formulation is equally significant in providing a limited, but important, bounding procedure for use with a solution algorithm. In addition, the dual problem is also used as part of the active set strategy, by which the full range of design constraints are reduced to those potentially active at the solution point.

3. STRUCTURAL OPTIMIZATION ALGORITHMS

3.1 Stress-ratioing

The basis of all structural optimization algorithms is to satisfy the optimality conditions KTC. However, these are complex in character and efforst have been made to simplify the equations by satisfying one part of the complete set. The earliest attempt at simplification involved structures subject only to stress constraints and assumed that equations KTC could be reduced to

$$\left. \begin{array}{c} \mu_i^*(\phi_i^2(z^*) - \bar{\phi}_i^2) = 0 \\[1em] \mu_i^* > 0 \end{array} \right\} \qquad i = 1,\ldots,n$$

If the further assumption is made that each constraint is related to one and only one design variable and that each of these variables can be scaled independently, then a solution algorithm is given by

$$z_i^* = \frac{\bar{\phi}^i}{\phi_i} z_i$$

where z^* represents an estimate of the optimizing vector of the inverse design variables. It is readily observed that this is equivalent to an assumption of static determinacy.

Although these assumptions give rise to an algorithm of limited value, it is, nevertheless, a powerful tool in the analyst's armory, when he is confronted with optimal design of large scale structures. Even for more general design problems, involving a range of constraints that includes stress constraints, the stress-ratioing algorithm can be used effectively for those design variables that are controlled by the stress constraints. However, the user must guard against the possibility that the algorithm may not converge to the optimum design.

3.2 Optimality Criteria

Stress-ratioing assumes that constraints are active and ignores the remaining parts of the Kuhn-Tucker conditions. Optimization techniques described by the title of optimality criterion are based on the assumption that the first set of equations in KTC apply. Employing slightly more compact notation in which all constraints are grouped under the term g_i, $i = 1,\ldots,p$, the relevant parts of KTC become

$$-\frac{\omega_k}{z_k^{*2}} + \sum_{i=1}^{p} \lambda_i^* \frac{\partial g_i}{\partial z_k} = 0, \qquad k = 1,\ldots,n$$

and the optimization algorithm is given by

$$\frac{1}{z_k^{*2}} = x_k^{*2} = \frac{1}{\omega_k} \sum_{i=1}^{p} \lambda_j^* \frac{\partial g_j}{\partial z_k}, \qquad k = 1,\ldots,n$$

Traditionally, this algorithm has been used for displacement constrained problems [7]. In the case of the design problem MWP, the algorithm becomes

$$\frac{1}{z_k^{*2}} = x_k^* = \frac{1}{\omega_k} \sum_{j=1}^{m} \lambda_j^* \left\{ \tilde{u}_j^k \right\}^t \left\{ \frac{k_k^k}{z_k} \right\} \{u^k\}, \quad k = 1,\ldots,n$$

Gauge constraints can easily be incorporated into the formulation to provide practical lower bounds on the design variables.

Although these formulas are simple in character, the algorithm requires accurate estimates of the optimizing Lagrange multipliers λ^*. In the case of a single displacement constraint, these equations can be solved to give estimates of the optimum design variable values. For multiple constraints the equations cannot be solved, so recourse is made to heuristic arguments to obtain a solution procedure. In particular, each constraint is considered to be independent of all other constraints. The single-constraint solution formula can then be applied to each constraint, in turn thereby creating as many estimates for the optimum design variables as there are constraints. In order to improve the convergence of this 'envelope' technique, the update formulas are raised to the power of an exponent selected a priori by the user. Other acceleration techniques have been suggested to improve the performance of this type of solution technique, but none remove the basic impediment created by inability to estimate the optimizing values for the Lagrange multipliers.

3.3   Newton Methods

Although the preceding algorithms are simple and extremely effective in certain design situations, they cannot be regarded as satisfactory for general applications. In order to achieve this generality, one must consider the Kuhn-Tucker conditions in their entirety. However, one must select from the inequality constraints those which, at each iteration, are considered to be active, i.e., have become equalities. Using the notation introduced earlier and denoting the set of active constraints as g_i, $i = 1,\ldots,p$, the KTC equations become

$$\nabla_z L(z^*,\lambda^*) = \nabla W(z^*) + G^t \lambda^* = 0$$

$$\lambda_i^* g_i(z^*) = 0 , \qquad\qquad i = 1,\ldots,p$$

If one starts from a feasible point, with known design variables z and Lagrange multipliers λ, and seeks to make a step to the optimum point z*, λ*, then

$$z^* = z + \delta z$$

$$\lambda^* = \lambda + \delta\lambda$$

If the optimum is found at z*, λ*, then a Taylor expansion from the starting point gives

$$\nabla_z L(z^*,\lambda^*) = \nabla W(z) + G^t\lambda + H\delta z + G^t\delta\lambda = 0$$

$$g(z^*) = g(z) + G\delta z = 0$$

where ∇W is the vector of derivatives of the objective function, G is the derivative matrix of the active constraints g_i, i = 1,...,p, and H is the hessian of the objective function. If these expansions are manipulated, then

$$\begin{bmatrix} H & G^t \\ G & 0 \end{bmatrix} \begin{Bmatrix} \delta z \\ \lambda^* \end{Bmatrix} = \begin{Bmatrix} -\nabla W(z) \\ - g(z) \end{Bmatrix}$$

Interpreting δz as a step-length in an iteration history, then

$$\delta z = z^{(k+1)} - z^{(k)}$$

$$\lambda^* = \lambda^{(k+1)}$$

and the iteration formulas can be obtained by solving these matrix equations, to give

$$\lambda^{(k+1)} = \left[G\left(z^{(k)}\right) H^{-1}\left(z^{(k)}\right) G^t\left(z^{(k)}\right) \right]^{-1}$$

$$\times \left[g(z) - G\left(z^{(k)}\right) H^{-1}\left(z^{(k)}\right) \nabla W\left(z^{(k)}\right) \right]$$

$$z^{(k+1)} = z^{(k)} + H^{-1}\left(z^{(k)}\right)\left[G^t\left(z^{(k)}\right)\lambda^{(k+1)} - \nabla W\left(z^{(k)}\right) \right]$$

These formulas can be considered as a Newton step towards the unconstrained optimum, which has been projected onto the active constraints. Because one is dealing with a Newton method, in essence the objective function is being approximated by a quadratic function. In order to compensate for errors in the curvature of the objective function, it is common in the allied subject of mathematical programming to consider the estimate

$z^{(k+1)} - z^{(k)}$ as a direction along which a search is made for a minimizing point. The efficiency of this approach in structural optimization is open to some debate and is discussed in some detail by Fleury [9].

Possibly the major difficulty of the solution algorithm is the assumed linearity of the constraints. The transformation to inverse variables improves the quality of the approximation, but cannot cope with the situation in which the curvature of the constraints is greater than that of the objective function. Bartholomew [8] has suggested that one solution to this difficulty would be to invoke a Wilson 'Solver' type of algorithm. Second derivatives of the constraints in a specified direction are available and the use of such second order algorithms is attractive, but largely untried at the present time.

3.4   Duality, Convergence, and Active Set Strategy

In the solution of a real structural optimization problem there are several distinct phases. Initially, for a large scale design it is usual to operate with one of the more simple algorithms, for example stress-ratioing, and change to a complex method when necessary. The change-over point may be selected based on a variety of criterion, but an important requirement dictates that such algorithms should be abandoned when potentially no active constraints are being enforced. Such constraints can be detected by the use of the dual problem DP which can effectively indicate non-active constraints through the sign of the Lagrange multipliers.

Duality has an important role to play during the terminal phase of a solution, where the design engineer needs a criterion for stopping the solution algorithm. The most useful criterion is to compare a primal feasible weight with a dual feasible value of W^2/F obtained by solving the dual problem DP. When this duality gap is less than a prescribed value, the algorithm can be terminated.

A solution of the complete dual problem is often as difficult to achieve as the original optimum design problem, but the two roles demanded above of the dual problem can be played by a linearized form. This is achieved by putting the values of the design variables attained at the end of a given iteration into DP as fixed values. The resulting problem can then be solved by the application of a standard linear programming algorithm, to provide both a lower bound on the optimum value and to indicate the activity level of the constraints.

As already indicated, the advanced algorithm of the Newton type require some form of active set strategy. The dual problem DP can be used, but the active set of strategies is more complex and, to a limited extent, the exact procedure followed is a question of taste. At the end of each iteration the violated constraints may be added to the active set, while those associated with zero or negative Lagrangian multipliers may be deleted. However, in order to avoid zig-zagging, it is customary to use a conservative constraint detection policy. For example, in the case of algorithms that develop their own Lagrangian multipliers, such as the Newton method, one may delete a specific constraint at each iteration identified by the most negative multiplier. An alternative philosophy used in the RAE system performs a sensitivity analysis on the set of negative multipliers and identifies the constraint to be deleted through changes in the Lagrangian function.

4. ILLUSTRATIVE EXAMPLES

Although there is no intention to present examples of practical designs achieved through the application of structural optimization, it is convenient to use specific problems to highlight how some of the algorithms of section 3 can be combined to produce a practical solution method. In addition, part of the role played by duality theory can be demonstrated.

For large scale statics problems, the designer is confronted with a large number of stress constraints and possibly a much smaller number of displacement constraints. It would be clearly very time-consuming and thus, expensive to immediately employ a complex optimization method that takes account of all the constraints, starting from an arbitrary feasible initial design. Intuitively, the design engineer will want to employ a simple method for sorting. Traditionally one uses the stress-ratioing method on the stress constraints until either a fully-stressed design is reached or the technique becomes inappropriate, due to the effect of the remaining constraints. At this point in the iteration history, the designer must change to one of the more comprehensive optimization methods, in order to converge to the required optimum.

The methods outlined in this paper have been used to solve a variety of large scale structural optimization problems. For details of implementation and examples, the reader is referred to the companion paper by Fleury [9].

5   CONCLUSIONS

The previous sections show that the solution of finite
dimensional structural optimization problems, using the finite
element technique, requires the application of the same
optimization theory as that employed in the infinite dimensional
case. For simple finite elements, however, the optimality and
duality equations takes a particular form. This form can be
exploited to develop a range of solution algorithms which can be
tailored by the design engineer to meet specific requirements.

REFERENCES

1.   Prager, W. and Taylor, J.E., "Problems of optimal structural
      design," J. Appl. Mech., Vol. 35, 1968, 102-106.
2.   Schield, R.T. and Prager, W., "Optimal structural design
      for given deflection," ZAMP, Vol. 21, No. 4, 1970, 513-523.
3.   Hemp, W.S., Optimum structures, Oxford University Press,
      1973.
4.   Morris, A.J., Bartholomew, P., and Dennis, J., "A computer
      based system for structural design, analysis and optimiza-
      tion," Proceedings AGARD Flight Mechanics Panel Symposium
      on 'The Use of Computers as a Design Tool', Neubiberg,
      Germany, September 1979.
5.   Chern,J-M. and Prager, W., "Minimum-weight design of
      statically determinate trusses subject to multiple con-
      straints," Int. J. Solids & Struct., Vol. 7, 1971, 931-940.
6.   Bartholomew, P., "A dual bound used for monitoring
      structural optimization programs," Engineering Optimization,
      Vol. 4, 1979, 45-50.
7.   Gellatly, R.A. and Berke, L., Optimal structural design,
      AFFDL-TR-70-165, 1971.
8.   Bartholomew, P., Private communication.
9.   Fleury, C., "Optimization of large flexural finite element
      systems," Optimization of Distributed Parameter Structures
      (Ed. E.J. Haug and J. Cea), Sijthoff & Nordhoff, Alphen
      aan den Rijn, 1980.

OPTIMIZATION OF LARGE FLEXURAL FINITE ELEMENT SYSTEMS

Claude Fleury

Aerospace Laboratory
The University of Liege
Rue Ernest Solvay, 21
B-4000 Liege, Belgium

ABSTRACT

Modern numerical methods for the optimization of large dis-cretized systems are now well developed and very efficient, in the case of thin-walled structures modeled by finite elements. However, this is not yet true for structures whose components are subject to bending loads. In this paper the idea of Generalized Optimality Criterion (GOC), set forth in previous work for bar, membrane, and shear panel elements is extended to deal with beam, plate, and flat shell elements. The modifications brought to the GOC method result in explicit approximation for the behavior constraints that are correct up to the first order, but that exhibit a more complex algebraic form. Indeed these explicit expressions are no longer merely linear in the reciprocal design variables. However, they continue to be additively separable and, therefore, dual methods remain fully applicable, just as in the original statement of the GOC approach. Numerical examples are offered to demonstrate the efficiency of the method presented.

1. INTRODUCTION

The structural optimization problem considered in this paper consists of weight minimization of a finite element model of the structure with fixed geometry and material properties. The design variables are taken as the transverse sizes of the structural

---

*This research has been sponsored by the U.S. Air Force Office of Scientific Research under grant AFOSR-80-0060.

members, namely the cross-sectional areas of bar and beam elements and the thicknesses of membrane, plate, and flat shell elements. The mathematical programming problem to be solved exhibits the following form: minimize

$$W = \sum_{i=1}^{n} \ell_i a_i \qquad (1.1)$$

subject to the constraints

$$h_j(a) \geq 0 \quad , \qquad j = 1,\ldots,m \qquad (1.2)$$

$$\bar{a}_i \geq a_i \geq \underline{a}_i \quad , \qquad i = 1,\ldots,n$$

where the a_i's denote the n design variables, which correspond to transverse sizes of either individual finite elements or, more often, of groups of finite elements (design variable linking). The structural weight of Eq. 1.1 is a linear objective function, because the ℓ_i's are constant coefficients (for example, specific weight times the length of a beam element or specific weight times the area of a plate element). The inequalities of Eqs. 1.2 represent behavior constraints, which impose limitations on quantities describing the structural response, e.g. stresses and displacements under multiple loading, natural frequencies, buckling loads, etc. The design variables are also subjected to side constraints of Eq. 1.3 where $\underline{a}_i$ and $\bar{a}_i$ are lower and upper limits that reflect fabrication and analysis validity considerations.

Standard minimization techniques can be applied to the non-linear programming problem of Eqs. 1.1 to 1.3. However this problem exhibits some characteristics that make it complicated when practical structural design applications are considered. The essential difficulty arises from the implicit nature of the behavior constraints of Eq. 1.2, in that their precise numerical evaluation for each particular design requires a complete finite element analysis. Since the solution scheme is iterative, it involves a large number of structural reanalyses. Therefore, the computational cost often becomes prohibitive when large structural systems are dealt with. However, in the case of thin-walled structures modeled by bar and membrane elements, a powerful design procedure has now emerged, which is based upon high quality explicit approximations of the behavior constraints [1].

For simplicity, this section and the following one are restricted to problems involving constraints on static stresses and displacements, in which case the behavior constraints of Eq. 1.2 can be written in form

$$h_j(a) \equiv \bar{u}_j - u_j(a) \geq 0 \qquad (1.4)$$

where $\bar{u}_j$ denotes an upper bound to a static response quantity $u_j(a)$ (stress, nodal displacement, relative displacement, etc.). Explicit approximations of these constraints can be generated, either by using virtual load considerations (optimality criteria approaches [2]) or by resorting to first order Taylor series expansion in terms of reciprocal design variables (mathematical programming approaches [3]). These approximations are of the form

$$\tilde{h}_j(a) \equiv \bar{u}_j - \sum_{i=1}^{n} \frac{c_{ij}}{a_i} \geq 0 \qquad (1.5)$$

In these expressions, the coefficients c_{ij} are related to virtual energy densities in the structural members. They can also be defined as the first derivatives of the response quantities with respect to the reciprocal design variables $1/a_i$ [4]. At each stage in the optimization process, the c_{ij}'s are assumed to be constant. By replacing the behavior constraints of Eq. 1.2 with the approximate forms of Eq. 1.5, an explicit problem is thus generated. Solving this explicit problem yields new values for the design variables. A structural reanalysis is then performed and the c_{ij}'s are updated. The whole process is repeated until convergence is achieved. Note that in the case of a statically determinate structure, the c_{ij}'s are, in fact, constant quantities, so only one structural analysis is needed to obtain the optimum design.

It is important to mention that this basic approach of transforming the initial problem into a sequence of explicit subproblems is now widely recognized [5] and is routinely employed for large scale industrial applications [6,7]. With respect to the solution of each subproblem, an attractive strategy is to solve it partially, using a primal solution scheme, before reanalyzing the structure and updating the approximate problem statement. This process facilitates generation of a sequence of steadily improved feasible designs. An alternative approach is to recognize that the explicit but approximate problem statement is of such high quality that it can be solved exactly, rather than partially, after each structural reanalysis. Adopting this viewpoint leads naturally to consideration of dual methods for solving the explicit problem and therefore to the concept of Generalized Optimality Criterion (GOC). In fact, the whole process of combining the linearization of the behavior constraints with respect to the reciprocal design variables and a dual solution scheme can be viewed as a rigorous generalization of conventional optimality criteria techniques [4,5].

This paper is concerned with structural models that are capable of carrying flexural forces. For beams and plates in pure bending, adequate intermediate variables can be selected, in terms of which high quality explicit approximations for the behavior constraints can still be constructed. The idea of GOC remains fully valid and it retains its interpretation in terms of energy densities in the structural members. The essential problem addressed in this paper is the establishment of the GOC approach in the general case where the structural members work both in extension and flexion (beam and flat shell elements). It is no longer possible to select a suitable intermediate variable for the linearization process. However, the virtual load procedure permits obtention of high quality, first order, explicit approximations of the behavior constraints.

2. PURE BENDING ELEMENTS

In this section, attention is focused on discretized models made up of pure beam and plate elements that are subjected to flexural loads only. The stiffness matrix of such a bending element is usually not merely proportional to its transverse size, as for bar and membrane elements. Therefore the optimization strategy reviewed in the previous section must be modified. The way to deal with a beam element subject to uniaxial bending depends upon the relationship between the principal moment of inertia I and the cross-sectional area a. A wide variety of situations is taken into consideration by adopting the relation

$$I = c\, a^p \tag{2.1}$$

where c is a constant that depends only on the shape of the beam cross-section and p is a positive number. Most often, p is taken as an integer, equal to 1, 2, or 3. For example, the case p = 2 is that of beams whose cross-sectional shape is kept invariant during redesign (dilatation or contraction). The flexural rigidity is proportional to the moment of inertia and therefore, in a finite element context, the structural stiffness matrix exhibits the following explicit form in terms of the cross-sectional areas:

$$K = \sum_{i=1}^{n} K_i = \sum_{i=1}^{n} a_i^p \bar{K}_i \,, \qquad p > 0 \tag{2.2}$$

where each matrix $\bar{K}_i$ is independent of the design variables a_i. Turning to plate elements subject to pure bending, since the stiffness is proportional to the cube of the thickness, it is apparent that Eq. 2.2 must be adopted with p = 3. Note that thin-walled structures modeled by bar and membrane elements, as well as sandwich beams and sandwich plates, are taken into consideration in the forthcoming developments by choosing p = 1.

The GOC method can be derived, just as in the case of thin-walled structures, by adopting a change of variables that tends to reduce the nonlinear character of the constraints

$$x_i = \frac{1}{a_i^p} \tag{2.3}$$

The next step is to linearize the constraints with respect to the new variables x_i,

$$\tilde{h}_j(x) \equiv \bar{u}_j - \left[u_j^0 + \sum_{i=1}^{n} \left(\frac{\partial u_j}{\partial x_i} \right)^0 (x_i - x_i^0) \right] \geq 0 \tag{2.4}$$

where the superscript "0" denotes quantities evaluated at the actual design point x^0, where the structural analysis is performed. Note that the finite element analysis must include auxiliary sensitivity analyses for evaluating the first partial derivatives of the response quantities. Restricting again the discussion to stress and displacement constraints, it is easily shown (see Ref. 8) that the first order Taylor series expansions of Eq. 2.4 reduce to the following form, when written in terms of the direct design variables a_i:

$$\tilde{h}_j(a) \equiv \bar{u}_j - \sum_{i=1}^{n} \frac{c_{ij}}{a_i^p} \geq 0 \tag{2.5}$$

The coefficients c_{ij} are the gradients of the response quantities u_j with respect to the intermediate variables x_i, defined in Eq. 2.3, but they can also be interpreted as virtual energy densities in the structural members. In this connection, it should be recognized that the virtual load procedure could be directly employed to derive the explicit approximations of Eq. 2.5, instead of resorting to first order Taylor series (see Section 3).

The GOC approach consists of replacing the behavior constraints of Eq. 1.2 with their approximate forms of Eq. 2.5 and applying a dual solution scheme to the resulting explicit problem. The optimality criterion equations can be obtained by writing the Kuhn-Tucker conditions for the explicit problem (see Ref. 8, p. 77). These equations, as well as the physical interpretation of the GOC method are very similar to those developed for thin-walled structures. For example, in the special case in which only one displacement constraint is specified, the optimality criterion states that the virtual strain energy density must be the same in each member. In this simple case, it is possible to analytically solve the explicit problem and to derive explicit redesign relations, in terms of known quantities. The active design variables can be shown to be given by

$$
a_i = \left[\frac{1}{\bar{u} - u_0} \sum_k \left(\ell_k^{\frac{p}{p+1}} \ c_k^{\frac{1}{p+1}} \right) \right]^{\frac{1}{p}} \left(\frac{c_i}{\ell_i} \right)^{\frac{1}{p+1}} \tag{2.6}
$$

while the remaining passive variables are fixed to an upper or a lower limit. The symbol u_0 denotes the contribution of these passive variables to the constraint $u < \bar{u}$. It is worthwhile mentioning that Eq. 2.6 is well suited to the design of plates in bending, with a single displacement constraint. Since $p = 3$, the redesign relation of Eq. 2.6 involves the fourth root of the coefficients c_i, rather than the third root, as employed in Ref. 9, on an intuitive basis. Note also that by taking $p = 1$ in Eq. 2.6, conventional redesign relations are recovered [2].

It can be concluded that the idea of GOC can easily be extended to deal with pure bending elements. High quality explicit approximations of the behavior constraints can still be generated by using first order Taylor series expansions, provided that adequate intermediate variables are selected. The resulting GOC retains its interpretation in terms of energy densities in the structural members. Determining the design variables that satisfy the GOC equations at each redesign stage can still be achieved efficiently by recoursing to dual methods, because the explicit approximate problem remains separable and strictly convex, when expressed in the intermediate design variables.

3. FLEXION-EXTENSION ELEMENTS

When flexure and extension forces act simultaneously, with comparable intensity at the element level, the definition of Eq. 2.2 of the stiffness matrix can no longer characterize the structural model with sufficient accuracy. To help fix ideas, consider a flat shell element made up of a membrane and a plate, stacked together. The stiffness matrix of this element exhibits the form

$$
K_i = a_i \ K_i^{(1)} + a_i^3 \ K_i^{(3)} \tag{3.1}
$$

where $K_i^{(1)}$ and $K_i^{(3)}$ are constant matrices. As a result, in the GOC approach, if the constraints are linearized with respect to the reciprocal design variables $1/a_i$, their first order explicit approximations, given by expressions similar to Eq. 1.5, will be of high quality only if the structural members behave mainly in extension. On the other hand, if the bending behavior is dominant, it is better to adopt the change of variables of Eq. 2.3, yielding first order explicit approximations of the form of Eq. 2.5 (with p = 3). As a matter of fact, the true situation is usually a combination of extension and bending. In a practical structure, some members work mainly in extension, some in flexure, and others, both in flexure and extension. This leads to the idea of using the following explicit approximations, which should be valid in any situation:

$$\tilde{h}_j(a) \equiv \bar{u}_j - \sum_{i=1}^{n} \left(\frac{c_{ij}^{(1)}}{a_i} + \frac{c_{ij}^{(3)}}{a_i^3} \right) \geq 0 \qquad (3.2)$$

where the coefficients $c_{ij}^{(1)}$ and $c_{ij}^{(3)}$ are considered constant throughout the redesign phase.

Because it is no longer possible to select appropriate intermediate design variables, the expressions of Eq. 3.2 cannot be constructed merely by using first order Taylor series expansion, as in the case of pure bending elements [see Eq. 2.4]. However, an essential requirement is that these explicit approximations remain correct up to the first order, despite the fact that they do not result from a strict linearization process. In other words, the following equality must hold:

$$\left. \frac{\partial \tilde{h}_j}{\partial a_i} \right|_{a^0} = \left. \frac{\partial h_j}{\partial a_i} \right|_{a^0} = \frac{c_{ij}^{(1)}}{\left(a_i^0\right)^2} + 3 \frac{c_{ij}^{(3)}}{\left(a_i^0\right)^4} \qquad (3.3)$$

This condition insures that, at the optimum, the solution to the explicit approximate problem satisfies the (first order) optimality conditions of the real problem, that is, the approximate and real restraint surfaces have the same tangent plane. As a result, the GOC approach should converge to a true minimum weight design. It will be shown hereafter how such first order explicit approximations can be obtained for various types of behavior constraints and structural models.

4.   STRESS AND DISPLACEMENT CONSTRAINTS

The key idea in constructing explicit approximations of the stress and displacement constraints is to come back to the virtual load procedure, which permits decomposing any static response quantity into the contributions of each element. Introducing a virtual load vector conjugated to the response quantity u_j (e.g., unit load for a nodal displacement), u_j can be expressed as follows:

$$u_j = q^T K q_j = \sum_{i=1}^{n} q^T K_i q_j \tag{4.1}$$

where q and q_j are respectively the real and virtual displacement vectors and K_i is the stiffness matrix of the ith element. On the other hand, it can be proved that the gradient of u_j is given by (see Ref. 8, page 75)

$$\frac{\partial u_j}{\partial a_i} = -q^T \frac{\partial K}{\partial a_i} q_j = -q^T \frac{\partial K_i}{\partial a_i} q_j \tag{4.2}$$

Now, for a rather general class of structural models, each element stiffness matrix can be assumed to have the following explicit form, in terms of its design variable [10],

$$K_i = \sum_{p=1}^{3} a_i^p K_i^{(p)} \tag{4.3}$$

where the matrices $K_i^{(p)}$ are independent of the design variables. Note that, most often, at least one of the $K_i^{(p)}$, p = 1,2, or 3, is zero (for example $K_i^{(2)}$ is zero in the stiffness matrix of Eq. 3.1 for a flat shell element). Introducing Eq. 4.3 into Eq. 4.1, it appears that a convenient explicit approximation of a stress or displacement constraint of Eq. 1.4 is

$$\tilde{h}_j(a) \equiv \bar{u}_j - \sum_{i=1}^{n} \sum_{p=1}^{3} \frac{c_{ij}^{(p)}}{a_i^p} \geq 0 \tag{4.4}$$

where the coefficients

$$c_{ij}^{(p)} = (q^T K_i^{(p)} q_j) a_i^{2p} \tag{4.5}$$

are assumed to be constant during the current stage. The gradient of this explicit approximate constraint is

$$\frac{\partial \tilde{h}_j}{\partial a_i} = \sum_{p=1}^{3} \frac{p \, c_{ij}^{(p)}}{a_i^{p+1}} \tag{4.6}$$

On the other hand, differentiating Eq. 4.3 and inserting the result into Eq. 4.2 shows that

$$\frac{\partial u_j}{\partial a_i} = - \sum_{p=1}^{3} p \, a_i^{p-1} \, q^T \, K_i^{(p)} \, q_j \tag{4.7}$$

Therefore, it can be concluded that the expressions of Eq. 4.4 represent first order explicit approximations, in that they constitute the exact values of the constraints and their first partial derivatives at the design point a^0 where the structural analysis is made:

$$\left\{ \begin{array}{l} \tilde{h}_j(a^0) = h_j(a^0) = \bar{u}_j - \sum\limits_{i=1}^{n} \sum\limits_{p=1}^{3} \frac{c_{ij}^{(p)}}{(a_i^0)^p} \tag{4.8} \\[20pt] \left.\frac{\partial h_j}{\partial a_i}\right|_{a^0} = \left.\frac{\partial h_j}{\partial a_i}\right|_{a^0} = \sum\limits_{p=1}^{3} \frac{p \, c_{ij}^{(p)}}{(a_i^0)^{p+1}} \tag{4.9} \end{array} \right.$$

5. FREQUENCY CONSTRAINTS

Constraints on natural frequencies usually consist of imposing lower limits

$$h_j \equiv \omega_j^2 - \underline{\omega}_j^2 \geq 0 \quad , \qquad j = 1,\ldots,m \tag{5.1}$$

They are directly written in terms of the squares of the frequencies, because these quantities naturally appear in the eigenproblem characterizing the structural modal analysis,

$$K \, q_j - \omega_j^2 \, M \, q_j = 0 \tag{5.2}$$

In this equation K and M represent the stiffness and mass matrices and q_j, $j = 1,\ldots,m$ are the modal displacements, i.e. the eigenvector solutions of Eq. 5.2, associated with eigenvalues ω_j^2. The structural mass matrix has a linear form in terms of the design variables,

$$M = \sum_{i=1}^{n} M_i + M_c = \sum_{i=1}^{n} a_i \, \bar{M}_i + M_c \tag{5.3}$$

where $\bar{M}_i$ and M_c are independent of the design variables. The matrix $\bar{M}_i$ denotes the mass matrix of the ith element, when $a_i = 1$.

The matrix M_c represents the contribution of the nonstructural masses, such as equipment, fuel, etc. It is well known that the first derivatives of the frequencies, with respect to the design variables, are given by

$$\frac{\partial \omega_j^2}{\partial a_i} = \frac{1}{\mu_j} q_j^T \left[\frac{\partial K_i}{\partial a_i} - \omega_j^2 \frac{\partial M_i}{\partial a_i} \right] q_j \tag{5.4}$$

where μ_j is the generalized mass of the jth mode

$$\mu_j = q_j^T M q_j \tag{5.5}$$

The best method to derive first order explicit approximation of the frequency constraints is less apparent than for stress and displacement constraints. In this paper, the following decomposition of the eigenvalues, in terms of the stiffness and mass contributions of each element, is used:

$$\omega_j^2 \equiv \omega_j^2 \left(1 + \frac{\bar{m}_j}{\mu_j} \right) - \frac{1}{\mu_j} \sum_{i=1}^{n} q_j^T (K_i - \omega_j^2 M_i) q_j \tag{5.6}$$

where

$$\bar{m}_j = q_j^T M_c q_j = \mu_j - \sum_{i=1}^{n} q_j^T M_i q_j \tag{5.7}$$

By taking account of the explicit definitions of Eq. 4.3 of the stiffness matrices K_i and of Eq. 5.3 for the mass matrices M_i, the high quality explicit approximations of the frequency constrains take the form of Eq. 4.4, with

$$\bar{u}_j = \omega_j^2 \left(1 + \frac{\bar{m}_j}{\mu_j} \right) - \omega_{-j}^2 \tag{5.8}$$

$$c_{ij}^{(1)} = \frac{q_j^T \left(K_i^{(1)} - \omega_j^2 \bar{M}_i \right) q_u}{\mu_j} a_i^2 \tag{5.9}$$

$$c_{ij}^{(p)} = \frac{q_j^T K_i^{(p)} q_j}{\mu_j} a_i^{2p} , \qquad p = 2,3 \tag{5.10}$$

The coefficients $c_{ij}^{(p)}$ and the modified limits $\bar{u}_j$ are frozen to their values at the current design point. Just as for the stress and displacement constraints, it is easily verified that Eq. 4.4 remain first order explicit approximations, satisfying Eqs. 4.8 and 4.9

6. SOLUTION OF THE EXPLICIT PROBLEM

From the foregoing developments, it appears that, at each stage of the optimization process, the following problem must be solved: minimize

$$W = \sum_{i=1}^{n} \ell_i \, a_i \tag{6.1}$$

subject to constraints

$$\sum_{i=1}^{n} \sum_{p=1}^{3} \frac{c_{ij}^{(p)}}{a_i^p} \leq \bar{u}_j \quad , \qquad j = 1,\ldots,m \tag{6.2}$$

$$\underline{a}_i \leq a_i \leq \bar{a}_i \quad , \qquad i = 1,\ldots,n \tag{6.3}$$

It is no longer possible to find intermediate variables, in terms of which the explicit constraints of Eq. 5.2 would be linear, so the primal solution of problem of Eqs. 6.1 to 6.3 is more diffi- cult to achieve if a gradient projection type algorithm is em- ployed, as in the mixed method developed in a previous work for thin walled structures [4]. The expressions of Eq. 6.2 are still explicit and they continue to exhibit a simple algebraic form. Therefore, a general optimization algorithm such as NEWSUMT [3] could be easily adapted to take the constraints of Eq. 6.2 into account. However, because they are still additively separable, resorting to dual methods remains probably the best strategy, just as in the case of thin-walled structures [4,5,6].

The minimization problem of Eqs. 6.1 to 6.3 can be solved efficiently as an auxiliary maximization problem in the m Lagrange multipliers r_j associated with the explicit behavior constraints of Eq. 6.2. This dual problem is as follows [11]: maximize

$$\ell(r) = \sum_{i=1}^{n} \ell_i \, a_i(r) + \sum_{j=1}^{m} r_j \, g_j(r) \tag{6.4}$$

subject to

$$r_j \geq 0 \quad , \qquad j = 1,\ldots,m \tag{6.5}$$

where $g_j(r)$ denote the components of the dual function gradient, which are equal to the values of the primal constraints,

$$g_j(r) = \frac{\partial \ell}{\partial r_j}(r) = \sum_{i=1}^{n} \sum_{p=1}^{3} \frac{c_{ij}^{(p)}}{a_i^p} - \bar{u}_j \tag{6.6}$$

The primal variables $a_i(r)$ are related to the dual variables r_j through the following one-dimensional minimization problems:

$$\min_{\underline{a}_i \leq a_i \leq \bar{a}_i} \left[\sum_{i=1}^{n} \ell_i a_i + \sum_{j=1}^{m} r_j \left(\sum_{i=1}^{n} \sum_{p=1}^{3} \frac{c_{ij}^{(p)}}{a_i^p} - \bar{u}_j \right) \right] \tag{6.7}$$

The $\tilde{n}$ active design variables can be obtained by solving the nonlinear algebraic equations

$$\sum_{p=1}^{3} \frac{c_i^{(p)}}{a_i^{p+1}} = \ell_i \quad , \qquad i = 1,\ldots,\tilde{n} \tag{6.8}$$

with

$$c_i^{(p)} = p \sum_{j=1}^{m} r_j \, c_{ij}^{(p)} \tag{6.9}$$

Assuming that each function to be minimized in Eq. 6.7 is unimodal, the subdivision between passive and active design variables can be decided as follows:

$$a_i = \underline{a}_i \quad , \qquad \text{if} \qquad \sum_{p=1}^{3} \frac{c_i^{(p)}}{\underline{a}_i^{p+1}} \geq \ell_i \tag{6.10}$$

$$a_i = \bar{a}_i \quad , \qquad \text{if} \qquad \sum_{p=1}^{3} \frac{c_i^{(p)}}{\bar{a}_i^{p+1}} \leq \ell_i \tag{6.11}$$

and a_i is active otherwise. Thus, the dual function of Eq. 6.4 can be considered as a function of the dual variables only.

The dual problem of Eqs. 6.4 and 6.5 exhibits an attractive feature. Namely, it is a quasi-unconstrained problem, because taking care of the nonnegativity constraints of Eq. 6.5 on the dual variables is straightforward. Given some nonnegative dual variables, the corresponding primal variables are computed from Eqs. 6.8 to 6.11 and the primal constraints are evaluated by using Eq. 5.6. The dual function of Eq. 6.4 and its gradient in Eq. 6.6 are directly known and a feasible ascent direction can therefore be determined. In a second order algorithm, the Hessian matrix of the dual function has to be evaluated as

$$H_{jk} = \frac{\partial^2 \ell}{\partial r_j \partial r_k} = \frac{\partial g_k}{\partial r_j} = - \sum_{i=1}^{\tilde{n}} \frac{\left(\sum_{p=1}^{3} p \, c_{ij}^{(p)} \, a_i^{1-p} \right) \left(\sum_{p=1}^{3} p \, c_{ik}^{(p)} \, a_i^{1-p} \right)}{\sum_{p=1}^{3} (p+1) \, c_i^{(p)} \, a_i^{2-p}}$$

$$(6.12)$$

where the summation is restricted to the $\tilde{n}$ active design variables (see Eqs. 6.8 to 6.11). Knowing the gradient in Eq. 6.6 and the Hessian matrix in Eq. 6.12 furnishes the Newton search direction

$$z = -H^{-1} g \qquad (6.13)$$

The next dual point is then given by

$$r^* = r + \tau z \qquad (6.14)$$

where τ is the step length taken along the direction z. Most often, a regular Newton method unit step [$\tau = 1$ in Eq. 6.14] is selected. However, the value of τ must sometimes be lowered to prevent one of the dual variables from becoming negative [5].

The second order dual optimizer implemented in SAMCEF [12] has been especially devised so that it seeks the maximum of the dual function by operating in a sequence of dual subspaces, with gradually increasing dimension. In this way, the effective dimensionality of the maximization problem never exceeds the number of active behavior constraints, which correspond to nonzero dual variables. Past experience with thin-walled structures indicates that this number is relatively small in practice, which explains the remarkable efficiency of the dual method approach [4,5]. The essential difficulty introduced in the foregoing dual formulation lies in the necessity to numerically solve the nonlinear algebraic equations of Eq. 6.8, by using, e.g. the Newton-Raphson method. A closed form solution is generally not possible, in contrast to the case of thin-walled structures, where explicit redesign relations are available [11]. However, in some special cases,

Eqs. 6.8 are still amenable to a closed form solution. For example, in structural models made up of flat shell elements, the terms corresponding to p = 2 are absent from the explicit form of the stiffness matrices of Eq. 3.1. Consequently the fourth order algebraic equations can be written

$$\ell_i \, a_i^4 - c_i^{(1)} \, a_i^2 - c_i^{(3)} = 0 \tag{6.15}$$

and they can be solved analytically.

7. NUMERICAL EXAMPLES

The two examples presented involve rather sophisticated flat shell elements that are characterized by a displacement field that is cubic in extension and quintic in flexure (hybrid quadrangular flat shell) [12]. The first example is taken from Ref. 13 and is concerned with minimum weight design of a simply supported square plate, subject to a natural frequency constraint. The dimensions of the plate are 10 in. by 10 in. and its material properties are E = 30 × 10^6 lb/in.^{2} and ν = 0.3. In Ref. 13, it is stated that the uniform plate with $\rho \omega^2$ = 1400 is taken as the initial estimate for the optimization problem. Therefore in the present study, assuming steel material with ρ = 0.283 lb/in.^{3}, the minimum frequency was chosen as $\nu = \omega/2\pi$ = 11.2 Hz. Iteration history data are presented in Table 7.1, for three cases, differing by the minimum thickness constraint ($\underline{a}$ = 0.1, 0.05, and 0.001 in.). The initial design in cases 1 and 3 corresponds to a uniform thickness a^0 = 0.2 in., while in case 2, a^0 equals 0.12 in. Final designs are given in Fig. 7.1. By symmetry only a quadrant of the plate has to be analyzed and designed. It can be observed that the design obtained in Ref. 13 is different from the design achieved in this study (case 1), however both designs have about the same weight (0.765 lb and 0.752 lb).

Attention is now directed to the I-beam structure shown in Fig. 7.2. The problem consists of minimizing the weight of the beam, while imposing lower bounds on the frequencies of the first three eigenmodes; flange flexure, torsion, and web flexure. Detailed data can be found in Ref. 8. In a first optimization exercise, a pure membrane model was employed. It involves 35 second degree displacement elements, including 10 fictitious diaphragms (without mass). These dummy members are introduced to obtain a satisfactory representation of the torsional mode. Only 5 analyses are sufficient to generate an optimum design for this membrane model. However when this final design was analyzed by using a more accurate model, made up of flat shell elements, the torsional frequency (mode 2) was seen to be violated by 10%. Therefore the problem was again solved with this new model, by

TABLE 7.1   ITERATION HISTORY DATA FOR SIMPLY SUPPORTED PLATE

Iter-ation	a = 0.1 in.		a = 0.05 in.		a = 0.001 in.	
	Weight (1b)	Fre-quency (Hz)	Weight (1b)	Fre-quency (Hz)	Weight (1b)	Fre-quency (Hz)
1	1.415	19.6	0.849	11.8	1.415	19.5
2	1.097	16.0	0.786	11.5	1.097	16.0
3	0.913	13.6	0.750	11.5	0.902	13.7
4	0.812	12.2	0.708	11.7	0.778	12.4
5	0.771	11.5	0.658	11.6	0.696	11.9
6	0.757	11.3	0.631	11.4	0.647	11.5
7	0.753	11.2	0.617	11.3	0.622	11.4
8	0.752	11.2	0.605	11.3	0.604	11.4
9	0.752	11.2	0.596	11.3	0.588	11.4
10			0.589	11.3	0.574	11.3
11			0.583	11.3	0.559	11.4
12			0.577	11.3	0.542	11.4
13					0.521	11.3

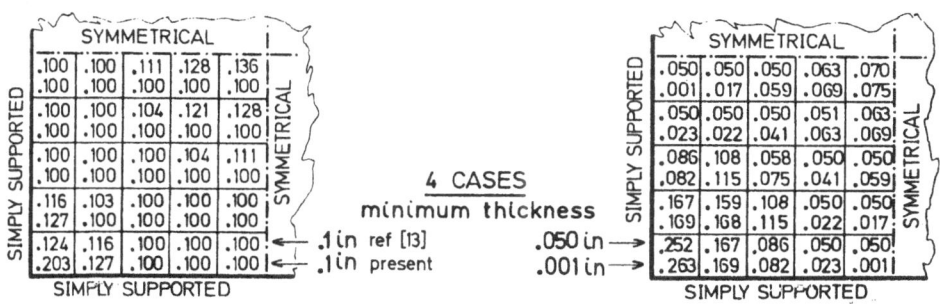

Figure 7.1   Final Designs for Simply Supported Plate

resorting to the theory proposed in the foregoing sections.   Iteration history data are illustrated in Fig. 7.2 and the final designs are given in Table 7.2, for both finite element models of the I-beam.

8.   CONCLUSIONS

It is now widely recognized that a powerful and rather general approach to structural optimization is achieved by replacing the original problem with a sequence of explicit approximate problems and solving them by using dual algorithms. This approach, which was initially conceived for thin-walled structures modelled by bar and membrane finite elements, has been extended in this paper to deal with structural systems made up of beam, plate, and flat shell elements. Although the present study was restricted to stress, displacement, and frequency constraints, it is worth noticing that linear buckling loads can be easily incorporated into the mathematical programming problem statement (see Ref. 8, Section 5). Because the method presented offers convergence properties that are independent of the number of design variables, large structural design problems can be treated at the expense of only a few finite element analyses.

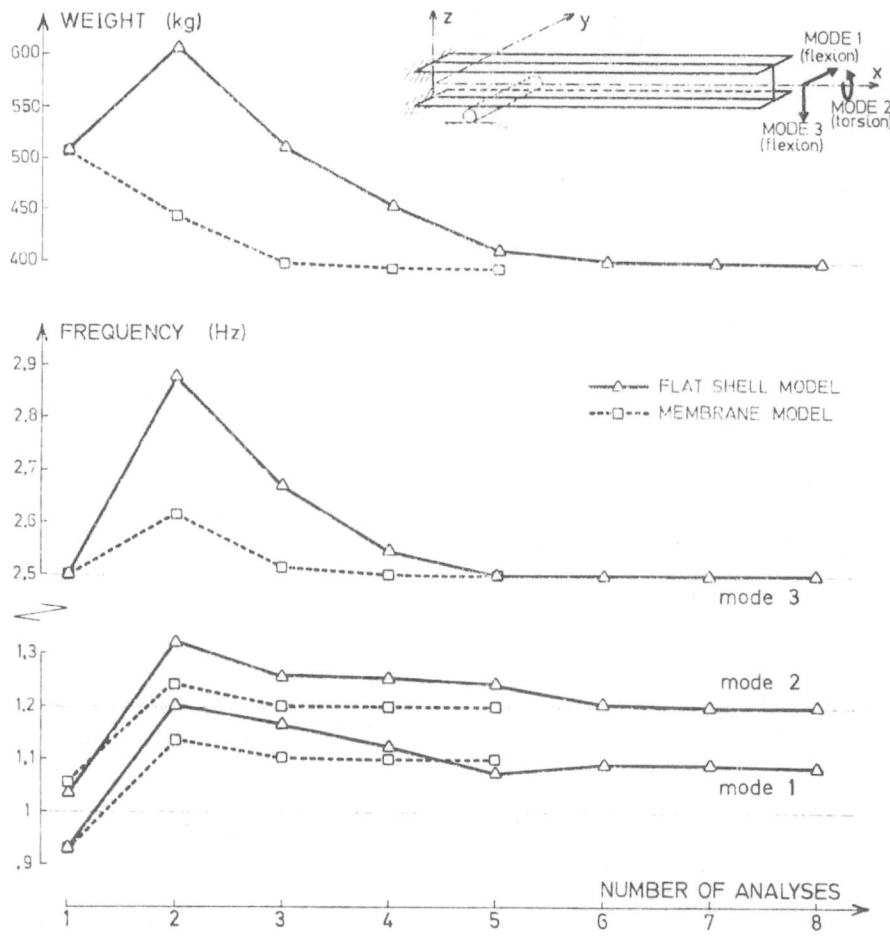

Figure 7.2   Iteration History for I-Beam

TABLE 7.2   FINAL DESIGN FOR I-BEAM MODEL

thickness (mm) $\begin{cases} \text{membrane model} \\ \text{flat shell model} \end{cases}$

Upper flange

| 15.80 | 17.30 | 11.67 | 6.143 | 1.815 |
| 16.59 | 15.09 | 10.08 | 5.090 | 1.363 |

Web

hinged

| 5.101 | 3.602 | 3.329 | 3.294 | 1.997 |
| 2.056 | 4.173 | 3.965 | 3.936 | 2.313 |

free

↑
supported

Lower flange

| 15.40 | 17.01 | 11.55 | 6.074 | 1.792 |
| 20.51 | 17.98 | 12.21 | 6.480 | 1.910 |

REFERENCES

1. Fleury, C., " A Unified Approach to Structural Weight Minimization," Comp. Meth. Appl. Mech. Eng., Vol. 20, No. 1, 1979, pp. 17-38.
2. Berke, L. and Khot, N.S., "Use of Optimality Criteria Methods for Large Scale Systems," AGARD-LS-70, 1974, pp. 1-29.
3. Schmit, L.A. and Miura, H., Approximation Concepts for Efficient Structural Synthesis, NASA-CR-2552, 1976.
4. Fleury, C. and Sander, G., Structural Optimization by Finite Elements, Final Scientific Report, Grant AFOSR-77-3118, LTAS, Univ. of Liège, 1978.
5. Fleury, C. and Schmit, L.A., Dual Methods and Approximation Concepts in Structural Synthesis, NASA-CR-3226, 1980.
6. Morris, A.J., Bartholomew, P. and Dennis, J., "A Computer Based System for Structural Design, Analysis and Optimization," AGARD-CP-280, 1980, paper 20.

756

7. Petiau, C. and Lecina, G., "Eléments Finis et Optimisation des Structures Aéronautiques," AGARD-CP-280, 1980, paper 23.
8. Fleury, C. and Sander, G., Generalized Optimality Criterion for Frequency Constraints, Buckling Constraints and Bending Elements, Final Scientific Report, Grant AFOSR-78-3652, Univ. of Liège, 1979.
9. Armand, J.L. and Lodier, B., "Optimal Design of Bending Elements," Int. J. Num. Methods Engng., Vol. 13, No. 2, 1978, pp. 573-584.
10. Kiusalaas, J. and Shaw, R.C.J., "An Algorithm for Optimal Structural Design with Frequency Constraints," Int. J. Num. Methods Engng., Vol. 13, No. 2, 1978, pp. 283-295.
11. Fleury, C., "Structural Weight Optimization by Dual Methods of Convex Programming," Int. J. Num. Methods Engng., Vol. 14, 1979, pp. 1761-1783.
12. SAMCEF, Système d'Analyse des Milieux Continus par Eléments Finis, LTAS, University of Liège
13. Haug, E.J., Pan, K.C. and Streeter, T.D., "A Computational Method for Optimal Structural Design. I: Piecewise Uniform Structures," Int. J. Num. Methods Engng., Vol. 5, 1972, pp. 171-184.

THE INTEGRATED APPROACH OF FEM-SLP FOR
SOLVING PROBLEMS OF OPTIMAL DESIGN

Pauli Pedersen

Department of Solid Mechanics
The Technical University of Denmark
DK-2800 Lyngby, Denmark

ABSTRACT

Today, the most efficient tool for solving boundary-value
problems is the finite element method (FEM); and probably among
the most efficient tools for solving optimization problems is the
Simplex method of linear programming, with nonlinear problems
solved as a sequence of linear problems (SLP). The integrated
approach of FEM-SLP forms a practical and reliable tool, easy to
use for research as well as for industrial optimal design.

These methods are not only intuitively understandable from a
physical/geometrical point of view, but they also rest on a well
established mathematical basis. In spite of this, one often feels
the reluctance of some research workers to use these methods, and
the integrated approach deserves a much wider attention. Problems
of inefficiency with respect to computer time are reported, but
these problems are diminished if the methods are not treated as
black boxes.

The present paper aims at clearing up these black boxes, and
concentrates on the critical points of the FEM-SLP approach, i.e.
the selection of an initial design, the analysis of the response-
sensitivities, and the selection of a sequence of move limits.
As applications to continuous distribution problems, two dimensional shape design for Min. Max. effective stress and one dimensional shape design of beams with multiple eigenvalue constraints
are treated.

1.   INTRODUCTION

As stated in a review by Niordson and the author [1], problems in optimal design may be formulated as follows: Find a design T such that

$$f_k(T) = 0 \quad , \qquad \text{for } k = 1, 2, \dots, K$$

$$h_j(T) \leq 0 \quad , \qquad \text{for } j = 1, 2, \dots, J \qquad \qquad (1.1)$$

$$\Phi(T) \quad , \qquad \text{is a minimum}$$

where the functionals f_k and h_j describe the constraints and the objective functional is denoted by Φ.   A necessary condition for optimality is given by the Kuhn-Tucker condition,

$$-\nabla\Phi(T) = \sum_{k=1}^{K} \lambda_k \nabla f_k(T) + \sum_{\text{active } j} \gamma_j \nabla h_j(T) \qquad (1.2)$$

where ∇ is the gradient operator, λ_k is an unknown set of real numbers, and γ_j is an unknown set of nonnegative real numbers. The condition of Eq. 1.2 simply states that no infinitesimal move from T, within the feasible design space, will decrease the object functional.

It is convenient to group the design variables into separate categories specifying size, geometry and topology. Most available results on optimal design deal only with sizes, i.e. areas of individual bars, height or width of beam sections, or thickness of disc, plate, and shell. Design variables dealing with geometry are joint positions of trusses or frames, positions of stiffeners, positions of supports, and the shape of two- or three-dimensional continua. Normally, one gains much more with geometrical variables than with size variables. Topological design, such as selection of the actual members of a structure or the structural model itself, is still very important. However, these problems include noncontinuous changes of the design and are therefore difficult to solve. For statically determinate structures such as trusses, solutions are obtainable [2], but in general, very few of these important problems are solved.

In the above-mentioned review [1], an objective point of view was taken covering variational methods and different methods of mathematical programming and optimization of structural elements as well as structural systems. The present paper is based on a subjective point of view, and will deal with what the author regards as the most valuable formulation and solution procedure. This approach is a combination of the finite element method (FEM)

and the Simplex method for solving a sequence of linear programming problems (SLP). Treated as an integrated approach, one may
call it the FEM-SLP approach.

Some arguments for the subjective point of view may be
appropriate:

(1) Although the methods of FEM and SLP rest on a well established mathematical basis, some research workers are
reluctant to use these methods. Except for the most
simple boundary-value problems, the FEM is a necessary
tool of analysis. How can one believe that he can solve
problems of optimal design without it? Much experience
in treating specific problems of analysis is available,
and if one does not use this experience in optimal design,
he must naturally expect difficulties, such as numerical
instabilities.

(2) About ten years ago, the trend in research on optimal design turned towards methods based on optimality conditions/
criteria (OC). The reason for this was that the mathematical programming approach (MP) was found to be too
costly in computer time. Today, one may ask:

(a) What is gained by the OC approach?

(b) Is it correct that the MP approach is costly?

The author would like to answer these questions as follows:

(a) Many important results are obtained for a variety
of idealized problems, mostly dealing with structural
elements and a single constraint. Many of these results may be used in more advanced programs to obtain
good initial designs, before an FEM-SLP optimization
is started. However, it is important to note the
following statement in Ref. 3: "There are no accurate, economical methods for optimizing practical-
sized redundant structures subject to multiple constraints on strength (stress), stiffness (displacement), and fabrication limits (minimum member sizes)."
This conclusion relates to the fact that OC are not
suited to multiple constraints.

(b) As to the second question, consider the SLP approach.
It is the author's experience that the computer time
required to obtain an optimum design is less than
ten times the computer time of the analysis that is
always required. Furthermore, the most valuable
design iterations are the first ones. The designer

may stop the design iterations when he gains less than the cost of one step of iteration. Problems of inefficiency are diminished if FEM and SLP are not treated as black boxes. The present paper aims at clearing up these boxes, and concentrates on the critical points of the FEM-SLP approach.

(3) One can obtain optimality conditions by variational methods, by the Kuhn-Tucker condition, by control theory, and even sometimes by the minimum principles of mechanics. It is important to realize that, with this, one has not solved the problem. The OC-based methods therefore include redesign algorithms, most of which are parallel to the classic, fully stressed (constant specific energy) iterations. For other than the most simple problems these algorithms are more or less ad hoc methods, with the absence of an objective, and the results approximate the optimal design to a degree that is difficult to assess. This is the author's main objection to the OC-based method. Why resort to ad hoc methods, when well documented iteration methods exist?

(4) A nice characteristic of the SLP approach is that increasing the number of constraints actually makes the solution easier to obtain. Among the methods of MP, many algorithms could probably be chosen. The reason for focusing on SLP is that with this approach the physical quantities of the problem are preserved as parameters and the method is therefore more direct to use for a designer, without a detailed knowledge of MP. It may be stated that the FEM-SLP approach is just computerizing the engineering way of thinking.

In the next section the optimality conditions are discussed in more detail. The formulation for two families of problems, which — from the point of view of OC — are difficult, are then presented. In section three, FEM-SLP is applied to two-dimensional shape design for min.-max. reference stress. This problem is difficult to treat with OC, because the objective is a local quantity with an unknown location. In section four, FEM-SLP is applied for one-dimensional shape design of beams, with multiple eigenvalue constraints. This problem is difficult to treat with OC, because it has a multiplicity of constraints. The paper aims at minimizing the mathematics involved.

2.   OPTIMALITY CONDITIONS/CRITERIA

As mentioned, many papers within the last decade have put forward the optimality conditions/criteria. Among the papers reviewed in Ref. 1, the reader may refer to papers by Prager and Taylor [4] and Masur [5] for general discussions of OC. Recent papers on OC are mostly related to its application. In general the OC state that a necessary condition for optimality is given by

Constant specific internal energy (2.1)

with the term specific related to the actual problem.

To see the agreement with the Kuhn-Tucker condition of Eq. 1.2, one may start with the simplest problem of a single equality constraint, for which Eq. 1.2 gives

$$-\nabla\phi(T) = \lambda\nabla f(T) \tag{2.2}$$

which states that the gradient of the cost ϕ should have the same direction as the gradient of the constraint. Assume furthermore that the cost is the sum of costs of the individual design sections and that each section depends only on one design variable, i.e.

$$\phi(T) = \sum_i \phi_i \quad \text{with} \quad \phi_i \equiv \phi_i(t_i) \tag{2.3}$$

The quantity of the constraint should also be determined as the sum over the design sections

$$f = \sum_i u_i - U \tag{2.4}$$

with U a constant. Normally, the quantity u_i depends explicitly on t_i, and only implicitly on the other design variables. Taking only the explicit dependence, as argued by virtual work principles, the assumptions of Eqs. 2.2, 2.3, and 2.4 give

$$-\phi_{i,t_i} = \lambda u_{i,t_i} \tag{2.5}$$

which, with u_i being an energy, gives the OC of Eq. 2.1, because λ is the same for all design sections i.

The three assumptions of Eqs. 2.2, 2.3, and 2.4 may seem rather restrictive, but they make the treatment of many different design problems possible. Let the constant of the equality constraint be

U = total internal complementary energy, or

U = total internal energy, or

U = external work, or $\qquad$ (2.6)

U = virtual external work,

then one may cover the design problems termed:

$$\left.\begin{array}{l}\text{Minimum cost for given stiffness,}\\\text{compliance, or displacement}\end{array}\right\} \qquad (2.7)$$

With d'Alemberts principle, one may also include dynamic problems and problems of stability. Thus the design problems may be extended to

$$\left.\begin{array}{l}\text{Minimum cost for given eigenfrequency}\\\text{or stability load}\end{array}\right\} \qquad (2.8)$$

Most papers on OC are based on the assumption of linear elasticity, but nonlinear elastic, visco-elastic, plastic, and nonconservative loads have also been treated. However, a paper that clearly classifies the necessary assumptions for Eq. 2.1 would be most valuable.

The extension of the problems to include absolute limits on the design variables

$$t_i^{min} \leq t_i \leq t_i^{max} \qquad (2.9)$$

follows directly from Eq. 1.2, with $h_i = t_i^{min} - t_i$ or $h_i = t_i - t_i^{max}$, and Eq. 2.5 is then modified to

$$\left.\begin{array}{ll} -\phi_{i,t_i} = \lambda u_{i,t_i} + \gamma_i(-1) \quad, & \text{for } t_i = t_i^{min} \\ \\ -\phi_{i,t_i} = \lambda u_{i,t_i} + \gamma_i(1) \quad, & \text{for } t_i = t_i^{max} \end{array}\right\} \qquad (2.10)$$

which, by $\gamma_i \geq 0$, shows that for $t_i = t_i^{min}$, the specific internal energy is often less than the constant value corresponding to $t_i^{min} < t_i < t_i^{max}$ and that for $t_i = t_i^{max}$, it is greater than this value. The practical problem is that one does not know where $t = t_i^{min}$ and where $t = t_i^{max}$.

Having more than one constraint of the type of Eq. 2.4 (possibly due to several loading cases), the OC of Eq. 2.5 will, from Eq. 1.2 and the assumptions of Eqs. 2.3 and 2.4, be

$$-\phi_i, t_i = \sum_k \lambda_k u_{ik}, t_i \tag{2.11}$$

The practical problem is that one does not know the linear combination factors λ_k. Note that a nonunique eigenfunction corresponds to a multiple constraint problem, cf. Olhoff and Rasmussen [6] and Masur and Mróz [7].

The simplicity and elegance of the OC are inherent, and a few critical comments may therefore be added to those in the introduction:

(1) The existence of designs that satisfy the conditions is not guaranteed. Therefore, the corresponding theorems should be read: "If there exists a design with constant specific energy, then...".

(2) The optimality conditions/criteria do not specify how a design that satisfies a specific condition can be obtained.

(3) With more than one constraint, one does not know the linear combination factors nor does one know whether the individual constraints are active or not.

3. SHAPE DESIGN FOR MIN.-MAX. REFERENCE STRESS

In a FEM stress analysis one first determines the nodal displacements {D} by solving

$$[S]\{D\} = \{A\} \tag{3.1}$$

where [S] is the total stiffness matrix and {A} are the nodal actions corresponding to {D}. These actions may include different equivalents for nondirect actions. Having solved the linear equations of Eq. 3.1, one gets the stresses $\{\sigma^e\}$ at a given point of element e from

$$\{\sigma^e\} = [Q^e]\{D^e\} \tag{3.2}$$

where the stress matrix $[Q^e]$ depends on the element geometry and material characteristics. For plane problems, $\{\sigma^e\}^T = \{\sigma_{11}, \sigma_{22}, \sigma_{12}\}^T$ ({ } is the symbol for a column vector and { }^{T} a transposed column vector, i.e. a row vector). Taking the squared von Mises' stress F as reference, one has for plane problems

$$F = \sigma_{11}^2 + \sigma_{22}^2 - \sigma_{11}\sigma_{22} + 3\sigma_{12}^2 \tag{3.3}$$

Now, for setting up the LP (linear programming) problem one needs the sensitivities of F to the design variables {T}. The effectiveness of the FEM stress-sensitivity analysis is very important. Although one can always obtain the necessary quantities as differences of reanalyses (to be sure, take central differences), a more detailed analysis is described here, in order to show that the sensitivity analysis does not necessitate any reanalysis.

Let t_i be a specific design parameter of {T}. Using the notation of comma for partial differentiation, one gets from Eq. 3.1,

$$[S]\{D\}_{,t_i} = \{A\}_{,t_i} - [S]_{,t_i}\{D\} \tag{3.4}$$

This equation states that, defining $\{A\}_{,t_i} - [S]_{,t_i}\{D\}$ as the action equivalent to a unit design change $\Delta t_i = 1$, the effort of computing $\{D\}_{,t_i}$ corresponds to one additional loading case and is thus normally inexpensive. The change in the real action $\{A\}_{,t_i}$ is often zero and, even when it is different from zero, it is easy to determine. Then the critical point for effectiveness (and numerical stability) is brought down to the evaluation of $[S]_{,t_i}\{D\}$. From the construction of the total stiffness matrix, it follows that

$$[S]_{,t_i}\{D\} \qquad \text{is assembled from} \qquad \sum [S^e]_{,t_i}\{D^e\} \tag{3.5}$$

for elements depending upon t_i. Furthermore, $[S^e]_{,t_i}$ may sometimes be evaluated analytically, as shown in Ref. 8, using the linear transformations from Ref. 9. Parallel transformations for tetrahedral elements are given in Ref. 10 and it would not be difficult to find optimal designs of three-dimensional shapes.

Having determined $\{D\}_{,t_i}$, one may determine $F_{,t_i}$ by central differences, as is done in Ref. 8. Alternatively, one may proceed on the analytical level. From Eq. 3.2, one gets

$$\{\sigma^e\}_{,t_i} = [Q^e]_{,t_i}\{D^e\} + [Q^e]\{D^e\}_{,t_i} \tag{3.6}$$

and defining $\{\tau\}^T$ by

$$\{\tau\}^T = \left\{2\sigma_{11} - \sigma_{22}, \ 2\sigma_{22} - \sigma_{11}, \ 6\sigma_{12}\right\}^T \tag{3.7}$$

one sees from Eq. 3.3 that the final result is given by

$$F_{,t_i} = \{\tau\}^T \{\sigma^e\}_{,t_i} \tag{3.8}$$

The next step is the formulation of the linear programming problem at a given stage of design. This may be done in the same manner as in optimal design of trusses [2,11,12]. In Section 4, a parallel formulation is given for eigenvalue problems. First, one puts some absolute limits on the design variables,

$$\{T\}^{min} \leq \{T\} \leq \{T\}^{max} \tag{3.9}$$

where $\{T\}^{min}$ and $\{T\}^{max}$ are user-given bounds. Next, one puts some move-limits on the change of the design variables,

$$\{0\} \geq \{\Delta T\}^{min} \text{ at } n \leq \{\Delta T\}^n \leq \{\Delta T\}^{max} \text{ at } n \geq \{0\} \tag{3.10}$$

and these move-limits may change with the design iteration n. With the move-limits, one can immediately eliminate the absolute limits of Eq. 3.9. If a design parameter is t_i^n before the design iteration n, then

$$\left. \begin{array}{l} t_i^{min} - t_i^n > \Delta t_i^{min} \text{ at } n \Rightarrow \Delta t_i^{min} \text{ at } n = t_i^{min} - t_i^n \\[2mm] t_i^{max} - t_i^n > \Delta t_i^{max} \text{ at } n \Rightarrow \Delta t_i^{max} \text{ at } n = t_i^{max} - t_i^n \end{array} \right\} \tag{3.11}$$

The move-limits in the LP formulation have many further advantages. In Ref. 12 there is a detailed discussion of how to choose the sequence of move-limits. In interactive optimal design, it is natural to let this be the most valuable controlling parameters for the designer. It is not difficult for an inexperienced user to specify the move-limits, because they are directly related to the design parameters and not abstract coefficients as in other formulations. Generally, the sequence of design iterations starts with rather large move-limits and should end up with move-limits related to the accuracy of a produced design. Absolute move-limits are preferred and both automatic increase and automatic decrease are often valuable.

The object of the optimization is to

Minimize F_{max} $\qquad$ (3.12)

subject to the constraint that the stress level everywhere, for all loading conditions m, is limited by F_{max}, i.e.

$$F_m(x) + \sum_{i=1}^{I} F_m(x)_{,t_i} \Delta t_i - F_{max} \leq 0 \qquad (3.13)$$

for m = 1,...,M and all x in the space of the design. This may seem to constitute many equations. However, in actual design there are normally only a few critical points and corresponding critical loading cases, and these are localized by the FEM analysis. The term critical here means close to the F_{max}. If one deals with constraints that are not active within the actual move-limits, then some unnecessary computing is done, but correct results are obtained. If one omits some active constraints, the next step of design iteration will cure the nonfeasibility. Thus, the selection of critical points is by no means critical to the success of the method.

One may assemble the critical stress points (possibly corresponding to different loads) in the vector {F}. Then Eq. 3.13 is rewritten as

$$F + [F]_{,\{T\}}\{\Delta T\}^n - \{U\}F_{max} \leq \{0\} \qquad (3.14)$$

where the column no. i of $[F]_{,\{T\}}$ is to be read as $\{F\}_{,t_i}$. All elements of the column vector {U} are equal to 1.

Going further into the LP approach, one again sees the value of the move-limits, because the transformation of nonnegative variables $\{\Delta T\}^+$ is given by

$$\{\Delta T\}^{n+} = \{\Delta T\}^n - \{\Delta T\}^{min} \text{ at } n \qquad (3.15)$$

or

$$\{\Delta T\}^n = \{\Delta T\}^{n+} + \{\Delta T\}^{min} \text{ at } n \qquad (3.16)$$

Note that the variable F_{max} is always nonnegative. The inequality constraints of Eqs. 3.10 and 3.13 are converted into equalities by nonnegative slack variables $\{\Delta \tilde{T}\}$ and $\{\tilde{F}\}$ and the LP problem to be solved is then

$$\text{Minimize } \left\{ \{0\}^T, \ \{0\}^T, \ 1, \ \{0\}^T \right\}^T \left\{ \begin{array}{c} \{\tilde{F}\} \\ \{\tilde{\Delta T}\} \\ F_{max} \\ \{\Delta T\}^{n+} \end{array} \right\} \tag{3.17}$$

subject to the linear equations,

$$\begin{bmatrix} [I], & [0], & -\{U\}, & [F]_{,\{T\}} \\ [0], & [I], & \{0\}, & [I] \end{bmatrix} \left\{ \begin{array}{c} \{\tilde{F}\} \\ \{\tilde{\Delta T}\} \\ F_{max} \\ \{\Delta T\}^{n+} \end{array} \right\}$$

$$= \left\{ \begin{array}{c} -\{F\} - [F]_{,\{T\}} \ \{\Delta T\}^{min} \text{ at } n \\ \{\Delta T\}^{max} \text{ at } n - \{\Delta T\}^{min} \text{ at } n \end{array} \right\} \tag{3.18}$$

Corresponding to the number of negative elements in $\left[-\{F\} - [F]_{,\{T\}}\{\Delta T\}^{min} \text{ at } n \right]$, one changes the sign for the actual equations and introduces artificial variables, i.e. variables with a very large coefficient in the objective of Eq. 3.17. The first basic solution for the Simplex procedure is now obtained. The problem of Eqs. 3.17 and 3.18 is then solved by two small sub-routines, each of about 20 statements, one for locating the best change to another basic solution (including test for optimality) and the other for performing the necessary row operations.

Referring to the examples given in Ref. 8, one may finally note some important practical aspects of the method:

(1) It is necessary to make sure that the FEM model does not degenerate. This means that the model must change with the design iterations. However, automatic mesh generation is standard in most FEM programs, and therefore requires no additional effort.

(2) The shape to be designed is described by a few global functions, and not by the FEM nodal coordinates. In this way one ensures smoothness and desired boundary behavior.

Furthermore, the model of the design is then less sensitive to FEM errors. Experience shows that when orthogonal functions are chosen, about five design parameters are enough.

(3) The most simple finite elements of constant strain/stress are not used, because experience has shown that they can give rise to serious errors. Elements of linear strain/stress are found to be highly reliable.

(4) Substructuring or superstructuring may be the answer to a large-scale problem.

(5) The number of necessary design iterations and the total computer times correspond to a factor of ten, when compared with a simple FEM stress analysis.

4. OPTIMAL DESIGN WITH MULTIPLE EIGENVALUE CONSTRAINTS

The intention of this section is to present the integrated FEM-SLP formulation for design problems with multiple eigenvalue constraints. A recent paper by Levy and Chai [13] is based on OC, and in the final conclusions it is stated that "Abandoning optimality criterion methods in favor of the more rigorous mathematical programming approaches would have to be considered carefully." This is exactly in the spirit of the present paper.

As in the preceding section, attention is focused on analysis and sensitivity analysis more than on optimization. The unpublished results shown in Fig. 4.1 were obtained in 1967 [14], without any specific sensitivity analysis, i.e. simply based on differences of reanalyses. Especially the result of Fig. 4.1(c), where two eigenvalues are considered, shows that rather advanced problems can be treated in that way.

The FEM formulation for the structural eigenfrequencies is

$$[S]\{D\} = \bar{\lambda}[M]\{D\} \tag{4.1}$$

where $\bar{\lambda}$ and $\{D\}$ are the squared eigenfrequency and corresponding eigenmode and $[M]$ is the total mass matrix, established analogously to the total stiffness matrix $[S]$. The most efficient analysis procedure for this problem is termed the method of inverse iteration (power method, Stodola method, or successive iterations). To obtain higher order eigenfrequencies, one shifts and/or orthogonalizes. The method is improved by simultaneous (subspace) iterations, at the same time overcoming the inherent problem of close or even equal eigenvalues.

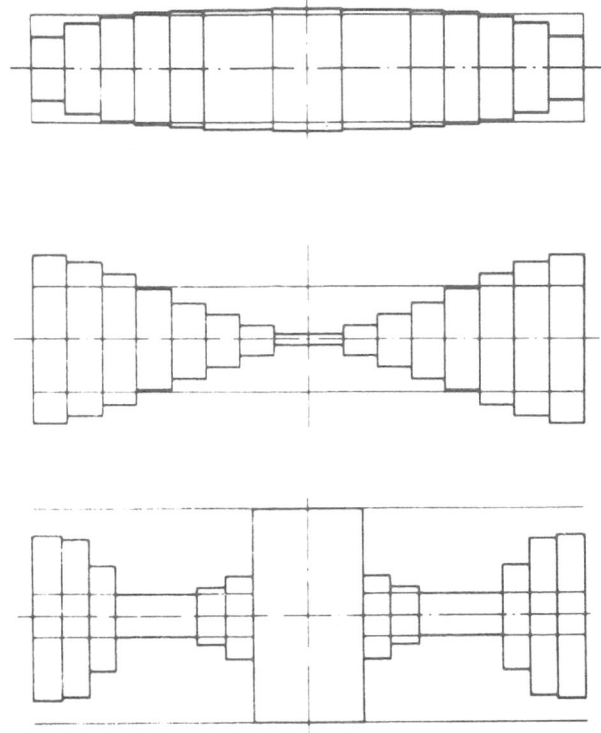

Figure 4.1   Optimal Design of Shafts for Maximum Critical
              Frequency. [(a) given volume and simply supported,
              (b) given volume and clamped, and (c) given torsional
              frequency and fixed sections]

Introducing the shift

$$\bar{\lambda} = \lambda_0 + \lambda \tag{4.2}$$

where λ_0 is a given shift value, Eq. 4.1 is transferred to

$$\left[[S] - \lambda_0[M] \right] \{D\} = \lambda[M]\{D\} \tag{4.3}$$

with λ and $\{D\}$ now being the eigenvalue and the eigenvector.
Premultiplying Eq. 4.3 by $\{D\}^T$, one gets the Rayleigh quotient

$$\lambda = \frac{\left[\{D\}^T[S]\{D\} - \lambda_0\{D\}^T[M]\{D\} \right]}{\{D\}^T[M]\{D\}} = \frac{\{D\}^T[S]\{D\}}{\{D\}^T[M]\{D\}} - \lambda_0 = \bar{\lambda} - \lambda_0 \tag{4.4}$$

Differentiating this with respect to the design variables t_i, one has

$$\bar{\lambda}_{,t_i} = \lambda_{,t_i} = \frac{\{D\}^T[S]_{,t_i}\{D\} - \bar{\lambda}\{D\}^T[M]_{,t_i}\{D\}}{\{D\}^T[M]\{D\}} \qquad (4.5)$$

The computational algorithm corresponding to the problem of Eq. 4.3 is

$$\Big[[S] - \lambda_0[M]\Big]\{D\}_n = [M]\{D\}_{n-1} \qquad (4.6)$$

with the convergence proven to be as follows:

$(d_j)_{n-1}/(d_j)_n \rightarrow \lambda$, which makes $\bar{\lambda}$ closest to λ_0

$\{D\}_n \rightarrow \{D\}$, the corresponding mode for $n \rightarrow \infty$ and j,

in principle, arbitrary . $\qquad (4.7)$

With the shift λ_0, one controls the eigenvalue to be found and the rate of convergence. As will be seen, Eqs. 4.4 and 4.5 are effectively evaluated at the element level.

Until now the structural model and the design parameters have not been specified. The examples that follow relate to a Timoshenko beam model, including rotational inertia. The cross-sectional areas are taken as design parameters. Introducing element nondimensional parameters, with L being some reference (total) length, one has for element e

length: $\ell_e = \xi_e L$

total volume: $V = \sum a_e \ell_e = L^3 \sum n_e^2 \xi_e$

area: $a_e = n_e^2 L^2$

moment of inertia: $I_e = \alpha^2 a_e^2 = \alpha^2 n_e^4 L^4$

mass density: $\rho_e = \rho$

Young's modulus: $E_e = E$

$\qquad (4.8)$

slenderness number: $\kappa_e = \alpha^2 \eta_e^2 / \xi_e^2$

coefficient of shear: $\dfrac{12EI_e}{\mu Ga_e \ell_e^2} = \dfrac{24(1 + \nu)}{\mu} \kappa_e = \beta_e$

and with these quantities the, four-by-four element stiffness matrix $[S^e]$ is

$$[S^e] = \frac{EL\alpha^2 \eta_e^4}{(1 + \beta_e)\xi_e^3}
\begin{bmatrix}
12 & 6L\xi_e & -12 & 6L\xi_e \\
6L\xi_e & L^2\xi_e^2(4 + \beta_e) & -6L\xi_e & L^2\xi_e^2(2 - \beta_e) \\
-12 & -6L\xi_e & 12 & -6L\xi_e \\
6L\xi_e & L^2\xi_e^2(2 - \beta_e) & -6L\xi_e & L^2\xi_e^2(4 + \beta_e)
\end{bmatrix}$$

(4.9)

The four-by-four consistent element mass matrix $[M^e]$ is

$$[M^e] = \frac{\rho L^3 \eta_e^2 \xi_e}{840(1 + \beta_e)^2}
\begin{bmatrix}
g_1 & g_3 L\xi_e & g_2 & -g_4 L\xi_e \\
g_3 L\xi_e & g_5 L^2\xi_e^2 & g_4 L\xi_e & -g_6 L^2\xi_e^2 \\
g_2 & g_4 L\xi_e & g_1 & -g_3 L\xi_e \\
-g_4 L\xi_e & -g_6 L^2\xi_e^2 & -g_3 L\xi_e & g_5 L^2\xi_e^2
\end{bmatrix}$$

(4.10)

where the nondimensional g functions are

$$g_1 = 312 + 588\beta_e + 280\beta_e^2 + 1008\kappa_e$$

$$g_2 = 108 + 252\beta_e + 140\beta_e^2 - 1008\kappa_e$$

$$g_3 = 44 + 77\beta_e + 35\beta_e^2 + (84 - 420\beta_e)\kappa_e$$

(4.11)

$$g_4 = 26 + 63\beta_e + 35\beta_e^2 + (-84 + 420\beta_e)\kappa_e$$

$$g_5 = 8 + 14\beta_e + 7\beta_e^2 + (112 + 140\beta_e + 280\beta_e^2)\kappa_e$$

$$g_6 = 6 + 14\beta_e + 7\beta_e^2 + (28 + 140\beta_e - 140\beta_e^2)\kappa_e$$

The results of Eqs. 4.9 and 4.10 may be found in Ref. 15.

From these results, it is seen that the sensitivity analysis is most simple when the areas are taken as design parameters

$$t_e = n_e^2 \tag{4.12}$$

When the influence of $\beta_{e,t_e} = \beta_{e,n_e^2}$ and $\kappa_{e,t_e} = \kappa_{e,n_e^2}$ are neglected, one gets

$$[S]_{,t_e} = 2[S^e]/n_e^2 \tag{4.13}$$

$$[M]_{,t_e} = [M^e]/n_e^2 \tag{4.14}$$

Inserting these results in Eq. 4.5 and choosing, for convenience of writing, a normalization $\{D\}^T[M]\{D\} = 1$, one gets

$$\lambda_{,t_e} = \left[2\{D^e\}^T[S^e]\{D^e\} - \bar{\lambda}\{D^e\}^T[M^e]\{D^e\} \right]/n_e^2 \tag{4.15}$$

These derivatives are determined without additional effort when the Rayleigh quotient is determined by addition of the element energies

$$\bar{\lambda} = \frac{\sum_e \{D^e\}^T[S^e]\{D^e\}}{\sum_e \{D^e\}^T[M^e]\{D^e\}} \tag{4.16}$$

Note the relations to the criteria of constant specific energy for the problem with a single eigenvalue constraint.

If one takes the eigenvalue of Eq. 4.1 to be a nondimensional frequency, it follows from Eqs. 4.9 and 4.10 that it should be defined by

$$\bar{\lambda} = \frac{\rho L^2}{840\alpha^2 E} \omega^2 \tag{4.17}$$

The problem of minimum volume with multiple eigenvalue constraints may now be formulated. The constraints are given by

$$\lambda_m + \sum_{i=1}^{I} \lambda_{m,t_i} \Delta t_i - \lambda_m^{min} \geq 0 \quad , \qquad m = 1,\ldots,M \tag{4.18}$$

Changing sign and introducing the nonnegative slack-variables $\{\tilde{\Gamma}\}$ one can, in parallel to Eq. 3.14, write

$$- \{\Gamma\} - [\Gamma]_{,\{T\}}\{\Delta T\}^n + \{\Gamma^{min}\} + \{\tilde{\Gamma}\} = \{0\} \tag{4.19}$$

With move-limits, the ready-to-solve LP program is then (cf. Eqs. 3.17 and 3.18),

$$\text{Minimize } \left\{ \{0\}^T, \ \{0\}^T, \ \{C\}^T \right\}^T \left\{ \begin{array}{c} \{\tilde{\Gamma}\} \\ \{\widetilde{\Delta T}\} \\ \{\Delta T\}^{n\dotplus} \end{array} \right\} \tag{4.20}$$

with, from Eq. 4.8, $c_i = \xi_i$ and $t_i = n_i^2$, subject to the linear equations

$$\begin{bmatrix} [I], & [0], & -[\Gamma]_{,\{T\}} \\ [0], & [I], & [I] \end{bmatrix} \left\{ \begin{array}{c} \{\tilde{\Gamma}\} \\ \{\widetilde{\Delta T}\} \\ \{\Delta T\}^{n+} \end{array} \right\}$$

$$= \left\{ \begin{array}{c} \{\Gamma\} - \{\Gamma\}^{min} + [\Gamma]_{,\{T\}} \{\Delta T\}^{min} \text{ at } n \\ \{\Delta T\}^{max} \text{ at } n - \{\Delta T\}^{min} \text{ at } n \end{array} \right\} \tag{4.21}$$

In the following, various results are shown, some of which are checked against known results, while others are new. These examples verify the effectiveness of the FEM-SLP integrated approach for dynamic problems. The total computer time for one of these designs was about 2 seconds, when the beam was modelled by 40 elements, using a Fortran H compiler and an IBM 3033 computer. The program constitutes less than 400 statements, in total, and no black boxes were used.

The results shown in Fig. 4.2 are for a simply supported beam, chosen very slender because four frequencies are encountered. Thus one wants the results to almost correspond to Bernoulli-Euler beam theory. However, taking shear deformations into account, zero area does not occur at the supports and no area constraints are active. The result shown in Fig. 4.2(a) agrees with known results (for the dual design problem, it corresponds to an increase in eigenfrequency of 6.5%). The results of Figs. 4.2(b), 4.2(c), and 4.2(d) are new and the trend in design changes are natural, bearing in mind the shape of the eigenmodes corresponding to the constraint frequencies.

The results shown in Fig. 4.3 are for a clamped-simply supported, slender beam. Again no area constraints are active. For the dual design problem corresponding to Fig. 4.3(a), an increase in eigenfrequency of 57% is reported, which corresponds to a reduction in volume of 60%. The designs of Figs. 4.3(b), 4.3(c), and 4.3(d) are more exotic, although the trend again is clear.

Very few results with flexible supports can be found in the many papers on optimal design. Therefore, in Fig. 4.4 results are shown for a flexible-clamped, simply supported beam. The optimum designs in Fig. 4.2 and 4.3 are quite different, and the optimum designs in Fig. 4.4 is the resulting balance.

The effects of slenderness are shown in Fig. 4.5. Especially at the supports, the influence of the slenderness is strong. Pierson [16] has given results for the cantilever case, so these results are omitted here. In the conclusions of Ref. 16, the practical restrictions for the optimal control approach are discussed. With the present FEM-SLP approach, these restrictions are not encountered. It is always difficult to compare computer time, but 131 sec are mentioned in Ref. 16 and here only about 2 sec were used, with four eigenvalue constraints.

5.   CONCLUSIONS

Optimal design is a highly nonlinear problem, which must be solved iteratively. Each iteration involves three steps: an analysis of the response of a design, a sensitivity analysis corresponding to possible changes in this design, and the decision of redesign.

For all kinds of response; static or dynamic, linear or nonlinear, global or local; the FEM is a unified approach, backed by a vast amount of experience. Treating sensitivity analysis as an integrated part of FEM, one obtains gradients of the response without too much cost. This information is valuable in itself, not only in relation to optimal design.

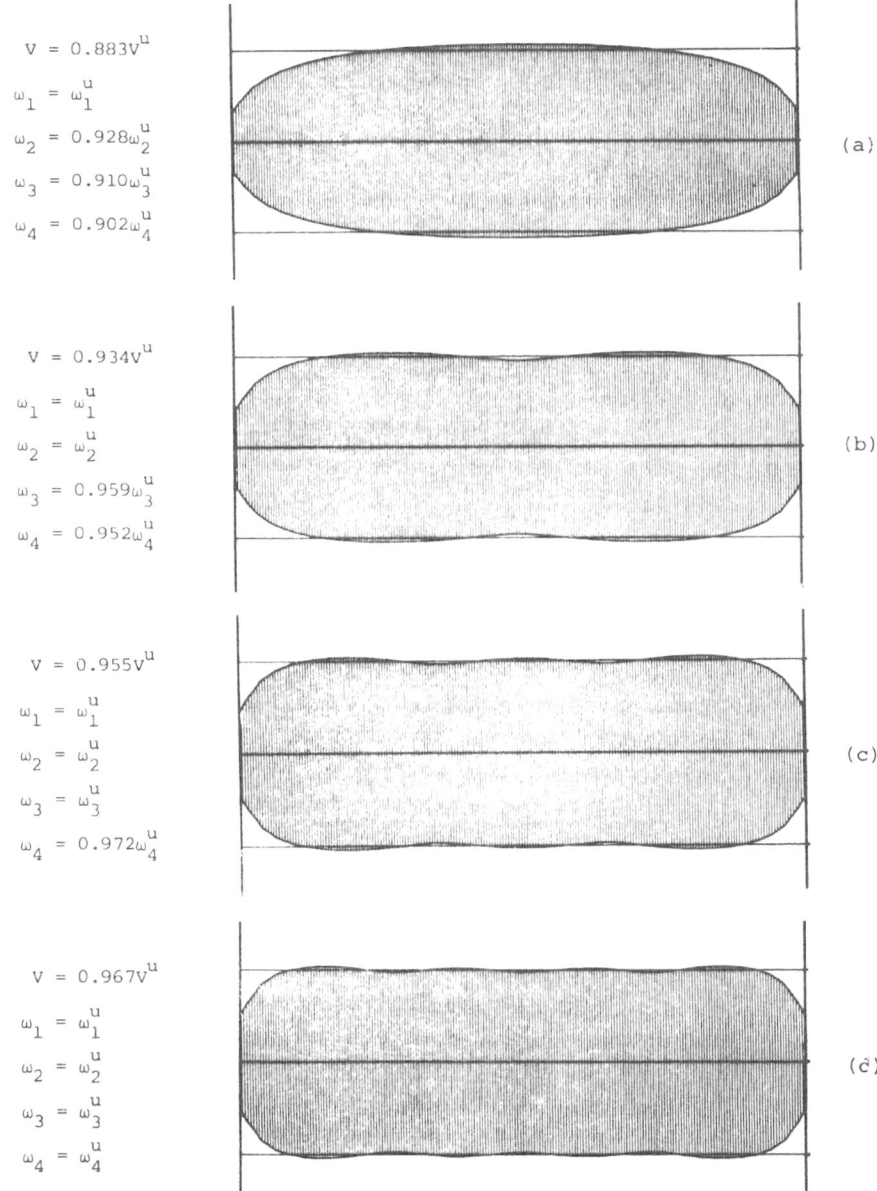

$V = 0.883V^u$

$\omega_1 = \omega_1^u$

$\omega_2 = 0.928\omega_2^u$

$\omega_3 = 0.910\omega_3^u$

$\omega_4 = 0.902\omega_4^u$

(a)

$V = 0.934V^u$

$\omega_1 = \omega_1^u$

$\omega_2 = \omega_2^u$

$\omega_3 = 0.959\omega_3^u$

$\omega_4 = 0.952\omega_4^u$

(b)

$V = 0.955V^u$

$\omega_1 = \omega_1^u$

$\omega_2 = \omega_2^u$

$\omega_3 = \omega_3^u$

$\omega_4 = 0.972\omega_4^u$

(c)

$V = 0.967V^u$

$\omega_1 = \omega_1^u$

$\omega_2 = \omega_2^u$

$\omega_3 = \omega_3^u$

$\omega_4 = \omega_4^u$

(d)

Figure 4.2   Optimal Design of Slender, Simply Supported Beams
for Minimum Volume, with Constraints on a Number
of Eigenfrequencies

776

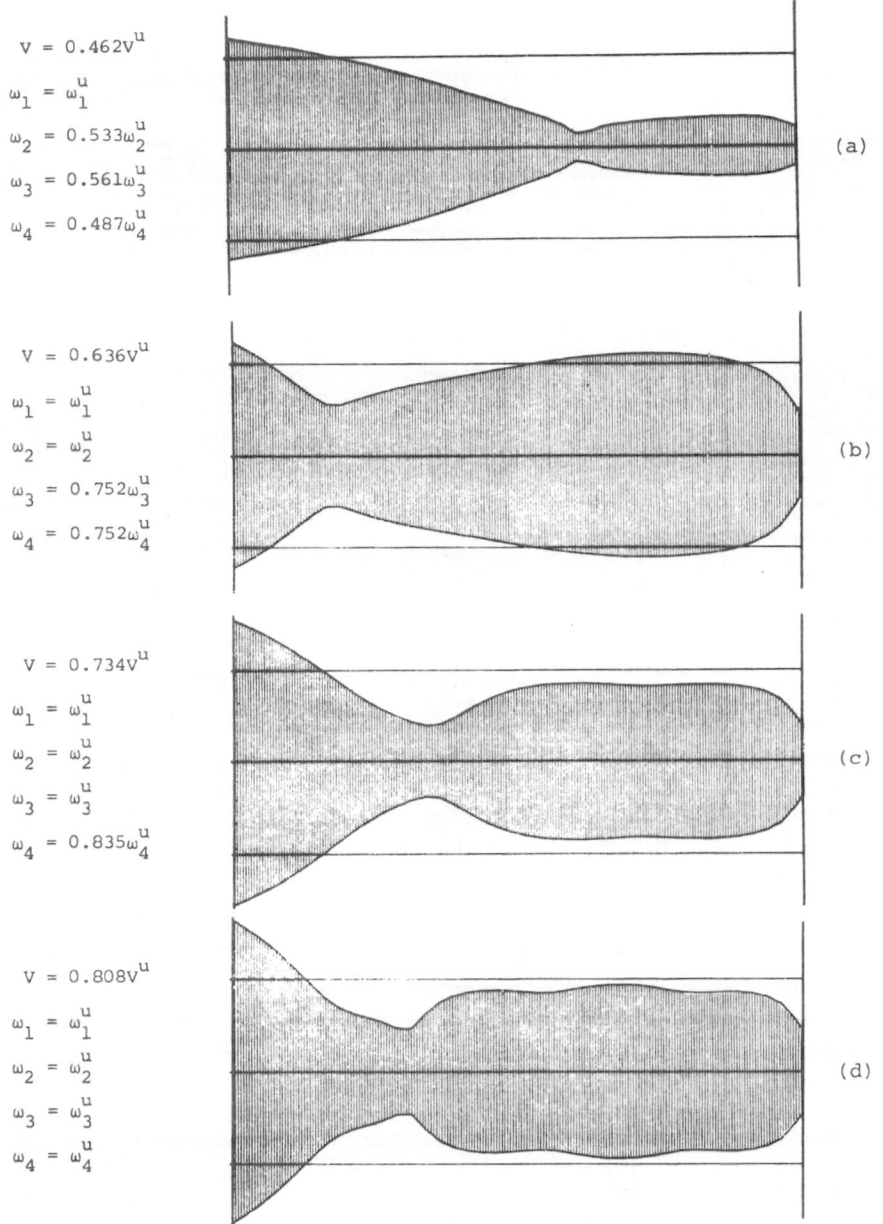

$V = 0.462V^u$

$\omega_1 = \omega_1^u$

$\omega_2 = 0.533\omega_2^u$

$\omega_3 = 0.561\omega_3^u$

$\omega_4 = 0.487\omega_4^u$

(a)

$V = 0.636V^u$

$\omega_1 = \omega_1^u$

$\omega_2 = \omega_2^u$

$\omega_3 = 0.752\omega_3^u$

$\omega_4 = 0.752\omega_4^u$

(b)

$V = 0.734V^u$

$\omega_1 = \omega_1^u$

$\omega_2 = \omega_2^u$

$\omega_3 = \omega_3^u$

$\omega_4 = 0.835\omega_4^u$

(c)

$V = 0.808V^u$

$\omega_1 = \omega_1^u$

$\omega_2 = \omega_2^u$

$\omega_3 = \omega_3^u$

$\omega_4 = \omega_4^u$

(d)

Figure 4.3   Optimal Design of Slender, Clamped, Simply Supported
             Beams for Minimum Volume, with Constraints on a
             Number of Eigenfrequencies [$\omega_1 = 2.44\omega_1$ when
             simply supported]

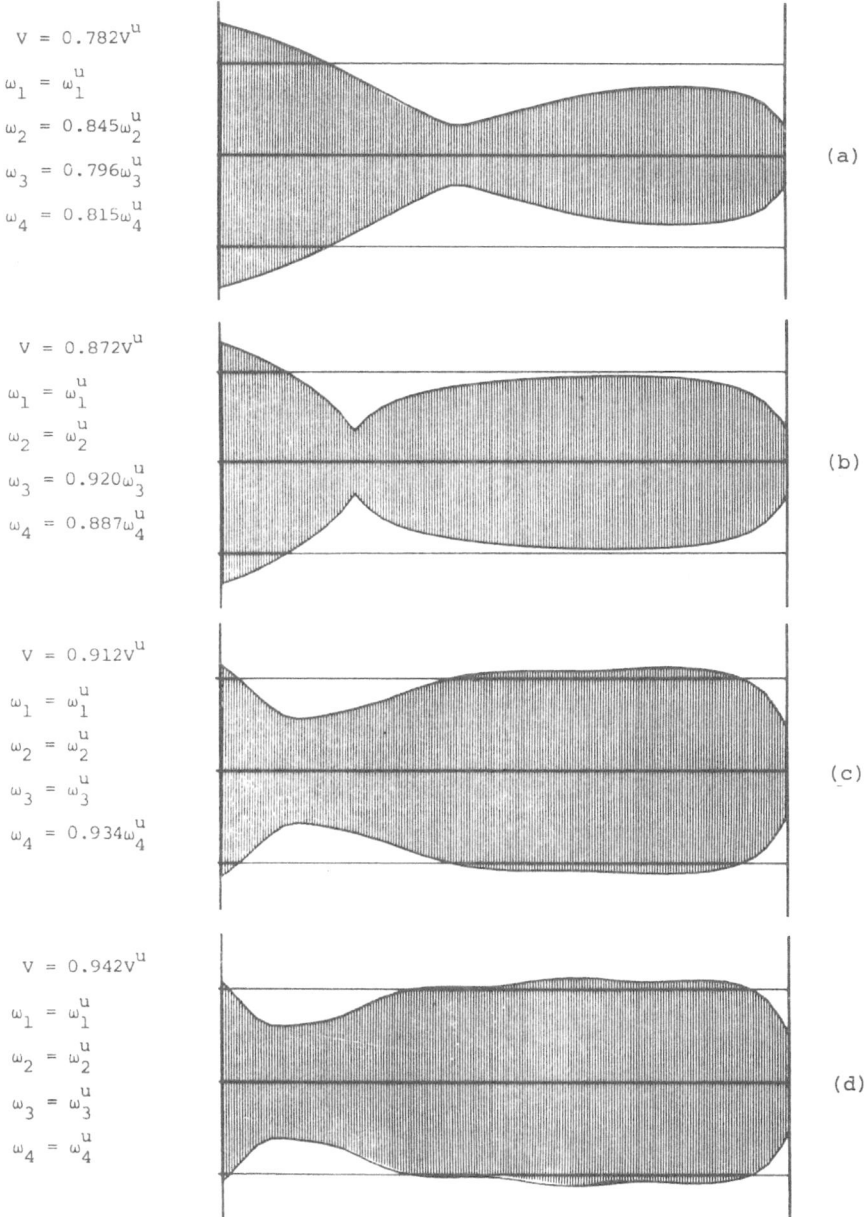

$V = 0.782V^u$
$\omega_1 = \omega_1^u$
$\omega_2 = 0.845\omega_2^u$
$\omega_3 = 0.796\omega_3^u$
$\omega_4 = 0.815\omega_4^u$

(a)

$V = 0.872V^u$
$\omega_1 = \omega_1^u$
$\omega_2 = \omega_2^u$
$\omega_3 = 0.920\omega_3^u$
$\omega_4 = 0.887\omega_4^u$

(b)

$V = 0.912V^u$
$\omega_1 = \omega_1^u$
$\omega_2 = \omega_2^u$
$\omega_3 = \omega_3^u$
$\omega_4 = 0.934\omega_4^u$

(c)

$V = 0.942V^u$
$\omega_1 = \omega_1^u$
$\omega_2 = \omega_2^u$
$\omega_3 = \omega_3^u$
$\omega_4 = \omega_4^u$

(d)

Figure 4.4   Optimal Design of Slender, Flexible-Clamped, Simply Supported Beams for Minimum Volume, with Constraints on a Number of Eigenfrequencies [$\omega_1 = 1.60\omega_1$ when simply supported]

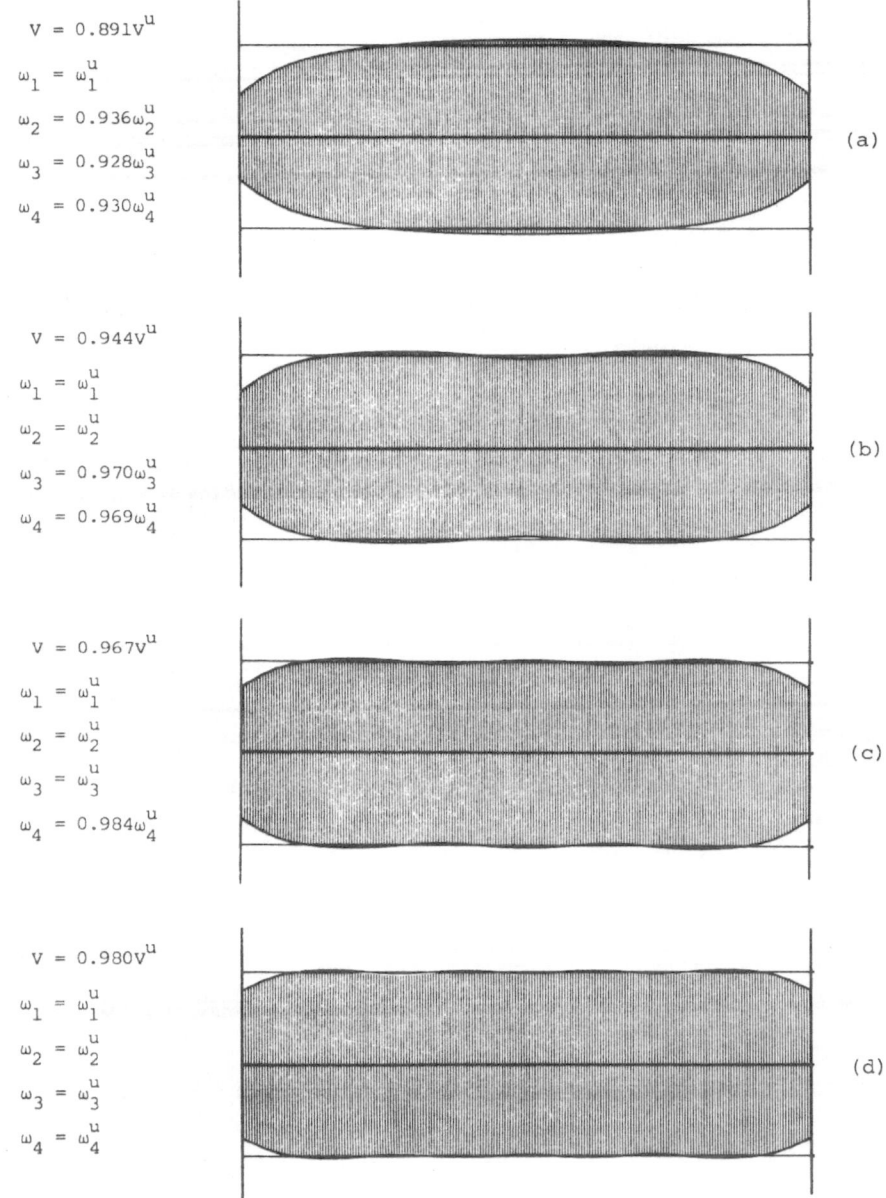

$V = 0.891V^u$

$\omega_1 = \omega_1^u$

$\omega_2 = 0.936\omega_2^u$

$\omega_3 = 0.928\omega_3^u$

$\omega_4 = 0.930\omega_4^u$

(a)

$V = 0.944V^u$

$\omega_1 = \omega_1^u$

$\omega_2 = \omega_2^u$

$\omega_3 = 0.970\omega_3^u$

$\omega_4 = 0.969\omega_4^u$

(b)

$V = 0.967V^u$

$\omega_1 = \omega_1^u$

$\omega_2 = \omega_2^u$

$\omega_3 = \omega_3^u$

$\omega_4 = 0.984\omega_4^u$

(c)

$V = 0.980V^u$

$\omega_1 = \omega_1^u$

$\omega_2 = \omega_2^u$

$\omega_3 = \omega_3^u$

$\omega_4 = \omega_4^u$

(d)

Figure 4.5   Optimal Design of Medium Slender ($\sqrt{(L^2 A)/I} = 33$),
Simply Supported Beams for Minimum Volume, with
Constraints on a Number of Eigenfrequencies
[$\omega_1 = 0.976\omega_1$ for Bernoulli-Euler Beam]

For each step of design iteration, the problem of redesign is a linear programming problem and can thus be solved with the Simplex algorithm. The introduction of move-limits to the formulation is of major importance, since they are the key to control for the designer. When setting up of the LP problem is also treated as an integral part of the FEM analysis, one obtains an approach that is both effective and reliable for optimal design of discrete as well as continuous systems.

REFERENCES

1. Niordson, F.I. and Pedersen, P., "A Review of Optimal Structural Design," Applied Mechanics (Eds. Becker and Mikhailov), Springer Verlag, Berlin, 1973, pp. 264-278.
2. Pedersen, P., "On the Minimum Mass Layout of Trusses," Proceedings No. 36, AGARD-CP-36-70, Symposium, Istanbul, 1969.
3. Gellatly, R.A., Helenbrook, R.G. and Kocher, L.H., "Multiple Constraints in Structural Optimization," Int. J. Num. Meth. in Engng., Vol. 13, 1978, pp. 297-309.
4. Prager, W. and Taylor, J.E., "Problems of Optimal Structural Design," J. of Appl Mech., Vol. 35, 1968, pp. 102-106.
5. Masur, E.F., "Optimum Stiffness and Strength of Elastic Structures," J. Engr. Mech. Div., ASCE, Vol. 95, No. EM5, 1970, pp. 621-640.
6. Olhoff, N. and Rasmussen, S.H., "On Single and Bimodal Optimum Buckling Loads of Clamped Columns," Int. J. Solids Structures, Vol. 13, 1977, pp. 605-614.
7. Masur, E.F. and Mróz, Z., "Non-Stationary Optimality Conditions in Structural Design," Int. J. Solids Structures, Vol. 15, 1979, pp. 503-512.
8. Kristensen, E.S. and Madsen, N.F., "On the Optimum Shape of Fillets in Plates Subjected to Multiple In-Plane Loading Cases," Int. J. Num. Meth. Engng., Vol. 10, 1976, pp. 1007-1019.
9. Pedersen, P., "Some Properties of Linear Strain Triangles and Optimal Finite Element Models," Int. J. Num. Meth. Engng., Vol. 7, 1973, pp. 415-429.
10. Pedersen, P., "On Computer-Aided Analytic Element Analysis and the Similarities of Tetrahedron Elements," Int. J. Num. Meth. Engng., Vol. 11, 1977, pp. 611-622.
11. Pedersen, P., "On the Optimal Layout of Multi-Purpose Trusses," Computers and Structures, Vol. 2, 1972, pp. 695-712.
12. Pedersen, P., "Optimal Joint Positions for Space Trusses," J. Struct. Div., ASCE, Vol. 99, No. ST12, 1973, pp. 2459-2476.
13. Levy, R. and Chai, K., "Implementation of Natural Frequency Analysis and Optimality Criterion Design," Computers and Structures, Vol. 10, 1979, pp. 277-282.

780

14. Pedersen, P., Optimal Design of Shafts (M.Sc. thesis, in Danish), 1967.
15. Przemieniecki, J.S., Theory of Matrix Structural Analysis, McGraw-Hill, 1968, p. 468.
16. Pierson, B.L., "An Optimal Control Approach to Minimum-Weight Vibrating Beam Design," J. Struct. Mech., Vol. 5, No. 2, 1977, pp. 147-178.

OPTIMUM DESIGN OF PORTAL FRAMES WITH TAPERED STEEL SECTIONS

David Anderson and M. Anwarul Islam

Department of Civil Engineering, Warwick University, Warwick, U.K.

Department of Civil Engineering, Manchester Polytechnic, Manchester, U.K.

ABSTRACT

This paper describes a computer program that determines suitable cross-sectional dimensions for the members of a steel portal frame, while minimizing weight. To achieve a practical minimum cost arrangement that does not entail excessive fabrication, the designer specifies positions at which plate thicknesses may change, while the program automatically ensures that the latter correspond to available sizes. The user initially provides a lower bound design which is analyzed by the computer. The dimensions of the design variables are automatically incremented by a specified factor and the frame is re-analyzed, to determine rates of change. The Simplex method is then used to alter the variables so that the permissible stresses and deflection limits of the British steelwork design code, B.S. 449, are not exceeded. Due to the non-linear nature of the design problem, specified move limits are imposed on the calculated changes, to avoid a non-optimal solution. Once the resulting modified values for the design variables have been determined, the frame is re-analyzed once more, and the procedure described is repeated. Iteration continues until re-analysis shows that the code's requirements are satisfied. The program is used to study the effect on a single bay frame of changes in the positions at which plate thicknesses may alter, and of changes to the move limits. To demonstrate the use of the program in practice, comparison is made with a well-optimized design obtained by experienced engineers using trial-and-error.

1. INTRODUCTION

Material economy can be achieved in single story, pitched roof steel building frames by using tapered sections, as shown in Fig. 1.1. In such frames, design variables are the widths and thicknesses of the individual plates that are welded together to form the structural members. Since doubly-symmetric I-sections are normally specified, the design variables at any cross-section are:

(1) the flange breadth, B
(2) the flange thickness, T
(3) the web depth, d
(4) the web thickness, t

Values must be assigned throughout the frame, so that permissible stresses and deflections are not exceeded when full service loading is applied.

The cost of fabricating such a structure from plate can only be justified if material weight is significantly lower than that for a similar frame composed mainly of prismatic members that are manufactured from standard rolled sections. A minimum weight approach to design is therefore both appropriate and desirable, provided that plate sizes are not changed with excessive frequency and plate lengths are convenient for fabrication. Since members can be tapered to conform approximately to the shape of the elastic bending moment diagram, little advantage can be gained from the plastic design method, in which redistribution of moments is allowed once yielding has commenced. Severe stability problems can also occur when plastic design is employed. The method proposed here is, therefore, based on elastic theory, assuming linear material behavior.

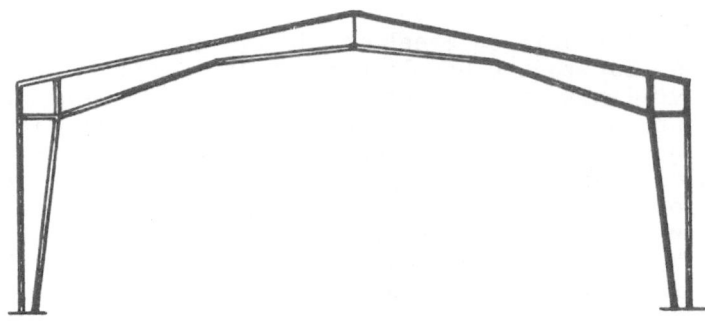

Figure 1.1   Tapered Single Store Frame.

2. STRUCTURAL EQUILIBRIUM EQUATIONS

In order to represent structural behavior, it is necessary to formulate equations relating the applied loading to the resulting displacements of the frame. This can be achieved most conveniently in a computer-based approach by employing the well-known matrix displacement method, in which stiffness matrices for individual members are assembled to represent the behavior of the structure as a whole. Trapezoidal web plates are usually adopted in tapered frames, since they can be obtained from rectangular plates by one cut. However, as continuously tapered flanges would require at least two such cuts to preserve symmetry, flange breadth is only varied by a step change in width at the junction between two plate lengths. A stiffness matrix is therefore required for a web-tapered member, such as that shown in Fig. 1.2.

Although it is usual to assume that conventional Bernoulli-Euler bending theory continues to hold for such members [1], it is not possible to obtain a stiffness matrix for major axis bending by exact algebraic integration of the governing differential equation. However, if the presence of the web is ignored, then algebraic integration can be easily performed [1]. Among other variables, the terms of the matrix will include the end depths D_0 and D_1. These are defined as the distances between the centroids of the flanges, at the shallow end and the deep end of the member. Significant errors can result from neglect of the web, but the present authors have found that these can be avoided by using an equivalent effective depth when calculating the terms of the stiffness matrix.

Consider the two cross-sections shown in Fig. 1.3. It can be shown that the moment of inertia of the section that consists

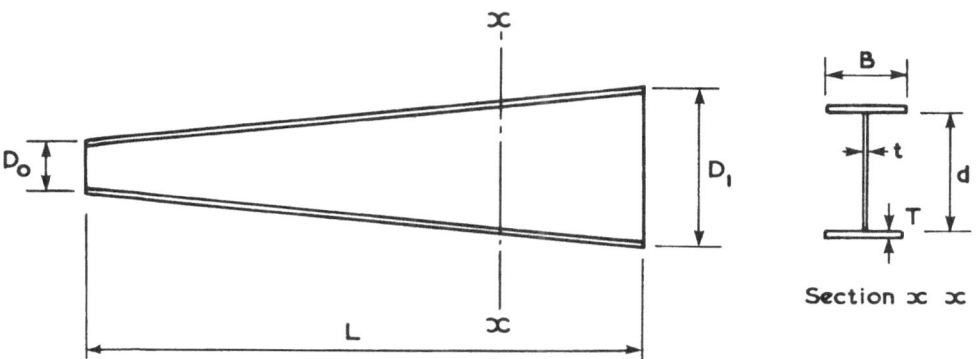

Figure 1.2 Web Tapered Member.

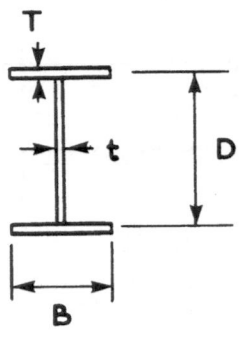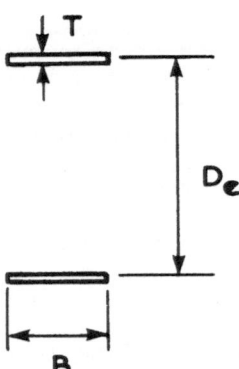

Figure 1.3 Equivalent Sections.

only of two flange plates equals that of the I-section, providing that the depth D_e is given by

$$D_e = [(D-T)^3 t/(6BT) + D^2]^{1/2} \tag{2.1}$$

where D_e is termed the equivalent effective depth. In evaluating the stiffness matrix, therefore, the true end depths D_0 and D_1 are replaced wherever they occur by the equivalent effective depths at the shallow end and at the deep end of the member, respectively. The derivation of Eq. 2.1 and investigations concerning the accuracy of this procedure are described elsewhere [2].

The analysis just described could be used to design a tapered frame by a process of trial-and-error. Usually, however, only a small number of designs can be considered, due to the limited time available, and the optimum design may not be found. It is there-fore proposed that an automatic iterative procedure, which begins with an infeasible lower-bound solution, be adopted. Subsequent increments in the cross-sectional dimensions are obtained by a mathematical programming technique, for minimization of weight, in order to achieve the optimum solution.

3.   DESIGN CONSTRAINTS

The program written to implement the proposed method designs to B.S. 449, the present British standard for the use of struc-tural steel in buildings [3]. This requires that the following condition be satisfied throughout the frame:

$$f_c/p_c + f_{bc}/p_{bc} \leq 1.0 \tag{3.1}$$

where

f_c is the calculated axial compressive stress

p_c is the allowable stress in axially loaded struts

f_{bc} is the calculated compressive stress due to bending

p_{bc} is the allowable compressive stress for members subject to bending

The permissible axial stress p_c is determined from the empirical Perry-Robertson strut formula [4], and depends on the ratio of the effective length ℓ to the radius of gyration r of the section, about the critical axis for buckling as a strut. If the member is also susceptible to lateral-torsional buckling as a beam, the permissible stress p_{bc} depends upon the ratio ℓ/r about the minor axis of the cross-section, and on the ratio of the overall depth to the flange thickness T [5]. Otherwise, p_{bc} is simply a proportion of the yield stress. In calculating increments in the cross-sectional dimensions to determine the final design, it is therefore necessary to ensure that the combined stress index on the left hand side of Eq. 3.1, at a location i, does not exceed unity. The stress index will be denoted f_i.

The code B.S. 449 also makes recommendations concerning maximum values of vertical and horizontal deflection. Although the recommended values are usually regarded as somewhat conservative for buildings with modern forms of cladding, such as plastic-coated steel, it is still desirable that the designer ensures that the frame is reasonably stiff. It is usual, therefore to impose limits on horizontal sway deflections at the eaves and sometimes also on the downward movement at the ridge. The limiting values should, however, be chosen by the designer, with regard for the particular nature and function of the building.

In order to prevent premature failure by local buckling, maximum limits must be placed on flange outstands, in relation to thickness T. The code B.S. 449 also restricts the depth to thickness ratio, d/t, for unstiffened webs. This restriction is much more severe than that which is applied in North America, but is necessary since the British code does not consider the possibility of web buckling due to shear. For structural mild steel, d/t is limited to 85. This restriction has been included in the computer program, in order to satisfy checking authorities. However, no calculation is made of shear stress in the web, since experience has shown that such stresses are invariably low in single story portal frames. The use of slender webs with stiffeners has not been considered, since they are generally uneconomic due to the extra fabrication required.

4. INITIAL DESIGN

The initial cross-sectional dimensions are specified by the designer. In order to ensure an economical final design, the initial values of the variables should be lower bound that usually correspond to an infeasible design. The following criteria are helpful at this stage:

(1) Plate width should not be less than the minimum value needed to ensure adequate connections to the flanges, for such items as purlins and sheeting rails.

(2) When the flange breadths take their initial value(s), plate thicknesses should be chosen so that the permissible outstand to thickness ratio is not exceeded. Thicknesses should also satisfy the minimum value required to counteract corrosive influences.

(3) Minimum cross-sectional dimensions can also be calculated by treating columns as members subject only to axial load.

If material waste is to be minimized, plate lengths should not be trimmed to simply avoid a short length of under-stressed plate. To aid in fabrication, it is also desirable that flange and web plates do not change size at the same point along a member. It is therefore proposed that the designer should specify those positions at which change of flange or web plate may be made. Loading is specified as a series of point forces (including moment, if required) along the length of the members.

5. OPTIMIZATION PROCEDURE

Consider an initial design that has been analyzed and let the combined stress index at a position i be f_i. As the initial design is usually infeasible, the value of f_i is likely to be greater than 1. In order to find the influence of the design variables on f_i, each is incremented in turn and the frame is re-analyzed, after which the increment is removed. A new set of values for stresses and deflections are given by each analysis, and the rate of change of the stress index f_i with variable v_j can therefore be expressed as

$$\frac{\partial f_i}{\partial v_j} = \frac{f_{ij} - f_i}{\Delta v_j} \tag{5.1}$$

where f_{ij} is the value of the stress index at position i, with the value of the design variable v_j incremented by Δv_j, and the frame reanalyzed. The design constraint for the stress index at location i is therefore

$$f_i + \frac{\partial f_i}{\partial v_1}\, \delta v_1 + \frac{\partial f_i}{\partial v_2}\, \delta v_2 + \ldots + \frac{\partial f_i}{\partial v_m}\, \delta v_m \leq 1 \qquad (5.2)$$

where δv_1, δv_2, ..., δv_m are the increases that should be made in the m design variables v_1, v_2, ..., v_m in order to reduce the stress index f_i to unity. Similar constraints apply to the stress indices at other positions around the frame and to the restricted deflections. Stress indices must be calculated at all load points around the frame and at points at which plate sizes may change. Clearly, for the latter case, and also where moments are applied as external loads, two stress indices must be calculated at each position, to ensure that the final design is not overstressed.

If weight is to be minimized, an expression for the objective Z is

$$Z = k_1 \delta v_1 + \ldots + k_j \delta v_j + \ldots + k_m \delta v_m \qquad (5.3)$$

where the general term k_j is a known constant that gives the increase in volume of material in the frame for an increase δv_j in the design variable v_j. However, if plate width and thickness are taken as separate variables, as proposed earlier, errors can arise in the value of Z if both the width and the thickness of a particular plate are to be increased. Hence for the flanges the cross-sectional area is treated as the design variable. The neglect of the small change in overall depth that may arise from possible changes in flange thickness leads to insignificant errors in calculation of the rates of change $\partial f_i / \partial v_j$. Increase of flange area is preferably achieved by increasing the width of the flange, since this significantly reduces the minor axis slenderness ratio ℓ/r_y and therefore leads to increases in the permissible stresses. Flange thickness is increased if the upper limit on the flange outstand to thickness ratio is reached.

Web thickness has also been discarded as a design variable in the optimization, because with low shear, the thickness will only be increased to satisfy the restriction on d/t. This could have a serious effect on the validity of the objective function, if large changes in web depth were permitted. However, such changes do not occur, for reasons described in the following.

The simplex method of linear programming is used to calculate the changes in design variables that are required to satisfy the stress and deflection constraints, which have the same general form as Eq. 5.2, while minimizing the objection function of Eq. 5.3. However, since deflections and both calculated and permissible stresses vary non-linearly with changes in cross-sectional dimensions, the optimum design may not be obtained by simply adding the changes in the design variables δv_j, given by the simplex calculation, to the initial dimensions. Indeed, it is unlikely that the resulting design would even be feasible. To overcome this problem, only a proportion of the increase calculated by the simplex method should be added to the initial value of each variable. The permitted proportion, or move limit, has been taken as a small percentage of the initial value of the variable. As a result, only small errors generally arise in the value of the objective function from the neglect of web thickness as a design variable, but iteration of the optimization procedure becomes necessary.

Once the permitted design changes δv_j have been added to the initial values of the design variables, the optimization procedure is repeated, with the new values replacing those initially chosen. The first stage of the cycle is to analyze the frame with the new values for the plate dimensions, web thicknesses having been modified as necessary to ensure that the restriction on d/t is not exceeded. Since the equilibrium equations are based on the center-line geometry of the frame, the positions of member center-lines are recalculated automatically, to take account of the new cross-sectional depths. Iteration continues until all the design constraints are shown to be satisfied by analysis of a newly revised design. A flow chart for the complete process is shown in Fig. 5.1.

6.   INITIAL TESTING OF THE PROGRAM

A number of tests were carried out by designing the idealized frame shown in Fig. 6.1. Fixed bases were adopted in these tests and no attempt was made to choose particular plate sizes from rolling mill lists because such rounding off could distort the comparisons that are to be made.

The frame was first designed using the following initial dimensions: flange breadth and web depth = 150 mm, flange thickness = 6 mm, and web thickness = 4 mm. In the final design, which satisfied the specified limits on stress and deflection, all cross-sectional dimensions had been increased, except for the retention of the 4 mm web thickness over the middle 1000 mm of the column and the minimum flange and web thicknesses over a 2000 mm length (measured horizontally) in each rafter. To test the optimality of the final design, material was redistributed while keeping the total weight constant, and the frame was re-analyzed. This was done by, for example, reducing the web depths in the eaves regions by 10% and increasing the flange dimensions, and vice-versa. It was found that one or more design constraints were then violated.

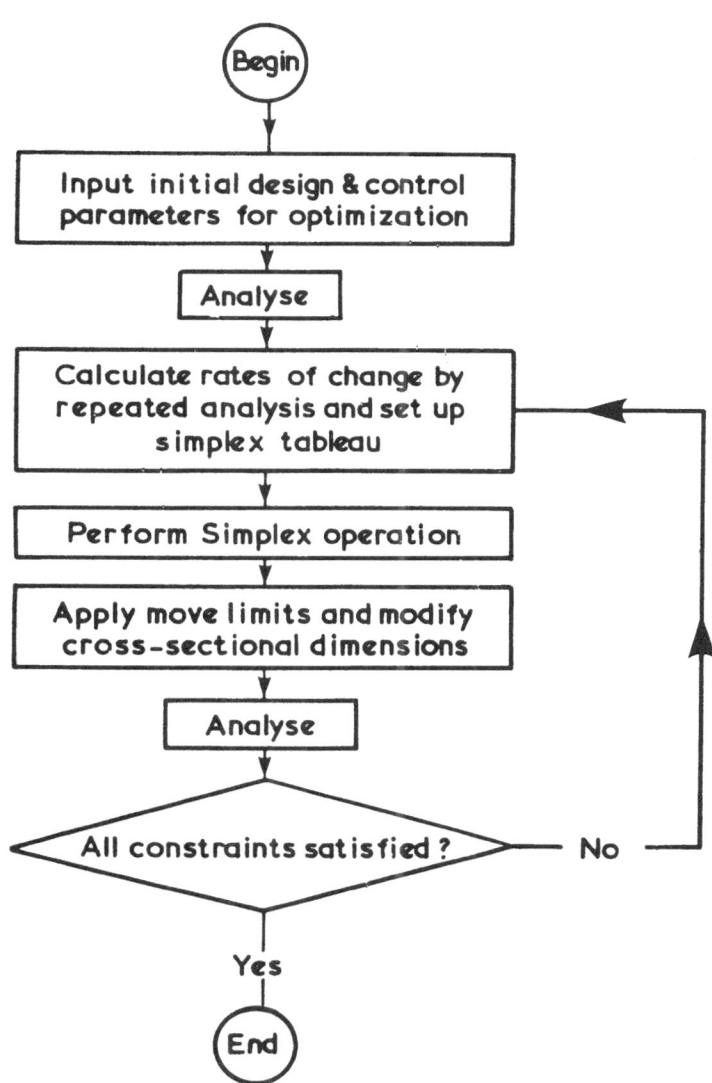

Figure 5.1   Flow Chart of Design Procedure.

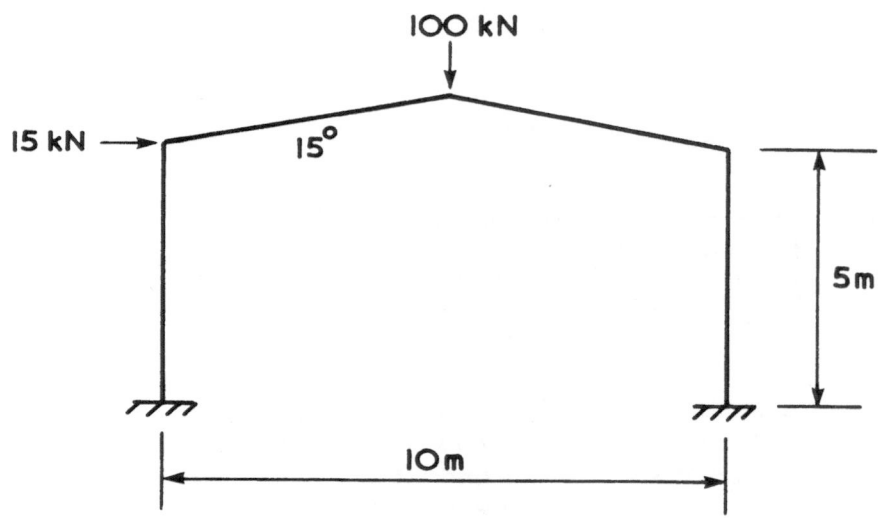

Figure 6.1   Idealized Frame for Testing of Program.

To examine the effect of the initial section dimensions on
the final solution, the frame was re-designed with the initial web
depth increased to 190 mm at the bases, eaves, and ridge.   The
final dimensions differed only very slightly from those obtained
earlier.   For example, the stanchion depth at the eaves increased
by 2%.   The weights of the two designs differed by only 0.3%.

Move limits have generally been taken as 10% of the current
value of a variable.   To investigate the effect of changing this
figure, the frame was redesigned using move limits of 15% and 5%.
The weights only differed by 0.5%, although the CPU time on a
Burroughs 6700 increased from 261 seconds for a 15% limit, to 382
seconds for a 10% limit, and to 659 seconds for a 5% limit.   However, when the frame was designed once more without any move
limits being imposed, the weight of the final design increased by
5%, although the design process was complete after only 4 iterations, corresponding to 95 seconds of CPU time.

In all the above cases, plate sizes for both flange and web
and angle of web taper, could be changed if required at 1000 mm
intervals (measured horizontally in the case of the rafters).

7.   20 m SPAN PINNED BASE FRAME

The frame shown in Fig. 7.1 was used to investigate the
effect of changing the number of positions at which plate sizes

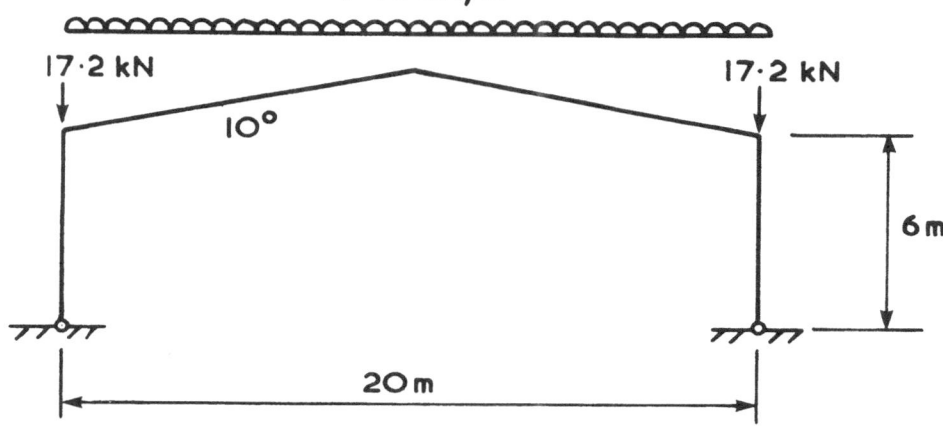

Figure 7.1   Pinned Base Frame.

and angle of taper could be changed. At first, no change was
permitted along the length of the stanchion, but changes could be
made at one-third points along each rafter from eaves to ridge.
In the second case, changes could be made at one-third points up
each stanchion and at four intermediate points along each rafter.
In both cases, however, the flange breadth was held constant and
uniform, throughout the frame. It is possible that this could be
an architectural requirement for certain structures. The results
showed that the weight could be reduced by 10.3%, if the increased
number of change points was adopted. Such information would enable
a designer to choose the more economical arrangement, after considering material and fabrication costs.

To achieve a practical design, it is also necessary to
choose particular plate sizes from manufacturers' lists. This was
done after the final theoretical sizes had been calculated. The
following strategy was adopted in the program: The thicker of
two consecutive standard plates was chosen, when the required
flange thickness was greater than the halfway value. In other
cases, the thinner plate was selected. The thicker plate was
always selected for webs, so that the restriction on d/t could be
satisfied. When this was done for the second of the above designs,
the weight increased by 4.7% and re-analysis confirmed that all
the design constraints were still satisfied. If this were not the
case, one or two alterations by hand would be sufficient to produce a feasible design.

Vertical loading almost always dominates the design of pitched roof portal frames, so it is proposed that this load condition be considered first. The resulting design can then provide initial sizes for other load cases. The first design for the 20 m frame was analyzed under typical wind loading, combined with dead load. It was found to be satisfactory.

8. COMPARISON WITH EXISTING DESIGN

A design had been prepared for the frame shown in Fig. 8.1 by an organization specializing in the fabrication of tapered frames. In view of their expertise, it was expected that the design would be very economical. The resulting sizes are shown in Fig. 8.2. Flanges had been selected from the following list of available rolled flats: 150 × 6, 170 × 6, 192 × 6, 192 × 8, 212 × 8, 234 × 8, 256 × 8, 256 × 10, 276 × 10, 298 × 10, 320 × 10, and 320 × 12. Possible web thicknesses were 4, 5, 6, 8, 10, 12, and 14 mm. The locations at which plates were changed were chosen to provide convenience in fabrication.

To enable the program to select from the list of available flats, it was decided that if, during any iteration, the simplex calculation gives an increase of flange area that is more than half the increase between two consecutive standard sizes, then the next higher is to be chosen. A move limit of 10% was imposed on increases in web depth, with a thicker web plate being chosen as soon as increased depth led to the violation of the restriction on d/t. In both designs, restrictions of (span/250) and (eave height/325) were imposed on vertical and horizontal deflection, respectively. Permissible stresses were based on an effective length of 2250 mm in the stanchions and 1500 mm along the rafter. Their calculation necessitated the selection of flat sizes for the flanges at the end of each iteration, as just described, rather than leaving such selection to the end of the design process. This effectively meant that move limits were not applied to flange area, and was the cause of some concern. However, removal of the limits would reduce CPU time, and the early tests of the program had shown that removal of all move limits only resulted in a small increase of weight, compared with a more accurate optimization.

The resulting design given by the program is shown in Fig. 8.3. This design is 4.8% heavier than the existing design, but is regarded as an encouraging result. It will be observed from Figs. 8.2 and 8.3 that the program has generally given greater depths to the members than those possessed by the existing design. Since the indirect increase in web thickness to satisfy the restriction on d/t is not accounted for in the objective function, the simplex operation initially finds increase of web depth to be

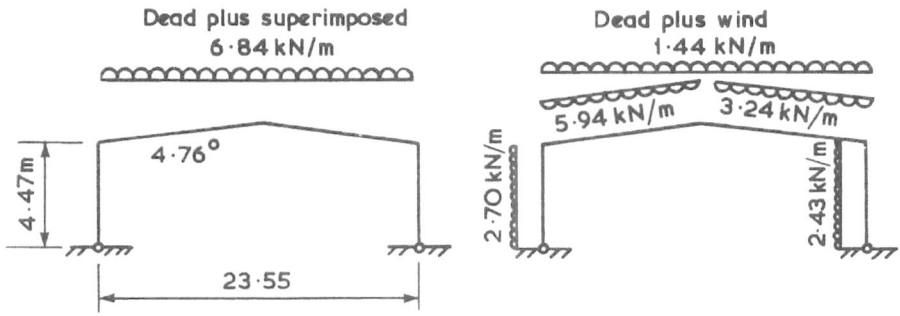

Figure 8.1   Frame for Comparison of Design Methods.

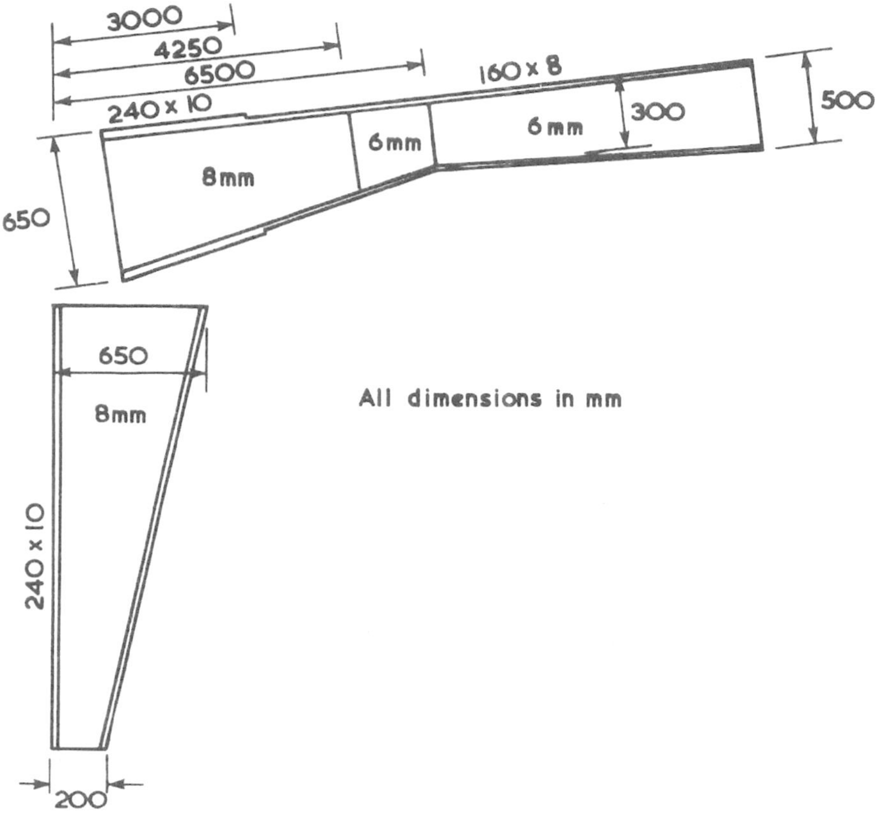

Figure 8.2   Existing Design.

Figure 8.3   Design by Proposed Method.

more cost effective than it actually is.   The greater web depths
and slightly heavier weight of the design shown in Fig. 8.3 are
not surprising, therefore.   It should be noted that the effective
removal of move limits from flange area has not resulted in such
variables being overemphasized in this example.

Finally, designs that proved satisfactory under dead plus
superimposed loading were also adequate for the dead load plus
wind load condition.

9.   CONCLUSION

An automatic optimal design program for tapered portal frames
has been developed, using the matrix displacement method for

analysis and the simplex procedure of linear programming for optimization. Design solutions can be obtained to satisfy both strength and deflection constraints. Permissible stresses and minimum plate thicknesses satisfy current British requirements.

The non-linear constraints usually associated with such optimization problems have been avoided by using repeated analysis and by following a multi-linear path to reach the optimum design. The objective function has been linearized by excluding the thickness of the web from the variables for optimization and by considering the area of the flange as a whole, instead of separate inclusion of breadth and thickness.

Before the method can be used in practice, a procedure must be devised to account for the indirect increase in web thickness, resulting from the restriction on d/t. When thicknesses are to correspond to available plate sizes, relatively large increases in web area can suddenly occur, but these are not accounted for in the objective function. One possible remedy, shortly to be investigated, would be to repeat the simplex operation immediately, with change of depth not to exceed that permitted without an increase in thickness. The two sets of changes in design variables could then be compared, to determine the more economical strategy. Alternatively, no attempt should be made to pick from available plate sizes for webs, until the iterative design procedure has been completed.

REFERENCES

1. Lee, G.C., Morrell, M.L., and Ketter, R.L., "Design of Tapered Members", Bulletin No. 173, Welding Research Council, New York, N.Y., 1972.
2. Islam, M.A., and Anderson, D., "Tapering I-Section Frames", Journal of the Structural Division, ASCE, Vol. 106, No. ST6, 1980.
3. B.S. 449: The use of structural steel in building. British Standards Institution, London, 1969.
4. Godfrey, G.B., "The Allowable Stresses In Axially Loaded Steel Struts". The Structural Engineer, Vol. 40, No. 3, 1962, p. 97.
5. Kerensky, O.A., Flint, A.R., and Brown, W.C. "The basis for design of beams and plate girders in the revised British Standard 153". Proc. Inst. Civ. Engrs., Part III, Vol. 5, No. 2, Aug. 1956, p. 396.

COMPUTER ORIENTED ALGORITHMS FOR SOLVING STRUCTURAL OPTIMIZATION PROBLEMS WITH DISCRETE PROGRAMMING TECHNIQUES

M. Rifat Saglam

Department of Computer Science, Bogazici University, Istanbul, Turkey

ABSTRACT

Design in many types of structural engineering problems involves the selection or the design of components from a discrete set of available or fabricated components. Mathematical models of these problems involve non-linear sets of equations. Recent research efforts in structural optimization have concentrated largely on finding continuous solutions to the non-linear model. This paper describes a computer oriented approach of linearization of the non-linear model and the discrete optimization of the problem.

1. NONLINEAR PROGRAMMING

1.1. General Formulation of Nonlinear Programming

The nonlinear programming problem has three fundamental ingredients:

(1) Finite number of real variables

(2) Finite number of constraints that the variables must satisfy

(3) Function of the variables that must be minimized or maximized

Mathematically, the problem can be stated as follows: Find $x_1, \ldots, x_n$ to satisfy inequality constraints;

$$g_i(x_1,\ldots,x_n) \le 0 , \qquad i=1,\ldots,m$$

equality constraints;

$$h_j(x_1,\ldots,x_n) = 0 , \qquad j=1,\ldots,k$$

and minimize an objective function;

$$z(x_1,\ldots,x_n)$$

Such problems may be visualized graphically by Fig. 1.1.

1.2. Linearization of Nonlinear Programming Problems

<u>Cutting Plane Method</u>. The cutting plane method is based on the useful property of convex problems that the linearized constraints are always entirely outside the feasible region

$$R = \{x : g_j(x) \le 0 ; \quad \text{all } j\}$$

One approximates the region R by linearized envelopes produced by the first terms of Taylor's series expansions of the constraint

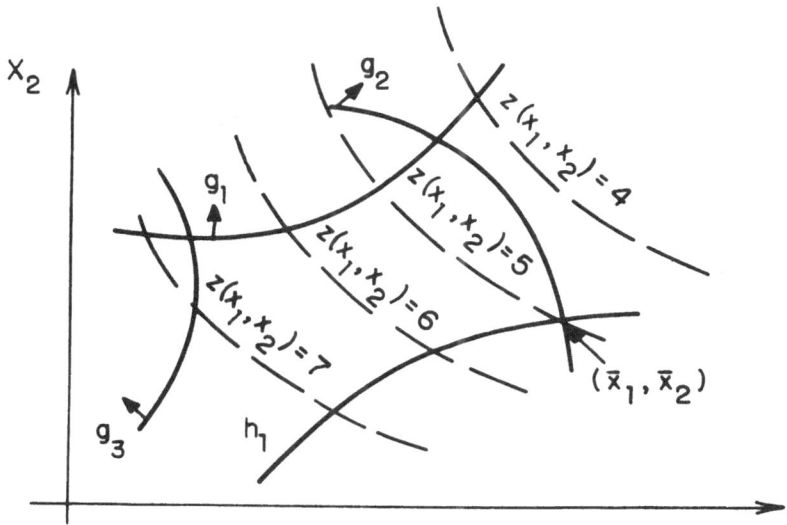

Figure 1.1   Typical Nonlinear Programming Problem in Two
             Variables

functions and solves this approximate problem by the LP (Linear Programming) Simplex Method.

The cutting plane procedure is as follows:

(1)   Constraints are linearized in the neighborhood of a starting point.

(2)   The resulting linear problem is solved by the LP method.

(3)   The solution is substituted in the original nonlinear constraints and the most violated constraint is detected.

(4)   This constraint is linearized about the optimum point of the previous problem and the resulting linear constraint is added to the linearized problem.

(5)   Return to Step 2, repeating Steps 2, 3, and 4 until all nonlinear constraints are satisfied to a desired degree of accuracy.

This process is illustrated in block diagram form in Fig. 1.2.

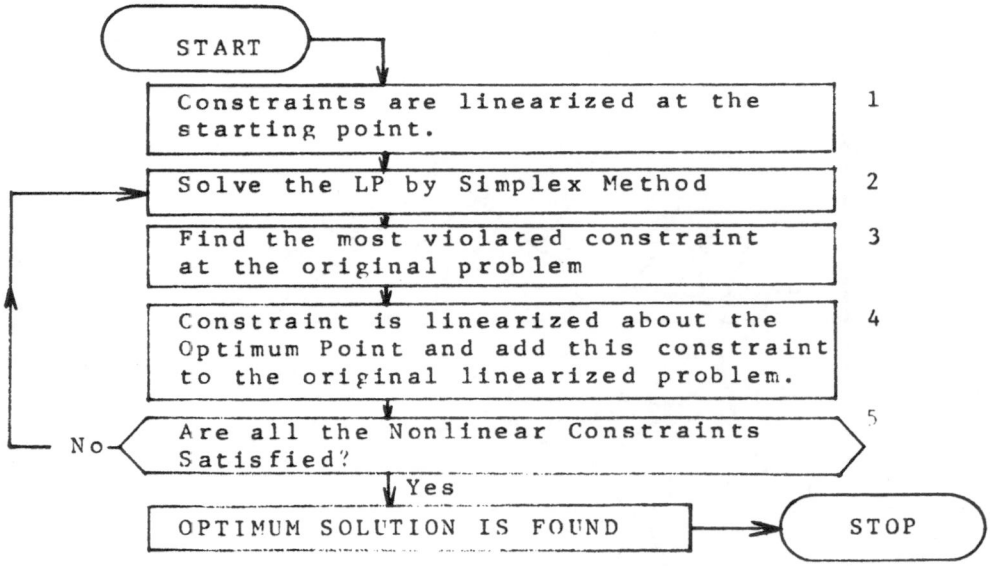

Figure 1.2   Nonlinear Optimization - Cutting Plane Method

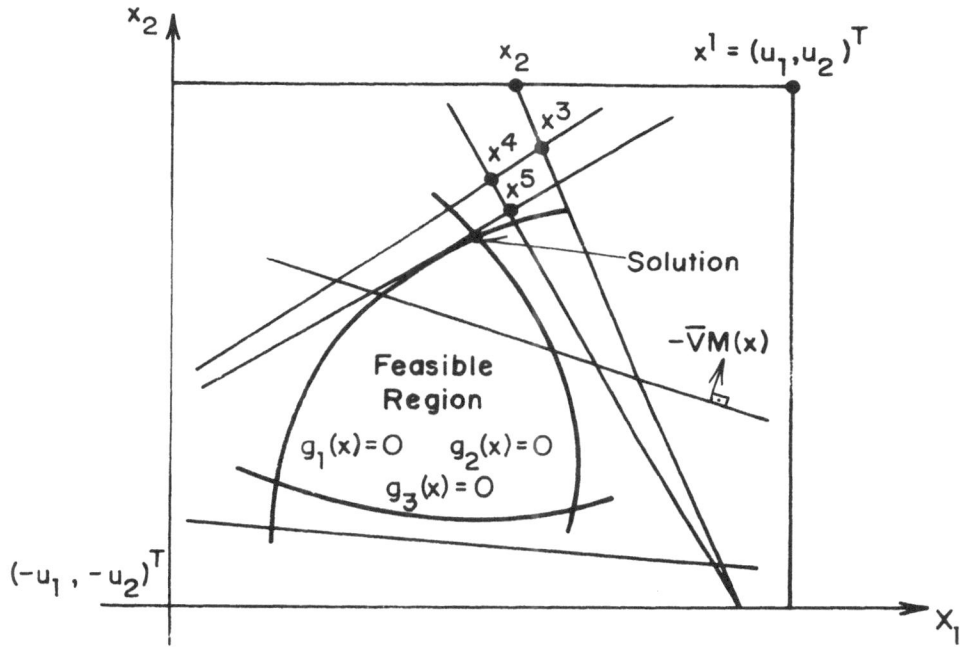

Figure 1.3   Cutting Plane Method

Several observations may be made concerning this method.

(1)   Every new linearization cuts off the old optimum, but not any portion of the convex set R (see Fig. 1.3).

(2)   The sequence of approximate x^k are, in general, all infeasible and give non-increasing values of z.

(3)   This method is efficient in solving nearly linear convex problems.

(4)   One of the computational advantages is that the linear sub-problem changes very little from step to step, so dual simplex methods can be used efficiently.

(5)   The method is restricted to convex problems.

(6)   If the optimum solution does not coincide with a vertex of the feasible solution, round-off errors may lead to oscillation.

(7)   These techniques can be applied to the nonlinear integer problems.

Method of Approximate Programming (MAP). This method can be applied to nonlinear and nonconvex problems. Taylor's expansion of the first terms are used again to linearize the objective function and nonlinear constraints. In this method, one linearizes all the nonlinear constraints at each iteration and no part of the preceding linear sub-problem is retained.

The MAP procedure is as follows:

(1)   Initialize, setting k = 0; x° is given.

(2)   Solve the linear sub-problem;

$$\min \ |\nabla z(x^k)|^T x$$

subject to

$$g_j(x^k) + |\nabla g_j(x^k)|^T (x-x^k) \geq 0, \ j=1,\ldots,m$$

and

$$|x_i - x_i^k| \leq U_i^k , \ \text{all } i.$$

where $U_i^k > 0$ are small predetermined numbers limiting steps. The resulting point is called x^{k+1}.

(3)   Set k = k+1 and repeat from Step 2, with gradually decreasing U_i^k until the following two conditions are satisfied;

$$|z(x^k) - z(x^{k+1})| < T_1$$

and

$$g_j(x^{k+1}) \geq - T_2$$

where $T_1 > 0$ and $T_2 > 0$ are preset tolerances.

This procedure is illustrated graphically by Fig. 1.4.

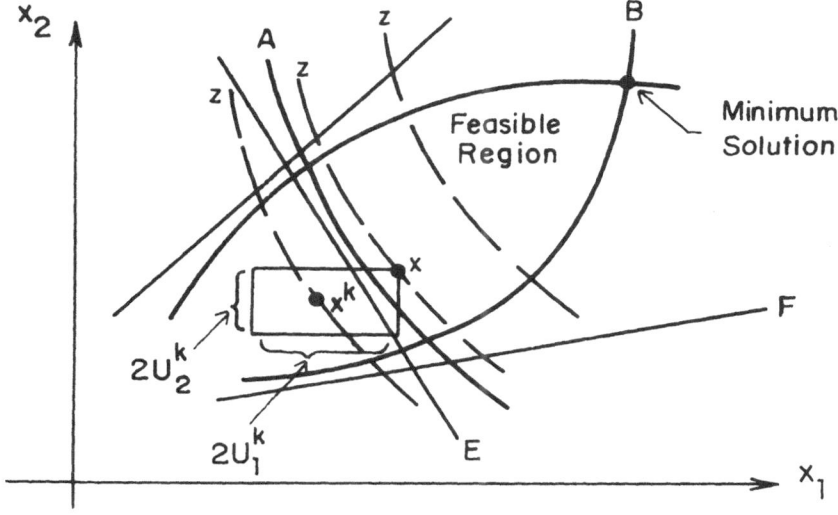

A,B,C - Nonlinear constraints

D,E,F - Linearized constraints

Z - Isocontours of the Objective Function

Figure 1.4   Method Approximate Programming

Important properties of this method are as follows:

(1)   This method is applicable to nonconvex problems and produces feasible, or nearly feasible, intermediate solutions with good accuracy.

(2)   Linear Programming is used in each iteration.

(3)   Experience must be used in the choice of the step size bounds U_i^k with respect to the nature of the problem.

(4)   Convergence of the method can be greatly improved if a good starting point is selected as an initial start x°.

2.   INTEGER (DISCRETE) PROGRAMMING

Integer programming deals with the class of mathematical programming problems in which some or all of the variables are required to be integers. Consider the case in which both the objective function and the constraints are linear, so that the general model can be formulated as follows:

Max.
or $z(X,Y) = c_1 X + c_2 Y;$ $(X,Y) \in s$
Min.

where

$$s = \{(X,Y)\; A_1 X + A_2 Y = b_1;\quad x \geq 0 \text{ integer}, y \geq 0\}$$

and $A_1 (m \times n_1)$, $A_2 (m \times n_2)$, $c_1 (1 \times n_1)$, $c_2 (1 \times n_2)$, and $b(m \times 1)$ are given. The variables $X(n_1 \times 1)$ and $Y(n_2 \times 1)$ are integer and continuous variables, respectively.

This problem is called mixed integer linear problems (MILP). Special cases occur when $n_2 = 0$, Integer Linear Program (ILP), and when $n_1 = 0$, Linear Program (LP). As can be seen from these definitions, MILP and ILP are special cases of the linear programming problem.

If $s = \emptyset$ (empty set), the problem is said to be infeasible. If $s \neq \emptyset$, any $(X,Y) \in s$ is called a feasible solution.

If there exists a least upper (or greatest lower) bound z^* such that

$$c_1 X + c_2 Y \leq z^* \quad \text{for all } (X,Y) \in s$$

the problem is said to be bounded. In this case, there exists $(X^\circ, Y^\circ) \in s$ such that $c_1 X^\circ + c_2 Y^\circ = z^*$ and (X°, Y°) is called an optimal solution. If no such z^* exists, the problem is said to be unbounded.

Attention will be focused on the ILP problem. If one simplifies the notation using summation notation, the problem may be written

$$z = \begin{matrix} \text{Min.} \\ \text{or} \\ \text{Max.} \end{matrix} \left\{ \sum_{j=1}^{n} c_j X_j \;:\; \sum_{j=1}^{n} A_{ij} X_j = b_i, \; i = 1,\ldots, m, \right.$$

$$\left. X_j \geq 0, \text{ integers}, \; j = 1,\ldots,n \right\}$$

2.1   Transformation to Discrete Variables

Design variables are to be chosen from a given discrete set for each optimization problem. To accommodate this, one has to modify the ILP problem. This can be done by defining additional constraints to the original problem. For instance, consider the constraint:

$$X_j \in s_j = \{s_{1j}, \ldots, s_{pj}\}$$

This is equivalent to the constraint set

$$X_j = \sum_{k=1}^{p} s_{kj} \delta_{kj}$$

$$\sum_{k=1}^{p} \delta_{kj} = 1, \quad \delta_{kj} = 0 \text{ or } 1, \quad k = 1,\ldots,p$$

where δ_{kj} are given discrete values for the specific problem considered.

If one substitutes these relations into the original set of equations

$$z = \text{Min.} \sum_{j=1}^{n} c_j \sum_{k=1}^{p} s_{kj} \delta_{kj}$$

subject to

$$\sum_{j=1}^{n} \sum_{k=1}^{p} a_{ij} s_{kj} \delta_{kj} = b_i, \qquad i=1,\ldots,m$$

$$\sum_{k=1}^{p} \delta_{kj} = 1, \qquad\qquad j=1,\ldots,n$$

$$\delta_{kj} = 0 \text{ or } 1, \qquad\qquad k=1,\ldots,p.$$

This set of equations represents the zero-one problem, a particular case of general integer programming problem in which each variable can take on only the values zero or one.

804

2.2 Cutting Plane Method

The cutting plane method is the reconstruction of the feasible solution space by imposing constraints on the original space, such that the required optimum feasible point is expressed as a proper extreme point of the modified solution space. The general idea is that these additional constraints systematically cut off a portion of the solution space such that no feasible points are ever excluded, as illustrated in Fig. 2.1.

<u>Derivative of the Gomory's Fractional Cut</u>. Suppose that the optimum solution of the linear program is given by

$$\text{Max.} \quad z = \bar{c}_o - \Sigma \, \bar{c}_j x_j$$

subject to

$$X_i = X_i^* - \sum_{j \in NB} |a_{ij}| X_i \, , \qquad i=1,\ldots,m$$

$$X_i, X_j \geq 0 \, ; \qquad j \in NB$$

where $\bar{c}_j = z_j - c_j$ is the reduced cost. The continuous solution is given by $X_i = X_i^* \geq 0$, $i=1,\ldots,m$, $X_j = 0$, $j \in NB$ and $z = \bar{c}_o$.

Consider any of the constraint equations for which $X_i^* \neq 0$, $i=1,\ldots,m$. That is, X_i does not have an integer value. Let

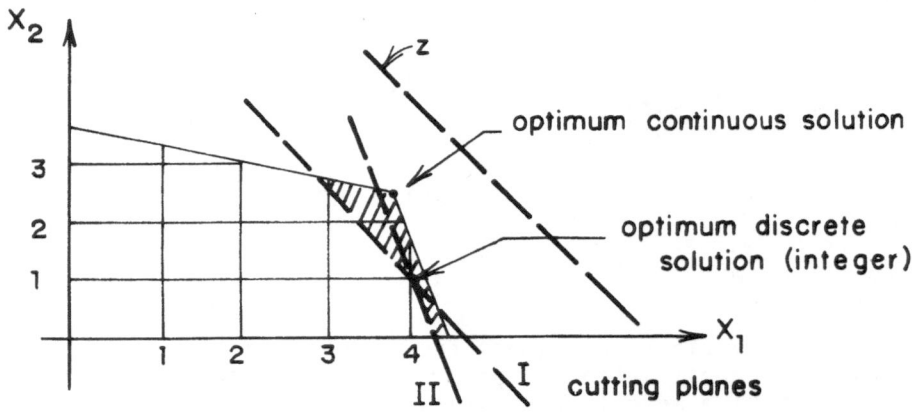

Figure 2.1   Cutting Plane Method

the selected equation be associated with X_k. This will be referred to as the source row. The X_k equation can be written as

$$X_k = |X_k^*| + f_k - \sum_{j \in NB} (|A_{kj}| + f_{kj})X_j$$

or

$$X_k - |X_k^*| + \sum_{j \in NB} |A_{kj}|X_j = f_k - \sum_{j \in NB} f_{kj}X_j$$

where

f_k = fractional part of basis (b_k)

f_{kj} = fractional part of A_{kj}.

Now, in order for X_k and X_j to be integers, the right hand side of the last equation must also be an integer. This implies that

$$f_k - \sum_{j \in NB} f_{kj}X_j \equiv 0$$

$$f_k - \sum_{j \in NB} f_{kj}X_j \leq f_k \leq 1$$

The necessary condition for integrality becomes

$$f_k - \sum_{j \in NB} f_{kj}X_j \leq 0, \text{ or in the form of equality:}$$

$$s - \sum_{j \in NB} f_{kj}X_j = - f_k$$

where $s \geq 0$ (Nonnegative slack variable).

The last equation is called the Gomory's fractional cut constraint. This constraint should be augmented to the LP tableau, from which it is derived and the resulting new LP should be solved. If the resulting optimum solution is an integer, the process ends. Otherwise, a new cut is constructed from the new LP tableau and the process is repeated.

Attention has to be paid to the selection of the source row. A number of different cuts can be generated from the current continuous optimum tableau. From the computational standpoint, the strongest cut is the one that cuts the deepest in the solution space, without eliminating any feasible integer point. One way of expressing this mathematically is as follows:

The f_{1_1} - cut is stronger than the f_1^2 - cut if $f_{kj}^1 \leq f_{kj}^2$ for all j and $f_k^1 \geq f_2^2$, with strict inequality holding throughout.

An algorithmic approach to formulate the above requirements may not be feasible for a large problem. But there is an empirical rule to select the source row that has

$$\max_i \{f_i | i=1,\ldots,m\} ,$$

or

$$\max_i \{f_i / \sum_{j \in NB} f_{ij} | i=1,\ldots,m\}$$

An iterative algorithm implementing this method is as follows:

(1) Relax the integer requirements and solve the corresponding LP problem.

(2) If the solution is integer, stop. Otherwise go to Step 3.

(3) Generate a Gomorian cut constraint from the equation and add this constraint to the existing tableau, as a source row that satisfies the following condition;

$$\max._i \{f_i | i=1,\ldots,m\} \text{ to utilize a deep-cut.}$$

(4) Use the dual simplex method to solve the new problem. Then go to Step 2.

A block diagram of this algorithm is given in Fig. 2.2.

2.3 Zero-One Implicit Enumeration

In general, the solution space of an integer program (IP) can be assumed to possess a finite number of possible feasible points. A straight forward method of solving integer problems is to exhaustively (or explicitly) enumerate all such points. In this case, the optimal solution is determined by the points that yields the best value of the object function.

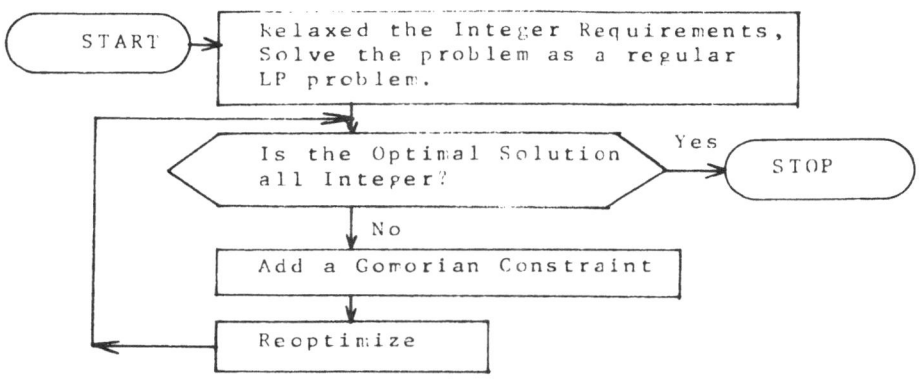

Figure 2.2   Gomory's Fractional Cutting Plane Method

The obvious drawback of this method is that the number of solution points may become too large to compute in reasonable computation time.   Then the idea of implicit (or partial) solution may be considered as an alternative method.   The implicit enumeration method considers only a portion of all possible solution points, while automatically discarding the remaining ones as nonpromising.   To illustrate this, consider determining (computing) all the feasible solutions for the following inequality equality:

$$3x_1 - 8x_2 + 5x_3 \leq -6 \quad x_j = (0,1) \quad j=1,2,3$$

The solution, obtained as shown in Fig. 2.3, is $x_1 = 0$, $x_2 = 1$, and $x_3 = 0$.

In the above example, it is easy to keep track of the enumerated solutions, because there are only eight of them.   In the general case, an efficient and flexible book-keeping method would be required for keeping track of all the solutions that have been considered (either implicitly or explicitly) and for generating the remaining ones in a non-redundant fashion. Efficiency here indicates that the scheme should not tax the computer memory and flexibility means that it should store and retrieve information easily.

Consider the following ILP:

max. $Z = CX$
$AX = \leq b$
$X \to binary$

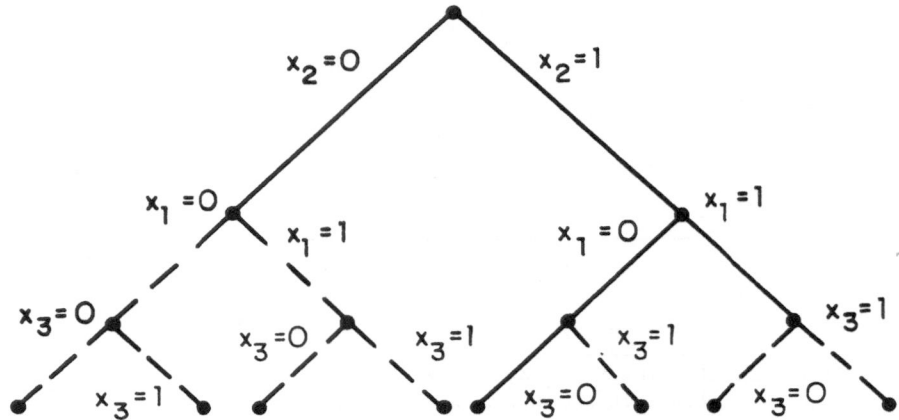

Figure 2.3   Enumeration Tree

Given $S_o = \{X \mid AX \leq b,\ X \text{ binary}\}$

the separation at V_k is determined by choosing a particular variable X_j, not chosen previously along the path P_k from v_o to v_k, and letting (partitioning equation)

$$S_k^* = \{S_k \cap \{X \mid X_j = 0\}\ ,\ S_k\ \{X \mid X_j = 1\}$$

The path P_k corresponds to an assignment of binary values to a subset of variables. Such an assignment is called a partial solution. Denote the index set of the assigned variables by W_k and let;

$$S_k^+ = \{j \mid j \epsilon W_k \text{ and } X_k = 1\}$$

$$S_k^- = \{j \mid j \epsilon W_k \text{ and } X_j = 0\}$$

$$F_k = \{j \mid j \epsilon W_k\}$$

A completion of W_k is an assignment of binary values to the free variables specified by the index set F_k. The partitioning equation satisfies

$$S_k \in S_j^* \to |H_k| < |H_j|$$

where

$$H_o = \{X|X \text{ binary}\} \quad \text{and} \quad H_k = \{X|X_j = 0,1. \quad j\epsilon F_k\}$$

Finiteness of an enumeration algorithm based on the partitioning equation is thus guaranteed. For ILP, total enumeration would terminate in 2^n steps.

The problem considered at V_k is

$$\text{max. } Z_k = \sum_{j\epsilon F_k} C_j X_j + \sum_{j\epsilon S_k^+} C_j$$

$$\sum_{j\epsilon F_k} a_{ij} X_j \leq b_i - \sum_{j\epsilon S_k^+} a_{ij} = S_i, \quad i=1,\ldots,m$$

$$X_j = 0 \text{ or } 1 \quad j\epsilon F_k$$

Let $T_k = H_k$. Since $C_j \leq 0$, $X^o(k)$ is obtained by setting $X_j = 0$, $j\epsilon F_k$. Thus $Z_k = Z_k^o = \sum_{j\epsilon S_k^+} C_j$.

If, in addition, $X = (S_1,\ldots,S_m) \geq 0$, then $X^o(k)$ is feasible and $Z_k = Z_k^o$.

Again, the fathoming cases are

(a) $\overline{Z}_k = \underline{Z_k}$ (b) $\overline{Z}_k \leq \underline{Z_o}$

Using the bound derived above, note that (a) occurs when $X^o(k)$ is feasible. A simple sufficient condition for (b) is also available. Suppose that some i:

$$t_i = \sum_{j\epsilon F_k} \min \{0, a_{ij}\} \geq S_i$$

In this case no completion of W_k can satisfy constraint i, $\overline{Z}_k = -\infty$, and v_k is fathomed.

For example, consider the constraint

$$\sum_{j \epsilon F_k} a_{ij} X_j = 3x_1 - 4x_2 + 3x_3 - 5x_4 \leq -10 = S_i$$

Vertex k is fathomed, since $t_i = -9 > S_i = -10$.

Let $Q_k = \{i : S_i < 0\}$. If $Q_k = \emptyset$, then v_k is fathomed. Since $X^\circ(k)$ is feasible, if $Q_k = \emptyset$, let

$$R_k = \{j : J\epsilon F_k \quad \text{and} \quad a_{ij} < 0 \text{ for some } i\epsilon Q_k\}$$

One partitions on some X_j, $j\epsilon R_k$, and then branches to the successor vertex corresponding to $X_j = 1$. The following rule selects a $j\epsilon R_k$ in an attempt to drive toward feasibility: Define

$$I_k = \sum_{i=1}^{m} \max. \{0, - S_i\} = - \sum_{i\epsilon Q_k} S_i$$

By choosing X_j, infeasibility at the successor vertex is

$$I_k(j) = \sum_{i=1}^{m} \max \{0, - S_i + a_{ij}\}$$

and X_p is selected such that

$$I_k(p) = \min_{j\epsilon R_k} I_k(j)$$

Implicit Enumeration Procedure.

(1) (Initialization) At $v_o, F_o = \{1,\ldots,n\}$, $\bar{Z}_o = \infty$ $\underline{Z}_o = - \infty$. Go to Step 2.

(2) (Calculating Bounds) At v_k, let $\bar{Z}_k = \sum_{j\epsilon S_k^+} C_j$. If $S \geq 0$, let $\underline{Z}_k = Z_k^\circ = \bar{Z}_k$ and let $\underline{Z}_o = \max\{\underline{Z}_o, \underline{Z}_k\}$. Go to Step 3.

(3) (Fathoming) If $t_i > S_i$ for any i, or if $\bar{Z}_k = \underline{Z}_k$, or or $f\bar{Z}_k \leq \underline{Z}_o$, v_k is fathomed and one goes to Step 4. If v_k is live, go to Step 5.

(4) (Backtracking) If no live vertex exists, go to Step 6, otherwise branch to the newest live vertex and go to Step 2.

(5) (Partitioning and branching) Partition on X_p, where $I_k(p) = \min_{j \in R_k} (j)$. Branch to the $X_p = 1$ vertex. Go to Step 2.

(6) (Termination) If $\underline{Z_o} = -\infty$, there is no feasible solution. If $\underline{Z_o} > -\infty$, that feasible solution which yielded $\underline{Z_o}$ is optimal.

3. DISCUSSION OF THE CONVERGENCY OF THESE METHODS

The nature of the engineering problems, as a mathematical model, and how it behaves during the optimization process is one of the significant factors effecting rate of convergence of algorithms under consideration. In each specific engineering problem, there are certain characteristics that directly or indirectly effect the rate of the optimization process, if they are used before or during the optimization. If these engineering and mathematical techniques are developed as an algorithmic technique and used properly, they will help to overcome some of the difficulties and speed convergence of the optimization algorithm. Several of the most significant factors are discussed here.

3.1 Starting Points

From the nature of engineering design, it is possible to expect that good or near optimal starting should accelerate the convergence to the optimum. This can be done several ways. Two important ways are:

(a) Initial design can start from the feasible region. Then the problem is solved as a linear continuous optimization problem. At the end of the linear optimization, a continuous solution or an immediate neighborhood of a continuous solution can be taken as a starting point for the discrete optimization. This method cuts down the iteration process considerably, because it is faster to solve the linear optimization cycles than the integer linear programming.

812

(b) In linear programming, if one considers only some of the constraints to reach the approximate optimal point, computation time is reduced.

3.2 Move Limits

In order to reduce the linearization error in the iterated LP Method, it is possible to introduce limitations between subsequent iteration points. This limitation is called either step-sizes or move limits. It is possible to take these limits larger at the early stage of the optimization and when the optimization gets closer to the continuous optimum point, these limits could be smaller so that the linearization error can be ignored.

3.3 Scaling

The procedure of scaling the design at the end of each cycle of the linear optimization process to maintain feasibility aids convergence, especially if the cutting plane method is used for the linearization process. (See Figs. 3.1 and 3.2).

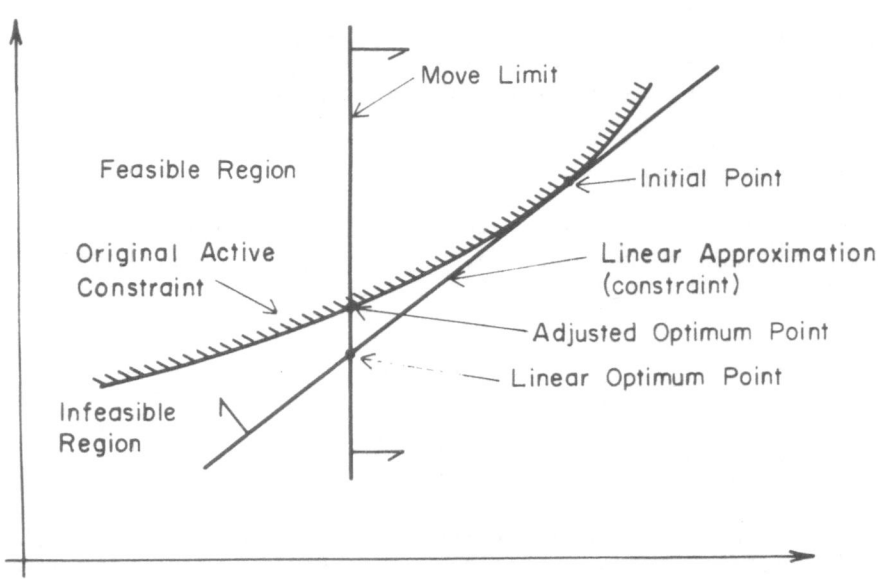

Figure 3.1 Scaling Procedure for the Convex Feasible Region

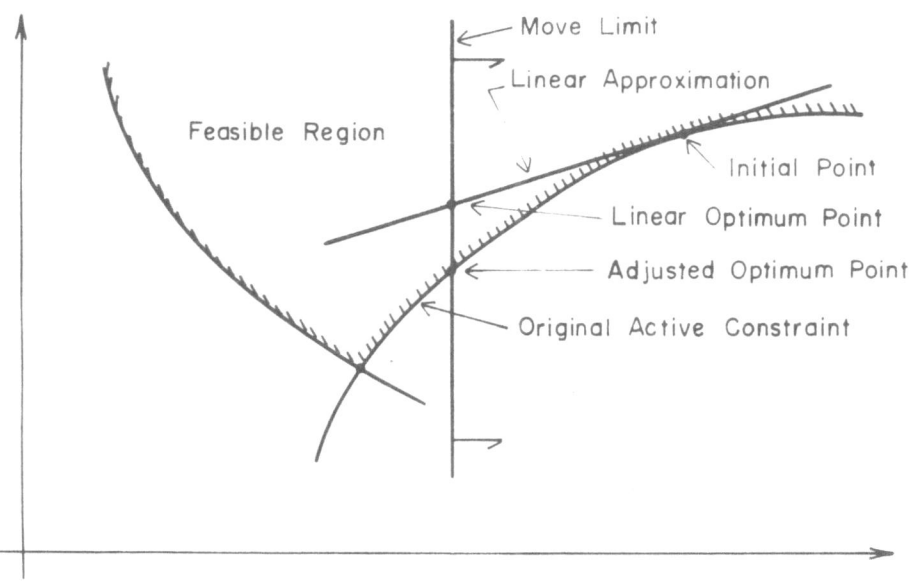

Figure 3.2   Scaling Procedure for the Non-Convex Feasible
             Region

3.4   Starting From Non-Empty Solution for a Discrete Programming Problem

At the end of the optimization cycle, the optimum continuous solution is known. This solution vector can be transformed to the discrete solution set at the neighborhood of the continuous solution, satisfying the feasibility conditions. This set of points can be taken as an initial partial solution. This process automatically puts the partial solution into the feasible region, as a first step, which in turn substantially improves convergence of the algorithm.

In some cases, certain variables take on certain values at the optimum solution, as in engineering design requirements. Partial solution can be taken as a permutation of these values at the initial stage of discrete algorithm.

3.5   Narrowing the Candidate Solution as Much as Possible

One of the advantages of knowing the optimum continuous solution is that during the discrete optimization, rather than considering all discrete points in the feasible region, it is possible to bound the feasible region to the immediate neighborhood of the continuous solution. Size of the candidate discrete solution set directly affects the size of the problem and consequently the computation time of the discrete algorithm. Therefore, if the candidate discrete solution set is taken as small as possible, the rate of convergence can be greatly increased.

3.6 Globol Versus Local Minimum

In the case of a non-convex feasible region, existence of local minima is more difficult to handle. Considering the difficulty, if not impossibility, of verifying a global optimum, such a favorable result seems impossible to guarantee. Also, the procedure indicated in selecting good starting points would enable one to reach the global optimum. If one has to test the global optimality condition, it is possible to start the same optimization procedure with an entirely different set of initial values. If the same optimum values are found, then one can reach the conclusion that the set of points that were found at the first optimization phase is the global optimum. Otherwise, the same iteration can be continued until the condition is satisfied.

For further details on these methods and their application to structural optimization, the reader is referred to Refs. 1 to 8.

REFERENCES

1. Balas, E., "An Additive Algorithm for Sovling Linear Programs with Zero-One Variables", Operation Research, Vol. 13, 1965, pp. 517-546.
2. Balas, E., "Discrete Programming by the Filter Method", Operation Research, Vol. 15, 1967, pp. 915-957.
3. Aldo, C. and Logeher, R.D., "Automated Optimum Design Discrete Components", Journal of Structural Division, ASCE, Jan. 1971.
4. Dorn, W.S., Gomory, R.E. and Greenberg, H.J., "Automatic Design of Optimal Structures", Journal de Mechanique, Vol. 3, No. 1, March, 1964.
5. Geoffrin, A.M., "Improved Implicit Enumeration Approach for Integer Programming", Operation Research, Vol. 17, 1969, pp. 437-454.
6. Gomory, R.E., "An Algorithm for Integer Solution to Linear Problems", Recent Advances in Mathematical Programming, New York, McGraw-Hill, 1963.

7. Kelly, J.E., "The Cutting-Plane Method for Solving Convex Programs", _J. Soc. Indust. Appl. Math._, Vol. 8, No. 4, December, 1960.

8. Sağlam, M. Rifat, "The Application of Discrete Programming Techniques in the Structural Optimization", Ph.D. Thesis, The University of Wisconsin, 1976.

THE SIZING OF STRUCTURES USING DYNAMIC RELAXATION[*]

B.H.V. Topping

Lecturer, Department of Civil Engineering and
Building Science, University of Edinburgh,
The King's Buildings, Edinburgh, U.K.

ABSTRACT

This paper presents a technique for automatic design of pin-jointed structures, based on parallel elastic and elasto-plastic analyses using the Dynamic Relaxation Analysis method. The procedure is developed from the Fully Stressed Stress-Ratio criteria, but is modified to account for maximum and minimum member area sizes and deflection constraints. Example structures are desired and then compared with nonlinear programming weight optimization solutions. The results are shown to be of comparable weight and computer solution time. It is also shown, with examples, that the Dynamic Relaxation analysis technique is particularly suitable for the interactive design of structures that includes cable elements.

1.   INTRODUCTION

Computer design of large structures idealized as pin-jointed assemblies can be accomplished automatically with the use of Fully Stressed Design Techniques [1]. Although Fully Stressed Design

---

*Most of the work included in this paper was undertaken in the Department of Civil Engineering, The City University, London, while the author was the recipient of an S.R.C. research studentship. The author wishes to thank Dr. M. R. Barnes, Lecturer in the Department of Civil Engineering, The City University, for his advice.

Solutions are not necessarily of minimum weight, they are always buildable and often have weights near that of the optimum solution. Hence these techniques are usually preferable to Nonlinear Optimization Algorithms, because of their comparatively high speed of convergence. In addition, there is no assurance that, for large structures with many design variables and constraints, nonlinear programming algorithms will always converge to a global optimum solution. Fully Stressed Design Methods also have an intuitive appeal to engineers, being akin to manual design processes. In Ref. 2, these methods were applied to the problem of finding the optimum layout of modular space structures. In this paper, the application of modified Fully Stressed Design Methods to structures of fixed topology is discussed.

The commonest Fully Stressed Technique is the Stress-Ratio Method, which consists of resizing each member according to its maximum stress level at the last reanalysis. Many structures may have constraints on the maximum and minimum permitted member area sizes, together with deflection constraints at many or all of the joints. Modifications to a Fully Stressed Design may be required if these constraints are not to be violated. In the case of minimum member sizes, allowance can readily be made by setting a member's cross sectional area to its minimum size whenever the stress-ratio resizing leads to an area less than the minimum. In this case the member will be understressed and the stress constraint for the member will become slack. However, a similar procedure cannot be adopted with maximum member area size constraints, because it would result in the member being overstressed and the stress constraint being violated. The areas of other members may usually be increased to reduce this overstress. It may also be necessary to increase some member areas to ensure that the structure does not violate any of the deflection constraints. Razani [3] suggested that an optimal search method might be used to allocate the required increases in member areas for deflection constraints, but concluded that this was difficult in practice. The method of allocation of member areas presented in this paper is based on the use of parallel elastic and elasto-plastic analyses, using the Dynamic Relaxation Solution Method. The structure is continually resized to comply with modified Fully Stressed Design Criteria that cater for maximum and minimum member area sizes and deflection constraints.

The Dynamic Relaxation analysis technique can easily be modified to account for cable members, with and without pretensioning. The method of sizing is applied to a bridging structure that is subjected to four loading cases and stress constraints. The problems of accounting for the effects of prestress in cable members is investigated. The Dynamic Relaxation (DR) analysis method and its application for form finding is discussed in detail in Refs. 2, 4, 5, 6, and 7.

2. THE SIZING OF PIN-JOINTED STRUCTURES USING
    MODIFIED FULLY STRESSED DESIGN CRITERIA

For structures subject to multiple load cases, the design
problem depends on the loading level in each loading case and
therefore concurrent vectors for each case must be considered
simultaneously. Each of these load cases will require additional
vectors for deflections, nodal residuals, and velocities.

2.1 Fully Stressed Design Criteria with Stress Constraints Only

Member areas of a structure that is subject to stress constraints only can be modified after every few iterations of the
integration procedure, according to the simple stress ratio
criteria:

$$A_{c+1} = A_c (\sigma_m / \sigma_{pm})_{max} \qquad (2.1)$$

where (σ_m / σ_{pm}) is the ratio of the current stress in member m and
the tensile/compressive permissible stress in member m. For
multiple loading cases, the maximum value of the ratio (σ_m / σ_{pm})
under all loading cases is used to ensure safety.

Before any modifications can be made to the structure, it is
important that the members are significantly stressed so that
their areas are not immediately cut to very small values. An
efficient damping procedure was developed [2], which is a combined Kinetic and Viscous Damping scheme in which the analysis is
initially undamped and each loading case is run until it reaches
a kinetic energy peak. When all cases have reached a peak, the
nodal velocities are set to zero and the iteration is restarted,
using a viscous damping factor based on the number of iterations
required to reach the first energy peak. The damping factor is
then set equal to:

$$C' = 4\pi f M_i = (\pi M_i)/(N_p \cdot \Delta t) \qquad (2.2)$$

where N_p is the number of iterations to the first kinetic energy
peak, for any loading case (assumed 1/4 of the fundamental period).
In this way, the damping factor insures that the analyses are
heavily damped, unlikely to diverge, and convergence is steady.

A simple three-bar truss, subject to two loading cases, was
analyzed and member areas resized according to equation (2.1)
after every 10 time intervals following the energy peak. This
structure, shown in Fig. 2.1 was first studied by Schmit [8],
using a steepest descent alternate mode algorithm, and subsequently by Razani [3], using the fully stressed design method.

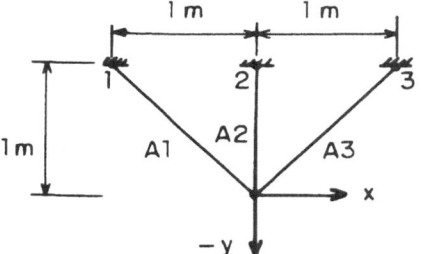

Load Case 1:
4x = 15.0N   4y = -25.9808N
Load Case 2:
4x = -20.0N   4y = 0.0N
Permissible Stresses:
Tension = 20.0 N/m²
Compression = -15.0 N/m²
Young's Modulus = 10⁵ N/m²

Figure 2.1   Three Member Truss

For comparison, the problem was designed using:

(a) The Stress-Ratio Fully Stressed Design (F.S.D.) Method
     (Matrix Analysis).

(b) A particular form of the steepest descent alternate mode
     algorithm (N.L.P.) which caters for maximum and minimum area,
     stress and deflection constraints developed by Elliott [9].

The solutions for each method, without constraints on member
sizes, are given in Table 2.1.   These analyses agree with the
work by Razani, Schmit and the D.R. solution confirms that the
F.S.D. solution for this problem is an optimum.

TABLE 2.1   SOLUTION OF THREE MEMBER TRUSS

| Member | D.R./F.S.D. Solutions | | | N.L.P. Solution | | |
	Areas m²	Load Case 1 Forces N	Load Case 2 Forces N	Areas m²	Load Case 1 Forces N	Load Case 2 Forces N
1	1.071	21.29	-16.06	1.070	21.28	-16.05
2	0.544	10.87	2.72	0.544	10.89	2.69
3	0.611	0.075	12.22	0.614	.0671	12.33

Volume = 2.92 m³   Volume = 2.92 m³

D.R. Solution Time = 0.096   Solution Time = 0.259

CDC seconds   CDC seconds

This problem was studied using the DR method, with a series
of initial member areas and damping factors. Generally it appears

that if the initial member areas are set to a large value
(usually their maximum), and the damping factor given as in
Eq. 2.2, a fast rate of convergence will be ensured.

2.2 Maximum and Minimum Member Area Size Constraints

On studying the results of the three-bar truss problem, it
can easily be seen that if a constraint on the maximum area is to
be set below 1.071 m^2, the most likely member to be overstressed
is member 1, where the compression force in load case 2 will be
most critical. It was hoped that the relative deflections of the
nodes of member 1 could be reduced to lower this compression
force. This would, in effect, correspond to pretensioning member
1 by shortening it. However, for structures subject to multiple
loading cases, pretension effects would obviously be best
arranged differently for each loading case. The best distribution
of prestress can therefore only be determined by an optimal
search method. It was thus decided to make the analysis elasto-
plastic, by setting the maximum force in any member that is at its
maximum size equal to the maximum area multiplied by the permissible stress. This can be accomplished efficiently in the explicit D.R. method by setting the maximum force sustained by any mem-
ber to $\sigma_T A_{max}$ in tension or $\sigma_c A_{max}$ in compression when calculating the residuals. The results for this design, using a maximum area
of 0.95 m^2, are given in Table 2.2, with the results of a subsequent elastic analysis for the structure. The elastic analysis

TABLE 2.2   D.R. ELASTO-PLASTIC RESULTS FOR THREE BAR TRUSS

Member	Area m^2	D.R. Elasto-Plastic Solution		Elastic Analysis	
		Load Case 1 Forces N	Load Case 2 Forces N	Load Case 1 Forces N	Load Case 2 Forces N
1	0.950	19.000(P)	-14.25(P)	19.70	-15.32
2	0.705	14.108	0.15	13.11	1.661
3	0.702	-2.208	14.034	-1.509	12.92

Volume = 3.041 m^3

Solution Time = 0.081

CDC seconds

shows that the areas of members 2 and 3 have not been increased enough to ensure that member 1 is not overstressed. It was noted that if the plastic force carried by the members of maximum size was reduced, then even more load was transferred to the elastic members, thereby producing an increase in their member areas on resizing. The extent to which the plastic force should be reduced to avoid overstress in the elastic analysis can be determined by running an elastic analysis simultaneously with the elasto-plastic analysis. The plastic force in members of maximum area is therefore defined as

$$\text{Plastic Force} = \text{Area}_{max} \times \text{Permissible Stress} \times AC_{m\ell c} \qquad (2.3)$$

where $AC_{m\ell c}$ is a factor for member m, under the action of load case ℓ at modification stage c.

The factor $AC_{m\ell c}$ has an initial value of 1.0. It must always remain less than or equal to 1.0 and is always reset equal to 1.0 if the area reduces to a value less than the maximum permitted area. The value of the factor is periodically amended throughout the procedure, to ensure that the plastic members are not overstressed in the elastic case. The $AC_{m\ell c}$ factor can be updated in the following way

$$AC_{m\ell c+1} = AC_{m\ell c} \times \left| \frac{\text{Permissible Stress of member m}}{\text{Elastic Stress of member m under load case}} \right|$$

$$(2.4)$$

Areas are modified using Eq. 2.1 and by considering member stresses under the action of all loading cases in both the elastic and elasto-plastic analyses. At convergence, all the areas and $AC_{m\ell c}$ factors must be constant.

The above technique ensures that any members that are overstressed and of maximum area are relieved of their overstress by increasing other members. In some cases in which alternative load paths are not available, or strain compatibility requirements might prohibit solution, convergence may not be possible. For sizing of three-dimensional, highly redundant, space frames this should not generally be a problem. The technique was applied to the three-bar truss problem, modifying member areas and the $AC_{m\ell c}$ factors after every 10 time intervals. Results for this method are given in Table 2.3. A comparative solution derived by using the nonlinear programming algorithm is given in Table 2.4.

It is interesting to note that despite the large change in volume, due to the constraint on maximum member size, the results for the two methods are the same. The D.R. method, however, was at least twice as fast as the N.L.P. algorithm.

TABLE 2.3   MODIFIED D.R. ELASTO-PLASTIC DESIGN
FOR THREE BAR TRUSS

| Member | Areas m^2 | Modified D.R. Elasto-Plastic Design | | Elastic Analysis | |
		Load Case 1 Forces N	Load Case 2 Forces N	Load Case 1 Forces N	Load Case 2 Forces N
1	0.950	19.00(P)	-14.25(P)	19.00	-14.25
2	0.8013	16.02	-6.11	14.10	0.159
3	0.9232	-3.56	18.46	-2.21	14.032
	Volume = 3.451 m^3		Solution Time = 0.354 CDC seconds		

TABLE 2.4   N.L.P. SOLUTION OF REFINED THREE BAR TRUSS

| | N.L.P. Solution | | |
Member	Areas m^2	Load Case 1.N	Load Case 2.N
1	0.950	19.00	-14.24
2	0.8014	14.11	0.136
3	0.927	-2.21	14.05
	Volume = 3.455 m^3	Solution Time = 1.970 CDC seconds	

2.3   Deflection Constraints

In addition to stress constraints and restrictions on maximum and minimum area sizes, many structures are designed so that deflections at certain or all nodes are constrained. The method for designing structures, accounting for deflection constraints presented in this section, was developed by further consideration of the three-bar truss problem. It was noted that if the deflection of a node was excessive this could be reduced by increasing the nodal stiffness in the appropriate direction. The extent of the increase in stiffness can be determined by applying a fictitious force in the elasto-plastic analysis which, on resizing, will result in an appropriate increase in stiffness. The magnitude of the fictitious force can be estimated from the current values of nodal elastic stiffness and deflection in the elastic analysis.

The fictitious forces of node i, load case ℓ, in the elasto-plastic analysis are defined as

$$FF_{ix\ell c+1} = FF_{ix\ell c} \left[\frac{S_{ix}}{2.0} (DC_{ix} - \Delta x_{i\ell}) \right] \qquad (2.5)$$

where S_{ix} is the stiffness of node i in the x direction (this has already been calculated to condition the fictitious masses); $DC_{ix\ell}$ is the constrained value of node i in the x direction for load case ℓ (the sign of the deflection is set the same as Δx_i); $\Delta x_{i\ell}$ is the deflection of node i, in the x direction, under load case ℓ, in the elastic analysis; and a $FF_{ix\ell c+1}$ is the fictitious force at modification stage c + 1, for node i, in the x direction, under load case ℓ.

The factor of 2.0 in Eq. 2.5 ensures that changes are not too rapid and that instability is avoided. All fictitious forces are initially set to zero and each time they are modified checks are made to ensure that:

 a) if the deflection is greater than 0 and the fictitious force is negative, the fictitious force is set equal to 0.

 b) if the deflection is less than 0 and the fictitious force is positive, the fictitious force is set equal to 0.

These checks ensure that if the deflection is less than the constrained value, the fictitious forces are either set to zero or ultimately reduced to zero.

When the three-bar truss was fixed without any maximum or minimum area constraints, the horizontal deflections were for load case 1, +0.00020 m and for load case 2, -0.00035 m. The results for a D.R. solution for a N.L.P. solution with the horizontal deflection constrained at 0.0003 m are given in Table 2.5. With the D.R. method the fictitious forces and the $AC_{m\ell c}$ factors were reset after every 10 iterations, when the member areas were resized. Although these results are different, the weights and solution times are similar.

2.4 A Further Comparison with the Nonlinear Programming Method

The structure considered for additional comparisons is shown in Fig. 2.2. The permissible stresses used are σ_T = 21.6 N/m^2 and σ_c = 10.0 N/m^2.

TABLE 2.5. SOLUTIONS OF THREE BAR TRUSS
WITH DEFLECTION CONSTRAINT

	D.R. Solution			N.L.P. Solution		
Max/Min Sizes	2.0 m^2	0.0 m^2		2.0 m^2	0.0 m^2	
Member	Areas	Load Case 1.N	Load Case 2.N	Areas	Load Case 1.N	Load Case 2.N
1	1.210	22.350	-15.561	1.131	21.79	-15.24
2	0.469	9.372	2.009	0.508	10.16	1.55
3	0.742	1.139	12.724	0.789	0.578	13.05
Deflection node 4 ×		0.00017	-0.0030		0.00019	-0.00030
	Volume = 3.230 m^3 Solution Time = 0.460 CDC seconds			Volume = 3.224 m^3 Solution Time = 0.319 CDC seconds		

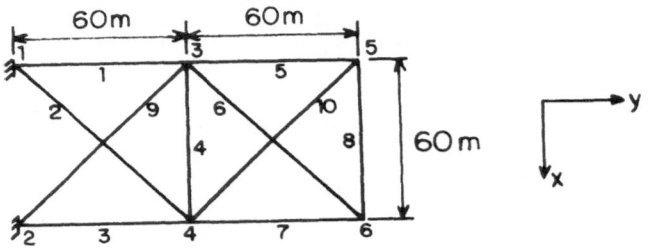

Figure 2.2   Ten Bar Truss

For this example, it was noted that if maximum and minimum size constraints were active, and a solution was possible, the D.R. iteration oscillated about a mean point. However, if a solution was not possible, without at least slightly over-stressing one of the members at a maximum area size, then convergence to the solution was rapid. This oscillating instability resulted because the $AC_{m\ell c}$ factors were modified at the same time as the member areas. This means that plasticity effects were readjusted before the effects of the last adjustment had been

taken into account by area modifications. If these area modifications had been made, they would have resulted in changes in the
stress levels in both the elastic and elasto-plastic analyses.
It was therefore decided to adjust the areas after every 10 time
intervals and the $AC_{m\ell c}$ factors after every 50 time intervals.
To ensure that the plasticity effects were not too rapid the
$AC_{m\ell c}$ factors were updated using the following expression:

$$AC_{m\ell c+1} = \frac{AC_{m\ell c}}{2.0} \left\{ 1 + \left| \frac{\text{Permissible Stress of member } m}{\text{Elastic Stress of member } m \text{ under load case } \ell} \right| \right\} \quad (2.6)$$

Masses were assigned using Equation 8 of Ref. 4 and an additional
mass component of 0.005 in each coordinate direction.

The structure was designed using both the D.R. procedure
and the nonlinear programming algorithm. Loading conditions for
the structure were as follows:

Load Case 1: $5x = 0.0$ N $5y = -10.0$ N

             $6x = 0.0$ N $6y = -10.0$ N

Load Case 2: $3x = 10.0$ N $3y = 0.0$ N

             $5x = 10.0$ N $5y = 0.0$ N

Comparative deflection constrained designs were made using the
D.R. procedure and the nonlinear programming algorithm. A constraint of 0.5 m on x and y deflections was applied at nodes 5
and 6. The maximum and minimum member areas were set at 2.4 m^2
and 0.5 m^2. The results for the solutions are given in Table 2.6.
The maximum area used in the N.L.P. solution was less than 2.0 m^2
and, although the solution was different from the D.R. solution,
the volumes were within 0.63%. The D.R. solution time was less
than the N.L.P. solution time.

3.   APPLICATION OF FULLY STRESSED DESIGN
     TO STRUCTURES WITH CABLE MEMBERS

Dynamic Relaxation is an explicit analysis method that is
ideally suited to consider on-off, nonlinear effects during the
analysis. The slackening of cables can therefore be easily
accounted for [6]. Using the D.R. method to analyze a structure
with cable elements, one modification is required to account for
the possible nonlinear effects. A one-dimensional array called
NS is established and used to keep a record of the cable members,
where $NS(m) = 1.0$ for bar members and $NS(m) = 0.0$ for cable
cable members.

TABLE 2.6   SOLUTIONS OF TEN BAR TRUSS

Max/Min Sizes	D.R. Solution			N.L.P. Solution		
	2.4 m^2		0.5 m^2	2.4 m^2		0.5 m^2
Members	Areas m^2	Load Case 1 N	Load Case 2 N	Areas m^2	Load Case 1 N	Load Case 2 N
1	1.461	-9.19	24.55	1.293	-8.511	21.552
2	0.500	-1.17	7.59	0.996	-2.106	11.947
3	2.205	-9.15	-15.31	1.849	-8.511	-18.448
4	0.500	2.40	1.69	0.679	2.979	-1.775
5	0.843	-8.43	7.09	1.096	-8.510	6.673
6	0.500	-2.24	4.09	0.758	-2.107	4.705
7	0.842	-8.42	-2.88	1.071	-8.510	-3.327
8	0.500	1.58	-2.91	0.728	1.490	-3.327
9	2.400	-1.16	-20.60	1.640	-2.106	-16.377
10	1.447	-2.22	-9.99	1.136	-2.107	-9.437
Deflec-tions						
5x		-0.0575	0.5001		-0.0340	0.5000
5y		-0.0978	0.1503		-0.0861	0.1366
6x		-0.0385	0.4649		-0.0217	0.4726
6y		-0.0849	-0.0631		-0.0753	-0.0784
	Volume = 792.31 m^3 Solution Time = 6.396 CDC seconds			Volume = 787.310 m^3 Solution Time = 7.152 CDC seconds		

Cable members obviously cannot sustain compression forces and this must be accounted for when calculating the residuals. All member forces that are negative (i.e. compressive) are modified as follows:

$$Tm' = Tm \times NS(m)$$

Cable members must also be accounted for at the Fully Stressed Design resize stage. This ensures that any resizing of the cable members only considers tension forces.

The use of prestress in computer-aided design and optimization appears not to have been considered by previous workers. Hofmeister and Felton [10] investigated prestress effects, but did not consider the extra weight of cables required to provide the optimized prestress distribution.

The Dynamic Relaxation analysis method can readily account for prestress effects [7]. For design with concurrent loading vectors, it is convenient to refer prestress modifications to initial, or slack, member lengths as follows:

$$Tm^{t+\Delta t} = \frac{EAm}{Lms} (Lmc - Lms)$$

where

$Tm^{t+\Delta t}$ = current force in member m

Lmc = current length of member m

Lms = initial or slack length of member m

EAm = elastic modulus multiplied by member area

During the design process, the prestress levels may be controlled interactively by the engineer or automatically adjusted to ensure that the final structure conforms with given deflection criteria. For example, the bridging structure shown in Fig. 3.1 could be designed to ensure that under dead load conditions the vertical deflection of nodes along the deck is zero.

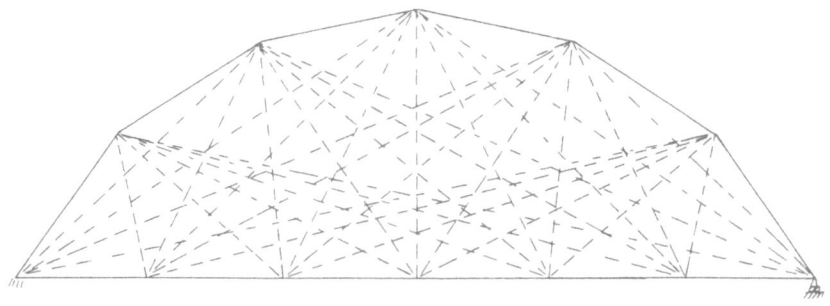

Figure 3.1   Bridging Structure

The structure was designed under four loading conditions. All members were initially assumed to be of the bar type and the structure was modified after every ten time intervals. The solution converged after 172 modifications. The structure was redesigned, assuming the dotted members shown in Fig. 3.1 to be cables. Members were modified after every fifteen time intervals. The solution converged after only 84 modifications. The structure was redesigned once more, using pretensioning to ensure that with one of the loading cases the vertical deflections of all nodes along the deck were zero. This was done by continually recalculating the slack lengths (Ls) of the cable members and by assuming that the initial deck coordinates were moved as far in the y direction as they were deflected down in the loading case. The initial lengths of the bar members were kept constant.

To avoid quasi-instability effects, the mass components derived using equation 8 of Ref. 4 were multiplied by 4 and an additional mass component of 0.005 was used. The structure was modified after every 10 time intervals and the member lengths adjusted after every 6 time intervals. The solution converged after 125 structure modifications. In each of these designs, many cable members were removed after their member areas reduced to zero during the design procedure. The problem was therefore one of form finding, as well as sizing. It should be pointed out that this structure becomes a mechanism once the applied loads have been removed and the cables are no longer in tension. This structure should be reconsidered with the deck and arch idealized using the rigidly jointed elements incorporated into D.R. analysis by Wakefield [11].

4.   CONCLUSIONS

D.R. solution times for problems with member area size and stress constraints are generally faster than the N.L.P. times. With the addition of deflection constraints the N.L.P. solutions are generally slower. In all cases, however, solution times for the two methods are of a similar order. Computer solution times generally indicate that the larger the problem, the more the D.R. method is to be favored [2].

The volume of solutions derived using the D.R. method is sometimes less than the N.L.P. solution volume [2]. This is because the N.L.P. algorithm frequently derives local optimum solutions and fails to find the globally optimum solution. However, the solutions from N.L.P. do give typical solutions, as would be expected from a nonlinear algorithm and, as such, represent a suitable yardstick for comparison.

In the problems studied, different but constant permissible stresses were considered for tension and compression members. Approximate expressions relating compressive permissible stresses and member areas can be developed [9] and these relationships could readily be incorporated into the D.R. method, to allow for the effects of buckling. It is also possible to incorporate deflection constraints that are different for positive and negative coordinate directions and, if required, self weight can also be accounted for during the process.

The final section of this paper illustrates that the D.R. method is particularly suitable for the design of structures idealized as rigid and cable elements.

REFERENCES

1. Gallagher, R.H., and Zienkiewicz, O.C., Optimum Structural Design - Theory and Applications, John Wiley & Sons, London, 1973.
2. Topping, B.H.V., "The Application of Dynamic Relaxation to the Design of Modular Space Structures," Ph.D. Thesis, Dept. of Civil Eng., The City University, London, September, 1978.
3. Razani, R., "The Behaviour of the Fully Stressed Design of Structures and its Relationship to Minimum Weight Design," AIAA Journal, Vol. 3, No. 12, 1965, pp. 2262-2268.
4. Topping, B.H.V., "Modified Fully Stressed Design Techniques using a Parallel Dynamic Relaxation Solution Method," Proc. World Congress on Shell and Spatial Structures, IASS, Madrid September, 1979.
5. Barnes, M.R., Topping, B.H.V., and Wakefield, D.S., "Aspects of Formfinding by Dynamic Relaxation," Proc. Int. Conf. on the Behaviour of Slender Structures, London, September, 1977.
6. Barnes, M.R., "Applications of Dynamic Relaxation to the Topological Design and Analysis of Cable, Membrane and Pneumatic Structures," 2nd Int. Conf. on Space Structures, Guilford, September, 1975.
7. Barnes, M.R., "Form Finding and Analysis of Tension Space Structures by Dynamic Relaxation," A thesis submitted on the basis of published papers for the Degree of Doctor of Philosophy, Dept. of Civil Eng., The City University, September, 1978.
8. Schmit, L.A., "Structural Design by Systematic Synthesis," Proc. 2nd Nat. Conf. on Electronic Computation, ASCE, Pittsburgh, September, 1960, pp. 105-132.
9. Elliott, D.W.C., "Structural Optimisation," Ph.D. Thesis, Dept. of Civil Eng., The University of Aston, Birmingham, 1971.

10. Hofmeister, L.D.,and Felton, L.P., "Prestressing in Structural Synthesis," _AIAA Journal_, Vol. 8, No. 2, February,1970 pp. 363-364.
11. Wakefield, D.S., "Dynamic Relaxation Analysis of Pretensioned Networks with Flexible Boundaries," _Proc. World Congress on Shell and Spatial Structures, IASS_, Madrid, September, 1979.

AUTOMATIC DESIGN OF FRAMES WITH TAPERED TUBULAR MEMBERS

C. Wayne Martin

Department of Engineering Mechanics, University of
Nebraska, Lincoln, Nebraska

ABSTRACT

Frames and guyed poles made from hollow tapered tubes are
frequently used to support electrical transmission lines and
lighting. This paper relates experience from developing two programs for automatic design of these structures and discusses some
techniques being considered for inclusion in a third-generation
program.

1. INTRODUCTION

Typical structures that have been designed by the programs
described in this paper are shown in Figure 1.1. The first
example shown in Fig. 1.1(a) supports an electrical transmission
line at a river crossing. It has six horizontal members, and
tapered vertical members. The second example shown in Fig. 1.1(b)
is a 140 ft. tapered pole, supporting lighting. Under maximum
loading, the top deflection of this pole is about 13 percent of
its height. The third example shown in Fig. 1.1(c) is a guyed
pole supporting a transmission line. The last example shown in
Fig. 1.1(d) is a dead-end structure, used where a transmission
line enters a power plant.

832

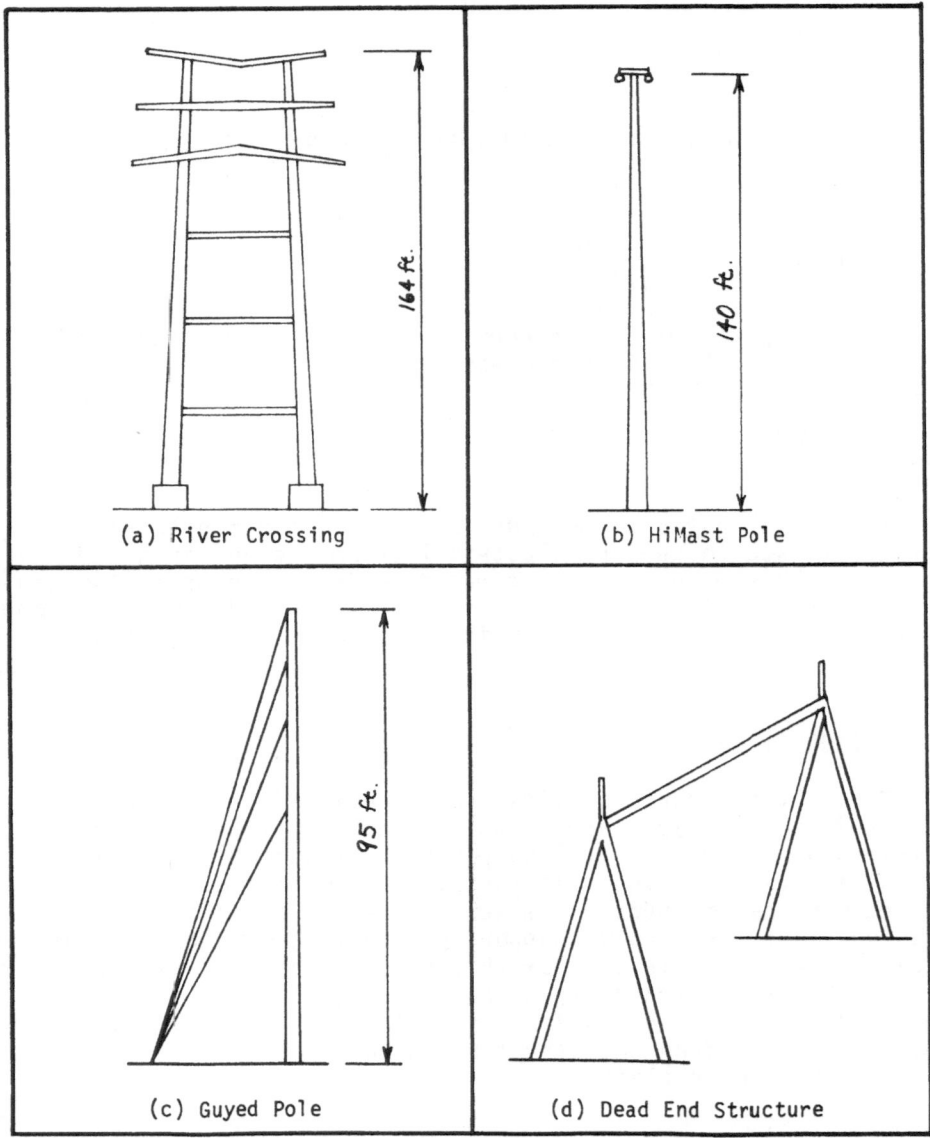

(a) River Crossing

(b) HiMast Pole

(c) Guyed Pole

(d) Dead End Structure

Figure 1.1   Typical Structures Designed

2. DESIGN PROBLEM CHARACTERISTICS

2.1 Analysis

Since the structures considered may have loads and deflections in any direction, the analysis techniques used must be appropriate to three-dimensional frames of arbitrary configuration, which are statically indeterminate. Members of the frames may be hollow, tapered tubes (beam-columns), truss elements, or guys. A large-deflection (nonlinear) analysis with multiple load conditions is required. Each load condition has a different deflected shape, and forces required for equilibrium must be found in the deformed geometry for each load condition. Consequently, much more computing time is required than for a similar linear problem.

Some trial designs (or even some final designs) may be controlled by buckling. Development of the programs described here was begun with the expectation that bifurcation would not be encountered. It was soon found that considerable user skill was required to describe the loading conditions so that bifurcation or very abrupt nonlinearity was avoided.

2.2 Design

Each element of the structure could have as many as three independent design variables, which are wall thickness, taper, and the radius at one end. However, in most practical structures, there are many constraints that greatly reduce the number of independent design variables. For example, the frame of Fig. 1.1(a) must be symmetrical. There can be no discontinuities in the radius of the tapered tubes, but discontinuities in thickness and taper are required for minimum weight.

The weight of the structure, which is to be minimized, is a nonlinear function that can be written explicitly and its gradient can be calculated. There are constraints on the maximum and minimum values of all thicknesses, radii, and tapers. As previously mentioned, there may be many equality constraints and indirect dependencies. Stress constraints are the most difficult constraints to evaluate. They are highly nonlinear and cannot be written explicitly. Their gradients could be calculated numerically, but at considerable expense in computing time. Constraints on radius-to-thickness ratios are required to avoid local buckling. Stress-dependent forms of these constraints are allowed by the codes, and sometimes result in significantly lighter structures.

3. PROGRAM APPROACH

3.1 Analysis

The analysis method used is the finite element stiffness method. An exact stiffness matrix for tapered tubes was developed. A new method was developed for calculating residuals (out-of-balance loads) for the tapered members, including the influence of axial forces on bending [1,2]. A modified Newton-Raphson method is used to solve the nonlinear equations for deflections [3]. The first program developed allowed stepwise application of loading. However, this did not appear to improve efficiency, so the second program applies the entire load in one step.

3.2 Design

The design algorithm is a nested, two-stage process. The outer loop is called design iteration. The inner loop is called optimization iteration. Complete analysis is performed only for each design iteration.

One of the most important concepts in the programs implementing the algorithms is definition of a design space and an analysis space. The design variable vector contains only the independent design variables, which are the minimum needed to uniquely define allowable variations in the structure. This may be written symbolically as

$$a_i = T_1 v_i$$

where the a_i are all of the radii, thicknesses, etc. required for finite element analysis, and v_i is the minimum set of independent design variables that are allowed to change in the design process. The transformation is, in these programs, nonlinear. It is not actually performed as a matrix multiplication. This transformation process allows definition of strings of tapered finite elements, in which the radius at the terminal end of one element matches the radius at the initial end of the following element. It also allows prescribed discontinuities in radius, as at a slip joint. Tapers and thicknesses within a string of elements can be either independent, or constant throughout the string. Diameters of cross members can be made dependent on diameters of vertical members, etc.

A modified Box-complex algorithm is used in the search for minimum weight [4,5]. The optimization iterations consists of moving the points in the complex. The best results have been obtained when the points in the complex are partly rational (to insure that motion is possible in all design variable-coordinate directions) and partly random. When a trial point move violates a constraint, the backward move is based on the degree of constraint violation. A penalty function has been added to the weight to encourage selection of high radius-to-thickness ratios, which tend to be more resistent to general buckling. During the optimization iterations, the location of the stress constraint is estimated using the element moments and forces from the complete analysis performed in the previous design iteration.

The nonlinear system stiffness equations may be written as

$$K(X,a_i)X = P$$

in which K is the system stiffness matrix, X represents displacements, a_i represents the radii, thicknesses, etc. and P represents external loads. The displacements of an element in the local x-coordinates are given by

$$x = T_2X$$

and the element forces p are given by

$$p = k(x,a_i)x$$

which is a nonlinear relation, because the influence of axial force on bending is included. The stresses σ are obtained from the element forces as

$$\sigma = H(x,a_i)p$$

As before, the design variables v_i are related to the analysis variables a_i by

$$a_i = T_1v_i$$

In these programs, during the optimization iteration, stresses are estimated by

$$\sigma + \Delta\sigma \approx H(x,T_1(v_i + \delta v_i))k(x,a_i) \, T_2 \, K^{-1}(X,a_i) \, P$$

This estimate of stress is exact for linear, small deflection analysis of determinate structures, but significant errors may occur because of redistribution of load within an indeterminate structure or because of nonlinearity. Both of these effects may be reduced by including an additional constraint that limits the allowable stiffness change within a design cycle.

At the end of the optimization cycle, a complete analysis is performed, which sometimes finds that the new design is over-stressed or that it would fail by buckling. In this case, the programs calculate a point on the hyperline in the design space between the last good design point and the new (overstressed) design point. Another analysis is then performed to confirm that the revised new design point does not violate any constraints.

An additional feature of the design process is a direct move to the stress or buckling constraint, maintaining equal relative stiffnesses. This can save considerable computing time in the early stages of the design process.

4. DESIGN EXAMPLES

The design process has been quite successful for designs that are not controlled by buckling and in which nonlinearities are not very abrupt. Figure 4.1 shows the design history for a 140 ft HiMast lighting pole. The deflection at the top of this pole under maximum design loading is about 13 percent of its height. There were 5 independent design variables in this case; the top diameter, the taper (constant for the entire pole), and 3 thick-nesses (one for each of three sections in which it would be fabricated). The analysis involved 6 elements, 7 nodes, 36 Degrees-of-Freedom, and a bandwidth of 12. Design required 88 seconds of CPU time on an IBM 360/65. The final design is on the constraints for minimum thickness, diameter, and taper and is a minimum weight design for the prescribed conditions.

Figure 4.2 shows the design history for a guyed pole with two loading conditions. This problem was formulated so that design was not controlled by buckling. It has two loading con-ditions. There were 7 active design variables, consisting of the top diameter, one taper, and 5 thicknesses. The program used an automatic start (which is taken as a very heavy pole, to insure that it is a feasible design) at 91,000 lb. The final design is on the constraints for minimum top diameter, maximum stress, and at 91 percent of the maximum radius-to-thickness ratio. Its weight of 5618 lb is 231 lb. lighter than the design provided as a test case by the sponsor of the work, but it has not been proven to be a true minimum weight design. This design process required 7 minutes of CPU time on an IBM 360/65.

The river-crossing structure of Fig. 1.1(a) is a third example. Its design history is somewhat erratic, because the design could fail by buckling. This case had 14 active design variables consisting of one taper, 5 diameters, and 8 thick-nesses. The analysis involved 19 elements, 16 nodes, 84 degrees-of-freedom, and one loading condition. The initial starting

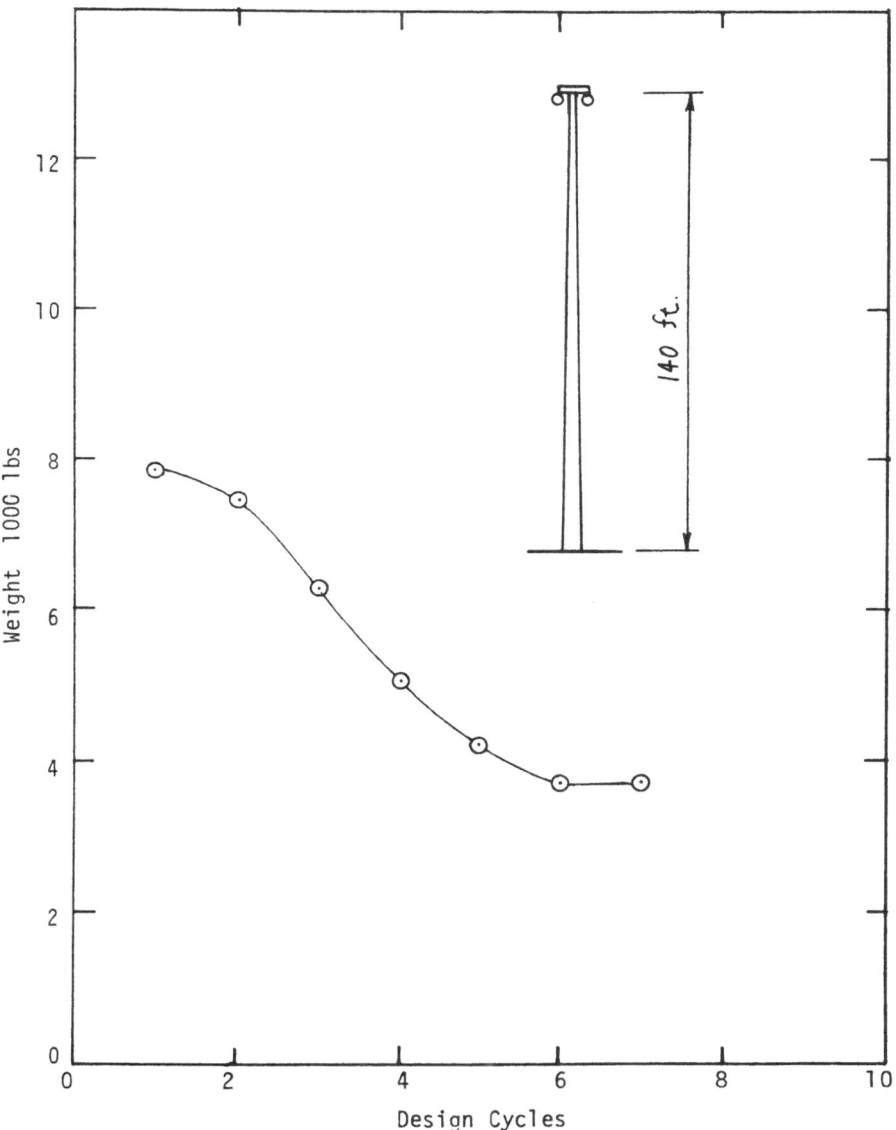

Figure 4.1   Design History of 140 ft HiMast Pole

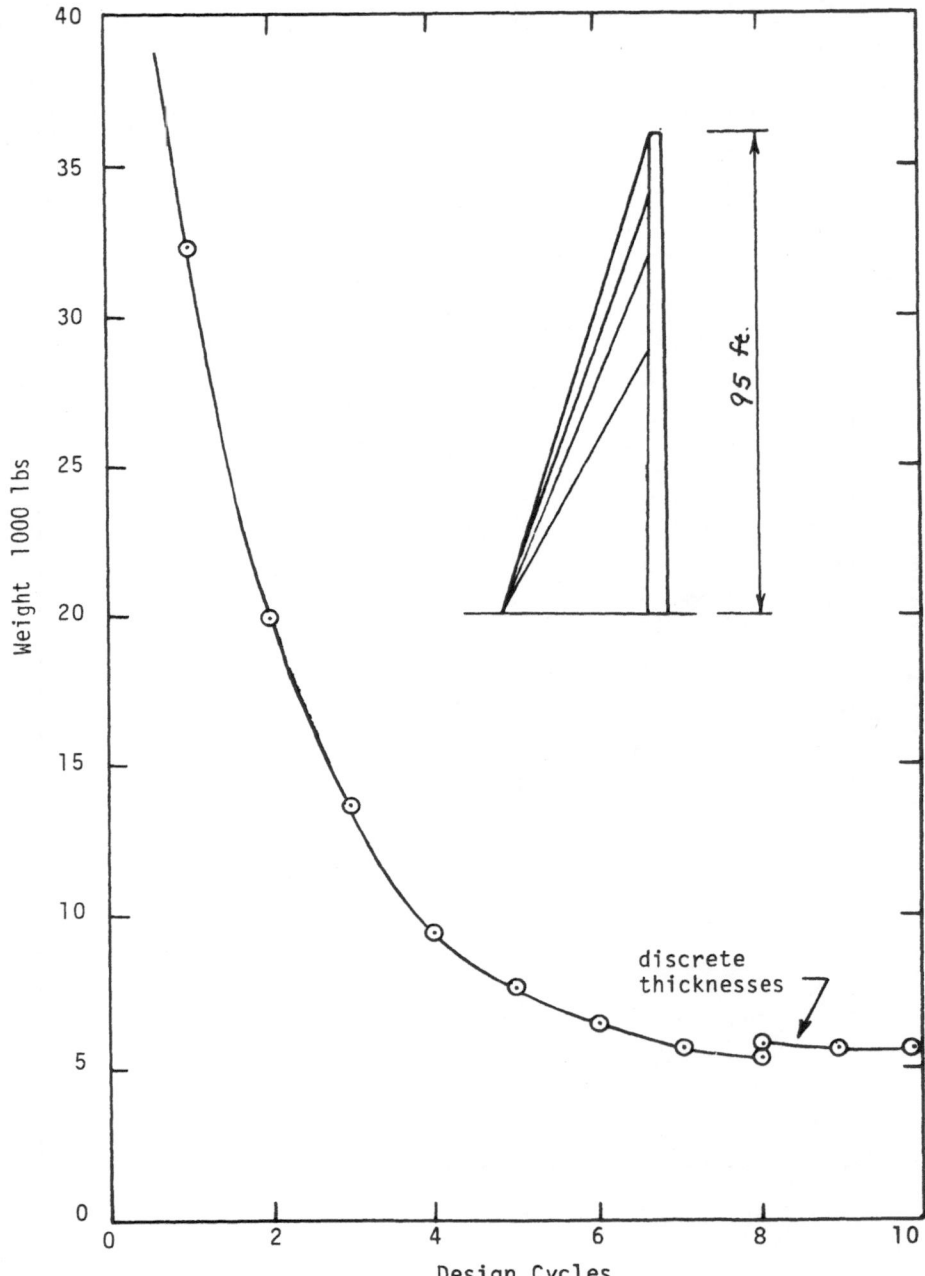

Figure 4.2   Design History of Guyed Pole

design was actually over-stressed, and the program ratioed up to a heavier design at 48,193 lb. The final design weight of 37950 lb. was 7 percent less than that for a design-by-hand, which took several man-weeks and weighed 40,760 lb. CPU time on the IBM 360/65 was 7 minutes 58 seconds.

5. POSSIBLE PROGRAM IMPROVEMENTS

The most significant improvement that could be made to these programs is addition of a true buckling constraint. While the second program has the ability to recover from a trial design that buckles, the process is not very efficient and the design history is often erratic. The essential ingredient appears to be calculation of a gradient to the buckling constraint, without excessive increase in computing time.

The Box - complex is a simple and robust algorithm that has performed well. However, it is believed that significant improvements in efficiency can be made by a good, rational method of distributing points in the initial complex.

There are a number of user-convenience features that can be very helpful. Input data for design or analysis should be identical, except for designation of which is to be done. It should be possible for the user to input his best guess at the design. Analysis of the initial input design should start with a buckling calculation and the program should be able to ratio us the initial input design, if necessary, to one that satisfies all constraints, before starting the automatic design. A history of the design process should be kept on disk or tape, and the program should have a restart capability. An iteractive front end for the program that could run on a microcomputer might be very helpful. Editing the input data and re-running the program should be easy.

6. CONCLUSIONS AND REMARKS

The nested, two-stage design process described herein has been demonstrated to be effective in design of frames with tapered members and moderate nonlinearity. A buckling constraint is needed for efficient design of structures with abrupt non-linearities, or where the design is actually controlled by bifurcation.

The feasibility of automatic design will increase because of continuing reduction in the cost of computing. Computing hardware costs have been decreasing by about one-half every two years since 1940 and there is no end in sight. With the advent of the

840

6809 and 8086 micro-processors, programs of the class discussed in this paper (around 5000 lines of FORTRAN) can run on computer systems whose total cost is less than $10,000. These computers can be dedicated to running only one program at a time, so inter-action between the user and computer is greatly simplified. Execution times are expected to be 2 to 10 times longer than the IBM 360/65.

Numerical failures are very disturbing to the average practicioner and must be avoided. Stable numerical techniques are desirable, even if slower in execution. The rapid advance in computing speed and continuing reduction in computing costs will reduce the advantage of techniques that are fast but not always stable.

REFERENCES

1. Lin, B., Nonlinear Space Frames With Tapered Members, M.S. Thesis, University of Nebraska, 1977.
2. Martin, C.W., Andersen, G.E., Lin, B. and Davy, D.T., "Non-linear Analysis of Space Frames with Tapered Tubular Members," Proceedings, Sixth Canadian Congress of Applied Mechanics, (V.J. Modi ed.), University of British Columbia, Vancouver, B.C., June, 1977.
3. Tezcan, S.S. and Ovunc, B., "An Iteration Method for the Non-Linear Buckling of Framed Structures," Space Structures, (R.M. Davies, ed.), Wiley, New York, 1967.
4. Kuester, J.L. and Mize, J.H., Optimization Techniques With Fortran, McGraw-Hill, New York, 1973.
5. Box, M.J., Davies, D. and Swann, W.H., Non-Linear Optimiza-tion Techniques, Imperial Chemical Industries Monograph No. 5, Oliver and Boyd, Edinburgh, 1969.